CONTRIBUTIONS TO
PROBABILITY AND STATISTICS

STANFORD STUDIES IN
MATHEMATICS AND STATISTICS, II

CONTRIBUTIONS TO PROBABILITY AND STATISTICS

Essays in Honor of Harold Hotelling

EDITED BY

INGRAM OLKIN SUDHISH G. GHURYE

WASSILY HOEFFDING WILLIAM G. MADOW

HENRY B. MANN

STANFORD UNIVERSITY PRESS

STANFORD, CALIFORNIA 1960

STANFORD UNIVERSITY PRESS
STANFORD, CALIFORNIA

Library of Congress Catalog Card Number: 60-5385

PRINTED IN THE UNITED STATES OF AMERICA

PREFACE

Work on this book began at the national meeting of the Institute of Mathematical Statistics held in Atlantic City in September, 1957. There Sudhish G. Ghurye and Ingram Olkin invited the other three editors to discuss the possibility of preparing a volume of essays to be dedicated to Harold Hotelling and offered to him on his sixty-fifth birthday. All five of us had been closely associated with Hotelling, either as students or as colleagues, and felt that a dedicatory volume would best express a real appreciation of his contributions.

Obviously, the selection of those to be invited to contribute to such a volume was a very difficult task. There were so many who might have been invited that it was clear that it would be impossible to make a selection without hurting the feelings of many, yet the major purpose seemed so worthwhile that we decided to risk the inevitable criticism. As the best way to limit the number who might be considered, we decided to ask only those who, to our knowledge, had been most closely associated with Professor Hotelling. For such errors as we may have made on this point, we apologize to all concerned, and not least to Professor Hotelling himself. Certainly, this volume would have been much longer had we asked all who have had some connection with Hotelling. As it is, those we invited to contribute have responded enthusiastically; indeed, only a few found it necessary to decline because of other commitments. To all the contributors, who have made this volume possible, we offer our sincere thanks.

At the initial meeting, the question whether one of us should assume the role of chairman of the editorial group was discussed. It was agreed that there would be no chairman, but that the work would be equally distributed. In the early stages of the book — the initial contacts with the potential authors, the review of the papers, and the initial decisions with respect to publication — all the editors shared equally. After the selection of the Stanford University Press as publisher, however, the heaviest load, of necessity, shifted to the one person who would be in contact with the publisher. Fortunately, during the academic year 1958–59, Ingram Olkin was a visiting professor of statistics at Stanford University, and he generously gave of his time to the many tasks connected with the editorial preparation of the volume. For this the other editors wish to express their appreciation and gratitude.

We also wish to express our thanks to Robert L. Ashenhurst, Patrick Billingsley, Meyer Dwass, and John Lamperti for the assistance they have given; and especially to the Statistics Branch, Mathematical Sciences Division, Office of Naval Research, for its generous support.

Finally, it is indeed a pleasure to thank the technical editor, Emlen T. Littell, for his extremely careful work on the volume.

The Editors

CONTENTS

PART I: AN APPRECIATION

PART II: CONTRIBUTIONS TO PROBABILITY AND STATISTICS

CONTRIBUTORS

R. L. Anderson
T. W. Anderson
Kenneth J. Arrow
R. R. Bahadur
M. S. Bartlett
V. P. Bhapkar
H. D. Block
R. C. Bose
Albert H. Bowker
R. A. Bradley
Kenneth A. Bush
Herman Chernoff
W. S. Connor
J. L. Doob
Meyer Dwass
Milton Friedman
S. G. Ghurye
B. G. Greenberg
Shanti S. Gupta
J. Hannan
Wassily Hoeffding
Gopinath Kallianpur
M. G. Kendall
Howard Levene
William G. Madow
Henry B. Mann
J. Marschak
Albert W. Marshall
M. V. Menon
Sigeiti Moriguti
Jerzy Neyman
George E. Nicholson, Jr.
Ingram Olkin
R. N. Pendergrass
Herbert Robbins
Joan Raup Rosenblatt
S. N. Roy
A. E. Sarhan
S. S. Shrikhande
Walter L. Smith
J. L. Snell
Milton Sobel
Herbert Solomon
Charles Stein
Milton E. Terry
John W. Tukey
S. S. Wilks
R. E. Williamson
J. Wolfowitz
Marvin Zelen

PART I

AN APPRECIATION

1 HAROLD HOTELLING

WILLIAM G. MADOW

Stanford Research Institute and Stanford University

Harold Hotelling has been the friend, and teacher or colleague, of all who have contributed to this volume. The volume is not, however, only a recognition of his friendship, or of his freely given help, or of the hours he has spent on others' problems, or of pleasant discussions of mutual interest. Rather, it is an acknowledgment that through his dedicated teaching, leadership in statistical research and organization, and vigorous advocacy of wide-ranging applications of modern statistical ideas and methods, Harold Hotelling has had a major role in the transition of statistics from its position in the early years of this century to the rapidly growing and vital field that it is today.

After taking his A.B. and M.S. degrees at the University of Washington, and his Ph.D. at Princeton University, he went to Stanford University, where he was Research Associate in the Food Research Institute from 1924 to 1927 and Associate Professor of Mathematics from 1927 to 1931. In 1929 he visited R. A. Fisher's laboratory at the Rothamsted Experiment Station.

During this same period, in addition to basic research in mathematics, mathematical economics, and applications of statistics, he made several pioneering contributions to statistical theory. He began the rigorous mathematical proofs of some fundamental results in statistical inference (1930 [2])[1]. The first demonstration that the standard error of estimate in regression analysis depended on the value of the independent variables and the use of the standard error to obtain the curved fiducial or confidence region for regression functions appeared in a paper written jointly with Holbrook Working (1929 [1]). The path-breaking paper (1931 [5]) in which he generalized Student's ratio for vector variates was the first in which the distribution of random vectors with dependent components was treated. The paper on differential equations with random coefficients (1927 [2]), although written with population estimates in mind, is of interest today in stochastic processes and in applications in the physical sciences and engineering.

In 1931, upon the retirement of Henry Ludwell Moore, a leading mathematical economist, Hotelling was called to Columbia University as Professor

[1] Dates and numbers refer to Hotelling's publications listed in the Bibliography, the fourth paper of Part I.

of Economics to develop the existing work in mathematical economics and to initiate a program in mathematical statistics. During the next fifteen years, under Hotelling's leadership, Columbia University became a major statistical center, not only for predoctoral but also for postdoctoral study and research both by statisticians and by those specializing in other areas. His influence on students and staff thus extended far beyond those majoring in statistics.

During these years, his interest in the status of statistics in the university began to be evident. (This aspect of Hotelling's activities as well as his development of students is considered further in the following article by J. Neyman.)

In 1942, at the outbreak of the Second World War, Hotelling organized at Columbia University the Statistical Research Group, which made important contributions, not only in the applications of statistics but also in the development of statistical theory.

Much of Hotelling's research while at Columbia University was in the area of multivariate statistical analysis. Among his papers may be especially cited those dealing with principal components (1933 [1] and 1936 [4]) and those dealing with relations between vector variates (1936 [5] and 1935 [2]).

Besides his contributions to multivariate statistical analysis, several other papers published during this period may be selected for particular mention. In the work that he did on the rank correlation coefficient with Margaret Richards Pabst, he made one of the early and still-important contributions to distribution-free tests (1936 [1]). With Lester R. Frankel (1938 [1]), he considered the general problem of transforming statistics to simplify their distributions. He published a basic paper (1939 [1]), in which he continued work begun earlier (1927 [1]) on the geometric method of approximating probabilities of the extreme tails of distributions of statistics. Further work in this area is indicated by an abstract published in 1947 (1947 [3]). He formulated and shed light on the difficult problem of the choice of one or more predictors from a set of predictors (1940 [1]). Problems of the determination of the maximum of a function that are related both to response surfaces and to stochastic approximation were also treated (1941 [1]). A paper (1944 [1]) on the design of weighing experiments has led to much further research. In this period, also, Hotelling published his first fundamental paper on the teaching of statistics (1940 [2]).

In 1946, Hotelling went to the University of North Carolina to serve as Associate Director of the university-wide Institute of Statistics and to organize the Department of Mathematical Statistics at Chapel Hill.

While at the University of North Carolina, Hotelling has published in the area of multivariate quality control (1947 [2], 1951 [3], 1955 [4]), in which he has considered, among other multivariate generalizations of the analysis of variance, the two-sample generalized T^2 problem. He has further developed his views on the teaching of statistics (1948 [1], [2]) and on the place of statistics in universities (1949 [4]; 1950 [2], [4]). His research on the correlation coefficient, done over many years, has been published (1953 [1], 1955

[2]). Problems of calculation involving matrices have also received his attention (1949 [3], 1950 [1]), as have the relationships between multivariate statistical analysis and factor analysis (1956 [1], 1957 [1]).

Hotelling has always been an inspired teacher who early motivates his students to read original research papers and to undertake their own research. He has set the highest standards, not only by his own research but also by his recommendations on the organization and applications of statistics. He has published in applied statistical research and has acted as consultant widely in industry and government in order to bring statistical research to the attention of those who might benefit from it.

He has, of course, contributed much to professional groups, both as a member of committees and as an executive. He has been an associate editor of several journals. During trips in Europe and India he has lectured at various national and international gatherings. He has received many honors, among which may be mentioned the honorary LL.D. from the University of Chicago, an honorary fellowship in the Royal Statistical Society, and the honorary presidency of the Indian Statistical Congress of 1940.

The impact of Hotelling's work has been continued and extended through his students who are active in statistical work in universities, industry, and government.

The years have seen the fruition of much of his work in research and in the teaching and organization of statistics throughout our nation. The number of departments of statistics and central statistical groups consisting of research workers in statistics grows steadily. Statistical research itself grows in volume and importance as the fields he initiated are further developed, and as additional statistical areas and models are formulated. More and more, statistical work in government and industry is being done by well-trained statisticians.

The contributors to this volume know they speak not only for themselves but also for Harold Hotelling's other students and fellow workers in offering this collection of papers to him on his sixty-fifth birthday, as a symbol of their devotion and gratitude.

2 HAROLD HOTELLING—A LEADER IN MATHEMATICAL STATISTICS

JERZY NEYMAN

Miller Institute for Basic Research in Science,
University of California, Berkeley

The occasion of the present jubilee volume for Professor Hotelling recalls a number of incidents occurring in the last two decades. When combined, these incidents form a chapter in the history of the development of mathematical statistics in the United States. What a change from 1938 to 1958! Volume 9 of the *Annals of Mathematical Statistics*, published in 1938, has only 313 pages. In comparison, Volume 29, published in 1958, has 1308 pages, more than four times as many. To my knowledge, twenty years ago there were no separate departments in American universities cultivating mathematical statistics and only in a few exceptional cases were there even modest programs of courses given in this discipline. Currently, there are a number of excellent departments in this country with well-developed systems of courses in probability and mathematical statistics. Also, there is an unprecedented demand for specialists in these domains on the part of many colleges and universities anxious to establish such programs of instruction.

Not only has the volume of statistical activity increased tremendously over the last twenty years, but also there has been an outstanding improvement in quality. The comparison of the current issues of the *Annals* with those of the late 1930's reveals that mathematical tools now commonly used by probabilists and statisticians were not generally available to the earlier authors. As a result, the problems studied now are frequently more delicate than those of two decades ago and, even where they are essentially the same, the current treatment is both more general and more precise.

How was all this progress achieved? One possible answer is that results of theoretical investigations have gradually percolated among the readers of the *Annals* and similar journals. However, this simple explanation is not sufficient because it presupposes the existence of a considerable number of individuals in the country who combined broad mathematical preparation with a sufficient interest in problems of statistics so that they would read the relevant journals and become actively involved in further research. Twenty years ago there were very few such persons. In order that the remarkable progress noted could become possible, something must have occurred to influence strongly the recruitment of persons capable of absorbing the material published in methodological research journals, capable of noticing such inadequacies as were present (that is, of noticing new problems), and capable of offering improvements. The phenomenon is somewhat akin to that in epidemiology, governed by the famous threshold theorem: unless the concentration of susceptibles is above a certain level, the introduction of infection will not cause a widespread epidemic.

Therefore, in trying to explain the remarkable growth of research in the theory of statistics, one must look for factors which created the "susceptibles." Undoubtedly, there were many such factors, including demands from more advanced scientific and industrial groups and the organization and activities of the Institute of Mathematical Statistics. However, two single factors seem to dominate all the others. They are the educational and organizational activity of Harold Hotelling (and of a few others, such as S. S. Wilks), marked by an outstanding event in 1940 [1], and the appearance in 1946 of an excellent book [2] by Harald Cramér.

In the 1930's, two major obstacles hampered the development of mathematical-statistical research. On the one hand, there was opposition on the "mathematical front," represented by the university departments of mathematics. On the other hand, resistance was encountered on the "applied front," from the university representatives of sciences in which statistical methods were frequently used.

The existence of the mathematical front was wholly natural. It resulted partly from the general character of the contemporary statistical literature and partly from the degree of information on the subject and attitudes among mathematicians. One has to admit that the bulk of the statistical literature of the time could not compare in precision and generality with that which was customary in pure mathematics. Thus, a specialist in abstract algebra or theory of functions, in differential geometry or theory of measure, when confronted with an average article on statistics, would immediately see its inadequacies. After a few similar experiences, he would conclude that the whole of statistical literature was no more than an accumulation of pompous verbosity produced by individuals whose mathematical education, not to speak of talent, lagged behind their academic ambitions. Unfortunately, by and large, opinions of this kind were justified, and the resistance exhibited by many first-class mathematics departments to the establishment of some sort of program in mathematical statistics should not be surprising.

The opposition on the applied front within the universities was based on entirely different grounds. The intensive publicizing of statistical methods by Karl Pearson and his school had firmly established the view that statistical methods are indispensable in a number of fields. However, the opinion prevailed that all a scientist needed to know in statistics was how to compute a few popular formulas; a "measure of central tendency," a standard deviation, a "Pearson r." All this could be learned without significant mathematical preparation. Also, once one had acquired the computational technique, one could successfully teach it to others. Thus, why should one bother about deeper studies involving elaborate mathematical developments?

With opposition on these two fronts, the situation was difficult. Also, it was not easy to remedy. For example, there was embarrassment in connection with the publication of papers on statistical theory. The statistical journals would not have them because they were "too mathematical." Also, the mathematical journals were not enthusiastic because the subject matter did not exactly fit the journals' programs and because it was under general suspicion.

Harold Hotelling was victorious on both fronts. His success on the mathematical front was due both to his own excellent research and to his unusual gift of identifying real talent and of attracting it to the general sphere of his interests. Hotelling's first success of this kind seems to have been J. L. Doob, then a young Ph.D. from Harvard, who, as a National Research Fellow, worked at Columbia and was attracted to the problem of consistency of maximum likelihood estimates. Incidentally, this occurrence was the more remarkable because of the somewhat paradoxical position of Hotelling, then Professor of Economics in the Faculty of Political Science.

After Doob there came many individuals whom Hotelling had identified as having talent. Frequently supported by grants from the Carnegie Foundation, these men gathered about him in the Faculty of Political Science and did excellent research in the mathematical theory of statistics: M. A. Girshick and W. G. Madow, Abraham Wald and Henry Mann, A. H. Bowker and Herbert Solomon. During the war the original group of talented individuals about Hotelling at Columbia partly dispersed. However, before very long, the National Defense Research Committee set up a new and much larger group, known as the Statistical Research Group, Columbia University, for which Hotelling selected the key figures, W. Allen Wallis and Jacob Wolfowitz.

When Hotelling departed from Columbia University in 1946 to assume his new position at the University of North Carolina, the ground was well prepared at Columbia for the creation of an outstanding Department of Mathematical Statistics. Abraham Wald became its first chairman and had no difficulty. In the meantime, Hotelling's ability to identify talent was again demonstrated by his selection of a staff for his new statistical research and teaching center at Chapel Hill, North Carolina. His selection included R. C. Bose, W. Hoeffding, W. G. Madow, G. E. Nicholson, Jr., H. E. Robbins, S. N. Roy, Walter L. Smith, and J. Wolfowitz. In addition, there is a long list of persons who began their allegiance to statistics either as students or under the influence of Hotelling and who have now achieved important positions or are on their way to achieving them. These include I. Blumen, R. A. Bradley, G. L. Burrows, U. Chand, W. L. Deemer, Jr., R. Dorfman, M. Dwass, M. Friedman, L. Frankel, S. G. Ghurye, D. M. Gilford, G. B. Kallianpur, W. H. Kruskal, J. Mandel, S. Moriguti, D. N. Nanda, I. Olkin,

W. R. Pabst, D. S. Penfold, W. Vickrey, J. W. Walker, W. A. Wallis, and W. J. Youden.

At the time of their first contact with Hotelling, many of these individuals were students or beginning scientists. Later they developed into outstanding personalities. As they dispersed into responsible teaching positions all over the country, they created the "density of susceptibles" necessary for the present grand-style epidemic-like development of mathematical-statistical research. The coincident publication of Cramér's book played an outstanding role in this process. This book presented a systematically organized collection of the results in algebra, analysis, and probability, basic for research in statistics; it established a novel standard of mathematical rigor and generality and clarified a number of ideas.

The development and dissemination of the excellent theoretical results of Hotelling himself, and of the many talented authors collected and inspired by him, contributed considerably to the establishment of statistics as a respectable mathematical discipline. An official mark of recognition of this respectability was the publication of a series of statistical papers by Hotelling, Doob, Madow, and Wald in the *Transactions of the American Mathematical Society*.

Hotelling's battle on the applied front was of a different character. This was a battle not for advancement in the scientific level of a young discipline or for recognition of its respectability, but for competence in teaching within American universities. It was a battle for progress in academic life. This battle was difficult and complicated because it involved the vested interests of the adversaries, the so-called applied statisticians. Many in number and located in many substantive departments, they taught an unbelievable number of "cookbook" courses. For them the suggestion of a program of instruction in statistical theory, of which they were ignorant, constituted a threat and an insult. It is not surprising that such suggestions met with formidable opposition when made in the various academic councils and committees.

In spite of all their strength within university committees, the applied statisticians had a weakness: the position they defended was intrinsically untenable and, therefore, they were open to public exposure and ridicule. Such an exposure came about at a meeting of the Institute of Mathematical Statistics held in September, 1940, when Harold Hotelling delivered his famous paper "The Teaching of Statistics." He described the contemporary system of statistical instruction and of recruitment of university teachers in this subject, and then presented his own views on what the system should be. In drawing the picture of the prevailing system, Professor Hotelling was being no more than frank and realistic. Yet, just because of these qualities, the article revealed all the striking extravagance of the situation. Hotelling's paper was received with enthusiasm and, by a unanimous vote, the audience decided to have it published in the *Annals* as an expression of IMS opinion on the matter.

The paper was published in Volume 11, 1940, and had a most salutary effect. Its moderate tone, combined with the extravagance of the picture drawn and the spontaneous support of a national scholarly society, made the authorities of several universities suspect that there was something unsound in their practices regarding the teaching of statistics and induced them to do some checking. In my opinion, the establishment of the excellent departments of statistics in such universities as Chicago, Columbia, Harvard, North Carolina and Stanford is due to a considerable extent to the publication of Hotelling's article.

In the preceding description of the "battle on the applied front," I used the past tense: "There was something unsound. . . ." This, of course, is as it should be; the past tense represents the truth. Unfortunately, it does not represent the whole truth. The whole truth is that the battle on the applied front is not yet finished and that, in spite of Hotelling's article and in spite of many subsequent relevant publications (see, for example, [3] and [4]), the current practices of offering statistical courses in substantive departments are only too often the same as those described and ridiculed by Hotelling. Frequently, this is true even in those universities in which there is an active department of statistics. In other words, at present the article "The Teaching of Statistics" published by Harold Hotelling in 1940, possibly with some minor changes in detail, is just as realistic, fresh, and useful as it was at the time of its first appearance. For this reason, the editors of the present volume agreed at my suggestion to republish this particular article, both as a tribute to its author and as a contribution to the further development of our discipline.

References

[1] Hotelling, H. The teaching of statistics. *Ann. Math. Stat.*, 1940, **11**, 457–70.

[2] Cramér, H. *Mathematical Methods of Statistics.* Princeton, N.J.: Princeton Univ. Press, 1946.

[3] National Research Council. Personnel and training problems created by the recent growth of the applied statistics in the United States. A report of the Committee on Applied Mathematical Statistics. *National Research Council Reprint and Circular Series*, No. 128, 1947.

[4] Institute of Mathematical Statistics. The teaching of statistics, A report of the IMS Committee on the Teaching of Statistics. *Ann. Math. Stat.*, 1948, **19**, 95–115.

3 THE TEACHING OF STATISTICS[1]

HAROLD HOTELLING

The very great increase in the teaching of statistics since the First World War has been associated on one hand with the development of statistical theory. This important series of discoveries has made available more and more powerful and accurate statistical methods, and has also acquired an intellectual interest of its own as embodying the modern version of the most important part of inductive logic and as providing scope for mathematical and logical ingenuity of high order. The increased teaching of statistics has also been associated with the rapidly growing applications of statistics in innumerable fields, made possible by the development of the theory, by the availability of persons having some knowledge of the theory, and by an increasing realization of the possibilities of application. Doubtless most students of statistics enter upon the subject, not for its intrinsic interest, but with the idea of applying statistical methods as a tool to some particular end. This object may be scientific research, or to fulfill a requirement for a degree, but is often connected with some purely practical pursuit offering the ready prospect of a remunerative job. But it would be a mistake to ignore those whose interest is more purely intellectual, who desire an insight into the peculiar problems of probable inference and the structure of empirical knowledge, who wish to get a fundamental acquaintance with one of the most fundamental of subjects, to see and understand fully the mathematical derivations underlying so much practical and scientific activity, and perhaps to make their own contributions.

Of the magnitude of the demand for statisticians there can be no doubt. The realization of what statistical methods can do in a multitude of fields has gradually led the administrators of government agencies, directors of scientific organizations and research institutes, and business men, to employ rapidly increasing numbers of persons with some knowledge of statistical methods, and to accord an unusual degree of recognition and promotion in many such cases. The uses of statistical methods, and especially of sampling

[1] Address at the meeting of the Institute of Mathematical Statistics at Hanover, N.H., September 10, 1940, while Professor Hotelling was a member of the faculty of Columbia University. Published in *Annals of Mathematical Statistics*, 1940, **11**, 457–70. Reprinted with permission of the Institute of Mathematical Statistics.

theory, are so varied that it is scarcely possible in a brief space to give any sort of survey of them. They enter, in one form or another, into the research work of the physicist, the chemist, the astronomer, the biologist, the psychologist, the anthropologist, the medical investigator, the economist, and the sociologist. Meteorology, which has lately acquired greatly increased importance, both civil and military, is with its masses of numerical observations very much a statistical matter. The engineer needs modern statistical methods both in the physical and in the economic aspects of his plans. The work of W. A. Shewhart has made clear the central importance of sampling theory in the economic control of quality of manufactured articles. Business men who use sampling surveys to test the markets for their products and the effectiveness of their advertising, who employ statisticians to make up index numbers and forecasts of business conditions, and whose manufacturing costs and quality are controlled with the help of recently devised statistical methods, are finding more and more uses for statisticians. Indeed, it seems as if the exploitation of the business and manufacturing possibilities of statistical methods has only begun, and that limitless further fields are coming into view. Insurance has of course always been essentially dependent on statistics.

But the most rapidly growing large class of positions for statisticians is at present in governmental activities. For some facts regarding the employment of statisticians by the federal government I am indebted to Dr. J. M. Thompson. It appears that it has about one hundred agencies using statistics, with almost eight hundred positions broadly classified as statistical or mathematical, in addition to more than six thousand generally classified as economists. The title "economist" covers many types of work, but much of it is largely statistical. The nature of the government's statistical work is varied and extensive. It includes such work as forecasting revenue from taxes, prices and production of agricultural commodities, general demand conditions, and weather. Some of the work consists in analyzing the effects of various taxes on other programs. In connection with proposed legislation, statisticians serving the lawmakers often attempt to outline the probable results of the legislation, as well as to assist in setting up definite formulae for carrying out the general policies aimed at in Acts of Congress. Administrators as well as lawmakers require statistical activities of a high order, exemplified in the Bureau of the Census, the Bureau of Agricultural Economics, and others. The scientific activities of the government, the work of the War Department, and many others that do not at first sight appear at all statistical, require the services of mathematical statisticians of high order. Even the judicial activities call for statistical theory of some of the most recently discovered kinds, as for instance in the investigation recently made of parole procedures. Cities and states, school and port authorities, employ numerous statisticians for other and widely diverse purposes.

The growing need, demand, and opportunity have confronted the educational system of the country with a series of problems regarding the teaching of statistics. Should statistics be taught in the department of agriculture, anthropology, astronomy, biology, business, economics, education, engineering, medicine, physics, political science, psychology, or sociology, or in all these departments? Should its teaching be entrusted to the department of mathematics, or to a separate department of statistics, and in either of these cases should other departments be prohibited from offering duplicating courses in statistics, as they are often inclined to do? To what students, and at what stage of their advancement, should a course in statistics be administered? Should there be mathematical or other prerequisites? How much of an investment in a statistical laboratory is warranted? Should courses be primarily theoretical, and mathematical, or should they be made as practical as possible, equipping the student in the shortest possible time for a job as statistician, or for statistical work in the field with which a particular department is concerned? What about degrees in statistics? Eclipsing all these in importance, though it seems to have received too little of the attention of college and university administrative officers, is the question: What sort of persons should be appointed to teach statistics?

To pressing practical problems answers are sure to be given either by considered policy or by processes of historical evolution. The latter are the more prominent in explaining the statistical teaching we have had. A synoptic picture of the origins, not many decades ago, of a good deal of it would perhaps be something like this. A university Department of X, where X stands for economics, psychology, or any one of numerous other fields, begins to note toward the end of the pre-statistical era that some of the outstanding work in its field involves statistics. The quantity and importance of such work are observed to increase, while at the same time its intelligibility seems to diminish. Evidently students turned out with degrees in the field of X who do not know something about statistics are going to be handicapped, and are not likely to reflect credit on Alma Mater. The department therefore resolves that its students must acquire at least an elementary knowledge of the fundamentals of statistics. To implement this principle, it perhaps inserts some acquaintance with statistics among the requirements for a degree. This situation naturally calls for the introduction of a course in statistics. Accordingly the head of the Department of X, in preparing the next Announcement of Courses, writes:

> "X 82. Elements of Statistics. An elementary but thorough course designed to acquaint students of X with the fundamental concepts of statistics and their applications in the field of X. The viewpoint will be practical throughout. Second semester, MWF at 10.
>
> "Instructor to be announced."

The problem now arises of finding someone to teach the new course. The few well-known statisticians in the country have positions elsewhere from which it would be impossible to dislodge them with the bait to be offered;

for though the department wishes to have statistics taught as an auxiliary to the study of X, it feels that there must be no question of the tail wagging the dog, and that economy is appropriate in this connection. The members of the department of professorial rank do not respond favorably to the suggestion that they should themselves undertake to teach the new and unfamiliar course. But every university department has a bright graduate student whose placement is an immediate problem. Young Jones has already demonstrated a quantitative turn of mind in the course on Money and Banking, or in the Ph.D. thesis on which he has already made substantial progress, dealing with The Proportion of Public School Yard Areas Surfaced with Gravel. He may even recall having had a high-school course in trigonometry. His personality is all that might be desired. He is a white, Protestant, native-born American. And so the "Instructor to be announced" materializes as Jones.

This earnest young scholar now finds that, in addition to completing his thesis, he must look up the literature of statistics and prepare a course in the subject. His attention is directed by older members of the department to some of the research papers in the field of X involving statistics. He pursues "statistics" through the library card catalog and the encyclopedias. He reads about census and vital statistics, price statistics, statistical mechanics. Perhaps he encounters probable errors. Eventually he learns that Karl Pearson is the great man of statistics, and that *Biometrika* is the central source of information. Unfortunately most of the papers in *Biometrika* and of Pearson's writings, while not lacking in vigor, trail off into mathematical discourse of a kind with which young Jones feels ill at ease. What he wants is a textbook, couched in simple language and omitting all mathematics, to make the subject clear to a beginner. Perhaps he finds the impressive books of Yule and Bowley, but decides that they are too abstruse. Elderton's "Frequency Curves and Correlation" is far too mathematical. Jones decides that a simple book on statistics must be written, and that he will do it if he can ever succeed in mastering the subject. In the meantime, he contents himself perforce with the less mathematical writings of Karl Pearson, with applied examples in the field of X, and with such nonmathematical textbooks as may have been written by other young men who have earlier trod the same path as that on which Jones is now beginning. Somehow or other he gets the class through the course. After doing this two or three times, Jones is an experienced teacher of statistics, and his services are much in demand. His course expands, takes on a settled form, and after a while crystallizes into a textbook. At the same time he may be getting out some research, consisting of studies in the field of X in which statistical methods play a part. His promotion is rapid. He becomes a Professor of Statistics, and perhaps an officer in a national association. His textbook has a large sale, and is used as a source by other young men writing textbooks on statistics.

The textbooks written in this way form an interesting literary cycle. Measures of "central tendency" and of dispersion are introduced, and the use of one as against another of these measures is debated on every ground except the criterion that modern research has shown to be the important one, the sampling stability. Sampling considerations, indeed, get little attention. The urge to simplify by leaving out the more difficult parts of the subject, and especially the mathematical parts, is accompanied by pride in the great number of examples drawn from real life, that is, actual data that have been collected.

But the most fascinating feature of this literary cycle is the opportunity it offers for research by the standard methods of literary investigation, tracing the influence of one author upon another through parallelism of passages, and so forth. This study is facilitated by the accumulation of errors with repeated copying. One outstanding example is in certain formulae connected with the rank correlation coefficient, derived originally by Karl Pearson in 1907 and copied from textbook to textbook without adequate checking back. As one error after another was introduced in this process, the formulae presented to students (and apparently made the basis of class exercises involving numerical substitution) became less and less like Pearson's original equations. Incidentally, in trying to check this original work of Pearson's, recent investigation has raised the suspicion that it is erroneous; at any rate, he does not give a fully adequate argument. Thus it may be that errors in copying, which are so useful in examining the history of statistics, never did any harm. The formulae in which the students were drilled may have been no worse than they would have been if all the copying had been done with more care.

While this process has been going on in the Department of X, the Y and Z Departments have likewise evolved the teaching of statistics. There is some interchange of ideas between the various statisticians on the campus, and there is a catholicity in the copying of textbooks. But by and large, statistics is regarded in the Economics Department as a branch of economics, in the Psychology Department as a part of psychology, and so forth. The astronomer is inclined to resent the suggestion that his students should be called upon to study their least squares with anyone but an astronomer. Medical and biological investigators suspect Economics and Psychology of charlatanry, and do not look with favor on the idea of turning their own students over to such departments for instruction in statistics. Most unthinkable of all would be putting the Department of Education in charge of an essential part of the training of scientific students. Thus the courses multiply.

The fact that it is essentially the same fundamental subject that is being taught under various names and with various kinds of notation in different departments is often concealed by including the teaching of statistical theory in a course whose title and prospectus are more suggestive of applications. A case in point is that of an economist of my acquaintance, not primarily engaged in teaching, who some years ago was invited to give a course in Price Forecasting in the Economics Department of a leading university. He carefully prepared a series of lectures on this subject, which had been the center of some extended research he had conducted. A large class enrolled for the course. But soon after beginning his series of lectures the economist noticed that the class was growing restive. Upon inquiring what was amiss, he learned that his discourse was unintelligible to many of them because he was using technical statistical terms and concepts with which they were not familiar. He thereupon undertook to use simpler language, and when this did not suffice to convey his meaning, to explain the statistical notions involved in his work on price forecasting. More and more his lectures came to deal with the elements of statistics, and less and less with price forecasting. At the end of the term he felt that he had given the students some elementary knowledge of statistical theory, for which they had not enrolled and for which he did not feel particularly well qualified, but had taught them virtually nothing about price forecasting. When the invitation was repeated the next year, the economist suggested imposing a course in statistics as a prerequisite for the course in Price Forecasting. This however was vetoed by the head of the Economics Department, who did not believe in prerequisites. The Price Forecasting course was not repeated.

This incident illustrates the evolution of a good deal of statistical teaching. At the beginning, the idea is to teach some application, but the teacher soon finds himself engaged at much more length than expected with the fundamentals of statistical theory and methods. In this way it has come about that a large number of persons are teaching theoretical statistics who initially had no intention of doing so, but were concerned with particular applications. The teaching of statistical theory has been undertaken belatedly and inexpertly because it was necessary to a discussion of some application originally in view. Thus it happens that a good deal of teaching of statistics, even of mathematical statistics, masquerades as something else.

The obvious inefficiency of overlapping and duplicating courses given independently in numerous departments by persons who are not really specialists in the subject leads to the suggestion that the whole matter be taken over by the Department of Mathematics. This is a promising solution, but it is doomed to failure if, as has sometimes happened, it means that the teaching of statistics is put under the jurisdiction of those who have no real interest in it. Moreover the teaching of statistics cannot be done appreciably better by mathematicians ignorant of the subject than by psychologists or agricultural experimenters ignorant of the subject. The latter indeed have a certain advantage in that the problems seem more real and definite to them; they can sense the difference between the important and the unimportant questions, even if they cannot express the questions in clear mathematical language, and can sometimes arrive intuitively at a correct result that leaves the mathematician puzzled. Also, they can understand more readily than can the mathematician the examples, drawn largely from biological material, which play so important a part in some of the leading expository work on statistics, such as R. A. Fisher's *Statistical Methods for Research Workers*. The pure mathematician has only one advantage over the non-mathematical worker in empirical fields: he is able to set about reading the serious literature of statistical theory. But he must still find this scattered literature, sort it out from a mass of rubbish, fallacies, and false starts, and trace it back historically until he can understand the notation and the presuppositions. He must also contend with the fact that a good deal that is important in statistics is still a matter of oral tradition, and some consists of laboratory techniques. In short, he needs a teacher before he himself sets out to teach the subject. When a Department of Mathematics calls in a young Ph.D., however brilliant, to teach statistics as a part or all of his program, the best thing it can do, if he has not already had a training in modern statistics, is to give him a furlough for a year or two to enable him to go where he can acquire such a training.

Qualifications of a good teacher of statistics include, first and foremost, a thorough knowledge of the subject. This statement seems trivial, but it has been ignored in such a way as to bring about the present unfortunate situation. Mathematicians and others, who deplore the tendency of Schools of Education to turn loose on the world teachers who have not specialized in the subjects they are to teach, would do well to consider their own tendency to entrust the teaching of statistics to persons who not only have not specialized in the subject, but have no sound knowledge of it whatever. A knowledge of theoretical statistics is not easy to obtain. There is no comprehensive treatise on the subject, starting from first principles and proceeding by sound deductions and well-chosen definitions to the methods that need to be used in practice. (I have been trying for years to write such a treatise, but it has turned out to be a bigger task than at first appeared. This is partly because some things formerly thought to have been proved turn out, on critical examination, not to be sound, and much new research has been necessary.) The literature is scattered through journals pertaining primarily to many kinds of applications, and it is only in recent years that any large proportion of the current contributions to statistical theory and methods have been gathered into a few periodicals devoted to statistical theory. On the other hand, the seeker after truth regarding statistical theory must make his way through or around an enormous amount of trash and downright error. The great accumulation of published writings on statistical theory and methods by authors who have not sufficiently studied the subject is even more dangerous than the classroom teaching by the same people.

A good teacher of statistics needs of course a mathematical background, including at least an acquaintance with the theory of functions and n-dimensional euclidean geometry. A good deal of additional algebra and analysis are likely to be helpful, as well as some differential geometry. But no amount of such mathematics constitutes by itself any approach to sufficiency in the qualifications of a teacher of statistics. The most essential thing is that the man shall know the theory of statistics itself thoroughly from the ground up, including the mathematical derivations of proper methods and a clear knowledge of how to apply them in various empirical fields. In addition to the pure mathematics and the knowledge of statistical theory, a competent statistician or teacher of statistics needs a really intimate acquaintance with the problems of one or more empirical subjects in which statistical methods are applied. This is quite important. Sometimes excellent mathematicians have wasted time and misled students through failure to get that feeling for applications that is necessary for proper statistical work.

The theory of statistics has been making advances so rapid and so fundamental that some of the first things that need to be said in an elementary course, even for prospective practical statisticians, are affected by some of the most recent researches. So elementary a question as "What definition is it wise to give to the term 'standard deviation'?," which must be faced by every teacher of Statistics 1, requires for an intelligent answer a rather thorough understanding of modern sampling theory and techniques. The answer, it now seems, is *not* the definition given in most textbooks. In the selection of a statistic to represent a parameter, for example in fitting frequency curves or in linkage estimation in genetics, the fundamental consideration is connected with the sampling distribution, as R. A. Fisher showed in founding the modern theory of estimation. This is ignored in most of the current teaching of statistics, with the result that innumerable students are sent out to waste the money and time of their employers by demanding larger samples than are necessary for the purposes in view, wasting costly information by calculating inefficient statistics and using tests that are not the most powerful. On the other hand, students of statistics who are taught rule-of-thumb methods without their derivations are never quite conscious of the exact limitations and assumptions involved, and may make unwarranted inferences from samples that are too small or in some way violate the conditions underlying the derivations of the formulae.

A good teacher of statistics must be thoroughly familiar with these recent advances. He must examine very critically textbook statements unsupported by full proofs. Even though the students are not capable of following the complete mathematical argument—indeed, especially if the students are not to examine it—the instructor needs to give it a critical study. The custom of omitting proofs, which would not be tolerated in pure mathematics beyond a very limited extent, is common in the teaching of statistics, and is excused on the ground that the students do not know enough mathematics to understand the proofs. Perhaps in some cases a better reason is that the teachers, and the authors of the textbooks, do not understand the proofs. In some instances no proofs exist, and in some instances no genuine proofs can exist, because the methods taught are demonstrably wrong. The custom prevalent in the teaching of mathematics of going over each proof carefully in the class is, among other things, a safeguard against infiltration of false propositions. This safeguard is missing from most of the teaching of statistics, and there has been an infiltration of errors. Since it is accepted that a great many students need to learn something about statistical methods without learning enough mathematics to understand the proofs, it follows that the elementary teaching of statistics to these students must, if the perpetuation of gross errors is to be avoided, be in the hands of really competent mathematical statisticians. This is perhaps the greatest reform needed in the teaching of statistics today. Until the *elementary* teaching of statistics is conducted by those with a thorough and critical knowledge of current research in statistical theory, of a sort that seems virtually inseparable from participation in that research, there is likely to be a continuation of the laborious drilling of thousands of students in methods that ought never to be used. Here, of all places, is the great need for participation of research workers in elementary teaching.

Teachers and textbook writers might well abandon the idea of telling what statistical methods are used, and say instead what methods ought to be used. But before they can do this with confidence they must have a very close acquaintance with the research of the last three decades in statistical theory.

How can an appointing officer know whether a prospective teacher of statistics knows his subject? This question requires no answer peculiar to statistics in distinction from other subjects. Publication of research, constituting a contribution to the particular field, has always been accepted as the best proof. A substantial contribution to fundamental statistical theory, which is to be distinguished from the mere application of known statistical methods to empirical data, is the best indication of the kind of scholarship appropriate to a teacher of statistics.

Participation in research is not novel as a criterion of what constitutes a good teacher of a college or university subject, if the subject is Greek literature, physics, chemistry, biology, or indeed any of those departments that have been long enough established to attain with respect to the organization of their teaching a state approximating equilibrium. The more reputable institutions of higher learning have long maintained the principle, though with occasional violations in practice, that the Ph.D. degree or its equivalent, representing among other things the completion of a piece of scholarly research, is a minimum condition for a regular faculty appointment. It has usually been maintained also that the Ph.D. thesis should be a new contribution of a strictly scholarly character to the field of the scholar's competence, and not merely a routine application of known methods to an extraneous field. Thus a thesis offered for the Ph.D. degree in mathematics would be judged by its contribution to mathematics, rather than to physics or accounting. Moreover the regard in which universities have held members of their faculties has been intimately connected with their output of scholarly research. Other criteria of excellence have not been ignored, but research has been recognized in a fairly consistent manner. Some say that there has been an over-emphasis on research, and that more attention ought to be given to other qualities related to teaching. However this may be, the facts remain that scholarly research is something capable of a reasonably objective evaluation by scholars in the field, that it offers the main hope of fundamental progress, and that familiarity with current research is a necessary, though not sufficient, condition for the most important teaching in institutions of higher learning.

A peculiarity of the teaching of statistics, of which in practice the theory of statistics is an essential even if unacknowledged part, is that a good deal of it has been conducted by persons engaged in research, not of a kind contributing to statistical theory, but consisting of the application of statistical methods and theory to something else. A similar situation would exist if the teaching of mathematics were in the hands of an assortment of various kinds of engineers, or if zoology and botany were taught by practicing physicians. The teaching of mathematics and of elementary biology might perhaps gain in liveliness and concreteness by such arrangements, with the accompanying emphasis on the particular applications of the fundamental sciences. Moreover the engineer might in the course of such teaching refresh his own knowledge of elementary mathematics, while the physician might gain by renewing his acquaintance with elementary biology. Such arrangements might occasionally be made with profit. But if they were the general rule the advantages of specialization would be lost; the fundamental sciences would not be developed in so well-rounded a manner as they are by specialists in them, while the special skills and knowledge of the physician and engineer could not be utilized to the full in their respective professions. Statistical theory is a big enough thing in itself to absorb the full-time attention of a specialist teaching it, without his going out into applications too freely. Some attention to applications is indeed valuable, and perhaps even indispensable as a stage in the training of a teacher of statistics and as a continuing interest. But particular applications should not dominate the teaching of the fundamental science, any more than particular diseases should dominate the teaching of anatomy and bacteriology to pre-medical students. These subjects are not ordinarily taught by practicing physicians, but by anatomists and bacteriologists respectively.

In medical education the principle has been accepted, after a long struggle, that a medical school should have full-time professors engaged primarily in teaching and research, and that such professors should not treat patients except in cases of unusual interest from the standpoint of the science or art of medicine. An analogous principle would be that an institution offering extensive instruction in statistics should have full-time professors engaged in the teaching of and research in statistical theory and methods, without spending time over applied statistical problems excepting insofar as such problems might present novel features calling for the development of new statistical methods or theoretical extensions having interest going beyond the immediate case. Sometimes the complaint is heard in medical schools that the teaching tends to become too theoretical on account of detachment from clinical practice, and a similar difficulty might conceivably develop in connection with statistics; but in neither case does the trouble seem to be beyond the ability of the personnel involved to cure if they have the right background.

A specialist in statistics on a university faculty has a threefold function. In addition to the usual duties of teaching and research, there is a need for him to advise his colleagues, and other research workers, regarding the statistical methods appropriate to their various investigations. The advisory function is a highly important one for the activities of the university as a whole, and should be taken into consideration in adjusting the teaching load. Probably every university statistician is visited from time to time by earnest research workers, deeply engrossed in their respective specialities, speaking technical jargons unfamiliar to the statistician, and seeking his advice on matters concerning which he has a sinking feeling of lack of comprehension. After some hours of psychoanalyzing his visitor the statistician may be able to ascertain what it is he *really* wants to know, and thereafter either refer him to some standard formula, or more often, undertake a piece of new mathematical research designed to fit the particular problem, and very possibly having value also for a more extended class of problems. The statistician is then very likely to find himself embarked on a co-operative research venture in a field that is new to him.

To function well in this third, the consultative or co-operative function, he must have an unusually large store of general information. No one stands in greater need than he of that knowledge of "something about everything and everything about something" that was once said to be the goal of a liberal education. In planning the education of statisticians and teachers of statistics these considerations point to a somewhat wider diffusion of studies among various fields than is customary in many institutions, especially in graduate work. The co-operation, and their other work, would also be facilitated if research workers in general were more strongly urged to get a training in mathematical statistics at an early stage in their careers.

The problem of departmental organization is secondary to that of getting men having the requisite qualities of extensive mathematical preparation, a thorough knowledge of modern theoretical statistics, an understanding of some fields at least in which statistical methods can be applied, and the type of inquiring mind sometimes described as a "research outlook." A Department of Mathematics may well handle the fundamental teaching in statistics, provided it has men properly qualified for such teaching. If it does not have such men, its teaching of statistics and its inability to provide the needed statistical advice will inevitably tempt the other departments to set up again their own duplicating courses in what amounts essentially to statistical theory and methods, and to repeat the mistakes of the past.

A separate Department of Statistics, if competently staffed, could very well provide advice for the whole institution as well as conducting elementary instruction in statistical methods and theory, both for students having calculus and for those without it, and should certainly carry on advanced teaching and research in statistical theory and methods. But for efficient functioning of the institution as a whole it should be agreed that the Department of Statistics or the Department of Mathematics should do *all* the elementary instruction in statistics, and that courses in statistics in other departments should be confined to applications of the basic theory. Normally such courses in applied statistics in the other departments should require as a prerequisite one or more of the basic courses in the Department of Statistics, or of Mathematics. The basic course to be required as a prerequisite to others should be the one which itself requires calculus as a prerequisite wherever this is practicable. It is practicable for students of engineering, physics, astronomy, and mathematical economics, since these students must have calculus anyhow. Moreover the value of the sequence consisting of calculus, statistical theory, and applied statistics, in this order, is so great that many other students are likely to avail themselves of it when it is once established and the true nature and value of statistics are more widely understood.

Exactly how far a Department of Statistics should go in particular applications would have to be decided anew from time to time by its members in the light of changing conditions and interests. It cannot teach everything that goes by the name of statistics. This problem may be exemplified by the case of population and vital statistics. This is a field with close connections with sociology, biology, medicine, and insurance. It is cultivated in conjunction with each of these subjects in various places. Some of its most interesting and important phases make use of quite advanced mathematics, as in the work of A. J. Lotka, and in addition there is extensive use, and more extensive need, of the statistical methods centered around sampling theory which are the appropriate domain of a Department of Statistics. Should the study of population and vital statistics be included in a Department of Statistics? I think not, except as a temporary arrangement, or in a small institution, in spite of the history of the word "statistics," which originated in connection with material of this kind, and in one of its meanings is still applied to it. (My use of the unqualified word "statistics" in this paper is in the sense of theory and methods, not in the sense of statistical facts such as those found by the census.) Medical, biological, and sociological considerations are prominent in the problems of vital statistics, and one of these departments might well handle the subject. But the vital statistician, like other research workers, should have acquired in the course of his training an intimate familiarity with the statistical theory and methods which are the appropriate province of a Department of Statistics. He also needs mathematics through integral equations, if he is to understand and extend the contributions of Lotka and Volterra. Students of vital statistics should have had an elementary course in statistical theory in the Department of Statistics, preferably the course requiring calculus.

A course in price statistics should be taught by an economist, presumably in the Department of Economics, but might well require as a prerequisite the same elementary courses in statistical theory and methods as would be required in psychology, medicine, and other fields. In addition, there are problems of time series analysis whose treatment calls for a mathematical statistician having some acquaintance with both economic and meteorological data. A course on the treatment of time series might appropriately be included in the Department of Statistics, requiring the general elementary course as a prerequisite, and itself serving as a prerequisite for courses in economic and meteorological statistics.

One of the chief obstacles to efficient organization of teaching is the habit of not prescribing prerequisites outside one's own department. But when once the elementary courses in statistics have become established in the hands of well-equipped specialists in statistical theory and methods, in whose competence general confidence can be reposed, the various departments of application will lose their motive for establishing their own duplicating courses, and will be able to cultivate more intensively their respective specialities.

The detection of biases and the details of practical statistical work vary greatly from one application to another. These, consequently, are matters for the departments concerned with applications rather than with the fundamentals of statistics, and should not be the chief features of a course in elementary statistical methods and theory. The work of a Department of Statistics should be concerned largely with sampling theory, and should emphasize the unity of statistical methods and theory, regardless of the field of application. It should deal with statistics as a coherent science of inductive inference, of the preparation of observations for inference, and of the planning of investigations so as to yield observations from which inferences can best be made.

The question what mathematical prerequisites should be established for the fundamental course in statistical theory must be answered by a compromise between the ideal and what is expedient at a particular time and place. In Europe a large number of students have had a year of calculus before coming to universities, that is, before reaching the age of eighteen. If a university were willing to restrict its entrants to such students (thus automatically solving the problem of overcrowding) it could give them another year of calculus, mixed perhaps with advanced algebra and geometry, and then in their sophomore year give them a thorough course in elementary statistics and probability, based on calculus. These students would then be ready to tackle advanced statistics in the third year in a really effective way. If the teaching of economic theory, physics, chemistry, and astronomy were geared to this program in such a way as to make real use of the calculus, the work in these subjects could be made far more efficient, in the sense that more material could be covered effectively in the allotted time, or an equivalent amount of material in less time. If, in addition, all the many departments in which statistical methods and theory are used required these statistical courses as prerequisites, and actually used the materials of these courses in their work, there would be a further huge gain in efficiency. The baccalaureate degree of such an institution would represent a far more thorough knowledge, and command of the tools of research, than is possible without an arrangement putting in this way the fundamentals first.

Institutions unwilling to undertake such a drastic improvement must face more or less delay and inadequacy in the acquisition by their students of the fundamentals of mathematics and of statistics. A division of the students into groups according to mathematical ability ought to be undertaken, and followed by a corresponding division of the elementary statistics course. Students having high mathematical ability could begin the study of statistics after completing calculus, and could look forward to rising ultimately to greater heights in pursuits involving mathematical or statistical knowledge than those of lesser mathematical talents. For these latter there would still be the possibility of acquiring, even without calculus, useful statistical tools; it is essential that this should be done under the guidance of instructors thoroughly familiar with the mathematics of statistics. The task of leading the blind must not be turned over to the blind. Students possessing the ability to master the calculus should be encouraged to begin the study of statistics with the course having calculus as a prerequisite, and should not be put into the necessarily slower group not having the calculus. I believe that these elementary courses should begin with the theory of probability, but should go on to the chief distribution functions used in practice, and should include applied problems and work on calculating machines.

Putting a sound program of statistical teaching into effect will take time, partly because of the scarcity of suitable teachers of statistics. Nevertheless, the process is well under way, and the prospects are good for substantial improvements in the teaching of statistics. A body of able young research men possessing the requisite knowledge of statistical fundamentals is now in existence and is growing. Some of the recent textbooks represent striking improvements. The Institute of Mathematical Statistics itself, with the *Annals of Mathematical Statistics*, is perhaps the best evidence of a changed view making for better things.

4 BIBLIOGRAPHY OF HAROLD HOTELLING

1. Publications Other than Reviews

1925

[1] Three-dimensional manifolds of states of motion. *Trans. Amer. Math. Soc.*, **27**, 329–44.

[2] A general mathematical theory of depreciation. *J. Amer. Stat. Assoc.*, **20**, 340–53.

[3] The distribution of correlation ratios calculated from random data. *Proc. Nat. Acad. Sci.*, **11**, 657–62.

1926

[1] Multiple-sheeted spaces and manifolds of states of motion. *Trans. Amer. Math. Soc.*, **28**, 479–90.

1927

[1] An application of analysis situs to statistics. *Bull. Amer. Math. Soc.*, **33**, 467–76.

[2] Differential equations subject to error, and population estimates. *J. Amer. Stat. Assoc.*, **22**, 283–314.

1928

[1] Spaces of statistics and their metrization. *Science*, **67**, 149–50.

[2] The physical state of protoplasm (with L. G. M. Baas Becking and Henriette v.d. Sande Bakhuysen). *Verhandel. Koninkl. Akad. Wetenschap. Amsterdam, Afdeeling Natuurkunde* (Tweedie Sect.), **25**(5), 28–31.

1929

[1] Applications of the theory of error to the interpretation of trends (with Holbrook Working). *J. Amer. Stat. Assoc.*, **24** (March suppl.), 73–85.

[2] Stability in competition. *Econ. J.*, **39**, 41–57.

[3] A determinant game. *Amer. Math. Month.*, **36**, 285.

1930

[1] British statistics and statisticians today. *J. Amer. Stat. Assoc.*, **25**, 186–90.

[2] The consistency and ultimate distribution of optimum statistics. *Trans. Amer. Math. Soc.*, **32**, 847–59.

1931

[1] Recent improvements in statistical inference. *J. Amer. Stat. Assoc.*, **26** (March suppl.), 79–89.
[2] The economics of exhaustible resources. *J. Polit. Econ.*, **39**, 137–75.
[3] Frequency distributions. In vol. 6, *Encyclopaedia of the Social Sciences*, New York: Macmillan, pp. 484–89.
[4] Causes of birth rate fluctuations (with Floy Hotelling). *J. Amer. Stat. Assoc.*, **26**, 135–49.
[5] The generalization of Student's ratio. *Ann. Math. Stat.*, **2**, 360–78.
[6] Outstanding problems in the theory of statistics. December 5, 1931 (duplicated).

1932

[1] The limits of a measure of skewness (with Leonard M. Solomons). *Ann. Math. Stat.*, **3**, 141–42.
[2] A new analysis of duration of pregnancy data (with Floy Hotelling). *Amer. J. Obst. Gyn.*, **23**, 643–57.
[3] Edgeworth's taxation paradox and the nature of demand and supply functions. *J. Polit. Econ.*, **40**, 577–616.

1933

[1] Analysis of a complex of statistical variables into principal components. *J. Educ. Psychol.*, **24**, 417–41 and 498–520.
[2] Note on Edgeworth's taxation paradox and Professor Garver's additional condition on demand functions. *Econometrica*, **1**, 408–9.
[3] Prosperity through increased production. Presented before the Econometric Society, December 27, 1933, at Philadelphia (duplicated).

1935

[1] Demand functions with limited budgets. *Econometrica*, **3**, 66–78.
[2] The most predictable criterion. *J. Educ. Psychol.*, **26**, 139–42.

1936

[1] Rank correlation and tests of significance involving no assumption of normality (with Margaret Richards Pabst). *Ann. Math. Stat.*, **7**, 29–43.
[2] Measurements and correlations on pelves of Indians of the Southwest (with W. W. Howells). *Amer. J. Phys. Anthropol.*, **21**, 91–106.
[3] Some little known applications of mathematics. *Math. Teacher*, **29**, 157–69.
[4] Simplified calculation of principal components. *Psychometrika*, **1**, 27–35.
[5] Relations between two sets of variates. *Biometrika*, **28**, 321–77.
[6] Curtailing production is anti-social. *Columbia Alumni News*, **28**(4), 3, 16.

1937

[1] Effects of pregnancy, mouth acidity and age on dental caries (with Daniel E. Ziskin). *J. Dent. Res.*, **16**, 507–19.

1938

[1] The transformation of statistics to simplify their distribution (with Lester R. Frankel). *Ann. Math. Stat.*, **9**, 87–96.
[2] The general welfare in relation to problems of taxation and of railway and utility rates. (Presidential address to the Econometric Society.) *Econometrica*, **6**, 242–69.

1939

[1] Tubes and spheres in n-spaces and a class of statistical problems. *Amer. J. Math.*, **61**, 440–60.
[2] The work of Henry Schultz. *Econometrica*, **7**, 97–103.
[3] The relation of prices to marginal cost in an optimum system. *Econometrica*, **7**, 151–55.
[4] A final note. *Econometrica*, **7**, 158–60.

1940

[1] The selection of variates for use in prediction with some comments on the general problem of nuisance parameters. *Ann. Math. Stat.*, **11**, 271–83.
[2] The teaching of statistics. *Ann. Math. Stat.*, **11**, 457–70.

1941

[1] Experimental determination of the maximum of a function. *Ann. Math. Stat.*, **12**, 20–45.
[2] Presidential Address (to Indian Statistical Congress at Madras). *Sankhyā*, **5**, 127–29.

1942

[1] Problems of prediction. *Amer. J. Sociol.*, **48**, 61–76. (Part of a symposium of four papers.)
[2] Rotations in psychology and the statistical revolution. *Science*, **95**, 504–7.
[3] Foreword to *Tables of Probability Functions*, Vol. II. National Bureau of Standards, Washington, D.C.: Government Printing Office.

1943

[1] Income-tax revision as proposed by Irving Fisher. *Econometrica*, **11**, 83–87.
[2] Some new methods in matrix calculation. *Ann. Math. Stat.*, **14**, 1–34.
[3] Dr. Peters' criticism of Fisher's statistics. *J. Educ. Res.*, **36**, 707–11.
[4] Further points on matrix calculation and simultaneous equations. *Ann. Math. Stat.*, **14**, 440–41.

1944

[1] Some improvements in weighing and other experimental techniques. *Ann. Math. Stat.*, **15**, 297–306.
[2] Note on a matric theorem of A. T. Craig. *Ann. Math. Stat.*, **15**, 427–29.

1945

[1] Graduate work in statistics at Columbia University. *Biometrics Bull.*, **1**, 22–23.

1947

[1] The training of statisticians. *Amer. Statistician*, **1**(3), 8–9.
[2] Multivariate quality control, illustrated by the air testing of sample bombsights. Chap. 3 in *Selected Techniques of Statistical Analysis*, ed. C. Eisenhart, M. W. Hastay, and W. A. Wallis, New York: McGraw-Hill.
[3] Effects of non-normality at high significance levels. *Ann. Math. Stat.*, **18**, 608–9 (abst.).

1948

[1] Points in the report of the Institute of Mathematical Statistics Committee on the teaching of statistics. *Math. Teacher*, **41**, 76–77.
[2] The teaching of statistics. A report of the Institute of Mathematical Statistics Committee on the Teaching of Statistics. Harold Hotelling, Chairman; Walter Bartky; W. Edwards Deming; Milton Friedman; Paul Hoel. *Ann. Math. Stat.*, **19**, 95–115.
[3] Mathematical statistics and econometrics at the University of North Carolina. *Econometrica*, **16**, 125–26.
[4] Discussion of paper of Georges Teissier. *Biometrics*, **4**, 50.
[5] Wake Forest for future leaders. *The News and Observer*, Raleigh, N.C., August 31, 1948, p. 4.
[6] Mathematics in our schools. *The News and Observer*, Raleigh, N.C., September 2, 1948, p. 4.

1949

[1] Discussion of paper of René Roy. *Econometrica*, **17** (July suppl.), 188–89.
[2] Stochastic processes. (Remarks by the chairman at the joint meeting of the Econometric Society and the Institute of Mathematical Statistics.) *Econometrica*, **17**, 66–68.
[3] Practical problems of matrix calculation. In *Proceedings of the Berkeley Symposium on Mathematical Statistics and Probability*. Berkeley, Calif.: Univ. California Press, pp. 275–94.
[4] The place of statistics in the university. In *Proceedings of the Berkeley Symposium on Mathematical Statistics and Probability*. Berkeley, Calif.: Univ. California Press, pp. 21–40.

1950

[1] Some computational devices. Chap. 10 in *Statistical Inference in Dynamic Economic Models*, ed. T. C. Koopmans, Cowles Commission Monograph No. 10, New York: Wiley.
[2] Needed steps in the mathematical training of social scientists. *Econometrica*, **18**, 198–99.

[3] Why we have the two-party system. *Amer. J. Econ. and Sociol.*, **10**, 13–15.

[4] Comments on the report, "Furtherance of statistical education." *Revue Inst. Internat. Statistique*, **18**, 69–70.

1951

[1] The impact of R. A. Fisher on statistics. *J. Amer. Stat. Assoc.*, **46**, 35–46.

[2] Abraham Wald. *Amer. Statistician*, **5**, 18–19.

[3] A generalized T test and measure of multivariate dispersion. In *Proceedings of the Second Berkeley Symposium on Mathematical Statistics and Probability*, ed. J. Neyman, Berkeley, Calif.: Univ. California Press, pp. 23–42.

[4] Foreword to "The Jacobians of certain matrix transformations useful in multivariate analysis." *Biometrika*, **38**, 345–46.

[5] Discussion of the paper of J. Neyman. *Proceedings of the International Statistical Conferences at Washington*, **III** A. The Hague: International Statistical Institute, p. 432.

1952

[1] Stability in competition. (Reprinted from *Econ. J.*, 1929, article listed above.) *Readings in Price Theory Selected by a Committee of the American Economic Association*. Homewood, Ill.: Irwin, pp. 467–84.

1953

[1] New light on the correlation coefficient and its transforms. *J. Roy. Stat. Soc.*, Ser. B, **15**, 193–225, with discussion, 225–32.

1954

[1] Multivariate analysis. Chap. 4 in *Statistics and Mathematics in Biology*, ed. O. Kempthorne, T. A. Bancroft, J. W. Gowen, and J. L. Lush. Ames, Iowa: Iowa State Coll. Press.

[2] Discussion of paper of J. Durbin (at the meeting of the Econometric Society at Innsbruck, Austria). *Econometrica*, **20**, 102, 105, 108–9.

1955

[1] Econometrics. In vol. 9, *Encyclopedia Americana*, Chicago: Americana Corp., p. 556.

[2] The correlation coefficient and its transforms—a new look. *Résumés des conférences faites au séminaire de statistique tenu à Rome, Septembre 1953*. The Hague: International Statistical Institute.

[3] The moments of the sample median (with John T. Chu). *Ann. Math. Stat.*, **26**, 593–606.

[4] Multivariate methods in testing complex equipment. *Papers Presented at the Colloquium in Statistical Design of Laboratory Experiments*. U. S. Naval Ordnance Laboratory, White Oak, Md. pp. 84–91.

1956

[1] Les rapports entre les méthodes statistiques récentes portant sur des variables multiples et l'analyse factorielle. In *L'analyse factorielle et ses applications*. Paris: Centre national de la recherche scientifique, pp. 107–19, with discussion, pp. 120–25.

1957

[1] The relations of the newer multivariate statistical methods to factor analysis. *Brit. J. Stat. Psychol.*, **10** (II), 69–79.

1958

[1] C. F. Roos, econometrician and mathematician. *Science*, **128**, 1194–95.
[2] The statistical method and the philosophy of science. *Amer. Statistician*, **12**(5), 9–14.

1959

[1] Summary (English) to "El método estadístico filosofía de la sciencia" (transl. of 1958 [2]). *Estadística*, **17**(62), 14–27.
[2] The general welfare in relation to problems of taxation and of railway and utility rates. Reprinted from *Econometrica*, 1938 [2], in *Readings in the Economics of Taxation Selected by a Committee of the American Economic Association*. Homewood, Ill.: Irwin, pp. 139–67.

2. Reviews

1926

[1] E. E. Day: Statistical Analysis. *J. Amer. Stat. Assoc.*, **21**, 360–63.
[2] L. I. Dublin and others: Population Problems in the United States and Canada. *J. Amer. Stat. Assoc.*, **21**, 503–5.

1927

[1] R. A. Fisher: Statistical Methods for Research Workers. *J. Amer. Stat. Assoc.*, **22**, 411–12.

1928

[1] W. P. Elderton: Frequency Curves and Correlation. *Bull. Amer. Math. Soc.*, **34**, 515–16.
[2] R. A. Fisher: Statistical Methods for Research Workers, 2d ed. *J. Amer. Stat. Assoc.*, **23**, 346.
[3] A. Cournot: Researches into the Mathematical Principles of the Theory of Wealth. *Amer. Math. Month.*, **35**, 439–40.

1930

[1] R. A. Fisher: Statistical Methods for Research Workers, 3d ed. *J. Amer. Stat. Assoc.*, **25**, 381–82.

1931

[1] G. C. Evans: Mathematical Introduction to Economics. *J. Polit. Econ.*, **39**, 107–9.

[2] G. C. Evans: Mathematical Introduction to Economics. *Amer. Math. Month.*, **38**, 101–3.

1932

[1] W. A. Shewhart: Economic Control of Quality of Manufactured Product. *J. Amer. Stat. Assoc.*, **27**, 215–17.

[2] Ragnar Frisch: New Methods of Measuring Marginal Utility. *J. Amer. Stat. Assoc.*, **27**, 451–52.

1933

[1] R. A. Fisher: Statistical Methods for Research Workers, 4th ed. *J. Amer. Stat. Assoc.*, **28**, 374–75.

[2] Horace Secrist: The Triumph of Mediocrity in Business. *J. Amer. Stat. Assoc.*, **28**, 463–65.

1934

[1] Reply to Secrist. *J. Amer. Stat. Assoc.*, **29**, 198–99.

1935

[1] J. G. Smith: Elementary Statistics, an Introduction to the Principles of Scientific Method. *Amer. Math. Month.*, **42**, 169–71.

[2] G. W. Snedecor: Calculation and Interpretation of Analysis of Variance and Covariance. *J. Amer. Stat. Assoc.*, **30**, 117–18.

[3] R. A. Fisher: Statistical Methods for Research Workers, 5th ed. *J. Amer. Stat. Assoc.*, **30**, 118.

[4] C. F. Roos: Dynamic Economics. *J. Amer. Stat. Assoc.*, **30**, 480–83.

[5] R. A. Fisher: The Design of Experiments. *J. Amer. Stat. Assoc.*, **30**, 771–72.

1937

[1] R. A. Fisher: Statistical Methods for Research Workers, 6th ed. *J. Amer. Stat. Assoc.*, **32**, 218–19.

[2] R. A. Fisher: The Design of Experiments, 2d ed. *J. Amer. Stat. Assoc.*, **32**, 580–82.

1938

[1] Henry Schultz: The Theory and Measurement of Demand. *J. Amer. Stat. Assoc.*, **33**, 744–47.

[2] R. A. Fisher and F. Yates: Statistical Tables for Biological, Agricultural and Medical Research. *Science*, **88**, 596–97.

1939

[1] R. G. D. Allen: Mathematical Analysis for Economists. *Ann. Amer. Acad. Polit. Soc. Sci.*, **205**, 154–55.

[2] R. A. Fisher: Statistical Methods for Research Workers, 7th ed. *J. Amer. Stat. Assoc.*, **34**, 423–24.

1941

[1] G. A. D. Preinreich: The Present Status of Renewal Theory. *J. Polit. Econ.*, **49**, 136–37.

1942

[1] K. L. Holzinger and H. H. Harman: Factor Analysis. *Science*, **95**, 504–7.

[2] Paul Horst and others: The Prediction of Personal Adjustment. *Amer. J. Sociol.*, **48**, 61–76.

[3] H. T. Davis: The Analysis of Economic Time Series. *Ann. Amer. Acad. Polit. Soc. Sci.*, **221**, 219–21.

1943

[1] E. C. Molina: Poisson's Exponential Binomial Limit. *J. Educ. Psychol.*, **34**, 124–25.

1944

[1] M. B. Nanavati and C. N. Vakil (eds.): India Speaking (from *Annals of the American Academy of Political and Social Science*). *Asia*, **47**, 382–84.

1945

[1] M. G. Kendall: The Advanced Theory of Statistics, Vol. 1. *Bull. Amer. Math. Soc.*, **51**, 214–16.

1947

[1] Harald Cramér: Mathematical Methods of Statistics. *J. Polit. Econ.*, **55**, 393–95.

1948

[1] M. G. Kendall: The Advanced Theory of Statistics, Vol. II. *Bull. Amer. Math. Soc.*, **54**, 863–68.

PART II

CONTRIBUTIONS TO PROBABILITY AND STATISTICS

5 SOME REMARKS ON THE DESIGN AND ANALYSIS OF FACTORIAL EXPERIMENTS

R. L. ANDERSON
Institute of Statistics, North Carolina State College

1. Introduction

It is a pleasure to contribute to this dedicatory volume to one of the pioneers in the fields of mathematical statistics and econometrics. In selecting a topic for this chapter I have been torn between revisiting the field of time-series analysis and publicizing some points of view on the use of least-squares methods. As is well known, Professor Hotelling has done considerable research in both fields and has been teaching courses on these two topics for many years. I have chosen to confine my remarks to a topic involving least squares: the analysis of factorial experiments. Some of this material is a compilation and revision of previous articles ([1], [2], [3], [4]).

2. Comparisons of Discrete and Continuous Models for Factorial Experiments with Continuous Factor Levels

Traditional analysis of factorial experiments is a simple extension of ordinary regression analysis, in which the regression variables take on the values 0 or 1. Hence the usual effects in a factorial experiment can be related in a simple manner to the regression coefficients in a polynomial regression model, if the factor levels are continuous. As a simple illustration, a two-factor experiment was chosen, each factor at three levels, e.g., three temperatures and three concentrations. Let us assume that each of the nine treatment combinations is applied to r experimental units. The expected average yield for the (ij) treatment combination, based on the discrete factorial model, is

$$E(\bar{Y}_{ij}) = \mu + \alpha_i + \beta_j + (\alpha\beta)_{ij} \mathrel{\hat{=}} \bar{Y} + a_i + b_j + (ab)_{ij}\,, \qquad i, j = 1, 2, 3\,,$$

where μ is the general mean; α_i the effect of the ith level of factor A; β_j the effect of the jth level of factor B; and $(\alpha\beta)_{ij}$ the additional effect, if any, of the (ij) combination—the interaction effect. It is assumed that $\sum_i \alpha_i = \sum_j \beta_j = \sum_i (\alpha\beta)_{ij} = \sum_j (\alpha\beta)_{ij} = 0$. The symbol $\hat{=}$ stands for "is estimated by." $\bar{Y}$ is the average of all $9r$ observations; $a_i = \bar{A}_i - \bar{Y}$, where $\bar{A}_i$ is the average yield of the $3r$ observations of the ith level of factor A; $b_j = \bar{B}_j - \bar{Y}$; $(ab)_{ij} = \bar{Y}_{ij} - \bar{A}_i - \bar{B}_j + \bar{Y}$.

Assuming that the factor levels are continuous and that the relationship is quadratic, one can use the regression model

$$E_1(\bar{Y}) = \theta_0' + \theta_1' X_1 + \theta_2' X_2 + \theta_{11}' X_1^2 + \theta_{22}' X_2^2 + \theta_{12}' X_1 X_2 ,$$

where X_1 is the amount of A used and X_2 the amount of B. For convenience, we will assume that the experimental levels of A and B are equally spaced; however, comparisons can be made (in a more complicated form) if this is not the case. For equal spacing, the results are simplified by using coded values (x_1 and x_2) of X_1 and X_2, obtained by subtracting the middle experimental level from each and dividing by the difference in successive levels; hence, the experimental values of x_1 and x_2 are $x_{11} = x_{21} = -1$, $x_{12} = x_{22} = 0$, $x_{13} = x_{23} = 1$. In order to simplify the least-squares equations, we will use the following model:

$$\begin{aligned} E_1(\bar{Y}) &= \theta_0 + \theta_1 x_1 + \theta_2 x_2 + \theta_{11} z_1 + \theta_{22} z_2 + \theta_{12} x_1 x_2 \\ &\triangleq \bar{Y} + t_1 x_1 + t_2 x_2 + t_{11} z_1 + t_{22} z_2 + t_{12} x_1 x_2 , \end{aligned}$$

where $\theta_0 = \mu$, $z_i = (x_i^2 - 2/3)$ (the average value of x_i^2 is 2/3). The t's are the usual least-squares estimates of the θ's, e.g., $t_1 = Sx_1\bar{Y}/Sx_1^2 = (\bar{A}_3 - \bar{A}_1)/2$.

Using these coded values of the X's, we relate the parameters of the two models as follows:

$$\theta_1 = (\alpha_3 - \alpha_1)/2 ; \qquad \theta_{11} = (\alpha_3 - 2\alpha_2 + \alpha_1)/2 ;$$

$$\theta_2 = (\beta_3 - \beta_1)/2 ; \qquad \theta_{22} = (\beta_3 - 2\beta_2 + \beta_1)/2 ;$$

$$\theta_{12} = [(\alpha\beta)_{11} - (\alpha\beta)_{13} - (\alpha\beta)_{31} + (\alpha\beta_{33})]/4 .$$

The parameter θ_{12} is often called the linear $\times$ linear component of the AB interaction. There are three other components of the AB interaction which can be estimated from a 3×3 experiment—the linear $\times$ quadratic, quadratic $\times$ linear, and quadratic $\times$ quadratic components, if the factor levels are continuous. These estimates can be used to test the adequacy of the quadratic model. In discussing these degree components in the remainder of this section, we will often use a short-cut notation which has become quite popular: ($hijk$), which indicates a four-factor component for the hth degree of the first factor, the ith of the second, the jth of the third, and the kth of the fourth. In the 3×3 experiment, (10) refers to θ_1, (01) to θ_2, (20) to θ_{11}, (02) to θ_{22}, and (11) to θ_{12}. The degree of the component is the sum, $h + i + j + k$.

It is interesting to compare these two models on the basis of expected values and variances of estimated average yields. If the true model is quadratic, both give the same expected values; in addition, the quadratic model provides a convenient method of interpolating between experimental points and even of extrapolating (if one is willing to risk assuming the true model is everywhere quadratic). Also the quadratic model gives more precise estimates of the average yield. The variance of the estimated average yield ($\bar{Y}_{ij}$) for every experimental point using the discrete model is σ^2/r, where σ^2 is the variance for individual yields. In using the quadratic model, $\bar{Y}_{ij}$ is replaced by the estimated value

$$\hat{Y}_1 = \bar{Y} + t_1 x_1 + t_2 x_2 + t_{11} z_1 + t_{22} z_2 + t_{12} x_1 x_2 .$$

The variance of $\hat{Y}_1$ ranges from $16\sigma^2/45r$ when $x_1^2 = x_2^2 = 0.4$ to $29\sigma^2/36r$ when $x_1^2 = x_2^2 = 1$.[1] Why then do we often prefer to use the less precise estimator $\bar{Y}_{ij}$ instead of $\hat{Y}_1$? The answer lies in the uncertainty of the true model. In general, one would not expect the response to be truly quadratic; however, it might be nearly so within the range of the experimental values. Hence the choice of whether to use the discrete or continuous model rests on a comparison of relative precision and relative bias. The quadratic estimator will generally be more precise but is probably biased to some extent. One can, of course, use the information on the three cubic and quartic components to test if there is a significant departure from a quadratic model; however, one is somewhat reluctant to use preliminary tests in formulating models. This writer feels that here is an important field for research, one which has received too little attention in the past.

Suppose the true model is $E_2(\bar{Y}) = E_1(\bar{Y}) + \theta_{122} x_1 z_2$. If we use $\hat{Y}_2 = \hat{Y}_1 + t_{122} x_1 z_2$, it is easy to show that t_{122} is uncorrelated with the other t's; hence, $\text{var}(\hat{Y}_2) = \text{var}(\hat{Y}_1) + x_1^2 z_2^2 \,\text{var}(t_{122})$, where $\text{var}(t_{122}) = 3\sigma^2/4r$. If $\hat{Y}_1$ is used, the estimator is biased by an amount $\theta_{122} x_1 z_2$. (The bias is zero only if x_1 or z_2 is 0.) The mean square error for $\hat{Y}_1$ is $\text{MSE}_1 = \text{var}(\hat{Y}_1) + \theta_{122}^2 x_1^2 z_2^2$. Except for the special case when x_1 or z_2 is 0, it is obvious that if $\text{var}(t_{122})$ is greater than θ_{122}^2, the variance of $\hat{Y}$ is increased more than the bias component by using the cubic term in the estimating equation. This is a general principle in all regression analysis; if the standard error of an estimated regression coefficient is larger than its expected value (in absolute terms), there seems to be good reason to omit the corresponding predictor from the regression equation. Of course, the difficulty is that one does not know the true values of $\text{var}(t_{122})$ and θ_{122}; hence, one must compare their estimators, $3s^2/4r$ and t_{122}. This is another interesting topic for research: how to formulate a decision rule regarding the addition of a new predictor to a regression model based only on sample values. Presumably the decision rule will be altered if one has several estimates of θ_{122} and σ^2, instead of only those from the current experiment.

One final calculation serves to indicate the complexity of this problem. In general it would seem that, if θ_{122} were in the model, θ_{112} should also be included, so that

$$E_3(\bar{Y}) = E_1(\bar{Y}) + \theta_{122} x_1 z_2 + \theta_{112} x_2 z_1 .$$

---

[1] These calculations are based on the fact that the t's are uncorrelated and $\text{var}(\bar{Y}) = \sigma^2/9r$, $\text{var}(t_i) = \sigma^2/6r$, $\text{var}(t_{ii}) = \sigma^2/2r$, and $\text{var}(t_{12}) = \sigma^2/4r$.

If we use

$$\hat{Y}_3 = \hat{Y}_1 + t_{122}x_1z_2 + t_{112}x_2z_1 ,$$
$$\operatorname{var}(\hat{Y}_3) = \operatorname{var}(\hat{Y}_1) + x_1^2z_2^2 \operatorname{var}(t_{122}) + x_2^2z_1^2 \operatorname{var}(t_{112}) .$$

If $\hat{Y}_2$ is used, we have the same conclusion regarding the inclusion of the second cubic term as for the first. However, since it seems reasonable to use either both cubic terms or neither of them, one would like to compare $\hat{Y}_1$ and $\hat{Y}_3$. The mean square error for $\hat{Y}_1$ is $\text{MSE}_1^* = \operatorname{var}(\hat{Y}_1) + (\theta_{122}x_1z_2 + \theta_{112}x_2z_1)^2$. This is smaller than $\operatorname{var}(\hat{Y}_3)$ if

$$(\theta_{122}x_1z_2 + \theta_{112}x_2z_1)^2 < x_1^2z_2^2 \operatorname{var}(t_{122}) + x_2^2z_1^2 \operatorname{var}(t_{112}) .$$

General statements are difficult here because of the introduction of the product term, $2\theta_{122}\theta_{112}x_1x_2z_1z_2$, in the left-hand side. Perhaps one could make statements regarding the average results, giving uniform weights to all x's in the range $(-1, +1)$; in this case the product term vanishes, and some conclusions can be based on the comparisons of expected values and variances, as before.

Finally, if there is a cubic response pattern, it seems reasonable that X_1^3 and X_2^3 should also be included in the model. But since their coefficients cannot be estimated from a 3×3 design, one would then require a new design. Here we enter into the problem of design construction which has been considered recently by Box and his associates. Many interesting facets of this problem are still worthy of research, especially the relationships between bias and precision as they affect predicted responses (especially optimal responses).

3. Use of Usual Confounded Arrangements when Levels Are Continuous

In the assignment of treatments to experimental units, we should take into account any previous knowledge of differences between the units; this may take the form of grouping the units into more or less homogeneous blocks. If each homogeneous block has the same number of experimental units as number of factor combinations, treatment comparisons can be made free of block effects by assigning each factor combination to one unit in each block. If the blocks are too small to accomplish this, certain or all of the expected values mentioned in Section 2 will contain block effects, i.e., the treatment and block effects are said to be *confounded*. One can usually estimate most or all of the treatment effects by adjusting for block effects, assuming there is an additive relationship between the two. However, the precision of such estimates could be considerably less than that of comparable estimates which did not involve confounding, if there were only a slight reduction in experimental variability resulting from the use of the smaller blocks.[2]

[2] This statement and subsequent remarks neglect interblock information. This writer is often dubious of the validity of assuming random block effects; granted this assumption is correct, he dislikes the use of estimated weights based on the few degrees of freedom usually available to estimate the block component.

To illustrate some of these ideas, consider Plan III in Cochran and Cox [5] for a 4×4 experiment in four blocks of four units each. The plan is given in the following tabulation:

Block			
a	b	c	d
33	31	32	30
20	22	21	23
11	13	10	12
02	00	03	01 .

The numbers in this table indicate the levels of the two factors, with 0 indicating the lowest level. Three orthogonal block contrasts are used (in terms of block constants, a, b, c, and d, where $a + b + c + d = 0$):

$$4C_1 = -a - b + c + d\,, \qquad 4C_2 = a - b - c + d\,, \qquad 4C_3 = -a + b - c + d\,.$$

The quadratic model becomes

$$\hat{Y} = \bar{Y} \pm C_1 \pm C_2 \pm C_3 + t_1 x_1 + t_2 x_2 + t_{11} z_1 + t_{22} z_2 + t_{12} x_1 x_2\,,$$

where the successive coded values for the X's are x_i: $(-3, -1, 1, 3)$; and $z_i = (x_i^2 - 5)/4$: $(1, -1, -1, 1)$. The coefficients of the C's in a given block are determined by the signs of the corresponding block constant. For example, the model for the yield of the plot with treatment *33* is

$$\hat{Y}_{33} = \bar{Y} - C_1 + C_2 - C_3 + 3t_1 + 3t_2 + t_{11} + t_{22} + 9t_{12}\,.$$

The nine least-squares equations are given in Table 1.

TABLE 1

Yield totals	Regression coefficients								
	$\bar{Y}$	C_1	C_2	C_3	t_1	t_2	t_{11}	t_{22}	t_{12}
SY	16	0	0	0	0	0	0	0	0
$B_1 = -B_a - B_b + B_c + B_d$	0	16	0	0	0	0	0	0	−16
$B_2 = B_a - B_b - B_c + B_d$	0	0	16	0	0	0	0	0	0
$B_3 = -B_a + B_b - B_c + B_d$	0	0	0	16	0	0	0	0	0
Sx_1Y	0	0	0	0	80	0	0	0	0
Sx_2Y	0	0	0	0	0	80	0	0	0
Sz_1Y	0	0	0	0	0	0	16	0	0
Sz_2Y	0	0	0	0	0	0	0	16	0
Sx_1x_2Y	0	−16	0	0	0	0	0	0	400

In this table, $B_a, B_b, \cdots$ are the block totals. The only confounding occurs with C_1 and $t_{12} = (11)$; each of the other constants can be computed from a single equation. We have

$$16C_1 - 16(11) = B_1\,,$$
$$-16C_1 + 400(11) = Sx_1x_2Y = T_{11}\,.$$

Hence we obtain $(11) = (T_{11} + B_1)/384$; $\text{var}\,(11) = \sigma_1^2/384$ as compared with $\sigma^2/400$ if there had been no confounding (no blocking), where σ_1^2 is the error variance when block effects are removed and σ^2 is the error variance with no blocking. The confounded design will produce a more precise estimator of t_{12} if $\sigma_1^2 < 384\sigma^2/400 = 0.96\sigma^2$. The quantity $100\% - 96\% = 4\%$ is called the percentage *loss of information* due to confounding; 96% is the *relative information* given by this design on (11). Note that the relative information is 100% for all other constants. All of these statements are based on the assumption that cubic and higher-degree components can be neglected. In the equations cited above, we have designated the coefficient of (11) in the first equation and of C_1 in the second, -16, as the *confounding constant*. These constants are presented for most confounded designs in a bulletin by Binet, Leslie, Weiner, and Anderson [4].

In this plan, the pure cubic components, (30) and (03), are not confounded with block effects; however, the other components, (21) and (12), are confounded. Fortunately they are not confounded with the same block effects as is (11). Using [4], we find that the least-squares equations for t_{112} or (21) and C_3 are

$$16C_3 - 32(21) = B_3\,, \qquad -32C_3 + 80(21) = T_{21}\,.$$

Hence $(21) = (T_{21} + 2B_3)/16$. There is an 80% loss of information on this component due to confounding. Similarly, (12) is confounded only with C_2. However, if one desires to estimate fourth-degree effects, adjusted for others, the problem becomes quite complicated, since both (13) and (31) are confounded with C_1. Using for the cubic components the coded values $(-1, 3, -3, 1)$, we have these equations:

$$\begin{aligned}
16C_1 - 16(11) - 32(13) - 32(31) &= B_1\,,\\
-16C_1 + 400(11) \qquad\qquad\qquad\qquad &= T_{11}\,,\\
-32C_1 \qquad\qquad + 400(13) \qquad\qquad &= T_{13}\,,\\
-32C_1 \qquad\qquad\qquad\qquad + 400(31) &= T_{31}\,.
\end{aligned}$$

The solution for (11) is $6400(11)/17 = (17T_{11} + 2T_{13} + 2T_{31} + 25B_1)/17$. Since the coefficient of (11) is reduced from 400 to 6400/17, there is a relative loss of information of 6%.

The usual procedure in estimating these components is to assume a quadratic model and test for goodness of fit by examining the contribution of the higher-degree components to the total variation. There are usually enough unconfounded high-degree components to enable the experimenter to make such an examination without excessive computing. However, one can determine the additional contribution of the high-degree components, even when mixed up with both blocks and lower-degree components, quite simply by use of the abbreviated Doolittle or a similar computing procedure, without inverting the coefficient matrix. For example, in the preceding problem the sum of squares attributable to (13) and (31), adjusted for C_1

and (11), is computed as follows: First adjust for C_1, to produce the equations

$$\begin{aligned} 384(11) - 32(13) - 32(31) &= T_{11} + B_1 = T'_{11} , \\ -32(11) + 336(13) - 64(31) &= T_{13} + 2B_1 = T'_{13} , \\ -32(11) - 64(13) + 336(31) &= T_{31} + 2B_1 = T'_{31} . \end{aligned}$$

The abbreviated Doolittle computations (for computing details, see [6]) proceed as follows:

384	-32	-32	T'_{11}	$= J_1$
1	$-1/12$	$-1/12$	$T'_{11}/384$	$= K_1$
	1000/3	$-200/3$	$(12T'_{13} + T'_{11})/12$	$= J_2$
	1	$-1/5$	$(12T'_{13} + T'_{11})/4000$	$= K_2$
		320	$(10T'_{31} + 2T'_{13} + T'_{11})/10$	$= J_3$
		1	$(10T'_{31} + 2T'_{13} + T'_{11})/3200$	$= K_3$.

The sum of squares attributable to (13) and (31), adjusted for C_1 and (11), is simply $J_2K_2 + J_3K_3$.

Another interesting example concerns the analysis of a $5 \times 3 \times 2$ experiment, using the following 3 blocks of 10 units each (see [4], pp. 49–50):

Block

a	*b*	*c*
001	000	010
020	011	021
100	110	101
111	121	120
210	201	200
221	220	211
301	300	310
320	311	321
400	410	401
411	421	420 .

In this case let D_1 and D_2 be the two block contrasts: $2D_1 = -a + c$; $6D_2 = a - 2b + c$, where $a + b + c = 0$.

We obtain these least-squares equations:

$$\begin{aligned} 20D_1 + 2(010) &= B_1 = -B_a + B_c , \\ 2D_1 + 20(010) &= T_{010} , \end{aligned}$$

$$\begin{aligned} 60D_2 - 6(011) + 6(020) - 6(110) &= B_2 = B_a - 2B_b + B_c\,, \\ -6D_2 + 20(011) \qquad\qquad\qquad &= T_{011}\,, \\ 6D_2 \qquad\quad + 60(020) \qquad\quad &= T_{020}\,, \\ -6D_2 \qquad\qquad\qquad + 40(110) &= T_{110}\,. \end{aligned}$$

It is easy to determine that $19.8(010) = T_{010} - 0.1B_1$; hence, there is a loss of 1% relative information on (010) due to confounding. After eliminating D_2 from the last three equations and inverting the coefficient matrix, we obtain

$$\begin{bmatrix} 13/252 & -1/1890 & 1/1260 \\ & 191/11{,}340 & -1/3780 \\ & & 8/315 \end{bmatrix}.$$

Hence the relative losses of information are as follows: (011): 8/260 = 2/65; (020): 120/11,460 = 2/191; (110): 5/320 = 1/64.

In factorial experiments, efforts are usually made to allocate treatment combinations to blocks in such a manner that most, if not all, of the confounding occurs with the higher-order interactions (or components of high degree for continuous factor levels). For example, with a $3 \times 3 \times 3$ experiment, it would be desirable to concentrate the confounding on the three-factor interactions or on the third- or higher-degree components. Extensive lists of confounded arrangements for traditional factorial experiments (based on the discrete model) have been published in the last twenty-five years, and are now readily available in various textbooks on design of experiments. For experiments in which all factors have the same number of levels and where the number of levels is the power of a prime, this confounding has been based on the use of the mathematics of Galois fields; particular reference is made to Bose [7] and Kempthorne [8]. Confounding for mixed factorials (differing numbers of levels) is much more complicated, but rules of procedure have been developed.

When the levels are continuous and the model is known to be of a certain degree, it is possible to construct designs which concentrate the confounding on higher-degree components. In many circumstances uncertainty regarding the correct model dictates that the confounding for continuous-factor-level experiments should also be concentrated on the higher-order interactions, regardless of degree. Recognizing that many multifactor experiments involving continuous factor levels would utilize traditional confounded designs, Binet, Leslie, Weiner, and Anderson [4] prepared their bulletin to accomplish three objectives:

1. To present short-cut computing procedures for estimating the parameters in polynomial models, adjusted for block effects.
2. To indicate the block-treatment confounding constants for these parameters for each design plan.

3. To construct some new confounded designs, especially for $3^2 \times 2^2$ experiments in blocks of two, three, or six and various mixed series involving one or more factors at four, five, or six levels.

This bulletin concentrated on singly replicated experiments in which the error variances must be estimated from high-degree contrasts.

For a given factorial experiment, a number of experimental plans are usually available. One of the striking results of this investigation was the discovery that the amount of information on a given parameter could vary quite widely from one plan to another. For example, three possible replications are given by Li [9] for a 4×2^2 experiment, each replication consisting of four blocks of four experimental units each. In Plan 2 of [9] the loss of information on (110) is 1/5, and in Plan 3 that on (101) is 4/5; whereas in Plan 1, 4/5 information on (110) and 1/5 information on (101) are lost in confounding. There is no other confounding of linear or quadratic components. The average loss of information over all three plans due to confounding is 1/3 for each of these quadratic components; i.e., one obtains balanced confounding if one uses all plans in a three-replicate experiment. However, it would appear more desirable to use only Plan 2, because there is full information on (101) and only a 20% loss on (110). Of course, the use of Plan 2 results in a greater loss on the cubic components, (111) and (201), than with the other plans. Another striking result concerns the 4×4 experiment discussed earlier. In Plan III (Cochran and Cox [5]), we have shown there is a loss of only 4% information on (11) due to confounding; in Plan II, this loss is 32% and in Plan I, 64%. The average loss due to confounding is 1/3, but it would be advisable to use only Plan III if interest is concentrated on the linear and quadratic components. Unfortunately, losses of information were not included in [4]; these were computed subsequently by this writer. Some of these results were included in the second edition of Cochran and Cox [5]. Table 2 in this paper presents a selection of the better plans for each factorial type; if more than one source of plans is available, plans from each source are included in this table. This table should be of some use in selecting plans which would lead to the loss of as little information as possible on linear and quadratic components but still furnish considerable information on higher-degree components. If higher-degree components are shown to be important, all components and block effects must be estimated simultaneously.

The designs for the following factorials have not been summarized in Table 2, because all plans have the same (or essentially the same) information on all linear and quadratic components: 3×2^3 in 6-unit blocks by Cochran and Cox; 4×3^2 in 12-unit blocks and 5×2^2 in 10-unit blocks by Li; $5 \times 3 \times 2$ in 10-unit blocks, 5×3^2 in 15-unit blocks, 6×5 in 15-unit blocks, 6^2 in 18-unit blocks, $6 \times 3 \times 2$ in 12-unit blocks and 7×2, 7×4, 7×2^2, 7×3, and 7×5—all in [4]. The following have no loss in information on linear and quadratic components: $3^3 \times 2$ in 18-unit blocks and 4×2^3

TABLE 2

LOSS OF RELATIVE INFORMATION ON LINEAR AND QUADRATIC COMPONENTS, USING STANDARD CONFOUNDING PLANS

Type of factorial	Units per block[a]	Plan[b]	Loss of information on linear and quadratic components[c]	Number of higher-degree components involved: Third, Unconf.	Third, Conf.	Others, Unconf.	Others, Conf.
3×2	3	B(1)	(01):1/9	0	1	0	0
3^2	3	Y	(11):1/2	0	2	0	1
3^2	4,5	B	(20), (02):1/9	2	0	0	1
3×2^2	6	C(II)	(011):1/9	3	0	0	1
3×2^2	6	C(I, III)	(011):1/9	2	1	0	1
$3^2 \times 2$	6	C	(110):1/8	2	3	0	4
$3^2 \times 2$	9	B	(001):1/81	3	2	3	1
3^3	9	Y, C	None	6	1	3	7
3^3	13,14	B	(200), (020), (002):1/89	7	0	6	4
$3^2 \times 2^2$	6	B	(0011): all, (1100):5/16	6	4	2	11
$3^2 \times 2^2$	12	L	(1100):1/8	8	2	8	5
$3^2 \times 2^2$	12	B(1-6)	(1100):1/32	6	4	2	11
$3^2 \times 2^2$	12	B(10)	(1010):1/12	7	3	5	8
$3^2 \times 2^2$	18	B(7)	(0011):1/81	10	0	10	3
4×2	4	B(2)	None	1	1	1	0
4×3	4	B(4)	(02): all	2	1	3	0
4×3	6	B(3)	(20):1/9	3	0	2	1
4×3	6	B(2)	(10):1/45	1	2	2	1
4^2	4	C(III)	(11):1/25	2	2	1	5
4^2	4	C(II)	(11):8/25	4	0	2	4
4^2	4	K, B(2)	(11):4/25	2	2	1	5
4^2	8	B(1)	None	4	0	5	1
4^2	8	B(2)	None	3	1	5	1
4×2^2	4	L(2)	(110):1/5	2	2	2	2
4×2^2	8	C(II)	None	4	0	3	1
$4 \times 3 \times 2$	12	L,C(IV–VI)	None	5	1	7	2
$4 \times 3 \times 2$	12	L,C(VIII)	(101):1/45	6	0	6	3
$4^2 \times 2$	8	L(3)	None	6	1	12	4
$4^2 \times 2$	16	L(9)	None	7	0	15	1
$4^2 \times 3$	12	L(II:4, 9)	(110):4/225	7	2	12	17
$4^2 \times 3$	12	L(I:2, 3)	(110):4/25	7	2	24	5
5×2	5	B(1, 3)	(01):1/25	1	1	2	1
5×3	5	B	(01):1/25, (02):2/47, (11):1/16	1	2	1	5
5×3	7,8	B	(20):5/189, (02):7/191	3	0	3	3
5×4	10	B(1, 2)	(02):1/25	4	0	8	2
5×4	10	B(3, 4)	(01):1/125	2	2	7	3
5^2	12,13	B(3)	(02), (20):1%	4	0	9	6
$5 \times 3 \times 2$	15	B(7, 8)	(001):1/225	4	2	12	3
6×2	6	B(1)	(11):1/105	2	0	3	2
6×3	9	B(1)	(10):1/945	1	2	6	3
6×4	12	B(2, 5)	None	3	1	12	2
6×4	12	B(3)	(11):1/525	4	0	9	5
$6 \times 3 \times 2$	18	B(9)	(101):1/945	6	0	16	5

[a] Plans with two numbers have two blocks of unequal size.

[b] References: *B*, Binet [4]; *C*, Cochran and Cox [5]; *K*, Kempthorne [8]; *L*, Li [9]; *Y*, Yates [10]. Plan numbers in original publication and in Binet *et al.* [4]; many plans given in *B* and *C* refer to Yates's methods.

[c] Assuming third- and higher-degree components negligible. The term (11) indicates linear × linear component, etc.

in 8-unit blocks in [9], and 4^3 in 16-unit blocks and 6×2^2 in 12-unit blocks in [4].

4. Effect of Unequal Initial Conditions in Experimental Units

Often the treatments used in factorial experiments contain elements which have similar (or even identical) counterparts in the experimental units; this is especially true for agricultural experiments in which different fertilizers are applied to the soil. The soil itself contains the same elements, perhaps in a different state, that are in the fertilizers. To illustrate some of the problems in the analysis of the resultant data, let us assume that a 2^2 experiment is performed, each treatment being applied to two experimental units. The treatments are designated as (00), (01), (10), and (11), where the numbers refer to units of elements A and B added to the soil. The eight experimental units contain different amounts of A and B, say (a_i, b_i) in each unit. Hence the amounts of each element actually available are (a_1, b_1), (a_2, b_2), (a_3, b_3+1), (a_4, b_4+1), (a_5+1, b_5), (a_6+1, b_6), (a_7+1, b_7+1), (a_8+1, b_8+1).

Let us assume that the true relationship between yield (Y) and total amount of A and B available (X_1 and X_2) is additive and quadratic, so that $E(Y) = \beta_0 + \beta_1 X_1 + \beta_{11} X_1^2 + \beta_2 X_2 + \beta_{22} X_2^2$. In many cases, β_0 will be zero. The usual main effect of A ($\boldsymbol{a}'$) and AB interaction ($\boldsymbol{ab}'$) will have the following expectations:

$$\begin{aligned}
E(8\boldsymbol{a}') &= E[-(00) - (01) + (10) + (11)] \\
&= [-a_1 - a_2 - a_3 - a_4 + a_5 + a_6 + a_7 + a_8 + 4]\beta_1 \\
&\quad + [-a_1^2 - a_2^2 - a_3^2 - a_4^2 + a_5^2 + a_6^2 + a_7^2 + a_8^2 + 2(a_5 + a_6 + a_7 + a_8) + 4]\beta_{11} \\
&\quad + [-b_1 - b_2 - b_3 - b_4 + b_5 + b_6 + b_7 + b_8]\beta_2 \\
&\quad + [-b_1^2 - b_2^2 - b_3^2 - b_4^2 + b_5^2 + b_6^2 + b_7^2 + b_8^2 + 2(-b_3 - b_4 + b_7 + b_8)]\beta_{22} ; \\
E(8\boldsymbol{ab}') &= E[(00) - (01) - (10) + (11)] \\
&= [a_1 + a_2 - a_3 - a_4 - a_5 - a_6 + a_7 + a_8]\beta_1 \\
&\quad + [a_1^2 + a_2^2 - a_3^2 - a_4^2 - a_5^2 - a_6^2 + a_7^2 + a_8^2 + 2(-a_5 - a_6 + a_7 + a_8)]\beta_{11} \\
&\quad + [b_1 + b_2 - b_3 - b_4 - b_5 - b_6 + b_7 + b_8]\beta_2 \\
&\quad + [b_1^2 + b_2^2 - b_3^2 - b_4^2 - b_5^2 - b_6^2 + b_7^2 + b_8^2 + 2(-b_3 - b_4 + b_7 + b_8)]\beta_{22} .
\end{aligned}$$

Throughout we have assumed that the average effect of other factors, including weather and experimental procedures, is zero for each (X_1, X_2) combination. If the results given are averaged over all allocations of the four treatments to the eight experimental units,

$$\text{ave}\,(8\boldsymbol{a}') = 4\beta_1 + (4 + \textstyle\sum a_i)\beta_{11}; \qquad \text{ave}\,(8\boldsymbol{ab}') = 0 .$$

Hence, if the response is quadratic, the usual estimator of the main effect of A is a mixture of the linear and quadratic coefficients.

In addition, it would appear that individual experiments often will indicate the presence of an AB interaction, sometimes positive and sometimes negative. Also, the values of $\boldsymbol{a}'$ and $\boldsymbol{b}'$ will vary considerably with changes in the initial conditions. This may account for some of the discordant results of similar experiments conducted at different places and times, if no adjustment is made for initial conditions. The experimenter is lulled into a false sense of security regarding the interpretation of his results, because the statistician tells him that valid inferences can be made if treatments are assigned at random to the available experimental units. This is not to say that randomization is not a useful device to protect against conscious and unconscious biases in the assignment of treatments to experimental units. However, it is inefficient and often misleading to ignore initial conditions if they can be measured. If the (a_i, b_i) are known, least-squares estimates of the β's can be obtained. The upper right-hand part of the matrix for the normal equations would be as follows:

β_0	β_1	β_2	β_{11}	β_{22}	Yield sums
8	SX_1	SX_2	SX_1^2	SX_2^2	SY
	SX_1^2	SX_1X_2	SX_1^3	$SX_1X_2^2$	SX_1Y
		SX_2^2	$SX_2X_1^2$	SX_2^3	SX_2Y
			SX_1^4	$SX_1^2X_2^2$	SX_1^2Y
				SX_2^4	SX_2^2Y ,

where $SX_1 = a_1 + a_2 + \cdots + (a_8 + 1)$; $SX_2 = b_1 + b_2 + \cdots + (b_8 + 1)$; and so on.

If the error variance is estimated from the squares of the differences $(Y_1 - Y_2, \cdots, Y_7 - Y_8)$, it also will be a function of the β's unless the (a_i, b_i) are the same for the two experimental units to which each treatment is applied. A study should be made of the effect of this result on the usual tests of main effects and interactions. If the (a_i, b_i) are known and the β's estimated by least squares, the mean square error (MSE) will no longer be a function of the β's.

If the (a_i, b_i) are known, there is an interesting problem in the best method of assigning treatments to the experimental units (and even in deciding what treatment combinations should be used) in order to obtain the best estimates of the β's and/or the optimal amount of A and B to be added to a given (a_i, b_i) soil. It is not clear to this writer if in this situation it would be desirable to divide the eight plots into two blocks of four plots each; perhaps blocking could be made on the basis of qualitative information on other factors. If quantitative information exists on the initial amounts of other nutrients, this should be utilized by inserting parameters in the model to reflect the effect of these nutrients on yield.

The obvious question to be asked is: Why are not most fertilizer experiments, for example, analyzed by the least-squares procedure? The answer is that adequate methods of determining pre-existing conditions usually are not available or they are too expensive and time-consuming; also there is the difficulty of calibrating the two sets of quantities: that in the soil (for example) and that added by the experimenter. These topics are receiving considerable attention in the analysis of fertilization studies at the present time. This writer suspects that there are many other types of experiments in which pre-existing conditions should be more adequately considered than simply by inserting block constants in our models.

One should mention that if the two factors A and B are compounds of various elements, some of them common to A and B, the analysis would indicate an interaction between the two factors. This is discussed at length in [1].

5. Alternative Models for Factorial Experiments

5.1 Introduction. This writer has been perturbed by the fact that the estimators of the main effects, for example, in the usual factorial model were not affected by any assumption regarding the presence or absence of interaction effects. As a matter of fact, experimenters have always been advised by this writer to analyze factorial experiments sequentially (see [3] and [6]) as follows:

1. First determine if interactions can be neglected. (This may take the form of an actual test of significance.)

2a. If interactions can be neglected, the usual main effects can be interpreted as being representative over various levels of the other factors.

2b. If interactions are important, one probably will want to consider the main effect of one factor for each combination of the other factors. In some population studies, main effects averaged over all levels of other factors may be desired, but it is necessary to be assured that the sample weights correspond to the population weights.

The writer has often considered reparametrizing the factorial model in order to have the parameters reflect main effects entirely free of interactions, two-factor interactions free of higher-order ones, and so forth. Then the estimators of the parameters are changed if the model is changed by making certain assumptions about the interactions. This is in line with estimates in traditional multiple regression analysis; it is recognized that the estimates of the regression coefficients will be affected by the inclusion or exclusion of other variables.

There is one serious drawback to such a new model: The estimated components will not be orthogonal to one another. Orthogonality is a useful property, because sums of squares are additive; hence, inferences can be made without reference to the other components. In subsequent sections, comparisons will be made of the traditional orthogonal model and the above-mentioned new model.

5.2 A Complete 2^3 Factorial Experiment. Let $\bar{P}$ represent the average yield of the r experimental units with all factors at the low level (often called the check plots); $\bar{A}$, average yield for those with only factor A at the high level and the other factors at the low level; $\overline{AB}$, average yield for those with only A and B at the high level; and so on. The standard model of Section 2 can be simplified considerably by considering orthogonal contrasts, such as

$$2\alpha' = \alpha_2 - \alpha_1 ,$$
$$4(\alpha\beta)' = [(\alpha\beta)_{11} - (\alpha\beta)_{12} - (\alpha\beta)_{21} + (\alpha\beta)_{22}] ,$$
$$8(\alpha\beta\gamma)' = [(\alpha\beta\gamma)_{222} + (\alpha\beta\gamma)_{112} + (\alpha\beta\gamma)_{121} + (\alpha\beta\gamma)_{211}] - [(\alpha\beta\gamma)_{111} + (\alpha\beta\gamma)_{122} + (\alpha\beta\gamma)_{212} + (\alpha\beta\gamma)_{221}] .$$

In this case $E(\bar{Y}_{ijk}) = \mu \pm \alpha' \pm \beta' \pm \gamma' \pm (\alpha\beta)' \pm \cdots \pm (\alpha\beta\gamma)'$, where the choice of + or − rests on the sign of the corresponding part for a given (ijk). Some examples are given below:

Mean yield	Signs for estimators μ	α'	β'	γ'	$(\alpha\beta)'$	$(\alpha\gamma)'$	$(\beta\gamma)'$	$(\alpha\beta\gamma)'$
$\bar{P}$	+	−	−	−	+	+	+	−
$\bar{A}$	+	+	−	−	−	−	+	+
$\overline{AB}$	+	+	+	−	+	−	−	−
$\overline{ABC}$	+	+	+	+	+	+	+	+

Examples of the estimators of these parameters are

$$8\alpha' \triangleq 8\boldsymbol{a}' = (\bar{A} + \overline{AB} + \overline{AC} + \overline{ABC}) - (\bar{P} + \bar{B} + \bar{C} + \overline{BC}) ,$$
$$8(\alpha\beta)' \triangleq 8\boldsymbol{ab}' = (\bar{P} + \bar{C} + \overline{AB} + \overline{ABC}) - (\bar{A} + \bar{B} + \overline{AC} + \overline{BC}) ,$$
$$8(\alpha\beta\gamma)' \triangleq 8\boldsymbol{abc}' = (\bar{A} + \bar{B} + \bar{C} + \overline{ABC}) - (\bar{P} + \overline{AB} + \overline{AC} + \overline{BC}) .$$

These estimators are uncorrelated; each has a variance of $\sigma^2/8r$, where σ^2 is the error variance. We note that $\boldsymbol{a}'$ is an unweighted average of four estimators of the effect of A: $(\bar{A} - \bar{P})/2$; $(\overline{AB} - \bar{B})/2$; $(\overline{AC} - \bar{C})/2$; $(\overline{ABC} - \overline{BC})/2$. In terms of main effects and interactions these four estimators have the same expectations only if the interactions vanish, as shown in the following:[3]

$$E(\bar{A} - \bar{P})/2 = \alpha' - (\alpha\beta)' - (\alpha\gamma)' + (\alpha\beta\gamma)' = \alpha/2 ,$$
$$E(\overline{AB} - \bar{B})/2 = \alpha' + (\alpha\beta)' - (\alpha\gamma)' - (\alpha\beta\gamma)' = [\alpha + (\alpha\beta)]/2 ,$$
$$E(\overline{AC} - \bar{C})/2 = \alpha' - (\alpha\beta)' + (\alpha\gamma)' - (\alpha\beta\gamma)' = [\alpha + (\alpha\gamma)]/2 ,$$
$$E(\overline{ABC} - \overline{BC})/2 = \alpha' + (\alpha\beta)' + (\alpha\gamma)' + (\alpha\beta\gamma)' = [\alpha + (\alpha\beta) + (\alpha\gamma) + (\alpha\beta\gamma)]/2 .$$

If one desires the average effect of A, averaged over these particular B and C treatments, $\boldsymbol{a}'$ is the proper estimator, because its average value is α' and it achieves minimum variance. However, the statistician has the obligation of informing the experimenter of the individual parts of this estimator. It is a mistake to recommend an over-all policy if separate recommendations under different combinations of the other variables are both feasible and profitable. It might happen that A would have no over-all effect,

[3] The constants on the extreme right will be explained later.

whereas it was harmful without B and beneficial with B. Experimenters recognize this situation; the statistician should provide an estimation procedure which takes account of such interactions when they exist. It would seem necessary first to determine if interactions could be neglected; if they cannot, one should then state the over-all results and individual results. Many texts on experimental designs emphasize the necessity of examining two-way and three-way tables of means. It might be desirable to have an operating procedure which tells one to compute such tables only when interactions are important.

A new model, which may more nearly coincide with the sequential nature of examining experimental results, is as follows:

$$\begin{aligned} E_3(\bar{P}) &= \pi , \\ E_3(\bar{A}) &= \pi + \alpha , \\ E_3(\overline{AB}) &= \pi + \alpha + \beta + (\alpha\beta) , \\ E_3(\overline{ABC}) &= \pi + \alpha + \beta + \gamma + (\alpha\beta) + (\alpha\gamma) + (\beta\gamma) + (\alpha\beta\gamma) . \end{aligned}$$

The symbol E_3 indicates a model with the three-factor interaction included. Note that the expectation of the yield, when two factors are together, always has an interaction added. The two sets of parameters are linearly related as follows:

$$\begin{aligned} (\alpha\beta\gamma) &= 8(\alpha\beta\gamma)' , \\ (\alpha\beta) &= 4[(\alpha\beta)' - (\alpha\beta\gamma)'] , \\ \alpha &= 2[\alpha' - (\alpha\beta)' - (\alpha\gamma)' + (\alpha\beta\gamma)'] , \\ \pi &= \mu - \alpha' - \beta' - \gamma' + (\alpha\beta)' + (\alpha\gamma)' + (\beta\gamma)' - (\alpha\beta\gamma)' . \end{aligned}$$

This indicates that the two-factor interactions will be the same (aside from a normalizing constant) only if the three-factor interaction is zero; also that main effects are comparable only when all interactions are zero.

The estimators for the new model are

$$\pi_3 \triangleq \boldsymbol{p}_3 = \bar{P} ; \qquad \boldsymbol{a}_3 = \bar{A} - \bar{P} ; \qquad \boldsymbol{ab}_3 = \overline{AB} - \bar{A} - \bar{B} + \bar{P} ;$$
$$\boldsymbol{abc}_3 = (\overline{ABC} + \bar{A} + \bar{B} + \bar{C}) - (\overline{AB} + \overline{AC} + \overline{BC} + \bar{P}) = 8\boldsymbol{abc}' .$$

The variances of these estimators are

$$\text{var}(\boldsymbol{p}_3) = \sigma^2/r ; \qquad \text{var}(\boldsymbol{a}_3) = 2\sigma^2/r ;$$
$$\text{var}(\boldsymbol{ab}_3) = 4\sigma^2/r ; \qquad \text{var}(\boldsymbol{abc}_3) = 8\sigma^2/r .$$

The subscript (3) indicates that this model includes the three-factor interaction.

Let us next assume that $(\alpha\beta\gamma)$ is negligible, so that this term can be dropped from $E(\overline{ABC})$; i.e., $E_2(\overline{ABC}) = \alpha + \beta + \gamma + (\alpha\beta) + (\alpha\gamma) + (\beta\gamma)$. In this case, we have eight equations to estimate seven unknowns; if least-squares procedures are used, the following matrix of coefficients is obtained:

p_2	a_2	b_2	c_2	ab_2	ac_2	bc_2	Yield sums
8	4	4	4	2	2	2	$S\bar{Y}_{ij} = \bar{G}$
4	4	2	2	2	2	1	$\bar{T}_a$
4	2	4	2	2	1	2	$\bar{T}_b$
4	2	2	4	1	2	2	$\bar{T}_c$
2	2	2	1	2	1	1	$\bar{T}_{ab}$
2	2	1	2	1	2	1	$\bar{T}_{ac}$
2	1	2	2	1	1	2	$\bar{T}_{bc}$

The sums $\bar{T}_a = \bar{A} + \overline{AB} + \overline{AC} + \overline{ABC}$; $\bar{T}_{ab} = \overline{AB} + \overline{ABC}$; and so forth.

The coefficient matrix is inverted quite easily by use of pattern matrix procedures (see [11]). The upper right-hand part of this symmetric matrix is

p_2	a_2	b_2	c_2	ab_2	ac_2	bc_2
7/8	−3/4	−3/4	−3/4	1/2	1/2	1/2
	3/2	1/2	1/2	−1	−1	0
		3/2	1/2	−1	0	−1
			3/2	0	−1	−1
				2	0	0
					2	0
						2

Hence

$$\begin{aligned} 4\boldsymbol{a}_2 &= -3\bar{G} + 6\bar{T}_a + 2\bar{T}_b + 2\bar{T}_c - 4\bar{T}_{ab} - 4\bar{T}_{ac} \\ &= 3(\bar{A} - \bar{P}) + [(\overline{AB} + \overline{AC} + \overline{BC}) - (\overline{ABC} + \bar{B} + \bar{C})]\,, \\ &= 4\boldsymbol{a}_3 - \boldsymbol{abc}_3\,, \end{aligned}$$

and

$$\begin{aligned} 2\boldsymbol{ab}_2 &= \bar{G} - 2\bar{T}_a - 2\bar{T}_b + 4\bar{T}_{ab} \\ &= (\overline{AB} - \bar{A} - \bar{B} + \bar{P}) + (\overline{ABC} - \overline{AC} - \overline{BC} + \bar{C}) \\ &= 2\boldsymbol{ab}_3 + \boldsymbol{abc}_3 = 8\boldsymbol{ab}'\,. \end{aligned}$$

The variances of these new estimators are var $(\boldsymbol{a}_2) = 3\sigma^2/2r$; var $(\boldsymbol{ab}_2) = 2\sigma^2/r$.

Thus if it can be safely assumed that $(\alpha\beta\gamma) = 0$, one can double one's information on the two-factor interaction effects and even increase that on the main effects by one-third. The gain in estimating $(\alpha\beta)$ is obtained because a second unbiased estimator is now available, $(\overline{ABC} - \overline{AC} - \overline{BC} + \bar{C})$, with the same variance as $\boldsymbol{ab}_3$; hence, the average of the two can be used.

And for α, we also have a second estimator, namely $(\overline{AB} + \overline{AC} + \overline{BC}) - (\overline{ABC} + \bar{B} + \bar{C})$. Since the variance of this estimator is $6\sigma^2/r$, as compared to $2\sigma^2/r$ for $\boldsymbol{a}_3$, the latter receives a triple weight in computing $\boldsymbol{a}_2$. As indicated in Section 2, this gain is to be balanced against bias in the estimates if $(\alpha\beta\gamma) \neq 0$. The expected values of the estimators, in this case, are

$$E(\boldsymbol{a}_2) = \alpha - (\alpha\beta\gamma)/4; \qquad E(\boldsymbol{ab}_2) = (\alpha\beta) + (\alpha\beta\gamma)/2 .$$

Hence the mean square errors are

$$\text{MSE}(\boldsymbol{a}_2) = 3\sigma^2/2r + (\alpha\beta\gamma)^2/16; \qquad \text{MSE}(\boldsymbol{ab}_2) = 2\sigma^2/r + (\alpha\beta\gamma)^2/4 .$$

If $(\alpha\beta\gamma)^2 < 8\sigma^2/r = \text{var}(\boldsymbol{abc}_3)$, an improvement is made by use of the biased estimator for α or for $(\alpha\beta)$. This is the same conclusion reached in Section 2. There is also a test of $H_0: (\alpha\beta\gamma) = 0$ by use of $\boldsymbol{abc}_3$ or by computing the mean square for $\boldsymbol{abc}_3$, $S\bar{Y}_{ij}^2 - (\boldsymbol{p}_2\bar{G} + \boldsymbol{a}_2\bar{T}_a + \cdots + \boldsymbol{bc}_2\bar{T}_{bc})$. This is compared with s^2/r, where s^2 is the estimate of σ^2. An over-all test for two-factor interactions can be made by comparing s^2/r with the mean square $[(\boldsymbol{ab}_2)^2 + (\boldsymbol{ac}_2)^2 + (\boldsymbol{bc}_2)^2]/6$. This test is simple, because the two-factor estimators are not correlated. If the lower right-hand corner of the inverse matrix (M, say) were not diagonal, the mean square would be $(\boldsymbol{b}'M^{-1}\boldsymbol{b})/3$, where $\boldsymbol{b}'$ is the row vector $(\boldsymbol{ab}_2, \boldsymbol{ac}_2, \boldsymbol{bc}_2)$.

If one could assume that all interactions were negligible, only the $\boldsymbol{p}$, $\boldsymbol{a}$, $\boldsymbol{b}$, and $\boldsymbol{c}$ columns and rows would be needed in the coefficient matrix. The resultant inverse is simply

$$\begin{array}{cccc} \boldsymbol{p}_1 & \boldsymbol{a}_1 & \boldsymbol{b}_1 & \boldsymbol{c}_1 \end{array}$$
$$\begin{bmatrix} 1/2 & -1/4 & -1/4 & -1/4 \\ & 1/2 & 0 & 0 \\ & & 1/2 & 0 \\ & & & 1/2 \end{bmatrix}.$$

Hence $4\boldsymbol{a}_1 = (\bar{A} - \bar{P}) + (\overline{AB} - \bar{B}) + (\overline{AC} - \bar{C}) + (\overline{ABC} - \overline{BC}) = 8\boldsymbol{a}'$; $\text{var}(\boldsymbol{a}_1) = \sigma^2/2r$. If the interaction effects are all zero, the information on main effects is quadrupled, because each of the estimators $(\bar{A} - \bar{P})$, $(\overline{AB} - \bar{B})$, $(\overline{AC} - \bar{C})$, and $(\overline{ABC} - \overline{BC})$ is an unbiased estimator of α and has the same variance, $2\sigma^2/r$. If these interactions are not zero, the mean square error is

$$\text{MSE}(\boldsymbol{a}_1) = \sigma^2/2r + [2(\alpha\beta) + 2(\alpha\gamma) + (\alpha\beta\gamma)]^2/16.$$

There is a gain if $[2(\alpha\beta) + 2(\alpha\gamma) + (\alpha\beta\gamma)]^2 < 24\sigma^2/r$. As expected, we have $\text{var}(2\boldsymbol{ab}_3 + 2\boldsymbol{ac}_3 + \boldsymbol{abc}_3) = 24\sigma^2/r$. Similarly, if $(\alpha\beta\gamma)$ is zero, but the two-factor interactions are not, we gain by ignoring these interactions (estimate only the main effects) if $[(\alpha\beta) + (\alpha\gamma)]^2 < 4\sigma^2/r$. Of course, $\text{var}(\boldsymbol{ab}_2 + \boldsymbol{ac}_2) = 4\sigma^2/r$. As before, one could test for goodness of fit by comparing s^2/r with the mean square for deviations from the main-effects model:

$$[S\bar{Y}_{ij}^2 - (\boldsymbol{p}_1\bar{G} + \boldsymbol{a}_1\bar{T}_a + \boldsymbol{b}_1\bar{T}_b + \boldsymbol{c}_1\bar{T}_c)]/4 .$$

This new type of model enables one to see at once the gain in efficiency to be realized on low-order effects if the model can be simplified, i.e., if high-order effects can be neglected. If previous experience does not indicate which, if any, of the interactions can be neglected, it would seem desirable to devise a sequential estimation procedure. One might first estimate only the main effects and test for goodness of fit; if this proved unsatisfactory, proceed to the model with two-factor interactions; then, if necessary, to the general model. Or perhaps one should start with the general model and study the effect of dropping interaction components from it. This can be accomplished by use of the procedure indicated previously for the two-factor components. There is a real need for operating rules for performing these preliminary tests in helping to formulate the model with the same data used to estimate the parameters in the model selected. A tentative procedure might be to conduct preliminary tests at, say, the 25% significance level. But we need some studies of the consequences of this or some other decision rule. Of course, as experience is accumulated in a given field of research, one may be able to formulate one's models more exactly, select factors which are more or less additive in their effects, or use new measurements for which an interaction-free model would be more nearly the correct one. It should be mentioned that it might be desirable to assume that certain two-factor interactions could be omitted from the model and others retained.

The best computing procedure to accomplish these objectives is somewhat uncertain. It is quite simple to compute the inverse matrix for each stage of estimation. In more complicated situations, one might prefer using only the forward solution of the abbreviated Doolittle or square-root computing procedure until one had decided on the best model. One could first compute the sum of squares attributable to main effects by considering only the main-effect columns and rows (plus that for π) and then compute the residual due to interactions; if this model proved unsatisfactory, one could proceed to the two-factor interaction columns and rows.

All that has been attempted here is to present a new method of looking at the results of factorial experiments. Whether the new parameters $[\alpha, (\alpha\beta), \cdots]$ are more easily interpreted than the standard orthogonal ones $[\alpha', (\alpha\beta)', \cdots]$ is a matter for further consideration. Perhaps standard analyses of factorial experiments, coupled with two-way and three-way tables of means, are sufficiently informative to enable the experimenter to make recommendations and plan future experiments. However, this writer would like a general model from which different estimates of parameters will be obtained if the model is changed. The new model has its greatest utility when the factor representatives (levels) can be ordered in some manner; the standard model of Section 2 is completely symmetrical as regards the factor representatives. Note that when interactions do exist, α represents the increase in yield of the second level over the first; hence, one needs to be able to designate which is first and which second—varietal trials may not fall in this category.

5.3 An Incomplete 2^3 Factorial Experiment. This new model was first introduced in the analysis of an incomplete factorial, in which the check plots could not be used because no yield would be forthcoming; the low level of each factor was absence of the factor. Hence only seven means are available, as $\bar{P} = 0$. Since there is a benchmark ($\pi = 0$), $\mu = \alpha' + \beta' + \gamma' - (\alpha\beta)' - (\alpha\gamma)' - (\beta\gamma)' + (\alpha\beta\gamma)'$. The expected values in terms of the standard orthogonal contrasts are as follows:

$$E(\bar{A}) = 2[\alpha' - (\alpha\beta)' - (\alpha\gamma)' + (\alpha\beta\gamma)'] ,$$

$$E(\overline{AB}) = 2[\alpha' + \beta' - (\alpha\gamma)' - (\beta\gamma)'] ,$$

$$E(\overline{ABC}) = 2[\alpha' + \beta' + \gamma' + (\alpha\beta\gamma)'] .$$

In this case the standard estimators also change when the model is simplified. When all interactions are included, we have the usual estimator (with $\bar{P} = 0$), e.g.,

$$8\boldsymbol{a}_3' = (\bar{A} + \overline{AB} + \overline{AC} + \overline{ABC}) - (\bar{B} + \bar{C} + \overline{BC}) ,$$

$$8\boldsymbol{ab}_3' = (\bar{C} + \overline{AB} + \overline{ABC}) - (\bar{A} + \bar{B} + \overline{AC} + \overline{BC}) ,$$

$$8\boldsymbol{abc}_3' = (\bar{A} + \bar{B} + \bar{C} + \overline{ABC}) - (\overline{AB} + \overline{AC} + \overline{BC}) .$$

Each estimator will have a variance of $7\sigma^2/64r$, but it will be correlated with each of the other estimators; the correlation coefficients are $\pm 1/7$.

If $(\alpha\beta\gamma)'$ is omitted from the model, the estimators are changed, e.g.,

$$28\boldsymbol{a}_2' = 4\bar{A} - 3\bar{B} - 3\bar{C} + 3\overline{AB} + 3\overline{AC} - 4\overline{BC} + 4\overline{ABC} .$$

Each estimator will have a variance of $3\sigma^2/28r$, and the correlation coefficients will be $\pm 1/6$. Finally, if all interactions are omitted from the model,

$$16\boldsymbol{a}_1' = 3\bar{A} - \bar{B} - C + 2\overline{AB} + 2\overline{AC} - 2\overline{BC} + \overline{ABC} .$$

Each estimator will have a variance of $3\sigma^2/32r$, and the correlation coefficients will be $-1/3$.

For this type of experiment, it is difficult to construct a realistic set of orthogonal contrasts; the standard components are not orthogonal, and they are not easily interpretable. However, the new model is the same as before, except that $\pi = 0$. Hence,

$$E(\bar{A}) = \alpha; \qquad E(\overline{AB}) = \alpha + \beta + (\alpha\beta); \cdots .$$

If all interactions are included, the estimators are

$$\boldsymbol{a}_3 = \bar{A}; \quad \boldsymbol{ab}_3 = \overline{AB} - \bar{A} - \bar{B}; \quad \boldsymbol{abc}_3 = \overline{ABC} - \overline{AB} - \overline{AC} - \overline{BC} + \bar{A} + \bar{B} + \bar{C} .$$

The variances of these estimators are

$$\text{var}(\boldsymbol{a}_3) = \sigma^2/r; \qquad \text{var}(\boldsymbol{ab}_3) = 3\sigma^2/r; \qquad \text{var}(\boldsymbol{abc}_3) = 7\sigma^2/r .$$

If $(\alpha\beta\gamma)$ can be omitted from the model, the upper right-hand side of the inverse matrix is

$$\frac{1}{7}\left[\begin{array}{ccc|ccc} \boldsymbol{a}_2 & \boldsymbol{b}_2 & \boldsymbol{c}_2 & \boldsymbol{ab}_2 & \boldsymbol{ac}_2 & \boldsymbol{bc}_2 \\ 6 & -1 & -1 & -4 & -4 & 3 \\ & 6 & -1 & -4 & 3 & -4 \\ & & 6 & 3 & -4 & -4 \\ \hline & & & 12 & -2 & -2 \\ & & & & 12 & -2 \\ & & & & & 12 \end{array}\right].$$

Hence

$$\begin{aligned} 7\boldsymbol{a}_2 &= 6\bar{T}_a - \bar{T}_b - \bar{T}_c - 4\bar{T}_{ab} - 4\bar{T}_{ac} + 3\bar{T}_{bc} \\ &= 6\bar{A} - \bar{B} - \bar{C} + \overline{AB} + \overline{AC} + \overline{BC} - \overline{ABC}\,. \end{aligned}$$

The variances have been reduced by 14% for main effects and by more than 40% for two-factor interactions. As indicated earlier, this gain is to be balanced against a bias in the estimates if $(\alpha\beta\gamma)$ is not zero. The mean square errors are

$$\begin{aligned} \text{MSE}\,(\boldsymbol{a}_2) &= 6\sigma^2/7r + (\alpha\beta\gamma)^2/49 \\ \text{MSE}\,(\boldsymbol{ab}_2) &= 12\sigma^2/7r + 9(\alpha\beta\gamma)^2/49\,. \end{aligned}$$

If $(\alpha\beta\gamma)^2 < 7\sigma^2/r = \text{var}\,(\boldsymbol{abc}_3)$, an improvement is made by use of the biased estimator for $\boldsymbol{\alpha}$ or for $(\alpha\beta)$. There is also a test of H_0: $(\alpha\beta\gamma)=0$, using the ratio of $\boldsymbol{abc}_3$ to its standard error; note that $\boldsymbol{abc}_3$ can be computed separately.

If all interactions can be neglected, the inverse matrix is

$$\frac{1}{8}\begin{bmatrix} 3 & -1 & -1 \\ -1 & 3 & -1 \\ -1 & -1 & 3 \end{bmatrix},$$

$8\boldsymbol{a}_1 = 3\bar{T}_a - \bar{T}_b - \bar{T}_c$, and $\text{var}\,(\boldsymbol{a}_1) = 3\sigma^2/8r$. If the interaction effects are all zero, one has almost tripled one's information on the main effects over the general case. If these interactions do not vanish, the mean square error is

$$\text{MSE}\,(\boldsymbol{a}_1) = \frac{1}{8}\left\{\frac{3\sigma^2}{r} + \frac{1}{8}\,[3(\alpha\beta + \alpha\gamma) - (\beta\gamma) + (\alpha\beta\gamma)]^2\right\}.$$

There is a gain if $[3(\alpha\beta + \alpha\gamma) - (\beta\gamma) + (\alpha\beta\gamma)]^2 < 40\sigma^2/r$. As expected, $\text{var}\,[3(\boldsymbol{ab}_3 + \boldsymbol{ac}_3) - \boldsymbol{bc}_3 + \boldsymbol{abc}_3] = 40\sigma^2/r$.

If $(\alpha\beta\gamma)$ is zero but the two-factor interactions do not vanish, we gain by ignoring these interactions (estimate only the main effects) if

$$[3(\alpha\beta + \alpha\gamma) - (\beta\gamma)]^2 < 216\sigma^2/7r; \text{ also } \text{var}\,[3(\boldsymbol{ab}_2 + \boldsymbol{ac}_2) - \boldsymbol{bc}_2] = 216\sigma^2/7r\,.$$

One could test for goodness of fit by comparing the mean square for deviations from the main-effects model, $S[\bar{Y}_{ij}^2 - (\boldsymbol{a}_1\bar{T}_a + \boldsymbol{b}_1\bar{T}_b + \boldsymbol{c}_1\bar{T}_c)]/4$, with s^2/r.

5.4 The New Model for an Incomplete 2^4 Factorial Experiment. We add $(D, AD, \cdots, ABCD)$ mean yields and have corresponding additions to the parameter and estimator space. If we assume the four-factor interaction can be neglected, the inverse matrix has the following elements for the $\boldsymbol{a}_3$, $\boldsymbol{ab}_3$, and $\boldsymbol{abc}_3$ rows (all divided by 15):

Row	a_3	b_3	c_3	d_3	ab_3	ac_3	ad_3	bc_3	bd_3	cd_3	abc_3	abd_3	acd_3	bcd_3
a_3	14	−1	−1	−1	−12	−12	−12	3	3	3	8	8	8	−7
ab_3	−12	−12	3	3	36	6	6	6	6	−9	−24	−24	6	6
abc_3	8	8	8	−7	−24	−24	6	−24	6	6	56	−4	−4	−4 .

Hence var $(\boldsymbol{a}_3) = 14\sigma^2/15r$, and so on.

If three-factor interactions can also be neglected, the $\boldsymbol{a}_2$ and $\boldsymbol{ab}_2$ rows in the inverse have the following elements (all divided by 44):

Row	a_2	b_2	c_2	d_2	ab_2	ac_2	ad_2	bc_2	bd_2	cd_2
a_2	28	−5	−5	−5	−14	−14	−14	8	8	8
ab_2	−14	−14	8	8	40	−4	−4	−4	−4	−4 .

Var $(\boldsymbol{ab}_2)$ is less than 40% of its value if three-factor interaction components must also be estimated.

Finally, if only main effects are of importance, var $(\boldsymbol{a}_1) = \sigma^2/5r$ and cov $(\boldsymbol{a}_1, \boldsymbol{b}_1) = -\sigma^2/20r$.

5.5 Concluding Remarks. It is interesting to note that the new model is essentially that used when matrix-inverting techniques are used with the model given in Section 2. Since the number of unknowns exceeds the number of equations, the matrix of coefficients in the normal equations will be singular. One method of reducing the number of coefficients has been the orthogonalizing device given in Section 5.2. Another is to equate certain of the constants to zero, e.g., for the 2^3 experiment, $0 = \alpha_1 = \beta_1 = \gamma_1 = (\alpha\beta)_{11} = (\alpha\beta)_{12} = (\alpha\beta)_{21} = \cdots = (\beta\gamma)_{21} = (\alpha\beta\gamma)_{111} = (\alpha\beta\gamma)_{112} = \cdots = (\alpha\beta\gamma)_{221} = 0$. Hence we would have μ replaced by π; α_2 by α; $\cdots$; and $(\alpha\beta\gamma)_{222}$ by $(\alpha\beta\gamma)$. Of course this is just another linear reparametrization. Similar procedures are followed for factors having more than two levels. For experiments with all factor combinations having the same number of replications (this excludes an incomplete factorial), the best computation procedure may be to compute the standard orthogonal estimates and then retransform these in accordance with the proper assumptions regarding the interactions. For incomplete factorials, the computations are somewhat tedious for any model; hence, it is probably best to set up the model first and estimate the parameters directly.

References

[1] Anderson, R. L. A comparison of discrete and continuous models in agricultural production analysis. Chap. 3 in E. L. Baum, E. O. Heady, and J. Blackmore (eds.), *Economic Analysis of Fertilizer Use Data.* Ames, Iowa: Iowa State College Press, 1956.

[2] Anderson, R. L. Some statistical problems in the analysis of fertilizer response data. Chap. 17 in E. L. Baum, E. O. Heady, J. T. Pesek, and C. G. Hildreth (eds.), *Fertilizer Innovations and Resource Use.* Ames, Iowa: Iowa State College Press, 1957.

[3] Anderson, R. L. Complete factorials, fractional factorials and confounding. Chap. 2 in Victor Chew (ed.), *Experimental Designs in Industry.* New York: Wiley, 1958.

[4] Binet, F. E., Leslie, R. T., Weiner, S., and Anderson, R. L. Analysis of confounded factorial experiments in single replications. North Carolina Agric. Exp. Sta. Tech. Bull. No. 113, NCAES, Raleigh, N.C., 1955.

[5] Cochran, W. G., and Cox, G. M. *Experimental Designs*, 2d ed. New York: Wiley, 1957.

[6] Anderson, R. L., and Bancroft, T. A. *Statistical Theory in Research.* New York: McGraw-Hill, 1952.

[7] Bose, R. C. Mathematical theory of the symmetrical factorial design. *Sankhyā*, 1947, **8**, 107–66.

[8] Kempthorne, O. *The Design and Analysis of Experiments.* New York: Wiley, 1952.

[9] Li, J. C. R. Design and statistical analysis of some confounded factorial experiments. Iowa State College Agric. Exp. Sta. Bull. No. 333, ISCAES, Ames, Iowa, 1944.

[10] Yates, F. The design and analysis of factorial experiments. Imperial Bureau of Soil Science Tech. Comm. No. 35, IBSS, Harpenden, Herts, England, 1937.

[11] Roy, S. N., and Sarhan, A. E. On inverting a class of patterned matrices. *Biometrika*, 1956, **43**, 227–31.

6 A LIMITATION OF THE OPTIMUM PROPERTY OF THE SEQUENTIAL PROBABILITY RATIO TEST[1]

T. W. ANDERSON, Columbia University

MILTON FRIEDMAN, University of Chicago

1. Introduction and Summary

The sequential probability ratio test is widely used to decide which of two hypotheses to accept or, more generally, to decide whether a parameter is large or small. A justification of its use is that the procedure possesses the optimum property of minimizing the expected sample size. However, as we shall show (Section 2), for the binomial distribution there exists a class of tests with a stronger optimum property.

This class of tests consists of curtailed single-sampling plans with acceptance number 0; in such plans, which have been widely used in sampling inspection, sample items are inspected one at a time until either an item is found to be defective, in which case the lot in question is rejected, or a designated number of items have been found to be nondefective, in which case the lot is accepted. Such a procedure never requires more observations for an individual sample sequence than any other test which gives the same probability of acceptance (between 0 and 1) at some value of p, the true percentage defective $(0 < p < 1)$, and which always leads to acceptance when $p = 0$; it requires fewer observations for some sample sequences $(0 < p < 1)$. This optimal property is clearly a strong one.

Peculiarly enough, the curtailed single-sampling plan is not necessarily a sequential probability ratio test (Section 3). There seems to be a contradiction: two different tests each of which can be proved to be optimal. The resolution of the contradiction is that, when it arises, there is no

[1] Research carried out at the Center for Advanced Study in the Behavioral Sciences, Stanford, California. The theorem of Section 2 (which stimulated this paper) was derived by Milton Friedman in 1945 while a member of the Statistical Research Group at Columbia University, of which Harold Hotelling was Official Investigator.

sequential probability ratio test that gives as low a "power" as the curtailed single-sampling plan; the sequential probability ratio test that comes closest will give greater "power," and it pays for this by requiring more observations on the average (Section 4).

The consideration of these two classes of tests reveals a limitation on the optimum property of the sequential probability ratio test as it is generally defined, at least for cases involving a discrete variate. That is, there can be more than one set of tests, not completely overlapping, for which similar optimum properties hold. We consider other limitations on the sequential probability ratio test (Section 5). The existence of such tests is related to the uniqueness of the tests; it is shown that only one sequential probability ratio test achieves a given pair of probabilities of error (Section 6). Finally, this study suggests that the usual way of defining these tests may not be the best way to satisfy the needs of the user of the tests (Section 7).

2. A Special Class of Sequential Procedures and a Strong Optimal Property

Consider sampling from a binomial distribution with parameter p; that is, let

$$\Pr\{X = 1\} = p , \qquad \Pr\{X = 0\} = 1 - p = q , \tag{2.1}$$

say. It may be convenient to think of sampling from a lot of items, each of which can be classified as a defective, $X = 1$, or a nondefective, $X = 0$. Then p is the proportion of defectives in the lot, which is assumed to be large enough so that it can be treated as an infinite population. If the proportion p is sufficiently small, we wish to accept the lot; if p is large, we wish to reject it.

A sequential sampling procedure consists in taking observations in sequence: $x_1, x_2, \cdots$, and at some point in the sequence stopping observation and deciding to accept or reject the lot. More precisely, at least one observation x_1 is taken. After m observations are taken ($m = 1, 2, \cdots$), there is a rule (based on $x_1, \cdots, x_m$) which indicates whether to stop sampling and accept the lot, to stop sampling and reject the lot, or to take one more observation and use the rule based on $m + 1$ observations. In this paper we consider only nonrandomized procedures.[2] We also assume that for each p the probability is 1 of coming to a decision; that is, the probability of stopping sampling before more than n observations are taken approaches 1 as n approaches infinity. For any specific procedure we designate by L_p the probability of accepting the lot when the proportion defective is p ($0 \leqq p \leqq 1$). (This is the operating characteristic.)

Let C be the curtailed single-sampling plan defined as follows: After taking m observations ($m = 1, \cdots, n - 1$) stop sampling and reject the lot

[2] Charles Stein (in unpublished work) has considered the effect of randomization on some of the problems treated in this paper.

if $x_m = 1$ and continue sampling if $x_m = 0$; after taking n observations stop sampling and reject the lot if $x_n = 1$ and accept the lot if $x_n = 0$. In other words, if no defective appears among the first n sample items, accept the lot; if a defective appears, stop sampling and reject the lot.

The plan has one constant to be adjusted, namely, the integer n. The probability of accepting the lot, L_p, is the probability of n nondefectives in a sample of n, namely

$$L_p(C) = (1 - p)^n = q^n \,. \tag{2.2}$$

It is evident that $L_0(C) = 1$, $L_p(C)$ decreases, and $L_1(C) = 0$. One way to fix n is to impose a requirement on $L_p(C)$ at some particular value of p, for example, the requirement that $L_{p_1}(C) \leqq \beta$ for some p_1 $(0 < p_1 < 1)$ and β $(0 < \beta < 1)$; that is, that the probability of accepting the lot should not exceed β if the proportion of defectives is as high as p_1. Then n is chosen as the smallest integer for which

$$(1 - p_1)^n \leqq \beta \,. \tag{2.3}$$

Theorem 1. *For any procedure E satisfying*

$$L_0(E) = 1 \,, \tag{2.4}$$

$$L_{p_1}(E) \leqq \beta \,, \tag{2.5}$$

where $0 < \beta < 1$, the number of observations required to reach a decision is for each sample sequence at least as large as the number for the curtailed single-sampling plan C. If the number is the same for every sample, then E is identical to C.

Proof. If $p = 0$, every observation is $x_i = 0$. Condition (2.4) implies that for any procedure E, the sequence of observations $0, 0, \cdots$ leads to acceptance. Since the probability must be 1 (for $p = 0$) of reaching a decision after a finite number of observations, there must be a number n^* such that the lot is accepted at the n^*th observation when $x_1 = 0, \cdots, x_{n^*} = 0$. Then $L_{p_1}(E)$ is at least equal to the probability of this sequence, and Condition (2.5) implies

$$(1 - p_1)^{n^*} \leqq \beta \,, \tag{2.6}$$

which implies $n^* \geqq n$. For any sequence with $x_1 = 0, \cdots, x_n = 0$, the number of observations required to reach a decision is, therefore, the same for E and for C if $n^* = n$ and larger for E than for C if $n^* > n$.

The preceding discussion implies that a sequence $x_1 = 0, \cdots, x_m = 0$ leads to continuation for E at the mth stage, $m = 1, \cdots, n - 1$. Now consider a sequence with $x_1 = 0, \cdots, x_{l-1} = 0, x_l = 1$, $l = 1, \cdots, n$, which under C leads to termination at the lth observation. Unless E has the same decision rule at the lth observation as C, such a sequence requires more observations to terminate under E than under C. ||

This optimum property is a strong one. For *every* sample sequence, C

is at least as good as any other procedure E in that it leads to a decision with no more observations. If E is different from C for some sequences of observations, C is better than E in this sense. This, of course, implies the weaker optimum property that the expected sample size under C is no greater than that under E for every value of p.

The result may be applied directly to problems in which it is desired solely to specify the risk of acceptance or rejection at a single point. For example, it might be desired solely to specify the risk of accepting a product with a stated percentage of defectives—the consumer's risk—and not to specify the producer's risk. For such a problem, a curtailed single-sampling plan C would be indicated.

The problem typically considered is that in which it is desired to set risks at two points, say p_0 and p_1, and to provide that at p_1 the risk of acceptance be less than or equal to β, and at p_0 the risk of rejection be less than or equal to α. If α and β happen to satisfy exactly the equations $1 - \alpha = (1 - p_0)^n$ and $\beta = (1 - p_1)^n$ for the same integral value of n, then it follows from Theorem 1 that the curtailed single-sampling plan C with maximum sample size n will never require a larger average number of observations than any other test for which $L_{p_1} \leqq \beta$ and $L_{p_0} \geqq 1 - \alpha$; hence it is a uniformly best plan for all values of p for such cases as well.

3. Is the Curtailed Single-sampling Procedure with Acceptance Number 0 a Sequential Probability Ratio Test?

The sequential probability ratio test has been shown by Wald and Wolfowitz [1] to have certain optimum properties. Since the curtailed single-sampling procedure C has an optimum property, it is of interest to compare the two.

Let $f_0(x)$ and $f_1(x)$ be two probability functions or two densities corresponding to hypotheses H_0 and H_1, respectively. The sequential probability ratio test for testing H_0 against H_1 is defined by two numbers[3] A and B $(0 < B < 1 < A < \infty)$ and the following rules: after m observations $(x_1, \cdots, x_m)$ have been taken $(m = 1, 2, \cdots)$, if

$$B < \prod_{i=1}^{m} \frac{f_1(x_i)}{f_0(x_i)} < A \,, \tag{3.1}$$

take another observation; if

$$\prod_{i=1}^{m} \frac{f_1(x_i)}{f_0(x_i)} \leqq B \,, \tag{3.2}$$

accept the hypothesis H_0 that sampling is from $f_0(x)$ [i.e., reject the hypothesis H_1 that sampling is from $f_1(x)$]; and if

$$A \leqq \prod_{i=1}^{m} \frac{f_1(x_i)}{f_0(x_i)} \,, \tag{3.3}$$

[3] The conditions $B < 1 < A$ are not necessary for the definition of the procedure but are necessary for the proof of the optimum property.

reject the hypothesis[4] H_0 (that is, accept H_1). Denote such a procedure by W (for Wald [2]). The optimum property is that for any other procedure E for which $L_{f_0}(E) \geqq L_{f_0}(W)$ and $L_{f_1}(E) \leqq L_{f_1}(W)$, the expected sample size is at least as large as for procedure W when sampling from $f_0(x)$ and $f_1(x)$, respectively.

The optimum property of the curtailed single-sampling plan is stronger than that of the sequential probability ratio test (since the former is in terms of every sample sequence), and the optimality is relative to a wider class of procedures (that is, procedures meeting only a requirement at p_1) than that of the sequential probability ratio test. These facts suggest that the curtailed single-sampling plan must always be a sequential probability ratio test. However, we must examine a little more closely what this means. There is one sense in which the curtailed single-sampling plan is always a special case of the Wald test, and another in which it may be a special case but need not be.

The essential point is that the definition of a sequential probability ratio test involves the specification of two parameter values, p_0 and p_1, as well as two probabilities of error (accepting H_0 when H_1 is true and accepting H_1 when H_0 is true). Construction of a curtailed single-sampling plan C, on the other hand, involves the specification of only one parameter value, p_1, and only one probability, the probability of accepting a lot when $p = p_1$. To put the curtailed single-sampling plan in terms comparable to the sequential probability ratio test we must associate some H_0 with our H_1, and there is no unique way to do so.

Under these circumstances what does it mean to ask whether a curtailed single-sampling plan is a special case of a sequential probability ratio test? One possible meaning is to ask whether there exists any H_0 which when combined with the H_1 of the curtailed single-sampling plan will yield a sequential probability ratio test identical with the curtailed single-sampling plan. The answer to this question is affirmative, as can readily be proved.

For our special binomial problem, H_0 is the hypothesis that $p = p_0$; H_1, that $p = p_1$; and we shall require that $p_0 < p_1$. In order for the curtailed single-sampling plan to be a sequential probability ratio test in the sense just defined, there must exist some $p_0 < p_1$ and some numbers A and B, with $B < 1 < A$, such that

$$B < \frac{(1-p_1)^m}{(1-p_0)^m} < A\,, \qquad m = 1, \cdots, n-1\,, \tag{3.4}$$

$$\frac{(1-p_1)^n}{(1-p_0)^n} \leqq B\,, \tag{3.5}$$

$$A \leqq \frac{(1-p_1)^{m'-1}}{(1-p_0)^{m'-1}} \cdot \frac{p_1}{p_0}\,, \qquad m' = 1, \cdots, n\,. \tag{3.6}$$

[4] A method of randomization is to randomize between accepting H_0 and continuing sampling when equality holds in (3.2) and to randomize between rejecting H_0 and continuing sampling when equality holds in (3.3).

These inequalities correspond to the inequalities (3.1), (3.2), and (3.3), for the binomial case where $f_i(x) = p_i^x(1-p_i)^{1-x}$, $i = 0, 1$, under the tentative assumption that the curtailed single-sampling plan is a sequential probability ratio test. Inequality (3.4) indicates continuation at the mth step (all 0's), (3.5) indicates accceptance of the lot and H_0 (n 0's), and (3.6) indicates rejection of the lot and H_0 (one 1).

If there exists a constant B to define the curtailed single-sampling plan as a sequential probability ratio test, it must satisfy (3.4) and (3.5); that is,

$$(3.7) \qquad \frac{(1-p_1)^n}{(1-p_0)^n} \leqq B < \frac{(1-p_1)^m}{(1-p_0)^m}, \qquad m = 1, \cdots, n-1 .$$

Since the right-hand side of (3.7) is smallest when m has its largest value, the limits (3.7) are tightest when $m = n - 1$. Clearly for any value of p_0 one can find a B to satisfy (3.7).

If there exists a constant A to define the curtailed single-sampling plan as a sequential probability ratio test, it must satisfy (3.4) and (3.6); that is,

$$(3.8) \qquad \frac{(1-p_1)^m}{(1-p_0)^m} < A \leqq \frac{(1-p_1)^{m'-1}}{(1-p_0)^{m'-1}} \cdot \frac{p_1}{p_0}, \qquad \begin{array}{l} m = 1, \cdots, n-1 , \\ m' = 1, \cdots, n . \end{array}$$

Since $(1-p_1)^m/(1-p_0)^m$ is a decreasing function of m, the limits in (3.8) are tightest when $m = 1$ and $m' = n$. However, we require $1 < A$. Hence, A must satisfy

$$(3.9) \qquad 1 < A \leqq \frac{(1-p_1)^{n-1}}{(1-p_0)^{n-1}} \cdot \frac{p_1}{p_0} .$$

The question then is to determine the set of p_0 for which the third expression in (3.9) is greater than the first expression. The function $(1-p_0)^{n-1}p_0$ has the derivative $(1-p_0)^{n-2}(1-np_0)$. Hence, it increases from a value of 0 at $p_0 = 0$ to a maximum at $p_0 = 1/n$; then it decreases to $p_0 = 1$. Thus if $p_1 \leqq 1/n$, the third expression in (3.9) is greater than 1 for every $p_0 < p_1$, and an A can be found that satisfies (3.9). If $p_1 > 1/n$, the third expression in (3.9) is greater than 1 for sufficiently small p_0 but not for all $p_0 < p_1$. Thus there exist values of p_0 for which an A can be found, but there will be some values of p_0 for which A cannot be found. The conclusion is that, given p_1, there always exists a p_0 such that the curtailed single-sampling plan is a sequential probability ratio test for p_1 and this p_0.

However, this is clearly not a fully satisfactory answer to the question whether a curtailed single-sampling plan C is a special case of the sequential probability ratio test. The proof of the optimum properties of the curtailed single-sampling plan holds for every null hypothesis H_0. Pick any such hypothesis, and let the values of α and β be those given by some curtailed single-sampling plan C (that is, some n). Then we have proved that this test requires fewer observations at both p_0 and p_1 than any other test yielding equal or greater power. For this proof to be a special case of the proof of the optimum properties of the sequential probability ratio test, the question whether a curtailed single-sampling plan is a special case of the sequential probability ratio test must be given a more demanding meaning, namely, whether for *every* H_0 that is relevant (namely, for which $p_0 < p_1$), there exists a sequential probability ratio test that is identical with the curtailed single-sampling plan.

The answer to this question is clear from the preceding analysis. If $p_1 \leqq 1/n$, then both (3.7) and (3.9) can be satisfied for every $p_0 < p_1$. If $p_1 > 1/n$, Inequality (3.7) can be satisfied for every $p_0 < p_1$, but there are always some values of $p_0 < p_1$ for which Inequality (3.9) cannot be satisfied. Hence, the curtailed single-sampling plan is not necessarily a sequential probability ratio test in this more demanding sense.

4. Relationship Between the Optimum Properties of the Tests

We seem to have a discrepancy. The curtailed single-sampling plan has a strong optimum property; the sequential probability ratio test has an optimum property; but there are cases when the curtailed single-sampling plan is not a sequential probability ratio test. How does this happen? The answer is that the theorem about the sequential probability ratio test asserts that for any H_0 and H_1 and any values of α and β *which can be achieved* by a sequential probability ratio test, there exists no other test yielding the same or lower values of α and β for which the average number of observations is smaller at either p_0 or p_1. But not all values of α and β can be achieved by a sequential probability ratio test defined for a particular H_0 and H_1. Whenever, for a given n, p_1, β, the value of p_0 is such that the sequential probability ratio test does not reduce to the curtailed single-sampling plan, then there is no sequential probability ratio test defined for these p_0 and p_1 that yields a probability of acceptance at p_0 as small as that yielded by the curtailed single-sampling plan.

THEOREM 2. *Given a curtailed single-sampling plan C with a probability of acceptance of β at p_1 and a probability of rejection of α at p_0 ($< p_1$), if this procedure is not a sequential probability ratio test defined in terms of p_0 and p_1, then there is no sequential probability ratio test that achieves these probabilities of errors.*

PROOF. Suppose a sequential probability ratio test is not the same as the curtailed single-sampling plan for a given n, p_0, and p_1, but yields the same value of $\beta = (1 - p_1)^n$. Then it must require continuation of sampling for a sample of n items having a value of zero. (Otherwise it would lead to acceptance for this and some other cases as well; hence the probability of acceptance when p_1 is true would be higher than β.) But this means that

$$B < \frac{(1 - p_1)^n}{(1 - p_0)^n}, \tag{4.1}$$

since the right-hand side is the probability ratio for such a sample.

A standard inequality for a sequential probability ratio test is given by Wald ([2], p. 41):

(4.2) $$L_{p_1}(W) \leqq BL_{p_0}(W) .$$

Hence

(4.3) $$L_{p_1}(W) < \frac{(1-p_1)^n}{(1-p_0)^n} L_{p_0}(W) .$$

Since

(4.4) $$L_{p_1}(W) = (1-p_1)^n = \beta ,$$

we have

(4.5) $$1 - \alpha = (1-p_0)^n < L_{p_0}(W) . \quad ||$$

The point of the theorem is that at some p_0 and p_1 a curtailed single-sampling plan can achieve an α and β that cannot be achieved by a sequential probability ratio test. The sequential probability ratio test for the given p_0, p_1, and β must achieve a smaller α, and this improvement is paid for by greater expected sample sizes at p_0 and p_1.

5. Probabilities of Errors for the Sequential Probability Ratio Test

It is evident that, at least for binomial distributions, sequential probability ratio tests cannot achieve all of the (α, β) points that can be achieved by (nonrandomized) sequential procedures. Consequently, it is possible for classes of tests other than the sequential probability ratio test to have similar optimal properties, as is the case for the curtailed single-sampling plan C. This raises questions about the set of (α, β) points that can be achieved by the sequential probability ratio test. We shall not make an exhaustive study of this subject, but simply note a few points.

For the binomial distribution, a sequential probability ratio test is defined for a given p_0 and p_1 by two constants B and A. At the nth step we continue sampling if

(5.1) $$B < \frac{p_1^m q_1^{n-m}}{p_0^m q_0^{n-m}} < A ,$$

where m is the number of defectives among the n observations. The inequalities can be rewritten as

(5.2) $$c_0 + sn < m < c_1 + sn ,$$

where

(5.3) $$s = \log \frac{q_0}{q_1} \Big/ \log \frac{p_1 q_0}{p_0 q_1} ,$$

(5.4) $$c_0 = \log B \Big/ \log \frac{p_1 q_0}{p_0 q_1} ,$$

(5.5) $$c_1 = \log A \Big/ \log \frac{p_1 q_0}{p_0 q_1} .$$

The conditions $B < 1 < A$ are equivalent to $c_0 < 0 < c_1$, and $s > 0$. In the plane of n and m, the region (5.2) lies between two lines, each with slope s and with intercepts c_0 and c_1, respectively. Each sample point can be plotted in this plane as a point (n, m) with integer-valued coordinates. Two procedures, defined by pairs of intercepts (c_0, c_1) and (c_0^*, c_1^*), are equivalent if there is no point (n, m), $n \geqq m \geqq 0$, between the lines $y = c_0 + sx$ and $y = c_0^* + sx$ or between $y = c_1 + sx$ and $y = c_1^* + sx$.

THEOREM 3. *If the slope s is rational, there is a denumerable number of sequential probability ratio tests. If the slope is irrational, there is a nondenumerable number.*

PROOF. Suppose $s = M/N$ (where M and N are relatively prime integers). Then a point (n, m) is on the line $y = c + sx$ for a value of $c = (Nm - Mn)/N$, that is, a rational value of c. The only lines needed in defining sequential probability ratio tests in this case are those with intercepts of the form shown. There is a denumerable number of such lines, and hence a denumerable number of pairs of such lines.

Now suppose s is irrational. Suppose the lower line of a sequential probability ratio test is $y = c_0 + sx$ for some $c_0 < -1$, and consider upper lines $y = c_1 + sx$ for $0 < c_1 < 1$. We shall show that any two values of c_1 define different tests and hence that the number of tests is nondenumerable. Consider the points (n, m) between $y = sx$ and $y = sx + 1$. (Because s is irrational, no point can fall on either line.) Let a_n be the distance of (n, m) above the line $y = sx$. Then $a_n = [sn] + 1 - sn$, and $1 - a_n = sn - [sn]$ is the fractional part of sn. (Read $[x]$ as the largest integer not greater than x.) It is known ([3], p. 71) that this sequence is uniformly distributed; that is,

$$\lim_{n\to\infty} \frac{\text{number of } (1 - a_j) \leqq z,\ j = 1, \cdots, n = z}{n} \quad (5.6)$$

This implies that between the lines $y = c_1^* + sx$ and $y = c_1^{**} + sx$ $(c_1^* < c_1^{**})$ there are an infinite number of points (n, m). Since $0 < s < 1$, some of these points (n, m) above $c_1^* + sx$ can be reached from points $(n - 1, m - 1)$ below $c_1^* + sx$ (continuation points). Hence, the two procedures are different. Thus the number of procedures is at least equal to the number of choices of c_1, which is nondenumerable. ||

The interesting problems considered in this paper occur not only for the binomial distribution, but also for other distributions. They clearly do so for other discrete distributions. What is less obvious is that they arise even when the distributions are continuous, if the distributions of the probability ratio are discrete. As an extreme example, consider rectangular distributions with unit range. Let

$$f_i(x) = \begin{cases} 1, & \text{if } \theta_i - \frac{1}{2} \leqq x \leqq \theta_i + \frac{1}{2}, \\ 0, & \text{otherwise}, \end{cases} \quad (5.7)$$

for $i = 0, 1$. Then if $\theta_0 < \theta_1$ and $\theta_1 - \theta_0 < 1$, the possible values of the probability ratio are

$$\frac{f_1(x)}{f_0(x)} = \begin{cases} 0, & \theta_0 - \frac{1}{2} \leqq x < \theta_1 - \frac{1}{2}, \\ 1, & \theta_1 - \frac{1}{2} \leqq x \leqq \theta_0 + \frac{1}{2}, \\ \infty, & \theta_0 + \frac{1}{2} < x \leqq \theta_1 + \frac{1}{2}, \end{cases} \tag{5.8}$$

where $\infty = 1/0$. Then if $B < 1 < A$ there is only one possible sequential probability ratio test. The procedure is to continue sampling as long as the observations fall in the interval $\theta_1 - \frac{1}{2} \leqq x \leqq \theta_0 + \frac{1}{2}$, to stop and accept H_0 as soon as an observation is less than $\theta_1 - \frac{1}{2}$, and to stop and reject H_0 as soon as an observation is greater than $\theta_0 + \frac{1}{2}$. The probability of each error is 0. The expected sample size under each hypothesis is $1/(\theta_1 - \theta_0)$. Clearly there are other procedures with smaller expected sample size. For example, one can take a single observation and accept H_0 if it is less than c ($\theta_1 - \frac{1}{2} \leqq c \leqq \theta_0 + \frac{1}{2}$) and reject H_0 if it is at least equal to c.[5]

Consider now distributions for which the probability ratio has a continuous distribution under each hypothesis. First consider the procedures based on the probability ratio $f_1(x)/f_0(x)$ of one observation; that is, procedures which reject H_0 if

$$\frac{f_1(x_1)}{f_0(x_1)} > k . \tag{5.9}$$

As k varies from 0 to ∞, α varies from 1 to 0, β varies from 0 to 1, and the set of (α, β) points forms a closed connected curve. As is well known, this curve is convex.

We can show that no sequential probability ratio test with $B < 1 < A$ can achieve an (α, β) point that is above and to the right of the set of (α, β) points achieved by use of one observation; that is, such that there is an (α, β) point based on one observation for which each coordinate is less than or equal to that of the point achieved by the sequential probability ratio test. If the sequential probability ratio test is not effectively based on one observation, the expected sample number under one or both hypotheses is greater than one. Then the existence of a single-observation test with both probabilities of error at least as small as those of the sequential probability ratio test would contradict the optimum-property result.

The result can be extended to cases where the conditions $B < 1 < A$ are not satisfied. Furthermore, it can be shown that each (α, β) point below the curve achieved by one-observation procedures can be achieved by a sequential probability ratio test. The argument is based on some continuity properties. The idea is that given an arbitrary (α, β) below the curve for one-observation procedures, there is a one-observation procedure with the same β and greater α; then the A can be increased and B can simultaneously be decreased while holding β fixed and decreasing α. Since Wijsman [4] has studied this situation, we shall not go into it further.

[5] More generally one could divide $(\theta_1 - \frac{1}{2}, \theta_0 + \frac{1}{2})$ into $(\theta_1 - \frac{1}{2}, c)$, (c, d), and $(d, \theta_0 + \frac{1}{2})$. If an observation falls in the first interval, accept H_0; if it falls in the second interval, take another observation; and if it falls in the third, reject H_0. This type of procedure (which may apply in other problems) gives a kind of randomization.

6. Uniqueness of a Sequential Probability Ratio Test

We have discussed some aspects of the existence of a sequential probability ratio test with specified probabilities of errors at the two hypotheses. A related question is the uniqueness of the sequential probability ratio test. That is, given two probability density functions or discrete probability functions, $f_0(x)$ and $f_1(x)$, is there more than one sequential probability ratio test that achieves a given α (probability of rejecting H_0 when H_0 is true) and β (probability of accepting H_0 when H_0 is not true)? The answer is that there is not. Weiss [5] has shown this for the case in which $f_1(x)/f_0(x)$ has a continuous distribution with positive probability on every interval in $(0, \infty)$. Here we give a proof that covers the general case and is simpler than the one given by Weiss.[6] In particular, the case of the binomial distribution is included.

THEOREM 4. *There is at most one sequential probability ratio test for testing H_0: $f(x) = f_0(x)$ against H_1: $f(x) = f_1(x)$, that achieves a given α and β.*

PROOF. Two tests are considered the same if they differ only on sample sequences that have probability 0 under each hypothesis. Suppose there is a pair of numbers A and B, $B < 1 < A$, such that the corresponding sequential probability ratio test achieves α and β. We shall show that any other pair A^* and B^* which gives the same probabilities of error defines the same test. First, suppose $B^* \leqq B$, $A \leqq A^*$. By the optimum property of the sequential probability ratio test [1] the expected sample sizes are the same for the two procedures, computed under each hypothesis. However, any sample sequence terminating on the nth step under the second procedure must terminate under the first procedure by at least the nth step. Hence

$$\sum_{m=1}^{n} p_i(m) \geqq \sum_{m=1}^{n} p_i^*(m) , \tag{6.1}$$

where $p_i(m)$ and $p_i^*(m)$ are the probabilities of terminating at exactly m steps under H_i. Since the expected sample size is the same for the two procedures, it follows that $p_i(m) = p_i^*(m)$. Thus for every sample sequence (except for a set of probability 0), the number of observations to reach a decision is the same under the two procedures. If a sample sequence leads to termination at the nth step and acceptance of H_0 under the second procedure (a probability ratio not greater than B^*), it must do so under the first (not greater than B). Similarly, a sample sequence leading to termination at the nth step and rejection of H_0 under the second procedure must do so under the first. Thus the two procedures are the same. The case of $B \leqq B^*$, $A^* \leqq A$ is treated the same way.

Now consider $B^* \leqq B$, $A^* \leqq A$. A sequence leading to acceptance at step m under the second procedure must lead to acceptance at least by step m under the first procedure because the sequence of probability ratios has not

[6] The proof in this paper is simpler than that of Weiss, but it uses a more powerful result, namely, the optimum property of the sequential probability ratio test, which is more difficult to prove.

reached A^* ($\leqq A$) by the mth step and at the mth step is less than or equal to B^* ($\leqq B$). Since the probabilities of acceptance for the two procedures are the same under each hypothesis, each sample sequence that leads to acceptance for one procedure must do so for the other. Similarly, any sequence leading to rejection under one procedure must do so under the other. It follows that the test with acceptance numbers B and A^* must yield the same decision for each sequence as would either (B^*, A^*) or (B, A). From the first case we see that all three of these procedures are the same.

The case $B \leqq B^*$, $A \leqq A^*$ is treated similarly. ||

7. Specification of Procedures

A statistician designing a test and accustomed to employing sequential probability ratio tests for this purpose will typically ask the user to specify two probabilities of error, α and β, at two parameter values termed H_0 and H_1. On this basis the statistician selects a sequential probability ratio test defined for this H_0 and H_1. However, as was shown above, he may not be able to find one that gives precisely[7] the specified α and β; if not, he is likely to choose one that comes as close as possible in the direction of yielding greater "power." In this event, he may be able to do better for the user. There may exist a sequential probability ratio test defined for a different pair of alternative hypotheses, H_0' and H_1', that nonetheless does better for the H_0 and H_1 specified by the user than the one chosen by the statistician; that is, which comes closer to yielding the specified α and β at H_0 and H_1 and has a smaller average number of observations at H_0 and H_1. Even if the statistician asks the user to specify values H_0, H_1, α, and β, there is no reason why he has to restrict himself to the subclass of tests defined for this H_0 and H_1. Yet the method of formulating the sequential probability ratio test suggests that he do so, and the usual "cookbook" rules tell him to do so.

We can make this point in a different way, particularly for the binomial case. As we have seen, if the number of defectives or values equal to 1 are plotted on the ordinate and the number of observations on the abscissa, a sequential probability ratio test is defined by a pair of parallel straight lines. It takes three parameters to describe any specific test—the slope of the lines and the two intercepts. Now it happens that the values of p_1 and p_0 specify the slope of the lines independently of the values of α and β [see (5.3)], but this limits more narrowly than is necessary the set of values of α and β that can be attained.[8] On the one hand, for any set of values

[7] We assume that he can compute exactly the operating-characteristic curve of the test, though in practice he may only be able to approximate it.

[8] This is another way of presenting the proof in Section 4. The curtailed single-sampling plan is defined by two parallel straight lines which have a slope less than $1/n$ and intercepts on the y axis, one positive, the other negative, that differ by less than unity. For some values of p_0 and p_1, the sequential probability ratio test specifies a slope greater than $1/n$, hence cannot be equivalent to the curtailed single-sampling plan.

of p_1, p_0, α, and β, there may be more than one triplet of slope and intercepts that will yield them. If one of these is the sequential probability ratio test, it will require no larger an average sample size than any of the others, but if none is, we cannot now in general say which one will have this property. On the other hand, for any triplet of slope and intercepts there are an indefinite number of combinations of p_1, p_0, α, and β for which that triplet is the sequential probability ratio test.

One direction of investigation is to reformulate the sequential tests in terms of specifications other than p_0, p_1, α, and β that may establish a closer connection between the questions asked of the user and the appropriate test; then to adapt the proofs of optimum properties accordingly. One reformulation that comes readily to mind is to specify a p^*, the probability of one decision at p^*, and the desired slope of the operating characteristic curve at p^*. However, this may not be the most suitable formulation.

More fundamentally, there is no reason why the definition of a class of tests needs to be in the same terms as the questions asked the user. The former should be chosen for mathematical and statistical convenience; the latter in light of the operating function to be served by the test. A fully satisfactory sampling-inspection theory would separate the two and require as information from the user only such items as the cost of inspection and the value or cost of rejecting lots of various characteristics. The rest would be the statistician's business.

References

[1] Wald, A., and Wolfowitz, J. Optimum character of the sequential probability ratio test, *Ann. Math. Stat.*, 1948, **19**, 326-39.

[2] Wald, A. *Sequential Analysis.* New York: Wiley, 1947.

[3] Pólya, G., and Szegö, G. *Aufgaben und Lehrsätze aus der Analysis*, Vol. 1. New York: Dover, 1945.

[4] Wijsman, R. A. On the existence and uniqueness of Wald's sequential test (unpublished paper).

[5] Weiss, L. On the uniqueness of Wald sequential tests. *Ann. Math. Stat.*, 1956, **27**, 1178-81.

7 DECISION THEORY AND THE CHOICE OF A LEVEL OF SIGNIFICANCE FOR THE *t*-TEST

KENNETH J. ARROW
Stanford University

1. Introduction

It is very remarkable that the rapid development of decision theory has had so little effect on statistical practice. Ever since the classic work of Neyman and Pearson [1], it has been apparent that, in the choice of a test for a hypothesis, the power of the test should play a role coordinate with the level of significance. Yet to this day the choice of a critical level for a test statistic is made in practice by an arbitrary conventional choice of the probability of a Type I error; hardly if ever is there explicit justification of the choice made by comparison with the power curve. Twenty-five years after his original paper on the subject, Neyman [2] felt again impelled to call attention to the disregard of the power concept in choosing a level of significance and a sample size in the testing of hypotheses.

It is true that the original Neyman-Pearson formulation, while making clear the importance of considering the two types of errors, gave no basis for balancing them. The problem is compounded for most practical cases, such as in tests for the mean of a normal distribution, by the multiplicity of the alternative hypotheses. To the single probability of a Type I error must be compared a whole power function. Each test thus defines a pair consisting of a number (the level of significance) and a function (the power function); a preference-ordering of the resulting space is by no means easy to define. It is not surprising that attention was largely concentrated on comparisons among tests with the same level of significance, so that only an ordering of power functions was needed, though even this was not a trivial matter when the alternative hypothesis was composite.

A decisive conceptual step forward was the introduction of the weight function by Wald [3]. In the case of testing a hypothesis about the mean of a normal distribution, Wald's theory may be sketched as follows: Let μ be the true mean, $\bar{\mu}$ the hypothetical mean, a the action of accepting the null hypothesis, r the action of rejecting. Let $w_a(\mu)$ be the loss incurred by accepting the null hypothesis when the true value of the mean is μ, and $w_r(\mu)$ the loss incurred by rejecting the null hypothesis when the true value of the mean is μ. It is assumed that the loss is zero if the action most appropriate to the true value of the mean is taken, so that we may assume that

$$w_a(\mu) \geqq 0, \quad w_r(\mu) \geqq 0, \quad \text{at least one equals zero.} \tag{1}$$

This assumes that the statistician is charged only with that portion of the loss which results from not knowing the true value of μ, a point especially stressed by L. J. Savage ([4], pp. 163–64, 169–71).

Let T be the rejection region in sample space, and $P(T|\mu)$ the probability of rejection if the true value of the mean is μ; we suppose for the moment that this is the only unknown parameter. Then $1 - P(T|\mu)$ is the probability of acceptance. The expected loss or *risk*, $R(\mu)$, is then given by

$$R(\mu) = w_a(\mu)[1 - P(T|\mu)] + w_r(\mu)P(T|\mu) . \tag{2}$$

Before simplifying and specializing, let us ask how useful is this formalism. It brings out explicitly the importance of the costs of different types of errors. For the statistician or his client to use this procedure, there are still two difficult questions to be answered. The first is the choice of the weight functions, $w_a(\mu)$ and $w_r(\mu)$. Knowledge of these implies considerable sophistication with respect to the implications of incorrect decisions. The second is the preference-ordering among risk functions, $R(\mu)$. For the latter purpose, Wald proposed the minimax principle; for each test, consider the greatest possible risk corresponding to some value of the parameter μ, that is, $\max_\mu R(\mu)$, and then choose that test for which this maximum risk is a minimum. Since the original formulation of the minimax principle, there has been considerable critical analysis of this and alternative principles for ordering risk functions (such as subjective *a priori* probabilities) as tested against intuitively plausible conditions. Perhaps the most nearly definitive treatment is that of Milnor [5], who showed in effect that *every* proposed ordering principle contradicts at least one reasonable axiom.

In this paper, we propose to use the minimax principle and to specialize the weight functions so that the value information demanded for their application is reduced to a reasonable minimum. Only tests for the mean of a normal population (with known or unknown variance) are considered. Instead of the present procedure, in which the test is determined by fixing a *statistical* level of significance, the proposed approach is to require fixing a difference in the parameter space which is significant in the more literal sense of the word, that is, a difference which is meaningful from the viewpoint of the losses associated with incorrect choices. The details of the tests for the one-sided and two-sided cases with known and unknown variances are given, along with some computational approximations for large samples.

2. Economically Significant Differences and the Equal-probability Test

It is necessary to have some information on the relative values of different errors in order to find an optimal test, but we wish to require as little as possible. Let us permit ourselves the following single question: For what value of the unknown mean do we consider the corresponding Type II error to be just as costly as a Type I error? We will call the difference between this value of the mean and the hypothetical value an *economically significant difference*, denoted by δ_0. Then the test proposed here has the requirement that the probability of a Type II error, when the difference between the true and hypothetical means is the economically significant difference, shall equal the probability of a Type I error. Such a test will be termed an *equal-probability test*.

In this section, the concepts will be developed for the case of one-sided alternatives to a hypothesis about the mean of a normal distribution with known variance. To begin with, assume that the hypothetical mean is 0, the known variance is 1, and the sample size is 1. The minimax test will be briefly recapitulated and shown to include the equal-probability test as a special case.

If x is the observation, then the optimal test will, of course, have the form of accepting when $x \leqq x_c$ and rejecting otherwise, for some *critical test-value* x_c. The cost of rejecting the null hypothesis will be assumed to fall toward zero as μ increases from zero, while the cost of accepting becomes greater as μ increases. In view of Equation (1), we may write

$$(3) \qquad w_r(\mu) = 0 \quad \text{for } \mu > \mu_0 , \qquad w_a(\mu) = 0 \quad \text{for } \mu < \mu_1 ,$$

where

$$(4) \qquad 0 < \mu_0 < \mu_1 .$$

Under these conditions it has been shown (Allen [6]; Karlin [7], p. 307) that the minimax critical test-value x_c is determined by the equation

$$(5) \qquad \max_{\mu \leqq \mu_0} w_r(\mu) P(x \geqq x_c | \mu) = \max_{\mu \geqq \mu_1} w_a(\mu) P(x \leqq x_c | \mu) .$$

By suitable choice of the units of measurement of losses, we can assume without loss of generality that

$$(6) \qquad w_r(0) = 1 ,$$

which is the cost of a Type I error. It is reasonable to assume that $w_r(\mu)$ falls off very rapidly for $\mu > 0$. Indeed, from a decision-theoretic viewpoint, there is no special parameter value which can be designated as the hypothetical value of the mean; all the information is already contained in the two weight functions. We may assume that the class of problems usually referred to as testing of hypotheses is characterized by the rapid fall of the function $w_r(\mu)$ beyond the value usually referred to as the hypothetical mean. To a first approximation, then, we may suppose that $w_r(\mu) = 0$ for $\mu > 0$, or

$$(7) \qquad \mu_0 = 0 .$$

There is great variety in the possible shapes of the function $w_a(\mu)$. As already remarked, it ought to be monotone-increasing; from general theoretical principles it ought to be bounded.[1] From the definition of the economically significant difference δ_0 and from the normalization (6),

$$w_a(\delta_0) = 1 \; . \tag{8}$$

The simplest function satisfying these conditions is the step-function

$$w_a(\mu) = \begin{cases} 0, & \text{for } \mu < \delta_0 \, , \\ 1, & \text{for } \mu \geqq \delta_0 \, . \end{cases} \tag{9}$$

One can of course argue that this class of functions is too simple; however, for practical purposes it is very useful to work with a one-parameter family of functions so that the particular weight function to be used in any given case can be determined by a single piece of information, such as a knowledge of δ_0.

If we assume the values given in Equations (7) and (9), then, for the normal distribution (and a great many others), Equation (5) reduces to

$$P(x \geqq x_c | \mu = 0) = P(x \leqq x_c | \mu = \delta_0) \; , \tag{10}$$

that is, the equal-probability test.[2]

In the case of the normal distribution, (10) leads immediately to the simple characterization

$$x_c = \delta_0/2 \; . \tag{11}$$

To examine the implications of this critical test-value as opposed to those implied by a conventional level of significance, let us first consider what values of δ_0 are reasonable. The most convenient way of judging this is to assume that we seek to characterize the distribution so as to make predictions about future values of the random variable. Suppose, for example, that we are interested in knowing the probability that $x \geqq 0$ (as in comparing a new proposed treatment with an existing one). Under the null hypothesis, the probability is, of course, .50. For values of $\mu = .25$, .50, 1.00, 2.00, and 3.00, the probability that $x \geqq 0$ becomes .60, .69, .84, .98, and 1.00 (to two decimal places), respectively. Even the smallest of these figures is considerably different from .50, the probability under the null hypothesis. It would appear that the Type II error corresponding to $\mu = 1.00$ would in almost any circumstances be more costly than the corresponding Type I error, so that δ_0 would be less than 1 and frequently less than .50.

Suppose, then, to illustrate the equal-probability test, we assume that $\delta_0 = .50$. Then the critical value x_c is .25, which corresponds to a level of significance in the classical sense of .40. This may seem outrageously high when compared with the usual level of .05 or less. But consider a Type II error corresponding to $\mu = .5$. Under the equal-probability test, the probability of this error is again .40, under the classical test with a .05 significance level, .87! A dim realization of the very low power of the classical

[1] Otherwise a form of the St. Petersburg paradox will occur; this was first observed by Menger [8], who, however, drew other inferences. See also Blackwell and Girshick ([9], pp. 104-11); and Arrow [10].

[2] A more general class of tests which includes the equal-probability test has been suggested by Lehmann ([11], p. 1171).

test for small samples is in the minds of those scientific workers who, after reporting absence of significant differences, add that the difference would have been significant if the sample size had been larger.

This simple example also shows that the equal-probability test will increase the real usefulness of the significance level, for it will now represent the probabilities of two errors whose meaning is easily understood, so that the statistician and the reader can immediately form a judgment about the reliability of the whole statistical procedure. Present practice does not provide any such figure, other than the uncertain information implied by the sample size.[3]

The formulation of the test for samples of any size and (known) standard deviation of any magnitude proceeds along obvious lines. It is worth noting, however, that the intuitive considerations on reasonable magnitudes of δ_0 remain valid if the difference of the means is measured in standard deviation units, not absolute units. From this point on, we will assume that *the economically significant difference is always measured in standard deviation units.* Let

$$z = \sqrt{N}\,(\bar{x} - \tilde{\mu})/\sigma\,, \tag{12}$$

where N is the sample size, $\bar{x}$ the sample arithmetic mean, $\tilde{\mu}$ the hypothetical mean, and σ the standard deviation of the underlying population. Then, of course, z has a normal distribution with unit variance and a mean which is zero under the null hypothesis. If δ_0 is the economically significant difference for x, $\sqrt{N}\,\delta_0$ is the economically significant difference for z, and the critical test-value is

$$z_c = \sqrt{N}\,\delta_0/2\,, \tag{13}$$

the null hypothesis being accepted for values of z smaller than z_c and rejected for larger ones. To illustrate, again with $\delta = .50$ and $\sigma = 1$ for a sample size of 9, the test with a .05 level of significance has a .56 probability of a Type II error at $\mu = .5$, while the equal-probability test has a level of significance (and a probability of Type II error at $\mu = .5$) of .23. The equal-probability and .05-level-of-significance tests coincide for a sample size of 43. For larger sample sizes, the constant-level-of-significance test provides excessive protection against Type II errors. With the same example, for a sample of 100, the .05-level-of-significance test gives a virtually zero probability of Type II error at $\mu = .5$, while the equal-probability test gives a probability of either error of less than .01. The slight loss of protection against Type II errors seems to be far more than made up by the great reduction in the probability of Type I errors.

Indeed, as N approaches infinity, the inadequacy of the constant-level-of-significance test becomes more glaring. The probabilities of all kinds of errors approach zero in the equal-probability test, while in the constant-

---

[3] Publication of the power of the test for at least one alternative hypothesis would be helpful even with the constant-level-of-significance test.

level-of-significance test the probability of a Type I error is maintained constant; all the gain in information due to larger sample size is taken out in the form of increased power. There is no logical foundation for this behavior.

3. Two-sided Test with Known Variance

In the two-sided case, the functions $w_a(\mu)$ and $w_r(\mu)$ are symmetric functions of μ about $\mu = \tilde{\mu}$ (the hypothetical mean). If we proceed as before, we assume that

$$w_r(\mu) = \begin{cases} 1, & \text{if } \mu = \tilde{\mu} , \\ 0, & \text{if } \mu \neq \tilde{\mu} , \end{cases} \tag{14}$$

$$w_a(\mu) = \begin{cases} 1, & \text{if } |\mu - \tilde{\mu}|/\sigma \geqq \delta_0 , \\ 0, & \text{if } |\mu - \tilde{\mu}|/\sigma < \delta_0 . \end{cases}$$

The analog of (10) is

$$P(|z| \geqq z_c | \mu = \tilde{\mu}) = P(|z| \leqq z_c | \mu = \tilde{\mu} + \delta_0 \sigma) . \tag{15}$$

Since z is normally distributed with mean $\sqrt{N}(\mu - \tilde{\mu})/\sigma$ and unit variance, Equation (15) is equivalent to

$$2 - 2\Phi(z_c) = \Phi(z_c - \sqrt{N}\,\delta_0) - \Phi(-z_c - \sqrt{N}\,\delta_0) , \tag{16}$$

which defines a relation between the critical test value z_c and the quantity $\sqrt{N}\,\delta_0$; here, $\Phi(x)$ is the cumulative normal distribution. The null hypothesis is rejected when $|z| \geqq z_c$ and accepted otherwise.

4. The Case of Unknown Variance

In the one-sided case, the previous considerations lead to the formula

$$P(t \geqq t_c | \mu = \tilde{\mu}) = P(t \leqq t_c | \mu = \tilde{\mu} + \delta_0 \sigma) , \tag{17}$$

where $t = \sqrt{N}(\bar{x} - \tilde{\mu})/s$, and s is the sample standard deviation. The statistic t has the cumulative non-central t-distribution with parameter $\lambda = (\mu - \tilde{\mu})/\sigma_{\bar{x}}$, equal to $0, \sqrt{N}\,\delta_0$, in evaluating left- and right-hand sides of (17), respectively; when $\lambda = 0$, the distribution is, of course, Student's distribution. The probabilities also depend on the number of degrees of freedom ($N - 1$ in the problem being studied), though this dependence is not indicated in the notation. Equation (17) can be solved by successive approximation with the aid of a table of the non-central t-distribution (Resnikoff and Lieberman [12]).

For samples of size greater than 30, it appears that $t_c = \sqrt{N}\,\delta_0/2$, the formula for the normal case, will be a good approximation. To see this, we use the remark of Wallis ([13], pp. 57–59) that $\bar{x} + ks$ is approximately normally distributed with mean $\mu + k\sigma$ and standard deviation $\sigma\sqrt{(2 + k^2)/2N}$. (Wallis gives this formula as $\sigma\sqrt{(1/N) + k^2/2(N-1)}$, but for N of any significant magnitude, the two should essentially coincide.) Then, if $t_0 = t_c/\sqrt{N}$ and $\psi(x) = 1 - \Phi(x) = \Phi(-x)$, we obtain

$$
\begin{aligned}
P(t \leqq t_c | \mu = \tilde{\mu} + \delta_0 \sigma) &= P(\bar{x} - \tilde{\mu} - t_0 s \leqq 0 | \mu = \tilde{\mu} + \delta_0 \sigma) \\
&= \Phi[(t_0 - \delta_0)/\sqrt{(2 + t_0^2)/2N}] \\
&= \psi[(\delta_0 - t_0)/\sqrt{(2 + t_0^2)/2N}] \, .
\end{aligned} \tag{18}
$$

Let p be the common value of the two sides of (17), and let k_p be the normal deviate exceeded with probability p. Then, from (18), we get

$$(\delta_0 - t_0)/\sqrt{(2 + t_0^2)/2N} = k_p \, . \tag{19}$$

Squaring both sides and simplifying yields

$$(1 - k_p^2/2N)t_0^2 - 2\delta^0 t^0 + (\delta_0^2 - k_p^2/N) = 0 \, . \tag{20}$$

It is also true that $P(t \leqq t_c | \mu = \tilde{\mu}) = 1 - p$. In Equation (20), set $\delta_0 = 0$, and replace k_p by $k_{1-p} = -k_p$:

$$(1 - k_p^2/2N)t_0^2 - k_p^2/N = 0 \, . \tag{21}$$

If we subtract (21) from (20), we find that $t_0 = \delta_0/2$, or $t_c = \sqrt{N}\,\delta_0/2$.

For the two-sided case, the critical test-value t_c is the solution of

$$P(|t| \geqq t_c | \mu = \tilde{\mu}) = P(|t| \leqq t_c | \mu = \tilde{\mu} + \delta_0 \sigma) \, , \tag{22}$$

which can in principle be solved with the aid of non-central t-tables. For sample size greater than 30, Wallis's procedure sketched above can be applied to yield another approximation. We can write

$$
\begin{aligned}
P(|t| \leqq t_c | \mu = \tilde{\mu} + \delta_0 \sigma) &= P(\bar{x} - \tilde{\mu} - t_0 s \leqq 0 | \mu = \tilde{\mu} + \delta_0 \sigma) \\
&\quad - P(\bar{x} - \tilde{\mu} + t_0 s \leqq 0 | \mu = \tilde{\mu} + \delta_0 \sigma) \\
&= \psi[(\delta_0 - t_0)/\sqrt{(2 + t_0^2)/2N}] \\
&\quad - \psi[(\delta_0 + t_0)/\sqrt{(2 + t_0^2)/2N}]
\end{aligned} \tag{23}
$$

with the aid of (20). Let c_2 and c_1 be the two expressions in brackets in (23), respectively, and let $2p$ be the common value of the two sides of (22). Then

$$
\begin{aligned}
2p &= P(|t| \geqq t_c | \mu = \tilde{\mu}) = 2P(t \geqq t_c | \mu = \tilde{\mu}) = 2P(\bar{x} - \tilde{\mu} - t_0 s \geqq 0 | \mu = \tilde{\mu}) \\
&= 2\psi[t_0/\sqrt{(2 + t_0^2)/2N}] \, ,
\end{aligned} \tag{24}
$$

so that

$$t_0/\sqrt{(2 + t_0^2)/2N} = k_p \, . \tag{25}$$

If we solve for t_0 from (25) and substitute into c_1 and c_2, we find that

$$c_1 = k_p + \delta_0\sqrt{(2N - k_p^2)/2} \, , \qquad c_2 = -k_p + \delta_0\sqrt{(2N - k_p^2)/2} \, . \tag{26}$$

Thus p has to satisfy the equation, from (23),

$$\psi(c_2) - \psi(c_1) = 2p \, , \tag{27}$$

which can be solved by successive approximations using only normal tables. Once $2p$ is found, t_c can be found from the relation $P(|t| \geqq t_c | \mu = \tilde{\mu}) = 2p$.

5. Tests of Regression Coefficients

Write a regression relation as

(28) $$\tilde{Y}_X = \mu_Y + \beta(X - \bar{X}) ,$$

where $\tilde{Y}_X$ is the conditional mean of Y given X, and $\bar{X}$ is the sample mean. If β_0 is a hypothetical value for the regression coefficient, then the null hypothesis asserts that the conditional mean is

(29) $$\tilde{Y}_{0X} = \mu_Y + \beta_0(X - \bar{X}) .$$

If the null hypothesis is false, an error is made concerning the conditional mean:

(30) $$\tilde{Y}_X - \tilde{Y}_{0X} = (\beta - \beta_0)(X - \bar{X}) .$$

It is this difference which is analogous to the error in the hypothetical mean in the univariate case if we know in advance for what value of X we are going to use the regression relation. More generally, we may imagine planning to use the regression relation for many different values of X. Let μ_X be the mean of these future values of X and let σ_X^2 be their variance. Then the expected mean square error in the conditional mean is

(31) $$L = |\beta - \beta_0| \, [\sigma_X^2 + (\mu_X - \bar{X})^2]^{\frac{1}{2}} .$$

As before, the economic significance of this loss will be judged relative to the standard deviation, in this case, the conditional standard deviation of Y given X, $\sigma_{Y \cdot X}$, which, in usual regression theory, we assume to be independent of X. Then the *economically significant difference* δ_0 will be a value of $L/\sigma_{Y \cdot X}$ such that the cost of a Type II error will equal that of a Type I error.

The test of significance is based, of course, on the statistic

(32) $$t = (b - \beta_0)/s_b ,$$

where b and s_b are the usual estimates of β and σ_b, respectively. The statistic t has, in general, the non-central t-distribution with parameter

(33) $$\lambda = (\beta - \beta_0)/\sigma_b .$$

If we find the value of $\beta > \beta_0$ for which $L/\sigma_{Y \cdot X} = \delta_0$ and then substitute into Equation (33), we find

(34) $$\lambda_0 = \{\delta_0/[\sigma_X^2 + (\mu_X - X)^2]^{\frac{1}{2}}\}(\sigma_{Y \cdot X}/\sigma_b) .$$

Notice that the ratio $\sigma_{Y \cdot X}/\sigma_b$ is equal to $\Sigma(X - \bar{X})^2$, the summation being over sample values, and so does not involve any unknown parameters.

In the one-sided case, the equal-probability test requires choosing t_c so that

(35) $$P(t \geqq t_c | \lambda = 0) = P(t \leqq t_c | \lambda = \lambda_0) ,$$

and in the two-sided case, so that

(36) $$P(|t| \geqq t_c | \lambda = 0) = P(|t| \leqq t_c | \lambda = \lambda_0) .$$

The approximations of Section 4 can be used, with $\sqrt{N}\,\delta_0$ replaced by λ_0.

References

[1] Neyman, J., and Pearson, E. S. On the problem of the most efficient tests of statistical hypotheses. *Phil. Trans. Roy. Soc. (London)*, Ser. A, 1933, **231**, 289-337.

[2] Neyman, J. The use of the concept of power in agricultural experimentation. *J. Indian Soc. Agric. Stat.*, 1958, **9**, 9-17.

[3] Wald, A. *Statistical Decision Functions*. New York: Wiley, 1950.

[4] Savage, L. J. *The Foundations of Statistics*. New York: Wiley, 1954.

[5] Milnor, J. Games against nature. Chap. 4 in R. M. Thrall, C. H. Coombs, and R. L. Davis (eds.), *Decision Processes*. New York: Wiley, 1954.

[6] Allen, S. G., Jr. A class of minimax tests for one-sided composite hypotheses. *Ann. Math. Stat.*, 1953, **24**, 295-98.

[7] Karlin, S. Pólya-type distributions, II, *Ann. Math. Stat.*, 1957, **28**, 281-308.

[8] Menger, K. Das Unsicherheitsmoment in der Wertlehre. *Z. Nationalökonomie*, 1933, **4**, 459-85.

[9] Blackwell, D., and Girshick, M. A. *Theory of Games and Statistical Decisions*. New York: Wiley, 1954.

[10] Arrow, K. J. Bernoulli utility indicators for distributions over arbitrary spaces. Technical Report No. 57, ONR Contract N6onr 25133, Stanford, Calif.: Department of Economics, Stanford University, July 30, 1958.

[11] Lehmann, E. Significance level and power. *Ann. Math. Stat.*, 1958, **29**, 1167-76.

[12] Resnikoff, G. J., and Lieberman, G. J. *Tables of the Non-Central t-Distribution*. Stanford, Calif.: Stanford Univ. Press, 1957.

[13] Wallis, W. A. Use of variables in acceptance inspection for percent defective. Chap. 1 in C. Eisenhart, M. Hastay, and W. A. Wallis (eds.), *Selected Techniques of Statistical Analysis*. New York: McGraw-Hill, 1947.

8 SIMULTANEOUS COMPARISON OF THE OPTIMUM AND SIGN TESTS OF A NORMAL MEAN

R. R. BAHADUR

Indian Statistical Institute, Calcutta

1. Introduction and Summary

Certain non-parametric tests are often used as preliminary tests in cases where superior tests based on normality or other assumptions are also applicable. This practice is sometimes, if not always, motivated by the idea that, in a given case and for given data, the non-parametric test provides an easily and quickly obtained estimate or prediction of the result of a subsequent, more refined analysis of the same data. It seems therefore of interest to study in such cases the joint distribution of the significance levels attained by the superior test and the non-parametric one. It may be added that the probability distribution of the level attained by a test is of interest apart from the present context of comparison, since it is usual to compute and record the level actually attained by the preferred test as a sort of sufficient statistic, independently of whether the hypothesis being tested is rejected in the particular case.

This paper presents in some detail an example of simultaneous comparison in the sense of the preceding paragraph, namely that of the optimum and sign tests of the mean of a univariate normal population, when the sample size is large. It would be interesting to carry out the parallel analyses for various other pairs of tests, e.g., for the two-sample t-test and the Wilcoxon-Mann-Whitney test, but the exact treatment of most other pairs presents certain technical difficulties. An inexact method, which can be applied to a wide class of pairs of tests, is described and exemplified in [1].

Turning now to the particular example of simultaneous comparison which is treated here, let $X_1, X_2, \cdots$ denote a sequence of independent observations from a normal population with mean μ and variance σ^2. It is desired to test $\mu = 0$ against $\mu > 0$. Suppose for the present that σ is known. Then, for any given $n = 1, 2, \cdots$, the level, $L_0(n)$ say, attained by the optimum test based on a sample of size n, is given by

$$L_0(n) = G(Y_n)\,, \tag{1}$$

where

(2) $$Y_n = \sqrt{n}\,\bar{X}_n/\sigma\,, \qquad \bar{X}_n = \sum_{i=1}^{n} X_i/n\,,$$

and

(3) $$G(x) = \int_x^\infty (2\pi)^{-\frac{1}{2}} \exp(-t^2/2)\,dt\,, \qquad -\infty < x < \infty\,.$$

Let

(4) $$R_n = \{\text{Total number of } X_i\text{'s} > 0,\ i = 1, 2, \cdots, n\}\,.$$

The significance level $L_s(n)$ attained by the sign test is given by

(5) $$L_s(n) = \sum_{r=R_n}^{n} \binom{n}{r}\left(\frac{1}{2}\right)^n .$$

The case when $\mu = 0$ is considered in Section 2. In this case, $L_0(n)$ is uniformly distributed in the interval (0, 1), and $L_s(n)$ tends to be so distributed as $n \to \infty$. The asymptotic joint distribution of $L_0(n)$ and $L_s(n)$ may be described by saying that

(6) $$(L_0(n), L_s(n)) \to (G(U), G(V))$$

in distribution, where G is given by Equation (3), and U and V are jointly normal random variables with $E(U) = E(V) = 0$, $E(U^2) = E(V^2) = 1$, and $E(UV) = (2/\pi)^{\frac{1}{2}}$ ($\doteq .8$). It thus appears that in large samples the joint distribution of L_0 and L_s is fairly well concentrated on the diagonal $L_0 = L_s$ of the unit square.

In the case when $\mu > 0$, both $L_0(n)$ and $L_s(n)$ tend to zero with probability one as $n \to \infty$. The respective rates of convergence are examined in Section 3. Let

(7) $$\theta = \mu/\sigma\,, \qquad p(\theta) = 1 - G(\theta)\,, \qquad q(\theta) = 1 - p(\theta)\,,$$

and

(8) $$\tau = (2p^p q^q)/\exp(\theta^2/2)\,.$$

It should be noted that p is the probability that a single X be positive. It is shown in Section 3 that

(9) $$\lim_{n\to\infty} [L_0(n)/L_s(n)]^{1/n} = \tau\,, \qquad \text{with probability one.}$$

Roughly speaking, one might therefore say that the asymptotic relation between the two attained levels is $L_0(n) = \tau^n L_s(n)$. It can be shown that τ is a strictly decreasing function of θ, varying from 1 to 0. Some values of τ are listed in Table 1.

TABLE 1

θ	0.0	0.5	1.0	1.5	2.0	2.5	3.0	3.5	4.0	∞
τ	1.00	.95	.78	.51	.24	.08	.02	.00	.00	.00

It follows from (9) and the stated properties of τ that (i) $L_0(n)$ is much smaller than $L_s(n)$ in large samples, i.e., there is considerable loss of information in using the sign test; (ii) the loss increases with increasing sample size; and (iii) roughly speaking, the loss for a given large n is an increasing function of θ. These conclusions are hardly detractions of the sign test as a preliminary test since, according to (i), the level attained by this test is almost certainly on the conservative side in large samples.

Section 4 of the paper concerns the question of how to estimate or to predict the actual value of L_0 in the given case, given only n and R_n. The tentative solutions considered suggest, in conformity with conclusions (ii) and (iii) above, that the data become more and more insufficient for the required prediction as n increases, and also as θ increases.

The estimates of L_0 and L_s given in Section 3 also afford an asymptotic evaluation of the relative sample sizes required by the sign test and the optimum test in order to achieve comparable values of L. It is shown in Section 5 that, if $\mu > 0$, then (roughly speaking) the sign test applied to a large sample of size n is equivalent to the optimum test applied to the first $n\varphi$ observations in the sample, where

$$\varphi(\theta) \doteq 2 \log_e (2p^p q^q)/\theta^2 , \qquad 0 < \varphi < 1 . \tag{10}$$

Consequently, φ could be regarded as the efficiency of the sign test when θ obtains. Some values of φ are given in Table 2.

TABLE 2

θ	0.0	0.5	1.0	1.5	2.0	2.5	3.0	3.5	4.0	∞
φ	.64	.60	.51	.40	.29	.21	.15	.11	.09	.00

It is interesting to note that

$$\lim_{\theta \to 0} \varphi(\theta) = 2/\pi . \tag{11}$$

This value for the asymptotic efficiency of the sign test in the neighborhood of $\theta = 0$ is the same as that obtained by Pitman [2] and by Lehmann and Hodges ([3], p. 333) by certain comparisons of power functions. The present interpretation of (11) is, however, different from those of [2] and [3], since φ is derived here by limiting operations on the sample space rather than by limiting operations on the probabilities of certain events. Further discussion of φ as an efficiency function is postponed to the last paragraphs of Section 5.

The case when σ is unknown is considered in Section 6. In this case, $L_s(n)$ is defined as before, but Y_n is replaced by the t-statistic, Y_n^* say, and the normal distribution is replaced by Student's distribution with $n - 1$ d.f., in the definition of $L_0(n)$ given by Equations (1), (2), and (3). Then the conclusion (6) remains valid in the case $\mu = 0$. It turns out, however, that in the non-null case τ and φ must both be increased by replacing θ^2 by $\log(1 + \theta^2)$ in (8) and (10). The resulting parametric functions, τ^* and φ^* say, are rather different from τ and φ. This may be seen by comparing the values of φ^* given in Table 3 with those for φ in Table 2.

TABLE 3

θ	0.0	0.5	1.0	1.5	2.0	2.5	3.0	3.5	4.0	∞
φ^*	.64	.67	.74	.76	.73	.66	.59	.49	.26	.00

The fact that φ^* exceeds φ for each $\theta > 0$ implies that the asymptotic efficiency of the sign test, as measured by simultaneous comparison, is higher when σ is unknown than when σ is known. In other words, in using the exact t-test, the loss of information due to ignorance of σ is not asymptotically negligible. It might be stated here that this "loss" is due not so much to the actual error in the estimate of σ (i.e., to the difference between Y_n and Y_n^*) as to the difference between the normal and Student's distributions. Indeed, if L_0 is defined by replacing Y_n by Y_n^* in (1), then the appropriate parametric functions τ and φ are again given by (8) and (10). This last definition of L_0 is inexact but very practical for large n. The general method of comparison of test levels which is described in [1] applies to levels which are inexact in the same sense.

2. The Null Case

For any given μ and σ, let us write

$$S_n = S_n(\mu, \sigma) = \sqrt{n}\,(\bar{X}_n - \mu)/\sigma\,, \tag{12}$$

and

$$T_n = T_n(p) = (R_n - np)/\sqrt{npq}\,, \tag{13}$$

for $n = 1, 2, \cdots$, where $\bar{X}_n$, R_n, and p are given by Equations (2), (4), and (8).

Suppose now that $\mu = 0$. Then $\theta = 0$, and $p = \frac{1}{2}$. For any n, let $G_n(x) = \Pr(T_n \geqq x \mid \theta = 0)$. Since $G_n(x) \to G(x)$ uniformly in x as $n \to \infty$, where G is given by (3), it follows easily from (5) and (13) that, for any sequence $X_1, X_2, \cdots$, $|L_s(n) - G(T_n)| \to 0$ as $n \to \infty$. Hence, by (1), the asymptotic joint distribution of $L_0(n)$ and $L_s(n)$ exists if and only if $G(S_n)$ and $G(T_n)$ have a limiting joint distribution, and the two distributions are then identical. Since G is a strictly monotonic function, the conclusion (6) is therefore plain from the following theorem.

Theorem 1. *For any given μ and σ, as $n \to \infty$ the joint distribution of S_n and T_n tends to the bivariate normal distribution with means zero, variances one, and covariance ρ, where*

$$\rho(\theta) = \exp(-\theta^2/2)/\sqrt{2\pi pq}\,\cdot \tag{14}$$

This theorem is stated and proved here for arbitrary μ and σ, rather than only for the case $\mu = 0$, because Equation (14) is required in Sections 3 and 4.

Proof. Write

(15) $$\xi(X) = (X-\mu)/\sigma\,, \qquad \eta(X) = \begin{cases} (q/p)^{\frac{1}{2}} & \text{if } X > 0\,, \\ -(p/q)^{\frac{1}{2}} & \text{if } X \leqq 0\,. \end{cases}$$

Then, with $\xi(X_i) = \xi_i$, $\eta(X_i) = \eta_i$, we have $S_n = \sum_{i=1}^{n} \xi_i/n^{\frac{1}{2}}$, $T_n = \sum_{i=1}^{n} \eta_i/n^{\frac{1}{2}}$, from (2), (4), (12), and (13). Consequently, by the central limit theorem for bivariate random variables (see, e.g., [4], p. 285), (S_n, T_n) tends in distribution to the normal distribution with means $E(\xi) = E(\eta) = 0$, variances $\operatorname{var}(\xi) = \operatorname{var}(\eta) = 1$, and covariance $= \operatorname{cov}(\xi, \eta)$. Now, η is a function of ξ; in fact $(pq)^{\frac{1}{2}}\eta + p$ is 0 or 1 according as $\xi \leqq -\theta$ or $\xi > -\theta$, where θ is given by (7). Hence

(16) $$\begin{aligned} \rho = \operatorname{cov}(\xi, \eta) &= (pq)^{-\frac{1}{2}} \operatorname{cov}[\xi, (pq)^{\frac{1}{2}}\eta + p] \\ &= (pq)^{-\frac{1}{2}} \int_{-\infty}^{\infty} \xi\,[(pq)^{\frac{1}{2}}\eta + p]\,(2\pi)^{-\frac{1}{2}} \exp(-\xi^2/2)\,d\xi \\ &= (pq)^{-\frac{1}{2}} \int_{-\theta}^{\infty} \xi\,(2\pi)^{-\frac{1}{2}} \exp(-\xi^2/2)\,d\xi \\ &= (2\pi pq)^{-\frac{1}{2}} \exp(-\theta^2/2)\,. \quad || \end{aligned}$$

Some values of ρ are given in Table 4.

TABLE 4

θ	0.0	0.5	1.0	1.5	2.0	2.5	3.0	3.5	4.0	∞
ρ	.80	.76	.66	.52	.36	.22	.12	.06	.02	.00

3. Comparison of Levels

The analysis in the non-null case depends on precise estimates of small-tail probabilities of the normal and binomial distributions. For the normal distribution (see [5], p. 166) we have

(17) $$G(x) = (1-\alpha)\,(2\pi x^2)^{-\frac{1}{2}} \exp(-x^2/2)\,, \qquad \text{for } x > 0\,,$$

where G is defined by (3), and $0 < \alpha(x) < x^{-2}$. Equation (17) evidently implies

(18) $$\log G(x) = -(x^2/2) - \log x - \tfrac{1}{2}\log(2\pi) + o(1)\,, \qquad \text{as } x \to \infty\,.$$

An analog of (17) for the binomial distribution is given in [6]. It follows from this analog (cf. also [3], p. 331) that if $k_1, k_2, \cdots$ is a sequence of integers such that

(19) $$\lim_{n\to\infty} k_n/n = p\,, \qquad \text{where } \tfrac{1}{2} < p < 1\,,$$

then

(20) $$\begin{aligned} \log\left[\sum_{r=k_n}^{n} \binom{n}{r} \left(\frac{1}{2}\right)^n\right] = -n \log[f(k_n/n)] - \tfrac{1}{2}\log n \\ - \tfrac{1}{2}\log[2\pi q(2p-1)^2/p] + o(1), \end{aligned}$$

where

$$f(y) = 2y^y(1-y)^{1-y}\,, \qquad \text{for } 0 < y < 1\,. \tag{21}$$

Now consider arbitrary but fixed $\mu > 0$ and σ, and let $X_{(\infty)} = (X_1, X_2, \cdots$ ad inf.) be an infinite sequence of independent normally distributed variables with mean μ and variance σ^2. It will be convenient to let $\{\varepsilon_n\}$ denote a sequence of random variables such that each ε_n is a function of $X_{(\infty)}$, and such that ε_n tends to zero with probability one. Similarly, we shall occasionally denote by $\{b_n\}$ a sequence of real-valued functions of $X_{(\infty)}$ such that the sequence is bounded with probability one. Thus, by the strong law of large numbers,

$$\bar{X}_n = \mu + \varepsilon_n\,, \quad \text{and} \quad R_n/n = p(\theta) + \varepsilon_n\,. \tag{22}$$

It follows from the law of the iterated logarithm (see [7]) that in fact

$$S_n = b_n (\log\log n)^{\frac{1}{2}}\,, \quad \text{and} \quad T_n = b_n (\log\log n)^{\frac{1}{2}}\,, \tag{23}$$

where S_n and T_n are defined by (12) and (13).

Since $\mu > 0$, it is clear from (2) and (22) that $Y_n \to \infty$ with probability one. Hence we may put $x = Y_n$ in (18), replacing $o(1)$ by ε_n. Since $Y_n = S_n + \sqrt{n}\,\theta$, it follows easily by referring to (1) that

$$\log L_0(n) = -\tfrac{1}{2}(S_n + \sqrt{n}\,\theta)^2 - \tfrac{1}{2}\log n - \tfrac{1}{2}\log(2\pi\theta^2) + \varepsilon_n\,. \tag{24}$$

According to (22), the random sequence $\{R_n\}$ satisfies (19) with probability one, so that (20) is valid with $k_n = R_n$ and $o(1) = \varepsilon_n$. Since $R_n/n = p + (pq/n)^{\frac{1}{2}} T_n$, we obtain

$$\log f(R_n/n) = \log f(p) + (pq/n)^{\frac{1}{2}} T_n \log(p/q) + (\tfrac{1}{2}n) T_n^2 + T_n^3 b_n / n^{\frac{3}{2}} \tag{25}$$

from (21) and (22). It follows from (5), (20), (23), and (25) that

$$\begin{aligned}\log L_s(n) = -\tfrac{1}{2}\{n[2\log f(p)] + 2(npq)^{\frac{1}{2}} T_n \log(p/q) + T_n^2\} \\ - \tfrac{1}{2}\log n - \tfrac{1}{2}\log[2\pi q(2p-1)^2/p] + \varepsilon_n\,.\end{aligned} \tag{26}$$

Now define

$$\gamma^2 = \theta^2 + pq[\log(p/q)]^2 - 2\theta(pq)^{\frac{1}{2}}[\log(p/q)]\rho\,, \tag{27}$$

where ρ is given by (14).

THEOREM 2. *Suppose that* $\mu > 0$. *Let* τ *be defined by* (7) *and* (8) *and* γ^2 *by* (27). *Let*

$$U_n = (n\gamma^2)^{-\frac{1}{2}} \log[\tau^n L_s(n)/L_0(n)]\,. \tag{28}$$

Then, (a) U_n *is asymptotically normally distributed with mean zero and variance one, and* (b)

$$\begin{aligned}\liminf_{n\to\infty}\{U_n/(2\log\log n)^{\frac{1}{2}}\} = -1\,, \\ \limsup_{n\to\infty}\{U_n/(2\log\log n)^{\frac{1}{2}}\} = 1\,,\end{aligned} \tag{29}$$

with probability one.

PROOF. It follows from (23), (24), and (26) that

$$n^{-\frac{1}{2}}[\log L_0(n) + n\theta^2/2] = -\theta S_n + \varepsilon_n\,, \tag{30}$$
$$n^{-\frac{1}{2}}[\log L_s(n) + n \log f(p)] = -[(pq)^{\frac{1}{2}} \log(p/q)]T_n + \varepsilon_n\,.$$

Write $\zeta(X) = \theta\xi(X) - [(pq)^{\frac{1}{2}} \log(p/q)]\eta(X)$, where ξ and η are given by (15). If we put $\zeta(X_i) = \zeta_i$, it follows easily from (30) that $U_n = (\sum_{i=1}^{n} \zeta_i)/(n\gamma^2)^{\frac{1}{2}} + \varepsilon_n$. Since $E(\zeta) = 0$ and $E(\zeta^2) = \gamma^2$, conclusions (a) and (b) are now immediate from the central limit theorem and the law of the iterated logarithm. It follows from (b), or otherwise directly from (30), that (9) holds with probability one. ||

4. Remarks on Predicting L_0 from L_s

In this section we suppose that $\mu > 0$ but that the value of μ is entirely unknown. The problem of predicting L_0 from L_s may be stated as follows: It is desired to construct a sequence $g_1, g_2, \cdots$ of functions on (0, 1) into (0, 1) such that, at least in large samples, the ratio $g_n(R_n/n)/L_0(n)$ is close to 1; one may require, for example, that the ratio tends to 1 in probability as $n \to \infty$. We have seen in the previous section that $g_n(R_n/n) = L_s(n)$ does not meet this requirement, and perhaps no sequence of functions g_n can. Indeed, it seems plausible that there exists no sequence g_n for which $\log[g_n/L_0(n)]$ has a limiting probability distribution.

Let us therefore consider the more modest requirement that, for some fixed $\delta > 0$,

$$U_n^* = (n\delta^2)^{-\frac{1}{2}} \log[g_n(R_n/n)/L_0(n)] \tag{31}$$

have exactly the same asymptotic properties as U_n defined by (28). This requirement can be met by taking $g_n(R_n/n) = [\tau(z_n)]^n L_s(n)$, where z_n is the estimate of θ based only on n and R_n; i.e., z_n is the solution of $p(\theta) = R_n/n$. The verification of this choice of g_n is straightforward from the estimates of L_s and L_0 given in Section 3, and is omitted. It follows from the omitted verification that the δ^2 required here is given by

$$\delta^2 = (\theta/\rho)^2(1 - \rho^2)\,, \tag{32}$$

where ρ is given by (14).

The functions g_n just discussed are suggested, of course, by Theorem 2. An apparently less sophisticated choice of g_n is to bypass the sign test and take $g_n = \exp(-nz_n^2/2)$. It turns out that this choice of g_n also meets the requirement of the preceding paragraph, and that the δ^2 required here is also given by (32).

Now, ρ can be shown to be a decreasing function of θ. Consequently, with either choice of g_n, the asymptotic variance of $\log[g_n/L_0(n)]$, namely $n\delta^2$, is a rapidly increasing function of θ.

5. Comparison of Equivalent Sample Sizes

Let l be a constant, $0 < l < 1$. Consider a given sequence of observations $X_{(\infty)} = (X_1, X_2, \cdots$ ad inf.) and for each $n = 1, 2, \cdots$ let $L_0(n)$ be defined by (1). Let $N_0^- =$ least integer m such that $L_0(m) \leqq l$, and $N_0^- = 1$ if no such

m exists. Let N_0^+ = least integer m such that $L_0(n) \leqq l$ for all $n \geqq m$, and $N_0^+ = \infty$ if no such m exists. Then N_0^- and N_0^+ are certainly upper and lower bounds, in the given case, to the number of observations required by the optimum test in order to achieve the level l. Let N_s^- and N_s^+ be defined by replacing L_0 by L_s in the definitions of N_0^- and N_0^+. It then follows that, for the given sequence $X_{(\infty)}$ and the given level l,

$$\frac{N_0^-}{N_s^+} \leqq \frac{\text{sample size required by the optimum test}}{\text{sample size required by sign test}} \leqq \frac{N_0^+}{N_s^-}\,, \tag{33}$$

provided that the first and third ratios in (33) are well defined. The following Theorem 3 states that, for almost all sequences $X_{(\infty)}$, the first and third ratios in (33) are well defined and tend to $\varphi(\theta)$ as $l \to 0$ (i.e., as the sample size required by either test tends to ∞), where φ is given by (10).

THEOREM 3. *Suppose that* $\mu > 0$. *For any sequence* $X_{(\infty)} = (X_1, X_2, \cdots \textit{ad inf.})$ *and any* l, $0 < l < 1$, *let* $N_0^-(l)$, $N_0^+(l)$, $N_s^-(l)$, *and* $N_s^+(l)$ *be determined as above by the level* l. *Then, with probability one,* (i) $N_0^-(l) \leqq N_0^+(l) < N_s^-(l) \leqq N_s^+(l) < \infty$, *for all sufficiently small* l, (ii) $\lim_{l\to 0} N_0^-(l) = \infty$, *and* (iii) $\lim_{l\to 0} N_0^-(l)/N_s^+(l) = \varphi(\theta) = \lim_{l\to 0} N_0^+(l)/N_s^-(l)$.

PROOF. It follows from (21), (23), and (30) that

$$\frac{1}{n}\log L_0(n) = -\theta^2/2 + \varepsilon_n\,, \qquad \frac{1}{n}\log L_s(n) = -\log\,[2p^p q^q] + \varepsilon_n\,. \tag{34}$$

Choose and fix a particular sequence $X_{(\infty)}$ for which the remainder terms ε_n in (34) do tend to zero as $n \to \infty$. It will suffice to establish (i), (ii), and (iii) for the chosen sequence.

Since, by (34), $L_0(m) \to 0$ as $m \to \infty$, for each l we have $L_0(m) < l$ for all sufficiently large m. Hence $N_0^-(l) \leqq N_0^+(l) < \infty$ for every l, by the definitions of N^- and N^+. We observe next that, since $L_0(m) > 0$ for each m by (1), we must have $N_0^-(l) \to \infty$ as $l \to 0$. The same arguments, with L_0 replaced by L_s, show that $N_s^-(l) \leqq N_s^+(l) < \infty$ for every l, and $N_s^-(l) \to \infty$ as $l \to 0$.

Now, by the definitions of $N_0^-(l)$ and $N_s^+(l)$, we have $L_0[N_0^-(l) - 1] > l \geqq L_0[N_0^-(l)]$ and $L_s[N_s^+(l)] \leqq l < L_s[N_s^+(l) - 1]$. Hence, for each l,

$$L_0[N_0^-(l) - 1] > L_s[N_s^+(l)]\,, \qquad L_0[N_0^-(l)] \leqq L_s[N_s^+(l) - 1]\,. \tag{35}$$

Since N_0^- and N_0^+ both tend to ∞ as $l \to 0$, by taking logarithms in (35) and applying (34), we easily have $N_0^-(l)/N_s^+(l) = \varphi + o(1)$, thus establishing the first half of (iii). The proof of the second half of (iii) is virtually the same and so is omitted. Part (ii) of the theorem has already been established. Part (i) follows from (iii) and the fact that $0 < \varphi < 1$. ||

In the preceding argument, φ is exhibited as the limiting ratio of the (random) sample sizes required by the optimum and sign tests in order to attain an arbitrarily small level l. The interpretation of φ as the asymptotic efficiency of the sign test thus afforded depends on the principle that, in a given case, the performance of test A is superior to, or equivalent to, or inferior to that of test B according as the level attained by A is less than, or equal to, or greater than the level attained by B. This "principle" seems to be consistent with, and is indeed suggested by, the manner in which tests are often used in practice, but whether it is a logical and self-sufficient theoretical basis is an open question.

It is of some interest that φ can also be interpreted in terms of power functions; specifically, φ can be exhibited as the limiting ratio of the (non-random) sample sizes required to attain an arbitrarily small error of Type 1 when the error of Type 2 is held fixed, as follows. Consider fixed $\mu > 0$ and σ, and let $\{k'_n\}$ be a sequence of constants such that $P(Y_n \geqq k'_n \mid \mu, \sigma) \to b_0$ as $n \to \infty$, where $0 < b_0 < 1$. For any l, $0 < l < 1$, let $M_0(l)$ be the least integer n such that $\Pr(Y_n \geqq k'_n \mid \mu = 0, \sigma) \leqq l$. Similarly, let $\{k''_n\}$ be a sequence of integers such that $\Pr(R_n \geqq k''_n \mid \mu, \sigma) \to b_s$, where $0 < b_s < 1$, and let $M_s(l)$ be the least n such that $\Pr(R_n \geqq k''_n \mid \mu = 0, \sigma) \leqq l$. It can then be shown that $\varphi = \lim_{l \to 0} M_0(l)/M_s(l)$. The proof of this theorem is omitted since it is not essential to the argument of this paper. It follows from this theorem that φ could be regarded as the dual of the efficiency function obtained by Hodges and Lehmann ([3], p. 333).

In concluding this discussion of φ, let us note a related question of some theoretical interest. The test of $\mu = 0$ based on Y_n is known to be optimum in terms of the Neyman-Pearson theory of testing hypotheses. Since Theorem 3 makes no explicit reference to this theory, Theorem 3 shows only that, in terms of simultaneous comparison, φ is the asymptotic efficiency of the test based on R_n relative to the test based on Y_n. It can be shown, however, that, as $n \to \infty$, the level attained by Y_n tends to zero at least as fast as the level attained by any other test statistic Z_n, so that φ serves in fact as the absolute efficiency of the sign test. This is a special case of a fairly general result which is given in [8].

6. The Case when σ Is Unknown

For each $n = 2, 3, \cdots$ let

$$Y^*_n = \sqrt{n}\,\bar{X}_n/V_n\,, \qquad \text{where } V^2_n = \sum_{i=1}^{n} (X_i - \bar{X}_n)^2/(n-1)\,. \tag{36}$$

Let $L^*_0(n)$ denote the level of the optimum test based on the first n observation when σ is unknown. Then

$$L^*_0(n) = G_{n-1}(Y^*_n)\,, \tag{37}$$

where, for any positive integer m, $G_m(x) = \int_x^\infty f_m(t)\,dt$, f_m being the density of Student's distribution with m degrees of freedom. It is easy to see that, if $\mu = 0$, $|L^*_0(n) - L_0(n)|$ tends to 0 in probability. Hence, in the null case, $L^*_0(n)$ and $L_s(n)$ have the same asymptotic joint distribution as $L_0(n)$ and $L_s(n)$.

To treat the non-null case, we note that $G_m(x)$ can be expressed as the expected value of $G(x\chi_m/\sqrt{m})$, where G is defined by (3) and χ^2_m is a chi-square with m d.f. By applying (17) to the integrand in this representation it follows that, for $m > 3$ and $x > 0$,

$$G_m(x) = \left\{\frac{m^{\frac{1}{2}}\Gamma[(m-1)/2]}{2^{\frac{1}{2}}\Gamma(m/2)}\right\}(2\pi x^2)^{-\frac{1}{2}}\left(1+\frac{x^2}{m}\right)^{-\frac{1}{2}(m-1)}(1-\beta)\,, \tag{38}$$

where $0 < \beta_m < (x^{-2} + m^{-1})[m/(m-3)]$. We omit the detailed calculation. It follows from (38) that

$$\log G_m(x) = -[(m-1)/2]\log(1 + x^2/m) - \log x - \tfrac{1}{2}\log(2\pi) + o(1)\,, \tag{39}$$

as m and x tend to ∞ in any manner.

If $\mu > 0$, we have $Y_n^* \rightarrow \infty$ with probability one. Consequently, (37) and (39) lead to an analog of (24). In particular, using the fact that

$$V_n^2/\sigma^2 = 1 + b_n(\log\log n/n)^{\frac{1}{2}}\,, \tag{40}$$

we have the following analog of (30):

$$n^{-\frac{1}{2}}\left[\log L_0^*(n) + \frac{n}{2}\log(1+\theta^2)\right] = -\frac{1}{1+\theta^2}\left\{\theta S_n - \frac{n^{\frac{1}{2}}\theta^2}{2}\left[\left(\frac{V_n}{\sigma}\right)^2 - 1\right]\right\} + \varepsilon_n\,. \tag{41}$$

It is easy to see from (41) that analogs of Theorems 2 and 3 are valid when L_0 is replaced by L_0^*, provided that τ and φ are replaced by τ^* and φ^*, where

$$\tau^* = (2p^pq^q)/(1+\theta^2)^{\frac{1}{2}}\,, \qquad \varphi^* = 2\log(2p^pq^q)/\log(1+\theta^2)\,. \tag{42}$$

7. Acknowledgments

The writer wishes to thank the editors for their comments and suggestions during the preparation of this paper. The writer also thanks Messrs. M. Chatterji and R. K. Naha of the Indian Statistical Institute for their careful preparation of tables on which Tables 1 through 4 are based.

References

[1] Bahadur, R. R. Stochastic comparison of tests. *Ann. Math. Stat.*, 1960, **31** (in press).

[2] Pitman, E. J. G. Nonparametric inference. (Unpublished notes of lectures at Columbia University, 1949.)

[3] Lehmann, E. L., and Hodges, J. L. The efficiency of some nonparametric competitors of the t-test. *Ann. Math. Stat.*, 1956, **27**, 324–35.

[4] Cramér, H. *Mathematical Methods of Statistics.* Princeton, N.J.: Princeton Univ. Press, 1946.

[5] Feller, W. *An Introduction to Probability Theory and Its Applications*, Vol. I, 2d ed. New York: Wiley, 1957.

[6] Bahadur, R. R. Some approximations to the binomial distribution function. *Ann. Math. Stat.*, 1960, **31** (in press).

[7] Hartman, P., and Wintner, A. On the law of the iterated logarithm. *Amer. J. Math.*, 1941, **63**, 169–76.

[8] Bahadur, R. R. On the asymptotic efficiency of tests and estimates. *Sankhyā*, 1960, **23** (in press).

9 SOME STOCHASTIC MODELS IN ECOLOGY AND EPIDEMIOLOGY

M. S. BARTLETT
University of Manchester

1. Introduction

Among Harold Hotelling's diverse publications, a brief report [1] on stochastic processes, given in 1949, testifies to his long interest in this subject; in that historical survey he mentioned some early work of his in 1928 on random walk (or random migration) theory applied in biology. I might also recall that when I visited Professor Hotelling's department at Chapel Hill in the fall of 1946 I lectured on stochastic processes [2]. In my contribution here I propose, therefore, to discuss some stochastic processes in biology; in particular, some stochastic population models in ecology and epidemiology.

Recent contributions of my own in these fields ([3], [4], [5], [6]) have already raised some theoretical problems which are difficult to solve exactly; nevertheless, approximate solutions are often adequate for applications, and some of these will be developed in this paper.[1]

In Section 2 I consider how far a stationary marginal distribution for the variable N representing population size in a certain class of stochastic models can still be said to exist when there is an "absorbing barrier," usually at the "extinction" state, $N = 0$. Provided the extinction probability is small enough, an effectively stationary process may be expected for quite a long interval of time, and I have suggested the phrase "quasi-stationary" for such a process.

In Section 3 a particular stochastic process which I have used as a model for measles epidemics is discussed by way of extension of these ideas to bivariate models; and approximate theoretical predictions, on the time to extinction of the number of infected persons, are compared briefly with official statistics of measles notifications.

[1] These results were first presented at the International Congress of Mathematicians held at Edinburgh in 1958, under the title: "Quasi-stationary behaviour of some non-linear biological models."

2. Population Processes

The term *population process* is used here for any homogeneous Markov process $N(t)$, where N can take the values $0, 1, 2, \cdots$, and, moreover, can change only by $+1$ or -1 with probabilities $\lambda_r dt$ and $\mu_r dt$ in the interval t, $t + dt$ when $N(t) = r$. $N(t)$ is often called a birth-and-death process, but the term "population process" will also be used, for example, for processes in two-integer variables $N_1(t)$, $N_2(t)$, where each variable changes only by one at a time (though they may simultaneously change, as in an epidemic model when a susceptible person becomes infected). In population models in *discrete* time, $N(t)$ would of course be allowed to change by more than one at a time, but we shall examine below only the theory of continuous-time models.

Considering first single-variable models, we assume that there are no barriers within the possible range of $N(t)$; i.e., λ_r and μ_r do not vanish inside, say, n and N ($>n$), although if these are the extreme values, $\lambda_N = 0$ and $\mu_n = 0$. The value N may be infinite, or n may equal 0, or both. In population models where no immigration into the population is envisaged, n is usually zero and $\lambda_0 = 0$, so that the population can become extinct. While N may be infinite, the death rate will be so high for large $N(t)$ that there is an effective ceiling for $N(t)$, and mathematical difficulties associated with what happens when $N(t)$ reaches infinity will not arise.

The "forward" differential equation for the probability $P\{N(t) = r\} \equiv P_t(r)$ is

$$dP_t(r)/dt = \lambda_{r-1} P_t(r-1) + \mu_{r+1} P_t(r+1) - (\lambda_r + \mu_r) P_t(r) , \tag{1}$$

for $r \geqq n$ [with $P_t(n-1)$ taken to be zero], and if a limiting distribution $P(r)$ exists, it will satisfy

$$\lambda_{r-1} P(r-1) + \mu_{r+1} P(r+1) = (\lambda_r + \mu_r) P(r) ,$$

whence

$$P(r) = \frac{\lambda_{r-1}}{\mu_r} P(r-1) , \tag{2}$$

$$P(r) = \frac{\lambda_0}{\mu_r} \prod_{s=1}^{r-1} \frac{\lambda_s}{\mu_s} P(0) , \qquad r > 1 , \tag{3}$$

if $n = 0$, $\lambda_0 \neq 0$, and $P(1) = \lambda_0 P(0)/\mu_1$.

Suppose, however, $n = 0$, $\lambda_0 = 0$. The chance of extinction at time t, if $N(0) = m$, $p_t(m)$ say, satisfies the "backward" differential equation

$$dp_t(m)/dt = \lambda_m p_t(m+1) + \mu_m p_t(m-1) - (\lambda_m + \mu_m) p_t(m) ,$$

and the limiting chance of extinction $p(m)$ is the minimal solution of

$$\lambda_m p(m+1) + \mu_m p(m-1) = (\lambda_m + \mu_m) p(m) , \tag{4}$$

whence, with $p(0) = 1$,

(5) $$p(m+1) - p(m) = \prod_{r=1}^{m} \frac{\mu_r}{\lambda_r} \{p(1) - 1\} ,$$

$$p(m) - 1 = \sum_{s=0}^{m-1} \left\{ \prod_{=1}^{s} \frac{\mu_r}{\lambda_r} \right\} \{p(1) - 1\} .$$

From this equation, as $p(m)$ is finite for all m, it follows that $p(1) = 1$ if

(6) $$\sum_{s=0}^{\infty} \prod_{r=1}^{s} \frac{\mu_r}{\lambda_r} = \infty ,$$

and hence that $p(m) = 1$ for all finite m (cf. [7], [8]).

In this case there can be no limiting distribution $P(r) \neq 0$. However, consider such a model as the logistic process (see [6])

$$\lambda_r = \max\{0, \alpha_1 r - \beta_1 r^2\} , \qquad \mu_r = \alpha_2 r + \beta_2 r^2 ,$$

all coefficients positive with $\alpha_1 > \alpha_2$. It is obvious that Equation (6) is satisfied (even if $\beta_1 = 0$), and hence that $p(m) = 1$. However, we have

$$dN(t) = \{(\alpha_1 - \alpha_2)N(t) - (\beta_1 + \beta_2)N^2(t)\}\, dt + dZ ,$$

where $E\{dZ\} = 0$, and $N(t)$ tends to approach the value $R = (\alpha_1 - \alpha_2)/(\beta_1 + \beta_2)$. The fact is that the passage time to zero may, when fluctuations about R are not too large, be so long that the result $p(m) = 1$ is irrelevant. It might be noticed that such a quasi-stationary process, as the writer has suggested calling it, becomes strictly stationary *on condition that there is no extinction*, but this merely means we replace μ_1 by 0, and have a new process with $n = 1$, $\lambda_1 \neq 0$; and we are more concerned with the properties of the process as it is.

The passage-time distribution is obviously determined by $p_t(m)$ if we can find this; in fact, $p_t(m)$ is the cumulative distribution function of the extinction time; but it is often more profitable to seek indirect information. Let us, recalling the stationary process given with $n = 1$, $\lambda_1 \neq 0$, consider a similar stationary process with $n = 0$, $\lambda_0 = \varepsilon \neq 0$, the other λ_r being as before. Then we can investigate the ultimate distribution (3) for this modified process, assuming of course that $\sum_{r=1}^{\infty} P(r) = 1 - P(0)$, so that if

$$T_r = \begin{cases} 1 , & r = 1 , \\ \dfrac{1}{\mu_r} \displaystyle\prod_{s=1}^{r-1} \frac{\lambda_s}{\mu_s} , & r > 1 , \end{cases}$$

(7) $$P(0) = 1/(1 + \varepsilon S) ,$$

(8) $$P(r) = \varepsilon T_r/(1 + \varepsilon S) , \qquad r > 0 ,$$

where $S \equiv \sum_{r=1}^{\infty} T_r < \infty$. Suppose the process is such that $P(r) \gg P(0)$ for a range of r, so that $\varepsilon T_r \gg 1$ for such r. Then $\varepsilon S \gg 1$, and

(9) $$P(0) \sim 1/(\varepsilon S) ,$$

(10) $$P(r) \sim T_r/S , \qquad r > 0 ,$$

where it will be noted that $P(r)$ is approximately independent of ε. In this sense it is legitimate to speak of an ultimate distribution ignoring ε. Moreover, for the original process with $\lambda_0 = 0$, the passage time to zero will be of the same order of magnitude as the recurrence time to zero for the above-modified process. Now the mean value of this recurrence time is

$$[1 - P(0)]/[\varepsilon P(0)] \sim 1/[\varepsilon P(0)] , \tag{11}$$

which from (9) is approximately S, again independent of ε. For example, as a further approximation to $P(r)$ when the mean value of r is large and the coefficient of variation small, we may replace the exact result (2) by

$$d \log P(r)/dr \sim [P(r+1) - P(r)]/P(r) = (\lambda_r/\mu_{r+1}) - 1 \sim (\lambda_r/\mu_r) - 1 ,$$

whence

$$P(r) \sim C \exp[-\tfrac{1}{2}(r - R)^2/\sigma^2] , \tag{12}$$

where R is the solution of $\lambda_r = \mu_r$, and

$$\sigma^2 = -1 \bigg/ \left[\frac{d(\lambda_r/\mu_r)}{dr}\right]_{r=R} . \tag{13}$$

Thus for the logistic model, where $R = (\alpha_1 - \alpha_2)/(\beta_1 + \beta_2)$, we find $\sigma^2 = (\alpha_1 - \beta_1 R)/(\beta_1 + \beta_2)$.

With this approximation we have

$$S \sim 1/P(1) \sim C^{-1} \exp[\tfrac{1}{2}(R - 1)^2/\sigma^2] , \tag{14}$$

which can be enormously large if $R \gg \sigma$. For the process $\alpha_1 = 0.8077$, $\alpha_2 = 0.1145$, $\beta_1 = 0.006932$, $\beta_2 = 0$; R is 100, $\sigma^2 = 16.52$, and S is of order 10^{130} (values corresponding to one of the numerical examples considered by Leslie [9]).

3. An Epidemic Model

The preceding arguments can be extended when relevant to two-variable models of the type that occur in ecology and epidemiology. As an illustration, we shall discuss the situation that arises in the stochastic model we have used for measles epidemics. This model is conveniently represented by means of the partial differential equation for the probability-generating function $\Pi_t(z, w) \equiv \sum_{i,s} P_t(i, s) w^i z^s$, where $P_t(i, s)$ is the probability at time t that the number I_t of infective persons is i, and the number S_t of susceptible persons is s. The possible transitions for this model in time dt are as follows:

Transition	Probability
$S_t \to S_t + 1, \quad I_t \to I_t$	$\nu\, dt$
$S_t \to S_t - 1, \quad I_t \to I_t + 1$	$\lambda I_t S_t\, dt$
$S_t \to S_t, \quad I_t \to I_t - 1$	$\mu I_t\, dt$
$S_t \to S_t, \quad I_t \to I_t + 1$	$\varepsilon\, dt$

The last type of transition will be absent if $\varepsilon = 0$, and the interesting question is whether in this model I_t becomes zero if $\varepsilon = 0$. The equation for $\Pi_t(z, w)$ is

$$(15)\qquad \frac{\partial \Pi_t}{\partial t} = \lambda(w^2 - zw)\frac{\partial^2 \Pi_t}{\partial z\,\partial w} + \mu(1-w)\frac{\partial \Pi_t}{\partial w} + \nu(z-1)\Pi_t + \varepsilon(w-1)\Pi_t\,.$$

As in the one-variable case treated in Section 2, a limiting distribution for S, I will exist if $\varepsilon \neq 0$, and consequently will satisfy the equation

$$(16)\qquad \lambda(w^2 - zw)\frac{\partial^2 \Pi}{\partial z\,\partial w} - \mu(1-w)\frac{\partial \Pi}{\partial w} + \nu(z-1)\Pi + \varepsilon(w-1)\Pi = 0\,.$$

We are particularly interested in the distribution of I, and this is given by $[\Pi]_{z=1}$. Putting $z = 1$ in (16), we obtain, after dividing by the factor $w-1$,

$$\lambda w\left[\frac{\partial^2 \Pi}{\partial z\,\partial w}\right]_{z=1} - \mu\left[\frac{\partial \Pi}{\partial w}\right]_{z=1} + \varepsilon[\Pi]_{z=1} = 0\,,$$

which in terms of $P(i, s)$ is

$$\lambda \sum_{i,s} is\, P(i, s)w^i - \mu \sum_{i,s} iP(i, s)w^{i-1} + \varepsilon \sum_{i,s} P(i, s)w^i = 0\,.$$

With

$$P(i) = \sum_s P(i, s), \qquad \bar{s}_i P(i) = \sum_s sP(i, s)\,,$$

this equation gives

$$(17)\qquad \mu P(1) = \varepsilon P(0)\,,$$

and

$$(18)\qquad \lambda i \bar{s}_i P(i) - \mu(i+1)P(i+1) + \varepsilon P(i) = 0, \qquad i > 0\,.$$

Without repeating the sort of justification given in Section 2, we shall neglect ε in this last equation, so that, if $R(i) \equiv iP(i)$,

$$(19)\qquad R(i+1) = \lambda \bar{s}_i R(i)/\mu\,.$$

While unfortunately this equation depends on $\bar{s}_i$, which is still unknown, it is possible to insert approximations for $\bar{s}_i$ in (19) and hence to solve for $R(i)$. Thus it has been shown [4] that, for the model discussed above, we have (in the sense of an ultimate distribution as discussed in Section 2)

$$E\{S\} \sim n = \mu/\lambda, \qquad E\{I\} = m = u/\mu\,,$$

$$\sigma_S^2/n \sim \sigma_I^2/m \sim 1 + n/m, \qquad \operatorname{cov}\{I, S\} \sim -n\,.$$

We might note also that $E\{IS\} = n = \operatorname{cov}\{I, S\} + mE\{S\}$, so that $E\{S\} \sim n + n/m$ to a better approximation. To this order of approximation,

$$(20)\qquad \bar{s}_i \sim n + \frac{n}{m} - \frac{n(i-m)}{n+m}\,,$$

and if this expression is substituted in (19), $R(i)$ may be found. We have

$$R(i) = R(1)\prod_{j=1}^{i}\left(1 + \frac{1}{m} - \frac{j-m}{n+m}\right),$$

or

$$\log R(i) - \log R(1) = \sum_{j=1}^{i} \log\left(1 + \frac{1}{m} - \frac{j-m}{n+m}\right)$$
$$= \sum_{j=1}^{i}\left(\frac{1}{m} - \frac{j-m}{n+m}\right) + \cdots$$
$$= \frac{i}{m} - \frac{i^2}{2(n+m)} + \frac{mi}{n+m} + \cdots,$$

whence

(21) $$R(i) \sim C \exp\left\{-\frac{1}{2}\frac{[i-m-(n+m)/m]^2}{n+m}\right\},$$
$$P(i) \sim \frac{C}{i} \exp\left\{-\frac{1}{2}\frac{[i-m-(n+m)/m]^2}{n+m}\right\}, \qquad i > 0,$$

where by integration $C \sim m/\sqrt{[2\pi(n+m)]}$.

Employing an argument similar to that used in Section 2, we have for the mean recurrence time to zero,

$$\frac{1-P(0)}{\varepsilon P(0)} \sim \frac{1}{\mu P(1)},$$

the natural logarithm of which is approximately

(22) $$-\log \mu + \tfrac{1}{2}\log(2\pi) + \tfrac{1}{2}\log(1 + n/m) - \tfrac{1}{2}\log m + \tfrac{1}{2}m/[1 + (n/m)] + 1 + \tfrac{1}{2}[1 + (n/m)]/m.$$

In the measles model, we shall, following Soper [10], take $\mu = \frac{1}{2}$ (when the unit of time is one week) and $n/m = 34.1$, so that, transforming to logarithms to the base 10, we find that (22) becomes (in this case)

(23) $$1.907 - \tfrac{1}{2}\log_{10} m + 0.00619m + 7.62/m.$$

Under conditions for which the approximation for $\bar{s}_i$ is valid, namely, m large, the mean recurrence time will be long and will also indicate the order of magnitude of the passage time to zero. The values obtained from formula (23), starting with large m, for which the approximation should be best, are given in Table 1.

TABLE 1

m	Time (years)	m	Time (weeks)
500	90	200	108
400	24	150	63
300	7	100	40
		50	33

It will be seen that the critical value of m suggested by these figures, for which the extinction time is of the order of the epidemic period (74 weeks in this model), is about 150 to 200. For higher values of m, the extinction time rapidly increases.

It is possible to list various complicating factors that may affect the extinction time in practice, such as (i) the effect of a fairly well-defined incubation period for measles and a limit to the persistence of infectivity from any one individual, (ii) the effect of spatial spread and diffusion of infection over an area, (iii) the effect of seasonal variation in incidence (probably largely due to summer dispersal and autumn reassembly of school children). The effect of (i) has been investigated by Monte Carlo work with a "discrete-time" model [11], the results (not yet complete) suggesting that some lowering of the critical value of m obtained from Table 1 should ensue. The effect of (ii) seems less important to this order of approximation. The theoretical effect of (iii) has not yet been examined in any detail, but might be expected to increase the critical size somewhat, to avoid fade-out in the summer months.

With such complications to consider, it is rather remarkable how well the observed statistics bear out the provisional prediction. English towns of critical size appear to be Bristol (average weekly notifications 67, *total* population about 415,000) and Hull (notifications 57, population about 269,000). Multiplying the notifications by two to give the average number m leads to a figure of 125. A more recent examination of American figures (to be reported somewhat more fully elsewhere) suggests that Rochester, Providence, and Akron are of the required critical size. For these towns, however, while their populations are comparable to those of the English towns mentioned, their average weekly notifications (for the period 1921–40, from monthly figures kindly supplied to me by Jane Worcester) are rather lower, being 24, 43, and 18 respectively. It is pertinent here to recall Hedrich's findings [12] for Baltimore, that the official notification figures appear to be only about one-third of their true value, so that the true mean value of m for these towns is probably of the order of 170 rather than one-third of this figure. This was checked by reference to birth and infant-mortality statistics for these towns. These figures also suggest that failure to notify is not so extreme a factor in England, and a further check on this point indicated that official notifications probably reach at least one-half of the true value.

References

[1] Hotelling, H. Stochastic processes. (Historical summary of the problem.) *Econometrica*, 1949, **17**, 66–68.

[2] Bartlett, M. S. Stochastic processes. (Duplicated notes of a course given at the University of North Carolina, fall quarter, 1946.)

[3] Bartlett, M. S. *An Introduction to Stochastic Processes*. Cambridge: Cambridge Univ. Press, 1955.

[4] Bartlett, M. S. Deterministic and stochastic models for recurrent epidemics. In vol. 4, J. Neyman (ed.), *Proceedings of the Third Berkeley Symposium on Mathematical Statistics and Probability*. Berkeley, Calif.: Univ. California Press, 1956.

[5] Bartlett, M. S. Measles periodicity and community size. *J. Roy. Stat. Soc.*, Ser. A, 1957, **120**, 48-60.

[6] Bartlett, M. S. On theoretical models for competitive and predatory biological systems. *Biometrika*, 1957, **44**, 27-42.

[7] Karlin, S., and McGregor, J. The classification of birth and death processes. *Trans. Amer. Math. Soc.*, 1957, **86**, 366-400.

[8] Kendall, D. G., and Reuter, G. E. H. The calculation of the ergodic projection for Markov chains and processes with a countable infinity of states. *Acta Math.*, 1957, **97**, 103-44.

[9] Leslie, P. H. A stochastic model for studying the properties of certain biological systems by numerical methods. *Biometrika*, 1958, **45**, 16-31.

[10] Soper, H. E. The interpretation of periodicity in disease prevalence. *J. Roy. Stat. Soc.*, Ser. A, 1929, **92**, 34-61.

[11] Bailey, N. T. J. *The Mathematical Theory of Epidemics*. London: Griffin, 1957.

[12] Hedrich, A. W. Monthly estimates of the child population "susceptible" to measles, 1900-1931, Baltimore, Md. *Amer. J. Hygiene*, 1933, **17**, 613-36.

10 RANDOM ORDERINGS AND STOCHASTIC THEORIES OF RESPONSES[1]

H. D. BLOCK, Cornell University

J. MARSCHAK, Yale University

1. The Problem

In interpreting human behavior there is a need to substitute "stochastic consistency of choices" for "absolute consistency of choices." The latter is usually assumed in economic theory, but is not well supported by experience. It is, in fact, not assumed in empirical econometrics and psychology.

The stochastic approach brings out the affinity between the phenomenon of choice and the more general psychological phenomenon of response to physical stimuli or, for that matter, to questionnaires, or of the action of "judges" who compare the performances of individuals.[2]

Let A be the set of all **a**lternatives (or **a**ctions) ever to be considered by a given subject. Let F be a subset of A. Let $(a; F)$ denote a "multiple choice" (also called "first choice" out of a set), i.e., the following observed fact: When forced to choose one element of F, the subject chose a. In the theory of choice, F is called the **f**easible set. When F consists of $2, 3, \cdots$ elements, the multiple choice is called binary, ternary, $\cdots$.

[1] Research carried out mostly under contracts of the Office of Naval Research with the Cowles Foundation for Research in Economics at Yale University and with Columbia University. The paper was completed while one of the authors was Ford Visiting Research Professor at the Graduate School of Industrial Administration, Carnegie Institute of Technology. We owe much to discussions with D. Davidson, G. Debreu, M. DeGroot, L. Hurwicz, and R. D. Luce.

[2] While the stochastic approach uses measurable *probabilities*, another approach, also promising to give realism and unity to the understanding of choices and responses, makes use of the measurable *time-delay* of a choice or response. The shorter the delay, the larger is the difference between certain values said to be attached by the subject to two alternative responses. A very long delay reveals a state of (almost) indifference, the "conflict-situation": Hamlet took a very long time to decide whether to kill his uncle. The logic of this approach and its combination with stochastic theories was treated by Cartwright and Festinger [1].

In another language, the set F is associated with a stimulus and consists of all possible responses of the subject to that stimulus; A is then interpreted as the set of all possible responses to all stimuli. In experiments on so-called preferences and attitudes, the stimulus consists in offering a list of alternative menus, alternative political candidates, alternative answers to an item in the questionnaire. This list is then the feasible set F. When a "judge" makes paired comparisons, the set F is a pair of individuals. In experiments on perceptional discrimination, physical stimuli—sounds, lights, or weights—impinge upon the subject; his response consists in stating that one of those objects is louder, or brighter, or heavier than the other(s). The set F is then the set of the n possible responses (or $n + 1$ if "I don't know" statements are permitted).

Another example of F is the set of all family budgets open to a family with a fixed amount of money, given the prices of goods (see Hotelling [2]). Each element of F is a different way of allocating the money among different goods.

In this paper—subject only to some attempted generalizations in Section 7—we shall call the observed choice, or response, $(a; F)$ the *basic* or *direct observation*. These observations can be counted, and the constraints on (i.e., the relations among) the resulting numbers (which may possibly be only 1's or 0's) we shall call *conditions*. Some of these conditions we shall call *directly testable*. Only those conditions are directly testable which involve the basic observations only. We shall encounter conditions that are not directly testable because they involve certain entities (constructs) that are not directly observable. An important class of these constructs will be called *utilities* (identical with *sensations* in psychophysics, *values* in anthropology, ethics, and older economics). These will be defined in a variety of ways, but will always denote certain real numbers, constant or random, associated with the elements of the set A.

However, to involve basic observations only is a necessary but not a sufficient property of a directly testable condition. For example, if the set A of alternatives is finite but of a size unknown, or too large to be exhausted by experiment, a constraint involving all alternatives simultaneously—such as Condition (o) in Section 4 and Condition (P_n) in Section 5—cannot be tested.

The purpose of this paper is to study the logical relations between various conditions describing consistency of responses, and in particular the logical relations between certain conditions that are not directly testable and others that are. This will tell whether a given class of observational results would or would not justify the acceptance (or the rejection) of a theory.

Our particular way of defining the class of basic observations and, correspondingly, of the directly testable conditions is to some extent arbitrary. Depending on the range of possible experiments and other observations, it may be preferable to define the class more narrowly, e.g., by including binary choices only. Or we might define this class more broadly. Following the practice of psychologists, we might admit the ranking, by the subject, of three or more objects as an observable fact, although the subject's observed action consists in this case of a verbal statement. (In the case of two objects, ranking and choosing is the same thing.) We might even admit as observable the subject's verbal statements of the relative "intensity" of his preferences.

Such variations of the domain of testability will be tentatively discussed in Section 7. In the rest of the paper, by using a particular demarcation of the class of directly testable conditions (the one most closely corresponding to the nature of economic observations), we are able to carry out a reasonably complete analysis of the relevant logical relations. The study may thus serve as a start when similar attempts are made under another definition of basic observations.

In particular, our operational approach seems to be unable to handle the following distinction that appears natural on grounds of common sense and may be important for predictions. If out of the pair $F = (a, b)$ of desirable objects a man chooses sometimes a and sometimes b, our introspection tells us that we may ascribe this to either or both of two different "causes":

1. He may have difficulty in perceiving all the relevant characteristics of the objects, as when a and b are two 10-ton carloads of the same merchandise but the exact quantity (in pounds, net of package) or, for that matter, all the differences in quality cannot be ascertained.
2. Even if he knew exactly the differences in the characteristics of the two objects, he might find them almost equally "desirable" or containing the promise of equal "satisfaction," and he will vacillate as a result.

To disentangle the two "causes"—call them "perceptibility" and "desirability" (anticipated "satisfaction") — may be important if one wants to predict how people will act if perceptibility is kept constant while desirability varies, or vice versa. Economists in particular have mostly considered men's decisions under the assumption of perfect perceptibility, also called perfect information. In the more recent developments of economic decision theory, imperfect information is introduced, but a clear, and probably fruitful, distinction is kept between varying the nature of (so-called) information and varying the satisfaction (the "payoff") attainable through the choice of action. Economists[3] use the term "utility" as interchangeable with desirability (satisfaction), and thus independent of perceptibility. Our concept is coarser. All the various definitions of utility given in this paper will be related to the empirical entities, called "alternatives." Each of these is identified precisely but combines the two aspects, information and desirability, in some unknown though presumably not-too-changeable fashion. Thus when two carloads are offered again, and, say, the firm's quality-control facilities (or, in a similar example, a housewife's abilities to discriminate between cuts of meat) have not changed, predictions from previous behavior observations can be made. If discrimination is learned, through experience, in some more-or-less-determined manner, the model has to be modified, of course, but we have not attempted to bring in learning.

[3] As pointed out by James Tobin in a private communication.

Stochastic theory introduces the probability $p(a; F)$ of the basic observation $(a; F)$, and it is assumed that inferences about constraints on these probabilities can be made from a finite number of basic observations. These constraints may be described with the help of parameters of certain distributions, and some of these parameters have been called, a century ago, sensations (Fechner [3], see especially pp. 70–103 of the 1889 edition). We shall call them *constant utilities*. Various types of constant utilities have been proposed. They will be studied in Section 2. In addition, there emerge naturally the concepts of *random utilities* and *random orderings* to be treated in Section 3. Sections 4 and 5 continue the study of directly testable conditions necessary and/or sufficient for the existence of constant and random utilities (and random orderings), respectively. Section 6 sums up the main logical relations between the random and the various forms of constant utilities, and includes a very incomplete attempt to relate our various models to the work of other authors. As already mentioned, a generalized experimental situation—the combination of ranking and choosing—will be formulated in Section 7. Section 8 discusses the memory effect and other difficulties of experiments on choice and ranking, difficulties perhaps encountered less in psychophysics proper. We shall postpone the description of Section 9 until the end of the present introductory section. Section 10 deals tentatively with a special statistical problem: how to test statistically our directly testable conditions — i.e., the properties of the probabilities $p(a; F)$—when (to avoid the memory effect and to avoid assuming that all subjects have identical distribution properties) each set F can be offered only a few times, possibly only once.

Clearly, non-stochastic theory is a special, strong form of stochastic theory. And whenever a theory of choices or responses is being submitted to a statistical test there is, explicitly or not, an underlying model of stochastic behavior of the subject (or possibly of the error-making observer).

We shall conclude this section with a brief examination of the non-stochastic, or absolute, theory of choice. We shall formulate it in terms of our "basic observations," and not, as usual, in terms of the subject's verbal preference or indifference statements. Define[4] for all a, b in A:

(1.1) "$a \succ b$" (a preferred to b) means "never $(b; F)$ if F contains a,"
"$a \succsim b$" means "not: $b \succ a$,"
"$a \sim b$" (indifference) means "$a \succsim b$ and $b \succsim a$."

Thus, if the subject forced to choose between a and b chooses b nine (but not ten) times out of ten, the non-stochastic theory calls him indifferent. We have always $a \succsim b$ or $b \succsim a$ since the subject is forced to choose. Therefore the relation "$\succsim$" is said to induce complete weak ordering on A provided the following condition (testable by basic observations) is satisfied:

[4] Preferences thus defined are "independent of irrelevant alternatives," a principle pointed out by Hotelling's pupil Arrow [4] and much used ever since. A better name for Arrow's principle is "irrelevance of added alternatives."

(1.2) if $a \gtrsim b$ and $b \gtrsim c$, then $a \gtrsim c$ (transitivity).

If, in addition, the set A is finite, $A = (a_1, \cdots, a_n)$, we can associate with each of its elements a_i an integer r_i, $1 \leqq r_i \leqq n$, called *rank*, such that

(1.3) $$r_i \leqq r_j \quad \text{if} \quad a_i \gtrsim a_j .$$

The vector $r = (r_1, \cdots, r_n)$, which can also be regarded as an integer-valued function on the set of integers $N = (1, \cdots, n)$, is called a ranking on N. If by some arbitrary convention ties are excluded, all r_i's are different integers and r is a permutation. Clearly any strictly decreasing monotone function on the integers $r_1, \cdots, r_n$ induces a real-valued function ω on A which is order-preserving in the sense that

(1.4) $$\omega(a_i) \geqq \omega(a_j) \quad \textit{if and only if } a_i \gtrsim a_j .$$

The function ω, called the ("*ordinal*") *utility function*, is unique up to increasing monotone transformations.

If the set A of alternatives is not finite, an order-preserving function ω on A need not exist. However, Debreu [5] proved that an order-preserving function exists if A and the ordering relation "$\gtrsim$" satisfy a certain rather weak condition[5] that may justify the assumption of ordinal utility functions over the space of commodity-bundles.[6]

In the stochastic models that will follow, the (testable) transitivity condition (1.2) and the "ordinal" utility function ω made possible by it will be suitably generalized. But, in addition, some stronger testable conditions and, correspondingly, more strictly measurable utility functions will arise naturally.

A final remark: The case when the set A includes wagers so that choices are, in general, made under uncertainty, is more general than that of choices among sure alternatives. This case has been often treated, ever since Daniel Bernoulli [8] and, for that matter, Marshall (Note IX, p. 843, [9]), by ordering the wagers according to their "expected utility." This leads to a non-stochastic utility that is more strictly measurable than ω: It is unique up to increasing linear transformations. This model, too, lends itself to stochastic generalizations, as will be briefly discussed in Section 9.

[5] The condition is, *A is perfectly separable and is ordered by the relation "$\gtrsim$"; and for every a' in A the sets $(a|a' \gtrsim a)$ and $(a|a \gtrsim a')$ are closed.* The condition is trivially fulfilled by every finite set, for it is perfectly separable and all its subsets are closed. Also, every subset of a finite-dimensional Euclidean space is perfectly separable, and it is reasonable to assume that the sets $(a|a' \gtrsim a)$ and $(a|a \gtrsim a')$ are closed when "$\gtrsim$" means a consumer's preference over the space of a finite number of commodities.

[6] Thus Abraham Wald [6] has suggested a method to evaluate the consumer's ordinal utility function ω from the basic observations $(a; F)$—purchases made as F varies with incomes and prices—assuming a non-stochastic model and not attempting a statistical estimate. More recently, H. Theil and H. Neudecker [7] gave a stochastic generalization of the same model, specifying, as did Wald, ω to be quadratic in the quantities of the goods consumed.

2. Stochastic Concepts of Constant Utilities

In general, $A = (a, b, \cdots)$ will continue to denote the set of distinct alternatives, and F the feasible subset. For mathematical ease, we shall assume A to be finite, unless otherwise stated, and identify it with $N = (1, \cdots, n)$. A feasible subset will be $M \subseteq N$. The probability that the subject forced to choose an element of M chooses i, denoted previously by $p(i; M)$, can be written more briefly thus: $i(M)$. Clearly

$$i(M) \geqq 0; \qquad \sum_{i \in M} i(M) = 1 \,. \tag{2.1}$$

When $M=(i, j, k, \cdots)$, $i(M)=i((i, j, k, \cdots))$ will be written simply $i(i, j, k, \cdots)$. Then $i(M)$ will be called the binary, ternary, $\cdots$, probability when M consists of 2, 3, $\cdots$, distinct elements. The binary probability will be sometimes written in still shorter forms: $i(i, j) = ij$, $i \neq j$. It will prove convenient to define $ii = \frac{1}{2}$ so that, using (2.1), always

$$ij + ji = 1 \,, \tag{2.2}$$

whether i and j are distinct or not.

For easier reference, the various *conditions* will be labeled by (more-or-less suggestive) letters, thus: (x). *A theorem* is an implication-relation between conditions (the "hypothesis" and the "conclusion"); by using arrows several theorems can be combined into one. In addition to the usual signs $\rightarrow$ ("implies") and $\leftrightarrow$ ("implies and is implied by"), we shall also use $\mapsto$ ("implies but is *not* implied by"). When $(x) \mapsto (y)$, (x), is said to be stronger than (y); rejection of (y) forces rejection of (x), but acceptance of (y) is inconclusive. We shall also use the sign $\vdash\!\dashv$ for "does not imply nor is implied by."

Each of the following three conditions (w), (v), (u), arranged in a sequence of increasing strength, constrains the set of probabilities $i(M)$ by postulating the existence of some real vector of order n (a real-valued function on the set N), called utility vector (utility function), and denoted by w, v, u, respectively. The stronger the constraint, the more strict is the sense in which the utility vector in measurable; that is, the smaller is the group of transformations under which the vector remains indeterminate.

Condition (w). *There is a constant real vector* $w = (w_1, \cdots, w_n)$ *such that*

$$w_i \geqq w_j \quad \text{if and only if } ij \geqq \tfrac{1}{2} \,, \tag{2.3}$$

where w_i may be called the *weak* utility of i; and w the weak utility function on N.

Condition (v). *There is a constant real vector* $v = (v_1, \cdots, v_n)$ *and, associated with it, a distribution function* φ_v, *strictly increasing* [*except when its value is* 0 *or* 1][7], *such that*

$$\varphi_v(v_i - v_j) = ij; \qquad \varphi_v(0) = ii = \tfrac{1}{2} \,. \tag{2.4}$$

[7] Regarding the proviso in brackets, see Section 8.

Then v_i may be called a *strong* utility of i; and v, a strong utility function on N.

CONDITION (u). *There is a constant positive vector* $u = (u_1, \cdots, u_n)$ *such that for any* i, j *in* M

$$u_i/u_j = i(M)/j(M)\,, \qquad \text{all } M \subseteq N\,, \tag{2.5}$$

where u_i may be called the *strict* utility of i; and u the strict utility function on N.

Clearly the weak utility function w is *unique up to an increasing monotone transformation*. It is analogous to the function ω of the non-stochastic model (Section 1), with "$a_i \gtrsim a_j$" interpreted as "$ij \geqq \frac{1}{2}$." We can call the cases $ij > \frac{1}{2}$ and $ij = \frac{1}{2}$, stochastic preference and indifference, respectively. These concepts are implicitly used in experimental work as when, e.g., Mosteller and Nogee [10] define indifference as the case when the subject chooses one of the two offered alternatives half of the time. As in the non-stochastic model, (w) implies a testable

CONDITION (t) (transitivity).

$$\textit{If} \quad ij \geqq \tfrac{1}{2} \quad \textit{and} \quad jk \geqq \tfrac{1}{2}\,, \quad \textit{then} \quad ik \geqq \tfrac{1}{2}\,. \tag{2.6}$$

Since always $ij \geqq$ or $\leqq \frac{1}{2}$, transitivity guarantees a (complete weak) ordering on N: the alternative i is ranked above, below, or on the level of j according as $ij >, <,$ or $= \frac{1}{2}$. Since N is finite we, have (as in Section 1):

THEOREM 2.1. $(w) \longleftrightarrow (t)$ *provided the set of alternatives is finite.*

If we admit infinite sets of alternatives and redefine the condition (w) accordingly, we have $(w) \mapsto (t)$. But if we denote by (D) Debreu's Condition (footnote 5) and replace "$a \gtrsim a'$" and "$ij \geqq \frac{1}{2}$" by "$aa' \geqq \frac{1}{2}$," we can write

THEOREM 2.2. $[(D), t] \longmapsto (w) \longmapsto (t)\,.$

Condition (v), the existence of "strong" utilities, may be made plausible by a physical analogy. Because of random variations in the properties of the air, metals, and so forth, the lowest voltage needed to produce a spark in a given direction is random; assume then that the probability of the occurrence of the spark is the larger, the larger the voltage. Condition (v) has been used in psychophysics (and more recently in the scaling of attitudes) ever since Fechner [3]. It is stated in the psychologists' adage (Guilford, [11]): "*Equally often noticed differences* (on the "sensation" scale) *are equal* [*unless noticed always or never*]." A similar thought might unify the recent attempts to measure the "power" of person X over person Y: on the one hand, by the probability that Y obeys X; and on the other hand, by the difference in the payoff (utility) to Y in the case of his obeying, as compared with his disobeying, X.[8]

[8] These two points of view, the empirical and the game-theoretical, are represented in somewhat modified form by Dahl [12] and by Shapley and Shubik [13], respectively. If one accepts John Harsanyi's proposal to measure the power of X over Y by "how much difference it makes to Y between being X's friend or enemy," this difference between maxmax and maxmin payoffs (see, e.g., Luce and Raiffa [14]) might be related to the probability of Y's decision to be a friend.

If (v) is satisfied by some v, φ_v, then clearly for *any increasing linear transform* $v' = \alpha + \beta v$ there exists a $\varphi_{v'}$ satisfying that condition. (If the set of alternatives is continuous, and the distribution φ_v strictly monotone, these are the only admissible transformations.[9]) The scale unit for v may be chosen by setting, for example, the quartile $\varphi_v^{-1}(.25) = 1$; this or a similarly chosen unit is sometimes called the "just noticeable difference," we believe. Moreover, because of (2.4), φ_v may be (and, in the continuous case, must be) so chosen that

$$\varphi(\zeta) + \varphi(-\zeta) = 1 \tag{2.7}$$

for all real ζ. Then φ_v is anti-symmetrical about the point $(0, \frac{1}{2})$; the median and the mean (if it exists) are zero. This is satisfied if φ_v is normal, as assumed by Fechner, whose test of normality was, however, rather crude. Normality of φ_v is also assumed in most textbooks; we do not know whether a formal test of this assumption has been developed. The "logistic" distribution function $\varphi_v(\zeta) = 1/(1 + e^{-\zeta})$ which, as we shall presently see, is required by the next stronger condition (u), also has the property (2.7).

THEOREM 2.3. $(v) \longmapsto (w)$.

PROOF. *Sufficiency*. Assume (v) and choose i, j with $v_i \geqq v_j$; then

$$ij = \varphi_v(v_i - v_j) \geqq \varphi_v(0) = \tfrac{1}{2}\,;$$

hence (w) is satisfied, with $w = v$.

No necessity. Let $N = (1, 2, 3)$ and assume $12 > 13 > 23 > \frac{1}{2}$. Then (w) is satisfied, with $w_1 > w_2 > w_3$; but no linear function of w will satisfy (v); for, by (v), if $\frac{1}{2} < 23$ then $0 = v_1 - v_1 < v_2 - v_3$, $v_1 - v_2 < v_1 - v_3$; hence, by (v), $12 < 13$, which contradicts the assumption.

As to the "strict" constant utility condition (u), it has been postulated, and developed most fully, by R. D. Luce [15].[10] We shall now give some of his results. Because of (2.1), Condition (u) can be rewritten in the form

$$i(M) = u_i \Big/ \sum_{j \in M} u_j, \qquad \textit{all} \quad M \subseteq N\,. \tag{2.8}$$

It follows that the vector u is *unique up to a positive factor* $\lambda = u_i/i(N)$. With N fixed, a convenient normalization is $\lambda = 1 = \sum_{i \in N} i(N)$; $u_i = i(N)$.

Putting $M = (i, j)$, $i(M) = ij$, we see that the probability of a binary choice is related to the ratio u_i/u_j by

---

[9] For, suppose that $\varphi_{v'}(v'(x) - v'(y)) \equiv \varphi_v(v(x) - v(y))$. Then with $\psi = \varphi_{v'}^{-1}\varphi_v$ we have (i) $v'(x) - v'(y) = \psi(v(x) - v(y))$, (ii) $v'(y) - v'(z) = \psi(v(y) - v(z))$, and (iii) $v'(x) - v'(z) = \psi(v(x) - v(z))$. Adding (i) and (ii) and using (iii) we find, letting $a = v(x) - v(y)$, $b = v(y) - v(z)$, that $\psi(a + b) = \psi(a) + \psi(b)$. The only solution of this last equation which is not unbounded in every interval is, as is well known, $\psi(a) = \beta a$. Then from (i), with y fixed, we have $v'(x) = \beta v(x) + \alpha$.

[10] Occasionally (in his Section 2.D.3, entitled "A generalization to two or more alternatives") Luce combines Condition (u) with another condition which we shall call (u') (existence of strict *disutilities* of *last* choices) and which is not generally consistent with (u): see our Theorem 3.8 below.

$$\psi(u_i/u_j) = ij, \quad \text{where} \quad \psi(\zeta) = \begin{cases} \zeta/(1+\zeta), & \zeta > 0, \\ 0, & \zeta \leqq 0. \end{cases} \tag{2.9}$$

Clearly ψ is a distribution function, for it is increasing, and $\psi(-\infty) = 0$, $\psi(\infty) = 1$. It is analogous to φ_v of condition (v), with ratios replacing differences, but is restricted to a particular form. Moreover, by putting v_k equal to $\log u_k$, we recover condition (v) itself, with φ_v a "logistic curve":

$$\varphi_v(\zeta) = 1/(1 + e^{-\zeta}). \tag{2.10}$$

This proves

THEOREM 2.4. $\qquad (u) \longmapsto (v),$

for (v) imposes constraints on binary choices only.

We shall now introduce two *directly testable* conditions equivalent to (u). First consider

CONDITION $(p.r)$ (constancy of probability-ratios). *$i(M) > 0$ for every i in M; and for every i, j in N, the ratio $i(M)/j(M)$ is constant over all sets $M \subseteq N$ that contain i and j.*

Condition $(p.r)$ follows directly from (u); conversely, rewriting $(p.r)$ as

$$i(M)/i(N) = j(M)/j(N) = \cdots = \sum_{k \in M} k(M) \Big/ \sum_{k \in M} k(N), \tag{2.11}$$

putting $k(N) = u_k$ and using (2.1), we obtain (u). Now consider

CONDITION $(c.p)$ (probability of choice as a conditional probability).

$$0 < i(M) = i(N) \Big/ \sum_{k \in M} k(N) \qquad \textit{all } i \textit{ in } M \subseteq N. \tag{2.12}$$

The fraction on the right-hand side, is, of course, the conditional probability that the element i of N is chosen *when it is known that the element chosen belongs to the subset M of N.* On the other hand, the quantity on the left-hand side is the probability that the element i of N is chosen *when the subject is forced to make his choice from the subset M of N.* Clearly these two quantities are not identical; $(c.p)$ is a verifiable, empirical proposition which may or may not have intuitive appeal.[11] Condition $(c.p)$ follows directly from (u) written in the form (2.8); on the other hand, on replacing i by j in (2.12) and comparing the two equations, one obtains $(p.r)$ which we have seen to be equivalent to (u). We summarize this and the previous results of this section in the following theorem, placing the directly testable conditions in the lower line:

THEOREM 2.5.

$$\begin{array}{ccccc} (u) & \longmapsto & (v) & \longmapsto & (w) \\ \updownarrow & & & & \updownarrow \\ (p.r) & \longleftrightarrow & (c.p) & \longmapsto & (t), \end{array}$$

[11] Luce tries to convey this appeal by an analogy with Arrow's (non-stochastic) principle which we called before the "irrelevance of added alternatives."

provided that the Debreu condition is satisfied. (This includes the case of a finite set of alternatives.)

3. Random Orderings and Random Utilities

In the non-stochastic theory briefly reviewed in Section 1, the preference statement "$a_i \succ a_j$" assigns probability 1 to a certain class of choices. In the stochastic models of Section 2, the corresponding statement "$ij > \frac{1}{2}$" merely sets a lower bound on that probability and is therefore weaker; however, some of those models are weaker than others.

A different way of weakening the non-stochastic theory consists in making the ("ordinal") utility function ω a random one; i.e., by defining a probability measure on the space of all real-valued functions on A. In the case of the economic example of Section 1, Wald's problem of evaluating the (non-stochastic) utility function ω of consumers from their observed choices $(a; F)$ is replaced by the following more general one: Estimate the probability distribution P on the space of random utility functions U, using the probabilities $p(a; F)$ already estimated from the observations $(a; F)$, or using these observations directly. Any random utility function will be regarded as equivalent to any of its increasing monotone transforms.

However, the choice probabilities $p(a; F)$ may or may not be consistent with the existence of a random utility function. That is, the following condition (defining, in fact, that function) may or may not be fulfilled:

CONDITION (U) (existence of random utilities). *There is a random vector* $U = (U_1, \cdots, U_n)$, *unique up to an increasing monotone transformation and such that, for any* i *in* $M \subseteq N$,

$$\Pr\{U_i \geqq U_j, \quad \text{all } j \text{ in } M\} = i(M). \tag{3.1}$$

U_i is called *random utility* of i. Putting $M = (i, j)$,

$$\Pr\{U_i \geqq U_j\} = ij. \tag{3.2}$$

By (2.2), it follows that, for all $i \neq j$,

$$\Pr\{U_i = U_j\} = 0; \tag{3.3}$$

for if $\Pr\{U_i = U_j\} > 0$, then by (3.2) $ij + ji = \Pr\{U_i \geqq U_j\} + \Pr\{U_i \leqq U_j\} = \Pr\{U_i \geqq U_j\} + \Pr\{U_i < U_j\} + \Pr\{U_i = U_j\} > 1$, contradicting (2.2).

When the set of alternatives is finite, we can consider one particular monotone transform of U, the ranking $r = (r_1, \cdots, r_n)$ on the set $N = (1, \cdots, n)$. Because of (3.3), no ties will occur. Hence R, the set of all rankings on N, consists of the $n!$ permutations on N. We shall denote by i_r the element of N that has rank i when the ranking is r; i.e., $r_{i_r} = i$. Using this notation, we write a given ranking $r = (r_1, \cdots, r_n) = 1_r 2_r \cdots n_r$ (without commas and parentheses). Thus, if $n = 3$, the ranking $r = 312$ means that $r_1 = 2$, $r_2 = 3$, $r_3 = 1$, and is therefore identical with $r = (2, 3, 1)$.

Denote by R_{iM} the set of all rankings on N in which i is the first among all elements of $M \subseteq N$:

$$R_{iM} = \{r \mid r_i \leqq r_j, \text{ all } j \in M\} \qquad i \in M. \tag{3.4}$$

For example, if $N = (1, 2, 3)$, $M = (1, 2)$, $i = 1$, then $R_{iM} = (123, 132, 312)$. Clearly, for every $M \subseteq N$,

(3.5) R_{iM} *and* R_{jM} *are disjoint for all* $i \neq j$ *in* M;

and if we denote by R^M the set of all permutations on M, we have

(3.6) $$\bigcup_{i \in M} R_{iM} = R^M ; \qquad \bigcup_{i \in N} R_{iN} = R^N = R .$$

The following condition will be presently shown to be equivalent to (U).

CONDITION (P) (existence of a probability distribution of rankings consistent with probabilities $i(M)$). *There are* $n!$ *numbers* $P(r)$ *such that*

(3.7) $$P(r) \geqq 0, \qquad \sum_{R_{iM}} P(r) = i(M), \qquad i \in M \subseteq N .$$

It follows, using (3.6), (2.1), that $\sum_R P(r) = 1$. We can call $P(r)$ the *probability of the ranking* r.

THEOREM 3.1. $(U) \longleftrightarrow (P)$.

PROOF. If we assume Condition (U) and define $P(r) = \Pr\{U_{1_r} > U_{2_r} > \cdots > U_{n_r}\}$, we obtain Condition (P). Conversely, assume (P) and define the random vector U thus: For any non-random real vector $s = (s_i, \cdots, s_n)$ let

(3.8) $$\Pr\{U = -s\} = \begin{cases} P(s), & s \in R , \\ 0, & s \notin R , \end{cases}$$

where R is the set of permutations on N. Then Condition (U) is satisfied.

THEOREM 3.2. *The existence of random utilities* U *does not imply the existence of weak constant utilities* w, *let alone the existence of strong or strict utilities,* v *or* u.

PROOF. Let $N = (1, 2, 3)$ and $0 < \alpha < \frac{1}{6}$. Let the probabilities of rankings $P(r)$ be

$$P(123) = P(231) = P(312) = \tfrac{1}{6} + \alpha > 0 ,$$
$$P(321) = P(213) = P(132) = \tfrac{1}{6} - \alpha > 0 .$$

(Then $\sum P(r) = 1$.) Let the probabilities of choices $i(M)$, $M \subseteq N$, be

(3.9) $$1(N) = 2(N) = 3(N) = \tfrac{1}{3} ,$$
$$1(1, 2) = 2(2, 3) = 3(1, 3) = \alpha + \tfrac{1}{2} ,$$
$$2(1, 2) = 3(2, 3) = 1(1, 3) = -\alpha + \tfrac{1}{2} .$$

Then $1(N) = P(123) + P(132)$, $1(1, 2) = P(123) + P(132) + P(312)$, etc., and Condition (P)—or its equivalent, (U)—is satisfied. But Condition (t) is not: $1(1, 2) > \frac{1}{2}$, $2(2, 3) > \frac{1}{2}$, but $1(1, 3) < \frac{1}{2}$. This proves the present theorem, using Theorem 2.5.[12]

[12] This counterexample is due to C. Winsten. Another interesting proof is due to P. Halmos. Three dice, A, B, C, are loaded so as to turn up with the following probabilities:

Face no.		1	2	3	4	5	6
Die:	A	0	0	.5	.5	0	0
	B	0	.6	0	0	0	.4
	C	.4	0	0	0	.6	0

Let a, b, c (independent chance variables) be the number of spots turning up on A, B, C, respectively. Then (P) is satisfied: the probabilities of the six possible rankings of the numbers a, b, c according to magnitude, must be consistent with the probabilities that one of the numbers is the largest in a given pair or triple. But (t) is not satisfied: $\Pr\{a > b\} = .6 > \frac{1}{2}$; $\Pr\{b > c\} = .64 > \frac{1}{2}$; $\Pr\{a > c\} = .4 < \frac{1}{2}$.

We shall show that the following conditions are necessary for (P), and therefore for its equivalent, (U):

CONDITION (e) (effect of enlarging the feasible set). *If* $L \subseteq M \subseteq N$, *then*

$$i(M) \leqq i(L) . \tag{3.10}$$

CONDITION (e_3). *For any three elements* i, j, k *of* N,

$$i(i, j) \geqq i(i, j, k); \text{ or, equivalently,} \tag{3.11}$$

$$i(i, j, k) \leqq \min(ij, ik) . \tag{3.11'}$$

Clearly (e_3) is a special case of (e); it was exemplified numerically in (3.9), for $n = 3$. Applying to (P) the fact that $R_{iM} = \{r \mid r_i \leqq r_j, j \in L\} \cap \{r \mid r_i \leqq r_j, j \in M - L\} \subseteq R_{iL}$, we obtain

THEOREM 3.3. $(P) \longrightarrow (e) \longmapsto (e_3)$.[13]

THEOREM 3.4. *The existence of random utilities U is not implied by the existence of strong constant utilities v, let alone by the existence of weak constant utilities w.*

PROOF. Denote by x, y, z the distinct generic elements of $N = (1, 2, 3)$. There exist three numbers (for example, $v_1 = .2$, $v_2 = .1$, $v_3 = 0$) such that

$$0 < v_x - v_y + \tfrac{1}{2} = \beta_{xy} < 1 ,$$
$$0 < v_1 - v_2 = \gamma_1 < 1 ,$$
$$0 < v_2 - v_3 = \gamma_2 < 1 ,$$
$$0 < v_3 - v_1 + 1 = \gamma_3 < 1 .$$

Since $\beta_{xy} + \beta_{yx} = 1 = \sum_x \gamma_x$, we can put $x(x, y) = \beta_{xy}$, $x(x, y, z) = \gamma_x$. Then Condition (v) is satisfied, with $\varphi_v(\zeta) = \zeta + \frac{1}{2}$. But since $\gamma_3 - \beta_{31} = \frac{1}{2} > 0$, $3(1, 2, 3) > 3(1, 3)$, Condition (e) and therefore also the stronger Condition (P) and its equivalent (U) are contradicted.

In the work of Thurstone [16] the following condition seems to be used:

CONDITION $(s.n)$ (symmetrical normal). *There is a normal random vector* $U = (U_1, \cdots, U_n)$ *that satisfies Condition* (U) *and has all variances equal* ($\sigma_{ii} = \sigma^2$, *say*) *and all covariances equal* ($\sigma_{ij} = \rho\sigma^2$, *say*).

Clearly this implies that both (U) and (v) are satisfied: let $v_i = EU_i$ and let φ_v, at all points $\zeta = v_i - v_j$, be normal, with mean zero and variance $2\sigma^2(1 - \rho)$. Even the following, weaker condition is sufficient for the conjunction $[(U), (v)]$:

CONDITION (s) (symmetry of adjusted random utilities). *There is a random vector U satisfying Condition* (U), *and a constant vector* $v = (v_1, \cdots, v_n)$ *such that* $V = U - v$ *has a distribution function symmetric in its arguments.*

If U, v satisfy this condition, then any pair of increasing linear transforms $\alpha + \beta U$, $\alpha + \beta v$ (α, β scalars) will satisfy it. Clearly $(s.n) \mapsto (s)$. Moreover,

THEOREM 3.5. $(s.n) \longmapsto (s) \longmapsto [(U), (v)]$.

[13] Whether or not (e) implies (P) will be left open until Section 5.

PROOF. *Sufficiency of* (s). If (s) is true, so is (U); and $i(i, j) = \Pr\{U_i \geqq U_j\} = \Pr\{V_j - V_i \leqq v_i - v_j\}$ = the value of a distribution function at $v_i - v_j$; because of symmetry, this function is the same for all i, j and can be written as φ_v, satisfying (v).

No sufficiency of $[(U), (v)]$. Let $N = (1, 2, 3)$. Let $v_i = i + 3$, and let V_1, V_2, V_3 be independently distributed with $\Pr\{V_i = 3\} = p_i = 1 - \Pr\{V_i = -3\}$, where $p_1 = 1 - p_3 = p > \frac{1}{2}, p_2 = \frac{1}{2}$. If $U = v + V$, then $\Pr\{U_1 > U_3\} = p^2$; $\Pr\{U_1 > U_2\} = p/2 = \Pr\{U_2 > U_3\}$; $\Pr\{U_1 > U_2, U_2 > U_3\} = p^2/2$, etc. Hence U satisfies Condition (U), if $1(1, 2) = 2(2, 3) = p/2$; $1(1, 3) = p^2$; $1(1, 2, 3) = p^2/2$, etc. Then Condition (v) is also satisfied, with

$$\varphi_v(v_1 - v_2) = \varphi_v(v_2 - v_3) = \varphi_v(-1) = p/2 > p^2 = \varphi_v(-2) = \varphi_v(v_1 - v_3) .$$

Yet no linear transformation of v or monotone transformation of U can make the distribution of V symmetric, since the p_i are not all equal. If, on the other hand, the set of alternatives is continuous, let V have a joint normal distribution with zero mean and a constant variance τ^2 of differences $V_x - V_y$: $\tau^2 = \sigma_{xx} + \sigma_{yy} - 2\sigma_{xy}$, all x, y, where $\sigma_{11} = \frac{1}{2}$ and (for $x, y \neq 1$) $\sigma_{xx} = 1$, $\sigma_{x1} = 0$, $\sigma_{xy} = \frac{1}{4}$. If $U = V + EU$ satisfies condition (U), then (v) is also satisfied, with $v = EU$ and φ_v normal with zero mean and variance $\tau^2 = \frac{3}{2}$. But no translation can make the distribution function of V symmetric in its arguments.

Mosteller [17] dropped Thurstone's symmetry assumption but maintained normality. We do not know whether tests have been developed for the joint condition $[(U), (v)]$ in general, or for the special case of joint normality with constant variance of differences.

We shall now discuss another condition that is also stronger than the conjunction $[(U), (v)]$ and is of particular interest, being equivalent to Condition (u) of Section 2 (existence of strict constant utilities). For a given permutation r of the set N define

$$\pi(r) = \prod_{j=1}^{n-1} j_r(j_r, (j+1)_r, \cdots, n_r) . \tag{3.12}$$

For example, $\pi(312) = 3(1, 2, 3)\cdot 1(1, 2)$. Suppose this were the probability of the ranking 312. This would exemplify

CONDITION (π) (probability of ranking as the product of probabilities of successive first choices). *Condition (P) is true and, for every ranking r of N, $P(r) = \pi(r)$ as defined in* (3.12).

Substituting $\pi(r)$ for $P(r)$ in (3.7), we can put Condition (π) into a form involving the "basic" probabilities (those of "first choices") only:

$$i(M) = \sum_{r \in R_{iM}} \pi(r) , \qquad \text{for all } M \text{ and } i \in M . \tag{3.13}$$

THEOREM 3.6. $\quad (u) \longleftrightarrow (\pi)$.

PROOF (outline). *Sufficiency of* (u). Define

$$\Omega(r) = \prod_{j=1}^{n-1} u_{j_r} \Big/ \sum_{k=j}^{n} u_{k_r} . \tag{3.14}$$

To prove that (u) implies (π), or (2.8) implies (3.13), is to prove the identity

$$\sum_{r \in R_{iM}} \Omega(r) = u_i \Big/ \sum_{h \in M} u_h , \tag{3.15}$$

for all $i \in M$, all $M \subseteq N$, and any positive numbers $u_1, \cdots, u_n$.[14] Without loss of generality let $M = (1, \cdots, m)$, $m \leqq n$, and, as in (3.6), denote by R^M the set of all permutations of M. Then $r^M = (r_1^M, \cdots, r_m^M) \in R^M$. The set $R = R^N$ is partitioned into m subsets of the form

$$R_{iM}^N = (r^N \mid r_i^N \leqq r_j^N; \quad \text{all } j \in M) , \qquad i \in M ;$$

the set R^N is also partitioned into $m!$ subsets of the form

$$R^N(r^M) = (r^N \mid r_i^N \leqq r_j^N \text{ if } r_i^M \leqq r_j^M; \quad \text{all } i, j \in M) .$$

Similarly, R^M is partitioned into subsets $R^M(r^L), L \subseteq M$. One obtains, by induction on m, first the identities

$$\sum_{r^M \in R^M} \Omega(r^M) = 1, \qquad \sum_{r^M \in R^M(r^L)} \Omega(r^M) = \Omega(r^L) ,$$

and then (3.15).

Necessity of (u). To show that (3.13) implies (2.8), let $u_i = i(N)$, all $i \in N$. Because of the identity (3.15), all numbers $i(M)$ that satisfy (3.13) also satisfy the system of equations

$$i(M) = \sum_{r \in R_{iM}} u_{1_r} \prod_{j=2}^{n-1} j_r(j_r, \cdots, n_r) , \qquad \sum_{i \in M} i(M) = 1 . \tag{3.16}$$

One then proves the uniqueness of this solution: Starting with $M = N$, then taking all sets M with $n-1$ elements, then with $n-2$ elements and so on, one can solve (3.16) for all $i(M)$ and one finds that they are all given by (2.8).[15]

THEOREM 3.7. *If* (u) *is true, then* (P) *and* (U) *hold, with* $P(r) = \pi(r) = \Omega(r)$.

This is clearly a corollary to Theorem 3.6, using Theorem 3.1.

[14] Block and Marschak [18] searched for an appropriate urn model, and Debreu [19] found it, thus providing a shorter proof.

[15] In more detail: If $M = N$ or $M = N - (n)$, the system (3.16) is readily solved for the $i(M)$. Now suppose that $i(M)$ has a unique value satisfying (3.16) for all M with more than h elements, $h \leqq n-1$; and let $H = (1, \cdots, h)$, $J = N - H$. Then (3.16) would imply that

$$i(H) = u_i + \sum_{j \in J} u_i \cdot i(N-(j)) + \sum_{j_1 \in J} \sum_{j_2 \in J-(j_1)} u_{j_1} \cdot j_1(N-(j_2)) \cdot i(N-(j_1, j_2))$$
$$+ \cdots + \sum_{j_1 \in J} \sum_{j_2 \in J-(j_1)} \cdots \sum_{j_{n-h} \in J-(j_1, \cdots, j_{n-h-1})} [u_{j_1} \cdot j_2(N-(j_1)) \cdots j_{n-h}(H+(j_{n-h}))] i(H). \tag{3.17}$$

Except for the two $i(H)$'s, every probability appearing here is the probability of a first choice from a set of more than h elements and by the induction hypothesis is uniquely determined in terms of the u's. Furthermore, the quantity in brackets is equal to

$$\sum_{s \in R^H} j_1(N) \cdot j_2(N-(j_1)) \cdots k_{n-h}(H+(j_{n-h})) \cdot 1_s(H) \cdot 2_s(H-(1_s)) \cdots (h-1)_s((h-1)_s, h_s);$$

thus the multiplier of $i(H)$ on the right-hand side of (3.17) is equal to the probability, p, that the set J precedes the set H. Since $1(N) > 0$ and $1 \in H$, it follows that $p < 1$ and (3.17) can be solved for $i(H)$.

Some interest attaches to a condition that, in effect, appears in the work of Luce [15]. In the same way as Condition (u) treats the probabilities of *first* choices, $p(i; M) = i(M)$, this new condition treats the probabilities of *last* choices. We may denote these by $p'(i; M) = i'(M)$ and remark that last choices are not "basic observations" in our sense. (However, see Section 7 on considering the rankings themselves, and hence also the last choices, "observable"; see also Marschak [20], Section 11.)

CONDITION (u'). *There exists a constant positive vector* $u' = (u'_1, \cdots, u'_n)$ *such that for any* i, j *in* M, $u'_i/u'_j = i'(M)/j'(M)$.
It is possible, however, to prove

THEOREM 3.8. *If* $n > 2$, (u') *is inconsistent with* (u) *unless* $i(N) = i'(N) = 1/n$ *for all* $i \in N$.

PROOF (outline). Define, for each ranking $r = 1_r 2_r \cdots n_r$ the inverted ranking $r^* = n_r(n-1)_r \cdots 2_r 1_r$ (hence $r^{**} = r$); and define

$$\Omega'(r) = \prod_{j=1}^{n-1} u'_{j_r} \Big/ \sum_{k=j}^{n} u'_{k_r} .$$

By Theorem 3.7, if (u) is true then $P(r) = \Omega(r)$. By the same theorem, if (u') is true then $P(r^*) = \Omega'(r)$, and hence $P(r) = P(r^{**}) = \Omega'(r^*)$. Therefore if both (u) and (u') are true,

$$\Omega(r) = \Omega'(r^*) . \tag{3.18}$$

Define the ranking $s = 2_r 1_r 3_r \cdots (n-1)_r n_r$. Since $\sum u_i = 1 = \sum u'_i$ without loss of generality, we have $\Omega(r)/\Omega(s) = (1 - u_{2_r})/(1 - u_{1_r})$; $\Omega'(r^*)/\Omega'(s^*) = u'_{2_r}/u'_{1_r}$. Since $1_r, 2_r$ are arbitrary, we have by (3.18) $u'_i/u'_j = (1 - u_i)/(1 - u_j)$; and summing on i, we find $u_j = 1 - (n-1)u'_j$. Now define $t = 1_r 2_r \cdots (n-2)_r n_r (n-1)_r$; then by (3.18)

$$\Omega(r)/\Omega(t) = u_{(n-1)_r}/u_{n_r} = \Omega'(r^*)/\Omega'(t^*) = (1 - u'_{(n-1)_r})/(1 - u'_{n_r}) ;$$

and in general $u_i/u_j = (1 - u'_i)/(1 - u'_j) = [1 - (n-1)u'_i]/[1 - (n-1)u'_j]$; or $(n-2)u'_i = (n-2)u'_j$. Hence, for $n > 2$, we have $u'_i = u'_j$ and $u_i = u_j$; so that $i(N) = i'(N) = 1/n$ for all i.

THEOREM 3.9. $\qquad (u) \longmapsto [(U, v)] \longleftarrow\!\!\dashv (s).$

PROOF. We have $(u) \mapsto (U)$ by Theorem 3.6 and 3.1, since (π) is strictly stronger than (P). Moreover, $(u) \mapsto (v)$, and $(s) \mapsto (U, v)$ by Theorems 2.4 and 3.5. It remains to prove that (U, v) does not imply (u). Consider any normal distribution symmetric in its arguments. Then, by Theorem 3.5, (U) and (v) are both satisfied; and the corresponding φ_v is normal, contrary to the implication (2.10) of (u).

Some of the results of this and the preceding sections are worth summarizing in

THEOREM 3.10.

$$\begin{array}{ccccc} (u) & \longmapsto & [(U),(v)] & \longmapsto & (U) \\ & & \downarrow & \nearrow & \big| \\ & & (v) & \longmapsto & (w) . \end{array}$$

4. Further Directly Testable Conditions for the Existence of Constant Utilities

We shall deal here with "strong" utilities v only; directly testable conditions for the existence of "strict" utilities u and "weak" utilities w were given in Section 2. Directly testable conditions for (v)—like those for (w) and unlike those for (u)—can, of course, involve binary choices only.

Unless otherwise stated, the set of alternatives will be assumed finite: $N = (1, \cdots, n)$. It is easy to see that we can rewrite

Condition (v). *There is a real vector* $v = (v_1, \cdots, v_n)$ *such that*

$$\text{if } hi \geqq jk \text{ then } v_h - v_i \geqq v_j - v_k \,, \tag{4.1}$$

provided hi *and* jk *are not both* 0 *or* 1.

This implies that $hi \geqq \frac{1}{2}$ if, and only if, $v_h - v_i \geqq 0$. It is seen immediately that (v) implies each of the following conditions:

Condition (t) (transitivity). *If* $ij \geqq \frac{1}{2}$ *and* $jk \geqq \frac{1}{2}$, *then* $ik \geqq \frac{1}{2}$.

Condition (t_s) (strong transitivity). *If* $ij \geqq \frac{1}{2}$ *and* $jk \geqq \frac{1}{2}$, *then* $ik \geqq \max(ij, jk)$.

Condition (t_s^*) *If* $ij \geqq \frac{1}{2}$, *then* $ix \geqq jx$, *all* x *in* N.

Condition (t_s^{**}) *If there exists* k *such that* $ik \geqq jk$, *then* $ij \geqq \frac{1}{2}$.

Condition (6_w) (weak condition on **6**-tuples). *If* $i_1i_2 \geqq j_1j_2$ *and* $i_2i_3 \geqq j_2j_3$, *then* $i_1i_3 \geqq j_1j_3$.

Condition (m_w) (weak condition on **m**-tuples). *For any* $m \leqq n$, *if* $i_hi_{h+1} \geqq j_hj_{h+1}$, $h = 1, \cdots, m-1$, *then* $i_1i_m \geqq j_1j_m$.

Condition (q) (quadruple condition). *If* $hi \geqq jk$, *then* $hj \geqq ik$.

Condition (6_s) (strong condition on **6**-tuples). *If* $i_1i_2 \geqq j_2j_3$ *and* $i_2i_3 \geqq j_1j_2$, *then* $i_1i_3 \geqq j_1j_3$.

Theorem 4.1.

$$\begin{array}{ccccccccccc} (v) & \longmapsto & (6_s) & \longmapsto & (q) & \longmapsto & (6_w) & \longmapsto & (t_s) & \longmapsto & (t) \\ & & & & & & \updownarrow & & \updownarrow & & \\ & & & & & & (m_w) & & (t_s^*) & \longleftrightarrow & (t_s^{**}) \,. \end{array}$$

Proof. Obviously (v) implies all the other conditions, and obviously $(t_s) \rightarrow (t)$. To prove $(t_s) \rightarrow (t_s^*)$, let $ij \geqq \frac{1}{2}$ and apply (t_s) to three cases: (1) $jx \geqq \frac{1}{2}$; (2) $jx < \frac{1}{2}$, $xi \geqq \frac{1}{2}$; (3) $jx < \frac{1}{2}$, $xi < \frac{1}{2}$. To prove the converse, let $ij \geqq \frac{1}{2}$, $jk \geqq \frac{1}{2}$; by $(t_s^*) ix \geqq jx$, $jy \geqq ky$; put $x = k, y = i$. To prove $(t_s^*) \leftrightarrow (t_s^{**})$, apply a contradiction argument, using (2.2). To prove $(6_w) \rightarrow (t_s)$, let $ij \geqq \frac{1}{2}$, $jk \geqq \frac{1}{2}$ and apply (6_w) with $i = i_1$, $j = i_2 = j_2$, $k = i_3$; and (1) $j_1 = j$, $j_3 = k$; (2) $j_1 = i$, $j_3 = j$. Condition (m_w) implies, and is itself obtained by successive application of (6_w). To prove $(q) \rightarrow (6_w)$ is straightforward. To prove $(6_s) \rightarrow (q)$, let $hi \geqq jk$, $i_1 = h$, $j_1 = i_2 = i$, $i_3 = j_2 = j$, $j_3 = k$.

The absence of implications is proved by counterexamples:[16] (t) does not

[16] For methods to construct these counterexamples see Block and Marschak [21]. Georgescu-Roegen [22] and Chipman [23] obtained a condition intermediate in strength between (t_s) and (t) by substituting min for max in (t_s). See Marschak ([20], Section 7.)

imply (t_s) since, with $n = 3$ and $12 > 13 > \frac{1}{2}$ and $23 > \frac{1}{2}$, (t) but not (t_s) is satisfied. To prove that (t_s) does not imply (6_w), let $n = 4$ and $\frac{1}{2} < 43 < 32 < 21 < 31 < 42 < 41$. To prove that (6_w) does not imply (q), let $n = 4$ and $\frac{1}{2} < 32 < 43 < 21 < 31 < 42 < 41$. To prove that (q) does not imply (6_s), let $n = 5$ and $\frac{1}{2} < 21 < 54 < 32 < 43 < 53 < 31 < 42 < 41 < 52 < 51$. To prove that (6_s) does not imply (v), a counterexample (with $n = 9$) was constructed, but must be omitted here.

All the conditions given so far are necessary but not sufficient for the existence of v. With the exception of (m_w), they all involve binary choices among elements of subsets of fixed size m; $m = 3$ in (t) and (t_s); $m = 4$ in (q); $m = 6$ in (6_w) and (6_s). The exception (m_w) is only apparent as (m_w) is equivalent to (6_w). Therefore all those conditions can be applied even when the size of N (the number n of all alternatives) is unknown. The following two conditions can be applied only if the size of N is known.

CONDITION (m_s) (strong condition on **m**-tuples). *For any* $m \leqq n$, $i_1 i_m \geqq j_1 j_m$, *if* $i_h i_{1+h} \geqq j_{r_h} j_{1+r_h}$, $h = 1, \cdots, m-1$, *where* $r = (r_1, \cdots, r_{m-1})$ *is any permutation of the set* $(1, \cdots, m-1)$.

CONDITION (o) (ordering of the probabilities of binary choices). *Arrange all the* $ij \geqq \frac{1}{2}$ *in a sequence of* $n(n-1)/2$ *inequalities (involving only* n *distinct alternatives):* $i_1 i_2 \geqq i_3 i_4 \geqq \cdots \geqq \frac{1}{2}$; *then the corresponding system of* $n(n-1)/2$ *inequalities in* n *distinct real numbers*

$$v_{i_1} - v_{i_2} \geqq v_{i_3} - v_{i_4} \geqq \cdots \geqq 0$$

must have a solution.

The following needs no proof.

THEOREM 4.2. $\quad (v) \longleftrightarrow (o) \longrightarrow (m_s) \longmapsto (6_s)$.

We have not investigated whether (m_s) implies (o).

Thus, for a finite set N of alternatives we have the condition (o) [perhaps also (m_s)?] sufficient and necessary for the existence of a strong utility vector v, and testable if N has known size. For finite N of unknown size our testable conditions are necessary only.

A summary of the main results of this section, assuming the set of alternatives finite, is given by

THEOREM 4.3.

$$\begin{array}{ccccccccccc} (v) & & & & & & & & & & (w) \\ \updownarrow & & & & & & & & & & \updownarrow \\ (o) & \longrightarrow & (m_s) & \longmapsto & (q) & \longmapsto & (6_w) & \longmapsto & (t_s) & \longmapsto & (t) . \end{array}$$

If the set of alternatives is ascribed some continuity properties (to be defined presently), some of the necessary testable conditions become also sufficient. We give two examples. Suppose one postulates

CONDITION $(s.c)$ (stochastic continuity). *For any elements* a, b, c *of the set* A *of alternatives such that* $ab < \lambda < ac$, *there is an element* d *in* A *with* $ad = \lambda$.

Debreu [24] has proved

THEOREM 4.4. $\quad [(s.c), (q)] \longmapsto (v)$.

Or one may postulate a stronger

CONDITION $(d.b.p)$ (continuous **d**ifferentiability of **b**inary **p**robabilities). *The set, A, of alternatives is representable by a real interval, and $p(a, b) = p(a; (a, b))$ is continuously differentiable with $\partial p/\partial b < 0, \partial p/\partial a > 0$.*

In Block and Marschak [21] proof was given of

THEOREM 4.5. $[(d.b.p), (6_w)] \longmapsto (v)$.

To show this, one first proves a

LEMMA. *Let $x, y \in [0, 1]$ and let $f(x, y)$ be continuously differentiable with $\partial f/\partial y < 0, \partial f/\partial x > 0$. Then, in order that f admit the representation $f(x, y) = h(u(x) - v(y))$, where h, u, v are monotone increasing functions of one variable, it is necessary and sufficient that there exist functions $\alpha(x), \beta(y)$ such that $(\partial f/\partial x)/(\partial f/\partial y) = \alpha(x)/\beta(y)$.*

5. Further Directly Testable Conditions for the Existence of Random Orderings

Directly testable sufficient conditions for (U), the existence of random utilities (or, equivalently, for (P), the existence of random orderings), were given in Section 3. It was shown, for example, that the probabilities $P(r)$ exist if the ratios $i(M)/j(M)$ do not depend on M. We have also shown that this condition is not necessary. On the other hand, the directly testable condition (e) is necessary: if the $P(r)$ exist, then $i(M) \leqq i(L)$ for any $i \in L \subseteq M \subseteq N$. We shall presently show that this condition is not sufficient. We have not found directly testable conditions that are both sufficient and necessary for the existence of $P(r)$ on N. The following partial results may, however, prove useful. They still use finite sets of alternatives only, of known or unknown size. The case of continuous sets has not been explored.

Condition (e_3) defined in Section 3 (3.11), as a special case of Condition (e), can also be conveniently denoted by (P_3), being the first link in a sequence of Conditions (P_m), $m = 3, \cdots, n$, which we shall state presently. Let us first state

THEOREM 5.1. *If $n = 3$, then* $(P_3) \longleftrightarrow (P)$.

PROOF. *Sufficiency of* (P_3). Denote by x, y, z the generic distinct elements of $N = (1, 2, 3)$, define $Q(zxy) = x(x, y) - x(x, y, z)$, and assume (P_3) or its equivalent, (e_3). Then by (3.11), $Q(zxy) \geqq 0$; and by (2.2), (2.1), the sum of the six numbers $Q(zxy)$ is equal to $3 - 2 = 1$; hence (P) is satisfied, with $P(xyz) = Q(xyz)$.

Necessity of (P_3). By Theorem 3.3.

CONDITION (P_4). *For any four distinct elements w, x, y, z of N,*

$$w(w, x, y) \geqq w(w, x, y, z) \; ; \tag{5.1}$$

$$w(w, x) - w(w, x, y) \geqq w(w, x, z) - w(w, x, y, z) \; . \tag{5.2}$$

THEOREM 5.2. $(P) \longrightarrow (P_4) \longmapsto (P_3)$.

PROOF. Add (5.1) and (5.2) to obtain (P_3) ($= e_3$) from (P_4). The converse is clearly not true. Moreover, if (P) is true, then

$$w(w, x, y) - w(w, x, y, z) = P(zwxy) + P(zwyx) , \tag{5.3}$$

$$w(w, x) - [w(w, x, y) + w(w, x, z)] - w(w, x, y, z) = P(zywx) + P(yzwx) ; \tag{5.4}$$

since the expressions on the right-hand side are non-negative, (P_4) is satisfied. The converse is not true, since (P) may involve subsets with more than four elements.

THEOREM 5.3. *If* $n = 4$, $(P_4) \longleftrightarrow (P)$.

PROOF. *Sufficiency of* (P_4). If x, y, z, w are generic distinct elements of $N = (1, 2, 3, 4)$, then (5.3), (5.4) is a system of 24 equations in as many unknowns, $P(r)$, non-negative and adding up to 1; the knowns being the $x(M)$, $M \subseteq N$. Each permutation r occurs exactly once in the set (5.3) and exactly once in the set (5.4). Thus the system breaks into six sets of four equations; each set involves exactly four permutations r (none of which appears in the other five sets) and is of the form

$$\begin{aligned} \alpha &= P(r^{(1)}) + P(r^{(2)}) , \\ \beta &= P(r^{(2)}) + P(r^{(3)}) , \\ \gamma &= P(r^{(3)}) + P(r^{(4)}) , \\ \delta &= P(r^{(4)}) + P(r^{(1)}) , \end{aligned} \tag{5.5}$$

where the known $\alpha, \beta, \gamma, \delta$ are non-negative. They obey the restriction $\alpha + \gamma = \beta + \delta$, so that the four equations are linearly dependent. A non-negative solution for the four $P(r)$ is obtained by putting $0 = P(r^{(4)})$ or $0 = P(r^{(2)})$ according as $\beta > \gamma$ or not. One then verifies (P) for all $x(M)$ and obtains $\sum_R P(r) = 1$.

Necessity of (P_4). By Theorem 5.2.

THEOREM 5.4. *When* $n = 4$, $(P_4) \longmapsto (e)$.

PROOF. *Sufficiency*. By Theorems 5.3 and 3.3.

No necessity. (P_4) implies

$$w(w, x) - w(w, x, y) \geqq w(w, x, z) - w(w, x, y, z) \geqq 0 ; \tag{5.6}$$

when $n = 4$, (e) is the pair of inequalities

$$w(w, x) - w(w, x, y) \geqq 0 , \tag{5.7}$$

$$w(w, x, z) - w(w, x, y, z) \geqq 0 , \tag{5.8}$$

which are implied by, but do not imply, (5.6).

Conditions (P_4) and (P_3) are special cases of

CONDITION (P_m). *Let* $y, z, x_1, \cdots, x_{m-2}$ *denote* m *generic distinct elements of a set* M *consisting of* m *elements,* $3 \leqq m \leqq n$. *Then for any* h, $3 \leqq h \leqq m$,

$$\begin{aligned} S_h = y(y, z) &- [y(y, z, x_1) + y(y, z, x_2) + \cdots + y(y, z, x_{h-2})] \\ &+ [y(y, z, x_1, x_2) + y(y, z, x_1, x_3) + \cdots + y(y, z, x_{h-3}, x_{h-2})] \\ &\cdots \pm y(y, z, x_1, \cdots, x_{h-2}) \geqq 0 . \end{aligned} \tag{5.9}$$

THEOREM 5.5. $(P) \longrightarrow (P_n)$, *with*

$$S_m = \sum_r P(1_r 2_r \cdots (m-2)_r yz)\,, \qquad m = 3, \cdots, n\,; \tag{5.10}$$

the sum being taken over all permutations r of the m-element-set M which end in yz.

PROOF. Assume (P) and count the number of times, with sign, that any particular permutation is included in S_m. A term of the type $\cdots y \cdots z \cdots$, where $k \geqq 1$ elements besides z follow y, has the coefficient

$$1 - \binom{k}{1} + \binom{k}{2} - \cdots + \binom{k}{k} = (1-1)^k = 0\,, \tag{5.11}$$

and so the only remaining terms will be as specified on the right-hand side of (5.10).

We have seen that, in addition, $(P_n) \rightarrow (P)$ for $n \leqq 4$. We do not know whether this is true for $n > 4$. And we note that (P_n) is not testable if n is unknown.

An interesting condition involving binary choices only was formulated independently by several authors (perhaps first by Guilbaud [25]):

CONDITION (c_3) (a condition on **c**ycles of **3** elements).[17] *For any three distinct elements i, j, k of N,*

$$1 \leqq ij + jk + ki \leqq 2\,. \tag{5.12}$$

THEOREM 5.6. $(e_3) \longmapsto (c_3)$.

PROOF. *Sufficiency*. Let $M = (i, j, k) \subseteq N$. By (e_3), $ij \geqq i(M)$, $jk \geqq j(M)$, $ki \geqq k(M)$; add; apply (2.1), (2.2).

No necessity. Use the example of Theorem 3.4, where $ij + jk + ki = \frac{3}{2}$, so that the binary probabilities obey (c_3); but one of the ternary probabilities contradicts (e_3). We shall now prove

THEOREM 5.7. *Condition* (c_3) *is sufficient and necessary for the binary probabilities to be consistent with Condition* (e_3).

PROOF. *Sufficiency*. Denote by x, y, z the generic distinct elements of $M = (i, j, k) \subseteq N$. We have to show that if the six binary probabilities xy satisfy (c_3), then there exist three numbers γ_x that have the properties required by (e_3) of the ternary probabilities, viz., $\sum_x \gamma_x = 1, 0 \leqq \gamma_x \leqq \min(xy, xz)$ for all x. Clearly such numbers exist if

$$s = \min(ij, jk) + \min(ji, jk) + \min(ki, kj) \geqq 1\,. \tag{5.13}$$

Let ij be minimal, and hence, by (2.2), ji maximal, among the six xy. If $kj < ki$, then, by (2.2), $s = ij + 1 \geqq 1$; if $kj \geqq ki$, then (c_3) implies $s \geqq 1$. Hence, (c_3) guarantees the existence of the desired numbers γ_x (in general not unique).

Necessity. By Theorem 5.6.

[17] Since (5.12) can be rewritten $ij + jk \geqq ik$, this condition has also been called "triangular." Its further properties are studied in Marschak [20].

Theorem 5.7 may be of importance if binary probabilities are the only observable ones (see Section 7). For the same reason, the following theorems, 5.8 and 5.9, will be of interest. We shall first state the relations between (e_3), (c_3), and the two transitivity conditions (t_s) and (t) of Section 4:

THEOREM 5.8.

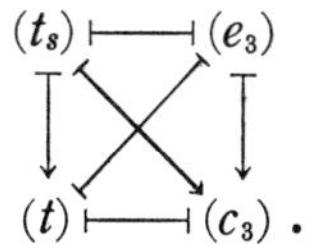

PROOF. $(t_s) \mapsto (t)$, $(e_3) \mapsto (c_3)$ by Theorems 4.1, 5.6. To show that (t_s) implies (c_3) assume, for some i, j, k in N, $ij + jk + ki < 1$, contradicting (c_3); then $ij + jk < 1$, and by (2.2) $ij < kj$; so that, if (t_s) [and therefore, by 4.1, (t_s^{**})] holds, then $ik < \frac{1}{2}$, $ki > \frac{1}{2}$, and by symmetry $ij > \frac{1}{2}$, $jk > \frac{1}{2}$, hence $ij + jk + ki > \frac{3}{2}$, contradicting the assumption. To show that (c_3) does not imply (t), let alone (t_s), consider $N = (i, j, k)$, and make $ij = jk = ki = .4$, thus satisfying (c_3) but not (t). On the other hand, the case $N = (i, j, k)$, $ij = jk = .1$, $kj = .6$ shows that (t) does not imply (c_3). Finally, to see that $(t_s) \mapsfrom\!\!\!\!\mapsto (e_3)$ and $(t) \mapsfrom\!\!\!\!\mapsto (e_3)$, note that (t_s) and (t) constrain only binary probabilities while (e_3) constrains ternary as well as binary ones, and its constraint on the latter is identical with (c_3), which was just shown not to imply (t_s) or (t).

The following condition on binary probabilities, due to Georgescu-Roegen [22], generalizes (c_3):

CONDITION (c_m). *For any* $x_1, \cdots, x_m$ *in* N,

$$1 \leqq x_1x_2 + x_2x_3 + \cdots + x_{m-1}x_m + x_mx_1 \leqq m - 1 . \tag{5.14}$$

A proposition stronger than (c_m) and—as will be proved presently—equivalent to (c_3) is

CONDITION (c^*). (c_m) *holds for all* $m \leqq n$.

THEOREM 5.9. $(c_3) \longleftrightarrow (c^*) \longmapsto (c_m)$.

PROOF. We have only to prove that (c_3) implies (c^*) and is not implied by (c_m), $m > 3$. First show that the conjunction $[(c_3), (c_{m-1})] \to (c_m)$: by (c_{m-1}),

$$1 \leqq x_1x_2 + x_2x_3 + \cdots + x_{m-1}x_1 \leqq m - 2 ; \tag{5.15}$$

and, by (c_3)

$$1 \leqq x_{m-1}x_m + x_mx_1 + x_1x_{m-1} \leqq 2 ,$$

or, by (2.2)

$$0 \leqq x_{m-1}x_m + x_mx_1 - x_{m-1}x_1 \leqq 1 ; \tag{5.16}$$

adding (5.15) and (5.16), we obtain (c_m); now put $m = 4, 5, \cdots$. To prove that (c_m) does not imply (c_3) let $m = n = 4$; if the matrix $[x_i\, x_j]$ is

$$\begin{bmatrix} .5 & .8 & .2 & .5 \\ .2 & .5 & .8 & .6 \\ .8 & .2 & .5 & .6 \\ .5 & .4 & .4 & .5 \end{bmatrix},$$

then (c_4) is satisfied, but (c_3) is not, for $x_1x_2 + x_2x_3 + x_3x_1 = 2.4 > 2$.

The main results of this section can be summarized in

THEOREM 5.10. *If* $n \geqq m > 3$, *then*

$$\begin{array}{ccccccccccc}
(P) & \longrightarrow & (P_n) & \longmapsto & (P_m) & \longmapsto & (e_3) & \longleftrightarrow & (P_3) & & \\
\big\downarrow & & & & & & \big\downarrow & & & & \\
(e) & \longmapsto & (c_m) & \longleftarrow\!\!\dashv & (c^*) & \longleftrightarrow & (c_3) & \longleftarrow\!\!\dashv & (t_s) & \longmapsto & (t)\,;
\end{array}$$

if $n = 3$ *or* 4, *then* $(P) \longleftrightarrow (P_n)$.

6. A Survey

Theorem 3.10 summarized the relations between conditions postulating the existence of random orderings, and of constant utilities of various degrees of strength. It is also useful to summarize the relations between Conditions (U), (u), (v), and (w) and some of the testable conditions, mainly from Sections 4 and 5:

THEOREM 6.1. *For A finite,*

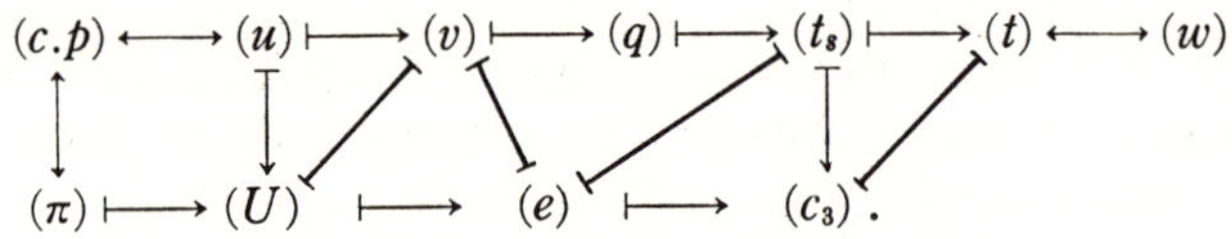

The following attempt to relate our various conditions to the hypotheses stated by earlier authors is very incomplete. Not only could we not do justice to the work of many psychologists; we have also failed to study thoroughly some of the statisticians' work.

In Section 2, the ideas of R. D. Luce, imposing a very strong constraint on the probabilities of multiple choices of any order, were identified with Condition (u); it can also be found in an early unpublished paper of Törnqvist [26]; and was stated orally (but taken back as unrealistic) by Debreu. R. A. Bradley and M. E. Terry [27] and L. R. Ford [28] applied this condition to binary choices only; in this form it is studied by J. Marschak [20] in some detail. A weaker constraint on the probabilities of binary choices is that of Fechner [3], identified with Condition (v). This condition, the still weaker Condition (w), and an intermediate case [the "strong transitivity", (t_s)], was also studied by C. Coombs [29]; H. Gulliksen [30]; the late S. Valavanis-Vail [31], to whom we owe the important Condition (t_s); W. Edwards [32]; A. G. Papandreou *et al.* [33]; K. May [34]; J. M. Davis [35]; and D. Davidson (in collaboration with one of the present authors) [36].

In Section 3, we identified an idea of Thurstone's with our Condition $(s.n)$ (somewhat weakened in the subsequent work of Mosteller): a special form of combining the model (v) of strong constant utilities with the model (U) of random orderings. Conditions which can be identified with (U) were studied by G. Th. Gilbaud [25], N. Georgescu-Roegen ([22], [37]), D. Rosenblatt [38], R. D. Luce [15], J. Marschak [39], and others.

M. G. Kendall [40], G. L. Mallows [41], I. R. Savage [42], E. L. Lehmann [43], H. D. Brunk [44] and other statisticians (we owe these references to Lehmann) have studied rank-order statistics, "ranking models," and paired comparisons, starting with problems of statistical inference. For example, given a sample of binary choices, estimate the "underlying" (constant) ranking (perhaps a vector-parameter of the distribution of our U?); or test the hypothesis that a certain parameter is higher with one than with another of a pair of parent populations. To our regret, we have not had the opportunity to do the important job of stating the relations (if any) between our network of conditions and the stochastic models that may underlie the work of these authors. We shall briefly deal with statistical inference in Section 10.

7. Varying the Domain of Testability

As remarked in Section 1, the definition of the class of basic observations, and therefore also of the class of testable conditions, depends on the range of possible experiments and other observations. Moreover, Condition (o) of Section 4 exemplifies the case of a constraint that involves all elements of the set of alternatives simultaneously; if this set is finite and of unknown size, such a condition is not testable, whatever the definition of the basic observation.

In this paper, the basic observations have been defined as $(a; F)$, the actual *choice* of a single element a out of a feasible subset F of alternatives. It was suspected that the subject's verbal statements about how he *ranks* the elements of a given subset of three or more alternatives were less reliable, for the characterization and prediction of actual behavior, than his actual choices. But this suspicion may have been unnecessary. Psychologists do ask their subjects to rank three or more objects, according to the subject's intensity of perception, or his preferences, etc. They also ask "judges" to rank individuals according to some characteristics. Presumably some consistency is assumed to prevail among all the rankings of any offered subsets of alternatives, and also between these rankings and the actual choices from any offered subset. In a stochastic model involving a finite set of alternatives $N = (1, 2, 3, 4)$, this would mean that, for example, a subject who, *at a given time*, ranks N in the order 1234, *would*, if asked to rank the subset $(2, 3, 4)$, produce the ranking 234; and if asked to choose from the subset $(2, 3, 4)$, *would* choose 2. The experimenter's faith in such consistency may be increased if the subject is told that the verbal statement about ranking will commit his actual choice when he is later presented with a pair of alternatives (a procedure suggested in a non-stochastic context by W. A. Wallis as quoted by L. J. Savage [45]), or, for that matter, with any subset of alternatives.[18] With the rankings (on the set N of all alternatives) admitted as basic observations, our condition (P) can be reformulated as a testable one, with $P(r)$ indicating the probability of an *observable* ranking

[18] On the other hand, the following procedure will not do for our present purposes (it was suggested in an earlier paper of Marschak [39]): to construct the subject's ranking of a set by asking him to make a choice from it; then removing the alternative chosen, and asking him to make a choice from the remaining subset; and so on. In the stochastic case, the subject may shift during the procedure from one ranking to another ranking of the same set.

r; that is, P is a distribution about which inferences can be made from experiments on rankings.

TESTABLE CONDITION (P). *The numbers* $P(r)$ *have the property* (3.7).

Since ranking experiments can be made on any subset $K \subseteq N$, we may, denoting by $P(r^K)$ the probability of a ranking r^K of K, generalize (P) into the following

TESTABLE CONDITION (P^K). *The numbers* $P(r^K)$ *have the property* (3.7) *with* K *replacing* N, *and with* $M \subseteq K \subseteq N$.

Note that when the subset K consists of two elements, the choosing from the subset and its ranking are the same thing. But the observation of ranking within triples and larger subsets does provide additional testable constraints not covered in the previous sections.

Testability is extended still further if the following procedure is permitted. The subject is given a set $M \subseteq N$, where M consists of m objects, and is told to select l of them ($l \leq m$) and to rank these l objects. The ranking is denoted by $r^{L,M} = (r_{a_1} \cdots r_{a_l})$ where $L = \{a_1, \cdots, a_l\}$. The condition ($P_{l,m}$) is said to hold if there exist $n!$ non-negative numbers $P(r^{N,N}) = P(r)$ whose sum is unity such that, for each set M having at least l, but not more than m elements, and each subset L of M with l elements

$$P(r^{L,M}) = \sum P(r)\,, \tag{7.1}$$

the summation being over all rankings r such that $r_i \leq r_j$ if either (1) $i \in L$ and $j \in M - L$, or (2) $i \in L$, $j \in L$, and $r_i^{L,M} \leq r_j^{L,M}$.

Note: with $l = 1$, ($P_{1,n}$) *becomes the condition* (P) of Section 3, since (7.1) then becomes

$$i(M) = \sum_{r | r_i \leq r_j, j \in M} P(r)\,.$$

Thus $(P) \longleftrightarrow (P_{1,n})$.

Clearly: $(P_{l,m}) \mapsto (P_{l,m-1})$, $(l < m)$; $(P_{m-1,m}) \leftrightarrow (P_{m,m})$. Moreover:

1. If $L = (i)$, then $P(r^{L,M}) = i(M)$.
2. If $L = (i), M = (i, j)$, then $P(r^{L,M}) = P(ij)$.
3. If $L = (i, j), M = (i, j)$, then $P(r^{L,M}) = P(ij)$.
4. If $L = N$ or $N - 1$, then $P(r^{L,N}) = P(r) = P(1_r 2_r \cdots n_r)$.

So far, we have discussed the possibilitity of extending the domain of testable conditions beyond the definition given in Section 1. On the other hand, the observational possibilities may be such as to make even that definition too wide.

For example, they may be such as to make binary but not ternary, etc. choices observable. As remarked orally by H. Simon, to replace the experiment—in which the subject names the heaviest of two objects placed in his two hands—by a similar experiment with three objects (taken in successive pairs, or weighed on three fingers, etc.) may change experimental conditions so drastically as to invite inconsistencies. On the other hand, no such inconsistencies need be expected in experiments where the subject tells which of the two, or three, or possibly four lights is brighter. In the economics of the actual market, binary choices are often not observable where ternary or higher ones are; for example, a speculator will (1) buy, (2) sell, or (3) do neither of these.

When the testability domain is extended, some sufficient existence conditions become irrelevant. As a simple example let N consist of three elements and first assume that both binary and ternary choices are observable but rankings are not. Theorem 5.1 gives a necessary and sufficient condition for the existence of probabilities of rankings as defined in Condition (P) of Section 3. But it does not say that these probabilities will be equal to the numbers $P(r)$ of the Testable Condition (P) of the present section, numbers about which inferences could be made if rankings were observable.

On the other hand, suppose that only binary choices among the three alternatives in N are observable, but neither ternary choice nor rankings are. Then Condition (c_3) ($1 \leqq xy + yz + zx \leqq 2$) becomes important; for, by Theorem 5.7, this condition is sufficient for the existence of the three ternary choice probabilities $x(x, y, z)$ and of the six probabilities of rankings $P(r)$; and nothing stronger can be stated with the observations at hand.

8. The Boundary Case: Perfect Discrimination or Preference

The case when the binary probability $p(a; F)$, with $F = (a, b)$, takes the boundary values 1 or 0 is called, in psychology of perceptions, perfect (as distinct from imperfect) discrimination. In the language of the theory of choice there is then perfect (as distinct from imperfect, or stochastic) preference. The term can be extended to the case when the feasible set F contains any number of elements.

To take account of the boundary case, special provisos are needed to make conditions such as (t_s), (v), (u) consistent with empirical observations. To take the weakest of those conditions, (t_s) or its equivalent (t_s^{**}): if there exists an element k of the set N of alternatives, such that $ik = jk$, then $ij = \frac{1}{2}$. Now, if three bodies i, j, k weigh, respectively, 1000, 10, and 6 grams, and xy is the probability that the subject says that x is heavier than y, then the observations will very likely show $ij = ik$; but they will very unlikely show $jk = \frac{1}{2}$. The conditions (t) and (w), both weaker than (t_s), are not endangered by this fact. But the stronger conditions (v) and (u) are. Without the proviso about boundary cases, made in Section 2 and again in Section 4, Condition (v) would imply, in our case of three bodies, $v_j = v_k$ for each of the lighter ones, because of the existence of the very heavy body i. Similarly, in Condition (u) all the (strict) utilities u_i had to be assumed positive to prevent any of the probabilities $i(M)$ from being 0 or 1.

Perfect discrimination is circumvented experimentally by presenting to the subject, at each trial, objects not too strongly differing from each other in the relevant physical dimension; and by keeping the subject uninformed of the physical dimension—for example, through blindfolding.

This approach is not sufficient, for experiments on choice, to the extent to which these are intended to reveal "desirabilities," with "perceptibility" kept more or less constant (see Section 1). Ideally, one would like to give the subject all the physical information, making the choice dependent only on his "desires." Suppose, then, that two alternatives differ from each other, however slightly, with respect to a single characteristic which is easily identified by the subject and is completely ordered with respect to the subject's choices. He will then presumably show perfect preference. Thus, in particular, when a and b are quantities of an economic "good" and a exceeds b, he will prefer a with probability 1, even if the difference in quantities is very small. (In fact, this may be regarded as the definition of a "homogeneous" economic good.) Also, if a and b are two vectors of quantities of the same m goods, $a = (a_1, \cdots, a_m)$, $b = (b_1, \cdots, b_m)$, and if a dominates b, i.e., if $a_i \geqq b_i$ for all i and $a_j > b_j$ for some j, then a will be perfectly preferred to b. The same is true if the vectors a and b are interpreted as two money wagers, (where a_i and b_i are, respectively, the monetary gains awarded when event E_i happens) and a dominates b. On the other hand, if a and b are two vectors neither of which dominates the other, stochastic instead of perfect preference may occur. Accordingly, the pairs of commodity-bundles used in the experiments by Papandreou and others [33] and the pairs of wagers used in the experiments of Davidson and Marschak [36] were such as to exclude domination.

The vacillation of the chooser between the alternatives has thus been ascribed by Simon [46] and by Quandt [47] to the presence, in each alternative, of several relevant characteristics that themselves are completely ordered, but to which the subject pays attention only one at a time. In general, we need not assume that every alternative is so reduced (albeit "unconsciously") to a bundle of characteristics, each of them an ordered one, so that the businessman decomposes, as it were, the job-applicant into his "personality factors," or considers a prospective plant site as a vector of numerical attributes. Yet in some important cases such vectorial representation is indeed manifest. This gives rise to disturbing discontinuities in applying our stochastic models. For example, denote by $a = (a_1, a_2)$, $b = (b_1, b_2)$, and $b' = (b_1', b_2')$, three pairs of "homogeneous commodities"—apples and oranges, not factories and vice-presidents; let ε_1 and ε_2 be two small positive numbers, and let $b_1 = b_1' = a_1 + \varepsilon_1$, $b_2 = a_2$, $b_2' = a_2 - \varepsilon_2$. Then, very likely, the binary probability ba is 1; while $b'a$ may be quite close to $\frac{1}{2}$; this big jump in probabilities would result from a very small physical change in the alternatives offered.

For such reasons, the psychophysicist's proviso that the "differences" should not be "noticed always or never" is doubly important in designing and interpreting experiments on economic choices,[19] if one wants such experiments to throw light on desirabilities, as separated from the perceptibility aspect of the choice phenomenon.

9. Remarks on Choice under Uncertainty

As visualized at the end of Section 1, we shall now outline possible stochastic reformulations of the theory of choice under uncertainty, weakening in particular the "expected utility" condition.

A wager associates each alternative (in the sense defined previously) $i = 1, \cdots, n$ with an event. Denote by $q = (q_1, \cdots, q_n)$ any wager in which

[19] The case of absolute preferences is discussed by N. Georgescu-Roegen [22] in the context of the economic theory of demand (for homogeneous commodities). J. S. Chipman [23] drew attention to important anomalies in the case of statistical decisions.

the event giving rise to alternative i has probability q_i (possibly zero). In its non-stochastic form, the expected utility condition says that there is a constant vector $\bar{\omega} = (\bar{\omega}_1, \cdots, \bar{\omega}_n)$, unique up to an increasing linear transformation, and such that

$$\sum \bar{\omega}_i q_i \geqq \sum \bar{\omega}_i q'_i \quad \text{if and only if } q \gtrsim q' , \tag{9.1}$$

where $\gtrsim$ means "preferred or indifferent." Since the observer cannot be sure that two subjects behave as if they ascribed the same probability to any given event, not only the "utility-numbers" $\bar{\omega}_i$ but also the probabilities q_i have to be interpreted in a subjective sense, thus: For a given subject, there exist a vector $\bar{\omega}$ and, associated with each wager, a vector q such that (9.1) is satisfied.

It has been shown[20] that this condition is implied by other conditions (called "axioms") that apparently appeal more immediately to what our intuition recommends as rational, or consistent. As we remarked at the beginning, consistency in an absolute sense is likely to be rejected by experience.[21] But it can be replaced by more general conditions, of stochastic consistency; these imply, or are implied by, or both, statistically testable hypotheses. In particular, each of the Conditions (w), (v), (u), (U) of our Sections 2 and 3 can be combined with the expected utility principle to yield a stronger condition to be denoted by $(\bar{w})$, $(\bar{v})$, $(\bar{u})$, $(\bar{U})$, respectively. In formulating these conditions, we shall denote by Q the set of all wagers (i.e., the set of all vectors q of order n, with $\sum q_i = 1$, $q_i \geqq 0$, all i); the feasible set of wagers will be denoted by $F \subseteq Q$; the probabilities of choices will be denoted as before, e.g., $q(F) = p(q; F)$; $qq' = p((q, q'))$.

CONDITION $(\bar{w})$. *There is a constant vector* $\bar{w} = (\bar{w}, \cdots, \bar{w}_n)$ *such that*

$$\sum \bar{w}_i q_i \geqq \sum \bar{w}_i q'_i , \qquad \textit{if } qq' \geqq \tfrac{1}{2} . \tag{9.2}$$

CONDITION $(\bar{v})$. *There is a constant vector* $\bar{v} = (\bar{v}_i, \cdots, \bar{v}_n)$ *and, associated with it, a distribution function* $\varphi_{\bar{v}}$*, strictly increasing except at its values* 1 *and* 0, *such that*

$$\varphi_{\bar{v}}(\sum \bar{v}_i q_i - \sum \bar{v}_i q'_i) = qq' . \tag{9.3}$$

CONDITION $(\bar{u})$. *There is a constant positive vector* $\bar{u} = (\bar{u}_1, \cdots, \bar{u}_n)$ *such that, for every* $F \subseteq Q$ *and every* q, q' *in* F,

$$\frac{\sum q_i \bar{u}_i}{\sum q'_i \bar{u}_i} = \frac{q(F)}{q'(F)} . \tag{9.4}$$

---

[20] This was done by J. von Neumann and O. Morganstern [48], with q interpreted "objectively"; their "axioms" and proof were subsequently simplified by various writers, most concisely by I. N. Herstein and J. Milnor [49]. The "subjective" interpretation of q was introduced by F. P. Ramsey [50] and made precise by L. J. Savage [45]. See also W. Edwards [32]. R. D. Luce [51] generalized the stochastic theory of choices under uncertainty by introducing random subjective probabilities.

[21] See the examples of actual behavior found by M. Allais [52] and discussed also by L. J. Savage ([45], pp. 101–4.)

Condition ($\bar{U}$). *There is a random vector* $\bar{U} = (\bar{U}_1, \cdots, \bar{U}_n)$ *such that for every* $F \subseteq Q$ *and every* q *in* F,

$$q(F) = \Pr\{\textstyle\sum (q_i - q'_i)\bar{U}_i \geqq 0; \quad \text{all } q' \text{ in } F\} . \tag{9.5}$$

Clearly $\bar{w}$ and $\bar{U}$ (unlike w and U) are unique up to an increasing linear transformation. So is $\bar{v}$ (by virtue of the continuity of the set Q; see footnote 9). And $\bar{u}$ is unique up to a positive multiplier (as is u).

The directly testable conditions of the previous sections—e.g., Conditions (t), (t_s), (q), etc.—can be appropriately reformulated, and some new ones added. A case most easily accessible to observations is that of even-chance wagers: i.e., the case when the subjective probabilities are $\frac{1}{2}$ for each of a pair of alternatives. Let iEj be the wager promising i if event E happens, and j otherwise; then the subjective probability of E (and therefore also each of the subjective probabilities q_i and q_j) is said to be equal to $\frac{1}{2}$ if the subject is indifferent between iEj and jEi. Or, using the stochastic interpretation and denoting by $(iEj \cdot jEi)$ the probability of choosing the first out of that pair of wagers, $q_i = q_j = \frac{1}{2}$ if $(iEj \cdot jEi) = \frac{1}{2}$. In principle this property is operationally ascertainable.[22] Consider an even-chance wager, with $q_1 = q_2 = \frac{1}{2}$, and another, $q'_3 = q'_4 = \frac{1}{2}$; the expressions $\sum q_i\bar{v}_i$ and $\sum q'_i\bar{v}_i$ in Condition ($\bar{v}$) become, respectively, $(\bar{v}_1 + \bar{v}_2)/2$ and $(\bar{v}_3 + \bar{v}_4)/2$; and so with similar expressions in ($\bar{u}$), ($\bar{w}$) and ($\bar{U}$). We can denote the even-chance wager aEb simply by $\overline{ab}$ (hence $\overline{aa} = a$), and the probability of choice between two such wagers by $(\overline{ab} \cdot \overline{cd})$. Then Condition ($\bar{v}$) can be rewritten—analogous to relation (4.1) and with the proviso that the considered probabilities of choice are not 1 or 0—thus:

(9.6) *if* $$(\overline{a_1a_2} \cdot \overline{b_1b_2}) \geqq (\overline{b_3b_4} \cdot \overline{a_3a_4}) ,$$

then $$v_{a_1} + v_{a_2} + v_{a_3} + v_{a_4} \geqq v_{b_1} + v_{b_2} + v_{b_3} + v_{b_4} .$$

This clearly implies the following *directly testable* set of conditions:

Condition (8_E) (8-tuple condition on even-chance wagers). *If* $(\overline{a_1a_2} \cdot \overline{b_1b_2}) \geqq (\overline{b_3b_4} \cdot \overline{a_3a_4})$, *then* $(\overline{a_{r_1}a_{r_2}} \cdot \overline{b_{s_1}b_{s_2}}) \geqq (\overline{b_{s_3}b_{s_4}} \cdot \overline{a_{r_3}a_{r_4}})$, *where* (r_1, r_2, r_3, r_4) *and* (s_1, s_2, s_3, s_4) *are arbitrary permutations of* $(1, 2, 3, 4)$.

Theorem 9.1. $(\bar{v}) \longrightarrow (8_E)$.

It can be shown that the set (8_E) of conditions is not only necessary but also sufficient for (9.6) provided the stochastic continuity condition ($s.c$) of Section 4 is satisfied with respect to the considered set of even-chance wagers.

If some of the eight alternatives in Condition (8_E) are made identical,

[22] For still greater experimental ease, Davidson and Marschak [36], following the device used by Davidson, Suppes and Siegel [53], and guided by considerations of continuity, assumed that, when the alternatives i, j, i', j' are money amounts, then the even-chance event E has also the following testable property:

$$(iEj \cdot j'\,Ei) = 1 = (jEi \cdot i'\,Ej) ,$$

provided $i - i' = j - j' = \varepsilon$, a small positive amount.

weaker directly testable conditions arise, involving a smaller number of distinct alternatives, and necessary for Condition ($\bar{v}$). In certain conjunctions, and assuming stochastic continuity, they are also sufficient for ($\bar{v}$), over the considered set of even-chance wagers. They were studied by Davidson and Marschak [36]; Debreu's paper [54], though dealing with even-chance wagers in a non-stochastic context, is also relevant.

10. Testing Stochastic Constraints when Replications Are Few

All the considered conditions have been constraints on probabilities of certain responses of the subject to varying external conditions. More precisely, these probabilities have been in the form $p(a; F)$ although in Section 7 more general forms were briefly discussed. These probabilities define the coordinates of the parameter space, and each directly testable condition ("hypothesis") defines a certain region of this space. The test consists in deciding, on the basis of observations, whether the parameter point—the set of probabilities $p(a; F)$ for all F—falls into that region.

A trial is easily replicated if the same feasible set F—the same set of stimuli—can be presented to a random sample of subjects *provided* it is reasonable to assume that each member of the parent population is characterized by the same relevant probability distribution of responses. This assumption has been made, in effect (with respect to choices of foods, birthday gifts, etc.) by Jones, Peryam, and Thurstone [55], Thurstone and Jones [56], and Gulliksen [57].

If this assumption of identical distributions is not justified, the hypothesis must be tested separately for each subject to find the proportion of the observed subjects that fail to satisfy certain strong conditions—such as (q) (see Section 4) and therefore (v); the proportion of subjects failing to satisfy weaker conditions—such as (t) and therefore (w); and so forth. One can thus estimate the frequency distribution of the various "degrees of stochastic consistency" among the parent population, even though the probabilities ($a; F$) are not the same for two persons with the same degree of consistency. Two persons may both obey (w) and both fail to satisfy (v), even though one of them tends to vegetarianism more than the other. In psychophysics, two persons may obey the Fechnerian condition (v) and yet have different functions φ_v, e.g., have different quartiles or "just noticeable" differences (see Section 2). A frequency distribution of degrees of (stochastic) consistency, in a given culture, is a useful thing to know. One may be interested in training people towards greater consistency, because of its importance for decision-making and leadership. This is not the same thing as inculcating identical tastes.

The number of replications, *for the same subject*, of the same set F of stimuli is limited by the fact that, in the course of successive replications, the subject's characteristic probabilities $p(a; F)$ for a given F may change. Inasmuch as this may be due to fatigue, psychophysicists do not seem to have found it necessary to curtail replications too much; they regard relative frequencies with which given responses a are elicited by a fixed F (e.g., a fixed pair of sounds) as workable estimates of the probabilities $p(a; F)$.

True, in psychophysical experiments the subject must be prevented from identifying the stimuli and remembering them. For example, he must be blindfolded. If he were asked to choose the heavier of two objects (or the string producing the higher pitch) but were not blindfolded, the visual stimulus, added to that of the weight (or sound), might help him to identify and remember the objects and thus possibly influence him when the same object is presented again.

For a similar reason, successive replications may lack independence when subjects have to respond to verbal stimuli, as in questionnaires on attitudes or on voting preferences. Students of attitudes try therefore to repeat a question in a different form, preferably after a lapse of a little time. The same lack of serial independence is likely to occur in the case of choices between verbally stated economic alternatives such as commodities and wagers. It might not be true of "blindfolded" choices between wines. But, as pointed out in Section 1, the economic theorist wants to separate preferences from difficulties of perception; his ideal is to observe the decisions of people while giving them as complete information as possible.

Accordingly, experimenters on choices have used only a small number of repetitions of the same pair of alternatives (Papandreou *et al.* [33]) or have avoided such repetitions altogether (Davidson and Marschak [36]). This creates special problems in devising statistical tests of significance, and at any rate makes such tests relatively weak. To bring out the logical nature of the problem, we shall concentrate on the mathematically easiest (though empirically weakest) case when each set F is presented once only; and we shall limit ourselves to the simplest of our hypotheses, the weak stochastic transitivity condition (t).

During the course of experiment the subject is asked to choose one from each of the sets (a, b), (b, c), (a, c). Denote by p_a, p_b, p_c the binary probabilities ab, bc, ca. Then $p = (p_a, p_b, p_c)$ is a point in $[0, 1]^3$. Assume that the subject has a certain probability distribution $\mathscr{P}$ of choosing points in $[0, 1]^3$; the random choice of triples (a, b, c) leads to a random choice from $[0, 1]^3$ with the distribution $\mathscr{P}$ on it. Let $q_i = p_i - \frac{1}{2}$, $i = a, b, c$. The region of $[0, 1]^3$ where condition (t) is *not* satisfied (call it T) is characterized by the fact that all three q_i are $\geqq 0$ or $\leqq 0$ and are not all $= 0$. That is, apart from a set of measure 0, all three q_i have the same sign. It is desired to estimate $\mathscr{P}(T) = \Theta$.

It may be useful to think of an analogous one-dimensional problem. A coin-making machine is characterized by an unknown probability distribution of the chance variable p (probability of a coin falling heads). One is permitted to toss coins, *each only a few times* (*perhaps only once*), in order to get evidence about the distribution of p; for example, one wants to test the hypothesis that the proportion of coins with $p > \frac{1}{2}$ (i.e., biased in favor of heads) is not more than a preassigned number. The region "$p > \frac{1}{2}$" clearly corresponds to our region T, and can be so denoted; the preassigned number is an upper bound Θ' on $\mathscr{P}(T) = \Theta$. Now, an infinite number of coins of which exactly 10% are biased for heads, would, if each is tossed once, show more than 5% (and not more than 55%) heads; hence, if out of a very large number of coins tossed, exactly $\bar{Z} = 5\%$ have fallen heads, all we can say is that the proportion of coins biased for heads is less than $\Theta' = 10\% = 2\bar{Z}$. Thus, *regardless of the number of coins* we are able to toss (each once), we cannot in general make statements involving a pre-assigned upper bound Θ' on the proportion of coins belonging to a given region. If Θ' had been 15%, then we could assert $\Theta < \Theta'$, but if Θ' had been 7%, then we could not assert $\Theta < \Theta'$. Thus, if we want to be sure that we can make an assertion, we can choose Θ' only after the tossings have been observed; if we are willing to accept the possibility of making no assertion, we can preassign Θ'. Similarly, in testing the stochastic transitivity condition (t), we shall be able to post-assign but not always to pre-assign an upper bound Θ' on Θ; and this regardless of the number of triples (a, b, c) we can present—each once—to the subject. The situation there is further complicated by the fact that we have only a finite sample and we introduce other parameters t^0, t' to analyze the confidence level.

Furthermore, if each coin could be tossed twice, we would still be unable to guarantee an assertion with a pre-assigned upper bound to Θ, but the ratio $\Theta'/\bar{Z}$ of the post-assigned bound to the observed proportion of heads would become smaller. And a similar statement can be made about the consequences of duplicating the presentation of each triple (a, b, c) to the subject.

We shall now develop a test of Condition (t). Let

$$f(p) = p_a p_b p_c + (1 - p_a)(1 - p_b)(1 - p_c) = \tfrac{1}{4} + q_a q_b + q_b q_c + q_c q_a \,, \tag{10.1}$$

where $q_i = p_i - \frac{1}{2}$. If $p \in T$ then $f(p) \geqq \frac{1}{4}$. If $p \notin T$ then $f(p) \leqq \frac{1}{2}$, for if $q_a = -\alpha$, $q_b = \beta$, $q_c = \gamma$ (α, β, γ non-negative), $f(p) = \frac{1}{4} + \beta\gamma - \alpha(\beta + \gamma) \leqq \frac{1}{4} + \beta\gamma \leqq \frac{1}{2}$, and the other possibilities are covered by symmetry. Define three random variables X_a, X_b, X_c: $X_a = 1$ if a is chosen from (a, b) and $X_a = 0$ otherwise; $X_b = 1$ if b is chosen from (b, c) and $X_b = 0$ otherwise; $X_c = 1$ if c is chosen from (a, c) and $X_c = 0$ otherwise. Define

$$Z = X_a X_b X_c + (1 - X_a)(1 - X_b)(1 - X_c) \,. \tag{10.2}$$

For a given vector p, $\Pr\{Z = 1\} = f(p)$. Now $\Pr\{Z = 1\} = \Pr\{Z = 1 \mid p \in T\} \cdot \mathscr{P}(p \in T) + \Pr\{Z = 1 \mid p \notin T\}\ \mathscr{P}(p \notin T)$. Hence $\Pr\{Z = 1\} \geqq (\Theta/4) + 0$ and $\Pr\{Z = 1\} \leqq \Theta + \frac{1}{2}(1 - \Theta) = \frac{1}{2} + (\Theta/2)$. Thus Z is a binomial chance variable with mean $\mu = \Pr\{Z = 1\}$ where

$$(\Theta/4) \leqq \mu \leqq \tfrac{1}{2} + (\Theta/2) \,. \tag{10.3}$$

Let $Z_1, \cdots, Z_m$ be independent observations on Z, each based on a different triple (a, b, c). Let $\bar{Z} = \sum_1^m Z_k/m$. Our first test is based on the the following considerations:

1. Let $0 < \Theta' \leqq 1$. If $\Theta \geqq \Theta'$ then $\mu \geqq \Theta'/4$. Let V' be a binomial chance

variable with $\Pr\{V' = 1\} = \Theta'/4$. Then $\Pr\{Z_k = 0\} \leqq \Pr\{V' = 0\}$. Let $t' \leqq \Theta'/4$. Then $\Pr\{\bar{Z} \leqq t'\} \leqq \Pr\{\bar{V}' \leqq t'\}$ where $\bar{V}'$ is the mean of m values of V'. If a confidence level α' is specified, one can choose m large enough so that $\Pr\{\bar{V}' \leqq t'\} \leqq \alpha'$. Hence:

$$\text{if } \Theta \geqq \Theta' , \Pr\{\bar{Z} \leqq t'\} \leqq \alpha' . \tag{10.4'}$$

2. Similarly, consider Θ^0, with $0 \leqq \Theta^0 < \frac{1}{2}$. If $\Theta \leqq \Theta^0$ then $\mu \leqq \frac{1}{2} + \Theta^0/2$. Let V^0 be a binomial random variable with $\Pr\{V^0 = 1\} = \frac{1}{2} + \Theta^0/2$. Then $\Pr\{Z_k = 1\} \leqq \Pr\{V^0 = 1\}$. Let $t^0 \geqq \frac{1}{2} + \Theta^0/2$. Then $\Pr\{\bar{Z} \geqq t^0\} \leqq \Pr\{\bar{V}^0 \geqq t^0\}$ where $\bar{V}^0$ is the mean of m values of V^0. Again, with a specified confidence level α^0, we can choose large enough m to make $\Pr\{\bar{V}^0 \geqq t^0\} \leqq \alpha^0$. Hence

$$\text{if } \Theta \leqq \Theta^0 , \Pr\{\bar{Z} \geqq t^0\} \leqq \alpha^0 . \tag{10.4^0}$$

Summarizing: If $\bar{Z} \geqq t^0$, we assert $\Theta > \Theta^0$; if $\bar{Z} \leqq t'$, we assert $\Theta < \Theta'$; if $t' < \bar{Z} < t^0$, we assert nothing. The consequences of this are shown in the following table of probabilities of occurrence:

Assertion	True state	
	$\Theta \leqq \Theta^0$	$\Theta \geqq \Theta'$
$\Theta > \Theta^0$	$<\alpha^0$	---
$\Theta < \Theta'$	---	$<\alpha'$
Nothing	---	---

,

where the dashes indicate probabilities that we have not estimated. This procedure protects us against extreme errors of classification, but leaves open the possibility of coming to no decision.

If m is large enough so that the normal approximation to the binomial may be used, then the constants involved in the test are determined by the following equations:

$$F(s^0) = 1 - \alpha^0 , \qquad F(s') = 1 - \alpha' , \tag{10.5}$$

where F is the normal distribution function with zero mean and unit variance, and

$$s' = (\Theta' - 4t')\sqrt{m}/2\sqrt{\Theta'(1 - \Theta'/4)} , \tag{10.6'}$$

$$s^0 = (2t^0 - 1 - \Theta^0)\sqrt{m}/\sqrt{(1 + \Theta^0)(1 - \Theta^0)} . \tag{10.6^0}$$

The experimenter may specify the parameters Θ^0, Θ', t^0, t' either before or after the experiment. Let us first discuss the case where the choice is made *after* he has the data. Suppose $\bar{Z}$ turns out to be quite high. He is then seeking to make an assertion $\Theta > \Theta^0$. He can do this with confidence level $1 - \alpha^0$ where α^0 is determined from (10.6^0), provided that $t^0 \leqq \bar{Z}$ and $\Theta^0 \leqq 2t^0 - 1$. He would like Θ^0 to be as high as possible and at the same time $1 - \alpha^0$ to be as high as possible. As he takes Θ^0 closer to $2t^0 - 1$, s^0 tends to zero and α^0 tends to $\frac{1}{2}$, which is not a convincing confidence level.

Thus to keep the confidence level $1-\alpha^0$ high he must take Θ^0 much less than $2t^0-1$; on the other hand, the smaller he takes Θ^0, the weaker the impact of his assertion $\Theta > \Theta^0$. Thus it is clear that he should choose t^0 as large as possible, namely $\bar{Z}$. Now if the experimenter wants to work with a fixed confidence level $1-\alpha^0$, he should adjust Θ^0 low enough so that he achieves, by (10.6^0), his desired confidence level. If, on the other hand, there is a fixed "tolerance level" Θ^0 which the experimenter regards as the significant indicator of inconsistency, then he would use this fixed Θ^0 and again with $t^0=\bar{Z}$ find his confidence level by (10.6^0). More generally, he can strike what balance he wishes, pushing Θ^0 up while driving $1-\alpha^0$ down, keeping of course $t^0=\bar{Z}$. If, instead, $\bar{Z}$ is quite low, then he wishes to assert $\Theta < \Theta'$. Here he would like Θ' as small as possible and $1-\alpha'$ as large as possible. Considerations similar to those above, applied now to (10.6$'$), show that he should choose $t'=\bar{Z}$ and similarly make an arrangement between Θ' and $1-\alpha'$.

It may be, however, that the experimenter is doing many experiments and would like to have a fixed procedure, regardless of the data. He can specify t^0 and Θ^0 in advance, subject to $\Theta^0 \leqq 2t^0-1$, and know that his confidence level is given by (10.6^0). (The larger $2t^0-1-\Theta^0$, the larger the confidence level, but the larger he takes t^0, the more likely he will have nothing to assert, while the smaller he takes Θ^0, the weaker are the assertions he does make; this balance he settles in advance.) Similarly he selects Θ', t' in advance. Now the data from many experiments can be handled by a fixed procedure; if $\bar{Z} \geqq t^0$ or $\bar{Z} \leqq t'$, an assertion is made, while if $t' < \bar{Z} < t^0$, no assertion is made. This procedure will be more convenient, but clearly it involves a loss of one or more of the following: confidence level, strength of assertion, and likelihood of making an assertion, as compared with the method of selecting the parameters after the experiment. However, if the experimenter is testing a large number of individuals in order to select a subset for further testing, he might find it better, as far as the data-processing is concerned, to fix the parameters in advance.

Our second method is to find a confidence interval for Θ. Let $0<\alpha<1$ and $0<\delta<1$ be given. Then it follows that for m sufficiently large $P(-\delta<\bar{Z}-\mu<\delta)>1-\alpha$; i.e., $P(\bar{Z}-\mu<\delta$ and $\bar{Z}-\mu>-\delta)>1-\alpha$, or $P(\bar{Z}<\delta+\mu$ and $\bar{Z}>\mu-\delta)>1-\alpha$; from (10.3) it follows that $P(\bar{Z}<\delta+\frac{1}{2}+(\Theta/2)$ and $\bar{Z}>(\Theta/4)-\delta)>1-\alpha$. Thus $P(2\bar{Z}-2\delta-1<\Theta<4\bar{Z}+4\delta)>1-\alpha$.

While it is true that the length of the confidence interval $(2\bar{Z}-2\delta-1, 4\bar{Z}+4\delta)$ is $2\bar{Z}+6\delta+1>1$, it is not centered at $\frac{1}{2}$. Hence, if $\bar{Z}$ is near zero or unity, the effective length of the interval may be quite small; e.g., if $\bar{Z}=0$, the conclusion is $-2\delta-1<\Theta<4\delta$, which has an effective length of only 4δ; while if $\bar{Z}=1$, the conclusion is $1-2\delta<\Theta<4+4\delta$, which has an effective length of 2δ. Since for a prescribed significance level $1-\alpha$ we do not know δ explicitly, we can use the estimate $\delta=k(\alpha)\sigma=k(\alpha)\cdot(pq/m)^{\frac{1}{2}} \leqq k(\alpha)/2(m)^{\frac{1}{2}}$. Thus

$$P\left(2\bar{Z}-\frac{k(\alpha)}{\sqrt{m}}-1<\Theta<4\bar{Z}+\frac{2k(\alpha)}{\sqrt{m}}\right)>1-\alpha\,.$$

If the normal approximation for the binomial is used, $k(\alpha)$ is determined from the equation

$$F(k(\alpha)) = 1 - \alpha/2\,, \tag{10.7}$$

where F is the normal distribution with zero mean and unit variance.

The confidence interval may also be used for making decisions; e.g., let $0 < \Theta^* \leqq \Theta^{**} < 1$. If the confidence interval is contained in the interval $(0, \Theta^*)$, assert that $\Theta \leqq \Theta^*$; if the confidence interval is contained in the interval $(\Theta^{**}, 1)$ assert that $\Theta \geqq \Theta^{**}$, otherwise assert nothing. In other words, if

$$\bar{Z} < \frac{\Theta^*}{4} - \frac{k(\alpha)}{2\sqrt{m}}\,,$$

assert that $\Theta \leqq \Theta^*$; if

$$\bar{Z} > \frac{1}{2}\left(\Theta^{**} + 1 + \frac{k(\alpha)}{\sqrt{m}}\right),$$

assert that $\Theta \geqq \Theta^{**}$; otherwise assert nothing. We then get the following table of probabilities:[22]

Assertion	True state		
	$0 \leqq \Theta < \Theta^*$	$\Theta^* \leqq \Theta \leqq \Theta^{**}$	$\Theta^{**} < \Theta \leqq 1$
$\Theta \leqq \Theta^*$	---	$<\alpha$	$<\alpha$
$\Theta \geqq \Theta^{**}$	$<\alpha$	$<\alpha$	---
Nothing	---	---	---

References

[1] Cartwright, D., and Festinger, L. A quantitative theory of decision. *Psychol. Rev.*, 1943, **50**, 595–621.

[2] Hotelling, H. Demand functions with limited budgets. *Econometrica*, 1935, **3**, 66–78.

[3] Fechner, G. Th. *Elemente der Psychophysik*. Leipzig: Breitkopf & Härtel, 1859.

[4] Arrow, K. J. *Social Choice and Individual Values*. New York: Wiley, 1951.

[5] Debreu, G. Representation of a preference ordering by a numerical function. Chap. 11 in R. M. Thrall, C. H. Coombs and R. L. Davis (eds.), *Decision Processes*. New York: Wiley, 1954.

[6] Wald, A. The approximate determination of indifference surfaces by means of Engel curves. *Econometrica*, 1940, **8**, 144–75.

[7] Theil, H., and Neudecker, H. Substitution, complementarity and the residual variation around Engel curves. *Rev. Econ. Studies*, 1957, **25**, 114–23.

[22] Naturally, much stronger tests are obtained if some *a priori* restrictions are admitted. Thus, in the Davidson and Marschak paper [36] two alternative hypotheses were formulated to test a stochastic constraint, viz., the transitivity condition (t) or (t_s): (1) The probability-triples (ac, bc, ca) are distributed uniformly over $[0, 1]^3$; (2) they are distributed over the region defined by Condition (t) or Condition (t_s), respectively. Mathematical simplicity is about the only serious claim in favor of these restrictions unless one invokes Laplace's "principle of ignorance."

[8] Bernoulli, D. Exposition of a new theory on the measurement of risk. (Translated by Dr. Louise Sommer.) *Econometrica*, 1954, **22**, 23-36.

[9] Marshall, A. *Principles of Economics*, 8th ed. London: Macmillan, 1920.

[10] Mosteller, F., and Nogee, P. An experimental measurement of utility. *J. Polit. Econ.*, 1951, **59**, 371-404.

[11] Guilford, J. P. *Psychometric Methods*, 2d ed. New York: McGraw-Hill, 1954.

[12] Dahl, R. A. The concept of power. *Behav. Sci.*, 1957, **2**, 201-15.

[13] Shapley, L., and Shubik, M. A method for evaluating the distribution of power in a committee system. *Amer. Polit. Sci. Rev.*, 1954, **48**, 787-92.

[14] Luce, R. D., and Raiffa, H. *Games and Decisions*. New York: Wiley, 1957.

[15] Luce, R. D. *Individual Choice Behavior*. New York: Wiley, 1959.

[16] Thurstone, L. L. A law of comparative judgment. *Psychol. Rev.*, 1927, **34**, 273-86.

[17] Mosteller, F. Remarks on the method of paired comparisons. *Psychometrika*, 1951, **26**, 3-9, 203-6, 207-18.

[18] Block, H. D., and Marschak, J. An identity in arithmetic. *Bull. Amer. Math. Soc.*, 1959, **65**, 123-24.

[19] Debreu, G. On "An identity in arithmetic." *Proc. Amer. Math. Soc.*, 1960, **11** (in press).

[20] Marschak, J. Binary choice constraints on random utility indicators. In *Stanford Symposium on Mathematical Methods in Social Sciences*. Stanford, Calif., Stanford Univ. Press, 1960.

[21] Block, H. D., and Marschak, J. Random orderings. Cowles Foundation Discussion Paper No. 42. New Haven, Conn.: Cowles Foundation, 1957 (duplicated).

[22] Georgescu-Roegen, N. Threshold in choice and the theory of demand. *Econometrica*, 1958, **26**, 157-68.

[23] Chipman, J. S. Stochastic choice and subjective probability. Technical Report No. 2, School of Business Administration. Minneapolis, Minn.: Univ. of Minn., 1958 (duplicated).

[24] Debreu, G. Stochastic choice and cardinal utility. *Econometrica*, 1958, **26**, 440-44.

[25] Guilbaud, G. Th. Sur une difficulté de la théorie du risque. *Colloques internationaux du centre national de la recherche scientifique (économétrie)*, 1953, **40**, 19-25.

[26] Törnqvist, L. A model for stochastic decision making. Cowles Commission Discussion Paper, Economics 2100. Chicago: Cowles Commission, 1954 (duplicated).

[27] Bradley, R. A., and Terry, M. E. Rank analysis of incomplete block designs, I. The method of paired comparisons. *Biometrika*, 1952, **39**, 324-44.

[28] Ford, L. R. Solution of a ranking problem from binary comparisons. *Amer. Math. Month.*, 1957, **64**(Aug. suppl.), 28-33.

[29] Coombs, C. On the use of inconsistency of preferences in psychological measurement. *J. Exp. Psychol.*, 1958, **55**, 1-7.

[30] Gulliksen, H. Paired comparisons and the logic of measurement. *Psychol. Rev.*, 1946, **53**, 199-213.

[31] Valavanis-Vail, S. A stochastic model for utilities. Seminar on the application of mathematics to social sciences. Ann Arbor: Univ. Michigan, 1957 (duplicated).

[32] Edwards, W. Probability preferences in gambling. *Amer. J. Psychol.*, 1955, **66** 349-64.

[33] Papandreou, A. G. (with the collaboration of O. H. Sauerlender, O. H. Brownlee, L. Hurwicz, W. Franklin). A test of a stochastic theory of choice, *Univ. Calif. Publ. in Econ.*, 1957, **16**, 1-18.

[34] May, K. Intransitivity, utility and the aggregation of preference patterns. *Econometrica*, 1954, **22**, 1-13.

[35] Davis, J. M. The transitivity of preferences. *Behav. Sci.*, 1958, **3**, 26-33.

[36] Davidson, D., and Marschak, J. Experimental tests of stochastic decision theories. Chap. 13 in C. W. Churchman and Ph. Ratoosh (eds.), *Measurement: Definitions and Theories*. New York: Wiley, 1959.

[37] Georgescu-Roegen, N. The pure theory of consumer behavior. *Quart. J. Econ.*, 1936, **50**, 545–93.

[38] Rosenblatt, D. On some stochastic process formulations of individual preference and consumer behavior (abst.). *Econometrica*, 1956, **24**, 347–48.

[39] Marschak, J. Norms and habits of decision making under certainty. In *Mathematical Models of Human Behavior* (a symposium). Stamford, Conn.: Dunlap and Associates, Inc., 1951, pp. 44-53.

[40] Kendall, M. G. Further contributions to the theory of paired comparisons. *Biometrics*, 1955, **11**, 43–67.

[41] Mallows, G. L. Non-null ranking models, I. *Biometrika*, 1957, **44**, 114–30.

[42] Savage, I. R. Contributions to the theory of rank orders. *Ann. Math. Stat.*, 1956, **27**, 590-615; 1957, **28**, 968–77.

[43] Lehmann, E. L. Ordered families of distributions. *Ann. Math. Stat.*, 1955, **26**, 399–419.

[44] Brunk, H. D. Mathematical models for ranking from paired comparisons. Technical Report No. 11, Air Force Office of Scientific Research. Columbia, Mo.: Department of Mathematics, University of Missouri, Nov. 1958 (duplicated).

[45] Savage, L. J. *The Foundations of Statistics*. New York: Wiley, 1954.

[46] Simon, H. A. A behavioral model of rational choice. *Quart. J. Econ.*, 1955, **69**, 99-118.

[47] Quandt, R. E. A probabilistic theory of consumer behavior. *Quart. J. Econ.*, 1956, **70**, 507-36.

[48] Neumann, J. von, and Morgenstern, O. *Theory of Games and Economic Behavior*, 2d ed. Princeton N.J.: Princeton Univ. Press, 1947.

[49] Herstein, I. N., and Milnor, J. An axiomatic approach to measurable utility. *Econometrica*, 1953, **21**, 291–97.

[50] Ramsey, F. P. *The Foundations of Mathematics and Other Logical Essays*. London: Kegan Paul, Trench, Tubner, 1931.

[51] Luce, R. D. A probabilistic theory of utility. *Econometrica*, 1958, **26**, 193-224.

[52] Allais, M. Le comportement de l'homme rationnel devant le risque: critique des postulats et axiomes de l'école américaine. *Econometrica*, 1953, **21**, 503–46.

[53] Davidson, D., and Suppes, P. (in collaboration with S. Siegel). *Decision Making: An Experimental Approach*. Stanford, Calif.: Stanford Univ. Press, 1957.

[54] Debreu, G. Cardinal utility for even-chance mixtures of pairs of sure prospects. *Rev. Econ. Studies*, 1958/9, **26**, 174–77.

[55] Jones, L. V., Peryam, D. R., and Thurstone, L. L. Development of a scale for measuring soldiers' food preferences. *Food Research*, 1955, **20**, 512-20.

[56] Thurstone, L. L., and Jones, L. The rational origin for measuring subjective values. *J. Amer. Stat. Assoc.*, 1957, **52**, 458-71.

[57] Gulliksen, H. Measurement of subjective values. *Psychometrika*, 1956, **21**, 229-44.

11 ON A METHOD OF CONSTRUCTING STEINER'S TRIPLE SYSTEMS

R. C. BOSE
University of North Carolina

1. Introduction

Steiner [1] in 1853 proposed the problem of arranging v objects in triplets such that every pair occurs in one and only one triplet. (The three objects in a triplet are all different.) Such a system may be called a Steiner's triple system. A Steiner's triple system is a special case of a balanced incomplete block (b.i.b.) design, which is defined as an arrangement of v objects into b blocks such that (i) each block contains k different objects, (ii) each object occurs in r blocks, and (iii) each pair of objects occurs in exactly λ blocks. Thus a Steiner's triple system is a b.i.b. design for which $k = 3$, $\lambda = 1$, each triplet being considered as a block. It is readily seen that for such a system

$$r = \frac{v-1}{2}, \quad b = \frac{v(v-1)}{6}. \tag{1}$$

Since r and b are both integral, v must be of the form $6t+1$ or $6t+3$. In a colloquium lecture at the Indian Statistical Institute, Fisher discussed a method of constructing a Steiner's triple system for the case $v = 21$, and the author modified his method to obtain a solution for the case $v = 19$. These solutions, though obtained in 1938, have never been published, possibly because it is difficult to generalize the method so as to obtain b.i.b. designs which are not triple systems. Even for the construction of triple systems some ingenuity is needed for resolving the cycles of irregular pairs. (This will be illustrated by the examples given later.) A discussion of these solutions seemed an appropriate subject for this dedicatory volume, owing to the interest shown by Professor Hotelling in this work when he visited the Indian Statistical Institute in 1939. Fisher's solution for $v = 21$ is given in Section 3, and the author's solution for the case $v = 19$ is given in Section 4. An excellent account of triple systems appears in Netto's book [2]. Other papers relating to the same subject are [3], [4], [5], and [6]. General methods of constructing b.i.b. designs are given in [7]. These in particular apply to Steiner's triple systems.

2. Some Preliminary Considerations

Consider a triple system with v objects. We have already seen that $v = 6t + 3$ or $6t + 1$. We shall say that the triple system is of the first kind if $v = 6t + 3$ and of the second kind if $v = 6t + 1$. In any case, v is odd.

Let the v objects be serially denoted by the numbers

$$0,\ 1,\ 2,\ \cdots,\ v - 1\,. \tag{2}$$

We shall define the object b to be the *successor* of the object a if

$$\text{(i)}\quad a \neq b\,, \qquad\qquad \text{(ii)}\quad a + 2b \equiv 0\ (\mathrm{mod}\ v)\,. \tag{3}$$

Given a, it follows from (ii) that if b is a successor of a, then

$$b = \frac{v - a}{2} \qquad \text{or} \qquad \frac{2v - a}{2}\,, \tag{4}$$

according as a is odd or even. This shows that when a is odd, then b will equal a if $a = v/3$, and when a is even, then b will equal a if $a = 0$ or $2v/3$. Hence if $v = 6t + 3$, then each object has a unique successor except the objects 0, $v/3$, $2v/3$, which have no successors. Similarly if $v = 6t + 1$, then each object has a unique successor except the object 0, which has no successor. If b is the successor of a, then a may be called the predecessor of b. Any object b ($b \neq 0$, $v/3$ or $2v/3$ when $v = 6t + 3$, and $b \neq 0$ when $v = 6t + 1$) has a unique predecessor. The predecessor of b is

$$a = v - 2b \qquad \text{or} \qquad a = 2v - 2b\,. \tag{5}$$

If we start with a given object and go on taking successors, then we must come back to the object. Hence, objects, other than those which have no predecessors or successors, must fall into a number of cycles such that any two consecutive members of a cycle stand in the predecessor–successor relationship.

Example 1. Let $v = 21$. Then the 18 objects other than 0, 7, and 14 have predecessors and successors. The objects fall into three cycles shown below:

$$\begin{array}{llllllll} \text{Cycle I} & 1 \longrightarrow 10 \longrightarrow 16 \longrightarrow 13 \longrightarrow 4 \longrightarrow 19 \longrightarrow 1\,, \\ \text{Cycle II} & 2 \longrightarrow 20 \longrightarrow 11 \longrightarrow 5 \longrightarrow 8 \longrightarrow 17 \longrightarrow 2\,, \\ \text{Cycle III} & 3 \longrightarrow 9 \longrightarrow 6 \longrightarrow 18 \longrightarrow 12 \longrightarrow 15 \longrightarrow 3\,. \end{array} \tag{6}$$

Example 2. Let $v = 19$. Then the 18 objects other than 0 have predecessors and successors. The objects fall into two cycles:

$$\begin{array}{ll} \text{Cycle I} & 1 \to 9 \to 5 \to 7 \to 6 \to 16 \to 11 \to 4 \to 17 \to 1\,, \\ \text{Cycle II} & 2 \to 18 \to 10 \to 14 \to 12 \to 13 \to 3 \to 8 \to 15 \to 2\,. \end{array} \tag{7}$$

Any two different objects a and b may be said to form a regular pair (a, b) if b is neither the successor nor the predecessor of a. Two different objects a and b will be said to form an irregular pair if a and b do not form a regular pair. (Note: The pairs considered here are all unordered pairs.)

If $v = 6t + 3$, then each of the $v - 3$ objects other than 0, $v/3$ or $2v/3$ is a member of two irregular pairs, viz., the pair which it forms with its successor and the pair which it forms with its predecessor. Hence the number of irregular pairs is $v - 3$. Since the total number of pairs is $v(v - 1)/2$, the number of regular pairs is $(v^2 - 3v + 6)/2$. For example, if $v = 21$, then we have 18 irregular pairs:

$$\begin{array}{llllll} (1,\ 10), & (10,\ 16), & (16,\ 13), & (13,\ \ 4), & (4,\ 19), & (19,\ 1)\ , \\ (2,\ 20), & (20,\ 11), & (11,\ \ 5), & (5,\ \ 8), & (8,\ 17), & (17,\ 2)\ , \\ (3,\ \ 9), & (9,\ \ 6), & (6,\ 18), & (18,\ 12), & (12,\ 15), & (15,\ 3)\ , \end{array} \tag{8}$$

and 192 regular pairs.

Similarly, when $v = 6t + 1$, each of the $v - 1$ objects other than 0 is a member of two irregular pairs. Hence the number of irregular pairs is $v - 1$, and the number of regular pairs is $(v^2 - 3v + 2)/2$. For example, if $v = 19$, then we have 18 irregular pairs:

$$\begin{array}{lllllllll} (1,\ 9), & (9,\ 5), & (5,\ 7), & (7,\ 6), & (6, 16), & (16, 11), & (11,\ 4), & (4,\ 17), & (17,\ 1)\ , \\ (2, 18), & (18, 10), & (10, 14), & (14, 12), & (12, 13), & (13,\ 3), & (3,\ 8), & (8,\ 15), & (15,\ 2)\ , \end{array} \tag{9}$$

and 153 regular pairs.

In general, let the objects a and b form a regular pair, and let c be determined by

$$a + b + c \equiv 0 \pmod{v}\ , \qquad 0 \leqq c < v\ ;$$

then (a, b, c) is called a regular triplet. It is readily shown that the pairs (a, c) and (b, c) are regular. If (a, c) is irregular, then either $a + 2c \equiv 0 \pmod{v}$ or $(2a + c) \equiv 0 \pmod{v}$. In the former case, $2(a + b + c) - (a + 2c) \equiv 0 \pmod{v}$ or $a + 2b \equiv 0 \pmod{v}$, which makes (a, b) an irregular pair contrary to the hypothesis. In the latter case, $(a + b + c) - (2a + c) \equiv 0 \pmod{v}$, or $b - a \equiv 0 \pmod{v}$, which gives $a = b$ contrary to the fact that a and b are different. In the same way, we show that the pair (b, c) is regular. It is clear that we would have gotten the same regular triplet if we had started with the regular pair (a, c) or (b, c) instead of (a, b). Hence, when $v = 6t + 3$, the $(v^2 - 3v + 6)/2$ regular pairs give rise to $(v^2 - 3v + 6)/6$ regular triplets, such that each triplet contains three different objects any two of which form a regular pair. Each regular triplet accounts for three regular pairs. Since every regular pair occurs exactly once among the regular triplets, and since each of the objects other than 0, $v/3$, $2v/3$ is a member of two irregular and $v - 3$ regular pairs, each such object occurs $(v - 3)/2$ times among the regular triplets (since each occurrence accounts for two regular pairs). Similarly, each of the objects 0, $v/3$, and $2v/3$, is a member of $v - 1$ regular pairs, and therefore occurs $(v - 1)/2$ times among the regular triplets. Thus, when $v = 21$, there are 64 regular triplets accounting for the 192 regular pairs. The objects 0, 7, and 14 occur ten times and the other objects occur nine times among the regular triplets.

Similar considerations apply to the case when $v = 6t + 1$. The $(v^2-3v+2)/2$

regular pairs give rise to $(v^2 - 3v + 2)/6$ regular triplets. Each regular pair occurs exactly once among the regular triplets. Each object other than 0 occurs $(v - 3)/2$ times, and 0 occurs $(v - 1)/2$ times among the regular triplets. For example, when $v = 19$, the 153 regular pairs give rise to 51 regular triplets, among which each of the 18 objects other than 0 occur eight times, and 0 occurs nine times.

3. Steiner's Triple System with $v = 21$

In this section we shall give Fisher's method of obtaining a Steiner's triple system with $v = 21$. It follows from (1) that such a system consists of 70 triplets, in which each of the 21 objects appears ten times. By definition each of the 210 pairs occurs exactly once, with each triplet accounting for three pairs.

Of the 210 pairs, 192 are regular pairs and 18 are irregular pairs given by (8). The regular pairs give rise to 64 regular triplets given in Table 1.

TABLE 1

(0, 1, 20)	(1, 2, 18)	(2, 5, 14)	(3, 19, 20)	(6, 7, 8)	(9, 14, 19)	(11, 15, 16)
(0, 2, 19)	(1, 3, 17)	(2, 6, 13)	(4, 5, 12)	(6, 16, 20)	(9, 15, 18)	(12, 13, 17)
(0, 3, 18)	*(1, 4, 16)*	(2, 7, 12)	(4, 6, 11)	(6, 17, 19)	(9, 16, 17)	(12, 14, 16)
(0, 4, 17)	(1, 5, 15)	*(2, 8, 11)*	(4, 7, 10)	(7, 15, 20)	(10, 12, 20)	(13, 14, 15)
(0, 5, 16)	(1, 6, 14)	(2, 9, 10)	(4, 8, 9)	(7, 16, 19)	(10, 13, 19)	
(0, 6, 15)	(1, 7, 13)	(3, 4, 14)	(4, 18, 20)	(7, 17, 18)	(10, 14, 18)	
(0, 7, 14)	(1, 8, 12)	(3, 5, 13)	(5, 6, 10)	(8, 14, 20)	(10, 15, 17)	
(0, 8, 13)	(1, 9, 11)	*(3, 6, 12)*	(5, 7, 9)	(8, 15, 19)	(11, 12, 19)	
(0, 9, 12)	(2, 3, 16)	(3, 7, 11)	(5, 17, 20)	(8, 16, 18)	(11, 13, 18)	
(0, 10, 11)	(2, 4, 15)	(3, 8, 10)	(5, 18, 19)	(9, 13, 20)	(11, 14, 17)	

The 64 regular triplets contain the 192 regular pairs. The objects 0, 7, and 21 occur ten times, and the other 18 objects occur nine times among the regular triplets. If the 18 objects other than 0, 7, and 14 could be grouped into six triplets such that among these triplets each of these objects occurs once, and each irregular pair occurs once, we would have a solution of our problem by adding these six triplets to the 64 regular triplets. But a little consideration shows that this is impossible to do. The three cycles given in (6) can be geometrically exhibited as in Fig. 1*a*.

The irregular pairs consist of the consecutive vertices of the three hexagons shown in Fig. 1*a*. It is clear that we cannot take a triplet, all three pairs of which are irregular. Hence a simple addition of six triplets to the 64 regular triplets cannot give us the solution of our problem. The question arises: Can we build a solution which retains a major part of the 64 regular triplets? We note that the alternate vertices of each of the three hexagons shown in Fig. 1*a* form a regular triplet. The cycles can be resolved by breaking up three of these regular triplets, say

$$(1, 4, 16), \quad (2, 8, 11), \quad (3, 6, 12)\,, \tag{10}$$

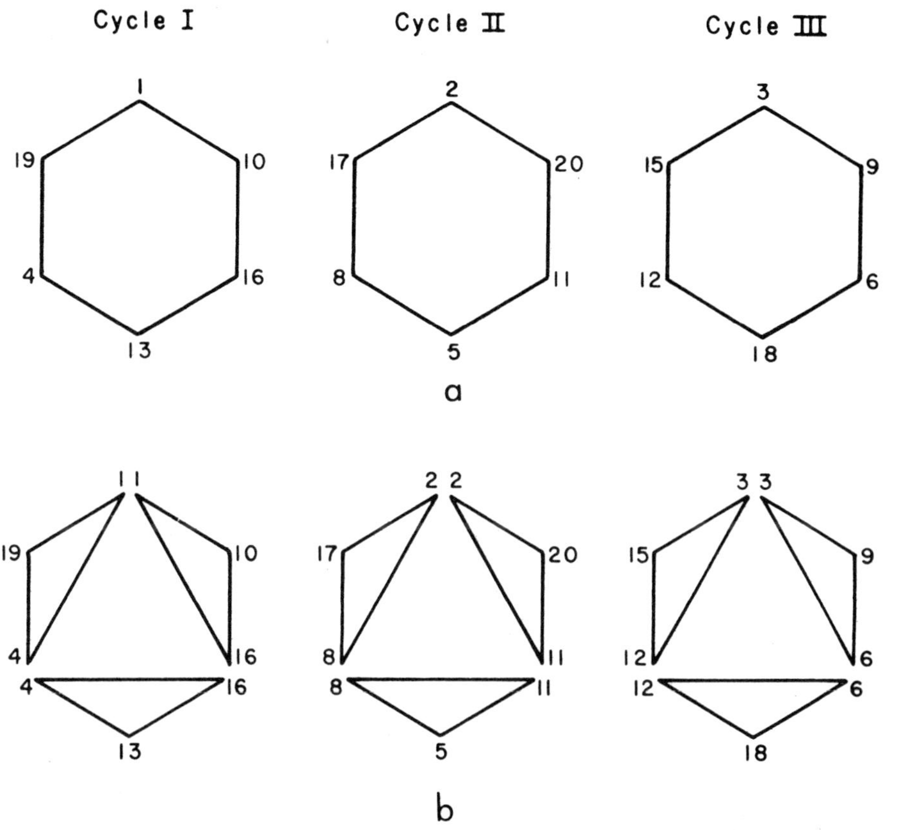

FIG. 1.

and using the objects to resolve the cycles as shown in Fig. 1*b*.

Thus a Steiner's triple system is obtained by deleting from the 64 triplets in Table 1 the three triplets in italic figures and adding the nine triplets shown in Table 2.

TABLE 2

(1, 10, 16)	(16, 13, 4)	(4, 19, 1)
(2, 20, 11)	(11, 5, 8)	(8, 17, 2)
(3, 9, 6)	(6, 18, 12)	(12, 15, 3)

The newly added triplets contain all the 18 irregular pairs once, and the nine regular pairs lost in breaking up the triplets in (10). Also they contain all the objects other than 0, 7, and 14 at least once, with the objects occurring in (10) contained twice. Thus the 61 regular triplets remaining after the italicized triplets are deleted from Table 1, together with the nine triplets of Table 2, give a system of 70 triplets in which each of the 210 pairs

(192 regular and 18 irregular) occur exactly once, and each object occurs ten times. It is clear that to resolve the first cycle we could have alternatively broken up the regular triplet (10, 13, 19) and used these three objects, obtaining the triplets (1, 10, 19), (10, 16, 13), (13, 4, 19) which contain the six irregular pairs of the first cycle and the three regular pairs of the triplet broken up. Similar remarks apply to the other cycles. Thus we can obtain eight different solutions of our problem, corresponding to two different ways of resolving each cycle.

4. Steiner's Triple System with $v = 19$

We shall now consider the construction of a Steiner's triple system with $v = 19$. It follows from (1) that such a system consists of 57 triplets, in which each of the 19 objects appears nine times. By definition, each of the 171 pairs occurs exactly once, with each triplet accounting for three pairs.

There are now 153 regular pairs, giving rise to 51 regular triplets, shown in the columns headed (R) in Table 3.

The 51 regular triplets contain the 153 regular pairs. The object 0 occurs nine times among the regular triplets, and the other 18 objects each occur eight times. If the 18 objects other than 0 could be grouped into six triplets such that among these triplets each of these objects occurs once and each irregular triplet occurs once, we would reach a solution of our problem by adding these triplets to the 53 regular triplets. This, however, is impossible to do, for reasons similar to those discussed in Section 3, in

TABLE 3

R	T	R	T	R	T
(0, 1, 18)	(6, 1, 18)	*(2, 4, 13)*	(2, 4, 11)	**(6, 14, 18)**	(10, 14, 18)
(0, 2, 17)		(2, 5, 12)		(6, 15, 17)	
(0, 3, 16)		*(2, 6, 11)*	(16, 6, 11)	(7, 13, 18)	
(0, 4, 15)	(14, 4, 15)	(2, 7, 10)		(7, 14, 17)	
(0, 5, 14)		**(2, 8, 9)**	(2, 8, 15)	(7, 15, 16)	
(0, 6, 13)	(2, 6, 13)	**(3, 4, 12)**	(3, 13, 12)	(8, 12, 18)	
(0, 7, 12)	(4, 7, 12)	(3, 5, 11)		(8, 13, 17)	
(0, 8, 11)	(9, 8, 11)	*(3, 6, 10)*	(3, 6, 7)	(8, 14, 16)	
(0, 9, 10)	(3, 9, 10)	*(3, 7, 9)*	(5, 7, 9)	**(9, 11, 18)**	(9, 2, 18)
(1, 2, 16)		(3, 17, 18)		(9, 12, 17)	
(1, 3, 15)		(4, 5, 10)		(9, 13, 16)	
(1, 4, 14)	(1, 4, 17)	(4, 6, 9)		*(9, 14, 15)*	(9, 14, 1)
(1, 5, 13)		**(4, 7, 8)**	(4, 3, 8)	(10, 11, 17)	
(1, 6, 12)	(6, 14, 12)	(4, 16, 18)		(10, 12, 16)	
(1, 7, 11)		(5, 6, 8)		(10, 13, 15)	
(1, 8, 10)		(5, 15, 18)		(11, 12, 15)	
(2, 3, 14)		(5, 16, 17)		(11, 13, 14)	

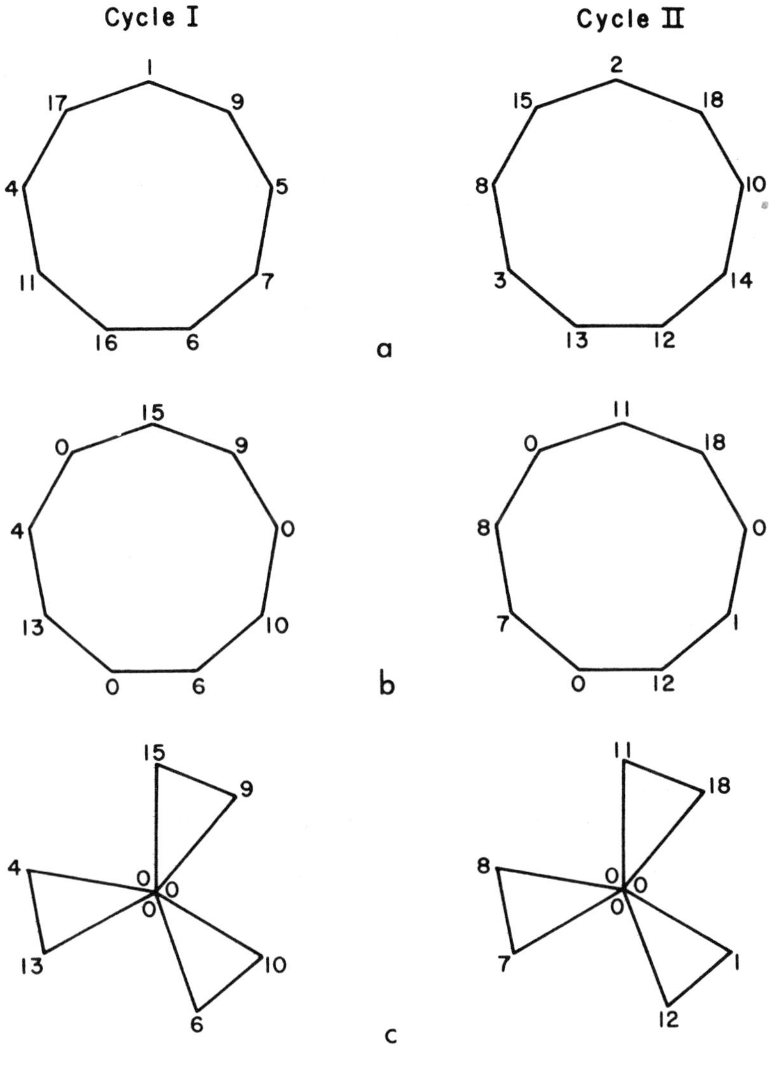

FIG. 2.

constructing the Steiner's triple system with $v = 21$. The two cycles given by (7) can be geometrically exhibited as in Fig. $2a$.

The 18 irregular pairs consist of the consecutive vertices of the two 9-gons shown in Fig. $2a$. These together with the 153 regular pairs make up the total of 171 pairs formed by the 19 objects. To resolve these cycles we now proceed as follows.

Suppose we modify Cycle I (7), represented by the first 9-gon in Fig. $2a$, by interchanging the object 5 with the object 3 of the regular triplet (3, 7, 9). The first cycle is now transformed to

$$1 \longrightarrow 9 \longrightarrow 3 \longrightarrow 7 \longrightarrow 6 \longrightarrow 16 \longrightarrow 11 \longrightarrow 4 \longrightarrow 16 \longrightarrow 1\,,$$

and the triplet (3, 7, 9) becomes triplet (5, 7, 9), which is shown in the column (T) of Table 3, to the right of (3, 7, 9). The regular triplet (3, 7, 9) accounted for three pairs (3, 7), (3, 9), and (7, 9); and the consecutive vertices 9, 5, 7 of the cycle accounted for the two irregular pairs (9, 5) and (5, 7). The interchange made keeps these pairs unchanged. Now the pairs (5, 7), (7, 9), and (9, 5) are given by the transformed triplet, and the pairs (9, 3) and (3, 7) are given by the consecutive vertices 9, 3, 7 of the cycle. Likewise, we further interchange the object 16 of the cycle with the object 2 of the regular triplet (2, 6, 11) and the object 17 of the cycle with the object 14 of the regular triplet (1, 4, 14). The transformations made on the first cycle and the regular triplets can be represented by

$$\begin{array}{ccccccccccccccccccc}
1 & \longrightarrow & 9 & \longrightarrow & 5 & \longrightarrow & 7 & \longrightarrow & 6 & \longrightarrow & 16 & \longrightarrow & 11 & \longrightarrow & 4 & \longrightarrow & 17 & \longrightarrow & 1 \\
 & & & & \Downarrow & & & & & & \Downarrow & & & & & & \Downarrow & & \\
1 & \longrightarrow & 9 & \longrightarrow & 3 & \longrightarrow & 7 & \longrightarrow & 6 & \longrightarrow & 2 & \longrightarrow & 11 & \longrightarrow & 4 & \longrightarrow & 14 & \longrightarrow & 1\,,
\end{array}$$

and

$$\begin{aligned}
(3,\ 7,\ \ 9) &\Rightarrow (5,\ 7,\ \ 9)\,, \\
(2,\ 6,\ 11) &\Rightarrow (16,\ 6,\ 11)\,, \\
(1,\ 4,\ 14) &\Rightarrow (1,\ 4,\ 17)\,.
\end{aligned}$$

These transformations have kept the pairs invariant. We now further make the following sets of three transformations:

$$\begin{array}{ccccccccccccccccccc}
1 & \longrightarrow & 9 & \longrightarrow & 3 & \longrightarrow & 7 & \longrightarrow & 6 & \longrightarrow & 2 & \longrightarrow & 11 & \longrightarrow & 4 & \longrightarrow & 14 & \longrightarrow & 1 \\
\Downarrow & & & & & & \Downarrow & & & & & & \Downarrow & & & & & & \Downarrow \\
15 & \longrightarrow & 9 & \longrightarrow & 3 & \longrightarrow & 10 & \longrightarrow & 6 & \longrightarrow & 2 & \longrightarrow & 13 & \longrightarrow & 4 & \longrightarrow & 14 & \longrightarrow & 15 \\
 & & & & \Downarrow & & & & & & \Downarrow & & & & & & \Downarrow & & \\
15 & \longrightarrow & 9 & \longrightarrow & 0 & \longrightarrow & 10 & \longrightarrow & 6 & \longrightarrow & 0 & \longrightarrow & 13 & \longrightarrow & 4 & \longrightarrow & 0 & \longrightarrow & 15\,,
\end{array}$$

and

$$\begin{array}{ll}
(3,\ \ 6,\ 10) \Rightarrow (3,\ \ 6,\ \ 7), & (0,\ 9,\ 10) \Rightarrow (3,\ 9,\ 10)\,, \\
(2,\ \ 4,\ 13) \Rightarrow (2,\ \ 4,\ 11), & (0,\ 6,\ 13) \Rightarrow (2,\ 6,\ 13)\,, \\
(9,\ 14,\ 15) \Rightarrow (9,\ 14,\ \ 1), & (0,\ 4,\ 15) \Rightarrow (14,\ 4,\ 15)\,.
\end{array}$$

The nine triplets in italic type in Table 3 have now been replaced by adjacent triplets to the right in column (T), while Cycle I of Fig. 2*a* is replaced by Cycle I in Fig. 2*b*. The pairs have remained invariant.

Similarly, we transform Cycle II and the corresponding regular triplets, by proper interchanges, represented by the scheme on the next page:

$$
\begin{array}{ccccccccccccccccccc}
2 & \rightarrow & 18 & \rightarrow & 10 & \rightarrow & 14 & \rightarrow & 12 & \rightarrow & 13 & \rightarrow & 3 & \rightarrow & 8 & \rightarrow & 15 & \rightarrow & 2 \\
 & & & & \Downarrow & & & & & & \Downarrow & & & & & & \Downarrow & & \\
2 & \rightarrow & 18 & \rightarrow & 6 & \rightarrow & 14 & \rightarrow & 12 & \rightarrow & 4 & \rightarrow & 3 & \rightarrow & 8 & \rightarrow & 9 & \rightarrow & 2 \\
\Downarrow & & & & & & \Downarrow & & & & & & \Downarrow & & & & & & \Downarrow \\
11 & \rightarrow & 18 & \rightarrow & 6 & \rightarrow & 1 & \rightarrow & 12 & \rightarrow & 4 & \rightarrow & 7 & \rightarrow & 8 & \rightarrow & 9 & \rightarrow & 11 \\
 & & & & \Downarrow & & & & & & \Downarrow & & & & & & \Downarrow & & \\
11 & \rightarrow & 18 & \rightarrow & 0 & \rightarrow & 1 & \rightarrow & 12 & \rightarrow & 0 & \rightarrow & 7 & \rightarrow & 8 & \rightarrow & 0 & \rightarrow & 11\,,
\end{array}
$$

$$
\begin{array}{lll}
(6,14,18) \Rightarrow (10,14,18), & (1,6,12) \Rightarrow (6,14,12), & (0,1,18) \Rightarrow (6,1,18)\,, \\
(3,4,12) \Rightarrow (3,13,12), & (4,7,8) \Rightarrow (4,3,8), & (0,7,12) \Rightarrow (4,7,12)\,, \\
(2,8,9) \Rightarrow (2,8,15), & (9,11,18) \Rightarrow (9,2,18), & (0,8,11) \Rightarrow (9,8,11)\,.
\end{array}
$$

The nine triplets in heavy type in Table 3 are replaced by adjacent triplets to the right in column (T) in the transformations involved in reducing Cycle II to the form shown in Fig. 2*b*.

As the result of both sets of transformations, we are left with 33 regular triplets, (neither in italics nor in heavy type and occurring in column (R) in Table 3), 18 transformed triplets occurring in column (T) in Table 3, and the two cycles shown in Fig. 2*b*. The 153 pairs occurring in the 51 plain triplets of Table 3, together with the 18 pairs given by the consecutive vertices of the two 9-gons in Fig. 2*b* constitute all the possible 171 different pairs that can be formed by the 19 objects. Our problem is therefore solved if we can find six triplets which contain exactly the 18 pairs corresponding to the two transformed cycles. This, however, can be immediately done by taking the triplets in Table 4. The resolution of the cycles is exhibited in Fig. 2*c*.

TABLE 4

(0, 15, 9)	(0, 11, 18)
(0, 10, 6)	(0, 1, 12)
(0, 13, 4)	(0, 7, 8)

The 51 triplets of Table 3, in plain type, together with the six triplets of Table 4, constitute a Steiner's triple system with $v = 19$. Each object appears in exactly nine of these triplets, and each of the 171 pairs occurs exactly once.

References

[1] Steiner, J. Combinatorische Aufgabe, *J. reine u. angew. Math.*, 1853, **45**, 181–82.

[2] Netto, E. *Lehrbuch der Combinatorik*. Leipzig und Berlin: B. G. Teubner, 1927.

[3] Heffter, L. Ueber Triplesysteme. *Math. Ann.*, 1897, **49**, 101-12.

[4] Moore, E. H. Concerning triple systems. *Math. Ann.*, 1893, **43**, 271–85.

[5] Netto, E. Zur Theorie der Triplesysteme. *Math. Ann.*, 1893, **42**, 143-52.

[6] Reiss, [n. i.]. Ueber eine Steinersche combinatorische Aufgabe. *J. reine u. angew. Math.*, 1859, **56**, 326-44.

[7] Bose, R. C. On the construction of balanced incomplete block designs. *Ann. Eugenics*, 1939, **9**, 353-99.

12 A REPRESENTATION OF HOTELLING'S T^2 AND ANDERSON'S CLASSIFICATION STATISTIC W IN TERMS OF SIMPLE STATISTICS[1]

ALBERT H. BOWKER
Stanford University

1. Introduction

It has been shown by Hsu [1] that the distribution of Hotelling's T^2 in the non-central case is equivalent to that of a non-central F-statistic. This paper gives a direct representation of T^2 in terms of the ratio of two independent chi-square variables, the numerator being non-central. The basic method is extended to obtain a representation of Anderson's classification statistic W as a function of the elements of two 2×2 independent Wishart matrices, the one being non-central.

Since the exact distribution of W is known to be exceedingly complicated [2], the present representation should prove extremely useful in finding percentage points for the distribution, either through an asymptotic expansion of the distribution function or by empirical sampling of the Wishart variables on high-speed computers.

2. Two Lemmas

Throughout the paper, we denote the joint density function of the elements of a p-dimensional Wishart matrix B, with n degrees of freedom and covariance matrix Σ, by $h(B|\Sigma, p, n)$. [The usual notation is $W(B|\Sigma, p, n)$. The notation $h(B|\Sigma, p, n)$ is used here to avoid confusion with the classification statistic W.] The symbol I_p is used for the p-dimensional identity matrix, and $\operatorname{tr} B$ denotes the trace of the matrix B.

We now prove two lemmas which justify the results obtained.

LEMMA 1. *Let B be a $p \times p$ symmetric random matrix whose elements have probability density function $f(B)$. Let G be a random $p \times p$ orthogonal matrix whose elements are distributed independently of the elements of B with density function $g(G)$.*

[1] Work performed under Contract No. AF 18(600)-941 with the USAF School of Aviation Medicine, Randolph Air Force Base, Texas.

(a) *Let O be any $p \times p$ orthogonal matrix, and define $E = O'BO$. Then, if the function f has the property that*

$$f(B) = f(OEO') = f(E)\ , \tag{2.1}$$

the matrix $B^* = GBG'$ is a $p \times p$ symmetric random matrix whose elements have probability density $f(B^*)$ and are distributed independently of G.

(b) *The Wishart density $h(B|I_p, p, n)$ has the property* (2.1).

PROOF OF (a). The joint probability density of the elements of G and B is $g(G) \cdot f(B)$. Let $B^* = GBG'$. The Jacobian of the transformation is $|GG'|^{(p+1)/2} = 1$. The joint density function of G and B^* is then $g(G)f(G'B^*G) = g(G)f(B^*)$, and part (a) follows.

PROOF OF (b). The Wishart density is

$$h(B|I_p, p, n) = C(p, n)\,|B|^{(n-p-1)/2} \exp\left(-\tfrac{1}{2} \operatorname{tr} B\right)\ , \tag{2.2}$$

where

$$C(p, n) = 2^{pn/2}\pi^{p(p-1)/4} \prod_{i=1}^{p} [\Gamma \tfrac{1}{2}(n - i + 1)]^{-1}\ .$$

Let O be any $p \times p$ orthogonal matrix and define $E = O'BO$. Then

$$\begin{aligned} h(B|I_p, p, n) &= h(OEO'|I_p, p, n) \\ &= h(E|I_p, p, n)\ , \end{aligned} \tag{2.3}$$

since $|OEO'| = |E|$ and $\operatorname{tr} OEO' = \operatorname{tr} EO'O = \operatorname{tr} E$. ||

It will be noted that the condition (2.1) is equivalent to requiring that f be a function only of D_β where D_β is a $p \times p$ diagonal matrix, whose diagonal elements are the characteristic roots of B.

LEMMA 2. *Let B be a $p \times p$ symmetric random matrix whose elements have probability density function $h(B|I_p, p, n)$. Partition B and B^{-1} as follows:*

$$B = \begin{bmatrix} B_{11} & B_{12} \\ B'_{12} & B_{22} \end{bmatrix},$$

and

$$B^{-1} = \begin{bmatrix} B^{11} & B^{12} \\ B^{12\prime} & B^{22} \end{bmatrix},$$

where B_{11} and B^{11} are the $k \times k$ matrices in the upper left-hand corner of B and B^{-1}, respectively. Let $(B)^{11-1} = B_{\cdot k+1, \ldots, p}$. Then $B_{\cdot k+1, \ldots, p}$ is a $k \times k$ symmetric random matrix whose elements have density function

$$h(B_{\cdot k+1, \ldots, p}|I_k, k, n - p + k)\ .$$

PROOF. We need two results from the theory of partitioned matrices (see, for example, [3], p. 363, and [4], p. 238):

$$|B| = |B_{22}|\,|B_{11} - B_{12}B_{22}^{-1}B'_{12}|\ , \tag{2.4}$$

and

$$(B^{11}) = (B_{11} - B_{12}B_{22}^{-1}B'_{12})^{-1}\ . \tag{2.5}$$

It follows that

$$B_{\cdot k+1,\ldots,p} = (B_{11} - B_{12} B_{22}^{-1} B_{12}') . \tag{2.6}$$

The distribution of B can be written

$$\begin{aligned} h(B|I_p, p, n) &= C(p, n) \, |B|^{(n-p-1)/2} \exp(-\tfrac{1}{2} \operatorname{tr} B) \\ &= C(p, n) \, |B_{22}|^{(n-p-1)/2} \exp(\tfrac{1}{2} \operatorname{tr} B_{22}) \\ &\quad \cdot |B_{11} - B_{12}B_{22}^{-1}B_{12}'|^{(n-p-1)/2} \exp(-\tfrac{1}{2} \operatorname{tr} B_{11}) . \end{aligned} \tag{2.7}$$

If we make the transformation from B_{11} to $B_{\cdot k+1,\ldots,p} = B_{11} - B_{12}B_{22}^{-1}B_{12}'$, the joint distribution of $B_{\cdot k+1,\ldots,p}$, B_{12}, and B_{22} is given by

$$\begin{aligned} & C(p, n) \, |B_{22}|^{(n-p-1)/2} \exp(-\tfrac{1}{2} \operatorname{tr} B_{22} - \tfrac{1}{2} \operatorname{tr} B_{12}B_{22}^{-1}B_{12}') \\ &\quad \cdot |B_{\cdot k+1,\ldots,p}|^{(n-p-1)/2} \exp(-\tfrac{1}{2} \operatorname{tr} B_{\cdot k+1,\ldots,p}) \\ &= \frac{C(p, n)}{C(k, n-p+k)} |B_{22}|^{(n-p-1)/2} \exp[-\tfrac{1}{2} \operatorname{tr} B_{22} - \tfrac{1}{2} \operatorname{tr} B_{12}B_{22}^{-1}B_{12}'] \\ &\quad \cdot h(B_{\cdot k+1,\ldots,p} \, | I_k, k, n-p+k) , \end{aligned} \tag{2.8}$$

and the lemma follows. ||

This result is well known in normal regression analysis, but an explicit statement seems to be elusive. (A related theorem is proved by Anderson in [5], Chapter 4.) If the Wishart matrix is assumed to be generated by a random sample of n p-dimensional normal vectors, the elements of the $k \times k$ matrix $B_{\cdot k+1,\ldots,p}$ are obtained from the sums of squares and cross products of the residuals from the regression of the first k variables on the remaining $p-k$. The distribution of the sums of squares and cross products of the residuals is again a Wishart distribution with a loss of one degree of freedom for each of the variables held constant. The notation $B_{\cdot k+1,\ldots,p}$ was adopted to show the relation to regression theory.

3. Representation of Hotelling's T^2

Hsu [1] has shown that Hotelling's T^2 can be written

$$T^2 = n \, t' \, U^{-1} t , \tag{3.1}$$

where the elements of $t' = (t_1, \cdots, t_p)$ are independently, normally distributed with unit variances, $E(t_i) = \tau_i$, $i = 1, 2, \cdots, p$, while U is a p-dimensional Wishart matrix, distributed independently of t, with density function $h(U|I_p, p, n)$. We can find an orthogonal matrix Γ whose first row is given by

$$\gamma_{1i} = \frac{\tau_i}{\sqrt{\sum \tau_i^2}} , \qquad i = 1, 2, \cdots, p .$$

If we make the transformation $Y = \Gamma t$, $A = \Gamma U \Gamma'$, then

$$T^2 = n \, Y' \Gamma \Gamma' A^{-1} \Gamma \Gamma' Y = n \, Y' A^{-1} Y , \tag{3.2}$$

where the elements of $Y' = (y_1, y_2, \cdots, y_p)$ are independently, normally distributed with unit variance, $E(y_{11}) = \sqrt{\sum_{i=1}^{p} \tau_i^2} = \mu$, say; $E(y_j) = 0$, $j = 2, \cdots, p$. The matrix A is again a p-dimensional Wishart matrix, whose elements are distributed independently of Y, with density function $h(A|I_p, p, n)$.

For any set of values of Y (except the null vector which has probability zero) we can construct an orthogonal matrix G whose first row is given by

$$g_j = \frac{y_j}{\sqrt{b_{11}}}, \qquad j = 1, 2, \cdots, p,$$

where $b_{11} = \sum_{i=1}^{p} y_i^2$. Then

$$\begin{aligned} T^2 &= n\,Y'\,G'\,G\,A^{-1}G'G\,Y \\ &= n\,(G\,Y)'\,(G\,A\,G')^{-1}\,G\,Y \\ &= n\,(\sqrt{b_{11}}, 0, \cdots, 0)\,A^{*-1}\,(\sqrt{b_{11}}, 0, \cdots, 0)' \\ &= n\sqrt{b_{11}}\,a^{*11}\sqrt{b_{11}} \\ &= n\frac{b_{11}}{a^{*}_{11\cdot 2\cdots p}}. \end{aligned} \tag{3.3}$$

By Lemma 1, the elements of A^* are distributed independently of the elements of G (and consequently of b_{11}) with a Wishart distribution $h(A^*|I_p, p, n)$. By Lemma 2, $a^*_{11\cdot 2\cdots p}$ has a chi-square distribution with $(n - p + 1)$ degrees of freedom. Since $b_{11} = \sum_{j=1}^{p} y_j^2$ is the sum of squares of p independent normal variables with unit variance, the first one being non-central, b_{11} has a non-central chi-square distribution with p degrees of freedom. We have a representation of T^2/n as the ratio of two independent chi-square variables, the numerator being non-central.

4. Representation of Anderson's Classification Statistic W

Anderson's classification statistic W [6] was proposed for the following problem: We have observed $N_1 + N_2 + 1$ independent p-dimensional chance vectors. The first N_1 vectors come from a population Π_1, the following N_2 vectors come from a population Π_2, and the last vector comes from a population Π where Π is either Π_1 or Π_2. The probability distribution in both Π_1 and Π_2 is assumed to be multivariate normal with the same covariance matrix Σ; the vector of expected values is $\mu^{(1)}$ in Π_1 and $\mu^{(2)}$ in Π_2. The values of $\mu^{(1)}$, $\mu^{(2)}$, and Σ are not known. On the basis of X, the $p\cdot(N_1 + N_2 + 1)$ matrix of observations, we want to classify the last observation as coming from either Π_1 or Π_2.

For purposes of classification, Anderson proposed the statistic

$$W = X'_{N_1+N_2+1} S^{-1}(X^{(1)} - X^{(2)}) - \tfrac{1}{2}(\bar{X}^{(1)} + \bar{X}^{(2)})\,S^{-1}(\bar{X}^{(1)} - \bar{X}^{(2)}), \tag{4.1}$$

where

$$\bar{X}^{(1)} = \frac{1}{N_1}\sum_{t=1}^{N_1} X_t, \qquad \bar{X}^{(2)} = \frac{1}{N_2}\sum_{t=N_1+1}^{N_1+N_2} X_t,$$

and

$$S = \frac{1}{N_1 + N_2 - 2}\left[\sum_{t=1}^{N_1}(X_t - \bar{X}^{(1)})(X_t - \bar{X}^{(1)})' + \sum_{t=N_1+1}^{N_1+N_2}(X_t - \bar{X}^{(2)})(X_t - \bar{X}^{(2)})'\right].$$

The last observation is classified as belonging to Π_1 if the observed value of W is greater than a pre-assigned cutoff point ω_0, and to Π_2 if $W \leqq \omega_0$.

It was pointed out in [7] that if we write

$$A_\Sigma = (N_1 + N_2 - 2)S ,$$

$$Z = \frac{(N_1 N_2)^{\frac{1}{2}}}{(N_1 + N_2)^{\frac{1}{2}}}(\bar{X}^{(1)} - \bar{X}^{(2)}) ,$$

$$Z^* = \frac{(N_1 + N_2)^{\frac{1}{2}}}{(N_1 + N_2 + 1)^{\frac{1}{2}}}\left[X_{N_1+N_2+1} - \frac{1}{(N_1 + N_2)}(N_1\bar{X}^{(1)} + N_2\bar{X}^{(2)})\right],$$

we can write

$$W = (N_1 + N_2 - 2)\left[\frac{(N_1 + N_2 + 1)^{\frac{1}{2}}}{(N_1 N_2)^{\frac{1}{2}}} Z^{*\prime} A_\Sigma^{-1} Z + \frac{(N_1 - N_2)}{2N_1N_2} Z' A_\Sigma^{-1} Z\right]. \tag{4.2}$$

Under either alternative, the three sets of variables are distributed independently of each other, the matrix A_Σ has a Wishart distribution with $n = N_1 + N_2 - 2$ degrees of freedom and covariance matrix Σ, the vector Z has a normal distribution with expected value $[(N_1N_2)^{\frac{1}{2}}][(N_1 + N_2)^{\frac{1}{2}}](\mu^{(1)} - \mu^{(2)})$ and covariance matrix Σ, and the vector Z^* has a normal distribution with covariance matrix Σ. Under the hypothesis that $X_{N_1+N_2+1}$ comes from Π_1, the expected value of Z^* is

$$\frac{N_2}{(N_1 + N_2)^{\frac{1}{2}}(N_1 + N_2 + 1)^{\frac{1}{2}}}(\mu^{(1)} - \mu^{(2)}),$$

while, under the alternative hypothesis, the expected value of Z^* is

$$\frac{-N_1}{(N_1 + N_2)^{\frac{1}{2}}(N_1 + N_2 - 1)^{\frac{1}{2}}}(\mu^{(1)} - \mu^{(2)}) .$$

There is a non-singular matrix Ψ (e.g., see [3]) such that $\Psi \Sigma \Psi' = I_p$ and $(\mu^{(1)} - \mu^{(2)})' \Psi' = (2\lambda, 0, \cdots, 0)$ with $\lambda \geqq 0$, $\lambda^2 = \frac{1}{4}(\mu^{(1)} - \mu^{(2)})' \Sigma^{-1}(\mu^{(1)} - \mu^{(2)})$. If we make the transformation $Y_1 = \Psi Z$, $Y_2 = \Psi Z^*$, $A = \Psi A_\Sigma \Psi'$, we have

$$\begin{aligned} W &= (N_1 + N_2 - 2)\left[\frac{(N_1 + N_2 + 1)^{\frac{1}{2}}}{(N_1N_2)^{\frac{1}{2}}} Y_1' \Psi'^{-1} \Psi' A^{-1} \Psi \Psi^{-1} Y_2 \right. \\ &\qquad \left. + \frac{(N_1 - N_2)}{2N_1N_2} Y_1' \Psi'^{-1} \Psi' A^{-1} \Psi \Psi^{-1} Y_1\right] \\ &= (N_1 + N_2 - 2)\left[\frac{(N_1 + N_2 + 1)^{\frac{1}{2}}}{(N_1N_2)^{\frac{1}{2}}} Y_1' A^{-1} Y_2 + \frac{(N_1 - N_2)}{2N_1N_2} Y_1' A^{-1} Y_1\right] \\ &= n\,[k_1 Y_1' A^{-1} Y_2 + k_2 Y_1' A^{-1} Y_1] , \quad \text{say.} \end{aligned} \tag{4.3}$$

The elements of Y_1 are independently, normally distributed with unit variance,

$$E(y_{11}) = \frac{2(N_1N_2)^{\frac{1}{2}}}{(N_1 + N_2)^{\frac{1}{2}}}\lambda = \mu ,$$

say, $E(y_{1j}) = 0$, $j = 2, \cdots, p$. The elements of Y_2 are also independently, normally distributed with unit variance, and $E(y_{21}) = c\mu$, $E(y_{2j}) = 0$, $j = 2, \cdots, p$, where

$$c = \left[\frac{N_2}{N_1(N_1 + N_2 + 1)}\right]^{\frac{1}{2}} \quad \text{or} \quad c = -\left[\frac{N_1}{N_2(N_1 + N_2 + 1)}\right]^{\frac{1}{2}},$$

depending on whether $\Pi = \Pi_1$, or $\Pi = \Pi_2$. The matrix A is a p-dimensional Wishart matrix whose elements are distributed independently of both Y_1 and Y_2 with density function $h(A|I_p, p, n)$.

Now let $Y_1'Y_1 = \sum_{i=1}^{p} y_{1i}^2 = b_{11}$, $Y_2'Y_2 = \sum_{i=1}^{p} y_{2i}^2 = b_{22}$, $Y_1'Y_2 = \sum_{i=1}^{p} y_{1i}y_{2i} = b_{12}$. We can construct an orthogonal matrix G such that

$$\begin{bmatrix} Y_1' \\ \\ Y_2' \end{bmatrix} G' = \begin{bmatrix} \sqrt{b_{11}} & 0 & 0 \cdots 0 \\ \\ \dfrac{b_{12}}{\sqrt{b_{11}}} & \sqrt{\dfrac{b_{11}b_{22} - b_{12}^2}{b_{11}}} & 0 \cdots 0 \end{bmatrix},$$

where $g_{1j} = y_{1j}/\sqrt{b_{11}}$, $j = 1, 2, \cdots, p$, and

$$g_{2j} = \frac{y_{2j} - (b_{12}/b_{11})\, y_{1j}}{\sqrt{\dfrac{b_{11}b_{22} - b_{12}^2}{b_{11}}}}, \qquad j = 1, 2, \cdots, p.$$

Since G is orthogonal, W can be written as

$$\begin{aligned}(4.4) \qquad W &= n(k_1 Y_1' G' G A^{-1} G' G Y_2 + k_2 Y_1' G' G A^{-1} G' G Y_1) \\ &= nk_1(\sqrt{b_{11}}, 0, \cdots, 0)\, A^{*-1}\left(\frac{b_{12}}{\sqrt{b_{11}}}, \sqrt{\frac{b_{11}b_{22} - b_{12}^2}{b_{11}}}, 0, \cdots, 0\right)' \\ &\quad + nk_2(\sqrt{b_{11}}, 0, \cdots, 0)\, A^{*-1}(\sqrt{b_{11}}, 0, \cdots, 0)'.\end{aligned}$$

By Lemma 1, the elements of A^* have a Wishart distribution independent of b_{11}, b_{12}, and b_{22}. If we write

$$A^{*-1} = \begin{bmatrix} A^{*11} & A^{*12} \\ A^{*12'} & A^{*22} \end{bmatrix},$$

where A^{*11} is the 2×2 matrix in the upper left-hand corner of A^{*-1}, then (4.4) reduces to

$$\begin{aligned}(4.5) \qquad W = nk_1(\sqrt{b_{11}}, 0)\, A^{*11}\left(\frac{b_{12}}{\sqrt{b_{11}}}, \sqrt{\frac{b_{11}b_{22} - b_{12}^2}{b_{11}}}\right)' \\ + nk_2(\sqrt{b_{11}}, 0)\, A^{*11}(\sqrt{b_{11}}; 0)'.\end{aligned}$$

By Lemma 2, the elements of the 2×2 matrix $(A^{*11})^{-1}$ have a Wishart distribution: $h(A_{\cdot 3 \ldots p}|I_2, 2, n - p + 2)$. We shall denote the elements

$$(A^{*11})^{-1} = A_{\cdot 3 \ldots p} = \begin{bmatrix} c_{11} & c_{12} \\ c_{12} & c_{22} \end{bmatrix}.$$

Thus, W can be expressed in terms of the c_{ij}'s and b_{ij}'s as

$$(4.6) \qquad W = nk_1 \frac{(c_{22}b_{12} - c_{12}\sqrt{b_{11}b_{22} - b_{12}^2})}{(c_{11}c_{22} - c_{12}^2)} + nk_2 \frac{b_{11}c_{22}}{(c_{11}c_{22} - c_{12}^2)}.$$

This expression is the basic result of the paper. Essentially we have here expressed W in terms of the elements of two independent 2×2 Wishart matrices with identity covariance matrix. The c_{ij}'s are central with $n-p+2$ degrees of freedom, and the b_{ij}'s have a non-central Wishart distribution with p degrees of freedom, given by the expression (e.g., see [8])

$$p(B) = \frac{|B|^{\frac{1}{2}(p-3)} \exp\left[-\frac{1}{2}(\mu^2(1+c^2) + b_{11} + b_{22})\right]}{2^p \pi^{\frac{1}{2}} \Gamma(\frac{1}{2}(p-1))} \cdot \sum_{j=0}^{\infty} \frac{[\frac{1}{2}(b_{11} + 2cb_{12} + c^2 b_{22})]^j (\frac{1}{2}\mu^2)^j}{\Gamma(\frac{1}{2}p + j)\, j!} . \tag{4.7}$$

The expression (4.6) may be rewritten in various ways. The most important case in applications arises when $N_1 = N_2 = N$, and $k_2 = 0$. In this case, we have

$$W = nk_1 \frac{b_{12} - (c_{12}/c_{22})(b_{11}b_{22} - b_{12}^2)^{\frac{1}{2}}}{c_{11} - (c_{12}^2/c_{22})} . \tag{4.8}$$

Note that $\sqrt{n-p+2}\,(c_{11}/c_{22})$ has Student's t-distribution with $(n-p+2)$ degrees of freedom and that $c_{11} - (c_{12}^2/c_{22})$ has an independent chi-square distribution with $(n-p+1)$ degrees of freedom. Thus W can be written as

$$\frac{\sqrt{2N+1}}{N} \frac{\left[b_{12} - \frac{t_{2N-p}}{\sqrt{2N-p}}(b_{11}b_{22} - b_{12}^2)^{\frac{1}{2}}\right]}{\frac{X^2_{2N-p-1}}{2N-2}} . \tag{4.9}$$

Note also that in this case $\mu = \sqrt{2N}\lambda$, while c is $1/\sqrt{2N+1}$ if $\Pi = \Pi_1$, or $-1/\sqrt{2N+1}$ if $\Pi = \Pi_2$. Thus, as $N \to \infty$,

$$c\mu \to \begin{cases} \lambda, & \text{if } \Pi = \Pi_1 , \\ -\lambda, & \text{if } \Pi = \Pi_2 . \end{cases}$$

The manipulations necessary to justify this reduction are heavy and may seem artificial. The geometric idea is very simple. Y_1 and Y_2 are p-dimensional vectors; clearly they will only have two non-zero components if the axes are chosen to be along Y_1 and perpendicular to Y_1 in the plane determined by Y_1 and Y_2. This rotation is accomplished by an orthogonal transformation whose elements depend on the components of Y_1 and Y_2. However, this transformation leaves the value of W invariant and since the Y's are distributed independently of the elements of A, the elements of the transformed matrix A^* have a Wishart distribution.

References

[1] Hsu, P. L. Notes on Hotelling's generalized T. *Ann. Math. Stat.*, 1938, **9**, 231-43.

[2] Sitgreaves, R. Some results on the distribution of the W classification statistic. USAF SAM Series in Statistics, Report No. 58-3, Randolph AFB, Texas: School of Aviation Medicine, 1957.

[3] Deemer, W. L., and Olkin, I. The Jacobians of certain matrix transformations useful in multivariate analysis. *Biometrika*, 1951, **38**, 345-67.

[4] Dwyer, P. S. *Linear Computations*. New York: Wiley, 1951.

[5] Anderson, T. W. *Introduction to Multivariate Statistical Analysis*. New York: Wiley, 1958.

[6] Anderson, T. W. Classification by multivariate analysis. *Psychometrika*, 1951, **16**, 31-50.

[7] Sitgreaves, R. On the distribution of two random matrices used in classification procedures. *Ann. Math. Stat.*, 1952, **23**, 263-70.

[8] Anderson, T. W., and Girshick, M. A. Some extensions of the Wishart distribution. *Ann. Math. Stat.*, 1944, **15**, 345-57.

13 EULER SQUARES

KENNETH A. BUSH

University of Idaho

1. Introduction

An arrangement of the numbers $0, 1, 2, \cdots, (s-1)$ is said to form a Latin square if we can construct an $s \times s$ matrix such that each row and each column contains each of these numbers. Two Latin squares are called orthogonal if each of the s^2 ordered pairs (u_{ij}, v_{ij}) occurs once where u_{ij} is the element in the ith row and jth column of the first square, and v_{ij} is similarly defined for the second square. Euler conjectured that if $s \equiv 2(4)$ no two orthogonal squares exist. Doubtless his conjecture was based on the only case known to him, $s = 2$, and on his inability to construct orthogonal squares of higher order satisfying this condition. Indeed it was not until 1900 that Tarry [1] established the validity of Euler's conjecture for the case $s = 6$ by a process of exhaustive enumeration. Despite strenuous efforts, including the use of high-speed digital computers, no further progress has been made with this problem.

In this note we shall show that these squares are "almost" constructible. Specifically, we shall show that if $s \equiv 2(4)$, then there exist two Latin squares such that $s^2 - s$ of the ordered pairs arise.

2. The Construction of "Almost" Orthogonal Euler Squares

We construct the first square as follows: Numbering the rows and columns from 0 to $s-1$, in the (i, j) square put u where $i + j - 2 \equiv u(s)$ and i and j are odd; otherwise put u in the (i, j) square where $i + j \equiv u(s)$.

The second square is obtained from the first by permuting rows so that all rows beginning with an even number are written first in increasing order, followed by the rows of the first square beginning with odd integers written in increasing order.

Following these rules, one obtains squares for $s = 6$, which have the appearance:

$$
\begin{array}{cccccc}
0 & 1 & 2 & 3 & 4 & 5 \\
1 & 0 & 3 & 2 & 5 & 4 \\
2 & 3 & 4 & 5 & 0 & 1 \\
3 & 2 & 5 & 4 & 1 & 0 \\
4 & 5 & 0 & 1 & 2 & 3 \\
5 & 4 & 1 & 0 & 3 & 2
\end{array}
\qquad\qquad
\begin{array}{cccccc}
0 & 1 & 2 & 3 & 4 & 5 \\
2 & 3 & 4 & 5 & 0 & 1 \\
4 & 5 & 0 & 1 & 2 & 3 \\
1 & 0 & 3 & 2 & 5 & 4 \\
3 & 2 & 5 & 4 & 1 & 0 \\
5 & 4 & 1 & 0 & 3 & 2
\end{array}
$$

It is clear from our construction that reduplications in pairs occur in the pairs formed from the first rows and the last rows. We shall, however, prove that all other pairs are distinct.

Suppose, on the contrary, that in the ith row and the jth column we find the elements u and v, and again in the mth row and nth column we find these same elements. We distinguish three cases.

First, we suppose i and m are $< s/2$. Then we have:

$$\begin{aligned} i + j + \delta &\equiv u(s)\,, & 2i + j &\equiv v(s)\,, \\ m + n + \delta' &\equiv u(s)\,, & 2m + n &\equiv v(s)\,, \end{aligned} \tag{1}$$

where $\delta = -2$ if i and j are both odd, $\delta' = -2$ if m and n are both odd, and δ, δ' are zero otherwise. We therefore have:

$$i - \delta \equiv v - u(s)\,, \qquad m - \delta' \equiv v - u(s)\,.$$

so that $m - i + \delta - \delta' \equiv 0(s)$.

We observe that the two equations on the right of (1) upon subtraction yield $2(i - m) + j - n \equiv 0(s)$. Since s is even, we have in consequence:

$$j - n \equiv 0(2)\,. \tag{2}$$

A similar argument with the other two equations in (1) results in $i - m \equiv 0(2)$. If i and j are both odd, then n and m are both odd. Hence $\delta - \delta' = 0$, and $m - i \equiv 0(s)$. By our inequality on i and m, it follows that $m = i$. Similarly, we see that $j = n$ from (2).

Similarly, if i and $m \geqq s/2$, we can show a contradiction. Thus

$$\begin{aligned} i + j + \delta &\equiv u(s)\,, & 2(i - s/2) + 1 + j + \delta'' &\equiv v(s)\,, \\ m + n + \delta' &\equiv u(s)\,, & 2(m - s/2) + 1 + n + \delta''' &\equiv v(s)\,, \end{aligned}$$

where δ and δ' have their previous meanings, but $\delta'' = -2$ if j is odd, $\delta''' = -2$ if n is odd. Otherwise $\delta'' = \delta''' = 0$. Using identical steps to those employed above, we again secure $j - n \equiv 0(2)$, $i - m \equiv 0(2)$. Hence $\delta - \delta' = 0$ and $\delta'' - \delta''' = 0$.

The final case arises if $i < s/2$ and $m \geqq s/2$. Our congruences now become

$$\begin{aligned} i + j + \delta &\equiv u(s)\,, & 2i + j &\equiv v(s)\,, \\ m + n + \delta' &\equiv u(s)\,, & 2m + 1 + n + \delta'' &\equiv v(s)\,, \end{aligned}$$

which reduce to

$$i - m - 1 + \delta' - \delta'' - \delta \equiv 0(s)\,.$$

We now have $j \equiv n + 1(2)$, $i \equiv m + 1(2)$. Hence $\delta = -2$ implies $\delta' = \delta'' = 0$, and $\delta'' = -2$ implies $\delta = 0$. Hence $\Delta = \delta' - \delta'' - \delta$ is either 0 or 2. If $\Delta = 0$, we have $i = 0$, $m = s - 1$. If $\Delta = 2$, we have $i = s/2 - 1$, $m = s/2$. But in this case $i \equiv 0(2)$; hence $\delta = 0$. Thus $\delta'' = -2$, n is odd, and since also m is odd, we have $\delta' = -2$, contradicting $\Delta = 2$.

3. Acknowledgment

The author wishes to acknowledge remarks of H. B. Mann which led to material simplification in the proof of this theorem.

Note: Just prior to publication, S. S. Shrikhande sent the author a copy of his joint paper with R. C. Bose entitled: "On The Construction of Pairwise Orthogonal Sets of Latin Squares and the Falsity of a Conjecture of Euler, II," in which it is shown that such squares always exist for the Euler numbers which exceed 6. Based in part upon the work of E. T. Parker, these exciting results, not yet published in final form, are available in the University of North Carolina Institute of Statistics Mimeograph Series No. 225, prepared in May, 1959.

Reference

[1] Tarry, G. Le problème des 36 officiers, *Assoc. Française, Comptes Rendus du 29me Session*. Vol. 1, pp. 122-23, Vol. 2, pp. 170-203. Paris: Masson, 1901.

14 A COMPROMISE BETWEEN BIAS AND VARIANCE IN THE USE OF NONREPRESENTATIVE SAMPLES

HERMAN CHERNOFF

Stanford University and Center for Advanced Study in the Behavioral Sciences

1. Introduction

It is customary for some advertisers to evaluate the appeal of a new product or technique by sampling the audience at a show. If the audience is assumed to represent the over-all population of interest, the percentage of the audience who favor the product or technique can be taken as a reasonable estimate of the corresponding percentage for the over-all population. If, as frequently happens, the selection is such that the audience is quite different in composition from the over-all population with respect to certain relevant characteristics, then this sample will be described as a nonrepresentative sample. The term "representative sample" will be defined in Section 2.

A common method for disposing of the problem of nonrepresentative samples is the following. Subdivide the population into groups (cells), each member of which has the same combination of relevant characteristics, e.g., age, sex, color of hair, and color of eyes. For each cell, estimate the expected response (proportion of population in the cell who would favor the product or technique). Then, using the known proportion of the over-all population in a particular cell, derive an estimate of the over-all response, which is a weighted average of the cell estimates. In fact, stratified sampling may be regarded as a deliberate use of nonrepresentative samples and of this estimation techique. When the average response varies considerably from one stratum to another and the sample sizes in the various strata are suitably controlled, stratified sampling offers an increase in efficiency over ordinary random sampling.

However, unnecessary stratification leads to losses of efficiency. In our problem, the estimation technique described above leads to *poor estimates* if there are cells which are well represented in the over-all population but poorly represented in the sample. Then the variances of the estimates for these cells are large and contribute a large amount to the variance of the estimate of the over-all response. Even worse is the possibility that some important cells may be completely unrepresented in the sample.

In sampling theory it is customary to evaluate an estimate of a parameter by its mean square deviation from the parameter. This may be expressed as the sum of the variance and the square of the bias. In this paper we shall study a class of estimates which *reduce variance* at the expense of *introducing bias*. Some of these estimates may be desirable when *a priori* information gives reason to believe that the bias is small compared with the reduction in the standard deviation. These estimates are derived by assuming that changes in expected response are *locally negligible* or *locally linear* in the neighborhood of a given cell. The bias of such estimates depends on the extent to which the assumptions are wrong.

We shall find it expedient to write separate equations for the bias and variance for our estimates, rather than to combine them into one equation for the mean square error. This expediency derives mainly from the algebraic forms of these equations, which make it relatively easy to compare variances for several estimates and to compare biases for several estimates. Furthermore, in some applications, there may be available *a priori* information which permits canceling a good portion of the bias when the expression for the bias is known.

2. The Case of Ordered Cells: Simple Grouping

Most of the ideas and methods of this paper are best explained in the context of the simple problem where there is a natural ordering of the cells. Conceivably, this situation would arise if there were only one important characteristic, such as age, which influences response. Then the population may be divided into age groups or cells. Although this situation may be relatively rare, we shall find it convenient for exposition to devote most of the paper to it. The importance of the ordering derives mainly from the assumption to be made that expected response does not fluctuate wildly as one moves from a cell to its neighbor.

Now, let us suppose that there are r cells labeled from 1 to r. Let p_i be the probability that an individual in the ith cell will favor the product or technique. Suppose π_i represents the proportion of the population in the ith cell. Then the over-all expected response in the population is given by

$$P = \sum_{i=1}^{r} \pi_i p_i .$$

Let n_i be the number of individuals in the sample who are in the ith cell and let m_i be the number of those who favor the product or technique.

At first we shall make the following basic assumption:

Assumption 1 (*Cell Representativity*).[1] *For given n_i, m_i has a binomial distribution corresponding to probability p_i and sample size n_i.*

[1] A sample will be said to be *representative* of a population if the distribution of the data is precisely that of a random sample with replacement from the population. Our usage differs slightly from that of random sampling in that we talk about subsamples

The estimate discussed in the introduction is represented by

(2.1) $$\hat{P}_1 = \sum \pi_i \hat{p}_{1i} ,$$

where

(2.2) $$\hat{p}_{1i} = m_i/n_i ,$$

if no n_i is zero. If no n_i is zero, $\hat{P}_1$ is unbiased; i.e., the bias is given by

(2.3) $$B(\hat{P}_1) = E(\hat{P}_1) - P = 0 ,$$

and its variance is given by

(2.4) $$V(\hat{P}_1) = \sum_{i=1}^{r} \pi_i^2 p_i(1 - p_i)/n_i .$$

We shall represent an approximation to the variance by

(2.5) $$V(\hat{P}_1) \sim P(1 - P) \sum_{i=1}^{r} \pi_i^2/n_i .$$

This approximation is derived under the assumption that the p_i's are all relatively close to P. Actually, $p(1 - p)$ varies relatively little for p in the interval from .3 to .7. We shall use such approximations to the variance merely to facilitate comparisons of the variances of various estimates. This usage makes it impossible to observe the advantages of stratified sampling. However, stratification questions are not assumed to be an issue in this paper. The main issue is that the sample is not representative of the population.

We shall consider several estimates, each of which can be represented in the form

(2.6) $$\hat{P} = \sum \pi_i \hat{p}_i ,$$

where the $\hat{p}_i$'s are "derived" in different ways. An extreme case is

(2.7) $$\hat{P}_2 = \sum m_i/\sum n_i = \sum \pi_i \hat{p}_{2i} ,$$

where

(2.8) $$\hat{p}_{2i} = \sum m_i/\sum n_i$$

would be the natural estimate of p_i if we assumed that all p_i's are equal. We have

---

being representative of subpopulations. Thus Assumption 1 states that, for given n_i, the subsample corresponding to the ith cell is representative of the subpopulation of all individuals in the ith cell.

The assumption of cell representativity implies that the bias introduced in the sampling technique operates mainly through the effect of the sampling on the cell frequencies n_i. For this assumption to be valid, all characteristics which affect response and which are disproportionately affected by the sampling technique must be considered in the formation of cells.

Note that nothing specific is said about how the n_i are obtained. They may or may not be random variables. When they are random variables, our results on bias and variance will hold conditionally for the given n_i.

$$B(\hat{P}_2) = \sum \left(\frac{n_i}{N} - \pi_i\right) p_i = \sum \frac{n_i}{N}(p_i - P) = \sum_i \pi_i \left(\frac{\sum_j n_j p_j}{N} - p_i\right), \tag{2.9}$$

where $N = \sum n_i$ and

$$V(\hat{P}_2) = \sum n_i p_i (1 - p_i)/N^2 \sim P(1-P)/N . \tag{2.10}$$

It is clear that $1/N$ is less than or equal to $\sum \pi_i^2/n_i$, with equality only if the n_i's are proportional to the π_i's. $V(\hat{P}_2)$ will be considerably smaller than $V(\hat{P}_1)$ if some of the π_i/n_i's are large. On the other hand, $B(\hat{P}_2)$ may be considerable if the p_i's are spread far apart.

Considerable reduction in variance at a slight cost in terms of bias may sometimes be obtained by estimating p_i, using only a few of its neighboring cells. Let

$$\hat{P}_3 = \sum_{i=1}^{r} \pi_i \hat{p}_{3i} , \tag{2.11}$$

where

$$\hat{p}_{3i} = \sum_{i \in S_i} m_i / \sum_{i \in S_i} n_i , \tag{2.12}$$

and S_i, a subset of the integers from 1 to r, corresponds to some of the cells near the ith. This estimate could be "derived" by making

Assumption 2 (*Local Constancy*). *In the neighborhood of the* ith *cell,* p *is approximately constant, i.e.,* $p_j \sim p_i$ *for* $j \in S_i$.

If Assumption 2 were satisfied, $\hat{P}_3$ would be approximately unbiased. In general, we have

$$B(\hat{P}_3) = \sum_{i=1}^{r} \pi_i \left\{ \frac{1}{N_i} \left[\sum_{j \in S_i} n_j p_j \right] - p_i \right\} , \tag{2.13}$$

and

$$V(\hat{P}_3) \sim P(1-P) \sum \pi_i \pi_k N_{ik} / N_i N_k , \tag{2.14}$$

where

$$N_i = \sum_{j \in S_i} n_j , \tag{2.15}$$

and

$$N_{ik} = \sum_{j \in S_i S_k} n_j , \tag{2.16}$$

where the sum of (2.16) is taken over all cells which are in both S_i and S_k.

First, we observe that both $\hat{P}_1$ and $\hat{P}_2$ are special and extreme cases of $\hat{P}_3$, which we shall call the *simple grouping estimate*. The estimate $\hat{P}_1$ corresponds to the case $S_i = \{i\}$, and $\hat{P}_2$ corresponds to the case $S_i = \{1, 2, \cdots, r\}$. If S_i consists of a few integers in the neighborhood of i, and p_i does not change very much with small changes in i, one may expect $B(\hat{P}_3)$ to be smaller than $B(\hat{P}_2)$. On the other hand, it is often possible to make

$V(\hat{P}_3)$ almost as small as $V(\hat{P}_2)$, by using only a few cells in the neighborhood of the ith for each S_i.

The following are three hypothetical examples which illustrate the preceding assertion about the variance. At the end of this section we shall elaborate on these examples and consider the bias and the mean square error which is, in the final analysis, the important measure of how good the procedures are.

Example 1.

i	1	2	3	4	5	6	7	8	9	10
π_i	.1	.1	.1	.1	.1	.1	.1	.1	.1	.1
n_i/N	.19	.05	.17	.01	.07	.11	.13	.15	.03	.09
S_i	{1, 2}	{1, 2}	{3, 4}	{3, 4}	{5, 6}	{5, 6}	{7}	{8, 9}	{8, 9}	{10}

$$V(\hat{P}_1) \sim 2.13\,P(1-P)/N$$
$$V(\hat{P}_2) \sim P(1-P)/N$$
$$V(\hat{P}_3) \sim 1.02\,P(1-P)N\,.$$

Example 2.

i	1	2	3	4	5	6	7	8	9	10
π_1	.1	.1	.1	.1	.1	.1	.1	.1	.1	.1
n_i/N	.01	.03	.05	.07	.09	.11	.13	.15	.17	.19
S_i	{1, 2}	{2, 3}	{3, 4}	{4, 5}	{5, 6}	{6, 7}	{7, 8}	{8, 9}	{9, 10}	{9, 10}

$$V(\hat{P}_1) \sim 2.13\,P(1-P)/N$$
$$V(\hat{P}_2) \sim P(1-P)/N$$
$$V(\hat{P}_3) \sim 1.36\,P(1-P)/N$$

Example 3.

i	1	2	3	4	5	6	7	8	9	10
π_i	.19	.17	.15	.13	.11	.09	.07	.05	.03	.01
n_i/N	.01	.03	.05	.07	.09	.11	.13	.15	.17	.19
S_i	{1, 2, 3}	{2, 3, 4}	{3, 4}	{4, 5}	{5, 6}	{6, 7}	{7, 8}	{8, 9}	{9, 10}	{9, 10}

$$V(\hat{P}_1) \sim 5.26\,P(1-P)/N$$
$$V(\hat{P}_2) \sim P(1-P)/N$$
$$V(\hat{P}_3) \sim 2.50\,P(1-P)/N$$

For Example 1, the ten numbers for n_i/N were taken to be .01, .03, $\cdots$, .19 and randomly assigned to the ten cells. The simple-minded grouping of this example was almost perfectly effective in reducing the variance to a minimum. In more realistic applications, one would expect that the low n_i/N would tend to correspond to neighboring cells as in Example 2. A simple grouping, in which few neighboring cells are involved in each S_i, would not be expected to be as effective as for Example 1. Even so, a considerable reduction is readily attained.

In Examples 1 and 2, the low n_i/N's are not very low, and the corresponding π_i's are not very large. Thus $V(\hat{P}_1)/V(\hat{P}_2)$ is only 2.1333. To increase the effect of the low n_i/N, we increased the corresponding π_i in Example 3. Here $V(\hat{P}_1)/V(\hat{P}_2)$ is more than 5, and the grouping used reduces the variance to less than half of $V(\hat{P}_1)$. (This is effectively equivalent to doubling the sample size as far as variance alone is concerned.) The grouping of Example 2 would yield $V(\hat{P}_2) \sim 2.86\, P(1-P)/N$, which is almost as good as the grouping used.

For practical applications, one must compare the decrease in standard deviation with the increase in bias introduced by using simple grouping. *A priori* information concerning the p_i could be used with Equation (2.13) to guess roughly the extent of the bias. A slightly more refined *guess* at the bias may be derived by applying

Assumption 3 (*Local Linearity*). *In the neighborhood of the* ith *cell, p is approximately linear in the cell index, i.e.,*

$$p_i \sim \alpha_i + \beta_i j \qquad \text{for } j \in S_i\,. \tag{2.17}$$

The corresponding approximation to the bias is given by

$$B_L(\hat{P}_3) = \sum_{i=1}^{r} \pi_i \beta_i (\bar{j}_i - i)\,, \tag{2.18}$$

where

$$\bar{j}_i = \sum_{j \in S_i} j n_j / N_i\,. \tag{2.19}$$

It is an interesting consequence that the bias effects of a local linear trend in the p_i can be reduced considerably by selecting the S_i so that the average $\bar{j}_i$ is close to i. *A priori* information concerning the p_i can be used to estimate the extent of bias. Furthermore, if it is sufficiently specific and dependable, it could also be used to correct partially for the bias.

Let us return to our three examples. In each case $\hat{P}_1$ has no bias. The bias of $\hat{P}_2$ can be estimated if we assume linearity of the p_i. However, this assumption is quite strong, and small deviations from linearity may have a large cumulative effect on this estimate of bias. We shall not concern ourselves with it. For $\hat{P}_3$ we shall apply $B_L(\hat{P}_3)$, assuming that the β_i is close to some number β. To be specific, we compute V, B, and the mean square error MSE $= V + B^2$, for $P = \frac{1}{2}$, $N = 400$, and $\beta = .05$.

Example 1.

	V	B	MSE $= V + B^2$
$\hat{P}_1$	$2.13\, P(1-P)/N = .00133$	0	.00133
$\hat{P}_2$	$P(1-P)/N = .00063$		
$\hat{P}_3$	$1.02\, P(1-P)/N = .00064$	$-.19\beta = -.010$	.00074

Example 2.

	V	B	MSE $= V + B^2$
$\hat{P}_1$	$2.13\, P(1-P)/N = .00133$	0	.00133
$\hat{P}_2$	$P(1-P)/N = .00063$		
$\hat{P}_3$	$1.36\, P(1-P)/N = .00085$	$.47\beta = .024$	.00142

Example 3.

	V	B	MSE $= V + B^2$
$\hat{P}_1$	$5.06\, P(1-P)/N = .00328$	0	.00328
$\hat{P}_2$	$P(1-P)/N = .00063$		
$\hat{P}_3$	$2.50\, P(1-P)/N = .00156$	$.84\beta = .042$	.00332

Thus we see that, in Examples 2 and 3, the $\hat{P}_3$ estimates, which previously seemed quite effective, yield no improvement for these values of β, P, and N. For smaller values of β or N, they will be useful. However, these examples illustrate that, while reducing variance, we should simultaneously attempt to keep the bias effects small. Assuming local linearity, we should, as remarked above, select the S_i so that the $\bar{j}_i$'s are close to i. Alternatively, we could try to compensate for one cell where $\bar{j}_i$ is greater than i by having the $\bar{j}_i$ in the next cell be less than i by a comparable amount. An examination of Example 1 will show that this is the reason why $B_L(\hat{P}_3)$ is so small there. For Examples 2 and 3, consider the following simple-minded grouping:

i	1	2	3	4	5	6	7	8	9	10
S_i^*	$\{1,2\}$	$\{2,3\}$	$\{3,4\}$	$\{3,4\}$	$\{5\}$	$\{6\}$	$\{7\}$	$\{8\}$	$\{9\}$	$\{10\}$

Then we have

	V	B	MSE $= V + B^2$
Example 2			
$\hat{P}_3$	$1.44\, P(1-P)/N = .00090$	$.15\beta = .008$	.00096
Example 3			
$\hat{P}_3$	$2.93\, P(1-P)/N = .00183$	$.28\beta = .014$	.00208

Thus $\hat{P}_3$ using S_i^* is a considerable improvement over $\hat{P}_1$ in these examples. Presumably a careful study would yield more effective groupings.

3. The Case of Ordered Cells: Linear Estimates

If it is assumed that

$$p_i = \alpha + \beta i \,, \tag{3.1}$$

one would be led to a fourth estimate,

$$\hat{P}_4 = \sum_{i=1}^{r} \pi_i \hat{p}_{4i} \,, \tag{3.2}$$

where

$$\hat{p}_{4i} = a + bi \,, \tag{3.3}$$

and a and b are the solutions of the normal equations

$$a \sum n_i + b \sum i n_i = \sum m_i \,, \tag{3.4}$$

$$a \sum i n_i + b \sum i^2 n_i = \sum i m_i \,. \tag{3.5}$$

The bias can be expressed by

$$B(\hat{P}_4) = \sum \pi_i E[a + bi - p_i] \,, \tag{3.6}$$

which would be zero if (3.1) holds.

Now let us assume that, in fact, p_i can be expressed as a quadratic in i, i.e.,

$$p_i \sim \alpha + \beta i + \gamma(i - \bar{i})^2 \,, \tag{3.7}$$

where the bar over a function of i represents its average when using the weights n_i/N. Thus $\bar{i} = \sum i n_i/N$, $\overline{i^2} = \sum i^2 n_i/N$, etc. Then

$$E(a + bi - p_i) \sim \gamma\left[\frac{\mu_3}{\mu_2}(i - \bar{i}) + \mu_2 - (i - \bar{i})^2\right], \tag{3.8}$$

where

$$\mu_s = \sum (i - \bar{i})^s/N \,. \tag{3.9}$$

Then the corresponding value of the bias is given by

$$B_Q(\hat{P}_4) \sim \gamma\left[\frac{\mu_3}{\mu_2} \sum (i - \bar{i})\pi_i + \mu_2 - \sum (i - \bar{i})^2 \pi_i\right]. \tag{3.10}$$

The variance of $\hat{P}_4$ is given by

$$V(\hat{P}_4) \sim \left[1 + \frac{(t - \bar{i})^2}{\mu_2}\right] \frac{P(1 - P)}{N} \,, \tag{3.11}$$

where

$$t = \sum i \pi_i \,. \tag{3.12}$$

The estimate $\hat{P}_4$ is similar to $\hat{P}_2$, in that it is based on a "global" assumption about relations between all the cells. We produced $\hat{P}_3$ to account for the possibility that the relations were approximately valid in small neighborhoods. Similarly, we may introduce $\hat{P}_5$ to generalize $\hat{P}_4$ to the case where a linear relation is assumed to hold in a *neighborhood* of the ith cell. Then we have

(3.13) $$\hat{P}_5 = \sum \pi_i \hat{p}_{5i} ,$$

where

(3.14) $$\hat{p}_{5i} = a_i + b_i i ,$$

(3.15) $$a_i \sum_{j \in S_i} n_j + b_i \sum_{j \in S_i} j n_j = \sum_{j \in S_i} m_j ,$$

(3.16) $$a_i \sum_{j \in S_i} j n_j + b_i \sum_{j \in S_i} j^2 n_j = \sum_{j \in S_i} j m_j .$$

To estimate the bias of $\hat{P}_5$, we assume the local quadratic relation,

Assumption 4 (*Local Quadraticity*).

(3.17) $$p_j \sim \alpha_i + \beta_i j + \gamma_i (j - \bar{j}_i)^2 , \qquad \textit{for } j \in S_i ,$$

where $\bar{j}_i$ *represents the average cell index over* S_i.

In general,

(3.18) $$\overline{f(j)}_i = \sum_{j \in S_i} n_j f(j)/N_i .$$

Then

(3.19) $$B_Q(\hat{P}_5) = \sum \pi_i \gamma_i \left[\frac{\mu_{3i}}{\mu_{2i}} (i - \bar{j}_i) + \mu_{2i} - (i - \bar{j}_i)^2 \right] ,$$

where

(3.20) $$\mu_{si} = \sum_{j \in S_i} (i - \bar{j}_i)^s n_i / N_i ,$$

and

(3.21) $$V(\hat{P}_5) \sim P(1 - P) \sum_{i,k} \frac{\pi_i}{N_i} \frac{\pi_k}{N_k} A_{ik} ,$$

where

(3.22) $$A_{ik} = \sum_{j \in S_i S_k} n_j \left[1 + \frac{(j - \bar{j}_i)(i - \bar{j}_i)}{\mu_{2i}} \right] \left[1 + \frac{(j - \bar{j}_k)(k - \bar{j}_k)}{\mu_{2k}} \right] .$$

4. The Case of Ordered Cells: Nonrepresentative Cells

Up to now we have assumed that the individuals in the ith cell of our sample are representative of the ith cell in the population. (See Assumption 1 of Section 2.) In some cases this assumption does not apply. For example, suppose the probability of a positive response $p(x)$ is a function of the value of a relevant characteristic x; for example, age. If the population is subdivided into cells (C_i, C_{i+1}), the average probability for the ith cell is

$$p_i = \int_{C_i}^{C_{i+1}} p(x)\pi(x)\,dx \Big/ \int_{C_i}^{C_{i+1}} \pi(x)\,dx \,, \tag{4.1}$$

if $\pi(x)$ is the distribution of ages in the population. Suppose that the sample is drawn from a population with age density $n(x)$, and that a subsample of age x is representative of the subpopulation of age x; i.e., an individual of age x in the sample also has probability $p(x)$ of giving a positive response. Then the average probability of a positive response for sampled individuals in the ith cell is

$$\int_{C_i}^{C_{i+1}} p(x)n(x)\,dx \Big/ \int_{C_i}^{C_{i+1}} n(x)\,dx \,. \tag{4.2}$$

Thus, distortion or loss of cell representativity is caused by the combination of *coarse grouping* and the *difference between* $n(x)$ *and* $\pi(x)$. In such a case, $\hat{P}_1$ is a biased estimate of $P = \int \pi(x)p(x)\,dx$. In this section we shall study the effect of such bias on the estimates $\hat{P}_3$ and $\hat{P}_5$. In doing so, we replace Assumption 1 by

ASSUMPTION 1* (*Point Representativity*). *The subsample with value x for the relevant characteristic is representative of the subpopulation with the same value x.*[2]

To study $B(\hat{P}_3)$ we modify the local linearity assumption to

ASSUMPTION 3* (*Local Linearity*). *In the neighborhood of the* ith *cell,* $\pi, n,$ *and* p *may be approximated by*

$$\pi(x) \sim \alpha_{1i} + \beta_{1i}(x - x_i) \,, \tag{4.3}$$

$$n(x) \sim \alpha_{2i} + \beta_{2i}(x - x_i) \,, \tag{4.4}$$

$$p(x) \sim \alpha_{3i} + \beta_{3i}(x - x_i) \,, \tag{4.5}$$

Here, the neighborhood of the ith *cell refers to* $\{x: x \in j\text{th cell}, j \in S_i\}$, *and* x_i *is the midpoint of the* ith *cell.*

Let l_i be the length of the ith cell. In what follows, we assume that S_i consists of a few (not a large number) cells of small length. The bias (given the n_j) in estimating p_i by $\hat{p}_{3i}$ is approximated by

$$B_i^* \sim \beta_i \left\{ \left[(\overline{x_j})_i - x_i \right] + \frac{1}{12} \left[\frac{\beta_{2i}}{\alpha_{2i}} (\overline{l_j^2})_i - \frac{\beta_{1i}}{\alpha_{1i}} l_i^2 \right] \right\} , \tag{4.6}$$

[2] In general, the investigator will represent his individuals in terms of several characteristics. Then x may be a vector giving exact age, exact income, sex, and color of eyes. It is important for the investigator to include enough characteristics so that Assumption 1 or 1* is satisfied. Thus any characteristic which is affected by the sampling technique and has an effect on the response should be included. In this section we assume ordering; i.e., one characteristic suffices.

and the conditional bias (given the n_j) in estimating P by $\hat{P}_3$ is

$$B_L^*(\hat{P}_3) \sim \sum \pi_i B_i^* , \tag{4.7}$$

where

$$\pi_i = \int_{c_i}^{c_{i+1}} \pi(x)\, dx . \tag{4.8}$$

To study $B(\hat{P}_5)$ we shall make

ASSUMPTION 4* (*Local Quadraticity*). *In the neighborhood of the* ith *cell*, π, n, *and* p *may be approximated by*

$$\pi(x) \sim \alpha_{1i} + \beta_{1i}(x - x_i) + \gamma_{1i}(x - x_i)^2 , \tag{4.9}$$

$$n(x) \sim \alpha_{2i} + \beta_{2i}(x - x_i) + \gamma_{2i}(x - x_i)^2 , \tag{4.10}$$

$$p(x) \sim \alpha_{3i} + \beta_{3i}(x - x_i) + \gamma_{3i}(x - x_i)^2 . \tag{4.11}$$

It will be most convenient to generalize $\hat{P}_5$ to allow for the possibility of using x_i in place of i to represent the ith cell in the linear estimate. Thus we have

$$\hat{P}_6 = \sum \pi_i \hat{p}_{6i} , \tag{4.12}$$

where

$$\hat{p}_{6i} = a_i + b_i y_i , \tag{4.13}$$

$$a_i \sum_{j \in S_i} n_j + b_i \sum_{j \in S_i} y_j n_j = \sum_{j \in S_i} m_j , \tag{4.14}$$

$$a_i \sum_{j \in S_i} y_j n_j + b_i \sum_{j \in S_i} y_j^2 n_j = \sum_{j \in S_i} y_j m_j , \tag{4.15}$$

and the y_i are arbitrarily prescribed numbers. In the special case where $y_i = i$, $\hat{P}_6$ reduces to $\hat{P}_5$. In the special case where $y_i = x_i$, $\hat{P}_6$ becomes the above-mentioned variation of the linear estimate $\hat{P}_5$, with i replaced by x_i. In general, we obtain a complicated formula, two special cases of which are of interest. The bias of P_i given $n_1, n_2, \cdots, n_r$ is

$$B_Q^*(\hat{P}_6) \sim \sum \pi_i B_i^{**} , \tag{4.16}$$

where

$$\begin{aligned} B_i^{**} = \beta_{3i}\Big[(\overline{x_j})_i - x_i\Big] + \frac{\beta_{3i}}{12}\Big[\frac{\beta_{2i}}{\alpha_{2i}}(\overline{l_j^2})_i - \frac{\beta_{1i}}{\alpha_{1i}} l_i^2\Big] \\ + \gamma_{3i}\Big\{\overline{[(x_j - x_i)^2]}_i + \frac{(\overline{l_j^2})_i - l_i^2}{12}\Big\} \\ + \frac{[y_i - (\overline{y_j})_i]}{\mu_{2i}^*}\overline{\Big\{\Big[\beta_{3i}(x_j - x_i) + \frac{\beta_{3i}\beta_{2i}}{12\alpha_{2i}} l_j^2 + \gamma_{3i}[(x_j - x_i)^2 + l_j^2/12]\Big][y_j - (\overline{y_j})_i]\Big\}}_i , \end{aligned} \tag{4.17}$$

and

$$\mu_{2i}^* = \sum_{j \in S_i} n_j [y_j - (\overline{y_j})_i]^2 \Big/ \sum_{j \in S_i} n_j . \tag{4.18}$$

The variance of $\hat{P}_6$ is the simple extension of that of $\hat{P}_5$ and is given by

$$V(\hat{P}_6) \sim P(1-P) \sum \frac{\pi_i \pi_k}{N_i N_k} A_{ik}^* , \tag{4.19}$$

where

$$A_{ik}^* \sim \sum_{j \in S_i S_k} n_j \left\{ 1 + \frac{[y_j - (\overline{y_j})_i][y_i - (\overline{y_j})_i]}{\mu_{2i}^*} \right\} \left\{ 1 + \frac{[y_j - (\overline{y_j})_k][y_k - (\overline{y_j})_k]}{\mu_{2k}^*} \right\} . \tag{4.20}$$

One special case of interest is that where $y_i = x_i$ for all i. This is the linear estimate where each cell is represented by its midpoint. Then

$$\begin{aligned} B_i^{**} = {} & \frac{\beta_{3i}}{12} \left[\frac{\beta_{1i}}{\alpha_{2i}} (\overline{l_j^2})_i - \frac{\beta_{1i}}{\alpha_{1i}} l_i^2 \right] + \gamma_{3i} \left\{ \overline{[(x_j - x_i)^2]}_i + \frac{(\overline{l_j^2})_i - l_i^2}{12} \right\} \\ & + \frac{[x_i - (\overline{x_j})_i]}{\mu_{2i}^*} \overline{\left\{ \left[\frac{\beta_{3i}\beta_{2i}}{12_{\alpha_{2i}}} l_j^2 + \gamma_{3i} [(x_j - x_i)^2 + l_j^2/12] \right] [x_j - (\overline{x_j})_i] \right\}}_i . \end{aligned} \tag{4.21}$$

A more specialized case is that where the $y_i = x_i$ and the x_i's are equally spaced. Then we simultaneously have a special case of $\hat{P}_6$ and of $\hat{P}_5$. Here let L be the common value of the l_i. Then

$$B_i^{**} = L^2 \left\{ \frac{\beta_{3i}}{12} \left[\frac{\beta_{2i}}{\alpha_{1i}} - \frac{\beta_{1i}}{\alpha_{1i}} \right] + \gamma_{3i} \left[\mu_{2i} - (i - \bar{j}_i)^2 + (i - \bar{j}_i) \frac{\mu_{3i}}{\mu_{2i}} \right] \right\} . \tag{4.22}$$

Note how the last part compares with Equation (3.19).

Finally, we consider the following variation of the preceding estimates. Suppose that the actual values of x are recorded, and these are used to estimate p as a linear function of x for x in the intervals of S_i. Thus we estimate $p(x)$ by

$$\hat{p}_i(x) = a_i + b_i x , \tag{4.23}$$

where

$$a_i \sum{}^{(i)} 1 + b_i \sum{}^{(i)} X_u = \sum{}^{(i)} m_u , \tag{4.24}$$

$$a_i \sum{}^{(i)} X_u + b_i \sum{}^{(i)} X_u^2 = \sum{}^{(i)} X_u m_u , \tag{4.25}$$

where X_u is the uth observation on x, m_u is one or zero according as the response is positive or not, and $\sum^{(i)}$ represents the sum over all N_i observations where X_u is in one of the cells of S_i. Then we estimate p_i by

$$\hat{p}_{7i} = \int_{c_i}^{c_{i+1}} \hat{p}_i(x) \pi(x)\, dx \Big/ \int_{c_i}^{c_{i+1}} \pi(x)\, dx . \tag{4.26}$$

Under the local-quadraticity Assumption 4*, it follows that the bias of $\hat{P}_7$, given the X_u, is

$$B_Q^*(\hat{P}_7) \sim \sum \pi_i B_i^{***} , \tag{4.27}$$

where

$$B_i^{***} = \gamma_{3i}\left\{\left[\frac{\mu_{3i}^{**}}{\mu_{2i}^{**}}(x_i - \bar{X}_i) + \mu_{2i}^{**} - (x_i - \bar{X}_i)^2\right] + \frac{l_i^2}{12}\left[\frac{\beta_{1i}}{\alpha_{1i}}\left[\frac{u_{3i}}{\alpha_{2i}} - 2(x_i - \bar{X}_i)\right] - 1\right]\right\}, \tag{4.28}$$

$$\bar{X}_i = \textstyle\sum^{(i)} X_u / N_i , \tag{4.29}$$

and

$$\mu_{ri}^{**} = \textstyle\sum^{(i)} (X_u - \bar{X}_i)^r / N_i . \tag{4.30}$$

This result is quite complicated. On the other hand, it can be shown by a rather involved argument that, if S_i contains only the ith cell and if π, n, and p may be approximated by fourth-order polynomials in the ith cell, then the observations will tend to spread out so that

$$B_i^{***} \sim \frac{\gamma_{3i} l_i^4}{180}\left[\left(\frac{\gamma_{2i}}{\alpha_{2i}} - \frac{\gamma_{1i}}{\alpha_{1i}}\right) + \frac{\beta_{2i}}{\alpha_{2i}}\left(\frac{\beta_{2i}}{\alpha_{2i}} - \frac{\beta_{1i}}{\alpha_{1i}}\right)\right]. \tag{4.31}$$

In any case, we may approximate the variance of $\hat{P}_7$ by

$$V(\hat{P}_7) \sim P(1 - P) \sum \frac{\pi_i \pi_k}{N_i N_k} A_{ik}^{**} , \tag{4.32}$$

where

$$A_{ik}^{**} = \sum_u^{(i,k)} \left[1 + \frac{(x_i - \bar{X}_i)(X_u - \bar{X}_i)}{\mu_{2i}^{**}}\right]\left[1 + \frac{(x_k - \bar{X}_k)(X_u - \bar{X}_k)}{\mu_{2k}^{**}}\right], \tag{4.33}$$

and $\sum^{(i,k)}$ represents the sum over all u for which X_u is in a cell of both S_i and S_k.

5. Remarks

5.1 Non-binomial Random Variables. While the discussion of the preceding section was in terms of binomial random variables, these played no essential role. If, instead of a zero or one response, we had a random variable y for each individual, the same ideas would apply. The formulas would have to be modified, with p replaced by $E(y)$ and $P(1 - P)$ replaced by the variance of y as a function of the cell. This would merely complicate the formulas for the variances slightly, making it a bit more difficult to compare variances without detailed computation. Incidentally, the binomial random-variable case may be treated as a special one of the class of examples where each cell has either one or zero observations. If the n_i in the binomial random-variable case is greater then zero, we then regard the cell as having one observation $\hat{p}_{1i}$, with mean p_i and variance $P(1 - P)/n_i$.

5.2 The Case of Several Characteristics. In many applications, individuals will be identified according to several characteristics. Many of the ideas of the preceding sections are still applicable. Each individual is classified in k ways and falls in one of many cells. The cell $i = (i_1, i_2, \cdots, i_k)$ corresponds to the i_1st level on the first classification, i_2nd on the second, etc. The estimate $\hat{P}_3$ may be trivially generalized to cover this case. One merely selects S_i as a group of cells in the neighborhood of i where one feels p does not fluctuate wildly.. Equations (2.13) and (2.14) are still valid. The estimate $\hat{P}_5$ may also be generalized. Here the fitting of a linear function in k dimensions may be a computational nuisance, but in principle the method generalizes.

Another approach consists of collapsing one classification at a time. Suppose that the first classification is age. For each $(i_2, i_3, \cdots, i_k)$, we have a group of cells ordered according to age. Applying one of the methods of the preceding sections, we obtain an estimate of the probability of response for each $(i_2, i_3, \cdots, i_k)$. Associated with this estimate is a variance. Then, keeping $(i_3, i_4, \cdots, i_k)$ fixed, we have an ordering according to i_2. Applying the same ideas, keeping in mind the fact that each cell has one observation, an estimate which is not necessarily binomial, we have an estimate for each $(i_3, i_4, \cdots, i_k)$. Continuing in this fashion, one ultimately obtains an over-all estimate of P.

If this method of collapsing classifications is to be applied, one may inquire about which classifications should be collapsed first. Intuitively, if one notices (a) that the need for the bias-producing methods $\hat{P}_3$ and $\hat{P}_5$ ordinarily tends to be reduced after a classification has been collapsed, and (b) there are locally negligible or locally linear changes in average response with small changes in the level of this classification, then this classification would seem to be a good one to collapse first.

5.3 Surveys with a Non-response Problem. In many sample surveys carried out by mail, difficulty arises when a portion of the sample refuses to respond, and, to the extent that non-response is correlated with the character being studied, there is a bias introduced by ignoring the non-responding individuals. Several methods of attacking this problem exist. For example, one may sample the non-respondents, using more persuasive approaches. Clearly, the methods of this paper would also apply if the respondents were identified by the relevant characteristics which would lead to Assumption 1 or 1* being satisfied.

5.4 Smoothing. The relation between the estimate $\hat{P}_3$ and the method of smoothing by moving averages is clear and will not be elaborated here.

5.5 Conclusion. In brief, by assuming that changes in average response are locally negligible or locally linear, minor and simple variations of the usual unbiased estimates may reduce variation considerably at the expense of introducing a small bias. These methods are worth applying, if *a priori* information concerning the average reponse would indicate that the (voluntary) bias so introduced is small compared with the reduction in the standard deviation of the estimate. The ideas and methods are applicable to the non-response problem in sample surveys. This study also points out the (involuntary) bias resulting from the combination of coarse grouping and nonrepresentative samples. Finally, it may be feasible to use the estimates of bias to correct (to a certain extent) for the voluntary and involuntary bias.

5.6 Acknowledgments. The author wishes to acknowledge gratefully that Howard Raiffa introduced him to the problem of nonrepresentative sampling and that some of the implications of the proposed solution were suggested to him by William G. Madow. Some improvements over the first draft were due to critical remarks by Dr. Madow, who incidentally remarked that certain techniques in this paper had been considered by the Census Bureau. The main ideas are quite simple and must have been considered by many statisticians, although the author knows of no specific references in the literature.

15 CONSTRUCTION OF FRACTIONAL FACTORIAL DESIGNS OF THE MIXED 2^m 3^n SERIES[1]

W. S. CONNOR

National Bureau of Standards

1. Introduction

This paper is based on some work in progress at the National Bureau of Standards to construct a catalogue of fractional factorial designs for the 2^m3^n series. This series has m factors at two levels and n factors at three levels. The total number $(m + n)$ of factors in these designs ranges from five to ten. Specifically, designs have been constructed for $2^43^1, \cdots, 2^93^1$; $2^33^2, \cdots, 2^83^2$; $2^23^3, \cdots, 2^73^3$; $2^13^4, \cdots, 2^63^4$; $2^13^5, \cdots, 2^53^5$; $2^13^6, \cdots, 2^43^6$; $2^13^7, 2^23^7, 2^33^7$; $2^13^8, 2^23^8$; and 2^13^9, a total of 39 designs altogether.

In brief, the method of construction is as follows: using standard methods, we obtain t (not necessarily distinct) fractions of the 2^m complete factorial and t (not necessarily distinct) fractions of the 3^n complete factorial. We then associate each fraction from the 2^m with a fraction from the 3^n. The association is such that every treatment combination in the fraction from the 2^m is adjoined to every treatment combination in the fraction from the 3^n, thereby forming treatment combinations in the fraction from the 2^m3^n.

Factors, treatment combinations, responses, and expected responses are defined in Section 2. In Section 3 the expected responses are expressed as linear functions of the grand average and the main and two-factor interaction effects, but not of higher-order interaction effects.

In these equations, the coefficient of a two-factor interaction effect is the product of the coefficients of the main effects of the factors. This definition of the expected responses permits the effective use of the Kronecker product of matrices to estimate the effects [1]. Also, this definition is compatible with the usual definitions in applied statistics.

Section 4 contains a brief review of the standard method for obtaining fractional factorial designs from the 2^m complete factorial, and Section 5 contains the analog for the 3^n complete factorial.

Section 6 contains a description of the method of construction of the mixed fractionals and sufficient conditions for the least-squares estimates to be

[1] Research supported by the Bureau of Ships, Navy Department, through a contract with the National Bureau of Standards.

mutually orthogonal functions of the responses. The conditions are illustrated by simple examples.

In Section 7 three examples are given, one for which all of the estimates are mutually orthogonal, one for which a few of the estimates are correlated, and one for which some of the estimates are not least-squares estimates.

A numerical example of a 1/2 fraction from the $2^3 3^2$ complete factorial is discussed in [2]. The method of construction for that design is essentially the same as the method described herein.

The present method was first described in a talk [3], although special cases were discussed earlier [4].

2. Treatments and Mean Responses

Let there be m factors

$$A_1, A_2, \cdots, A_m, \tag{2.1}$$

each at two levels, and n factors

$$B_1, B_2, \cdots, B_n, \tag{2.2}$$

each at three levels. Then there are $2^m 3^n$ treatment combinations, or simply treatments, in the complete factorial. The symbol

$$a_1^{x_1} a_2^{x_2} \cdots a_m^{x_m} b_1^{z_1} b_2^{z_2} \cdots b_n^{z_n} \tag{2.3}$$

denotes the treatment in which the factors $A_1, A_2, \cdots, A_m$ occur at the levels

$$x_1, x_2, \cdots, x_m, \qquad x_i = 0 \text{ or } 1, \tag{2.4}$$

and the factors $B_1, B_2, \cdots, B_n$ occur at the levels

$$z_1, z_2, \cdots, z_n, \qquad z_i = 0, 1, \text{ or } 2. \tag{2.5}$$

In accordance with the standard convention, a treatment and the expected response to that treatment will be denoted by the same symbol. Thus if

$$y(x_1, x_2, \cdots, x_m, z_1, z_2, \cdots, z_n) \tag{2.6}$$

is the response corresponding to the treatment (2.3), then

$$E\{y(x_1, x_2, \cdots, x_m, z_1, z_2, \cdots, z_n)\} = a_1^{x_1} a_2^{x_2} \cdots a_m^{x_m} b_1^{z_1} b_2^{z_2} \cdots b_n^{z_n}, \tag{2.7}$$

where E denotes "expectation."

These remarks include the case when only the factors A_i are present and the case when only the factors B_j are present.

3. Equations of Expectation when Some Factors Have Two Levels and Other Factors Have Three Levels

The effects will be denoted by

$$A_1^{\lambda_1} A_2^{\lambda_2} \cdots A_m^{\lambda_m} B_1^{\mu_1} B_2^{\mu_2} \cdots B_n^{\mu_n}, \tag{3.1}$$

$\lambda_i = 0$ or 1 and $\mu_j = 0$, 1, or 2, where, because three-factor and higher-order interaction effects are not to be defined, not more than two exponents can be different from zero.

When all λ's and μ's are zero, we shall write I in place of $A_1^0 A_2^0 \cdots A_m^0 B_1^0 B_2^0 \cdots B_n^0$ and call I the "grand average."

The pure A effects are defined for one or two λ's equal to 1 and all μ's equal to zero. When only one λ is 1, the symbol (3.1) denotes a main effect, and when two λ's are 1, a two-factor interaction effect.

The pure B effects are defined for one or two μ's equal to 1 or 2 and all λ's equal to zero. When only one μ is different from zero, the symbol (3.1) denotes a main effect. It is a linear effect if $\mu = 1$, and a quadratic effect if $\mu = 2$. When two μ's are different from zero, the symbol (3.1) denotes a two-factor interaction effect. It is a linear $\times$ linear effect, a linear $\times$ quadratic effect, a quadratic $\times$ linear effect, or a quadratic $\times$ quadratic effect according as the couplet of non-zero μ's is (1, 1), (1, 2), (2, 1), or (2, 2).

The mixed AB effects are two-factor interaction effects, and are designated linear $\times$ linear and linear $\times$ quadratic according as the couplet (λ, μ) is (1, 1) or (1, 2).

The expected response to any treatment $a_1^{x_1} a_2^{x_2} \cdots a_m^{x_m} b_1^{z_1} b_2^{z_2} \cdots b_n^{z_n}$ will be expressed as a linear function of the main and two-factor interaction effects as follows:

$$(3.2)\quad a_1^{x_1} a_2^{x_2} \cdots a_m^{x_m} b_1^{z_1} b_2^{z_2} \cdots b_n^{z_n}$$
$$= \prod_{i,j} [(1 - A_i)^{1-x_i}(1 + A_i)^{x_i}(1 - B_j + B_j^2)^{f_0(z_j)}(1 - 2B_j^2)^{f_1(z_j)}(1 + B_j + B_j^2)^{f_2(z_j)}] ,$$

where $f_0(z_j) = \frac{1}{2}(z_j - 1)(z_j - 2)$, $f_1(z_j) = z_j(-z_j + 2)$, and $f_2(z_j) = \frac{1}{2}z_j(z_j - 1)$; and where the right-hand side is expanded algebraically and then all terms which contain more than two letters are deleted. Also, the symbol I is substituted for 1.

To illustrate, for $m = 2$ and $n = 2$,

$$(3.3)\qquad \begin{aligned} a_1 b_1 b_2^2 &= (1 + A_1)(1 - A_2)(1 - 2B_1^2)(1 + B_2 + B_2^2) \\ &= I + A_1 - A_2 - A_1A_2 - 2B_1^2 - 2A_1B_1^2 + 2A_2B_1^2 \\ &\quad + B_2 + A_1B_2 - A_2B_2 - 2B_1^2B_2 + B_2^2 + A_1B_2^2 - A_2B_2^2 - 2B_1^2B_2^2 , \end{aligned}$$

after deletion of the terms with more than two letters.

Equation (3.2) has the property that the coefficient of any interaction effect is the product of the coefficients of the corresponding main effects. This is readily seen to be true in the example (3.3).

Because there are $2^m 3^n$ treatment combinations in the complete factorial, but the expected responses, when defined as in Equation (3.2), are functions of only $1 + \frac{1}{2}m(m + 1) + 2n^2 + 2mn$ effects, we cannot in general solve uniquely for the effects in terms of the expected responses.

If the effects (3.1) had not been restricted by allowing at most two non-zero exponents, so that the expected responses were given by (3.2) without any deletions, then we could solve uniquely for the effects as functions of the expected responses. This has been done in [1]. The resulting expressions for the effects as functions of the expected responses are the same as those in standard references such as [5].

The expression for the typical effect $A_1^{x_1}A_2^{x_2}\cdots A_m^{x_m}B_1^{z_1}B_2^{z_2}\cdots B_n^{z_n}$ is

$$(3.4)\quad A_1^{x_1}A_2^{x_2}\cdots A_m^{x_m}B_1^{z_1}B_2^{z_2}\cdots B_n^{z_n}$$
$$= D\prod_{i,j}[(-1+a_i)^{x_i}(1+a_i)^{1-x_i}(1+b_j+b_j^2)^{f_0(z_j)}(-1+b_j^2)^{f_1(z_j)}(1-2b_j+b_j^2)^{f_2(z_j)}]\,,$$

where the divisor $D = (2^{m+n_1}3^{n_0}6^{n_2})^{-1}$, n_{μ_j} = the number of B factors with exponent μ_j, $n_0 + n_1 + n_2 = n$; $f_0(z_j), f_1(z_j)$, and $f_2(z_j)$ are defined below (3.2); and where the right-hand side is expanded algebraically.

For example, if $m = 1$ and $n = 2$,

$$(3.5)\qquad A_1B_1^2 = \frac{1}{36}(-1+a_1)(1-2b_1+b_1^2)(1+b_2+b_2^2)$$
$$= \frac{1}{36}(-1-b_2-b_2^2+2b_1+2b_1b_2+2b_1b_2^2-b_1^2-b_1^2b_2-b_1^2b_2^2$$
$$+a_1+a_1b_2+a_1b_2^2-2a_1b_1-2a_1b_1b_2-2a_1b_1b_2^2$$
$$+a_1b_1^2+a_1b_1^2b_2+a_1b_1^2b_2^2)\,.$$

Of course, in our context the functions given by (3.4) are not necessarily definitions of effects. We have introduced the main and two-factor interaction effects in the expected responses, given by (3.2), and, as has already been pointed out, we usually cannot solve uniquely for these effects as functions of the expected responses.

4. Fractional Replication when the Factors Have Two Levels

In this section we shall briefly review some of the standard theory concerning fractional replication of the 2^m complete factorial. For other discussions, one may read, for example, [6].

In the usual theory of fractional replication, a $1/2^p$ fraction of the 2^m complete factorial consists of the treatment combinations $a_1^{x_1}a_2^{x_2}\cdots a_m^{x_m}$ whose sets of exponents $(x_1, x_2, \cdots, x_m)$ satisfy p linearly independent equations

$$(4.1)\qquad l_i = \alpha_{i1}x_1 + \alpha_{i2}x_2 + \cdots + \alpha_{im}x_m = c_i\,, \qquad i = 1, \cdots, p$$

in $GF(2)$, where $(\alpha_{i1}, \cdots, \alpha_{im}) \neq (0, \cdots, 0)$.

Because each of the c_i can assume one of two values, 0 or 1, there are 2^p different fractions which can be produced in this way. Each fraction contains 2^{m-p} treatment combinations, and because different fractions do not have any treatment combinations in common, the fractions collectively contain all 2^m treatment combinations.

Corresponding to the form l_i is the interaction (possibly a main effect)

$G_i = A_1^{\alpha_{i1}} A_2^{\alpha_{i2}} \cdots A_m^{\alpha_{im}}$, by which is meant the comparison of the expected responses to those treatment combinations for which

$$\alpha_{i1}x_1 + \alpha_{i2}x_2 + \cdots + \alpha_{im}x_m = 0 , \tag{4.2}$$

with the expected responses to those for which

$$a_{i1}x_1 + \alpha_{i2}x_2 + \cdots + \alpha_{im}x_m = 1 . \tag{4.3}$$

Half of the sets $(x_1, \cdots, x_m)$ satisfy (4.2) and the others satisfy (4.3). If only one α differs from 0, the interaction is a main effect; if two α's differ from 0, it is a two-factor interaction, and so on.

The pure A effects defined by (3.4) correspond to the interactions G_i. In fact, for any set of α's, every comparison between (4.2) and (4.3) differs only by a multiplicative constant from the corresponding effect in (3.4).

We now refer back to (4.1), and observe that the sets $(x_1, \cdots, x_m)$ which satisfy the equations $l_i = c_i$ also satisfy the equations

$$d_1 l_1 + \cdots + d_p l_p = d_1 c_1 + \cdots + d_p c_p , \tag{4.4}$$

where the quantities $d_1, \cdots, d_p$ are 0 or 1, but are not all 0. The interaction which corresponds to (4.4) is the product $G_1^{d_1} G_2^{d_2} \cdots G_p^{d_p}$.

Associated with the fraction defined by (4.1) is the identity relationship

$$\begin{aligned} I = G_1 &= \cdots = G_p \\ &= G_1 G_2 = \cdots = G_{p-1} G_p \\ &= \cdots = G_1 G_2 \cdots G_p , \end{aligned} \tag{4.5}$$

which provides vital information about the fraction. By selecting from the identity relationship any p interactions, which are independent in the sense of (4.1), one may construct all of the 2^p different fractions.

In addition, (4.5) provides information about which functions can be estimated from the fraction. To see this, consider an arbitrary interaction, $A_1^{\alpha_1} A_2^{\alpha_2} \cdots A_m^{\alpha_m}$. It contrasts the expected responses to the treatment combinations whose exponents satisfy

$$l = \alpha_1 x_1 + \alpha_2 x_2 + \cdots + \alpha_m x_m = 0 , \tag{4.6}$$

with the expected responses to the treatment combinations whose exponents satisfy

$$l = \alpha_1 x_1 + \alpha_2 x_2 + \cdots + \alpha_m x_m = 1 . \tag{4.7}$$

Now from (4.1) and (4.6) it follows that

$$d_1 l_1 + d_2 l_2 + \cdots + d_p l_p + l = d_1 c_1 + d_2 c_2 + \cdots + d_p c_p , \tag{4.8}$$

and from (4.1) and (4.7) that

$$d_1 l_1 + d_2 l_2 + \cdots + d_p l_p + l = d_1 c_1 + d_2 c_2 + \cdots + d_p c_p + 1 , \tag{4.9}$$

where the right-hand sides of (4.8) and (4.9) are either 0 and 1 or 1 and 0.

Because (4.8) is a linear combination of (4.1) and (4.6), any set $(x_1, \cdots, x_m)$

which satisfies (4.1) and (4.6) also satisfies (4.8). Similarly, any set $(x_1, \cdots, x_m)$ which satisfies (4.1) and (4.7) also satisfies (4.9). Half of the treatment combinations in the fraction (4.1) have exponents satisfying (4.6) and (4.8), and the other half have exponents satisfying (4.7) and (4.9).

For the fraction defined by (4.1) it follows that the interaction $A_1^{\alpha'_1} A_2^{\alpha'_2} \cdots A_m^{\alpha'_m}$ where

$$
\begin{aligned}
\alpha'_1 &= d_1\alpha_{11} + d_2\alpha_{21} + \cdots + d_p\alpha_{p1} + \alpha_1 , \\
\alpha'_2 &= d_1\alpha_{12} + d_2\alpha_{22} + \cdots + d_p\alpha_{p2} + \alpha_2 , \\
&\ \ \vdots \\
\alpha'_m &= d_1\alpha_{1m} + d_2\alpha_{2m} + \cdots + d_p\alpha_{pm} + \alpha_m ,
\end{aligned} \tag{4.10}
$$

is a comparison between the same two sets of treatments, as is the interaction $A_1^{\alpha_1} A_2^{\alpha_2} \cdots A_m^{\alpha_m}$.

Because the d's in (4.8) and (4.9) may be either 0 or 1, but not all zero, there are $2^p - 1$ interactions represented by $A_1^{\alpha'_1} A_2^{\alpha'_2} \cdots A_m^{\alpha'_m}$. Thus, there is only one function of the expected responses, apart from a multiplicative constant, which corresponds to $A_1^{\alpha_1} A_2^{\alpha_2} \cdots A_m^{\alpha_m}$ and to the $2^p - 1$ interactions represented by $A_1^{\alpha'_1} A_2^{\alpha'_2} \cdots A_m^{\alpha'_m}$. These 2^p interactions are said to be "aliased" with each other.

We shall now consider the expected responses (3.2) for different fractions (4.1). If in (4.1) $c_i = 0$, then in the equations of expectation the column vector of coefficients of any effect is the same as the column vector for the effect with which it is aliased. If in (4.1) $c_i = 1$, then these column vectors of coefficients differ only in sign.

For example, suppose that $m = 3$, $n = 0$, and $p = 2$, and let the equations (4.1) be as follows:

$$
\begin{aligned}
x_1 + x_2 \qquad &= 0 , \\
x_1 \qquad + x_3 &= 1 .
\end{aligned} \tag{4.11}
$$

The only linear combination of these equations is $x_2 + x_3 = 1$. If $x_1 = 0$, then $x_2 = 0$ and $x_3 = 1$, and (3.2) becomes

$$(1 - A_1)(1 - A_2)(1 + A_3) . \tag{4.12}$$

If $x_1 = 1$, then $x_2 = 1$ and $x_3 = 0$, so that (3.2) becomes

$$(1 + A_1)(1 + A_2)(1 - A_3) . \tag{4.13}$$

In both (4.12) and (4.13), A_1 has the same coefficient as does A_2, but A_1 and A_3 have different coefficients.

The situation can be summarized by writing the identity relationship as

$$I = A_1A_2 = -A_1A_3 = -A_2A_3 , \tag{4.14}$$

and by indicating the sets of aliased interactions as follows:

$$I = A_1A_2 = -A_1A_3 = -A_2A_3 \tag{4.15}$$

and

$$A_1 = \quad A_2 = \quad -A_3 = -A_1A_2A_3 \,.$$

Because the effect of any interaction which is not in the identity relationship has coefficient 1 for half the treatment combinations in the fraction and coefficient -1 for the other half, the two functions $I + A_1A_2 - A_1A_3 - A_2A_3$, and $A_1 + A_2 - A_3 - A_1A_2A_3$ are estimable, and moreover, when estimated by the method of least squares, their estimates are orthogonal functions of the responses.

In general, we may write (4.5) as

$$\begin{aligned} I &= \pm G_1 = \cdots = \pm G_p = \pm G_1G_2 = \cdots = \pm G_{p-1}G_p \\ &= \cdots \\ &= \pm G_1 \cdots = G_p \,, \end{aligned} \tag{4.16}$$

where the sign of an independent interaction G_i is $+$ or $-$, according as the corresponding form l_i is equal to 0 or 1. The signs of the products of the interactions are the products of the associated signs. As in the above example, the identity relationship (4.16) may be used to identify the estimable functions.

If two interactions are not aliased with each other, then in the equations of expectation the column vector of coefficients of one interaction effect is orthogonal to the column vector of coefficients of the other interaction effect. We shall describe this by saying that the effects (or interactions) are orthogonal to each other.

5. Fractional Replication when the Factors Have Three Levels

In this section we shall summarize some of the standard theory concerning the fractional replication of the 3^n complete factorial. A suitable reference is [6].

Because of the close analog with the 2^m factorial, we shall be concise. A $1/3^q$ fraction of the 3^n complete factorial consists of the treatment combinations $b_1^{z_1}b_2^{z_2}\cdots b_m^{z_m}$ whose sets of exponents $(z_1, z_2, \cdots, z_m)$ satisfy q linearly independent equations

$$l'_j = \beta_{j1}z_1 + \beta_{j2}z_2 + \cdots + \beta_{jn}z_n = e_j \,, \qquad j = 1, \cdots, q \tag{5.1}$$

in $GF(3)$, where $(\beta_{j1}, \cdots, \beta_{jn}) \neq (0, \cdots, 0)$ and the first non-zero element is 1.

Because each e_j may be 0, 1, or 2, there are 3^q different fractions of 3^{n-q} treatment combinations each. They have no treatment combinations in common, and collectively they exhaust all 3^n treatment combinations.

Corresponding to the form l'_j is the interaction $G'_j = B_1^{\beta_{j1}}B_2^{\beta_{j2}}\cdots B_m^{\beta_{jm}}$, by which is meant the comparison in the complete factorial of the expected responses to the three sets of treatment combinations whose exponents satisfy

$$l'_j = 0, \qquad l'_j = 1, \qquad \text{or} \qquad l'_j = 2 \,. \tag{5.2}$$

Because there are three sets of expected responses to be compared, they may be compared in many ways. It is customary to choose two linearly independent comparisons, from which all other comparisons can be calculated. We shall define two such comparisons.

The linear effect $L(\beta_1, \cdots, \beta_n)$ of the interaction $B_1^{\beta_1} \cdots B_n^{\beta_n}$ is defined as $(2 \cdot 3^{n-1})^{-1}$ times the following quantity: the sum of the expected responses for which $(z_1, \cdots, z_n)$ satisfies

(5.3) $$\beta_1 z_1 + \cdots + \beta_n z_n = 2$$

minus those for which $(z_1, \cdots, z_n)$ satisfies

(5.4) $$\beta_1 z_1 + \cdots + \beta_n z_n = 0 .$$

The quadratic effect $Q(\beta_1, \cdots, \beta_n)$ is defined as $(2 \cdot 3^n)^{-1}$ times the following quantity: the sum of the expected responses for which $(z_1, \cdots, z_n)$ satisfies

(5.5) $$\beta_1 z_1 + \cdots + \beta_n z_n = 0 \text{ or } 2 ,$$

minus twice those for which it satisfies

(5.6) $$\beta_1 z_1 + \cdots + \beta_n z_n = 1 .$$

These definitions include the main effects. For any factor B, the linear and quadratic effects $L(1)$ and $Q(1)$ are the same as the definitions given in (3.4) for B and B^2, respectively.

However, for any two factors B_1 and B_2, the effects $L(11)$, $Q(11)$, $L(12)$, and $Q(12)$ are not contained among the effects (3.4). Instead, they are related to the effects B_1B_2, $B_1B_2^2$, $B_1^2B_2$, and $B_1^2B_2^2$ as follows:

(5.7) $$\begin{bmatrix} B_1B_2 \\ B_1B_2^2 \\ B_1^2B_2 \\ B_1^2B_2^2 \end{bmatrix} = \frac{1}{4} \begin{bmatrix} -3 & -3 & -3 & 3 \\ 1 & -3 & 1 & 3 \\ 1 & -3 & -1 & -3 \\ 1 & 1 & -1 & 1 \end{bmatrix} \begin{bmatrix} L(11) \\ Q(11) \\ L(12) \\ Q(12) \end{bmatrix} .$$

Referring now to the q equations (5.1), we observe that there are $(3^q - 1 - 2q)/2$ linearly independent linear combinations of the equations, and that any set $(z_1, \cdots, z_n)$ which satisfies the equations also satisfies the linear combinations of them.

In particular, we note that corresponding to any linear function $f_1 l_1' + \cdots + f_q l_q'$, where $f_1, \cdots, f_q$ are 0, 1, or 2 but are not all zero, is the interaction $(G_1')^{f_1} \cdots (G_q')^{f_q}$.

Associated with the fraction (5.1) is the identity relationship

(5.8) $$\begin{aligned} I' &= G_1' = \cdots = G_q' \\ &= G_1'G_2' = \cdots = G_{q-1}'G_q' \\ &= G_1'(G_2')^2 = \cdots = G_{q-1}'(G_q')^2 \\ &= \cdots \\ &= G_1'G_2' \cdots G_q' = \cdots = G_1'(G_2')^2 \cdots (G_q')^2 , \end{aligned}$$

which may be used to construct 3^q fractions, including the fraction (5.1), and also may be used to discover the interactions which are aliased with each other. The detailed argument is analogous to the case where the factors have two levels.

For example, suppose that $n = 3$ and $q = 2$, and that

$$I' = B_1B_2 = B_2B_3^2 = B_1B_2^2B_3^2 = B_1B_3 \,. \tag{5.9}$$

The nine fractions are the solutions to the following nine sets of simultaneous equations:

	Equations	Values of e's
(5.10)	$z_1 + z_2 = e_1$:	0 0 0 1 1 1 2 2 2
	$z_2 + 2z_3 = e_2$:	0 1 2 0 1 2 0 1 2 .

The first fraction consists of the treatment combinations $b_1^0b_2^0b_3^0$, $b_1^2b_2^1b_3^1$, and $b_1^1b_2^2b_3^2$; the second of $b_1^0b_2^0b_3^2$, $b_1^2b_2^1b_3^0$, and $b_1^1b_2^2b_3^1$; and so on.

The two sets of interactions which are aliased within the set are as follows:

$$I' = B_1B_2 = B_2B_3^2 = B_1B_2^2B_3^2 = B_1B_3, \tag{5.11}$$

and

$$\begin{aligned} B_1 &= B_1B_2^2 = B_1B_2B_3^2 = B_1B_2B_3 = B_1B_3^2 \\ &= B_2 \quad = B_1B_2^2B_3 = B_2B_3 = B_3 \,. \end{aligned}$$

For the general case, the identity relationship (5.8) may be used to identify the sets of mutually aliased interactions. The implications of such aliasing for the equations of expectation are as follows. The column vectors of coefficients of effects represented in the identity relationship I' form a vector space of rank 1, and the column vectors for effects in any other set form a vector space of rank 2. For every interaction in such a set, the linear and quadratic effects are orthogonal and therefore comprise a basis for the vector space.

Finally, if two interactions are not aliased, then the effects of one interaction are orthogonal to the effects of the other interaction.

6. Construction of Fractional Factorial Designs with Factors of Two and Three Levels

We may let the treatment combinations in a fraction of the 2^m complete factorial, as determined from (4.1), be denoted by the column vector

$$a^X = \begin{bmatrix} a_1^{x_{11}} & a_2^{x_{21}} & \cdots & a_m^{x_{m1}} \\ a_1^{x_{12}} & a_2^{x_{22}} & \cdots & a_m^{x_{m2}} \\ & & \vdots & \\ a_1^{x_{1u}} & a_2^{x_{2u}} & \cdots & a_m^{x_{mu}} \end{bmatrix}, \tag{6.1}$$

where $u = 2^{m-p}$. Similarly, we may let the treatment combinations in a fraction of the 3^n complete factorial, as determined from (5.1), be denoted by the column vector

(6.2)
$$b^Z = \begin{bmatrix} b_1^{z_{11}} & b_2^{z_{21}} & \cdots & b_n^{z_{n1}} \\ b_1^{z_{12}} & b_2^{z_{22}} & \cdots & b_n^{z_{n2}} \\ & & \vdots & \\ b_1^{z_{1v}} & b_2^{z_{2v}} & \cdots & b_n^{z_{nv}} \end{bmatrix},$$

where $v = 3^{n-q}$. Then the Kronecker product

(6.3)
$$a^X \otimes b^Z$$

is a column vector whose elements are the uv treatments obtained by taking the symbolic product of each treatment in (3.12) with each treatment in (4.12). The Kronecker product of matrices is discussed in [7] and has been used in the present context in [1].

For the design (6.3) the identity relationship is obtained from I and I' by multiplying the interactions in I by the interactions in I', and connecting the products by equality signs. Referring to (4.5) and (5.8), we find this identity relationship is

(6.4)
$$\begin{aligned} I'' &= G_1 = \cdots = G_p \\ &= G_1G_2 = \cdots = G_{p-1}G_p \\ &= \cdots \\ &= G_1G_2 \cdots G_p \\ &= G_1' = \cdots = G_q' \\ &= G_1'G_2' = \cdots = G_{q-1}'G_q' \\ &= G_1'(G_2')^2 = \cdots = G_{q-1}'(G_q')^2 \\ &= \cdots \\ &= G_1'G_2' \cdots G_q' = \cdots = G_1'(G_2')^2 \cdots (G_q')^2 \\ &= G_1G_1' = \cdots = G_1G_q' = \cdots = G_pG_1' = \cdots = G_pG_q' \\ &= \cdots \\ &= G_1G_2 \cdots G_pG_1'(G_2')^2 \cdots (G_q')^2 . \end{aligned}$$

The product GG' of any interaction G from the 2^m with any interaction G' from the 3^n is defined in the following way. Corresponding to the interaction G is the form l, defined by (4.1), and corresponding to the interaction G' is the form l', defined by (5.1). The interaction G is the contrast between $l = 0$ and $l = 1$, and the interaction GG' compares G for $l' = 0$, 1, and 2.

Equivalently, the interaction G' denotes contrasts among $l' = 0$, 1, and 2, and GG' compares these contrasts for $l = 0$ and $l = 1$. For our purpose we shall consider the two contrasts L and Q defined in Section 5. Using these contrasts, we get the mixed parameters defined by (3.4).

We may illustrate (6.4) by beginning with $I = A_1 = A_2 = A_1A_2$ and $I' = B_1B_2$. The identity relationship which corresponds to the design (6.3) is

$$(6.5) \qquad I'' = A_1 = A_2 = A_1A_2 = B_1B_2 = A_1B_1B_2 = A_2B_1B_2 = A_1A_2B_1B_2 \, .$$

If we limit our attention to the grand average, main effects, and two-factor interactions, it follows from (6.5) that the sets of aliased interactions are

$$(6.6) \qquad (I, A_1, A_2, A_1A_2, B_1B_2),$$

and

$$(B_1, A_1B_1, A_2B_1, B_1B_2^2, B_2, A_1B_2, A_2B_2) \, .$$

Of course, interactions which are not aliased with each other are orthogonal interactions.

We now are in a position to describe our general method of construction of mixed fractional factorials. We form a design by pairing t distinct fractions like (6.1) with t distinct fractions like (6.2). That is, the design consists of the treatment combinations

$$(6.7) \qquad a^{X_1} \otimes b^{Z_1}, a^{X_2} \otimes b^{Z_2}, \cdots a^{X_t} \otimes b^{Z_t} \, .$$

Of especial interest are designs for which the estimates of the effects are orthogonal functions of the responses. We shall give a set of sufficient conditions for achieving such orthogonality.

The first condition applies to the pure A effects, and will be stated as a condition on the identity relationship I used to generate the t distinct fractions.

Condition 1. Either (i) the lowest-order interaction in I is a five-factor or higher-order interaction, or (ii) the t fractions taken together form a larger fraction for which the lowest-order interaction in the identity relationship is a five-factor or higher-order interaction.

The second condition applies to the pure B effects and is analogous to the condition for the A effects.

Condition 2. Either (i) the lowest-order interaction in I' is a five-factor or higher-order interaction, or (ii) the t fractions taken together form a larger fraction for which the lowest-order interaction in the identity relationship is a five-factor or higher-order interaction.

Because the integers 2 and 3 are relatively prime, it is impossible to satisfy both Condition 1, part (ii) and Condition 2, part (ii) at the same time. However, the other three combinations are possible.

That these conditions are sufficient to insure orthogonality among the estimates of the pure A effects and among the estimates of the pure B effects follows from the discussions in Sections 4 and 5. For other pairs of estimates we shall consider the several cases separately.

In designs for which Condition 1, (i) and Condition 2, (i) are met, every interaction in I'' is at least a five-factor interaction, so that all of the estimates are mutually orthogonal.

In designs for which Condition 1, (i) and Condition 2, (ii) are met, I' contains some interactions for fewer than five factors; and it follows that there may be aliasing between pure A interactions and mixed A, B interactions.

However, for the t fractions taken together, such aliasing disappears, and the estimates are orthogonal.

The same considerations apply to designs for which Condition 1, (ii) and Condition 2, (i) are met. Some simple examples follow:

Condition 1, (i); Condition 2, (i): Let $m=n=5$ and choose $I=A_1A_2A_3A_4A_5$, $I'=B_1B_2B_3B_4B_5$ to produce a 1/6 fraction.

Condition 1, (i); Condition 2, (ii): Let $m=5$, $n=4$ and choose $I=A_1A_2A_3A_4A_5$, and $I'=B_1B_2B_3B_4$ to produce a 1/2 fraction. Corresponding to I are two half-replicates a^{X_1} and a^{X_2} from the 2^5, and corresponding to I' are three third-replicates b^{z_1}, b^{z_2}, and b^{z_3}. The design is

$$a^{X_1}\otimes b^{Z_1},\quad a^{X_2}\otimes b^{Z_2},\quad a^{X_2}\otimes b^{Z_3}\,.$$

Condition 1, (ii); Condition 2, (i): Let $m=4$, $n=5$ and choose $I=A_1A_2A_3A_4$, $I'=B_1B_2B_3B_4B_5$. The design is

$$a^{X_1}\otimes b^{Z_1},\quad a^{X_2}\otimes b^{Z_2}\,.$$

Designs may be constructed for which most of the estimates are orthogonal. The remaining estimates are correlated but are not aliased. Such designs will be illustrated in the next section.

One useful device for obtaining orthogonal estimates is to omit some of the treatment combinations from the estimates of certain effects. An example of this technique will be given below.

7. Some Examples

The first example will be of a design with completely orthogonal estimates. Let $m=8$, $n=2$. There are 77 effects to be estimated, and for this purpose we shall choose 128 treatment combinations out of the 2304 in the complete factorial. This is a 1/18 fraction.

Let the identity relationship for the fractions from the 2^8 factorial contain the generators

$$G_1=A_1A_2A_3A_4A_5A_6A_7A_8\,,\quad G_2=A_1A_2A_3A_4\,,$$
$$G_3=A_1A_2A_5A_6\,,\quad G_4=A_2A_3A_5A_7\,. \tag{7.1}$$

Then choose eight fractions $a^{X_1},\cdots,a^{X_8}$ such that the eight fractions taken together comprise a fraction having $I=G_1$ as its identity relationship. This can be done, for example, by choosing the constants c of (4.1) as follows:

(7.2)

G_1	G_2	G_3	G_4
0	0	0	0
0	0	0	1
0	0	1	0
0	0	1	1
0	1	0	0
0	1	0	1
0	1	1	0
0	1	1	1.

Let the sets $b^{z_1}, \cdots, b^{z_8}$ from the 3^2 be the complete factorial. Then the design is

$$a^{X_1} \otimes b^{Z_1}, \cdots, a^{X_8} \otimes b^{Z_8} . \tag{7.3}$$

Estimation by least squares results in very simple normal equations. The matrix of coefficients in the normal equations is a diagonal matrix, with elements along the diagonal expressed as functions of the number of treatment combinations f. Each element is the coefficient of an effect, as follows:

Effect	Element
$A_i, A_iA_{i'}$	f
A_iB_j, B_j	$\frac{2}{3}f$
$A_iB_j^2, B_j^2$	$2f$
$B_jB_j,$	$\frac{4}{9}f$
$B_jB_{j'}^2, B_j^2B_{j'}$	$\frac{4}{3}f$
$B_j^2B_{j'}^2$	$4f$.

These elements also occur along the diagonal when the design is not orthogonal.

The next example is a design for which some of the estimates are correlated. Let $m = 2$ and $n = 3$. Thirty-four effects will be estimated from 54 treatment combinations, a half-fraction.

Let the identity relationships be $I = A_1A_2$ and $I' = B_1B_2B_3$, and let the sets a^{X_1} and a^{X_2} be the half-replicates from the 2^2 and the sets b^{Z_1}, b^{Z_2}, and b^{Z_3} be the third-replicates from the 3^3. Then the design is

$$a^{X_1} \otimes b^{Z_1}, \quad a^{X_2} \otimes b^{Z_2}, \quad a^{X_2} \otimes b^{Z_3} . \tag{7.4}$$

The estimates of certain parameters can be expressed in terms of the matrix equation

$$\hat{\rho}_i = CY_i , \quad C = \frac{1}{96}\begin{bmatrix} 3 & 1 \\ 1 & 3 \end{bmatrix} , \tag{7.5}$$

where the ρ_i are as follows:

$$\begin{aligned} \rho_1' &= \tfrac{3}{2}(A_1, A_2) , & \rho_2' &= \tfrac{3}{2}(I, A_1A_2) , \\ \rho_3' &= (A_1B_1, A_2B_1) , & \rho_4' &= (A_1B_2, A_2B_2) , \\ \rho_5' &= (A_1B_3, A_2B_3) , & \rho_6' &= 3(A_1B_1^2, A_2B_1^2) , \\ \rho_7' &= 3(A_1B_2^2, A_2B_2^2) , & \rho_8' &= 3(A_1B_3^2, A_2B_3^2) , \end{aligned} \tag{7.6}$$

and there is a column vector Y_i for each ρ_i. The elements of Y_i are the functions of the responses which occur in the normal equations.

The remaining parameters are estimated by dividing their Y's by the appropriate coefficients from (7.3).

The third example is obtained from the second, and illustrates how to obtain orthogonal estimates by ignoring some of the responses. The estimates are obtained from the equations of expectation in the following way. The column vectors of coefficients of the pure A effects are modified by replacing the 18 elements which correspond to $a^{X_2} \otimes b^{Z_3}$ by zero. The effect of doing this is to make all of the column vectors mutually orthogonal. Then, taking the scalar product of the transpose of each vector with the column vector of responses produces a set of mutually orthogonal, unbiased estimates.

References

[1] Bose, R. C., and Connor, W. S. Analysis of fractionally replicated 2^m3^n designs. (To be published in the *Review of the International Statistical Institute*.)

[2] Connor, W. S. A fractional factorial experiment of the mixed 2^m3^n series. *Trans. Rochester Soc. for Quality Control*, 1958, 59-70.

[3] Connor, W. S., Some recent work on mixed fractional factorial designs. Paper read at the joint meeting of the American Statistical Association and the Biometric Society (ENAR), Chicago, Illinois, December 28, 1958.

[4] Morrison, M. Fractional replication for mixed series. *Biometrics*, 1956, **12**, 1-19.

[5] Cochran, G., and Cox, M. *Experimental Designs*, 2d ed. New York: Wiley, 1957.

[6] Kempthorne, O. *The Design and Analysis of Experiments*. New York: Wiley, 1952.

[7] Macduffee, C. C. *The Theory of Matrices*. New York: Chelsea, 1946.

16 APPLICATION OF BOUNDARY THEORY TO SUMS OF INDEPENDENT RANDOM VARIABLES

J. L. DOOB,[1] University of Illinois

J. L. SNELL[2] and R. E. WILLIAMSON,[2] Dartmouth College

1. Introduction

In [1] Doob has given a potential theory and a boundary theory for discrete-time Markov chains. This theory follows that of Martin [2] for classical Newtonian potentials.

The purpose of this paper is to investigate the nature of this boundary for Markov chains which arise as sums of independent random variables. We shall also see that the general theorems of [1] applied to our special chains give new limit theorems for sums of independent random variables. In Section 2 we summarize without proofs the results of [1] that we shall use. In Section 3 we apply these results to sums of independent random variables.

2. The Martin Boundary

2.1 Superregular and Regular Functions. Let R be a subset of the integers and p a matrix with entries $p(i, j)$, i, j in R. Let $p(i, A) = \sum_{j \in A} p(i, j)$. We say that p is a *substochastic transition matrix* if

$$\text{(i)} \quad 0 \leqq p(i, j) \leqq 1 , \qquad i, j \text{ in } R ,$$
$$\text{(ii)} \quad p(i, R) \leqq 1 , \qquad i \text{ in } R .$$

If (ii) is replaced by equality, p is called *stochastic*. We denote by $p^{(n)}$ the nth power of the matrix p. We define $p^{(0)}$ to be the identity matrix. If p is substochastic, it is often convenient to make it stochastic by adding an element ρ to R and define $p(\rho, \rho) = 1$, $p(i, \rho) = 1 - p(i, R)$.

Let p be stochastic. We denote by $\{X_{\xi n}, n \geqq 0\}$ a Markov process defined on Ω (a point of which is denoted by ω) with $X_{\xi 0} = \xi$ and transition matrix

[1] Research done in part under an Air Force contract.

[2] The work done by the second and third author was supported by a grant given to Dartmouth College by the National Science Foundation.

p. The n-step transition probabilities for this process are then given by $p^{(n)}$. The sample paths are called paths from ξ. If p is substochastic, it can be made stochastic as indicated above, and it then determines a Markov process.

If u is a function defined on R, we can represent u as a column vector. Then pu yields a new function on R.

Let u be a function defined on R satisfying the following conditions:

(a) $-\infty < u \leqq \infty$ on R.

(b) For each strictly positive integer n, $p^{(n)}u$ is finite on R and $pu \leqq u$.

Then u is called *superregular*. It is strictly superregular at a point of R if there is strict inequality in (b) at this point. It is *subregular* if its negative is superregular, and *regular* if it is both superregular and subregular. Regular functions play the role of harmonic functions and subregular the role of subharmonic functions in the classical theory.

If u is a positive (by which we shall always mean non-negative) regular function on R, which has the property that any positive regular function v on R with $v \leqq u$ is proportional to u, then u is called *minimal*.

2.2 Potentials. For a given substochastic matrix p, we define a matrix g by

$$g(i, j) = \sum_{n=0}^{\infty} p^{(n)}(i, j) .$$

Then $g(\xi, A)$ is the expected number of times that a path from ξ enters A. A subset A of R is called transient if $g(\cdot, A)$ is finite on R. The sets of interest for potential theory are the transient sets.

If v is a function defined on R, gv is called its *potential*. If u is the potential of a positive function, $pu \leqq u$. It is then superregular if $p^{(n)}u$ is finite for $n \geqq 1$.

If u is positive and superregular, then $u \geqq pu \geqq p^{(2)}u \geqq \cdots$. We denote by $p^{\infty}u$ the limit of this monotone sequence. Then $p^{\infty}u$ is regular and u may be represented in the form $u = p^{\infty}u + g(u - pu)$. This represents u as the sum of a regular function and a potential. This corresponds to the Riesz representation of a superharmonic function as a harmonic function plus the potential of a positive mass distribution.

2.3 The *h*-Process. Let h be a positive superregular function defined on R. Define a new transition function by

$$q(i, j) = \begin{cases} \dfrac{p(i, j)h(j)}{h(i)}, & \text{if } 0 < h(i) < \infty , \\ 0, & \text{otherwise} . \end{cases}$$

Probability paths corresponding to the process determined by this new transition function are called h-paths. Almost no h-path from a point where h is strictly positive and finite reaches a zero or an infinity of h. Hence we may exclude these from the state space of q. The n step transition probabilities for this new process are given by $q^{(n)}(i, j) = p^{(n)}(i, j)h(j)/h(i)$. Hence the function $g^h(i, j)$ for the h-process is given by $g^h(i, j) = g(i, j)h(j)/h(i)$.

If u is superregular (regular) for p, then u/h is superregular (regular) for q. Hence h is minimal for p if and only if the constant function 1 is minimal for q.

2.4 Definition of the Martin Boundary. We say that state j is *covered* by state i if $g(i, j) > 0$. We shall define a boundary for the chain obtained by restricting the state space to the set S of all states covered by a single state (taken for convenience to be the state 0).[3] Every state covered by a state of S is in S. The transient states of S are denoted by S_t. We are interested in positive superregular functions u defined on S with $0 < u(0) < \infty$. If $u(0)$ is finite, then u is necessarily finite on all of S.

Let $u_j(i)$ denote the probability that a path from i ever reaches j. Define K by

$$K(i, j) = \frac{u_j(i)}{u_j(0)}, \qquad i, j \text{ in } S.$$

Then

$$K(i, j) = \frac{g(i, j)}{g(0, j)}, \qquad \text{if } j \text{ is in } S_t.$$

The function $K(\cdot, j)$ is positive superregular on S with $K(0, j) = 1$. It is regular, except possibly at j, where it is regular if and only if j is persistent (recurrent). It is constant on each persistent class.

The set S is metrized by the distance function d defined by

$$d(\nu_1, \nu_2) = \sum_i | K(i, \nu_1) - K(i, \nu_2) | 2^{-i} u_i(1).$$

Then $d(\nu_1, \nu_2) = 0$ implies that $\nu_1 = \nu_2$ or that ν_1 and ν_2 are in the same persistent class.

The space S is completed by adding limit points and so completed is a compact metric space. Let S' be the set consisting of the points of metrized S corresponding to persistent classes and of the set of limit points of metrized S in the completed space. The set S' is called the *Martin exit boundary* of S in accordance with Martin's definition [2] for the case of Newtonian potential.

Points corresponding to persistent classes are called degenerate boundary points. For ξ a degenerate point $K(i, \xi)$ has been defined. If ξ is a non-degenerate boundary point, $K(i, \xi)$ is defined by $K(i, \xi) = \lim_{j \to \xi} K(i, j)$. The function $K(i, \cdot)$ is continuous on the space $S \cup S'$. For every ξ of this space, $K(\cdot, \xi)$ is a positive, finite-valued, superregular function on S with $K(0, \xi) = 1$. Different points ξ yield different superregular functions and conversely. Thus a point of the metric space can be identified with a su-

[3] In [1] this reference state is state 1.

perregular function. A boundary point ξ is called *minimal* if $K(\cdot, \xi)$ is a minimal regular function. The degenerate boundary points are minimal. The set of minimal boundary points is a Borel set.

THEOREM 1. *Let v be a positive superregular function defined on S, finite-valued, and $v(0) = 1$. Then v can be represented in the form*

$$v(i) = \int_{S \cup S'} K(i, \xi) V(d\xi) , \qquad V(S \cup S') = 1 .$$

Here V can be chosen to have the value 0 for the set of non-minimal boundary points and, with this choice, is uniquely determined by v.

The integral over the boundary gives the regular component, and the integral over S gives the potential component in the Riesz-type decomposition of v.

THEOREM 2. *If h is positive and superregular on S, finite and positive at* 0, *then almost every h-path from a point at which h is finite and strictly positive converges (in the sense of Martin topology) to a minimal boundary point or reaches ρ. (The alternative cannot occur if h is regular.) If h is minimal, almost all h-paths from i converge to the minimal boundary point corresponding to h.*

In particular, if h in Theorem 2 is taken to be the constant function 1, we see that the sample paths of the process itself converge to the boundary, if p is stochastic. Since $K(i, \cdot)$ is continuous, Theorem 3 follows from Theorem 2.

THEOREM 3. *Let $\{X_n, n \geqq 1\}$ be any Markov process with the given transition matrix and with initial states distributed in S. Then $\lim_{n\to\infty} K(i, X_n)$ exists with probability* 1. *The limit on a path converging to ξ is $K(i, \xi)$.*

The three theorems just stated are the key results from [1] that we shall use.

3. The Martin Boundary for Random Walks

3.1 Definition of a Random Walk. We assume we are given a random variable Y with values the lattice points in N-dimensional space. Let $p(j) = \Pr\{Y = j\}$. (Now i and j are N-dimensional vectors with integer components.) By a *random walk determined by Y* we shall mean a Markov chain with state space the lattice points of N-dimensional space and transition probabilities defined by $p(i, j) = p(j - i)$. We denote by $\{X_n, n \geqq 1\}$ the random walk obtained by starting at state 0. Then $X_n = Y_1 + Y_2 + \cdots + Y_n$ where $Y_1, Y_2, \cdots, Y_n$ are independent random variables each with the common distribution of Y. Then, by a theorem of Chung and Fuchs [3], either all states are transient or all are recurrent. In the recurrent case, the boundary theory is not interesting, and so we shall assume throughout that we are in the transient case. Thus the set of states which can be reached from 0 is also the set of states covered by 0. Let S be this set of states.

Our boundary theory will be a boundary for the set S. All of our results will apply to the chain obtained by restricting the random walk to S. We shall first show that random walks differ from the general Markov chain in that there can be no non-constant bounded positive regular functions. That is to say, the constant function 1 is minimal.

THEOREM 4. *For a random walk in N-dimensions the constant function* 1 *is minimal.*

PROOF. Assume that the constant function 1 is not minimal. Then there is a bounded regular function u, positive, and not a constant. We may assume that $u(0) = 1$. Let $\{X_n, n \geqq 1\}$ be the random walk from 0 and form the process $\{u(X_n), n \geqq 1\}$. Since u is regular, the resulting process is a martingale. Since it is positive and bounded, a standard martingale theorem states that $\lim u(X_n) = u_\infty$ exists with probability 1 and $E[u_\infty] = 1$. Consider the event $\{\omega | u_\infty(\omega) < k\}$. This event is an event determined by the sequence of independent random variables $Y_1, Y_2, \cdots$, where $Y_n = X_n - X_{n-1}$, $n \geqq 2$. It is invariant under a finite permutation of the coordinates of the sample sequences for this process. Hence, by a theorem of Hewitt and L. J. Savage [4], this event must have measure 0 or 1. Therefore u_∞ is constant with probability 1 and, since $E[u_\infty] = 1$, this constant must be 1. Assume now that i is in S and $u(i) \neq 1$. The same argument shows that $\Pr\{u_\infty = u(i) | X_n = i\} = 1$. This clearly contradicts the fact that $u_\infty = 1$ with probability 1. ||

3.2 Minimal Functions for a Random Walk. The following theorem characterizes the minimal functions for random walks in N-dimensions.

THEOREM 5. *For a random walk in N-dimensions a function h, positive with* $h(0) = 1$, *is minimal if and only if it satisfies the following two conditions*:

$$\text{(a)} \quad \sum h(i)p(i) = 1 ,$$

$$\text{(b)} \quad h(i + j) = h(i)h(j) , \qquad i, j \text{ in } S .$$

PROOF. Assume that h is regular, positive and $h(0) = 1$. Then (a) is simply the condition that h is regular at 0. We need prove only that (b) is satisfied for the first half of the theorem. Let $u(i) = h(i + k)$ for i in S and k a fixed element of S. Then we shall show that u is regular. In fact, since h is regular at $i + k$,

$$\sum_j p(j)h(j + i + k) = h(i + k) ,$$

or

$$\sum_j p(j)u(i + j) = u(i) .$$

But this is the condition that u is regular at i.

Since h is regular at i, we have

$$h(i) = \sum_{S^+} h(i + j)p(j) ,$$

where S^+ is the set of all j with $h(j) > 0$. The restricted sum is possible because if $h(j) = 0$, then also $h(i + j) = 0$. We can write this last equality in the form

$$h(i) = \sum_{S^+} \frac{h(i + j)h(j)p(j)}{h(j)} .$$

Here $u(i) = h(i + j)/h(j)$ for fixed j is regular with $u(0) = 1$. Thus we have expressed h as a convex combination of regular functions, each with value 1 at 0. If h is minimal, this can happen only if $h(i + j)/h(j)$ does not depend on j. That is, if

$$h(i) = \frac{h(i + j)}{h(j)} ,$$

where $h(j) \neq 0$. But if $h(j) = 0$, then $h(i + j) = 0$ also, so that in any case we have (b).

Assume now, conversely, that h is positive, $h(0) = 1$, and h satisfies (a) and (b). We have then

$$\sum_j p(i, j)\frac{h(j)}{h(i)} = \sum_j p(j - i)h(j - i) = 1 ,$$

if $h(i) \neq 0$. If $h(i) = 0$, then

$$\sum p(i, j)h(j) = \sum p(j)h(j + i) = \sum p(j)h(j)h(i) = 0 .$$

Hence

$$\sum_j p(i, j)h(j) = h(i) ,$$

and h is regular. Consider next the h-process. The transition matrix for this process is

$$q(i, j) = \frac{p(i, j)h(j)}{h(i)} = p(j - i)h(j - i) .$$

Therefore, q leads to the random walk determined by Y with distribution $\Pr\{Y = j\} = p(j)h(j)$ on S^+. By Theorem 4 the constant function 1 is minimal for this process, and hence h is minimal for the original process. ||

The theorem just proved reduces the search for minimal functions to the study of positive solutions of the functional equation $u(i + j) = u(i)u(j)$ on a set S of lattice points in N dimensions closed under addition. Let u be such a solution and let S^+ be the set on which u is strictly positive and S^0 the set on which it is 0. Then S^+ is also closed under addition. We can extend u to the group generated by S^+ by defining $u(j - i) = u(j)/u(i)$. The extended function still satisfies the functional equation. Let $e_1, e_2, \cdots, e_r$ be a basis for this linear group. Then $r \leqq N$ and every element of S can be uniquely written in the form $i = i_1e_1 + i_2e_2 + \cdots + i_re_r$, where $i_1, i_2, \cdots, i_r$ are integers. Then, by the functional equation,

$$u(i) = u(e_1)^{i_1}u(e_2)^{i_2}\cdots u(e_r)^{i_r} .$$

The following results concerning the dimension of S^0 have been proved by G. Lewis and kindly communicated to us. If S and S^+ have the same dimension, then S^0 is empty. If S^+ has dimension less than S, then S^0 must have the same dimension as S. In the case $N = 2$, if S^+ is one-dimensional, then S^0 must lie entirely on one side of S^+.

3.3. The One-dimensional Random Walk. We consider now in detail the one-dimensional case; that is, the case where Y is numerical-valued. Then S is a subset of the integers. The only positive solutions of the functional equation $u(i + j) = u(i)u(j)$ are those of the form $u(i) = s^i$ for $s > 0$. For regularity we must also have $\sum p(j)s^j = 1$. Let $\Phi(s) = E[s^Y]$ be the moment-generating function of Y. Then $h(i) = s_0^i$ is minimal if and only if $\Phi(s_0) = 1$. One such s_0 is $s_0 = 1$. This gives the constant function 1. Since Φ is convex, there can be at most one other s_0 with $\Phi(s_0) = 1$. If Y takes on only positive values, then $\Phi(s) = 1$ has only the solution $s_0 = 1$. An example where two values occur is the following. Let $p(1)=p$ and $p(-1)=q$ where $p + q = 1$, and we suppose $p \neq \frac{1}{2}$, so that the states are transient. Then $\Phi(s) = ps + qs^{-1}$ and $\Phi(q/p) = 1$. Hence $h(i) = (q/p)^i$ is minimal. Thus there are two minimal points in the boundary.

By Theorem 1, if $\Phi(s) = 1$ has only one solution, there is only one positive regular function h with $h(0) = 1$, namely $h(i) = 1$. If there are two solutions $1, s_0$, then every such regular function is of the form $h(i) = a + (1 - a)s_0^i$ for $0 \leqq a \leqq 1$.

We consider next the question of approach to the boundary. Since the constant function is minimal, almost all paths of the random walk approach the boundary point corresponding to the constant function. Theorem 3 then states that $\lim_{n\to\infty} K(i, X_n) = 1$ with probability 1. The functions $g(i, j)$ and $u_j(i)$ depend only on the difference $j - i$. Hence

$$K(i, j) = \frac{g(0, j - i)}{g(0, j)} = \frac{u_{j-i}(0)}{u_j(0)} .$$

Assume now that Y has a mean $\mu > 0$. Then a renewal theorem [5] states that $\lim_{j\to\infty} g(0, j) = 1/\mu$, and hence

$$\lim_{j\to\infty} K(i, j) = 1 .$$

Thus any sequence approaching $+\infty$ tends to the Martin boundary point corresponding to the constant function. Note that in this case Theorem 2 is a consequence of the renewal theorem since almost all paths tend to infinity. In particular, if Y takes on only positive values, the Martin boundary must then have only one point.

Assume now that there is a positive $s_0 \neq 1$ with $\Phi(s_0) = 1$. Let $h(i) = s_0^i$. Then the h-process is a random walk determined by a random variable Y^h. If this has a mean μ_h, it is given by

$$\mu_h = \sum jp(j)\, s_0^j = s_0\Phi'(s_0) .$$

Since Φ is convex, $\mu = \Phi'(1) > 0$ if and only if $s_0 < 1$, and in that case $\mu_h < 0$.

Thus, again by the renewal theorem, if $0 < \mu < \infty$ and if $0 > \mu_h > -\infty$, $\lim_{j\to-\infty} K^h(i, j) = 1$. But since $K^h(i, j) = K(i, j)/h(i)$, we have

$$\lim_{j\to-\infty} K(i, j) = s_0^i .$$

Thus any sequence tending to $+\infty$, $[-\infty]$ approaches the Martin boundary point corresponding to the minimal function 1, $[s_0^i]$.

Assume now that Y takes on only positive values but has infinite mean. Then the renewal theorem states only that $\lim_{j\to\infty} g(0, j) = 0$, and hence we cannot conclude from this that $\lim_{j\to\infty} K(i, j) = 1$. An example showing in fact that this is not necessarily true has been given by Dyson. (The details of this example will appear in a forthcoming book by Chung [6].) For this example, let $n_i = 2^{i^2} - 1$ and $p(n_i) = 2^{-i-1}$, $i = 0, 1, 2, \cdots$. Consider the random walk determined by this distribution. For this walk to reach n_i, in fact on the first step, is relatively much more likely than that it ever reaches $n_i - 1$. In fact, it can be shown that for this random walk,

$$\lim_{i\to\infty} \frac{u_{n_i}(0)}{u_{n_i-1}(0)} = \infty .$$

Since $u(j) \leqq 1$, $K(1, j) = u_{j-1}(0)/u_j(0)$ cannot tend to a limit as j tends to ∞. Thus, for this example we cannot conclude that the Martin boundary has only one point, even though we know that it has only one minimal point. We do have, however, by Theorem 3, that $\lim_{n\to\infty} K(i, X_n) = 1$ with probability 1, even though almost all sample paths of X_n tend to ∞!

There is one class of positive-valued random variables with infinite mean for which we can conclude the stronger limit, and from this that the Martin boundary has only one point. These random variables arise as follows. Let Z be a lattice-valued random variable with mean 0 and $S_n = Z_1 + \cdots + Z_n$. Assume that 0 is a possible value for these partial sums. Then let Y be the time between occurrences of 0. Consider now the random walk $\{X_n, n \geqq 1\}$ determined by Y. This random walk may be considered to represent the times that the partial sums S_j equal 0. Hence, for this random walk $u_j(0) = \Pr\{S_j = 0\}$. By a theorem of Chung and Erdös [7],

$$\lim_{j\to\infty} \frac{\Pr\{S_{j-k} = 0\}}{\Pr\{S_j = 0\}} = 1 .$$

Hence

$$\lim_{j\to\infty} \frac{u_{j-i}(0)}{u_j(0)} = 1 ,$$

and $\lim_{j\to\infty} K(i, j) = 1$. Thus in this case the Martin boundary has only one point, corresponding to the constant function.

3.4 A Two-dimensional Random Walk. In the one-dimensional case we found that the Martin boundary can have at most two minimal points. In two dimensions there will in general be many more minimal points. We shall not treat the general two-dimensional random walk but rather a special case which arises naturally from the one-dimensional case.

Assume that for a one-dimensional random walk we form a new chain by taking as state the pair (n, j), where $n \geqq 0$ and j is an arbitrary integer. We define the transition probabilities by

$$p((m, i), (n, j)) = \begin{cases} p(j-i)\,, & \text{if} \quad n = m+1\,, \\ 0\,, & \text{otherwise}\,. \end{cases}$$

We take S to be the set of states which can be reached from $(0, 0)$. Then, for this chain to be in state (n, j), having started at $(0, 0)$ can be interpreted to mean that for the one-dimensional random walk $X_n = j$. We shall refer to this chain as the space-time random walk in one dimension. This random walk may be considered to be a random walk in two dimensions determined by the random variable Y with distribution

$$\Pr\{Y = (i, j)\} = \begin{cases} p(j), & i = 1\,, \\ 0\,, & \text{otherwise}\,. \end{cases}$$

The state space S_t consists of all (n, j) with $p^{(n)}(j) > 0$. For this chain a state is entered at most once. Hence the mean number of times in a state is simply the probability that the state is ever entered. Thus

$$g((m, i), (n, j)) = p^{(n-m)}(j-i)\,,$$

and

$$K((m, i), (n, j)) = \frac{p^{(n-m)}(j-i)}{p^{(n)}(j)}\,.$$

Let h be minimal for this chain, with $h(0, 0) = 1$. Then h must satisfy the functional equation (b) of Theorem 5 on the space-time state space S_t.

If h does not vanish, we have seen that it must have the form $h(n, i) = s_1^n s_2^j$, where, by (a) of Theorem 5,

$$\sum s_1 s_2^j p(j) = 1\,,$$

so that $\Phi(s_2)$ (defined in Section 3) must be finite and $s_1 = 1/\Phi(s_2)$. Thus

$$h(n, j) = \frac{s^j}{\Phi(s)^n}\,,$$

where we have written s for s_2. Conversely, if $\Phi(s) < \infty$, the function h defined on S_t in this way is minimal regular.

If h is minimal for this chain, with $h(0, 0) = 1$, it may vanish in S_t. We have already noted Lewis's theorem, that the set of positivity of h must lie on some line L, and that the set of zeros is then on one side of L. Excluding now and hereafter the trivial case $p(0) = 1$, we can assume, in view of the special nature of the walk in space-time, that the slope of the line L in (n, j) space is a finite number, say b, and that L contains a point of S_t besides the origin. Suppose first that S_t is on or below L. Then it is clear that Y has b as its maximum possible value and that h is given by

$$h(n, j) = \begin{cases} p(b)^{-n}, & \text{if } j = nb, \\ 0, & \text{otherwise}. \end{cases}$$

Conversely, if Y has b as its maximum value, this function is obviously regular, and is minimal according to Theorem 5. In this case, the interval on which Φ is defined stretches to 0. Thus there are no minimal regular functions which are strictly positive at (0, 0) but which have zeros, unless Y has a maximum or minimum, in which case the maximum, or minimum, corresponds to exactly one such function (aside from constant factors).

Let I be the interval on which Φ is defined. If I contains only the point 1, the positive constant functions are the only minimal functions on S_t. Otherwise, for every s in I there is the minimal non-vanishing function h we have defined above. Let Q be the set of minimal boundary points ξ whose corresponding minimal function $K(\cdot, \xi)$ is strictly positive. We map I in a one-to-one way onto Q by making ξ correspond to s if

$$K((n, j), \xi) = \frac{s^j}{\Phi(s)^n}. \tag{1}$$

The map is continuous, because the value at (n, j) of the minimal function corresponding to s is a continuous function of s. The map is therefore bicontinuous between every compact subinterval of I and the corresponding subset of Q, and is therefore bicontinuous between I and Q. If Y has a finite maximum, b, I is unbounded on the right, and the minimal function with zeros which we have made correspond to b above is the limit, $s \to \infty$, of the minimal function in (1). We now adjoin $+\infty$ to I, and this minimal function to Q, and, similarly, extend I to 0 if Y has a finite minimum. We denote the extension of I by I'; that of Q by Q'. If we use the usual topology of the extended line, we thus have a homeomorphism between I' (which may contain infinite endpoints) and Q', the class of minimal boundary points. A measure of Borel subsets of Q' induces a measure of Borel subsets of I'. By Theorem 1, if h is a positive regular function, with $h(0, 0) = 1$, there is a distribution function on I' such that

$$h(n, i) = \int_{I'} \frac{s^i}{\Phi(s)^n} dF(s),$$

with the obvious conventions if I' contains 0 or ∞, and F is determined uniquely by h. As an example, let X have the Poisson distribution with mean 1. Then $p(j) = e^{-1}/j!$ and $\Phi(s) = e^{s-1}$. If h is positive and regular with $h(0, 0) = 1$, it can be uniquely represented in the form

$$h(n, j) = e^n \int_0^\infty s^j e^{-ns} dF(s).$$

In this case, I' is the interval $[0, \infty)$, and h is strictly positive unless F assigns measure 1 to the point 0, in which case h is given by

$$h(n, j) = \begin{cases} e^n, & \text{if } j = 0, \\ 0, & \text{otherwise}. \end{cases}$$

In the one-dimensional case, we did not exhibit any non-minimal boundary points. In the two-dimensional space-time case, we shall now exhibit an example in which there is a boundary point which is not only non-minimal but does not even correspond to a regular function. Consider any case in which the right-hand endpoint $c\ (\leqq \infty)$ of I is not in I (or not in I' if $c = \infty$). The example just considered, in which Y has a Poisson distribution, is just such an example, with $c = \infty$. Then

$$\lim_{s \to c} \frac{s^j}{\Phi(s)^n} = \begin{cases} 1, & \text{if } n = j = 0, \\ 0, & \text{otherwise}. \end{cases}$$

(We note that $(0, 0)$ is the only point in S_t with $n = 0$.) The limit function h is positive and superregular, but is not regular. In fact, $h = K(\cdot, (0, 0))$. Since the boundary is compact, and since h is a limit of functions corresponding to boundary points, h itself corresponds to a boundary point. That is, $(0, 0)$ is a boundary point.

3.5 Approach to the Boundary for the Space-time Random Walk. In the one-dimensional case, we saw that the renewal theorem enabled us to give sequences which approach the boundary and which did not involve sample path ideas. Again, for the space-time random walk we know from Theorem 2 that almost all sample [illegible]hs tend to the boundary point corresponding to the constant function. For a non-probabilistic approach we need a more special theorem. We shall assume throughout this section that $\Phi(s)$ exists on a non-degenerate interval I and that there is no j with $p(j) = 1$. Let s be an interior point of I and form the h-process for $h(n, j) = s^j/\Phi(s)^n$. This is again a space-time random walk, but determined by the distribution $q(i) = p(i)s^i/\Phi(s)$. The mean of this distribution, which we denote by $\Psi(s)$, is

$$\Psi(s) = \sum_i ip(i)s^i/\Phi(s) = \frac{s\Phi'(s)}{\Phi(s)}.$$

The variance may be computed similarly and found to be $\Psi'(s)/s$. Since the variance is positive, Ψ' is positive and Ψ is increasing. We know from Theorem 2 that almost all sample paths of this h-process approach the point ξ corresponding to s. We know also from the law of large numbers that for almost all sample paths $X_n^h(\omega)/n \to \Psi(s)$. This suggests that, if $\{(n_k, i_k), k \geqq 1\}$ is a sequence such that $i_k/n_k \to \Psi(s)$, then this sequence should approach the boundary point corresponding to s.

THEOREM 6. *Assume that the generating function $\Phi(s)$ exists on a non-degenerate interval I and let s be an interior point of this interval. Let $\{(n_k, i_k), k \geqq 1\}$ be a sequence of points in S such that $i_k/n_k \to \Psi(s)$. Then this sequence approaches the boundary point corresponding to s.*

PROOF. Since Ψ is a monotone function, for k sufficiently large there is a unique s_k in I such that $\Psi(s_k) = i_k/n_k$. We apply the local central limit theorem (Chap. 9 in [8]), to the h_k process with $h_k(j) = s_k^j/\Phi(s_k)$. According to this theorem and the discussion preceding it, if w is the highest common factor of the class of differences between the values of X_j which have strictly positive probability, and $\sigma_k^2 = \Psi'(s_k)$, then

$$\frac{\sqrt{2\pi n}\,\sigma_k p^{(n)}(j)s_k^j}{\Phi(s_k)^n} - w\exp\left(\frac{[j - n\Psi(s_k)]^2 w^2}{2n\sigma_k^2}\right) \to 0\,,$$

where the limit is uniform as j varies and n tends to infinity. From the proof of the local limit theorem it can also be seen that the limit is uniform in k. Putting $n_k - m$ for n and $i_k - i$ for j gives

$$\frac{p^{(n_k-m)}(i_k - i)s_k^{-i}}{p^{(n_k)}(i_k)\Phi(s_k)^{-m}} \to 1\,.$$

This means that (n_k, i_k) tends to the Martin boundary point corresponding to s with approach in the Martin topology. A theorem similar to Theorem 6 has been proved by Kemeny [9].

3.6 A Fatou Theorem. Consider the general N-dimensional case. We have defined the classes of regular and superregular functions. These classes are defined as those whose functions have certain average properties, analogous to those enjoyed by the classes of harmonic and superharmonic functions in N-dimensional space. One of the most important properties of harmonic functions is embodied in the classical Fatou theorem and its generalizations. The most familiar version of this theorem states that, if u is a positive harmonic function in an N-dimensional solid sphere, and if ξ is a boundary point of the sphere, not in an exceptional set of zero $(N-1)$-dimensional "surface area," then u has a limit at ξ along non-tangential paths to ξ. The obvious analog of this theorem in the present context is that a positive regular function has a finite limit along appropriate paths to most Martin boundary points. In order to make this more definite, as well as true, one must define the "appropriate paths," that is, the analogs of the non-tangential paths to the boundary of a sphere, and one must define a suitable measure of sets of Martin boundary points, the analog of the measure of sets of boundary points of a sphere. In studying these problems, one soon sees that the appropriate Martin boundary measure is obtained as follows. One chooses a reference point on the N-dimensional lattice, the one in terms of which the Martin boundary is defined, and considers the probability paths, that is, the random walk paths, from this point. These almost all converge to points of the Martin boundary, as already remarked above, and the distribution of boundary points obtained in this way is the desired measure. Unfortunately, since the positive constant functions are minimal regular, the paths converge with probability 1 to a single Martin boundary point, that corresponding to the function 1. Thus the version of Fatou's theorem we are looking for, if obtained as the analog of the one stated above, is not very interesting, since the exceptional set of Martin boundary points may include all but one point! A recently proved, relativized Fatou theorem [10], that the quotient of two positive harmonic functions in a sphere has a limit along non-tangential paths to almost any boundary point (where "almost any" refers to a measure of boundary sets determined in a natural way by the denominator harmonic function), suggests a more interesting theorem in the present context to be stated and proved below. We shall consider in the present paper only the two-dimensional case, and in fact we shall consider only the space-time regular functions based on space dimensionality 1.

Accordingly, we shall discuss from now on a one-dimensional walk, in which Φ is defined on a non-degenerate open interval I, as well as (perhaps) on one or both endpoints. The domain of Φ will be denoted by I_1. We assume there is no j with $p(j) = 1$. The points of I are then minimal boundary points. Let s be a minimal boundary point, and let $\{(n_k, j_k), k \geqq 1\}$ be a sequence of points of S. By Theorem 6, the sequence converges to s if $j_k/n_k \to \Psi(s)$. We shall say that there is non-tangential convergence if, for some constant c, $|j_k/n_k - \Psi(s)| \leqq c n_k^{-\frac{1}{2}}$, and we shall say that a function defined on space-time has a non-tangential limit at s if it has a limit on every non-tangential sequence converging to s. The limit must then be independent of the sequence. We can now state the desired theorem.

THEOREM 7. *Let u and h be positive regular functions, determined by the monotone functions U and H on the Martin boundary in the manner of Theorem 1. Then u/h has a finite non-tangential limit at H almost every point of I. More specifically, if dU/dH exists and has the finite value a at a point s_0 of I, u/h has the non-tangential limit a at s_0.*

Here dU/dH at s_0 means the limit, which exists and is finite H almost everywhere, of the ratio of the U-measure of a closed interval containing s_0 divided by the H-measure of the interval, as the interval length approaches 0.

The proof will be carried through by means of several lemmas. Throughout the proof, U, H, s_0, and a have the properties described in the statement of the theorem and $\{(n_k, j_k\}, k \geqq 1\}$ is a sequence of space-time points converging non-tangentially to s_0. For sufficiently large k there is then a unique point s_k for which $\Psi(s_k) = j_k/n_k$; we can and shall suppose without loss of generality that s_k is actually defined for all k. Define $f_k(s) = s^{j_k}/\Phi(s)^{n_k}$.

LEMMA 1. *If $\gamma(s) = s^{\Psi(s_0)}/\Phi(s)$, γ has its maximum value at s_0 and is strictly monotone increasing (decreasing) when $s < s_0 (s > s_0)$.*

In fact,

$$\frac{d \log \gamma}{ds} = 0$$

at s_0 and

$$\frac{d^2 \log \gamma}{ds^2} = -\Psi'(s)/s < 0 .$$

LEMMA 2. *Under the hypotheses given above,*

$$\sup \frac{f_k(s_k)}{f_k(s_0)} < \infty .$$

To prove this lemma, define δ_k by

$$\delta_k = j_k/n_k - \Psi(s_0) = \Psi(s_k) - \Psi(s_0) = (s_k - s_0)\Psi'(s_0) + \cdots = O(n_k^{-\frac{1}{2}}) , \tag{2}$$

in the classical O notation, referring to $k \to \infty$. Hence

$$s_k - s_0 = \delta_k/\Psi'(s_0) + O(n_k^{-1}) , \tag{3}$$

so that

$$\begin{aligned} \log f_k(s_k) - \log f_k(s_0) &= n_k\left\{\frac{j_k}{n_k}\log\frac{s_k}{s_0} - \log\frac{\Phi(s_k)}{\Phi(s_0)}\right\} \\ &= n_k\left\{[\Psi(s_0)+\delta_k]\left[\frac{\delta_k}{s_0\Psi'(s_0)} + O(n_k^{-1})\right]\right. \\ &\qquad \left. -\delta_k\frac{\Psi(s_0)}{s_0\Psi'(s_0)} + O(n_k^{-1})\right\} \\ &= \Psi(s_0)O(1) + O(1) = O(1) , \end{aligned} \tag{4}$$

as was to be proved. ||

LEMMA 3. *If there is no open interval containing s_0 in which H is identically constant, then*

$$\lim_{k\to\infty} \frac{h(n_k, j_k)}{f_k(s)} = \infty \tag{5}$$

uniformly for s in any interval containing no point of some neighborhood of s_0.

In fact, if δ is a strictly positive number, and if k is so large that $|s_k - s_0| < \delta$, then applying Lemma 2, we find that

$$\begin{aligned} \frac{h(n_k, j_k)}{f_k(s)} &\geqq \int_{s_0-\delta}^{s_0+\delta} \frac{f_k(t)}{f_k(s)} H(dt) \\ &\geqq \min [f_k(s_0 \pm \delta)][H(s_0+\delta) - H(s_0-\delta)]/f_k(s) . \end{aligned} \tag{6}$$

If s lies in a set excluding the interval $(s_0 - 2\delta, s_0 + 2\delta)$, the right side becomes infinite geometrically fast when $k \to \infty$, in view of Lemma 1.

We now proceed to the proof of the theorem. We can suppose that U and H are right continuous and vanish at s_0. Since a is well-defined, the hypothesis of Lemma 3 is satisfied. By the definition of a,

$$U(s) - aH(s) = H(s)\,\varepsilon(s) , \tag{7}$$

where $\varepsilon(s)$ goes to 0 with $s - s_0$. To simplify the writing, we shall write u_k and h_k for $u(n_k, j_k)$, $h(n_k, j_k)$. Then

$$\frac{u_k}{h_k} = a + \int_{I_1} \frac{f_k(s)}{h_k}\, d[H(s)\,\varepsilon(s)] , \tag{8}$$

and we must prove that the integral is $o(1)$ when $k \to \infty$. According to Lemma 3, the integrand converges uniformly to 0 outside each neighborhood of s_0. Thus, if δ is strictly positive,

$$\frac{u_k}{h_k} = a + \int_{s_0-\delta}^{s_0+\delta} \frac{f_k(s)}{h_k}\, d[H(s)\,\varepsilon(s)] + o(1) \tag{9}$$

$$= \frac{1}{h_k}[f_k(s_0+\delta)\,H(s_0+\delta)\,\varepsilon(s_0+\delta) - f_k(s_0-\delta)H(s_0+\delta)\varepsilon(s_0-\delta)]$$

$$-\left(\frac{1}{h_k}\right)\int_{s_0-\delta}^{s_0+\delta} f_k'(s)H(s)\,\varepsilon(s)\,ds + o(1)\,.$$

The terms outside the integral are $o(1)$ by Lemma 3. We rewrite the integral in the form

$$\frac{1}{h_k}\int_{s_0-\delta}^{s_0+\delta} f_k'(s)\,[H(s) - H(s_k)]\,\varepsilon(s)\,ds + \frac{H(s_k)}{h_k}\int_{s_0-\delta}^{s_0+\delta} f_k'(s)\,\varepsilon(s)\,ds\,. \tag{10}$$

According to Lemma 1, the first integrand has the same sign as $\varepsilon(s)$, and the first integral is therefore at most (in absolute value)

$$\frac{\varepsilon'(\delta)}{h_k}\int_{s_0-\delta}^{s_0+\delta} f_k(s)\,[H(s) - H(s_k)]\,ds\,, \tag{11}$$

where $\varepsilon'(\delta) = \sup_{|s-s_0|\leqq\delta} |\varepsilon(s)|$. If we now integrate by parts in (11), we are reversing the previous integration by parts and again find that the terms outside the integral are $o(1)$, so that the expression in (11) can be written in the form

$$\frac{\varepsilon'(\delta)}{h_k}\int_{s_0-\delta}^{s_0+\delta} f_k(s)dH(s) + o(1) \leqq \varepsilon'(\delta) + o(1). \tag{12}$$

The second term in (10) is numerically at most

$$\varepsilon'(\delta)\,|\,H(s_k)\,|\,[2f_k(s_k) - f_k(s_0-\delta) - f_k(s_0+\delta)]/h_k\,. \tag{13}$$

Applying Lemma 3, we obtain for this

$$2\varepsilon'(\delta)\,|\,H(s_k)\,|\,f_k(s_k)/h_k + o(1)\,, \tag{14}$$

and, on applying Lemma 2, this expression is at most

$$\varepsilon'(\delta)O(1)\,|\,H(s_k)\,|\,f_k(s_0)/h_k\,. \tag{15}$$

Now we see from the integral expression for the function h that

$$h_k \geqq \left|\int_{s_k}^{s_0} f_k(s)dH(s)\right| \geqq f_k(s_0)\,|\,H(s_k)\,|. \tag{16}$$

Hence the expression in (15) is at most $\varepsilon'(\delta)O(1)$, and we have proved that the sum in (10) is at most $\varepsilon'(\delta)O(1)$ in absolute value. Thus, going back to (9), we have proved that

$$|\,u_k/h_k - a\,| \leqq \varepsilon'(\delta)O(1) + o(1)\,, \tag{17}$$

and this inequality implies the truth of the theorem. ||

The description of $\varepsilon(s)$ after (7) is valid only if H is continuous at s_0, and therefore the proof of Theorem 7 is valid only in this case. The other case is, however, still more easily treated directly.

References

[1] Doob, J. L. Discrete potential theory and boundaries. *J. Math. and Mech.*, 1959, **8**, 433-58.

[2] Martin, R. S. Minimal positive harmonic functions. *Trans. Amer. Math. Soc.*, 1941, **49**, 137-72.

[3] Chung, K. L., and Fuchs, W. H. J. On the distribution of values of sums of random variables. *Mem. Amer. Math. Soc., No. 6.* New York: AMS, 1951.

[4] Hewitt, E., and Savage, L. J. Symmetric measures on Cartesian products. *Trans. Amer. Math. Soc.*, 1955, **80**, 470-501.

[5] Chung, K. L., and Pollard, H. An extension of renewal theory. *Proc. Amer. Math. Soc.*, 1952, **3**, 303-9.

[6] Chung, K. L. (forthcoming book).

[7] Chung, K. L., and Erdös, P. Probability limit theorems assuming only the first moment, I. *Mem. Amer. Math. Soc., No. 6.* New York: AMS, 1951.

[8] Gnedenko, B. V., and Kolmogorov, A. N. Limit distributions for sums of independent random variables. (Translated and annotated by K. L. Chung.) Cambridge, Mass.: Addison Wesley, 1954.

[9] Kemeny, J. G. A limit theorem assuming no moments. *Proc. Amer. Math. Soc.*, 1959, **10**, 607-12.

[10] Doob, J. L. A relativized Fatou theorem. *Proc. Nat. Acad. Sci.*, 1959, **45**, 215-22.

17 SOME k-SAMPLE RANK-ORDER TESTS

MEYER DWASS
Northwestern University

1. Introduction

Recently, there has been an increasing interest in the use of rank-order procedures for the problem of comparing several populations. The purpose of this note is to point out several new ideas on this subject.

Suppose $X_1 = (X_{11}, \cdots, X_{1n_1}), \cdots, X_k = (X_{k1}, \cdots, X_{kn_k})$ are k independent vectors of independent observations from k possibly different populations. The components of each vector are identically distributed, and all random variables under discussion are assumed to have continuous c.d.f.'s. The null hypothesis is

H: The components of all k vectors follow the same distribution law.

For $k = 2$, there is a large variety of rank-order tests for H, the hypothesis that both populations are the same; the ensuing discussion revolves about tests in the case $k \geqq 2$, which are based on two-sample tests. Let $T(X_i; X_j)$ be a rank-order test statistic for testing the hypothesis that the ith and jth populations are the same. For example, T might be (a) the Wilcoxon statistic, or (b) the Hoeffding C_1 statistic, or (c) the Kolmogorov-Smirnov statistic.

There are, of course, many ways for combining two-sample tests to give a k-sample test. Two procedures are given here, special cases of which have been mentioned in the literature before. The reasons for selecting these two particular procedures are for ease in handling the null distribution or for ease in making certain confidence-interval estimates.

2. Procedure I

Let $R_j = (R_{j1}, \cdots, R_{jn_j}), \quad j = 2, \cdots, k,$ be random vectors defined as follows:

$$R_{ji} = \text{number of } X_{rs} \leqq X_{ji}, \qquad r = 1, \cdots, j; \; s = 1, \cdots, n_r .$$

In other words, R_j is the vector of the rank-order positions of the elements of X_j in the combined sample of the $n_1 + \cdots + n_j$ observations $X_1, \cdots, X_j$. The first procedure is based on the following theorem, which is of independent interest.

THEOREM 1. *The $k-1$ random vectors $R_2, R_3, \cdots, R_k$, defined above, are statistically independent, if the null hypothesis, H, holds.*

PROOF. The proof of the theorem is completely elementary, and the details are omitted.

The following notational convention is adopted: $T(X_1, \cdots, X_{j-1}; X_j)$ is a two-sample test statistic, as described earlier, where the first sample consists of the pooled sample of the $n_1 + \cdots + n_{j-1}$ observations which are the components of $X_1, \cdots, X_{j-1}$, and the second sample consists of the n_j observations which are the components of X_j. In terms of this notation, the following corollary is immediate.

COROLLARY. *Assume that the two-sample statistic T is invariant under permutations within each of the two samples. (In other words, $T(X_1, \cdots, X_{j-1}; X_j)$ is a function of R_j only.) Then the $k-1$ random variables $T(X_1; X_2)$, $T(X_1, X_2; X_3), \cdots, T(X_1, \cdots, X_{k-1}; X_k)$ are statistically independent when H holds.*

Remarks. (a) The theorem and the corollary are not true, in general, if H does not hold.

(b) The following special case of this theorem had been noticed earlier. Let $T(X_i; X_j) = \sup_y |F_i(y) - F_j(y)|$ where $F_i(y)$ is the empirical c.d.f. of the observations making up the X_i vector. Then it was proved in [1] that

$$(1) \qquad T(X_1; X_2) = \sup_y \left| \frac{n_1}{n_1 + n_2} F_1(y) - \frac{n_2}{n_1 + n_2} F_2(y) \right| \quad \text{and} \quad T(X_1, X_2; X_3)$$

(properly normed to have a limiting distribution) are *asymptotically independent* as the sample sizes approach infinity. Some similar results are also proved when k is greater than 3. The above corollary shows that this result is not restricted to the Kolmogorov-Smirnov statistic, nor is it asymptotic in nature. For $k = 3$, the independence of the two statistics (1) for finite sample size, but with $n_1 = n_2 = n_3$, was pointed out by David [2].

(c) As a second example, consider the Wilcoxon statistic, $U_{ij} = T(X_i; X_j) =$ (the number of times components of X_j are greater than or equal to components of X_i). It is easy to verify that, in this case, $T(X_1, \cdots, X_{j-1}; X_j) = U_{1j} + U_{2j} + \cdots + U_{j-1,j}$. Hence, the following $k-1$ linear combinations of Wilcoxon statistics,

$$\begin{array}{l} U_{12} \\ U_{13} + U_{23} \\ \vdots \\ U_{1k} + U_{2k} + \cdots + U_{k-1,k}, \end{array}$$

are statistically independent, under the null hypothesis H.

(d) The importance of the independence of the $k-1$ random variables of the corollary is that a test of H can be based upon them, and that only the distribution of the two-sample statistics is required for the determination of the level of significance. For example, consider the Wilcoxon statistics of example (c) above. A possible test of H is the following. Accept H only if

$$V = \max\left\{C_2\left|U_{12}-\frac{n_1n_2}{2}\right|, \cdots, C_k\left|U_{1k}-\frac{n_1n_k}{2}+\cdots+U_{k-1,k}-\frac{n_{k-1}n_k}{2}\right|\right\}<\lambda, \tag{2}$$

where $C_2, \cdots, C_k$ and λ are appropriately chosen constants.

A general analog of the test described above is that H be accepted only if

$$\max\{C_2T(X_1; X_2), \cdots, C_kT(X_1, \cdots, X_{k-1}; X_k)\} < \lambda .$$

A heuristic rationale of such a test is the following. Each of the $T(X_1, \cdots, X_{j-1}; X_j)$ statistics plays the role of the estimate of a "contrast" among the k populations. There are only $k-1$ "independent" contrasts and this test is based on one convenient choice of estimates of such a set of contrasts.

3. Procedure II

This procedure is similar to the first one, as it is based on the maximum of a set of "contrast" estimates. Specifically, the testing procedure is to accept H only if

$$W = \max_{i,j \text{ in } A} c_{ij}\, T(X_i; X_j) < \lambda , \tag{3}$$

where A is a set of pairs of indices from $(1, \cdots, k)$ and c_{ij} are suitably chosen constants. The set A may consist of all $\binom{k}{2}$ pairs of indices. This kind of test has been discussed by David [2], and Kiefer [3], when T is a Kolmogorov-Smirnov type of statistic. The distribution of W presents considerably more formidable difficulties than does that of V, since the maximum in (3) is not a maximum of independent random variables. A general inequality which may be useful here is the easily verified one,

$$P(W < \lambda) \geqq 1 - \sum_{(i,j) \text{ in } A} P[c_{ij}\, T(X_i; X_j) \geqq \lambda] . \tag{4}$$

An example of a case where the distribution problem can be dealt with, at least asymptotically, is that with the Wilcoxon statistic. Specifically, suppose that $n_1 = n_2 = \cdots = n_k = n$ and that $T(X_i; X_j) = |U_{ij} - n^2/2|$. Then we have the following theorem.

THEOREM 2. *Under the null hypothesis, H,*

$$\lim_{n\to\infty} P\left\{\sqrt{\frac{12}{n^2(2n+1)}} \max_{i,j=1,\dots,k} \left|U_{ij} - \frac{n^2}{2}\right| \leq t\right\} = G(t) ,$$

where G is the c.d.f. of the range of k independent $N(0, 1)$ random variables.

PROOF. Let $Y_1, \cdots, Y_k$ be k independent $N(0, 1)$ random variables, and let

Z be the random vector with $\binom{k}{2}$ components $Y_i - Y_j$; $i, j = 1, \cdots, k$; $(i < j)$. Let Z_n be the random vector with $\binom{k}{2}$ components

$$\sqrt{\frac{12}{n^2(2n+1)}}(U_{ij} - n^2/2) .$$

The proof follows from the fact that the limiting distribution of Z_n as n approaches infinity is that of Z. This last fact can be proved by a slight adaptation of the methods of [4], among other ways, and the details are omitted.

By similar techniques, it is not hard to show that if the distribution of the components of the vector X_i are $N(\theta_i, 1)$, with the θ_i "close" to each other, then the Pitman efficiency of the above-described test, compared with the one based on the range of sample averages, $\max_{ij} |\bar{X}_i - \bar{X}_j|$ is quite high.

4. Confidence-interval Procedures

Here the very special assumption is made that the k underlying distributions are all the same except for location parameters. That is, there are constants, $\theta_1, \cdots, \theta_k$, such that the $n_1 \cdots + n_k$ random variables $(X_{11} - \theta_1, \cdots, X_{1n_1} - \theta_1), \cdots, (X_{k1} - \theta_k, \cdots, X_{kn_k} - \theta_k)$ are identically distributed. The n_i-tuples with subtracted θ_i will be denoted simply as $X_i - \theta$.

Let $S(X_1, \cdots, X_k)$ be a test statistic such that H is accepted only if S lies in the set of real numbers $\mathcal{a}$. Let $1 - \alpha = P(S \in \mathcal{a} | H)$. It is easy to show, and well known, that a confidence-set statement for $(\theta_1, \cdots, \theta_k)$ can be made as follows: $\theta_1, \cdots, \theta_k$ lies in the set

$$C(X_1, \cdots, X_k) = \{\Delta_1, \cdots, \Delta_k : S(X_1 - \Delta_1, \cdots, X_k - \Delta_k) \in \mathcal{a}\} . \tag{5}$$

Then the probability that C covers the true point $\theta_1, \cdots, \theta_k$ is $1 - \alpha$.

Though Procedure I has certain null-distribution advantages, Procedure II seems easier to handle from the point of view of constructing confidence intervals, at least with some statistics. Suppose then that $T(X_i; X_j) = U_{ij}$ is the two-sample Wilcoxon statistic described before, and suppose also, for simplicity, that $n_1 = n_2 = \cdots = n_k = n$. Let λ be a number such that $P\{\max_{i,j} |U_{ij} - n^2/2| < \lambda | H\} \geqq 1 - \alpha$. Such a number λ can be determined from (4) or approximately for large n from Theorem 2. Then the following confidence statements are simultaneously true for all $i, j = 1, \cdots, k$ with probability greater than or equal to $1 - \alpha$: $\theta_i - \theta_j$ is covered by the set of Δ such that

$$\frac{n^2}{2} - \lambda < T(X_i - \Delta; X_j) < \frac{n^2}{2} + \lambda . \tag{6}$$

The set of confidence intervals described in (6) can be determined monographically as follows. Plot the $n_i X_i$ values on one graph and plot the $n_j X_j$ values on another graph of the same scale. Place the two graphs one above the other so that all the X_i values are to the right of all the X_j values.

Slide the X_i graph to the left until $[(n^2/2) - \lambda]$ interchanges between X_i and X_j values have taken place and just up to the point of the $[\{(n^2/2) - \lambda\} + 1]$th interchange. Read the distance δ_1 between the origins. Repeat the same procedure exchanging the words "left" and "right" in the above description, obtaining the number δ_2. Then the set of Δ's of (6) is the interval (δ_2, δ_1).

The same kind of sliding procedure to obtain confidence intervals can easily be done with some other statistics, for example with the Kolmogorov-Smirnov statistic. In this case, however, the distribution of the random variable W of (3) is not known explicitly, not even asymptotically for large sample size.

Finally it should be pointed out that there is nothing special about the case of location parameters. Scale parameter differences and differences described in terms of more general groups of transformations could be handled in a similar way.

References

[1] Fisz, M. A limit theorem for empirical distribution functions. *Bull. Acad. Polon. Sci.*, Cl. III, 1957, **5**, 695-98.

[2] David, H. T. A three-sample Kolmogorov-Smirnov test. *Ann. Math. Stat.*, 1958, **29**, 842-51.

[3] Kiefer, J. Distance tests with good power for the nonparametric k-sample problem (abst.). *Ann. Math. Stat.*, 1955, **26**, 775.

[4] Dwass, M. On the asymptotic normality of certain rank order statistics. *Ann. Math. Stat.*, 1953, **24**, 303-6.

18 CHARACTERIZATION OF SOME LOCATION AND SCALE PARAMETER FAMILIES OF DISTRIBUTIONS

S. G. GHURYE

Northwestern University

1. Introduction

This paper originated from a problem discussed by Zinger [1], as follows: Let $X_1, \cdots, X_n$ be independent and identically distributed random variables; let $t(X) = (1/n) \sum X_i$, $s(X) = [\sum X_i^2 - nt^2(X)]^{\frac{1}{2}}$ and $Y_i = [X_i - t(X)]/s(X)$. If the X-distribution is normal, then the Y_i are uniformly distributed on the $(n-2)$-dimensional sphere $\{\sum y_i = 0,\ \sum y_i^2 = 1\}$. Conversely, is the normal distribution characterized by this property? Zinger proves this to be so if $n \geqq 6$.[1]

It is natural to question whether such a characterization holds for other distributions, with a suitable choice of s and t. In the above problem, the statistics s and t have the property that $y_i = ax_i + b$, $a > 0$, implies $s(y) = as(x)$ and $t(y) = at(x) + b$. Consequently, the distribution of Y_i is invariant under change of location and scale in the X-distribution. This suggests more generally the following characterization of a location and scale parameter family of c.d.f.'s $F[(x-\theta)/\sigma]$: for each $n \geqq 3$ we have a pair of functions $[s_n(x), t_n(x)]$ of n variables and an n-dimensional distribution P_n such that

(i) $y_i = ax_i + b$, $a > 0$, implies $s_n(y) = as_n(x)$ and $t_n(y) = at_n(x) + b$,

(ii) P_n is degenerate and allocates probability 1 to the surface $\{s_n(x) = 1, t_n(x) = 0\}$,

and such that if $X_1, \cdots, X_n$ are independent, identically distributed random variables, the joint distribution of $Y_i = [X_i - t_n(X)]/s_n(X)$, $i = 1, 2, \cdots, n$, is P_n if and only if the common c.d.f. of the X_i is $F[x - \theta/\sigma]$ for some $\sigma > 0$ and θ.

Zinger's result is one special case of this type, except that it has been proved only for $n \geqq 6$. Other such special results are easily obtained. It would seem natural to try as a candidate for (s, t) in each case the appropriate sufficient statistic. Since the parameter and the statistic each have

[1] Although Zinger states his theorem in this generality, his proof seems to assume the existence of a continuous probability density. Also, in the implicit assumption that the Y_i are determinate with probability 1, there is inherent an assumption that the distribution is continuous.

two coordinates, one expects to have to deal with complete sufficient statistics. This consideration helps to discover other distributions characterized in this way, but the hope of deriving a general theory has not materialized. The foregoing discussion also indicates the possibility of characterizing one-parameter families $F(x/\sigma)$, $F(x-\theta)$ in a similar way.

Another possible line of extension is one to multidimensional distributions. I have been able to include in this paper characterizations of the multivariate normal and Wishart distributions, which have played prominent roles in some of Professor Hotelling's work. In fact, just as Zinger's result has a close association with Student's t, so Theorem 4 below is closely associated with Hotelling's T^2.

The results obtained are stated in the next section, and proofs are given in Section 3. Part of the latter section deals with a functional equation which might be of some independent interest.

2. Statement of Results

Results concerning some location and scale parameter families are stated as Theorem 1 and those concerning only scale parameter families as Theorem 2. Theorems 3 and 4 deal with the Wishart and multivariate normal distributions. In proving these results, we have restricted the distributions to be of the continuous type. It would be nice to be able to prove them without such restrictions.

Definition. A distribution is said to be of the continuous type if it is absolutely continuous and possesses a probability density whose points of discontinuity have no finite limit points.

This agrees with Cramér's definition in the one-dimensional case ([2], p. 169), but is much more restrictive than his for more then one dimension.

Since the set of discontinuities of the density f is assumed to be denumerable, we may redefine f on this set without affecting the distribution. Hence if, as $x \to a$, $f(x)$ has a unique finite limit different from $f(a)$, we may redefine f at a so as to make it continuous. Consequently, we shall assume that, at each point of discontinuity, either the density does not have a unique limit or the limit is infinite.

THEOREM 1. *Let $X_1, \cdots, X_n$ be independent random variables with a common distribution of the continuous type, and let $Y_i = (X_i - t)/s$, where s and t are defined in each case below. Let the Y-vector be uniformly distributed on the set $\{s(y_1, \cdots, y_n) = 1,\ t(y_1, \cdots, y_n) = 0\}$. Then*

(a) *if* $n \geqq 6$, $t(x) = (1/n)\sum x_i$, $s(x) = [\sum x_i^2 - nt^2(x)]^{\frac{1}{2}}$, *the X-distribution is normal (Zinger* [1]);

(b) *if* $n \geqq 5$, $t(x) = \min(x_1, \cdots, x_n)$, $s(x) = \sum (x_i - t)$, *there exist* $\sigma > 0$ *and* θ *such that*

$$\Pr\{X_i > x\} = \begin{cases} 1, & \text{if } x \leqq \theta, \\ \exp[-(x-\theta)/\sigma], & \text{if } x > \theta; \end{cases}$$

(c) *if* $n \geqq 4$, $t(x) = \min(x_1, \cdots, x_n)$, *and* $s(x) = \max(x_1, \cdots, x_n) - t$, *there exist* $\sigma > 0$ *and* θ *such that* X_i *is distributed uniformly on* $(\theta, \theta + \sigma)$ *with density* $1/\sigma$.

COROLLARY 1.1. *With the preceding notation, let* $Z_i = X_i - t(X)$ *and suppose the Z-vector has density* $g(z)$ *on the surface* $\{t(Z) = 0\}$. *Then*

(a) *if* $n \geqq 6$ *and* $g(z)$ *depends only on* $\sum z_i^2$, *the X-distribution is normal;*

(b) *if* $n \geqq 5$ *and* $g(z)$ *depends only on* $\sum z_i$, *the X-distribution is the same as in case (b) of the Theorem;*

(c) *if* $n \geqq 4$ *and* $g(z)$ *depends only on* $\max(z_1, \cdots, z_n)$, *the* X_i *have a uniform distribution.*

THEOREM 2. *Let* $X_1, \cdots, X_n$ *be independent random variables with a common distribution of the continuous type and let* $Y_i = X_i/s$, *where s is as defined in each case below. Let the Y-vector be uniformly distributed on the set S as defined in each case. Then*

(a) *if* $n \geqq 4$, $s(x) = [\sum x_i^2]^{\frac{1}{2}}$ *and* $S = \{s(y) = 1\}$, X_i *is normally distributed with mean* 0;

(b) *if* $n \geqq 4$, $s(x) = \sum x_i$, *and* $S(y) = \{s(y) = 1, y_i \geqq 0, i = 1, \cdots, n\}$, *there exists a* $\sigma > 0$ *such that the c.d.f. of* X_i *is either* $F(x/\sigma)$ *or* $1 - F(-x/\sigma)$, *where* $F(x) = 0$ *if* $x \leqq 0$, *and* $1 - e^{-x}$ *if* $x > 0$;

(c) *if* $n \geqq 3$, $s(x) = \max(x_1, \cdots, x_n)$, *and* $S = \{s(y) = 1, y_i \geqq 0, i = 1, \cdots, n\}$, *there exists a* $\sigma > 0$ *such that* X_i *is distributed uniformly on* $(0, \sigma)$ *with density* $1/\sigma$.

COROLLARY 2.1. *If* n, s, *and S are as in* 2(b), *and there exists an* $\alpha > -1$ *such that the Y-vector has a density proportional to* $(y_1 y_2 \cdots y_n)^\alpha$ *on S and zero elsewhere, then there exists a* $\sigma > 0$ *such that the c.d.f. of* X_i *is either* $F(x/\sigma)$ *or* $1 - F(-x/\sigma)$, *where*

$$F(x) = \begin{cases} 0, & \text{if } x \leqq 0, \\ [\Gamma(\alpha + 1)]^{-1} \int_0^x u^\alpha \exp(-u)\, du, & \text{if } x > 0. \end{cases} \tag{1}$$

COROLLARY 2.2. *Let* $f(x)$ *be a probability density function of the continuous type on the positive real axis with the property that there exist functions* $g(x)$, $h(x)$ *such that for some* $n \geqq 4$,

$$\int_0^\infty \prod_1^n f(sy_i) s^{n-1}\, ds = g(\sum y_i) \prod_1^n h(y_i) \tag{2}$$

for all $y_1, \cdots, y_n \geqq 0$. *Then f is the density of a gamma distribution of the type in Equation* (1).

THEOREM 3. *Let* $X_1, \cdots, X_n, n \geqq 4$, *be independent* $p \times p$ *symmetric random matrices which have a common distribution of the continuous type and are, with probability* 1, *positive definite. Let* $S = \sum X_i$, *and* $S^{\frac{1}{2}}$ *be a positive definite square root of S (which may be defined unambiguously as follows: if* $\lambda_1 \geqq \lambda_2 \geqq \cdots \geqq \lambda_p \geqq 0$ *are square roots of the characteristic roots of S, and* Ω *is an orthogonal matrix such that* $S = \Omega D_\lambda^2 \Omega'$, *then let* $S^{\frac{1}{2}} = \Omega D_\lambda \Omega'$, *where D denotes a diagonal matrix). Let* $Y_i = S^{-\frac{1}{2}} X_i S^{-\frac{1}{2}}$, *so that the* Y_i *are positive definite with probability* 1 *and satisfy* $\sum Y_i = I$. *If the joint distribution of the* Y_i

on this surface is absolutely continuous, with a density proportional to some power of $\Pi_1^n (\det Y_i)$, *then the* X_i *have a Wishart distribution.*

THEOREM 4. *Let* $X_1, \cdots, X_n, n \geq \max(p+1, 6)$, *be independent p-dimensional random row-vectors with a common distribution of the continuous type. Let* $t = (1/n) \sum X_i$ *and T be the unique triangular matrix, with non-negative diagonal elements and zeros below the diagonal, such that* $T'T = \sum (X_i - t)'(X_i - t)$. *Let* $\mathscr{P}$ *denote the joint distribution of* $Y_i = (X_i - t)T^{-1}$, $i = 1, 2, \cdots, n$, *when the* X_i *are normally distributed with zero means and covariance matrix I.*

Then $Y_1, \cdots, Y_n$ *have the distribution* $\mathscr{P}$ *if and only if the common distribution of the* X_i *is normal with some mean vector* μ *and some covariance matrix* Σ.

Remark. These results have been proved for different minimum values of n, which always exceed the minimum possible values for which the results would make sense. It is an open question how far the indicated bounds on n can be reduced by improving the method of proof. In Theorem 2, statements (a) and (b) are certainly not true for $n = 2$, since the determining distribution is essentially that of X_1/X_2, and we know that the normal and gamma distributions are not characterized by the distribution of the ratio, since the ratio of the reciprocals also has the same distribution. In fact, it is known [3] that the distribution of neither X_i nor $1/X_i$ need be normal in order for X_1/X_2 to have the Cauchy distribution. Thus it seems that, if some general theory in terms of sufficient statistics is possible, it will still be subject to restrictions on n.

3. Proofs

Many of the proofs are similar to each other; in such cases, the first proof is presented in detail, and the subsequent ones are sketched in outline. To start with, we discuss a functional equation which has been solved by Zinger under the assumption that the function is finitely integrable. A more general result is given in the following lemma and is extended in Lemma 2 to functions of several variables.

LEMMA 1. *Let* $f_1(x), f_2(x), f_3(x)$ *be measurable, complex-valued functions of real x, such that*

(i) $f_1(x)f_2(y)f_3(z)$ *is determinate, i.e., not of the form* $0 \cdot \infty$, *for all* (x, y, z) *and depends only on* $x + y + z$ *and* $x^2 + y^2 + z^2$,

(ii) *there exist real numbers* a_1, a_2, a_3, *not all equal, for which* $f_1(a_1)f_2(a_2)f_3(a_3) \neq 0$, *and similarly,* b_1, b_2, b_3, *for which* $|f_1(b_1)f_2(b_2)f_3(b_3)| < \infty$.

Then there exist finite complex numbers $\alpha, \beta, \gamma_1, \gamma_2, \gamma_3$, *with* $\gamma_1\gamma_2\gamma_3 \neq 0$, *such that, for all x,*

$$f_1(x)/\gamma_1 = f_2(x)/\gamma_2 = f_3(x)/\gamma_3 = \exp(\alpha x^2 + \beta x) . \tag{3}$$

PROOF. From property (i), it follows that if one of the functions has a zero, the other two are everywhere of finite modulus, since otherwise the product would be indeterminate of the form $0 \cdot \infty$ for some points. In fact, since $f_1(x)f_2(a_1)f_3(a_3) = f_1(a_1)f_2(x)f_3(a_3)$, we see immediately that $f_1(x) = 0$ implies $f_2(x) = 0$ and in the same way, also $f_3(x) = 0$, and hence $|f_i(y)| < \infty$, $i = 1, 2, 3$, for all y.

We shall now show that the functions are everywhere nonvanishing and of finite modulus. For this, suppose that they have a zero at $x = c$. Then they are everywhere of finite modulus. Also, if (i_1, i_2, i_3) be any permutation of $(1, 2, 3)$, we have $f_1(a_{i_1})f_2(a_{i_2})f_3(a_{i_3}) = f_1(a_1)f_2(a_2)f_3(a_3)$, so that $f_i(a_j) \neq 0$ for $i, j = 1, 2, 3$. Similarly, we get

$$f_1(x)\,f_2(a_3)\,f_3(a_3) = f_1(a_3)\,f_2(x)\,f_3(a_3) = f_1(a_3)\,f_2(a_3)\,f_3(x) = f(x)\,,\ \text{say.} \tag{4}$$

We know that $f(a_1)f(a_2)f(a_3) \neq 0$, $f(x)f(y)f(z)$ depends only on $x + y + z$ and $x^2 + y^2 + z^2$, $f(c) = 0$, and $|f(x)| < \infty$ for all x.

Let a, b denote a pair of unequal members of the triplet (a_1, a_2, a_3), and suppose for convenience that $a < b < c$ (the other cases are similar). Then for any x, if we can find y, z such that $2x + a = c + y + z$ and $2x^2 + a^2 = c^2 + y^2 + z^2$, we have $f(x)f(x)f(a) = f(c)f(y)f(z) = 0$; and since $f(a) \neq 0$, this means $f(x) = 0$. Now, such y and z exist if and only if $2(y^2 + z^2) = 2(2x^2 + a^2 - c^2) \geqq (y + z)^2 = (2x + a - c)^2$. Hence, $f(x) = 0$ for all $x \geqq c - (c - a)/4$. We can repeat the argument, using $c - (c - a)/4$ instead of c. Proceeding this way, we see that $f(x) = 0$ for all $x > a$, which contradicts the assumption that $f(b) \neq 0$. Thus the f_i have no zeros. From the fact that the functions $1/f_i(x)$ also have the properties assumed of the f_i, it follows that $|f_i(x)| < \infty$ everywhere.

The first two equalities of (3) are thus established by (4), and it remains to show that f has the postulated exponential form. We are given that f is everywhere of positive, but finite, modulus and that there exists a function $g(u, v)$, of $u > 0$ and v, such that

$$f(x)\,f(y)\,f(z) = g(x^2 + y^2 + z^2, x + y + z)\,. \tag{5}$$

From (5), we immediately have

$$f(0)\,f(-y)\,f(x + y) = g(x^2 + 2xy + 2y^2, x)\,,$$
$$f(-x)\,f(-y)\,f(x + y) = g(2x^2 + 2xy + 2y^2, 0)\,,$$
$$f^2(0)\,f(-x) = g(x^2, -x)\,,$$

and hence,

$$f^3(0)\,g(2x^2 + 2xy + 2y^2, 0) = g(x^2, -x)\,g(x^2 + 2xy + 2y^2, x)\,. \tag{6}$$

Now, to any $u \geqq 0$ and x, there corresponds a y such that $2y^2 + 2xy = u$; consequently, (6) implies that for all $u \geqq 0$ and x,

$$g(x^2, -x)\,g(x^2 + u, x) = f^3(0)\,g(2x^2 + u, 0)\,. \tag{7}$$

On the other hand, from (5) we also get

$$\begin{aligned} f^3(0)\,g(x^2 + 2y^2, x) &= f^3(0)\,f(x)\,f(y)\,f(-y) \\ &= f^2(0)\,f(x)\,f(0)\,f(y)\,f(-y) = g(x^2, x)\,g(2y^2, 0)\,, \end{aligned}$$

that is to say, for all $u \geqq 0$,

$$f^3(0)\, g(x^2 + u, x) = g(x^2, x)\, g(u, 0)\ . \tag{8}$$

Combining (7) and (8), we see that

$$\begin{aligned} f^6(0)\, g(2x^2 + u, 0) &= f^3(0)\, g(x^2 + u, x)\, g(x^2, -x) \\ &= g(u, 0)\, g(x^2, x)\, g(x^2, -x) = g(u, 0)\, g(2x^2, 0)\, f^3(0)\ , \end{aligned}$$

so that for all $y, z \geqq 0$,

$$g(y, 0)\, g(z, 0) = f^3(0)\, g(y + z, 0)\ . \tag{9}$$

This implies the existence of an α such that, for all $u \geqq 0$,

$$g(u, 0) = f^3(0) \exp(\alpha u)\ . \tag{10}$$

If a proof of this assertion were called for, we would refer to the elegant treatment of Cauchy's functional equation by Kac [4].

Finally, using (5) and (10) we find that

$$\begin{aligned} f^6(0) f(x) f(y) \exp[\alpha(2x^2 + 2xy + 2y^2)] \\ &= f^3(0) f(x) f(y) g(2x^2 + 2xy + 2y^2, 0) \\ &= f^3(0) f(x) f(y) f(-x) f(-y) f(x + y) \\ &= f(0) f(x) f(-x) f(0) f(y) f(-y) f(0) f(x + y) \\ &= g(2x^2, 0)\, g(2y^2, 0) f(0) f(x + y) \\ &= f^7(0) \exp[\alpha(2x^2 + 2y^2)] f(x + y)\ . \end{aligned} \tag{11}$$

In other words, $h(x) = f(x) \exp(-\alpha x^2)/f(0)$ satisfies the equation $h(x + y) = h(x) h(y)$; and hence, there exists a β such that $f(x) = f(0) \exp(\alpha x^2 + \beta x)$. ||

Lemma 2. *Let $f_1(\boldsymbol{x}), f_2(\boldsymbol{x}), f_3(\boldsymbol{x})$ be measurable, complex-valued functions of the p-dimensional, real row-vector $\boldsymbol{x} = (x_1, \cdots, x_p)$, $p > 1$, (also referred to as the point $\boldsymbol{x}$), such that*

(i) *$f_1(\boldsymbol{x}) f_2(\boldsymbol{y}) f_3(\boldsymbol{z})$ is determinate, i.e., not of the form $0 \cdot \infty$, for all $(\boldsymbol{x}, \boldsymbol{y}, \boldsymbol{z})$ and depends only on $\boldsymbol{x} + \boldsymbol{y} + \boldsymbol{z}$ and $\boldsymbol{x}'\boldsymbol{x} + \boldsymbol{y}'\boldsymbol{y} + \boldsymbol{z}'\boldsymbol{z}$;*

(ii) *there exist points $\boldsymbol{a}_0, \boldsymbol{a}_1, \cdots, \boldsymbol{a}_p$, not all lying in the same $(p-1)$-dimensional hyperplane, for which $f_1(\boldsymbol{a}_0) f_2(\boldsymbol{a}_1) f_3(\boldsymbol{a}_2) \neq 0$ and $|f_1(\boldsymbol{a}_i)| + |f_2(\boldsymbol{a}_i)| + |f_3(\boldsymbol{a}_i)| \neq 0$, $i = 0, 1, \cdots, p$ (i.e., at each $\boldsymbol{a}_i$ some $f_j \neq 0$, and each $f_j \neq 0$ at some $\boldsymbol{a}_i$), and similarly, points $\boldsymbol{b}_0, \boldsymbol{b}_1, \cdots, \boldsymbol{b}_p$, not all lying in one $(p-1)$-dimensional hyperplane, such that at each $\boldsymbol{b}_i$ some $|f_j| < \infty$ and each $|f_j| < \infty$ for some $\boldsymbol{b}_i$.*

Then there exist finite complex numbers $\gamma_i \neq 0$, $i = 1, 2, 3$, a linear form $l(\boldsymbol{x})$ and a quadratic form $q(\boldsymbol{x})$ such that, for all $\boldsymbol{x}$,

$$f_1(\boldsymbol{x})/\gamma_1 = f_2(\boldsymbol{x})/\gamma_2 = f_3(\boldsymbol{x})/\gamma_3 = \exp[l(\boldsymbol{x}) + q(\boldsymbol{x})]\ . \tag{12}$$

Proof. As before, we note that, if one of the functions is zero at a point, the others are too, and all three are everywhere of finite modulus. Hence, $f_i(\boldsymbol{a}_j) \neq 0$, $i = 1, 2, 3$, $j = 0, 1, \cdots, p$.

To show that the functions are nowhere zero, let us suppose the contrary. Then we know that $f(\boldsymbol{x})$, defined as in (4), is everywhere of finite modulus and satisfies the assumptions of the corollary with $f_1 = f_2 = f_3 = f$. Consequently, $f(\boldsymbol{a}) \neq 0 \neq f(\boldsymbol{b})$ implies that $f[\boldsymbol{a} + (\boldsymbol{b} - \boldsymbol{a})t]$ is a function of real t, which satisfies the assumptions of the previous lemma. Hence, if f is non-zero at two points, it is non-zero everywhere on the line joining them. Since $f \neq 0$ at $\boldsymbol{a}_0, \boldsymbol{a}_1, \cdots, \boldsymbol{a}_p$, and these points do not all belong to the same $(p-1)$-dimensional hyperplane, it follows that f is everywhere non-zero. Similarly, we see that it is everywhere of finite modulus.

It remains to show that f has the postulated exponential form. For this, we can use induction on p. Since the result is true for $p = 1$, let us assume that it holds for $p = 1, 2, \cdots, r-1$, and that $f(x_1, x_2, \cdots, x_r)$ satisfies the assumptions of the corollary with $p = r$. Then for fixed x_1, $f(x_1, x_2, \cdots, x_r)$ is a function of $x_2, \cdots, x_r$, which satisfies the assumptions and is, by the induction hypothesis, of the form of (12); that is to say, there exist functions $\alpha_{ij}(x_1)$, $\beta_i(x_1)$, $\gamma(x_1)$ such that

$$f(x_1, \cdots, x_r) = \gamma(x_1) \exp\left[\sum_{2 \leq i \leq j \leq r} \alpha_{ij}(x_1) x_i x_j + \sum_{2 \leq i \leq r} \beta_i(x_1) x_i\right]. \tag{13}$$

On the other hand, for fixed $x_2, \cdots, x_r$, f is a function of x_1 satisfying the assumptions of Lemma 2. It follows that $\gamma(x) = \gamma \exp(\alpha x^2 + \beta x)$, $\alpha_{ij}(x) = \alpha_{ij} + \alpha'_{ij} x + \alpha''_{ij} x^2$ and $\beta_i(x) = \beta_i + \beta'_i x + \beta''_i x^2$.

Finally, on putting $x_1 = t x_2$ in (13), we obtain on the left-hand side, for each t, a function of $r-1$ variables satisfying the assumptions of the corollary. Therefore, the exponent on the right-hand side is at most of degree 2, so that

$$\sum_{2 \leq i \leq j \leq r} (\alpha'_{ij} t x_2 + \alpha''_{ij} t^2 x_2^2) x_i x_j + \sum_{2 \leq i \leq r} \beta''_i t^2 x_2^2 x_i \equiv 0 .$$

Thus, the result is proved for $p = r$, and hence for all p. ||

PROOF OF THEOREM 1. (a) The equation satisfied by the density $f(x)$ of X_i is obtained by deriving the joint density of $Y_1, \cdots, Y_n$ on the sphere S and equating it to a constant. For this, we first obtain the joint density of $Y_1, \cdots, Y_{n-2}$. Since the transformation from $X_1, \cdots, X_n$ to $s, t, Y_1, \cdots, Y_{n-2}$ is not $1:1$, we divide the sample space of the X's into two regions, $X_n \geqq X_{n-1}$ and its complement, so that on each of these domains the transformation is $1:1$. To obtain the Jacobian of this transformation, we note that $\sum y_i = 0$ and $\sum y_i^2 = 1$ give us $\sum dy_i = 0$ and $\sum y_i dy_i = 0$, which can be solved for dy_{n-1} and dy_n. We thus find $\partial y_{n-1}/\partial y_j = (y_j - y_n)/(y_n - y_{n-1})$ and $\partial y_n/\partial y_j = (y_{n-1} - y_j)/(y_n - y_{n-1})$, $j = 1, 2, \cdots, n-2$. Consequently, the Jacobian is the absolute value of

$$s^{n-2} \begin{vmatrix} 1 & 0 & 0 & \cdots & 0 & 1 & y_1 \\ 0 & 1 & 0 & \cdots & 0 & 1 & y_2 \\ & & & \cdots & & & \\ 0 & 0 & 0 & \cdots & 1 & 1 & y_{n-2} \\ a_1 & a_2 & a_3 & \cdots & a_{n-2} & 1 & y_{n-1} \\ b_1 & b_2 & b_3 & \cdots & b_{n-2} & 1 & y_n \end{vmatrix},$$

where $sa_j = \partial x_{n-1}/\partial y_j = s\partial y_{n-1}/dy_j$, and $sb_j = \partial x_n/\partial y_j = s\partial y_n/\partial y_j$. Subtracting the sum of the first $n-2$ columns from the $(n-1)$th, and subtracting the sum of $y_j \times$ (jth column), $j = 1, \cdots, n-2$, from the last column, we see that the absolute value of the determinant is $ns^{n-2}/(y_n - y_{n-1})$. The density of $y_1, \cdots, y_{n-2}$ is obtained by integrating over $(0, \infty) \times (-\infty, \infty)$ with respect to (s, t), and finally the joint density of $y_1 y_2, \cdots, y_n$ on the sphere S is obtained by transforming an element of volume in the space of $(y_1, y_2, \cdots, y_{n-2})$ into its counterpart on S. The hypothesis of uniformity of the distribution on S then amounts to

$$I(y) \equiv \int_{-\infty}^{\infty} \int_0^{\infty} \prod_1^n f(t + sy_i) s^{n-2} ds\, dt = \text{constant} > 0 \,, \tag{14}$$

if $\sum y_i = 0$ and $\sum y_i^2 = 1$. From this it follows that, without any restriction on the y_i, $I(y)$ is a function of $\sum y_i$ and $\sum y_i^2$ only. Hence, for $n \geqq 6$, we may successively choose three sets of values $(y_1, y_2, \cdots, y_6, y_7, \cdots, y_n)$, $(y_1, y_2, y_3, y_1, y_2, y_3, y_7, \cdots, y_n)$ and $(y_4, y_5, y_6, y_4, y_5, y_6, y_7, \cdots, y_n)$, with $y_1 + y_2 + y_3 = y_4 + y_5 + y_6$ and $y_1^2 + y_2^2 + y_3^2 = y_4^2 + y_5^2 + y_6^2$. From these, we obtain

$$\int_{-\infty}^{\infty} \int_0^{\infty} (f_1 f_2 f_3 - f_4 f_5 f_6)^2 f_7 \cdots f_n s^{n-2} ds\, dt = 0 \,, \tag{15}$$

where $f_i = f(t + sy_i)$. If $n > 6$, we may integrate out $y_7, \cdots, y_n$, and have

$$\int_{-\infty}^{\infty} \int_0^{\infty} (f_1 f_2 f_3 - f_4 f_5 f_6)^2 s^4 ds\, dt = 0 \,. \tag{16}$$

Hence, $f(t + sy_1)f(t + sy_2)f(t + sy_3) = f(t + sy_4)f(t + sy_5)f(t + sy_6)$ for almost every (s, t). Consequently, if the equality fails to hold at (s_0, t_0), f must have a discontinuity at one of the points $t_0 + s_0 y_i$. We shall now see that this is not possible.

Suppose then that f is discontinuous at a point a_0. Since by assumption discontinuity points are isolated, given any continuity point a_1 we can find continuity points a_2, a_3, a_4, such that $a_0 + 2a_1 = a_2 + a_3 + a_4$ and $a_0^2 + 2a_1^2 = a_2^2 + a_3^2 + a_4^2$. Also, there exist intervals A_i containing respectively the points a_i, $i = 0, 1, 2, 3, 4$, such that f is continuous in $\cup A_i - \{a_0\}$, and such that to each $x_0 \in A_0 - \{a_0\}$ there corresponds $x_i \in A_i$, $i = 1, 2, 3, 4$, with the property that $x_0 + 2x_1 = x_2 + x_3 + x_4$, $x_0^2 + 2x_1^2 = x_2^2 + x_3^2 + x_4^2$, and $x_i \to a_i$, $i = 1, 2, 3, 4$, as $x_0 \to a_0$. Hence, $f(x_0)f^2(x_1) = f(x_2)f(x_3)f(x_4)$. Since, by the remark about distributions of the continuous type, $f(x_0)$ does not have a unique finite limit as x_0 converges to the discontinuity point a_0, we must have $f(a_1) = 0$. Thus the existence of a discontinuity implies that $f = 0$ at every continuity point. This, of course, contradicts the assumption that f is an almost-everywhere-continuous probability density. We conclude that f has no discontinuities and satisfies the equation $f(x_1)f(x_2)f(x_3) = f(y_1)f(y_2)f(y_3)$ whenever $x_1 + x_2 + x_3 = y_1 + y_2 + y_3$ and $x_1^2 + x_2^2 + x_3^2 = y_1^2 + y_2^2 + y_3^2$. Application of the lemma completes the proof. This proof is essentially Zinger's, but we have supplied some details that he did not present.

(b) Similarly, by considering the domains $\{\min(x_1, \cdots, x_n) = x_i\}$, $i = 1, 2, \cdots, n$, separately, we again get Equation (14) if $\min(y_1, \cdots, y_n) = 0$ and $\sum y_i = 1$. Now, for $n \geqq 5$, and any y_i such that $\min(y_1, \cdots, y_n) = 0$, if we write $z_i = y_i/\sum y_j$, then $I(z) = \text{const}$. Therefore, by changing from s to $s\sum y_i$, in this integral, we see that, for $\min(y) = 0$, $I(y)$ is a function of $\sum y_i$ only. Hence, for any $y_1, y_2, y_3, y_4 \geqq 0$ and such that $y_1 + y_2 = y_3 + y_4$, using the three sets $(y_1, y_2, y_3, y_4, 0, \cdots, 0)$, $(y_1, y_2, y_1, y_2, 0, \cdots, 0)$ and $(y_3, y_4, y_3, y_4, 0, \cdots, 0)$, we obtain

$$\int_{-\infty}^{\infty}\int_{0}^{\infty} f^{n-4}(t)\,(f_1 f_2 - f_3 f_4)^2\, s^{n-2}\, ds\, dt = 0\,. \tag{17}$$

Hence, $f(x) \neq 0$ implies that $f(x + y_1)f(x + y_2)$ depends on $y_1, y_2 > 0$ only through $y_1 + y_2$. Consequently, $f(x) > 0$ implies $f(x + y) = f(x)\exp(ay)$ for all $y > 0$. From this, the result follows.

(c) In the same way as before, considering the subset $\{\min(y) = y_{n-1} = 0, \max(y) = y_n = 1\}$, we have

$$\int_{-\infty}^{\infty}\int_{0}^{\infty} f(t)f(t+s)\prod_{1}^{n-2} f(t+sy_i)\, s^{n-2}\, ds\, dt = \text{constant} > 0\,, \tag{18}$$

provided $0 \leqq y_i \leqq 1$. Taking successively the sets of values $(y_1 = \cdots = y_{n-1} = 0, y_n = 1)$, $(y_1 = y, y_2 = \cdots = y_{n-1} = 0, y_n = 1)$ and $(y_1 = y_2 = y, y_3 = \cdots = y_{n-1} = 0, y_n = 1)$, where $y \in (0, 1)$, we get $f(t)f(t+s)[f(t) - f(t+sy)] = 0$. Hence, the positive range of f is an interval, and f is constant on it. ||

Proof of Corollary 1.1 is contained in the preceding proof.

Proof of Theorem 2. (a) By a procedure similar to that used above, we easily see that $\int_{-\infty}^{\infty} \Pi_1^n f(sy_i)\, s^{n-1}\, ds$ is a function of $\sum y_i^2$ only, so that for $n \geqq 4$, we may choose successively the values $(y_1, \cdots, y_n)$, $(y_1, y_2, y_1, y_2, y_5, \cdots y_n)$ and $(y_3, y_4, y_3, y_4, y_5, \cdots, y_n)$ with $y_1^2 + y_2^2 = y_3^2 + y_4^2$, and obtain

$$\int_{0}^{\infty} (f_1 f_2 - f_3 f_4)^2 f_5 \cdots f_n\, s^{n-1}\, ds = 0\,, \tag{19}$$

where $f_i = f(sy_i)$. Hence $f(x_1)f(x_2) = f(x_3)f(x_4)$ if $x_1^2 + x_2^2 = x_3^2 + x_4^2$. It follows from this, as in the lemma, that $f(x) = c\exp(ax^2)$.

(b) In this case, we observe that in the distribution of X_i there cannot be positive probability on both sides of the origin. Suppose then that there is no probability to the left of 0. Taking the values $(y_1, \cdots, y_n)$, $(y_1, y_2, y_1, y_2, y_5, \cdots, y_n)$, and $(y_3, y_4, y_3, y_4, y_5, \cdots, y_n)$, with $y_1 + y_2 = y_3 + y_4$, $y_i \geqq 0$, we get Equation (19). Thus for $x, y \geqq 0$, $f(x)f(y)$ is a function of $x + y$ only, so that $f(x) = c\exp(ax)$ for all $x > 0$.

(c) Considering the region $\{\max(x) = x_1\}$, we find that for $0 \leqq y_i \leqq y_1$, $\int_0^{\infty} \Pi_1^n f(sy_i) s^{n-1}\, ds$ is a function of y_1 only. Hence, using the values

$(y, y, y_3, y_4, \cdots, y_n)$, $(y, y_3, y_3, y_4, \cdots, y_n)$, and $(y, y, y, y_4, \cdots, y_n)$ with $0 \leqq y_i \leqq y$, we get

$$\int_0^\infty f(sy)\,[f(sy) - f(sy_3)]^2 f(sy_4) \cdots f(sy_n)\, s^{n-1}\, ds = 0 \,. \tag{20}$$

Consequently, $f(x) \neq 0$ implies $f(x') = f(x)$ for $0 \leqq x' \leqq x$. ||

PROOF OF COROLLARY 2.1. As before, we get the relation

$$\int_0^\infty \prod_1^n f(sy_i)\, s^{n-1} ds = (\text{constant})\ (y_1 \cdots y_n)^\alpha \,, \tag{21}$$

if $y_i \geqq 0$ and $\sum y_i = 1$. Hence, writing $g(x) = f(x)x^{-\alpha}$, we get

$$\int_0^\infty \prod_1^n g(sy_i)\, s^{\alpha n+n-1}\, ds = \text{a function of } \sum y_i \,, \tag{22}$$

if $y_i \geqq 0$. This is the same equation as that encountered in the proof of Theorem 2(b), and we thus get the result stated. ||

PROOF OF COROLLARY 2.2. It is enough to show that the right-hand side in Equation (2) reduces to that in (21) and then to apply the previous corollary. For this, note that changing y_i to uy_i, $i = 1, \cdots, n$, $u > 0$, in (2) gives us

$$u^n g(u \textstyle\sum y_i) \prod_1^n h(uy_i) = g(\sum y_i) \prod_1^n h(y_i) \,. \tag{23}$$

When we write $k_u(y) = uh(uy)/h(y)$, Equation (23) implies that $\Pi_1^n k_u(y_i)$ depends on the y's only through $\sum y_i$. Hence, for each $u > 0$, there exist numbers $a(u), b(u)$ such that, for all $y > 0$, $k_u(y) = a(u) \exp[b(u)y]$, i.e.,

$$h(uy) = u^{-1}h(y)\, a(u) \exp[b(u)y] = y^{-1}h(u)\, a(y) \exp[b(y)u] \,. \tag{24}$$

From (24), with $u = 1$, we obtain

$$h(y) = h(1)y^{-1}\, a(y) \exp[b(y)] \,, \tag{25}$$

and hence,

$$h(uy) = h(1)(uy)^{-1}a(uy) \exp[b(uy)] \,. \tag{26}$$

Using (25) in (24), we obtain

$$\begin{aligned} h(uy) &= h(1)(uy)^{-1}a(y)\, a(u) \exp[b(y) + b(u)y] \\ &= h(1)(uy)^{-1}a(y)\, a(u) \exp[b(u) + b(y)u] \,. \end{aligned} \tag{27}$$

Hence, there exists a constant b, such that $b(y) = (1 - y)b$. Using this fact in (26) and (27), we find that $a(uy) = a(u)\, a(y)$. Consequently, $a(u) = u^a$ for some a, and

$$h(y) = cy^{a-1} \exp(-by) \,. \tag{28}$$

Substitution in (23) establishes that $g(y) = g(1)e^{-b}y^{-an}\exp(by)$. Thus, the right-hand side of (2) reduces to the right-hand side of (21). ||

PROOF OF THEOREM 3. Denoting by $f(x)$ the probability density of X, where x is a $p \times p$ symmetric matrix, we are given that the positive range of f is contained in the set of all positive definite x, and we can show as before that f is continuous on its positive range. The joint density of $Y_1, \cdots, Y_n$ on the surface $\{y_i \text{ positive definite}, \sum y_i = I\}$ is found by finding the distribution of $S, Y_1, \cdots, Y_{n-1}$, integrating out S, and changing from the nonsingular distribution of $Y_1, \cdots, Y_{n-1}$ to that of $Y_1, \cdots, Y_n$. The last operation is simple in this case, because the only difference between the two densities is a multiplicative constant which we do not need to evaluate.

The Jacobian of the transformation from $(X_1, \cdots, X_n)$ to $(S, Y_1, \cdots, Y_{n-1})$ can be computed in two steps as the product of the Jacobians of the transformations $(X_1, \cdots, X_n) \to (Y_1, \cdots, Y_{n-1}, X_n)$, and $(Y_1, \cdots, Y_{n-1}, X_n) \to (Y_1, \cdots, Y_{n-1}, S)$. By Theorem 3.7 of Deemer and Olkin [5], the first Jacobian is $[\det S]^{(p+1)(n-1)/2}$, and the second is unity. Thus the hypothesis of the theorem states that

$$\int_{\mathscr{S}} \prod_1^n f(s^{\frac{1}{2}} y_i s^{\frac{1}{2}})(\det s)^{(p+1)(n-1)/2}\, ds = (\text{constant}) \prod_1^n (\det y_i)^a , \tag{29}$$

where $\mathscr{S}$ is the set of all positive definite $p \times p$ matrices. Writing $g(x) = f(x)(\det x)^{-a}$ if x is positive definite, Equation (29) becomes

$$\int_{\mathscr{S}} \prod_1^n g(s^{\frac{1}{2}} y_i s^{\frac{1}{2}})(\det s)^{\beta}\, ds = \text{constant} , \tag{30}$$

if y_i are positive definite and such that $\sum y_i = I$, with $\beta = an + (p+1)(n-1)/2$. By the procedure used in Theorem 2(b), we consequently find that x, y, x', y' positive definite and $x + y = x' + y'$ imply

$$g(x)g(y) = g(x')g(y') . \tag{31}$$

From this we see immediately that since $g(x) \not\equiv 0$, it cannot vanish for any positive definite x. For, suppose $g(a) = 0$; then $[g(2a/3)]^2 = g(a)g(a/3) = 0$, so that $g[(2/3)^n a] = 0$ for $n = 1, 2, \cdots$. Now, for any positive definite x, $2x - (2/3)^n a$ is also positive definite for sufficiently large n. Hence, $[g(x)]^2 = g((2/3)^n a)\, g(2x - (2/3)^n a) = 0$ for all positive definite x.

We can now extend the definition of g to all $p \times p$ symmetric matrices $\bar{x}$ by noting that every $\bar{x}$ can be expressed as $x_1 - x_2$, where x_1 and x_2 are positive definite, and that for all such pairs (x_1, x_2) corresponding to a given $\bar{x}$, $g(x_1)/g(x_2)$ has a unique value. Defining this to be $g(\bar{x})/g(0)$, it is obvious that g now satisfies (31) for all symmetric matrices. Consequently, if x_{ij}, $i \leqq j = 1, 2, \cdots, p$, denote the elements of x, then $g(x) = \Pi_{i \leqq j}\, g_{ij}(x_{ij})$ where the g_{ij} are functions of one variable satisfying an equation of the type of (31). Hence, $g(x) = c \exp(\sum \lambda_{ij} x_{ij})$, so that f is a Wishart density. ||

PROOF OF THEOREM 4. Since the procedure is, in principle, akin to that of the proof of Theorem 1(a), the whole of it will not be presented in detail.

Let X, Y denote $n \times p$ matrices whose rows are respectively $\boldsymbol{X}_i$, $\boldsymbol{Y}_i$, and let $\boldsymbol{e}$ denote the n-dimensional vector $(1, 1, \cdots, 1)$. Then we have a transformation from the np variables X to the $p[n+1+(p+1)/2]$ variables $(\boldsymbol{t}, T, Y)$ subject to $p[1+(p+1)/2]$ constraints:

$$X = \begin{pmatrix} \boldsymbol{t} \\ \vdots \\ \boldsymbol{t} \end{pmatrix} + YT\,, \quad \boldsymbol{e}Y = 0, \quad Y'Y = I\,. \tag{32}$$

This is a multiple-valued transformation, since, although $\boldsymbol{t}$ and T are uniquely determined by X, there are 2^p values of Y for each X ([6], Lemma 3). As before, we shall consider one specific branch of this transformation, on which it is 1 : 1. Let Y_I denote a uniquely identified set of $p[n-1-(p+1)/2]$ elements of Y which will be regarded as the independent variables. To evaluate the Jacobian $J(X; \boldsymbol{t}, T, Y_I)$, we make use of the work of Olkin and Roy [6] and also of the fact that the Jacobian of a chain of transformations is the product of the individual Jacobians.

Let Ω_0 be an $(n-1) \times p$ matrix such that $(\boldsymbol{e}'/\sqrt{n}, \Omega_0') = \Omega'$ is orthogonal, and let $U = \Omega_0 X$, $V = \Omega_0 Y$. Consider the chain of transformations $X \to (\boldsymbol{t}, U) \to (\boldsymbol{t}, T, V) \to (\boldsymbol{t}, T, Y)$. This gives us

$$J(X; \boldsymbol{t}, T, Y_I) = J(X; \boldsymbol{t}, U)\, J(\boldsymbol{t}, U; \boldsymbol{t}, T, V_I)\, J(\boldsymbol{t}, T, V_I; \boldsymbol{t}, T, Y_I)\,, \tag{33}$$

Note that $X'\Omega' = (\boldsymbol{t}'\sqrt{n}, U')$, $T'T = U'U$ and $U = VT$. Of the three terms on the right-hand side of (33), the first equals $n^{p/2}$, the second equals $2^p \Pi_1^p t_{ii}^{n-1-i}/J(V'V; V_D)$ by Equation (5) of [6], and the third is $J(V_I; Y_I)$. Hence,

$$J(X; \boldsymbol{t}, T, Y_I) = (4n)^{p/2} \prod_1^n t_{ii}^{n-1-i} J(Y)\,. \tag{34}$$

where $J(Y) = J(V_I; Y_I)/J(V'V; V_D)$ is a function only of Y, whose actual value does not concern us.

The joint density of Y_I is thus found to be

$$(4n)^{p/2} J(Y) \int_{\mathscr{R}_p \times \mathscr{T}} \left[\prod_1^n f(\boldsymbol{t} + T\boldsymbol{y}_i)\right] \prod_1^p t_{ii}^{n-1-i}\, dt\, dT\,, \tag{35}$$

where $\mathscr{T}$ is the set of all T (i.e., the set of all triangular matrices with non-negative diagonal elements and zeros below the diagonal). On substituting the $\mathscr{N}(0, I)$ density for f in (35), we see that the integral has a constant value. Hence, the hypothesis of the theorem states that f is a probability density of the continuous type, satisfying the equation

$$\int_{\mathscr{R}_p \times \mathscr{T}} \left[\prod_1^n f(\boldsymbol{t} + T\boldsymbol{y}_i)\right] \prod_1^p t_{ii}^{n-1-i}\, dt\, dT = \text{constant}\,, \tag{36}$$

for $\sum \boldsymbol{y}_i = \boldsymbol{0}$ and $\sum \boldsymbol{y}_i'\boldsymbol{y}_i = I$.

The solution of this equation proceeds along lines similar to those of the proof of Theorem 1(a), and is completed by the use of Lemma 2. ||

References

[1] Zinger, A. A. On a problem of A. N. Kolmogorov. *Vestnik Leningrad. Univ.*, 1956, **1**, 53-56.

[2] Cramér, H. *Mathematical Methods of Statistics*. Princeton, N.J.: Princeton Univ. Press, 1946.

[3] Steck, G. P. A uniqueness property not enjoyed by the normal distribution. *Ann. Math. Stat.*, 1958, **29**, 604-6.

[4] Kac, M. Une remarque sur les équations fonctionelles. *Comment. Math. Helv.*, 1936-37, **9**, 170-71.

[5] Deemer W. L., and Olkin, I. The Jacobians of certain matrix transformations useful in multivariate analysis. *Biometrika*, 1951, **38**, 345-67.

[6] Olkin, I., and Roy, S. N. On multivariate distribution theory. *Ann. Math. Stat.*, 1954, **25**, 329-39.

19 GENERALIZATION OF SOME RESULTS FOR INVERSION OF PARTITIONED MATRICES[1]

B. G. GREENBERG and A. E. SARHAN

University of North Carolina

1. Summary

Matrices which arise from the application of least-squares procedures to the analysis of variance and the estimation of response surfaces ordinarily have a patterned arrangement which facilitates inversion. These matrices usually have submatrices in the diagonal which are either diagonal matrices or reducible to linear combinations involving diagonal matrices. Such matrices can be precisely inverted by substituting unknowns in a generalized form and solving for the unknowns. Several typical patterns are considered and their inverses are given in terms of the known elements. The major arithmetic involves evaluation of a determinant whose order is equal to the order of the partitioned matrix. All elements of the inverse matrix involve this determinant and a smaller one, or both, as well as minors from these determinants.

2. Introduction

In a previous paper, Roy and Sarhan [1] presented methods of partitioning matrices whose elements were patterned so as to facilitate the inversion process. In Greenberg and Sarhan [2], numerical methods of inversion by partitioning patterned matrices arising in the analysis of variance and response-surface estimation were considered.

The principle involved in the numerical method was a simple and well-known one. By partitioning a patterned matrix and inferring the general form of its inverse, unknowns in the inverse are estimable by equating the cross multiplication of the two matrices to the identity matrix. The present paper extends some of the findings about partitioning previously reported, and now all of the observations reported in [1] can be generalized.

The notation adopted will be: $C(k \times m)$ is a matrix with k rows and m columns, and $J(k \times m)$ when all the elements of C are unity; I_p is the

[1] Sponsored by Office of Ordnance Research, U. S. Army.

identity matrix of order p, $e'_p = [1, 1, \cdots, 1]$, i.e., a row vector with all p elements unity.

3. Partitioned Matrices with Diagonal Matrices in the Diagonal Position

When a matrix can be partitioned so that the diagonal submatrices are themselves diagonal matrices, the diagonal submatrices in the inverse will have elements equal to the reciprocal of the elements of the diagonals.

For example, consider the simple matrix whose partitioning is as follows:

$$C = \begin{bmatrix} k_1 I_n & aJ \\ aJ & k_2 I_p \end{bmatrix}.$$

If the determinant is nonvanishing, the inverse exists and is

$$C^{-1} = \begin{bmatrix} \rho I_n + \alpha J & \beta J \\ \beta J & \mu I_p + \gamma J \end{bmatrix}.$$

By direct cross multiplying and equating the diagonal elements to the identity matrix, it is observable that $\rho = 1/k_1$ and $\mu = 1/k_2$.

To evaluate α, β, γ, the equating of CC^{-1} to the identity matrix gives $k_1\alpha + a\beta p = 0$, $a\beta n + k_2\gamma = 0$, $(a/k_1) + an\alpha + k_2\beta = 0$, and $k_1\beta + (a/k_2) + ap\gamma = 0$. These give:

$$\alpha = \frac{a^2 p}{k_1 \begin{vmatrix} k_1 & an \\ ap & k_2 \end{vmatrix}} = \frac{a^2 p}{k_1(k_1 k_2 - na^2 p)},$$

$$\gamma = \frac{a^2 n}{k_2 \begin{vmatrix} k_1 & an \\ ap & k_2 \end{vmatrix}} = \frac{a^2 n}{k_2(k_1 k_2 - na^2 p)},$$

$$\beta = \frac{-a}{\begin{vmatrix} k_1 & an \\ ap & k_2 \end{vmatrix}} = \frac{-a}{(k_1 k_2 - na^2 p)}.$$

A more general pattern of matrix that may occur in the analysis of variance is a structure as follows:

$$V = \begin{bmatrix} k_1 I_n & aJ & bJ \\ aJ & k_2 I_p & cJ \\ bJ & cJ & k_3 I_s \end{bmatrix}.$$

The inverse may be written down in symbolic form, and will be, as previously:

$$V^{-1} = \begin{bmatrix} \frac{1}{k_1} I_n + \alpha J & \beta J & \gamma J \\ \beta J & \frac{1}{k_2} I_p + \delta J & \theta J \\ \gamma J & \theta J & \frac{1}{k_3} I_s + \xi J \end{bmatrix}.$$

Following the usual cross multiplication procedure, we have results that are easily modified as follows:

$$I + k_1\alpha J + a\beta p J + b\gamma s J = I ,$$

$$\frac{a}{k_1} + an\alpha J + k_2\beta J + c\gamma s J = 0 ,$$

$$\frac{b}{k_1} + bn\alpha J + cp\beta J + k_3\gamma J = 0 .$$

These equations reduce to

$$\begin{aligned} k_1\alpha + ap\beta + bs\gamma &= 0 , \\ an\alpha + k_2\beta + cs\gamma &= -a/k_1 , \\ bn\alpha + cp\beta + k_3\gamma &= -b/k_1 . \end{aligned} \tag{3.1}$$

The pattern of the relationships is observable to permit direct extension to four, five, or n submatrices in a row or column. Solution of (3.1) yields the following values:

$$\alpha = \frac{a^2pk_3 + b^2k_2s - 2abcps}{k_1\Delta} ,$$

$$\beta = \frac{bcs - ak_3}{\Delta} ,$$

$$\gamma = \frac{acp - k_2b}{\Delta} ,$$

where Δ is the determinant of coefficients in (3.1).

The set of equations to replace (3.1) to estimate the parameters for the second column of submatrices is as follows:

$$\begin{aligned} k_1\beta + ap\delta + bs\theta &= -a/k_2 , \\ an\beta + k_2\delta + cs\theta &= 0 , \\ bn\beta + cp\delta + k_3\theta &= -c/k_2 . \end{aligned} \tag{3.2}$$

The determinant of the coefficients is identical in both (3.1) and (3.2). Solution of the equations yields

$$\delta = \frac{a^2nk_3 + c^2sk_1 - 2abcns}{k_2\Delta} , \qquad \theta = \frac{abn - k_1c}{\Delta} .$$

Finally, the last column of submatrices yields a check upon the values of γ and θ as well as the equation for ξ which provides

$$\xi = \frac{b^2k_2n + c^2pk_1 - 2abcnp}{k_3\Delta} .$$

Example. In [1], a special patterned matrix which frequently arises was shown as follows:

$$C = \begin{bmatrix} kI_k & J & J \\ J & kI_{k-1} & J \\ J & J & kI_{k-1} \end{bmatrix} .$$

This can be inverted by using the method of Section 3. Thus,

$\alpha = 2(k-1)/k^2$, $\delta = 1/k$, $\xi = 1/k$, $\beta = -1/k$, $\gamma = -1/k$, and $\theta = 0$.

4. A General Result

It can be shown that if the matrix C is non-singular,

$$C = \begin{bmatrix} k_1 I & a_{12}J & a_{13}J & \cdots & a_{1n}J \\ & k_2 I & a_{23}J & \cdots & a_{2n}J \\ & & \ddots & & \\ & & & & k_n I \end{bmatrix},$$

the inverse is of the form

$$C^{-1} = \begin{bmatrix} \frac{1}{k_1}I + \alpha J & \beta J & \gamma J & \cdots & \omega J \\ \beta J & \frac{1}{k_2}I + \delta J & \theta J & \cdots & \varphi J \\ \gamma J & \theta J & \frac{1}{k_3}I + \xi J & \cdots & \lambda J \\ \vdots & \vdots & \vdots & & \vdots \\ \omega J & \varphi J & \lambda J & \cdots & \frac{1}{k_n}I + \Sigma J \end{bmatrix},$$

where all unknowns can be evaluated as previously indicated.

5. Partitioned Matrices with Column and Row Vectors

In fitting response surfaces, a matrix frequently encountered is one which is of a form [1] such as

$$C((r+1)\times(r+1)) = \begin{bmatrix} a & b & b & b & \cdots & b \\ b & (k+c) & c & c & \cdots & c \\ b & c & (k+c) & c & \cdots & c \\ b & c & c & (k+c) & \cdots & c \\ \vdots & \vdots & \vdots & \vdots & & \vdots \\ b & c & c & c & \cdots & k+c \end{bmatrix}.$$

This can be symbolized as

$$C = \begin{bmatrix} a & be_r' \\ be_r & kI_r + cJ \end{bmatrix}.$$

After cross multiplication, the inverse is clearly of the form

$$C^{-1} = \begin{bmatrix} \alpha & \beta e_r' \\ \beta e_r & \frac{1}{k}I_r + \gamma J \end{bmatrix}.$$

This type of matrix can also be extended to the situation where there are more than 2×2 submatrices present. For example, a more useful case is as follows:

$$V = \begin{bmatrix} a & be' & ce' \\ be & k_1 I_r + dJ & k_2 I_r + gJ \\ ce & k_2 I_r + gJ & k_3 I_r + fJ \end{bmatrix},$$

and

$$V^{-1} = \begin{bmatrix} \alpha & \beta e' & \gamma e' \\ \beta e & \tau_1 I_r + \xi_1 J & \tau_2 I_r + \xi_2 J \\ \gamma e & \tau_2 I_r + \xi_2 J & \tau_3 I_r + \xi_3 J \end{bmatrix}.$$

There are nine unknowns expressed by the Greek letters $\alpha, \beta, \gamma, \tau_1, \tau_2, \tau_3$, and ξ_1, ξ_2, and ξ_3. A general solution of this can be obtained by writing the equations of the nine submatrices resulting from cross multiplication. Thus, the results are as follows:

$$a\alpha + be'\beta e + ce'\gamma e = 1I + 0J, \tag{5.1}$$

$$a\beta e' + be'(\tau_1 I + \xi_1 J) + ce'(\tau_2 I + \xi_2 J) = 0I + 0J, \tag{5.2}$$

$$a\gamma e' + be'(\tau_2 I + \xi_2 J) + ce'(\tau_3 I + \xi_3 J) = 0I + 0J, \tag{5.3}$$

$$be\alpha + (k_1 I + dJ)\beta e + (k_2 I + gJ)\gamma e = 0I + 0J, \tag{5.4}$$

$$be\beta e' + (k_1 I + dJ)(\tau_1 I + \xi_1 J) + (k_2 I + gJ)(\tau_2 I + \xi_2 J) = 1I + 0J, \tag{5.5}$$

$$be\gamma e' + (k_1 I + dJ)(\tau_2 I + \xi_2 J) + (k_2 I + gJ)(\tau_3 I + \xi_3 J) = 0I + 0J, \tag{5.6}$$

$$ce\alpha + (k_2 I + gJ)\beta e + (k_3 I + fJ)\gamma e = 0I + 0J, \tag{5.7}$$

$$ce\beta e' + (k_2 I + gJ)(\tau_1 I + \xi_1 J) + (k_3 I + fJ)(\tau_2 I + \xi_2 J) = 0I + 0J, \tag{5.8}$$

$$ce\gamma e' + (k_2 I + gJ)(\tau_2 I + \xi_2 J) + (k_3 I + fJ)(\tau_3 I + \xi_3 J) = 1I + 0J. \tag{5.9}$$

By expanding and rewriting (5.5), (5.6), and (5.9), one obtains

$$b\beta J + k_1\tau_1 I + (\xi_1 k_1 + \tau_1 d + dr\xi_1)J + k_2\tau_2 I + (k_2\xi_2 + g\tau_2 + g\xi_2 r)J = 1I + 0J, \tag{5.5$'$}$$

$$b\gamma J + k_1\tau_2 I + (\xi_2 k_1 + \tau_2 d + dr\xi_2)J + k_2\tau_3 I + (k_2\xi_3 + \tau_3 g + g\xi_3 r)J = 0I + 0J, \tag{5.6$'$}$$

$$c\gamma J + k_2\tau_2 I + (g\tau_2 + k_2\xi_2 + rg\xi_2)J + k_3\tau_3 I + (f\tau_3 + k_3\xi_3 + rf\xi_3)J = 1I + 0J. \tag{5.9$'$}$$

Equating the coefficients of I from (5.5′), (5.6′), and (5.9′), we have

$$\begin{aligned} k_1\tau_1 + k_2\tau_2 &= 1, \\ k_1\tau_2 + k_2\tau_3 &= 0, \\ k_2\tau_2 + k_3\tau_3 &= 1. \end{aligned}$$

This gives $\tau_1 = k_3/\Delta_1$, $\tau_2 = -k_2/\Delta_1$, $\tau_3 = k_1\Delta_1$, where

$$\Delta_1 = \begin{vmatrix} k_1 & k_2 \\ k_2 & k_3 \end{vmatrix}.$$

Solving for α, β, and γ involves the expansion and rewriting of Equations

(5.1), (5.4), and (5.7). Doing so and equating the coefficients of I, one obtains

$$
\begin{aligned}
a\alpha + br\beta \qquad\quad + cr\gamma \qquad\quad &= 1 , \\
b\alpha + (k_1 + dr)\beta + (k_2 + gr)\gamma &= 0 , \\
c\alpha + (k_2 + gr)\beta + (k_3 + fr)\gamma &= 0 .
\end{aligned} \tag{5.10}
$$

This gives

$$[\alpha\ \beta\ \gamma] = [1\ 0\ 0]\begin{bmatrix} a & b & c \\ br & k_1 + dr & k_2 + gr \\ cr & k_2 + gr & k_3 + fr \end{bmatrix}^{-1} .$$

Solving for ξ_1, ξ_2, and ξ_3 involves the expansion and rewriting of Equations (5.5) and any two of (5.3), (5.6), and (5.9). The solution yields

$$\xi_1 = \frac{\begin{vmatrix} a & cr \\ c & k_3 + fr \end{vmatrix}}{r\Delta_2} - \frac{k_3}{r\Delta_1} ,$$

$$\xi_2 = -\frac{\begin{vmatrix} a & br \\ c & k_2 + gr \end{vmatrix}}{r\Delta_2} + \frac{k_2}{r\Delta_1} ,$$

$$\xi_3 = \frac{\begin{vmatrix} a & br \\ b & k_1 + dr \end{vmatrix}}{r\Delta_2} - \frac{k_1}{r\Delta_1} ,$$

where Δ_2 is the determinant of the constants in (5.10).

In summary, the nine unknowns can be written at once by reference to the two determinants Δ_1 and Δ_2. Thus,

$$\alpha = \frac{\text{cofactor of } a}{\Delta_2} , \qquad \tau_1 = \frac{\text{cofactor of } k_1}{\Delta_1} ,$$

$$\beta = \frac{\text{cofactor of } br}{\Delta_2} , \qquad \tau_2 = \frac{\text{cofactor of } k_2}{\Delta_1} ,$$

$$\gamma = \frac{\text{cofactor of } cr}{\Delta_2} , \qquad \tau_3 = \frac{\text{cofactor of } k_3}{\Delta_1} ,$$

$$\xi_1 = \frac{\text{cofactor of } (k_1 + dr)}{r\Delta_2} - \frac{\text{cofactor of } k_1}{r\Delta_1} ,$$

$$\xi_2 = \frac{\text{cofactor of } (k_2 + gr)}{r\Delta_2} - \frac{\text{cofactor of } k_2}{r\Delta_1} ,$$

$$\xi_3 = \frac{\text{cofactor of } (k_3 + fr)}{r\Delta_2} - \frac{\text{cofactor of } k_3}{r\Delta_1} .$$

These results can be extended in a straightforward manner to invert a matrix with more than three diagonal submatrices, or one in which the submatrix $(k_3I + fJ)$ consists of s rows and columns instead of r rows and columns. The latter case, for instance, involves the substitution of s for r in the third column of Δ_2.

Example. In [2], a 10×10 matrix arising from an actual problem in response-surface estimation was inverted by numerical methods, using the principle of cross multiplication with unknowns. That illustration, however, did not take advantage of the direct, condensed procedures developed in this section. For purpose of comparison in highlighting the convenience, the same example is inverted by using the method of this section.

The matrix to be inverted is partitioned as follows:

$$V = \begin{bmatrix} 17 & 22e' & 3e' & 6e' \\ 22e & 19I_3 + 27J & 5I_3 + 3J & -5I_3 + 11J \\ 3e & 5I_3 + 3J & 19I_3 + 8J & -11I_3 + 11J \\ 6e & -5I_3 + 11J & -11I_3 + 11J & 19I_3 + 3J \end{bmatrix}.$$

The inverse is written as

$$V^{-1} = \begin{bmatrix} \alpha & \beta e' & \gamma e' & \delta e' \\ \beta e & aI + bJ & cI + dJ & eI + fJ \\ \gamma e & cI + dJ & gI + hJ & kI + tJ \\ \delta e & eI + fJ & kI + tJ & mI + nJ \end{bmatrix}.$$

To evaluate the unknowns, write

$$\Delta_2 = \begin{vmatrix} 17 & 66 & 9 & 18 \\ 22 & 100 & 14 & 28 \\ 3 & 14 & 43 & 22 \\ 6 & 28 & 22 & 28 \end{vmatrix} = 123{,}904 ,$$

and

$$\Delta_1 = \begin{vmatrix} 19 & 5 & -5 \\ 5 & 19 & -11 \\ -5 & -11 & 19 \end{vmatrix} = 4160 .$$

Then,

$$a = \frac{\begin{vmatrix} 19 & -11 \\ -11 & 19 \end{vmatrix}}{\Delta_1} = \frac{240}{4160} = \frac{3}{52} ,$$

$$c = -\frac{\begin{vmatrix} 5 & -11 \\ -5 & 19 \end{vmatrix}}{\Delta_1} = -\frac{40}{4160} = -\frac{1}{104} ,$$

$$e = \frac{\begin{vmatrix} 5 & 19 \\ -5 & -11 \end{vmatrix}}{\Delta_1} = \frac{40}{4160} = \frac{1}{104} ,$$

$$g = \frac{\begin{vmatrix} 19 & -5 \\ -5 & 19 \end{vmatrix}}{\Delta_1} = \frac{336}{4160} = \frac{21}{260} ,$$

$$k = -\frac{\begin{vmatrix} 19 & 5 \\ -5 & -11 \end{vmatrix}}{\Delta_1} = \frac{184}{4160} = \frac{23}{520} ,$$

$$m = \frac{\begin{vmatrix} 19 & 5 \\ 5 & 19 \end{vmatrix}}{\Delta_1} = \frac{336}{4160} = \frac{21}{260},$$

$$\alpha = \frac{\text{cofactor of 17 in } \Delta_2}{\Delta_2} = \frac{391}{968},$$

$$\beta = \frac{\text{cofactor of 66 in } \Delta_2}{\Delta_2} = -\frac{87}{968},$$

$$\gamma = \frac{\text{cofactor of 9 in } \Delta_2}{\Delta_2} = -\frac{1}{968},$$

$$\delta = \frac{\text{cofactor of 18 in } \Delta_2}{\Delta_2} = \frac{4}{968}.$$

Finally, to solve for the remaining unknowns:

$$b = \frac{\text{cofactor of 100}}{3\Delta_2} - \frac{\text{cofactor of 19}}{3\Delta_1} = \frac{35}{(121)(52)},$$

$$d = \frac{\text{cofactor of 14}}{3\Delta_2} - \frac{\text{cofactor of 5}}{3\Delta_1} = \frac{31}{(121)(52)},$$

$$f = \frac{\text{cofactor of 28}}{3\Delta_2} - \frac{\text{cofactor of } -5}{3\Delta_1} = -\frac{127}{(121)(52)(2)},$$

$$h = \frac{\text{cofactor of 43}}{3\Delta_2} - \frac{\text{cofactor of 19}}{3\Delta_1} = -\frac{849}{(121)(52)(10)},$$

$$t = \frac{\text{cofactor of 22}}{3\Delta_2} - \frac{\text{cofactor of } -11}{3\Delta_1} = -\frac{843}{(121)(52)(5)},$$

$$n = \frac{\text{cofactor of 28}}{3\Delta_2} - \frac{\text{cofactor of 19}}{3\Delta_1} = \frac{10{,}176}{3(123{,}904)} - \frac{336}{3(4160)}$$

$$= \frac{57}{(121)(52)(20)}.$$

References

[1] Roy, S. N., and Sarhan, A. E. On inverting a class of patterned matrices. *Biometrika*, 1956, **43**, 227-31.

[2] Greenberg, B. G., and Sarhan, A. E. Matrix inversion, its interest and application in analysis of data. Paper presented before 31st Session, International Statistical Institute in Brussels, 1958; also in *J. Amer. Stat. Assoc.*, 1959, **54**, 755-66.

20 SELECTING A SUBSET CONTAINING THE BEST OF SEVERAL BINOMIAL POPULATIONS

SHANTI S. GUPTA and MILTON SOBEL

Bell Telephone Laboratories

1. Summary

A procedure R is studied for selecting a subset of several given binomial populations which includes the "best" population with high probability. For a known number of observations from the k given binomial populations, the procedure R selects a subset which is small (the exact size depends on the observed results), never empty, and yet large enough to guarantee with preassigned probability that, regardless of the true (unknown) parameter values, it will include the best population, i.e., the one with the highest parameter value. Expressions for the probability of a correct selection with procedure R are derived and, in the case of a common number of observations, these are used to construct tables needed to carry out the procedure R. Formulas are obtained for the expected number of populations retained in the selected subset and, for selected cases, tables are given for the expected proportion of populations retained. The tables are used to determine the common number of observations required to control the expected size (or proportion) of the retained subset when the best population has a certain "distance" from the others. Alternative procedures based on the arc sine transformation, which require no new tables, are briefly discussed. Some additional mathematical results are obtained (particularly, for $k = 2$) dealing with the properties of monotonicity and symmetry for the probability of a correct selection and the expected number of populations retained. The application of the procedures and the method of using the tables are illustrated.

2. Introduction

It frequently happens in problems concerned with ranking and selection that, whatever the original formulation or purpose of the experiment, the actual outcome is the rejection of certain processes and the acceptance of the remaining processes as being superior with respect to a desired characteristic. We shall try to formalize this in the special case when the observations are of an attribute or binomial character. Each process is associated with a binomial distribution and hence also with a probability of success (success corresponding to a good unit) on a single trial. The problem is to select a subset of the k populations which is small, never empty, and always large enough so that, with preassigned probability P^*, it contains the best population (i.e., the one with the highest probability of success on a single trial). It is desirable that this be accomplished without the knowledge of any of the true probabilities of success on a single trial, i.e., without any knowlege of the true point in the parameter space. The number of populations to be retained in the selected subset is not fixed in advance but is determined in a simple and meaningful way by the observed results. It should be noted that the problem of selecting a subset containing the population with the *smallest* parameter is obviously equivalent to the problem treated here, and the appended tables are again applicable.

The corresponding problem for normal populations with unknown means is treated in [1] for the case of a common, known variance and in [1] and [2] for the case of a common, unknown variance. This "selecting a subset" formulation is to be distinguished from (and compared with) the "indifference zone" formulation which is treated in [3] and [4]. In the latter formulation, the "width" of an "indifference zone" in the parameter space is preassigned, the number of observations needed is tabulated, and the final decision is the selection of a *single* population which is asserted *to be the best population.* In the former or present formulation, the number of observations is given, constants needed for carrying out the procedure are tabulated, and the final decision is the selection of a *subset* of populations which is asserted *to contain the best population.*

The present formulation is more flexible in that the experimenter can make a decision on any given number of observations and can assert that a certain subset contains the best population, the size of the subset depending on the observed results. If the experimenter must subsequently choose one best population, he is free either to choose one from this subset on the basis of non-statistical considerations or he may choose to reduce the size of the retained subset on the basis of further experimentation in the future.

Clearly, a procedure which attempts to select a subset containing the best population can be regarded, from the point of view of the rejected populations, as an elimination or screening procedure. In fact, one joint confidence statement which can be made after experimentation with confidence level P^* is that the parameter values of the eliminated populations are all *less than or equal to* the largest of the parameter values of the retained populations.

In the present formulation, the expected size of the retained subset can be regarded as a criterion of the efficiency of the procedure and as a means of comparing the efficiencies of different procedures which give the same probability guarantee that the best population is included in the selected subset. The expected size of the retained subset is always at least one and, for the proposed procedure and a fixed number of populations, depends on the true, unknown configuration of probabilities of success on a single trial as well as on the sample size from each population. Hence, assuming a common sample size n from each population and assuming that n is not determined by other considerations, we can use this concept to determine the smallest value of n needed to "control" the expected size of the selected subset (i.e., to make it at most $1 + \varepsilon$ for some preassigned positive ε), at some *particular* alternative configuration in which the two largest parameter values are not equal. Alternatively, we can determine the smallest common sample size n needed to control the *maximum* expected size of the retained subset over all configurations in which the two largest parameters are a certain "distance" apart.

In Section 4 the basic procedure R with equal sample sizes per population is discussed and the corresponding normal approximation is given. Section 5 treats the same basic procedure R for unequal sample sizes and its normal approximation. Section 6 deals with the expected size of the selected subset and the problem of determining the smallest common sample size n needed to control the expected size of the selected subset when the best population has a probability of success on a single trial which is δ greater than that of all the other populations. Alternative procedures based on the arc sine transformation are briefly discussed in Section 7. Sections 8 and 9 give some relevant mathematical proofs for equal sample sizes and general sample sizes, respectively, which are referred to in the earlier sections. For equal sample sizes, tables of constants needed to carry out the procedure R and also tables of the expected proportion of the k populations included in the selected subset are appended for selected cases.

(The non-mathematical reader may wish at this point to refer to Section 4 where the procedure and the uses of the table are described and illustrated.)

3. Formal Statement of the Problems

Let $\Pi_1, \Pi_2, \cdots, \Pi_k$ be k given binomial populations with unknown probabilities of success on a single trial $p_1, p_2, \cdots, p_k$, respectively, where $0 \leqq p_i \leqq 1$, $i = 1, 2, \cdots, k$. The ranked probabilities are denoted by

$$p_{[1]} \leqq p_{[2]} \leqq \cdots \leqq p_{[k]}\,, \tag{3.1}$$

and it is assumed that there is no *a priori* information available about the correct pairing of the ordered $p_{[i]}$ and the k given populations from which observations are taken.

Any population whose parameter value equals $p_{[k]}$ will be defined as a best population. A correct selection (CS) is defined as a selection of any subset of the k given populations which contains at least one best population. In computing the probability of a correct selection $P\{CS\}$, it will clearly be conservative to assume that $p_{[k]} > p_{[k-1]}$ since the equality of $p_{[k]}$ and $p_{[k-1]}$ can only increase this probability. However, in finding the infimum of the $P\{CS\}$ with respect to the $p_{[i]}$, $i = 1, 2, \cdots, k$, it will be necessary to consider configurations with $p_{[k]} = p_{[k-1]} = \cdots = p_{[1]}$ $(= p$, say) as a limit of other configurations in which $p_{[k]} > p_{[k-1]}$. Hence, in such a limiting configuration with a common p, we shall regard one particular "tagged" population as being *the best* population in order to find the infimum of the $P\{CS\}$. In general, for $p_{[k]} = p_{[k-1]}$ the phrase "the best population" (instead of "a best population") will imply that a particular population whose parameter value is equal to $p_{[k]}$ has been tagged as the best population.

Let n_i denote the number of observations from Π_i, $i = 1, 2, \cdots, k$; the vector $\boldsymbol{n} = \{n_1, n_2, \cdots, n_k\}$ is assumed to be given in the primary problem below. Let P^* be a preassigned constant with $0 \leqq P^* \leqq 1$. Let $P\{CS; k, \boldsymbol{n}, \boldsymbol{p}, R\}$ denote the probability of a correct selection when the procedure R is used with the given k, $\boldsymbol{n}$, and when the true configuration of parameter values is $\boldsymbol{p} = \{p_1, p_2, \cdots p_k\}$; let the space of all possible values of $\boldsymbol{p}$ be denoted by Ω.

The problem of primary interest is to define a procedure R which selects a subset of the k given populations that is small, never empty, and yet large enough so that it contains the best population with probability at least P^*, regardless of the true configuration $\boldsymbol{p}$ in Ω, i.e., so that,

$$\inf_{\boldsymbol{p} \text{ in } \Omega} P\{CS; k, \boldsymbol{n}, \boldsymbol{p}, R\} \geqq P^* . \tag{3.2}$$

It is clear that we prefer procedures which make the selected subset as small as possible subject to (3.2), since one can always attain any specified P^* (even unity) by putting all the populations in the selected subset (without using the observations).

After having defined a particular procedure $R = R(k, \boldsymbol{n}, P^*)$ for each possible set of values of k, $\boldsymbol{n}$, and P^*, we consider in Section 6 two secondary problems within this class of proposed procedures. Let $E\{S; k, \boldsymbol{n}, \boldsymbol{p}, P^*, R\}$ denote the expected size of the selected subset when the proposed procedure R is used with the given k, $\boldsymbol{n}$, P^*, and when $\boldsymbol{p}$ is the true parameter configuration in Ω. Let $\boldsymbol{n}_0 = \{n, n, \cdots, n\}$ denote a k-vector all of whose components equal n. Regarding n as a variable, we write $R(\boldsymbol{n}_0)$, or simply $R(n)$, to denote the proposed procedure R [which satisfies (3.2) for the given k and P^*] which is being considered as a function of n. Let ε be a preassigned constant with $0 \leqq \varepsilon \leqq k-1$.

One secondary problem is to find the smallest common sample size n such that for some particular set of parameter values $\boldsymbol{p}_0$ in Ω

$$E\{S; k, \boldsymbol{n}_0, \boldsymbol{p}_0, P^*, R(n)\} \leqq 1 + \varepsilon . \tag{3.3}$$

An alternative secondary problem is to find the smallest common sample size n such that

$$\sup_{\boldsymbol{p} \text{ in } \Omega(\delta)} E\{S; k, \boldsymbol{n}_0, \boldsymbol{p}, P^*, R(n)\} \leqq 1 + \varepsilon , \tag{3.4}$$

where $\Omega(\delta)$ is the set of all configurations $\boldsymbol{p}$ in which $p_{[k]} - p_{[k-1]} = \delta$.

Numerical examples of (3.3) and (3.4) are given in Illustrations 3 and 4, respectively, in Section 6.

4. Procedure R for Equal Sample Sizes

Let x_i denote observed number of successes based on n observations from Π_i, $i = 1, 2, \cdots, k$, and let $x_{\max}$ denote the largest of the observed values of x_i, $i = 1, 2, \cdots, k$. Then the procedure R is

"Retain in the selected subset only those populations Π_i for which

$$x_i \geqq x_{\max} - d, \tag{4.1}$$

where $d = d(n, k, P^*)$ is a non-negative constant."

The constant d is the smallest non-negative integer that will satisfy the basic probability requirement (3.2) for all configurations $\mathbf{p} = \{p_1, p_2, \cdots, p_k\}$ and is given in the appended Table 2, Parts A, B, C, and D for selected values of n, k, and P^*.

Illustration 1. Fifty independent observations ($n = 50$) are to be taken from each of $k = 5$ binomial populations, and observations from different populations are also assumed to be independent. Before the observations are taken, the experimenter decides to guarantee a probability $P^* = .95$ of a correct selection regardless of the true p's, and he consults Table 2C to find that $d = d(50, 5, .95) = 11$. The number of successes in the five populations are observed to be 32, 28, 40, 35, and 25. Hence $x_{\max} - d = 29$, and according to procedure R only the three populations that gave 32, 35, and 40 successes are retained in the selected subset. Then the experimenter can assert with 95% confidence that one of the three retained populations is best or, equivalently, that with 95% confidence the two eliminated populations have p-values which are less than or equal to the maximum of the p-values of the three retained populations.

It will be shown in Section 9 that the probability of a correct selection $P\{CS\}$ assumes its minimum value when all components of the vector $\mathbf{p}$ are equal. Hence attention can be restricted to the cases $\mathbf{p} = \{p, p, \cdots, p\}$, and the integer d needed for (4.1) is the smallest, non-negative integer for which

$$\inf_{0 \leqq p \leqq 1} P\{CS \mid p, d\} \geqq P^*, \tag{4.2}$$

where

$$P\{CS \mid p, d\} = \sum_{\alpha=0}^{n} \binom{n}{\alpha} p^{\alpha} (1-p)^{n-\alpha} \left[\sum_{j=0}^{\alpha+d} \binom{n}{j} p^{j} (1-p)^{n-j} \right]^{k-1}, \tag{4.3}$$

and where the arguments n, k are dropped from the left-hand side of (4.2) and R is replaced by d. Such a value of d ($\leqq n$) must exist, since for $d = n$ (when all populations are retained in the selected subset) the probability (4.3) is identically unity for all p.

Normal Approximation. The inequality (4.1) which defines the procedure R can be written as

"Retain Π_i if and only if for all $j \neq i$

$$\frac{x_j - np}{\sqrt{npq}} \leqq \frac{x_i - np}{\sqrt{npq}} + \frac{d + \frac{1}{2}}{\sqrt{npq}}. \text{''} \tag{4.4}$$

Then the normal approximation to the binomial (with continuity correction already taken into account) gives the equation approximating d as

$$\inf_{0 \leqq p \leqq 1} \int_{-\infty}^{\infty} \left\{ F\left(x + \frac{\tilde{d} + \frac{1}{2}}{\sqrt{npq}} \right) \right\}^{k-1} f(x)\, dx = P^* , \tag{4.5}$$

where $F(x)$ and $f(x)$ are the standard normal c.d.f. and density functions, respectively, and $\tilde{d}$ need not be an integer. The left-hand member of (4.5) is minimized at $p = \frac{1}{2}$, and hence we obtain for moderate k and large n the approximating equation

$$\int_{-\infty}^{\infty} \{F(x+u)\}^{k-1} f(x)\, dx = P^* , \tag{4.6}$$

where

$$\tilde{d} = \tfrac{1}{2}(u\sqrt{n} - 1) , \tag{4.7}$$

and the recommended approximation d' is the smallest integer equal to or greater than $\tilde{d}$. The values of u satisfying (4.6) have been tabulated in [3] for $k = 2(1)\ 10$ and in [1] for $k = 2(1)\ 50$. It is interesting to point out that in all cases in which both exact and approximate values were computed, the recommended approximation d' was conservative (i.e., $d' \geqq d$) and never exceeded the exact value by more than unity (i.e., $d' \leqq d + 1$).

It is of some interest to note that if we define $d(p) = d(p \mid n, k, P^*)$ as the smallest integer for which $P\{CS \mid p, d\} \geqq P^*$, then $d(p)$ can be regarded as the P^*-percentage point of the distribution of the statistic

$$v = \max\ [x_1, x_2, \cdots, x_{k-1}] - x_0 , \tag{4.8}$$

where $x_0, x_1, \cdots, x_{k-1}$ are all independent chance variables having binomial distributions with a common parameter p and based on a common number n of observations. The cumulative distribution function of v, i.e., $P\{v \leqq d\}$, is given by (4.3). Since for fixed k and large n the minimum in (4.2) occurs near $p = \frac{1}{2}$, the quantity $d(\frac{1}{2})$ can be used as a first approximation to the exact value of d.

5. Procedure R for General Sample Sizes

Let $y_i = x_i/n_i$, where x_i is the number of successes and n_i is the number of observations in the ith population, $i = 1, 2, \cdots, k$, and let $y_{\max}$ denote the largest of the observed y_i. Let $\boldsymbol{n}$ denote the vector $\{n_1, n_2, \cdots, n_k\}$. Then the procedure R for n_i not necessarily equal takes the form

"Retain in the selected subset only those populations Π_i for which

$$y_i \geqq y_{\max} - c , \tag{5.1}$$

where $c = c(\boldsymbol{n}, k, P^*)$ is a non-negative constant that can be determined in advance of experimentation."

The constant c is the smallest non-negative number for which

$$\min_{i=1,2,\cdots,k} \left\{ \inf_{0 \leqq p \leqq 1} P\{CS; p, c, n_i\} \right\} \geqq P^* , \tag{5.2}$$

where

(5.3) $P\{CS; p, c, n_i\}$

$$= \sum_{\alpha=0}^{n_i} \binom{n_i}{\alpha} p^{\alpha} (1-p)^{n_i-\alpha} \prod_{\substack{j=1 \\ j\neq i}}^{k} \left\{ \sum_{\beta=0}^{\left[n_j\left(\frac{\alpha}{n_i}+c\right)\right]} \binom{n_j}{\beta} p^{\beta}(1-p)^{n_j-\beta} \right\}$$

is the probability that a single tagged population with n_i observations is retained in the selected subset and where $[x]$ denotes the largest integer less than or equal to x. The arguments $\boldsymbol{n}$, k, and R in the left-hand members of (5.2) and (5.3) are understood and not shown. The justification for using a common p-value in (5.2) and (5.3) is shown in Section 9. Such a value of c ($\leqq 1$) must clearly exist, since for $c = 1$ (when all populations are retained in the selected subset) the probability in (5.3) is identically unity for all p.

If the components of $\boldsymbol{n}$ are all large, then (5.3) can be approximated by

$$P\{CS; p, c, n_i\} \cong \int_{-\infty}^{\infty} f(x) \prod_{\substack{j=1 \\ j\neq i}}^{k} \left\{ F\left(x\sqrt{\frac{n_j}{n_i}} + c\sqrt{\frac{n_j}{pq}} \right) \right\} dx\,. \tag{5.4}$$

Equation (5.4) shows that the infimum in (5.2) is attained for large sample sizes by setting the common p equal to $\frac{1}{2}$ and tagging one of the populations as best. This result is quite useful if the sample sizes are not too small, as is shown in the following illustration. Equation (5.4) appears to indicate that the minimum in (5.2) is attained for large sample sizes by taking n_i to be the *largest* component in $\boldsymbol{n}$. Actually, it was empirically found, for fixed $\boldsymbol{n}$ with moderately large components, that this is generally true only for an interval of c-values of the form $[0, c_0]$ and that the reverse may be true for larger c-values (i.e., the minimum may be attained by taking n_i to be the *smallest* component in $\boldsymbol{n}$) as shown in the illustration below. With this information, it is not difficult to solve for c (by trial and error with a table of binomial probabilities) the equation obtained by setting the right-hand member of (5.3) equal to P^*.

It is interesting and useful to note that, for the special case $k = 2$, Equation (5.4) can be written in the simpler form

$$P\{CS; p, c, n_i\} \cong F\left(\frac{c}{\sqrt{pq\left(\frac{1}{n_1} + \frac{1}{n_2}\right)}} \right), \tag{5.5}$$

which is independent of i. In this case, if n_1 and n_2 are large enough, then (5.5) shows that both possible assignments of sample sizes give approximately the same probability of a correct selection and that the common "worst" value of p will be close to $\frac{1}{2}$. A stronger result, which shows that we need consider only one of the two possible assignments in this case for any n_1 and n_2, is proved in Section 9.

Illustration 2. An experimenter has $k = 3$ binomial populations from which he has taken $n_1 = 20$, $n_2 = 40$, and $n_3 = 40$ observations. Since the tables are not available, he wishes to compute the appropriate c-value needed to carry out the procedure R for selecting a subset containing the best population with a guarantee of at least $P^* = .75$ for the probability of a correct selection, regardless of the true configuration of p-values.

Since the sample sizes are not too small, he assumes that the infimum in (5.2) is attained when there is a common p close to $\frac{1}{2}$. A rough initial estimate of the required c is obtained by computing $\bar{n} = 100/3 = (20 + 40 + 40)/3$; assuming that all components of $\boldsymbol{n}$ are equal to $\bar{n}$, we enter Table 2A and find the rough estimate of c to be $4 \div (100/3) = .12$. From the form of (5.3) it is easily seen that we can assume that c is some multiple of 1/40 (since the probability of a correct selection does not change between multiples of 1/40). Hence we try $c = .100$ and $c = .125$ for $i = 1$ (i.e., $n_i = 20$) and $i = 2$ (i.e., $n_i = 40$). An exact computation based on (5.3) with $p = \frac{1}{2}$ and $i = 1$ gives .7001 and .7632 for $c = .100$ and $c = .125$, respectively. For $p = \frac{1}{2}$, $i = 2$, and $c = .125$, the result is .7730. Hence, the correct value is assumed to be $c = .125$ and the procedure R is carried out as in Illustration 1. As a partial check on this result, we computed (5.3) with $c = .125$ for $p = .45$, .48, and .49 and obtained .7639, .7630, and .7630, respectively; the four decimal places are purposely given here in order to show that the minimum (if it is a unique minimum) occurs between $p = .48$ and $p = .49$.

6. Expected Size of the Selected Subset for Procedure *R*

For the procedure R, the size S of the selected subset is a chance variable which can take on only integer values from 1 to k, inclusive. For any fixed values of $\boldsymbol{n}$, k, and P^* the expected size of the selected subset is a function of the true configuration $\boldsymbol{p} = \{p_1, p_2, \cdots, p_k\}$, and this function, in analogy with power functions associated with the usual tests of simple hypotheses, will be regarded as a criterion of the efficiency of any procedure which satisfies the basic probability requirement (3.2). Carrying on the analogy with power functions, we shall be concerned with finding the common sample size n necessary either to control $E\{S\}$ at some preassigned level for a particular alternative in the parameter space or to control the maximum $E\{S\}$ over a given subset of the parameter space.

Let Y_i denote a chance variable which equals 1 if Π_i is included in the selected subset and equals 0 otherwise. Then for any values of $\boldsymbol{n}$, $\boldsymbol{p}$, k, and P^*,

$$E\{S\} = E\left\{\sum_{i=1}^{k} Y_i\right\} = \sum_{i=1}^{k} E\{Y_i\} = \sum_{i=1}^{k} P\{\Pi_i \text{ is selected}\}\,. \tag{6.1}$$

Let x_i denote the number of successes observed from the population Π_i associated with p_i and based on n_i observations, $i = 1, 2, \cdots, k$. Then Π_i will be included in the selected subset if and only if the observed proportion of successes from each of the other populations does not exceed $(x_i/n_i)+c$ where c is defined by the basic probability requirement (3.2). Hence, dropping the subscript on x_i, we obtain from (6.1)

$$E\{S\} = \sum_{i=1}^{k}\sum_{x=0}^{n_i}\binom{n_i}{x}p_i^x(1-p_i)^{n_i-x}\prod_{\substack{\alpha=1\\ \alpha\neq i}}^{k}\left\{\sum_{j=0}^{\left[n_\alpha\left(\frac{x}{n_i}+c\right)\right]}\binom{n_\alpha}{j}p_\alpha^j(1-p_\alpha)^{n_\alpha-j}\right\}, \tag{6.2}$$

where $[y]$ is the largest integer less than or equal to y. For the special case $n_1 = n_2 = \cdots = n_k = n$, we replace $[nc]$ by d (or assume that c is of the form d/n where d is an integer) and obtain

$$(6.3) \qquad E\{S\} = \sum_{i=1}^{k} \sum_{x=0}^{n} \binom{n}{x} p_i^x (1-p_i)^{n-x} \prod_{\substack{\alpha=1 \\ \alpha \neq i}}^{k} \left\{ \sum_{j=0}^{x+d} \binom{n}{j} p_\alpha^j (1-p_\alpha)^{n-j} \right\},$$

For any d, k, and n, the expression above is symmetric in the unordered parameters p_i and can also be written, without any change in the form of the expression, in terms of the ordered $p_{[i]}$. In particular, for the special set of configurations

$$(6.4) \qquad p_{[k]} = p + \delta\,; \qquad p_{[k-1]} = p_{[k-2]} = \cdots = p_{[1]} = p\,,$$

where $0 \leqq \delta \leqq 1$ and $0 \leqq p \leqq 1 - \delta$, Equation (6.3) reduces to

$$(6.5) \qquad E\{S\} = \sum_{x=0}^{n} \binom{n}{x} (p+\delta)^x (q-\delta)^{n-x} \left\{ \sum_{j=0}^{x+d} \binom{n}{j} p^j q^{n-j} \right\}^{k-1}$$
$$+ (k-1) \sum_{x=0}^{n} \binom{n}{x} p^x q^{n-x} \left\{ \sum_{j=0}^{x+d} \binom{n}{j} p^j q^{n-j} \right\}^{k-2} \left\{ \sum_{=0}^{x+d} \binom{n}{j} (p+\delta)^j (q-\delta)^{n-j} \right\},$$

where $q = 1 - p$. For $\delta > 0$, the first of the two expressions in (6.5) is the $P\{CS\}$ for configuration (6.4).

It is interesting to note that for any fixed n, k, d, and δ the expression in (6.5) has the same value at $p = 0$ and at $p = 1 - \delta$, which is given by

$$(6.6) \qquad E\{S\} = 1 + (k-1) \sum_{j=0}^{d} \binom{n}{j} \delta^j (1-\delta)^{n-j} \,.$$

In fact, for $p = 1 - \delta$, the last summation in (6.5) is 0 if $x < n - d$ and is unity if $x \geqq n - d$ and, in the latter case, the next to the last summation also reduces to unity since $d \geqq 0$; the first two summations in (6.5) also reduce to unity for $p = 1 - \delta$, and the result (6.6) easily follows. For $p = 0$, the same result (6.6) is straightforward.

For the special case $k = 2$, it is indicated in Section 9 that $E\{S\}$ has certain symmetries; for the true configuration (6.4) with $n_1 = n_2$ ($= n$ say) this reduces to a symmetry about $p = (1-\delta)/2$ [or about $p + \delta = (1+\delta)/2$].

Values of the *proportion* of populations retained, $E\{S\}/k$, computed from (6.5) for the true configuration (6.4) are given in Table 3 for selected values of n, k, δ, p, and P^*; in each case the value of d used in (6.5) is the one to be found in Table 2*A*, *B*, *C*, or *D*, for the given values of n, k, and P^*.

It should be added that we can also obtain a normal approximation to $E\{S\}$. For the special case when the true configuration is given by (6.4) and $n_1 = n_2 = \cdots = n_k = n$, this reduces to

$$(6.7) \qquad E\{S\} \cong \int_{-\infty}^{\infty} \left\{ F\left(x \sqrt{\frac{(p+\delta)(q-\delta)}{pq}} + (c+\delta) \sqrt{\frac{n}{pq}} \right) \right\}^{k-1} f(x)\,dx$$
$$+ (k-1) \int_{-\infty}^{\infty} \left\{ F\left(x + c \sqrt{\frac{n}{pq}} \right) \right\}^{k-2}$$
$$\cdot F\left(x \sqrt{\frac{pq}{(p+\delta)(q-\delta)}} + (c-\delta) \sqrt{\frac{n}{(p+\delta)(q-\delta)}} \right) f(x)\,dx\,,$$

where $f(x)$ and $F(x)$ are the standard normal density and c.d.f., respectively For the special case $k = 2$, we find [either by evaluating the integrals in (6.7) or by using a normal approximation to the difference of two proportions] that (6.6) reduces to the simpler form

$$(6.8) \quad E\{S|k=2\} \cong F\left(\frac{(c-\delta)\sqrt{n}}{\sqrt{pq+(p+\delta)(q-\delta)}}\right) + F\left(\frac{(c+\delta)\sqrt{n}}{\sqrt{pq+(p+\delta)(q-\delta)}}\right).$$

It is interesting to note that for fixed c, δ, and n the expression in (6.8) is maximized by setting $p = (1-\delta)/2$ or $q = (1+\delta)/2 = p+\delta$. This can be numerically observed for $k = 2$ in Table 3 ($\delta = .50$) for the larger values of n and the smaller values of P^*.

It should also be noted for any k, any P^*, and under the configuration (6.4), that the limit of the expected proportion of populations retained, as the common number n of observations tends to infinity, equals $1/k$ for $\delta > 0$ and any p, and the limit equals P^* for $\delta = 0$, provided that $pq \neq 0$. (These results can be obtained from (6.7) and (6.1), respectively.) For unequal sample sizes, all tending to infinity in any manner, the result for $\delta > 0$ remains true for any k; the result for $\delta = 0$ remains true for $k = 2$ [since the expected proportion retained is asymptotically equal to the expression for $P\{CS\}$ in (5.5)], but for $k \geqq 3$ we can say only that the limit is greater than or equal to P^*. For the equal-sample-size case, it is interesting to compare these results with the values in the appended tables for the expected proportion of populations retained.

Illustration 3. An experimenter wishes to carry out procedure R for $k = 10$ and $P^* = .90$; the common sample size n to be taken from each population has not yet been determined. The experimenter decides that he would like to use the smallest n such that, if the true configuration is given by (6.4) with $\delta = .25$ and $p + \delta = .75$ then he can expect to eliminate at least three of the ten populations. From Table 3 with $\delta = .25$, we find that, corresponding to an expected proportion retained of at most .70, the smallest value of n required lies between 20 and 25; linear or graphical interpolation gives the estimated answer as $n = 24$. A check on this answer can be obtained, if desired, by computing (6.5) with $n = 24$ and the data of this problem; this gives an expected proportion retained of .694, which satisfies the requirement. Table 2B then shows that the appropriate d-value corresponding to $n = 24$ is $d = 7$, and the procedure R can now be carried out with the observed data as in Illustration 1.

Illustration 4. An experimenter wishes to carry out procedure R for $k = 5$ and $P^* = .75$; the common sample size n is not yet determined, as in Illustration 3. The experimenter decides that he would like to use the smallest value of n such that, if the true configuration is given by (6.4) with $\delta = .25$ and any p with $0 \leqq p \leqq 1 - \delta = .75$, then he can expect to eliminate two of the five populations regardless of the true p-value. Table 3 with $k = 5$, $\delta = .25$, and $P^* = .75$ shows that 20 observations are sufficient to accomplish the required goal (i.e., to keep the proportion retained under .60 for all values of p) and that $n = 18$ or $n = 19$ would be a reasonable estimate of the smallest value of n with this property. A check on the maximum proportion retained for $n = 18$ and $n = 19$ [which can be calculated from (6.5) with $p + \delta = 1.00$] gives .615 and .572, respectively, and so the accepted answer is $n = 19$. (These values, .615 and .572, must be regarded as conjectured maxima, since it has not been shown that for $0 \leqq \delta < \frac{1}{2}$ there is a unique maximum of $E\{S\}$ which is attained in the closed interval $\frac{1}{2} \leqq p + \delta \leqq 1$.)

By Table 2A, for $n = 19$, $k = 3$, and $P^* = .75$, we need a d-value of 4, and the procedure R can then be carried out with the observed data as in Illustration 1.

7. Other Procedures Based on Transformation of Variables

Another procedure of interest is based on the transformed variables

$$z_i = 2 \operatorname{arc} \sin \sqrt{x_i/n_i}\,, \qquad i = 1, 2, \cdots, k\,. \tag{7.1}$$

Let $z_{\max}$ denote the largest of the observed values of the z_i, $i = 1, 2, \cdots, k$. It is well known ([5], pp. 377-82) that for large n_i the statistic z_i is approximately normal with mean 2 arc sin $\sqrt{p_i}$ and variance $1/n_i$. Hence, for large n_i, we can use the following procedure R' for selecting the subset containing the best population.

"Retain in the selected subset only those populations Π_i for which

$$z_i \geqq z_{\max} - b\,, \tag{7.2}$$

where $b = b(\boldsymbol{n}, k, P^*)$ is a non-negative constant to be determined."

For large values of n_i we replace all the z_i in (7.2) by normal chance variables, and this converts the problem into the one involving location parameters of normal populations which has been solved in [1]. The equation determining b becomes

$$\min_{i=1,2,\ldots,k} \int_{-\infty}^{\infty} \prod_{\substack{j=1\\ j\neq i}}^{k} F\left[\sqrt{n_j}\left(\frac{x}{\sqrt{n_i}} + b\right)\right] f(x)\, dx = P^*\,. \tag{7.3}$$

For $n_1 = n_2 \cdots = n_k = n$ (say), this reduces to

$$\int_{-\infty}^{\infty} \{F(x + b\sqrt{n})\}^{k-1} f(x)\, dx = P^*\,, \tag{7.4}$$

which is solvable by using the tables in [1] or [3] as noted in Section 4. It is interesting to note that with the result given by equation (7.4) no new tables are needed.

Modifications of the arc sine transformation as suggested in [5] and [6] can also be used in place of (7.1) to improve the approach to normality.

It is possible to compute the exact probability of a correct selection by using (7.2) with z_i replaced by z_i' which is defined by

$$z_i' = \operatorname{arc} \sin \sqrt{\frac{x_i}{n_i + 1}} + \operatorname{arc} \sin \sqrt{\frac{x_i + 1}{n_i + 1}}\,, \tag{7.5}$$

as in [6] and by comparing this with the P^*-value which is a lower bound, under the assumption that the z_i' are exactly normally distributed. This investigation showed that the exact probability of a correct selection does occasionally dip slightly below the P^*-value. A brief table of this maximum error (or maximum dip below the P^*-values) over all values of $p\,(0 \leq p \leq 1)$ is given in Table 1.

TABLE 1

MAXIMUM VALUE OF $[P^* - P\{CS\}]$

[Under Rule (7.2) with z_i replaced by z_i' defined by (7.5).]

n \ P^*	0.75	0.90	0.95	0.99
5	0.008	0.000	0.005	0.001
10	0.012	0.010	0.008	0.004
15	0.042	0.013	0.000	0.006
20	0.014	0.014	0.004	0.008

8. Mathematical Results for Equal Sample Sizes

In order to prove for $k = 2$, for $n_1 = n_2$ ($= n$, say), and for any d with $0 \leqq d \leqq n-1$ that the probability of a correct selection approaches its minimum value as p_1 and p_2 (with $p_1 \neq p_2$) both approach $\frac{1}{2}$, it is sufficient to regard the best population as being "tagged" and show that for $p_1 = p_2$ ($= p$, say), the probability of selecting a subset containing the tagged population attains a minimum at $p = \frac{1}{2}$. To show the latter result we shall first prove a more general result about symmetric, unimodal distributions that will yield the desired result as a corollary.

Let $\mathscr{X}_i$ denote a set of equally spaced values z_i with a common spacing d; the value of d is assumed to be the same for all i, $i = 1, 2, \cdots, n$. Let Z_i, $i = 1, 2, \cdots, n$, denote a given set of n discrete, independent chance variables such that Z_i assigns probability 1 to $\mathscr{X}_i$ and such that the distribution of Z_i is symmetric and unimodal with respect to $\mathscr{X}_i$. We assume, without loss of generality, that $d = 1$ and that each Z_i is symmetric about zero. (If the Z_i had points of symmetry μ_i, not neccessarily the same, then they could be made symmetric about zero by translation, and the properties of symmetry and unimodality of sums of the Z_i are not affected by such translations.) In making the assumption of symmetry, we obtain two types (say Type I and Type II) according as the translated Z_i takes on only integer values or only half-integer values, viz., $\pm\frac{1}{2}, \pm\frac{3}{2}$, etc. Since Z_i is symmetric about zero, we define unimodality with respect to $\mathscr{X}_i$ simply by

$$r_i(j) \geqq r_i(j+1)\,, \qquad j = j_0, j_0 + 1, \cdots, \tag{8.1}$$

where $r_i(j)$ denotes $P\{Z_i = j\}$ and $j_0 = 0$ or $\frac{1}{2}$ according as Z_i is Type I or II. Strict unimodality is defined by requiring that (8.1) hold with strict inequality whenever either side is positive.

It now follows that Z_i takes on with positive probability a succession of integers 0, $\pm 1, \cdots, \pm t_i$ where $0 \leqq t_i \leqq \infty$, or a succession of half-integers $\pm\frac{1}{2}, \pm\frac{3}{2}, \cdots, \pm t_i$ where $\frac{1}{2} \leqq t_i \leqq \infty$.

Let $S_n = Z_1 + Z_2 + \cdots + Z_n$, let $a_n(j) = P\{S_n = j\}$, and let $A_n(d) = P\{S_n \leqq d\}$. In Theorem 1 we shall consider only Z_i of Type I. We shall regard $A_n(d)$ as a function only of $r_i(1), r_i(2), \cdots, r_i(t_i)$, making use of the assumed symmetry and the relation

$$r_i(0) = 1 - 2\sum_{j=1}^{t_i} r_i(j)\,. \tag{8.2}$$

We first require a lemma which proves the unimodality of S_n, in which the different Z_i need not be of the same type. Let $m(Z_i)$ denote the number of values that Z_i takes on with positive probability.

LEMMA 1. *Under the above assumptions on Z_i, $i = 1, 2, \cdots, n$, the distribution of S_n is also symmetric and unimodal; the set $\mathscr{X}$ associated with S_n depends on whether the number of Z_i of Type II is odd or even. If, in addition, at least one Z_i (say, Z_2) is strictly unimodal and $m(Z_2) \geqq \max_i[m(Z_i)-1]$, or if the Z_i are unimodal with finite t_i and such that $|m(Z_i) - m(Z_j)| \leqq 1$ for all pairs i, j, then S_n is strictly unimodal.*

PROOF. It is sufficient to consider the sum S_2 of Z_1 and Z_2, since the rest of the proof follows easily by induction. The fact that S_2 is symmetric (about zero) and takes on only integers or only half-integers is clear and the proof is omitted. Using the symmetry of Z_2, we have for any j

$$a_2(j) = r_2(t_2)\, r_1(j - t_2) + r_2(t_2 - 1)\, r_1(j - t_2 + 1) + \cdots + r_2(t_2)\, r_1(j + t_2)\,, \tag{8.3}$$

$$a_2(j) - a_2(j+1) = \sum_{\beta=0}^{[t_2]} \{r_2(t_2' + \beta) - r_2(t_2' + \beta + 1)\} \cdot \{r_1(j - t_2' - \beta) - r_1(j + 1 + t_2' + \beta)\}\,, \tag{8.4}$$

where $[t_2]$ denotes the largest integer not greater than t_2 (which may be infinite) and $t_2' = t_2 - [t_2] = 0$ if t_2 is an integer or $+\infty$ and $t_2' = \frac{1}{2}$ otherwise. The first factor is non-negative since Z_2 is unimodal and the second is also, since for $j \geqq 0$ and $\beta \geqq 0$ we have $|j - t_2' - \beta| \leqq j + 1 + t_2' + \beta$, and the unimodality of Z_1 can be applied.

If (say) Z_2 is given to be strictly unimodal and $t_2 \geqq t_1 - \frac{1}{2}$, then the first factor is positive for every β with $0 \leqq \beta \leqq [t_2]$, and it suffices to find any one positive second factor. Clearly we can assume that $t_1 > 0$. If $t_2 = \infty$, then for any j the first argument in the second factor of (8.4) runs from $j \geqq 0$ to $-\infty$, and there must exist a value of β such that $r_1(j - \beta) = r_1(\beta - j) > r_1(\beta - j + 1)$. If t_2 is finite, then t_1 is also. For $t_2 - t_1 \leqq j \leqq t_2 + t_1$, we need only take $\beta = [t_2]$ in order that

$$|j - t_2' - [t_2]| = |j - t_2| \leqq t_1 < 1 + t_2 \leqq 1 + j + t_2' + [t_2]\,, \tag{8.5}$$

and it follows from the definition of t_1 that the second factor is positive. For $0 \leqq j \leqq t_2 - t_1 - \frac{1}{2}$, we take $\beta = j - t_2' + t_1$ in order that

$$|j - t_2' - \beta| = t_1 < 1 + 2j + t_1 \leqq 1 + j + t_2' + \beta\,, \tag{8.6}$$

and the second factor is again positive as in (8.5).

For the remainder of the lemma we assume $t_1 \leqq t_2$. It follows from the assumption $|m(Z_1) - m(Z_2)| \leqq 1$ that $t_1 \leqq t_2 \leqq t_1 + \frac{1}{2}$. For $\beta = [t_2]$ in (8.4) the first factor is positive, and for $t_2 - t_1 \leqq j \leqq t_2 + t_1$ the argument in (8.5) again holds. We need consider only $0 \leqq j \leqq t_2 - t_1 - \frac{1}{2}$, which reduces to the single case $j = 0$. However, $j = 0$ is a value for S_2 only if Z_1 and Z_2 are of the same type, and then we must have $t_1 = t_2$, which was already considered. For n chance variables Z_i, we can start by convoluting two (say, Z_1 and Z_2), where Z_1 is chosen so that $m(Z_1) = \max_i m(Z_i)$. Then S_2 is strictly unimodal and $m(S_2) \geqq \max_{i \geqq 3} m(Z_i)$, and the previous part of the lemma can be used. This proves the lemma.

It is interesting to note that the last part of the lemma cannot be extended to infinite t_i as is shown by the following counterexample:

Z_i $(i = 1, 2)$	$\cdots$	-4	-3	-2	-1	0	1	2	3	4	$\cdots$
$P(Z_1)$	$\cdots$	$\frac{1}{32}$	$\frac{1}{32}$	$\frac{1}{16}$	$\frac{1}{16}$	$\frac{1}{2}$	$\frac{1}{16}$	$\frac{1}{16}$	$\frac{1}{32}$	$\frac{1}{32}$	$\cdots$
$P(Z_2)$	$\cdots$	$\frac{1}{64}$	$\frac{1}{32}$	$\frac{1}{32}$	$\frac{1}{4}$	$\frac{1}{4}$	$\frac{1}{4}$	$\frac{1}{32}$	$\frac{1}{32}$	$\frac{1}{64}$	$\cdots$.

Then, for $S_2 = Z_1 + Z_2$, we compute $P\{S_2 = 0\} = P\{S_2 = 1\} = 21/2^7$. (For the case of identically distributed chance variables the lemma also holds for infinite t_i but the proof is omitted.) It follows as a corollary from the lemma that S_n is strictly unimodal if all the Z_i are strictly unimodal or if they are all unimodal and identically distributed with a common finite range.

This lemma is related to a result of A. Wintner (see p. 255 of [7]), which proves the closure property under convolution for symmetric, unimodal distributions that possess densities.

It will also be useful when the t_i are all finite to define the non-negative integers $t_{\max} = \max t_i$, $i = 1, 2, \cdots, n$,

$$\underline{t} = \max\{0, t_{\max} - 1 - \bar{t}\}\,, \quad \text{and} \quad \bar{t} = \sum_{i=1}^{n} t_i - t_{\max}\,. \tag{8.7}$$

THEOREM 1. *If the Z_i, $i = 1, 2, \cdots, n$, are independent, symmetric about zero and unimodal with respect to the integers (i.e., of Type I), then $A_n(d) = P\{S_n \leqq d\}$ for $d \geqq 0$ is a non-increasing function of $r_i(j)$ for each i, $i = 1, 2, \cdots, n$, and each j, $j = 1, 2, \cdots, t_i$. If $\underline{t} \leqq \bar{t}$ and each Z_i is strictly unimodal then, for $d = \underline{t}, \underline{t} + 1, \cdots, \bar{t}$, $A_n(d)$ is a strictly decreasing function of each $r_i(j)$. (If the Z_i have points of symmetry μ_i, then the condition $d \geqq 0$ is replaced by $d \geqq \sum_{i=1}^{n} \mu_i$.)*

PROOF. It is assumed that at least one $t_i \geqq 1$, since otherwise there is nothing to prove. Regarding S_n for $n \geqq 2$ as a convolution of Z_n and S_{n-1}, we write

$$A_n(d) = \left[1 - 2\sum_{j=1}^{t_n} r_n(j)\right] A_{n-1}(d) + \sum_{\substack{j=-t_n \\ j \neq 0}}^{t_n} r_n(j)\, A_{n-1}(d - j)\,, \tag{8.8}$$

$$A_n(d) = A_{n-1}(d) + \sum_{j=1}^{t_n} r_n(j)\{[A_{n-1}(d+j) - A_{n-1}(d)] - [A_{n-1}(d) - A_{n-1}(d-j)]\}\,, \tag{8.9}$$

$$A_n(d) = A_{n-1}(d) - \sum_{j=1}^{t_n} r_n(j)\left\{\sum_{\alpha=1}^{j}\left[a_{n-1}(d-\alpha+1) - a_{n-1}(d+\alpha)\right]\right\}, \tag{8.10}$$

If $d \geqq 0$, then, since $\alpha \geqq 1$, we have the strict inequality

$$|d - \alpha + 1| < d + \alpha\,, \tag{8.11}$$

and hence by Lemma 1 above each square bracket in (8.10) is non-negative. This proves the first part of the theorem for each $r_n(j)$ and also shows that $A_n(d) \leqq A_{n-1}(d)$ for $d \geqq 0$.

For the second part we add to (8.11) the results that, since $1 \leqq \alpha \leqq t_n \leqq t_{\max}$,

$$d - \alpha + 1 \leqq \bar{t} \leqq \sum_{i=1}^{n-1} t_i\,, \qquad \text{for } d \leqq \bar{t}\,, \tag{8.12}$$

$$-(d - \alpha + 1) \leqq \bar{t} \leqq \sum_{i=1}^{n-1} t_i\,, \qquad \text{for } d \geqq t_{\max} - 1 - \bar{t}\,. \tag{8.13}$$

It follows from these results that every square bracket in (8.10) is strictly positive for $\underline{t} \leqq d \leqq \bar{t}$, and this proves the result for each $r_n(j)$ and also shows that $A_n(d) < A_{n-1}(d)$ for $\underline{t} \leqq d \leqq \bar{t}$.

A similar analysis of S_n as a convolution of Z_i and $Z_1 + \cdots + Z_{i-1} + Z_{i+1} + \cdots + Z_n$ gives the same result for each $r_i(j)$, since the bounds $\underline{t}$ and $\bar{t}$ are now independent of i. This proves the theorem for $n \geqq 2$.

For $n = 1$, the first result holds for all $d \geqq 0$; for the second result there is nothing to show unless $t_1 = 1$. In the latter case it need only be shown that the result holds for $d = 0$, and this is clear. This completes the proof of the theorem.

COROLLARY 1. *If the Z_i are identically distributed [so that we can write $t_i = t$ and $r_i(j) = r(j)$] and the common Z is strictly unimodal, then for $0 \leqq d \leqq (n-1)t$ the probability $A_n(d)$ is a strictly decreasing function of each $r(j)$, $j = 1, 2, \cdots, t$.*

PROOF. Differences along the equiangular line in n dimensions can be expressed as sums of differences on lines parallel to the coordinate axes. The sum of the differences is positive since, by Theorem 1, each difference is positive. Hence the corollary is proved.

COROLLARY 2. *If the Z_i are identically distributed (as in Corollary 1) with $t = 1$ and $r(1) = r$, say, and if the common Z is the difference of two independent and identically distributed (zero-one) binomial chance variables with common parameter p, then for $d = 0, 1, 2, \cdots, n-1$ the probability $A_n(d)$ is strictly decreasing in the interval $0 < p < \frac{1}{2}$, is strictly increasing in the interval $\frac{1}{2} < p < 1$, and is symmetric about $p = \frac{1}{2}$.*

PROOF. Here Z takes on the values -1, 0, 1 with probabilities $r = pq$, $1 - 2pq$, and pq, respectively. Hence, for $d = 0, 1, \cdots, n-1$ the function

$A_n(d)$ is a strictly decreasing function of $r = pq$ and [since $A_n(d)$ depends on p only through $r = pq$] this corollary follows from the fact that pq is strictly increasing for $0 < p < \frac{1}{2}$, is strictly decreasing for $\frac{1}{2} < p < 1$, and is symmetric about $p = \frac{1}{2}$.

The following elementary problem has an interesting connection with this paper, which will be explained below.

Problem. Some event S, which has probability p in a single throw of a single fair die, is called a success. A person continues to throw $2n$ fair dice, of which n are red and n are green, until he obtains a throw in which the observed number of red and green successes are equal and then he stops. Is the expected number of red successes in the last throw less than, equal to, or greater than np?

Mathematically, this question is easily shown to be equivalent to the question "Is the function

$$g(p) = \sum_{x=0}^{n} (x - np) \left[\binom{n}{x} p^x q^{n-x}\right]^2 \tag{8.14}$$

less than, equal to, or greater than zero?" To show the relation of this to the present paper we consider for $k = 2$, $n_1 = n_2 = n$, and any d the probability of a correct selection, the limit of which, as p_1, p_2 $(p_1 \neq p_2)$ both approach p, is

$$P = \tfrac{1}{2} + \tfrac{1}{2} \sum_{x=0}^{n} \left[\binom{n}{x} p^x q^{n-x}\right]^2 + \sum_{j=1}^{d} \sum_{x=j}^{n} \binom{n}{x} p^x q^{n-x} \binom{n}{x-j} p^{x-j} q^{n-x+j}. \tag{8.15}$$

For $d = 0$, the double summation in (8.15) vanishes. Setting $d = 0$ and differentiating with respect to p gives

$$\frac{dP}{dp} = \frac{1}{pq} \sum_{x=0}^{n} (x - np) \left[\binom{n}{x} p^x q^{n-x}\right]^2 = \frac{1}{pq} g(p). \tag{8.16}$$

For $d \geqq 0$, letting x_1 denote the number of successes from the tagged population, we have from (4.1)

$$P = P\{x_1 \geqq x_{\max} - d\} = P\{x_2 - x_1 \leqq d\} = P\left\{\sum_{i=1}^{n} Z_i \leqq d\right\} = A_n(d). \tag{8.17}$$

It now follows from Corollary 2 to Theorem 1 that dP/dp, and hence also $g(p)$, is negative for $0 < p < \frac{1}{2}$, positive for $\frac{1}{2} < p < 1$, and (by continuity) zero for $p = \frac{1}{2}$. For example, if a success S is the showing of a six on a die and $n = 6$, then the expected number of red sixes on the last throw is calculated to be $.7923 < np = 1$ since $p = \frac{1}{6} < \frac{1}{2}$.

9. Mathematical Results for General Sample Sizes

We shall now prove that the probability of a correct selection for the procedure R is minimized when all p's are equal; this will provide the justification needed for using a common p in the probability requirements (4.2) and (5.2).

Let $\boldsymbol{m} = \{m_1, m_2, \cdots, m_k\}$ be a reordering of the components of $\boldsymbol{n} = \{n_1, n_2, \cdots, n_k\}$ such that m_i is that particular one of the components $n_1, n_2, \cdots, n_k$

which is associated with $p_{[i]}$, $i = 1, 2, \cdots, k$. Then, for $p_{[k]} > p_{[k-1]}$, the probability of a correct selection is given by

$$(9.1) \qquad P\{CS; \mathbf{p}, c\} = \sum_{\alpha=0}^{m_k} \binom{m_k}{\alpha} p_{[k]}^{\alpha} (1 - p_{[k]})^{m_k - \alpha} \cdot \prod_{j=1}^{k-1} \left\{ \sum_{\beta=0}^{[m_j(\frac{\alpha}{m_k} + c)]} \binom{m_j}{\beta} p_{[j]}^{\beta} (1 - p_{[j]})^{m_j - \beta} \right\} .$$

Each of the binomial sums in the braces in (9.1) starts at $\beta = 0$ and is, therefore, a non-increasing function of $p_{[j]}$. (This can be more easily seen by writing it as an incomplete beta function, but we omit this detail.) Since $p_{[j]} \leqq p_{[k]}$, a (partial) minimum is attained by holding $p_{[k]}$ fixed and setting $p_{[j]} = p_{[k]}$ ($= p$, say) first for $j = k - 1$, then for $j = k - 2, \cdots$, and finally for $j = 1$. The product of the sums in braces in (9.1) is thus (partially) minimized by holding p fixed and minimizing the sums in braces one at a time in an ordered manner. Hence, if the number of observations m_k associated with the best population were known, then we would take the infimum of (5.3) with a common p ranging over the unit interval, with n_i replaced by m_k, and with all the other n_j $(j \neq i)$ replaced by m_j with any manner of pairing. Since the value of m_k is not known and since the value of p which minimizes (5.3) for each i is not known, it is necessary to use a double minimization as in (5.2) to obtain the over-all minimum. This completes the justification for (4.2) and (5.2).

For any fixed values of $\mathbf{n}$, k, and d, it has been observed in all the numerical computations that the expression in (4.3) approaches its infimum value when $p_{[k]} > p_{[k-1]}$ and all the $p_{[i]}$, $i = 1, 2, \cdots, k$, approach a common $p \leqq \frac{1}{2}$ and usually quite close to $\frac{1}{2}$ as shown in Illustration 2. It follows from the normal approximation (5.4) that, as the components of $\mathbf{n}$ all tend to infinity, this common "worst" value of p approaches $\frac{1}{2}$. For $k = 2$, $n_1 = n_2$ ($= n$, say) and any d, it was shown in the previous section that the infimum of (4.3) is attained when the common worst value of p is exactly equal to $\frac{1}{2}$.

For the special case $k = 2$ and any n_1, n_2, we now show that the right-hand member of (5.3) can be minimized (and hence the constant c required for procedure R can be found exactly) by considering either one of the two possible assignments, i.e., by disregarding the minimum over $i = 1, 2$ in (5.2).

Remark 1. For $k = 2$, any pair of positive integers n_1, n_2 and any c and any p with $0 \leqq c \leqq 1$ and $0 \leqq p \leqq 1$, we have

$$(9.2) \qquad P\{CS; p, c, n_1\} = P\{CS; 1 - p, c, n_2\} ,$$

where $P\{CS; p, c, n_i\}$ is defined by (5.3).

Proof. For $k = 2$, let $y_{(2)}$ and n_2 be associated with the population tagged as best, and let $y^*_{(i)} = 1 - y_{(i)}$, $i = 1, 2$. Then

$$(9.3) \qquad P\{CS; p, c, n_2\} = P\{y_{(2)} \geqq y_{(1)} - c)\} = P\{y^*_{(1)} \geqq y^*_{(2)} - c\} .$$

Since the last two expressions have the same form and since $y^*_{(i)}$ has the same distribution as $y_{(i)}$, $i = 1, 2$, except that p and q are interchanged, the result (9.2) follows.

Remark 1.1. For $k = 2$, any sample sizes n_1, n_2, and any c with $0 \leqq c \leqq 1$, the two functions $P\{CS; p, c, n_i\}$ for $i = 1, 2$ have the same infimum for $0 \leqq p \leqq 1$, the infimum being attained at points symmetrical about $p = \frac{1}{2}$.

Proof. By (9.2) the two functions are mirror images with respect to the line $p = \frac{1}{2}$ and the remark follows.

It follows that the infimum of the probability of a correct selection for $k = 2$ and any sample sizes n_1, n_2 can be obtained from either of the two functions in (5.3) and the minimum in (5.2) can be disregarded.

Corresponding to Remark 1 for the $P\{CS\}$ function, there is a similar result for the $E\{S\}$ function. For $k = 2$ and any c, let the right-hand member of (6.2) be denoted by $E\{S; p_1, p_2, n_1, n_2\}$.

Remark 2. For $k = 2$, any c with $0 \leqq c \leqq 1$, any sample sizes n_1, n_2, and any probabilities p_1, p_2 such that p_i and n_i are associated with the same population ($i = 1, 2$), we have

$$E\{S; p_1, p_2, n_1, n_2\} = E\{S; q_2, q_1, n_2, n_1\} . \tag{9.4}$$

The proof is similar to that of Remark 1 and is omitted.

Remark 2.1. For $k = 2$, any c with $0 \leqq c \leqq 1$, $n_1 = n_2$ ($=n$, say) and $p_2 = p_1 + \delta$ with $\delta \geqq 0$, the expectation $E\{S\}$, considered as a function of p_1 for $0 \leqq p_1 \leqq 1 - \delta$, is symmetric about $p_1 = (1 - \delta)/2$ [or about $p_1 + \delta = (1 + \delta)/2$].

Proof. For $n_1 = n_2$ we have symmetry about the line $p_1 = q_2$, and for $p_2 = p_1 + \delta$ with δ fixed the result follows. This symmetry about $p_1 + \delta = (1 + \delta)/2$ can be numerically observed for $k = 2$ and (say) $\delta = .50$ in the appended tables for $E\{S\}/k$.

10. Acknowledgments

The authors wish to thank Professor Wassily Hoeffding, University of North Carolina, for making some useful suggestions that led to the present form of Theorem 1 and its corollaries. The authors also wish to acknowledge the help of Miss Phyllis A. Groll and Miss Marilyn J. Huyett, both of the Bell Telephone Laboratories, who performed the tedious task of computing and checking all the tables of this paper.

[*References appear on page 248, following Table 3.*]

TABLE 2

d-VALUES

Values needed to guarantee a probability of a correct selection of at least P^* for any true configuration of parameters when procedure R is used with a common number n of observations from each of k binomial populations.

A. $P^* = .75$

k \ n	1	2	3	4	5	6	7	8	9	10	11	12	13	14	15	16	17	18	19	20	25	30	35	40	45	50	60	70	80	90	100	125	150	175	200	250	300	350	400	450	500
2	0	1	1	1	1	1	1	1	1	2	2	2	2	2	2	2	2	2	2	2	2	3	3	3	3	3	4	4	4	5	5	5	6	6	7	8	8	9	10	10	11
3	1	1	1	1	2	2	2	2	2	2	2	2	3	3	3	3	3	3	3	3	4	4	4	5	5	5	6	6	6	7	7	8	9	9	10	11	12	13	14	15	16
4	1	1	1	2	2	2	2	2	3	3	3	3	3	3	3	3	3	4	4	4	5	5	5	5	6	6	7	7	8	8	9	9	10	11	12	13	15	16	17	18	19
5	1	1	2	2	2	2	2	3	3	3	3	3	3	3	4	4	4	4	4	4	5	5	5	6	6	7	7	8	8	9	9	10	11	12	13	15	16	17	18	20	21
6	1	1	2	2	2	2	3	3	3	3	3	3	4	4	4	4	4	4	4	4	5	5	6	6	7	7	8	8	9	9	10	11	12	13	14	16	17	18	20	22	22
7	1	1	2	2	2	3	3	3	3	3	3	4	4	4	4	4	4	4	4	5	5	6	6	7	7	7	8	9	9	10	10	12	13	14	15	16	18	19	21	22	23
8	1	2	2	2	2	3	3	3	3	3	4	4	4	4	4	4	4	5	5	5	5	6	6	7	7	8	8	9	10	10	11	12	13	14	15	17	19	20	21	23	24
9	1	2	2	2	2	3	3	3	3	3	4	4	4	4	4	4	5	5	5	5	6	6	7	7	7	8	9	9	10	10	11	12	14	15	16	17	19	21	22	23	25
10	1	2	2	2	3	3	3	3	3	4	4	4	4	4	4	5	5	5	5	5	6	6	7	7	8	8	9	9	10	11	11	13	14	15	16	18	20	21	23	24	25
11	1	2	2	2	3	3	3	3	3	4	4	4	4	4	4	5	5	5	5	5	6	6	7	7	8	8	9	10	10	11	12	13	14	15	16	18	20	22	23	25	26
12	1	2	2	2	3	3	3	3	4	4	4	4	4	4	5	5	5	5	5	5	6	6	7	7	8	8	9	10	11	11	12	13	14	16	17	19	20	22	24	25	26
13	1	2	2	2	3	3	3	3	4	4	4	4	4	4	5	5	5	5	5	5	6	7	7	8	8	8	9	10	11	11	12	13	15	16	17	19	21	22	24	26	27
14	1	2	2	2	3	3	3	3	4	4	4	4	4	5	5	5	5	5	5	5	6	7	7	8	8	9	9	10	11	12	12	14	15	16	17	19	21	23	24	26	27
15	1	2	2	2	3	3	3	3	4	4	4	4	4	5	5	5	5	5	5	5	6	7	7	8	8	9	10	10	11	12	12	14	15	16	17	20	21	23	25	26	28
16	1	2	2	2	3	3	3	3	4	4	4	4	4	5	5	5	5	5	5	6	6	7	7	8	8	9	10	10	11	12	12	14	15	17	18	20	22	23	25	26	28
17	1	2	2	2	3	3	3	4	4	4	4	4	5	5	5	5	5	5	5	6	6	7	7	8	8	9	10	11	11	12	13	14	15	17	18	20	22	24	25	27	28
18	1	2	2	2	3	3	3	4	4	4	4	4	5	5	5	5	5	5	6	6	6	7	8	8	9	9	10	11	11	12	13	14	16	17	18	20	22	24	26	27	29
19	1	2	2	3	3	3	3	4	4	4	4	4	5	5	5	5	5	5	6	6	6	7	8	8	9	9	10	11	12	12	13	14	16	17	18	20	22	24	26	27	29
20	1	2	2	3	3	3	3	4	4	4	4	4	5	5	5	5	5	5	6	6	6	7	8	8	9	9	10	11	12	12	13	15	16	17	18	21	23	24	26	28	29
25	1	2	2	3	3	3	4	4	4	4	4	5	5	5	5	5	6	6	6	6	6	7	8	9	9	10	10	11	12	13	13	15	17	18	19	21	23	25	27	29	30
30	1	2	2	3	3	3	4	4	4	4	5	5	5	5	5	6	6	6	6	6	7	8	8	9	9	10	11	12	12	13	14	16	17	18	20	22	24	26	28	29	31
35	1	2	2	3	3	3	4	4	4	4	5	5	5	5	5	6	6	6	6	6	7	8	8	9	10	10	11	12	13	13	14	16	17	19	20	22	25	27	28	30	32
40	1	2	3	3	3	4	4	4	4	5	5	5	5	5	6	6	6	6	6	6	7	8	9	9	10	10	11	12	13	14	14	16	18	19	20	23	25	27	29	31	32
45	1	2	3	3	3	4	4	4	4	5	5	5	5	5	6	6	6	6	6	7	7	8	9	9	10	10	11	12	13	14	15	16	18	19	21	23	25	27	29	31	33
50	1	2	3	3	3	4	4	4	4	5	5	5	5	6	6	6	6	6	6	7	7	8	9	9	10	11	12	12	13	14	15	17	18	20	21	24	26	28	30	33	33

TABLE 2 (*Contd.*)

B. $P^* = .90$

k \ n	1	2	3	4	5	6	7	8	9	10	11	12	13	14	15	16	17	18	19	20	25	30	35	40	45	50	60	70	80	90	100	125	150	175	200	250	300	350	400	450	500
2	1	1	2	2	2	2	2	3	3	3	3	3	3	3	4	4	4	4	4	4	5	5	5	6	6	6	7	8	8	9	9	10	11	12	13	14	16	17	18	19	20
3	1	2	2	2	2	3	3	3	3	4	4	4	4	4	4	4	5	5	5	5	6	6	7	7	7	8	9	9	10	11	11	12	14	15	16	18	19	21	22	24	25
4	1	2	2	2	3	3	3	3	4	4	4	4	4	5	5	5	5	5	5	5	6	7	7	8	8	9	10	10	11	12	12	14	15	16	17	19	21	23	25	26	27
5	1	2	2	3	3	3	3	4	4	4	4	5	5	5	5	5	5	6	6	6	6	7	8	8	9	9	10	10	12	12	13	15	16	17	18	21	23	24	26	28	29
6	1	2	2	3	3	3	4	4	4	4	4	5	5	5	5	5	6	6	6	6	7	7	8	9	9	10	10	11	12	13	14	15	17	18	19	21	23	25	27	29	30
7	1	2	2	3	3	3	4	4	4	4	5	5	5	5	5	6	6	6	6	6	7	8	8	9	9	10	11	12	13	13	14	16	17	19	20	22	24	26	28	30	31
8	1	2	2	3	3	3	4	4	4	4	5	5	5	5	6	6	6	6	6	6	7	8	8	9	10	10	11	12	13	14	14	16	18	19	20	23	25	27	29	30	32
9	1	2	2	3	3	4	4	4	4	5	5	5	5	5	6	6	6	6	6	7	7	8	9	9	10	10	11	12	13	14	15	16	18	19	21	23	25	27	29	31	33
10	1	2	2	3	3	4	4	4	4	5	5	5	5	6	6	6	6	6	6	7	7	8	9	9	10	11	12	12	13	14	15	17	18	20	21	24	26	28	30	32	33
11	1	2	3	3	3	4	4	4	4	5	5	5	5	6	6	6	6	6	7	7	8	8	9	10	10	11	12	13	14	15	15	17	19	20	22	24	26	28	30	32	34
12	1	2	3	3	3	4	4	4	5	5	5	5	5	6	6	6	6	6	7	7	8	8	9	10	10	11	12	13	14	15	15	17	19	20	22	24	27	29	31	33	34
13	1	2	3	3	3	4	4	4	5	5	5	5	6	6	6	6	6	7	7	7	8	8	9	10	10	11	12	13	14	15	15	17	19	20	22	24	27	29	31	33	34
14	1	2	3	3	3	4	4	4	5	5	5	5	6	6	6	6	6	7	7	7	8	9	9	10	11	11	12	13	14	15	16	18	19	21	22	25	27	29	31	33	35
15	1	2	3	3	3	4	4	4	5	5	5	5	6	6	6	6	6	7	7	7	8	9	9	10	11	11	12	13	14	15	16	18	19	21	22	25	27	30	32	34	35
16	1	2	3	3	3	4	4	4	5	5	5	5	6	6	6	6	7	7	7	7	8	9	9	10	11	11	12	13	14	15	16	18	20	21	23	25	28	30	32	34	36
17	1	2	3	3	3	4	4	4	5	5	5	6	6	6	6	6	7	7	7	7	8	9	10	10	11	11	13	14	14	15	16	18	20	21	23	26	28	30	32	34	36
18	1	2	3	3	4	4	4	5	5	5	5	6	6	6	6	6	7	7	7	7	8	9	10	10	11	12	13	14	15	16	16	18	20	22	23	26	28	31	33	35	37
19	1	2	3	3	4	4	4	5	5	5	5	6	6	6	6	6	7	7	7	7	8	9	10	10	11	12	13	14	15	16	16	18	20	22	23	26	29	31	33	35	37
20	1	2	3	3	4	4	4	5	5	5	5	6	6	6	6	7	7	7	7	7	8	9	10	10	11	12	13	14	15	16	16	18	20	22	23	26	29	31	33	35	37
25	1	2	3	3	4	4	4	5	5	5	6	6	6	6	7	7	7	7	7	8	8	9	10	11	11	12	13	14	15	16	17	19	21	22	24	27	29	32	34	36	38
30	1	2	3	3	4	4	4	5	5	5	6	6	6	6	7	7	7	7	8	8	9	9	10	11	12	12	13	14	15	16	17	19	21	23	24	27	30	32	35	37	39
35	1	2	3	4	4	4	5	5	5	6	6	6	6	7	7	7	7	7	8	8	9	10	10	11	12	12	14	15	16	17	18	20	22	23	25	28	31	33	35	37	39
40	1	2	3	4	4	4	5	5	5	6	6	6	6	7	7	7	7	8	8	8	9	10	11	11	12	13	14	15	16	17	18	20	22	24	26	28	31	33	36	38	40
45	1	2	3	4	4	4	5	5	5	6	6	6	7	7	7	7	7	8	8	8	9	10	11	11	12	13	14	15	16	17	18	20	22	24	26	29	31	34	36	38	40
50	1	2	3	4	4	4	5	5	5	6	6	6	7	7	7	7	8	8	8	8	9	10	11	12	12	13	14	15	16	17	18	20	22	24	26	29	32	34	37	39	41

TABLE 2 (*Contd.*)

C. $P^* = .95$

k \ n	1	2	3	4	5	6	7	8	9	10	11	12	13	14	15	16	17	18	19	20	25	30	35	40	45	50	60	70	80	90	100	125	150	175	200	250	300	350	400	450	500
2	1	2	2	2	3	3	3	3	3	4	4	4	4	4	5	5	5	5	5	5	6	6	7	7	8	8	9	10	10	11	12	13	14	15	16	18	20	22	23	25	25
3	1	2	2	3	3	3	4	4	4	4	4	5	5	5	5	5	6	6	6	6	7	7	8	9	9	10	10	11	12	13	14	15	17	18	19	21	23	25	27	29	30
4	1	2	2	3	3	3	4	4	4	5	5	5	5	5	6	6	6	6	6	6	7	8	9	9	10	10	11	12	13	14	15	16	18	19	21	23	25	27	29	31	33
5	1	2	3	3	3	4	4	4	5	5	5	5	5	6	6	6	6	6	7	7	8	8	9	10	10	11	12	13	14	14	15	17	19	20	22	24	26	29	31	32	34
6	1	2	3	3	3	4	4	4	5	5	5	5	6	6	6	6	6	7	7	7	8	9	9	10	11	11	12	13	14	15	16	18	19	21	22	25	27	30	32	33	35
7	1	2	3	3	4	4	4	4	5	5	5	6	6	6	6	6	7	7	7	7	8	9	10	10	11	11	12	14	14	15	16	18	20	21	23	26	28	30	32	34	36
8	1	2	3	3	4	4	4	5	5	5	5	6	6	6	6	7	7	7	7	7	8	9	10	10	11	12	13	14	15	16	17	19	20	22	23	26	29	31	33	35	37
9	1	2	3	3	4	4	4	5	5	5	5	6	6	6	6	7	7	7	7	7	8	9	10	11	11	12	13	14	15	16	17	19	21	22	24	27	29	32	34	36	38
10	1	2	3	3	4	4	4	5	5	5	6	6	6	6	7	7	7	7	7	8	8	9	10	11	11	12	13	14	15	16	17	19	21	23	24	27	30	32	34	36	38
11	1	2	3	3	4	4	4	5	5	5	6	6	6	6	7	7	7	7	7	8	9	9	10	11	12	12	13	14	15	16	17	19	21	23	24	27	30	32	35	37	39
12	1	2	3	3	4	4	5	5	5	5	6	6	6	6	7	7	7	7	8	8	9	10	10	11	12	12	14	15	16	17	18	20	21	23	25	28	30	33	35	37	39
13	1	2	3	3	4	4	5	5	5	6	6	6	6	7	7	7	7	7	8	8	9	10	10	11	12	13	14	15	16	17	18	20	22	23	25	28	31	33	35	38	40
14	1	2	3	3	4	4	5	5	5	6	6	6	6	7	7	7	7	7	8	8	9	10	11	11	12	13	14	15	16	17	18	20	22	24	25	28	31	33	36	38	40
15	1	2	3	3	4	4	5	5	5	6	6	6	6	7	7	7	7	8	8	8	9	10	11	11	12	13	14	15	16	17	18	20	22	24	25	28	31	34	36	38	40
16	1	2	3	3	4	4	5	5	5	6	6	6	6	7	7	7	7	8	8	8	9	10	11	11	12	13	14	15	16	17	18	20	22	24	26	29	31	34	36	38	41
17	1	2	3	3	4	4	5	5	5	6	6	6	6	7	7	7	7	8	8	8	9	10	11	11	12	13	14	15	16	17	18	20	22	24	26	29	31	34	37	39	41
18	1	2	3	3	4	4	5	5	5	6	6	6	7	7	7	7	7	8	8	8	9	10	11	12	12	13	14	15	16	17	18	21	23	24	26	29	32	34	37	39	41
19	1	2	3	3	4	4	5	5	5	6	6	6	7	7	7	7	8	8	8	8	9	10	11	12	12	13	14	15	17	18	18	21	23	24	26	29	32	35	37	39	41
20	1	2	3	4	4	4	5	5	5	6	6	6	7	7	7	7	8	8	8	8	9	10	11	12	12	13	14	16	17	18	19	21	23	25	26	29	32	35	37	39	42
25	1	2	3	4	4	5	5	5	6	6	6	7	7	7	7	8	8	8	8	8	9	10	11	12	13	13	15	16	17	18	19	21	23	25	27	30	33	36	38	40	43
30	1	2	3	4	4	5	5	5	6	6	6	7	7	7	8	8	8	8	8	9	10	11	11	12	13	14	15	16	17	18	19	22	24	26	27	31	34	36	39	41	43
35	1	2	3	4	4	5	5	6	6	6	7	7	7	7	8	8	8	8	9	9	10	11	12	12	13	14	15	16	18	19	20	22	24	26	28	31	34	37	39	42	44
40	1	2	3	4	4	5	5	6	6	6	7	7	7	7	8	8	8	8	9	9	10	11	12	13	13	14	15	17	18	19	20	22	24	26	28	32	35	37	40	42	45
45	1	2	3	4	4	5	5	6	6	6	7	7	7	7	8	8	8	8	9	9	10	11	12	13	14	14	16	17	18	19	20	23	25	27	29	32	35	38	40	43	45
50	1	2	3	4	5	5	5	6	6	6	7	7	7	8	8	8	8	9	9	9	10	11	12	13	14	14	16	17	18	19	20	23	25	27	29	32	35	38	41	43	46

TABLE 2 (*Contd.*)

D. $P^* = .99$

k \ n	1	2	3	4	5	6	7	8	9	10	11	12	13	14	15	16	17	18	19	20	25	30	35	40	45	50	60	70	80	90	100	125	150	175	200	250	300	350	400	450	500
2	1	2	3	3	4	4	4	5	5	5	5	6	6	6	6	7	7	7	7	7	8	9	10	10	11	12	13	14	15	16	16	18	20	22	23	26	29	31	33	35	37
3	1	2	3	3	4	4	5	5	5	6	6	6	6	7	7	7	7	8	8	8	9	10	11	11	12	13	14	15	16	17	18	20	22	24	26	29	31	34	36	38	40
4	1	2	3	4	4	4	5	5	6	6	6	6	7	7	7	8	8	8	8	8	9	10	11	12	13	13	15	16	17	18	19	21	23	25	27	30	33	36	38	40	42
5	1	2	3	4	4	5	5	5	6	6	6	7	7	7	7	8	8	8	8	9	10	11	11	12	13	13	15	16	18	19	20	22	24	26	28	31	34	37	39	42	44
6	1	2	3	4	4	5	5	6	6	6	6	7	7	7	8	8	8	8	9	9	10	11	12	13	13	14	16	17	18	19	20	22	25	27	28	32	35	38	40	43	45
7	1	2	3	4	4	5	5	6	6	6	7	7	7	8	8	8	8	9	9	9	10	11	12	13	14	14	16	17	18	19	20	23	25	27	29	32	35	38	41	43	46
8	1	2	3	4	4	5	5	6	6	6	7	7	7	8	8	8	8	9	9	9	10	11	12	13	14	15	16	17	19	20	21	23	25	27	29	33	36	39	41	44	46
9	1	2	3	4	4	5	5	6	6	6	7	7	7	8	8	8	9	9	9	9	10	11	12	13	14	15	16	18	19	20	21	23	26	28	30	33	36	39	42	45	47
10	1	2	3	4	4	5	5	6	6	7	7	7	8	8	8	8	9	9	9	9	10	12	12	13	14	15	16	18	19	20	21	24	26	28	30	34	37	40	42	45	47
11	1	2	3	4	4	5	5	6	6	7	7	7	8	8	8	8	9	9	9	9	11	12	13	13	14	15	17	18	19	20	21	24	26	28	30	34	37	40	43	45	48
12	1	2	3	4	5	5	5	6	6	7	7	7	8	8	8	9	9	9	9	10	11	12	13	14	14	15	17	18	19	21	22	24	26	29	31	34	37	40	43	46	48
13	1	2	3	4	5	5	5	6	6	7	7	7	8	8	8	9	9	9	9	10	11	12	13	14	14	15	17	18	19	21	22	24	27	29	31	34	38	41	44	46	49
14	1	2	3	4	5	5	6	6	6	7	7	7	8	8	8	9	9	9	9	10	11	12	13	14	15	16	17	18	20	21	22	25	27	29	31	35	38	41	44	47	49
15	1	2	3	4	5	5	6	6	6	7	7	7	8	8	8	9	9	9	9	10	11	12	13	14	15	16	17	18	20	21	22	25	27	29	31	35	38	41	44	47	49
16	1	2	3	4	5	5	6	6	6	7	7	7	8	8	8	9	9	9	9	10	11	12	13	14	15	16	17	19	20	21	22	25	27	29	31	35	38	41	44	47	50
17	1	2	3	4	5	5	6	6	6	7	7	8	8	8	8	9	9	9	10	10	11	12	13	14	15	16	17	19	20	21	22	25	27	30	32	35	39	42	45	47	50
18	1	2	3	4	5	5	6	6	6	7	7	8	8	8	9	9	9	9	10	10	11	12	13	14	15	16	17	19	20	21	22	25	27	30	32	35	39	42	45	48	50
19	1	2	3	4	5	5	6	6	7	7	7	8	8	8	9	9	9	9	10	10	11	12	13	14	15	16	17	19	20	21	23	25	28	30	32	36	39	42	45	48	50
20	1	2	3	4	5	5	6	6	7	7	7	8	8	8	9	9	9	9	10	10	11	12	13	14	15	16	18	19	20	21	23	25	28	30	32	36	39	42	45	48	51
25	1	2	3	4	5	5	6	6	7	7	7	8	8	9	9	9	9	10	10	10	12	13	14	15	15	16	18	19	21	22	23	26	28	30	33	36	40	43	46	49	52
30	1	2	3	4	5	5	6	6	7	7	7	8	8	9	9	9	9	10	10	10	12	13	14	15	16	17	18	20	21	22	23	26	29	31	33	37	40	44	47	50	52
35	1	2	3	4	5	5	6	6	7	7	8	8	8	9	9	9	10	10	10	10	12	13	14	15	16	17	18	20	21	22	24	26	29	31	33	37	41	44	47	50	53
40	1	2	3	4	5	5	6	6	7	7	8	8	8	9	9	9	10	10	10	10	12	13	14	15	16	17	18	20	21	23	24	27	29	32	34	38	41	45	48	51	53
45	1	2	3	4	5	5	6	6	7	7	8	8	8	9	9	9	10	10	10	11	12	13	14	15	16	17	19	20	22	23	24	27	29	32	34	38	42	45	48	51	54
50	1	2	3	4	5	5	6	6	7	7	8	8	8	9	9	9	10	10	10	11	12	13	14	15	16	17	19	20	22	23	24	27	30	32	34	38	42	45	48	51	54

TABLE 3

EXPECTED PROPORTION OF POPULATIONS RETAINED IN THE SELECTED SUBSET BY PROCEDURE R

For the true configuration $p_{[1]} = p_{[2]} = \cdots = p_{[k-1]} = p$; $p_{[k]} = p + \delta$ for selected values of n, k, P^*, δ, and $p + \delta$.

		$P^* = .75$				$P^* = .90$				$P^* = .95$			
n	$p+\delta$ \ k	2	3	5	10	2	3	5	10	2	3	5	10
						$\delta = .00$							
5	.50	.828	.904	.845	.935	.945	.904	.964	.935	.989	.980	.964	.992
	.75	.868	.950	.926	.986	.969	.950	.989	.986	.996	.993	.989	.999
	.95	.982	.999	.999	1.000	.999	.999	1.000	1.000	1.000	1.000	1.000	1.000
	1.00	1.000	1.000	1.000	1.000	1.000	1.000	1.000	1.000	1.000	1.000	1.000	1.000
10	.50	.868	.785	.838	.886	.942	.962	.934	.961	.979	.962	.979	.961
	.75	.904	.846	.912	.958	.966	.984	.974	.990	.991	.984	.994	.990
	.95	.993	.990	.999	1.000	.999	1.000	1.000	1.000	1.000	1.000	1.000	1.000
	1.00	1.000	1.000	1.000	1.000	1.000	1.000	1.000	1.000	1.000	.000	1.000	1.000
15	.50	.819	.832	.858	.771	.951	.913	.932	.949	.979	.961	.972	.981
	.75	.855	.887	.923	.880	.972	.952	.971	.985	.990	.983	.991	.996
	.95	.981	.996	.999	.999	1.000	1.000	1.000	1.000	1.000	1.000	1.000	1.000
	1.00	1.000	1.000	1.000	1.000	1.000	1.000	1.000	1.000	1.000	1.000	1.000	1.000
20	.50	.785	.781	.792	.804	.923	.928	.939	.947	.960	.965	.972	.977
	.75	.820	.838	.869	.900	.951	.962	.974	.984	.978	.985	.991	.995
	.95	.966	.990	.998	1.000	.999	1.000	1.000	1.000	1.000	1.000	1.000	1.000
	1.00	1.000	1.000	1.000	1.000	1.000	1.000	1.000	1.000	1.000	1.000	1.000	1.000
25	.50	.760	.830	.833	.834	.941	.942	.902	.905	.968	.970	.974	.951
	.75	.794	.883	.901	.919	.964	.971	.951	.963	.984	.988	.992	.985
	.95	.951	.996	.999	1.000	1.000	1.000	1.000	1.000	1.000	1.000	1.000	1.000
	1.00	1.000	1.000	1.000	1.000	1.000	1.000	1.000	1.000	1.000	1.000	1.000	1.000
						$\delta = .10$							
5	.50	.817	.895	.835	.929	.938	.895	.960	.969	.987	.977	.960	.992
	.75	.838	.922	.881	.963	.952	.922	.976	.963	.991	.985	.976	.997
	.95	.928	.984	.979	.998	.989	.984	.998	.998	.999	.999	.998	1.000
	1.00	.959	.993	.993	1.000	.996	.994	1.000	1.000	1.000	1.000	1.000	1.000
10	.50	.846	.761	.819	.874	.925	.951	.923	.956	.970	.951	.975	.956
	.75	.864	.791	.857	.915	.939	.965	.946	.974	.978	.965	.984	.974
	.95	.937	.908	.967	.992	.983	.995	.994	.999	.997	.995	.999	.999
	1.00	.965	.953	.990	.999	.994	.999	.999	1.000	.999	.999	1.000	1.000
15	.50	.786	.797	.830	.747	.926	.885	.914	.939	.963	.943	.962	.977
	.75	.800	.823	.864	.801	.939	.906	.936	.960	.972	.957	.975	.986
	.95	.869	.923	.966	.955	.982	.975	.991	.997	.995	.993	.998	1.000
	1.00	.908	.963	.990	.989	.994	.992	.998	1.000	.999	.999	1.000	1.000
20	.50	.743	.733	.750	.774	.884	.894	.914	.934	.930	.941	.957	.970
	.75	.754	.756	.785	.821	.898	.913	.935	.955	.942	.955	.970	.981
	.95	.806	.856	.916	.959	.954	.976	.990	.997	.983	.993	.997	.999
	1.00	.838	.911	.965	.990	.978	.992	.998	1.000	.994	.998	1.000	1.000
25	.50	.711	.772	.783	.800	.896	.902	.861	.879	.934	.942	.955	.934
	.75	.719	.793	.814	.841	.909	.919	.887	.910	.946	.956	.968	.954
	.95	.752	.885	.931	.965	.962	.979	.973	.988	.985	.993	.997	.996
	1.00	.769	.935	.973	.991	.983	.994	.992	.998	.995	.998	1.000	1.000

TABLE 3 (*Contd.*)

n	k / $p+\delta$	$P^*=.75$				$P^*=.90$				$P^*=.95$			
		2	3	5	10	2	3	5	10	2	3	5	10
	$\delta=.25$												
5	.50	.762	.850	.796	.925	.896	.850	.946	.925	.971	.959	.946	.992
	.75	.762	.849	.794	.916	.896	.849	.941	.916	.971	.958	.941	.988
	.95	.798	.908	.883	.974	.934	.908	.978	.974	.988	.983	.978	.998
	1.00	.816	.931	.917	.986	.948	.931	.987	.986	.992	.990	.987	.999
10	.50	.742	.643	.715	.816	.838	.883	.853	.926	.915	.883	.939	.926
	.75	.742	.643	.713	.808	.838	.882	.849	.919	.915	.882	.936	.919
	.95	.754	.668	.783	.895	.870	.928	.912	.968	.947	.928	.973	.968
	1.00	.763	.684	.821	.930	.888	.948	.937	.982	.961	.948	.984	.982
15	.50	.651	.628	.671	.607	.805	.735	.792	.864	.875	.821	.884	.935
	.75	.651	.628	.670	.604	.805	.734	.789	.858	.875	.830	.882	.931
	.95	.632	.635	.719	.678	.827	.768	.850	.924	.908	.877	.934	.973
	1.00	.618	.641	.749	.718	.843	.791	.881	.949	.926	.901	.955	.984
20	.50	.596	.528	.528	.580	.714	.705	.747	.812	.781	.792	.837	.892
	.75	.596	.529	.527	.577	.714	.705	.745	.808	.781	.792	.836	.888
	.95	.566	.503	.531	.628	.709	.728	.798	.877	.796	.832	.892	.943
	1.00	.546	.483	.532	.655	.707	.745	.849	.908	.809	.857	.919	.963
25	.50	.562	.527	.516	.557	.703	.682	.615	.669	.763	.761	.798	.770
	.75	.562	.527	.516	.555	.703	.682	.614	.666	.763	.761	.797	.766
	.95	.533	.497	.509	.587	.694	.696	.634	.721	.772	.795	.851	.832
	1.00	.516	.476	.503	.605	.689	.707	.649	.754	.781	.818	.880	.866
	$\delta=.50$												
5	.50	.594	.667	.600	.831	.750	.667	.850	.831	.906	.875	.850	.972
	.75	.610	.646	.568	.769	.737	.646	.801	.769	.878	.836	.801	.948
	.95	.600	.657	.586	.806	.744	.657	.831	.806	.896	.860	.831	.964
	1.00	.594	.667	.600	.831	.750	.667	.850	.831	.906	.875	.850	.972
10	.50	.527	.370	.338	.439	.586	.585	.502	.661	.688	.585	.698	.661
	.75	.551	.400	.369	.440	.607	.588	.505	.625	.691	.588	.667	.625
	.95	.537	.383	.351	.439	.595	.585	.502	.645	.689	.585	.685	.645
	1.00	.527	.370	.338	.439	.586	.585	.502	.661	.688	.585	.698	.661
15	.50	.502	.345	.247	.153	.530	.373	.321	.373	.575	.434	.443	.550
	.75	.511	.367	.284	.194	.553	.404	.357	.393	.598	.464	.461	.537
	.95	.505	.353	.263	.170	.539	.385	.336	.381	.585	.447	.450	.543
	1.00	.502	.345	.247	.153	.530	.373	.321	.378	.575	.434	.443	.550
20	.50	.500	.334	.205	.119	.503	.347	.246	.218	.510	.372	.305	.327
	.75	.502	.341	.221	.149	.513	.370	.283	.261	.527	.402	.343	.356
	.95	.501	.336	.210	.130	.506	.356	.261	.236	.517	.384	.321	.339
	1.00	.500	.334	.205	.119	.503	.347	.246	.218	.510	.372	.305	.327
25	.50	.500	.334	.202	.107	.501	.338	.206	.119	.504	.348	.243	.148
	.75	.501	.338	.211	.126	.507	.352	.223	.150	.514	.370	.279	.188
	.95	.500	.335	.205	.113	.503	.343	.212	.131	.507	.356	.257	.165
	1.00	.500	.334	.202	.107	.501	.338	.206	.119	.504	.348	.243	.148

References

[1] Gupta, S. S. On a decision rule for a problem in ranking means. Institute of Statistics, Mimeograph Series Report No. 150, Chapel Hill, N.C.: University of North Carolina, May 1956.

[2] Gupta, S. S., and Sobel, M. On a statistic which arises in selection and ranking problems. *Ann. Math. Stat.*, 1957, **28**, 957–67.

[3] Bechhofer, R. E. A single-sample multiple decision procedure for ranking means of normal populations with known variances. *Ann. Math. Stat.*, 1954, **25**, 16–29.

[4] Sobel, M. and Huyett, M. J. Selecting the best one of several binomial populations. *Bell System Tech. J.*, 1957, **36**, 537–76.

[5] Eisenhart, C., Hastay, M. W., and Wallis, W. A. (eds.). *Techniques of Statistical Analysis*. New York: McGraw-Hill, 1947.

[6] Freeman, M. F., and Tukey, J. W. Transformations related to the angular and the square root. *Ann. Math. Stat.*, 1950, **21**, 607–11.

[7] Gnedenko, B. V., and Kolmogorov, A. N. *Limit Distributions for Sums of Independent Random Variables*. (Translated from the Russian and annotated by K. L. Chung.) Cambridge, Mass.: Addison Wesley, 1954.

21 CONSISTENCY OF MAXIMUM LIKELIHOOD ESTIMATION OF DISCRETE DISTRIBUTIONS[1]

J. HANNAN
Michigan State University

1. Summary

In dominated classes of discrete distributions, maximum likelihood (ML hereafter) estimation can be identified with a certain minimum "information" estimation and viewed as a particular minimum "distance" method. This viewpoint permits a direct extension of the Hotelling consistency proof [1] for finite discrete distributions under a near-minimal assumption and suggests several classes of examples displaying various degrees of inconsistency.

Properties of the "information" convergence topology are studied more generally in an early digression. Its domination of the $\mathscr{L}_p$ metric topologies is a sharpening of the standard non-negativity of information.

The consistency theorems obtained (but not their proofs) are particular cases of that of Doob [2] and of that of Kiefer and Wolfowitz [3]. A version of Doob's theorem is included in an appendix.

2. Introduction

To set the discrete case in perspective and to facilitate comparison with other work, definitions are introduced for dominated classes in general.

Let $(\mathscr{X}, \mathscr{A}, \mu)$ be a measure space with μ σ-finite and points x. Let $\mathscr{C}$ be a topological space of probability measures P dominated by μ. Let $p = dP/d\mu$, $g = \log p$ and, for independent identically distributed $x_1, x_2, \cdots$, let E_n be the expectation operator averaging over the first n. Then

$$E_n g = n^{-1} \log \prod_{1}^{n} p(x_j) , \tag{2.1}$$

and a sequence $\{P_n\}$, with P_n $\mathscr{A}^n$-measurable to $\mathscr{C}$, is called a (strong) asymptotic ML estimate[2] of P if, except on a set N with $P^\infty(N) \equiv 0$ on $\mathscr{C}$,

[1] Supported in part by National Science Foundation Grant No. G-5011, Department of Statistics, Michigan State University.

[2] Other definitions [3], [4], [5] have required that, for $c > 0$, the difference in (2.2) is bounded below by $n^{-1} \log c$.

(2.2) $$\sup_{\mathscr{C}} E_n g - E_n g_n \to 0 .$$

As a partial motivation for the digression of Section 3, temporarily view an ML estimate (2.2) as an infimizing sequence for the sequence of random functions of P' in $\mathscr{C}$:

(2.3) $$E_n(g - g') .$$

For each P' for which $E_P(g - g')$ exists [which turns out (3.4) to be no restriction],

(2.4) $$E_n(g - g') \to E_P(g - g') , \qquad \text{a. e. } (P^\infty) ,$$

by the strong law of large numbers (SLLN hereafter). It will follow from Lemma 1 that $E_P(g - g')$ has the unique minimizer P and thus that an ML estimate of the minimizer is a "minimizer" of an "estimate," in both cases, of $E_P(g - g')$.

3. Relations Between Topologies on Probability Measures

Let $(\mathscr{X}, \mathscr{A})$ be a measurable space, P and P' be probability measures on $\mathscr{A}$, p and p' their densities with respect to a dominating measure ν, $r = p^{\frac{1}{2}}$, and $r' = p'^{\frac{1}{2}}$.

Kraft [6] (see also Kakutani [7]) has noted the equivalence of the metric topologies on the probability measures on $\mathscr{A}$ based on the respective $\mathscr{L}_1$ and $\mathscr{L}_2$ metrics

(3.1) $$V(P, P') = E_\nu |p - p'| , \qquad d(P, P') = E_\nu^{\frac{1}{2}}(r - r')^2 .$$

Denoting the $\mathscr{L}_2$ scalar product $E_\nu rr'$ by ρ, $0 \leqq \rho \leqq 1$ and, from

(3.2) $$E_\nu^2 |r - r'|(r + r') \leqq E_\nu(r - r')^2 E_\nu(r + r')^2 \leqq E_\nu |p - p'| E_\nu(r + r')^2 ,$$

it follows that

(3.3) $$V^2 \leqq d^2 2(1 + \rho) \leqq V2(1 + \rho) .$$

Henceforth letting $g = \log p$, a log inequality yields

(3.4) $$E_P(g' - g)^+ \leqq E_P(p'/p - 1)^+ \leqq E_\nu(p' - p)^+ = V/2 .$$

Defining the "information" [8]:

(3.5) $$\Delta_P(P') = E_P(g - g') ,$$

the following lemma improves on the usual non-negativity result [2], [4], [8].

LEMMA 1. $$\Delta_P(P') \geqq d^2(P, P') .$$ [3]

PROOF. Integrating $p(g - g') = 2p \log (r/r') \geqq 2p(1 - r'/r)$ with respect to ν and identifying $E_\nu 2p(1 - r'/r) = 2(1 - \rho)$ as d^2 yields the required result.

The same proof shows that if p and p' are only supposed to be non-negative and summable (ν),

[3] Hoeffding and Wolfowitz [9] have sharpened this result by using Jensen's inequality before using the log inequality: $E_\nu 2p \log (r/r') \geqq -2 \log \rho \geqq 2(1 - \rho)$.

LEMMA 1′. $E_\nu p(g - g') \geqq E_\nu(p - p') + E_\nu(r - r')^2$.

In view of (3.3), Lemma 1 has the corollary,

$$\Delta_P(P') \geqq V^2(P, p')/4 , \tag{3.6}$$

while from (3.2) Lemma 1′ has the corresponding corollary,

$$E_\nu p(g - g') \geqq E_\nu(p - p') + E_\nu^2 |p - p'| /(E_\nu^{\frac{1}{2}} p + E_\nu^{\frac{1}{2}} p')^2 . \tag{3.6′}$$

A strengthening of (3.6), which clarifies the infimizers of Δ subject to fixed V, will be obtained by considering separately the positive and negative parts of Δ.

Let A also denote the indicator function of a measurable set A and define, for any p-measure Q with $q = dQ/d\nu$, Q_A to be the measure with $dQ_A/d\nu = qA/Q(A)$. Let $\delta = V(P, p')/2$, $S = [p \leqq p']$, $T = [p > p']$, $t = P(T)$, and define, for δ in $(0, 1)$ and t in $[\delta, 1]$,

$$\psi(t \mid \delta) = t\left[-\log\left(1 - \frac{\delta}{t}\right)\right] + (1 - t)\left[-\log\left(1 + \frac{\delta}{1 - t}\right)\right] . \tag{3.7}$$

Then

$$E_P(g - g')^+ = E_P T\left(-\log\frac{p'}{p}\right) = t\left[-\log\frac{t - \delta}{t}\right] + E_P T\left(-\log\frac{p'_T}{p_T}\right) ,$$

$$E_P[-(g' - g)^+] = E_P S\left(-\log\frac{p'}{p}\right) = (1 - t)\left(-\log\frac{1 - t + \delta}{1 - t}\right) + E_P S\left(-\log\frac{p'_S}{p_S}\right) ,$$

and hence

$$\Delta_P(P') = \psi(t \mid \delta) + t\Delta_{P_T}(P'_T) + (1 - t)\Delta_{P_S}(P'_S) . \tag{3.8}$$

Since, for each δ, ψ is strictly convex in t and attains its minimum at a unique $t(\delta)$ in $(\delta, 1)$, Equation (3.8) yields the following improvement of (3.6):

$$\Delta_P(P') \geqq \psi(t(\delta) \mid \delta) , \tag{3.9}$$

and equality holds if and only if

$$1 - P\left\{\frac{p'}{p} = 1 + \frac{\delta}{1 - t(\delta)}\right\} = t(\delta) = P\left\{\frac{p'}{p} = 1 - \frac{\delta}{t(\delta)}\right\} . \tag{3.10}$$

Moreover, since ψ is strictly increasing with respect to δ, the following lemma is immediate from (3.8).

LEMMA 2. *If, for η in $(0, 1)$, P, P' vary such that* $\liminf \delta \geqq \eta$, $\Delta_P(P') \to \psi(t(\eta) \mid \eta)$ *if and only if* $t \to t(\eta)$, $\delta \to \eta$, $\Delta_{P_T}(P'_T)$ *and* $\Delta_{P_S}(P'_S) \to 0$.

4. Equi-likelihood and Its Role in ML Consistency

Consistency of estimates satisfying (2.2) usually stems from the weaker equi-likelihood condition[4]

$$\text{(4.1)}\quad \text{for each } P \text{ in } \mathscr{C}, \qquad \limsup E_n(g - g_n) \leqq 0\,, \qquad \text{a.e. } (P^\infty)\,.$$

For example, if $\mathscr{C}$ is finite (but otherwise arbitrary), the SLLN insures that

$$\text{(4.2)}\qquad \limsup E_n(g - g_n) = \limsup \Delta_P(P_n)\,, \qquad \text{a.e. } (P^\infty)\,,$$

and $\Delta_P(P_n) \to 0$ on the intersection of the sets of (4.1) and (4.2). Further, for certain compact metric $\mathscr{C}$, it was proved by Wald [4] and emphasized by LeCam [11] that

$$\text{(4.3)}\qquad \limsup E_n(g - g_n) \geqq \limsup \Delta_P(P_n)\,, \qquad \text{a.e. } (P^\infty)\,,$$

which permits the same conclusion.

5. Consistency of Equi-likelihood Estimates in the Discrete case

In the discrete case, μ is counting measure on a denumerable set, E_μ will be abbreviated by Σ, and the ML definition will be phrased in terms of a "minimum information."

Letting P_n^* denote the empirical distribution of the first n of $x_1, x_2, \cdots, P_n^*$ is itself dominated by μ and, abbreviating $\Delta_{P_n^*}(P)$ to $\Delta_n(P)$,

$$\text{(5.1)}\qquad \Delta_n(P) = E_n(g_n^* - g) = \Sigma g_n^* p_n^* - E_n g\,.$$

The ML property of P_n, (2.2), is expressible,

$$\text{(5.2)}\qquad \Delta_n(P_n) - \inf_{\mathscr{C}} \Delta_n(P) \to 0\,, \quad \text{except on } N \text{ with } P^\infty(N) \equiv 0 \text{ on } \mathscr{C}\,,$$

and the equi-likelihood property (4.1),

$$\text{(5.3)}\quad \text{for each } P \text{ in } \mathscr{C}, \qquad \limsup(\Delta_n(P_n) - \Delta_n(P)) \leqq 0\,, \qquad \text{a.e. } (P^\infty)\,.$$

The discrete case is distinguished by the presence of a natural estimate of p,

$$\text{(5.4)}\qquad p_n^* \to p\,, \qquad \text{a.e. } (\mu), \quad \text{a.e. } (P^\infty)\,.$$

Moreover, the generally weaker $\mathscr{L}_1$ convergence of probability densities to a probability density here implies a.e. (μ) convergence, so that (5.4) is equivalent to

$$\text{(5.5)}\qquad V(P_n^*, P) \to 0\,, \qquad \text{a.e. } (P^\infty)\,.$$

The next lemma is basic for the consistency theorem which follows it.

LEMMA 3. $\qquad E_P(-g) < \infty \Rightarrow \Delta_n(P) \to 0\,, \qquad$ *a.e.* (P^∞).

[4] Le Cam [10] has exploited this in constructing estimates "uniformly consistent on compacts" from ML estimates, for a sequence of compacts increasing to $\mathscr{C}$, by a diagonalization process.

PROOF. For each sequence $\boldsymbol{x} = (x_1, x_2, \cdots)$ for which $E_n g \to E_P g$ and $p_n^* \to p$, a.e. (μ), hence a.e. (P^∞) by the SLLN and (5.4), it follows from (5.1) by Fatou's lemma that

$$\limsup \Delta_n(P) = \limsup \Sigma g_n^* p_n^* - E_P g \leqq \Sigma g p - E_p g = 0 .$$

THEOREM 1. *If P_n is an estimate satisfying* (5.3),

$$E_P(-g) < \infty \Rightarrow V(P_n, P) \to 0 , \qquad a.e.\ (P^\infty) .$$

PROOF. By the triangle inequality for V, by (3.3) and Lemma 1,

$$V(P_n, P) \leqq V(P_n^*, P) + 2\Delta_n^{\frac{1}{2}}(P_n) ,$$

and hence it $\to 0$, a.e. (P^∞), by (5.5) and Lemma 3 together with (5.3) for P.

Remark. The form of this proof is that used by Hotelling [1] in the finite case with Lemma 1, (3.3), Lemma 3, and (4.5) strengthening his main arguments. [An alternative form, which does not extend from the finite case, has been recently noted by Rao [12] who relies on $m_n = \max(p(x)/p_n^*(x)) \to 1$, a.e. ($P^\infty$), to show, as follows immediately from $\Delta_P(P_n) \leqq E_P g + m_n E_n(-g_n)$ and (5.3), that $\Delta_P(P_n) \to 0$, a.e. (P^∞).]

As will be seen in Section 6 (Lemma 4), the converse of Lemma 3 holds, and hence $E_P(-g) < \infty$ is necessary for the above proof. A slightly better theorem is possible and indeed is a particular case both of the consistency theorem of Doob [2] (which is discussed in the Appendix) and of the general consistency result[5] of Kiefer and Wolfowitz [3].

THEOREM 1′. *If P_n is an estimate satisfying* (5.3) *and* $H = -\inf_{P' \in \mathscr{C}}(g - g')$,

$$E_P H < \infty \Rightarrow V(P_n, P) \to 0 , \qquad a.e.\ (P^\infty) .$$

PROOF. Letting I denote a finite set and also its indicator function, Equation (3.6′) (with $p = Ip_n^*, p' = Ip_n$) gives

$$\Sigma^2 I|p_n^* - p_n|/4 \leqq P_n(I) - P_n^*(I) + E_n I(g_n^* - g_n) . \tag{5.6}$$

Bounding the right-hand member of (5.6) by $E_n(1 - I)(1 + H) + E_n I(g_n^* - g) + E_n(g - g_n)$, the greatest limit of the bound is $\leqq E_P(1 - I)(1 + H)$, a.e. (P^∞), by the SLLN, (5.5) and (5.3) for P. Since $\Sigma I|p_n^* - p|$ is monotone in I, it then follows from $E_P H < \infty$ that $\lim \Sigma I|p_n^* - p_n| = 0$, a.e. ($P^\infty$). Since the finite subsets are enumerable, $V(P_n^*, P_n) \to 0$, a.e. (P^∞), and the theorem follows from the triangle inequality for V and (5.5).

6. Inconsistency of ML Estimates in the Discrete Case

The following lemma and remark will be useful in the construction of $\mathscr{C}$ where ML estimates are not consistent. Here and throughout the rest

[5] A prominent special case of the Kiefer-Wolfowitz result furnishes consistency of ML estimates of the d.f. of a mixing distribution and includes the present theorems via the representation of discrete distributions as mixtures of degenerates.

of the paper $\boldsymbol{x}$ will continue to denote the sequence $(x_1, x_2, \cdots)$.

LEMMA 4. $$E_P(-g) = \infty \Rightarrow \limsup \Delta_n(P) = \infty\,, \qquad a.e.\ (P^\infty)\,.$$

PROOF. For $1 < M < \infty$, let $A_n = [\boldsymbol{x} \mid -g(x_n) \geqq nM]$ and $q_n = P^\infty(A_n)$. On A_n,

$$[g_n^*(x_n) - g(x_n)]p_n^*(x_n) \geqq [g_n^*(x_n) + nM]p_n^*(x_n) \geqq (\log n^{-1} + nM)/n > M - 1\,,$$

and hence, by (3.4), $\Delta_n(P) \geqq E_n(g_n^* - g)^+ - 1 > M - 2$. Since

$$\sum_n q_n = \sum_n P[-g \geqq nM] \geqq E_P(-g/M) - 1 = \infty\,,$$

the Borel-Cantelli lemma implies $P^\infty(A_n$ infinitely often$) = 1$, completing the proof.

Remark. Since $\Delta_n(P) > M - 2$ on $B_n = \bigcup_1^n [\boldsymbol{x} \mid -g(x_k) \geqq nM]$ and $P^\infty(B_n) = 1 - (1 - q_n)^n$,

$$nq_n \to \infty \Rightarrow P^\infty(B_n) \to 1 \Rightarrow \Delta_n(P) \to \infty\,, \qquad \text{i.m.}\ (P^\infty)\,,$$

and

$$\sum e^{-nq_n} < \infty \Rightarrow P^\infty\Big(\bigcap_{m>n} B_m\Big) \geqq 1 - \sum_{m>n} e^{-mq_m} \to 1 \Rightarrow \Delta_n(P) \to \infty\,, \qquad \text{a.e.}\ (P^\infty)\,.$$

Example 1. For the basic example of inconsistent ML estimates suggested by Lemma 4, take P such that $E_P(-g) = \infty$, take η in $(0, 1)$, and let

$$\mathscr{C}_1 = \{P\} \cup \{P' \mid V(P', P) \geqq 2\eta\}\,.$$

Noting that, for each P_n^* with $2\eta_n = 2\eta + V(P_n^*, P) < 2$, any P_n' with $p_n'(x) = (1 - \eta_n)p_n^*(x)$ on $[p_n^* > 0]$ has $V(P_n', P) \geqq V(P_n', P_n^*) - V(P_n^*, P) = 2\eta$ and $\Delta_n(P_n') = -\log(1 - \eta_n)$, it follows that, if P_n is an ML estimate (5.2), $\limsup \Delta_n(P_n) \leqq -\log(1 - \eta)$, a.e. (P^∞). By the structure of $\mathscr{C}_1$, Lemma 4 gives

$$\limsup V(P_n, P) \geqq 2\eta\,, \qquad \text{a.e.}\ (P^\infty)\,. \tag{6.1}$$

If P satisfies the stronger assumptions of the Remark on Lemma 4, the inconsistency (6.1) has corresponding stronger forms. For example, if $\sum_n e^{-nq_n} < \infty$, then we find

$$\liminf V(P_n, P) \geqq 2\eta\,, \qquad \text{a.e.}\ (P^\infty)\,. \tag{6.2}$$

The following variant on Example 1 uses Lemma 2 to exhibit convergence of inconsistent ML estimates.

Example 2. Specialize Example 1 by taking $\eta > \frac{1}{2}$ and taking P such that (6.2) holds for ML estimates P_n and, for some atom a, $P(a) = t(\eta)$ as defined in Section 3. Letting Q, with $V(Q, P) = 2\eta$, be defined by $q(x) = [1 + \eta/1 - P(a)]p(x)$ for $x \neq a$ [and hence $q(a) = P(a) - \eta$], it will be shown that

$$V(P_n, Q) \to 0\,, \qquad \text{a.e.}\ (P^\infty)\,. \tag{6.3}$$

An initial application of Lemma 2 (or direct calculation) shows that Q_n, defined for P_n^* such that $2\eta_n = 2\eta + V(P_n^*, P) < 2$ and $\eta_n \leqq P_n^*(a) < 1$ by

$$q_n(x) = \left(1 + \frac{\eta_n}{1 - P_n^*(a)}\right)p_n^*(x) \quad \text{on } x \neq a, \quad q_n(a) = P_n^*(a) - \eta_n ,$$

has $\Delta_n(Q_n) \to \psi(P(a) \mid \eta)$, a.e. (P^∞), while $V(Q_n, P) \geqq 2\eta$. In view of (5.5), a second application of Lemma 2 assures that P_n is ML if and only if $\Delta_n(P_n) \to \psi(P(a) \mid \eta)$, a.e. (P^∞), and characterizes the latter property. This example was constructed so that

$$t = P_n^*[p_n^* > p_n] \to t(\eta) = P(a) \qquad \text{only if } T = [p_n^* > p] \to \{a\} ,$$

and it may therefore be verified that the characterization of Lemma 2 entails (6.3).

The following example shows that the isolation of P is not necessary for the inconsistency phenomena of Examples 1 and 2.

Example 3. Modify the structure of the previous examples by adjoining to $\mathscr{C}_1$ a set $\mathscr{C}_h$ of all P' with $g' - g \leqq h$ for any h with $E_P h < \infty$. Since, for P' in $\mathscr{C}_h$,

$$\Delta_n(P') = \Delta_n(P) - \Sigma(g' - g)p_n^* \geqq \Delta_n(P) - E_n h ,$$

and $E_n h \to E_P h$, a.e. (P^∞), the adjunction will not affect the previous conclusions.

Finally, a slight modification of Example 3 results in composite inconsistency at least in the case of Example 1.

Example 4. Modify the structure of Example 1 by adjoining to $\mathscr{C}_1$ a set $\mathscr{C}_c$ of all P' with $|g' - g| \leqq c$ and $V(P', P) < 2\eta$ for any constant c with $0 < c < \infty$. Since, for each P' in $\mathscr{C}_c$, $E_{P'}(-g') \geqq e^{-c}E_P(-g') = \infty$, it follows by appropriate modification of the argument leading to (6.1) that, if P_n is an ML estimate,

$$\liminf V(P_n, P') > 0, \qquad \text{a.e. } (P'^\infty) \text{ for each } P' \text{ in } \mathscr{C}_c . \tag{6.4}$$

Appendix: DOOB'S CONSISTENCY THEOREM

This theorem (Theorem 5 of [2]) was based on Theorem 4 of [2] which was later (see Wald [4]) corrected by adding a domination condition to the hypothesis. Slightly more general one-sided versions, Theorems 4′ and 5′, are here presented in the notation of Section 2 specialized to the case where $\mathscr{X}$ is the real line. The consistency theorem of Section 5 will be obtained as the discrete case of Theorem 5′.

Theorem 4′ depends on the following result. Let H be a non-negative summable function of x and let I denote an arbitrary interval and also its indicator function. From the monotonicity of the interval function $E_P IH$, it follows that

$$\Lambda(H) = \bigcap_I [\boldsymbol{x} \mid E_n IH \to E_P IH]$$

is the intersection over those I with rational endpoints or degenerate but with $P(I) > 0$. That

$$P^{\infty}(\Lambda(H)) = 1 \tag{A.1}$$

then follows from the SLLN for each IH of the denumerable intersection. (Indeed $\Lambda(H) = [\boldsymbol{x} \mid \sup_x |E_n(-\infty, x]H - E_P(-\infty, x]H| \to 0]$, and (A.1) is an extended Glivenko-Cantelli theorem.)

If h is a uniform limit of step functions, $\sum_1^k c_j I_j$, and hence in particular if h is continuous and bounded, it is immediate that

$$E_n h \to E_P h \qquad \text{on } \Lambda(1)\,. \tag{A.2}$$

In contrast, continuous a.e. $(P)h$ are characterized by the existence, for each h and $\varepsilon > 0$, of a step function $\sum_1^k c_j I_j$ such that

$$|(\textstyle\sum I_j)h - \sum c_j I_j| < \varepsilon \quad \text{and} \quad P(\cup I_j) > 1 - \varepsilon\,, \tag{A.3}$$

and even if $E_P h$ exists, this structure does not suffice for (A.2).

Theorem 4′ is a positive result in this direction. Let

$$\mathscr{H} = \{h \mid h \text{ continuous a.e. } (P),\, h/H \text{ bounded below}\}\,,$$

and

$$\Lambda = \Lambda(1) \cap \Lambda(H)\,.$$

THEOREM 4′. *If $0 \leqq H$ and $E_P H < \infty$, then $P^{\infty}(\Lambda) = 1$ and, for $\boldsymbol{x}$ in Λ and h in $\mathscr{H}$,* $\liminf E_n h \geqq E_P h$.

PROOF. $P^{\infty}(\Lambda) = 1$ by (A.1) and, if $h/H \geqq -M > -\infty$, $h \geqq -M(1 - \sum I_j)H + \sum c_j I_j - \varepsilon$ by (A.3). Since $E_n(1 - \sum I_j)H \to E_P(1 - \sum I_j)H$ on $\Lambda(H)$ and $E_n \sum c_j I_j \to E_P \sum c_j I_j \geqq E_P(\sum I_j)h - \varepsilon$ on $\Lambda(1)$, the desired inequality follows from the facts that as $\varepsilon \to 0$, $E_P(1 - \sum I_j)H \to 0$ and $E_P(\sum I_j)h \to E_P h$.

Theorem 5′ will follow mainly from the following corollary to Theorem 4′. For an estimate P_n and $\boldsymbol{x}$ in Λ,

$$h_N = \inf_{j \geqq N} (g - g_{n_j}) \qquad \text{in } \mathscr{H} \text{ infinitely often} \tag{A.4}$$

implies

$$\liminf E_{n_j}(g - g_{n_j}) \geqq E_P \liminf (g - g_n)\,, \tag{A.5}$$

since the left-hand member $\geqq \liminf E_n h_N$ while $E_P h_N \uparrow E_P \lim h_N$.

THEOREM 5′. *If H satisfies the assumption of Theorem 4′ and P_n is (at P) an equi-likelihood estimate (4.1), then, for each $\boldsymbol{x}$ in Λ such that* $\limsup E_n(g - g_n) \leqq 0$, *every convergent, a.e. (μ) subsequence p_{n_j} satisfying* (A.4) *converges to p, a.e. (μ).*

PROOF. Since by Fatou's lemma $E_P \liminf (p_{n_j}/p) \leqq 1$, Theorem 5′ follows from the fact that, by (A.5) and Jensen's inequality,

$$0 \geqq E_P \liminf (g - g_{n_j}) > -\log E_P \limsup (p_{n_j}/p)\,,$$

unless $\limsup p_{n_j} = p$, a.e. (μ).

In the special case where μ is counting measure on a denumerable set without finite limit points, every function of x is equivalent to an everywhere-continuous function of x. Hence, if $E_P H < \infty$ for the H of Theorem 1′ in Section 5, (A.4) holds for every $\boldsymbol{x}$ and every subsequence p_{n_j}. Since, moreover, every sequence p_n is sequentially compact, Theorem 5′ specialized to this case yields Theorem 1′.

References

[1] Hotelling, H. The consistency and ultimate distribution of optimum statistics. *Trans. Amer. Math. Soc.*, 1930, **32**, 847-59.

[2] Doob, J. L. Probability and statistics. *Trans. Amer. Math. Soc.*, 1934, **36**, 759–75.

[3] Kiefer, J., and Wolfowitz, J. Consistency of the maximum likelihood estimator in the presence of infinitely many incidental parameters. *Ann. Math. Stat.*, 1956, **27**, 887–906.

[4] Wald, A. Note on the consistency of the maximum likelihood estimate. *Ann. Math. Stat.*, 1949, **20**, 595–601.

[5] Doob, J. L. *Stochastic Processes.* New York: Wiley, 1953.

[6] Kraft, C. Some conditions for consistency and uniform consistency of statistical procedures. *Univ. Calif. Publ. Stat.*, 1955, **2**, 125-42.

[7] Kakutani, S. On equivalence of infinite product measures. *Ann. Math.*, 1948, **49**, 214–24

[8] Kullback, S., and Leibler, R. A. Information and sufficiency. *Ann. Math. Stat.*, 1951, **22**, 79–86.

[9] Hoeffding, W., and Wolfowitz, J. Distinguishability of sets of distributions. *Ann. Math. Stat.*, 1958, **29**, 700–18.

[10] Le Cam, L. On the asymptotic theory of estimation and testing hypotheses. In J. Neyman (ed.), *Proceedings of the Third Berkeley Symposium on Mathematical Statistics and Probability*, vol. 1. Berkeley, Calif: Univ. California Press, 1955.

[11] Le Cam, L. On some asymptotic properties of maximum likelihood estimates and related Bayes' estimates. *Univ. Calif. Publ. Stat.*, 1953, **1**, 277-330.

[12] Rao, C. R. Maximum likelihood estimation for the multinomal distribution. *Sankhyā*, 1957, **18**, 139–48.

22 AN UPPER BOUND FOR THE VARIANCE OF KENDALL'S "TAU" AND OF RELATED STATISTICS[1]

WASSILY HOEFFDING

University of North Carolina

1. Results

Let $X_1, X_2, \cdots, X_n$ be independent and identically distributed random variables (real- or vector-valued). Let $f(X_1, X_2)$ denote a bounded function such that $f(X_1, X_2) = f(X_2, X_1)$. With no loss of generality we shall assume that the bounds are $0 \leqq f(X_1, X_2) \leqq 1$. Let

$$U = \frac{2}{n(n-1)} \sum_{1 \leqq i < j \leqq n} f(X_i, X_j) .$$

Examples of statistics of this form are given below. The mean of U is $E(U) = p$ and the variance is

$$\operatorname{var}(U) = \frac{2}{n(n-1)} \{2(n-2)(r - p^2) + s - p^2\} ,$$

where

$$p = Ef(X_1, X_2) , \qquad r = Ef(X_1, X_2) f(X_1, X_3) , \qquad s = Ef^2(X_1, X_2) .$$

As n tends to infinity, $\sqrt{n}(U - p)$ has a normal limiting distribution [1]. Hence, if we have an upper bound for the variance of U which depends only on p and n, we can obtain an approximate confidence region for p and a lower bound for the power of a test based on U when n is large (see [2]).

It is known [1] that $2(r - p^2) \leqq s - p^2$, and since obviously $s \leqq p$, we have $\operatorname{var}(U) \leqq 2p(1-p)/n$.

In this paper we shall show that, under the stated assumptions,

$$r - p^2 \leqq H(p) = \begin{cases} p^{\frac{3}{2}} - p^2 , & p \geqq \frac{1}{2} , \\ (1-p)^{\frac{3}{2}} - (1-p)^2 , & p \leqq \frac{1}{2} . \end{cases} \tag{1}$$

[1] This research was supported by the United States Air Force through the Air Force Office of Scientific Research, Air Research and Development Command, under Contract No. AF 18(600)-458. Reproduction in whole or in part is permitted for any purpose of the United States Government.

It is easily seen that the sign of equality holds in (1) if, with probability 1, $f(X_1, X_2) = g(X_1)g(X_2)$ (for $p \geqq \frac{1}{2}$) or $f(X_1, X_2) = 1 - g(X_1)g(X_2)$ (for $p \leqq \frac{1}{2}$), where $g(X)$ takes the values 0 and 1 only.

Inequality (1) implies that

$$\text{var}(U) \leqq \frac{2}{n(n-1)}[2(n-2)H(p) + p - p^2]\,. \tag{2}$$

An inequality analogous to (1) was conjectured by Daniels and Kendall [3] for the variance of the finite population analog of the statistic t defined in Example 1 below. A proof of this conjecture suggested by Sundrum [4] does not seem to be complete.

We now give three examples of statistics to which the present bound is applicable; in the first two examples the bound can be attained.

Example 1. Let $X_i = (Y_i, Z_i)$ be a random vector with two continuously distributed components, and $f(X_1, X_2) = 1$ or 0 according as $(Y_1 - Y_2)(Z_1 - Z_2)$ is positive or negative. In this case, $2U - 1$ is the rank correlation coefficient known as Kendall's tau. The condition for equality in (1) is satisfied if Z is a function of Y of a certain form, for instance, positive and decreasing for $Y < 0$ and negative and increasing for $Y > 0$ (if $p \geqq \frac{1}{2}$); or negative and increasing for $Y < 0$ and positive and decreasing for $Y > 0$ (if $p \leqq \frac{1}{2}$). For if we let $g(X) = 0$ or 1 according as $Y < 0$ or $Y > 0$, then, with probability 1, $f(X_1, X_2) = g(X_1)\,g(X_2)$ in the first case, and $f(X_1, X_2) = 1 - g(X_1)\,g(X_2)$ in the second case. These two cases correspond to the inverse canonical ranking and the canonical ranking as defined by Daniels and Kendall [3] for finite populations.

Example 2. Let X_i be a real-valued random variable, $f(X_1, X_2) = 0$ or 1 according as $X_1 + X_2 < 0$ or > 0. A statistic closely related to the corresponding U, namely, the number of pairs i, j, where $1 \leqq i \leqq j \leqq n$, such that $X_i + X_j > 0$, has been considered by Walsh [5] and shown by Tukey [6] to be essentially identical with Wilcoxon's signed-rank sum [7]. In this case, the sign of equality in (1) is attained, for instance, if X_i can take only two values a and b, such that either $a + b < 0 < b$ (if $p \geqq \frac{1}{2}$) or $a < 0 < a + b$ (if $p \leqq \frac{1}{2}$). Here we may take $g(X) = 0$ or 1 according as $X < 0$ or $X > 0$ in the first case, and $g(X) = 0$ or 1 according as $X > 0$ or $X < 0$ in the second case.

Example 3. Let X_i again be real-valued, and let

$$f(X_1, X_2) = 1 - 2\max[F_0(X_1),\ F_0(X_2)] + F_0^2(X_1) + F_0^2(X_2)\,,$$

where $F_0(x)$ is a continuous (cumulative) distribution function. Then, if X_i has the distribution function $F(x)$, $p = \frac{1}{3} + 2\int [F(x) - F_0(x)]^2\, dF_0(x)$, and $\frac{1}{2}(U - \frac{1}{3})$ differs in large samples negligibly little from the Cramér-von Mises goodness-of-fit criterion $\int [F_n(x) - F_0(x)]^2\, dF_0(x)$, where $nF_n(x)$ denotes the number of observations $\leqq x$. In this case, the condition for equality in (1) cannot be satisfied, and presumably the upper bound for the variance can be further improved.

2. Proof of Inequality (1)

We first assume that

(3) $$f(X_1, X_2) = 0 \text{ or } 1.$$

Let $X_1, X_2, \cdots, X_n$ be independent and identically distributed random variables,

$$f_{ij} = f(X_i, X_j)\,, \qquad i \neq j\,, \qquad f_{ii} = 0\,,$$

$$\hat{p} = n^{-2} \sum_{i=1}^{n} \sum_{j=1}^{n} f_{ij}\,, \qquad \hat{r} = n^{-3} \sum_{i=1}^{n} \sum_{j=1}^{n} \sum_{k=1}^{n} f_{ij} f_{ik}\,.$$

As $n \to \infty$, $\hat{p}$ and $\hat{r}$ converge in probability to p and r, respectively.

We shall show that

(4) $$\hat{r} - \hat{p}^2 \leqq H(\hat{p}) + \varepsilon_n\,,$$

where $H(p)$ is the function defined in (1) and ε_n are numbers which converge to 0 as $n \to \infty$. Since the function $H(p)$ is continuous, Inequality (4) easily implies Inequality (1).

Both $\hat{p}$ and $\hat{r}$ are functions of the $n \times n$ matrix $||f_{ij}||$ whose elements satisfy the conditions

(5) $$f_{ij} = 0 \text{ or } 1\,, \qquad f_{ii} = 0\,, \qquad f_{ij} = f_{ji}\,.$$

Let

$$F_i = \sum_{j=1}^{n} f_{ij}\,.$$

Then

$$n^2 \hat{p} = \sum_{i=1}^{n} F_i\,, \qquad n^3 \hat{r} = \sum_{i=1}^{n} F_i^2\,.$$

We first show that, in order to find an upper bound for $\hat{r}$ when $\hat{p}$ is fixed, we may assume that

(6) $$F_i \geqq F_j \text{ implies } f_{ik} \geqq f_{jk}\,, \qquad \text{for all } k \neq i\,.$$

Suppose that there are integers i, j, k such that $F_i \geqq F_j$, $k \neq i$, and $f_{ik} < f_{jk}$. Thus $f_{ik} = 0$, $f_{jk} = 1$. Let $||f'_{uv}||$ be the $n \times n$ matrix defined by $f'_{ik} = f'_{ki} = 1$, $f'_{jk} = f'_{kj} = 0$, $f'_{uv} = f_{uv}$ otherwise. The transformed matrix satisfies (5), and the value of $\hat{p}$ is not affected by the transformation. Also, by writing F'_u for the sum of the uth row in the new matrix, $F'_i = F_i + 1$, $F'_j = F_j - 1$, $F'_u = F_u$ otherwise. Therefore

$$\sum (F'_u)^2 - \sum F_u^2 = (F_i + 1)^2 - F_i^2 + (F_j - 1)^2 - F_j^2 = 2(F_i - F_j) + 2 > 0\,.$$

Thus the value of $\hat{r}$ is increased by the transformation. By repeated application of this transformation, we can obtain a matrix which satisfies (6), for instance as follows. We first take one of the rows with the largest sum to be the ith row, and for every $j \neq i$ we apply the transformation for each $k \neq i$ with $f_{ik} < f_{jk}$. Having exhausted all k and all j, we are left with a matrix such that every column different from the ith that has 0 in the ith place will consist of zeros only. From now on, this ith row and the corresponding column are held fixed. We next take one of the rows with the largest sum among the remaining $n-1$ rows and repeat the procedure described above, confining it to the now-available $n-1$ rows and $n-1$ columns, and so on. Eventually, we obtain a matrix which satisfies the conditions (5) and (6) without changing the value of $\hat{p}$ and without decreasing the value of $\hat{r}$.

With no loss of generality we may assume that, in addition to satisfying (5) and (6), the matrix $||f_{ij}||$ satisfies the condition

$$F_1 \geqq F_2 \geqq \cdots \geqq F_n , \tag{7}$$

for this can always be achieved by rearranging suitable rows and columns, which will not affect the values of $\hat{p}$ and $\hat{r}$. Thus we now restrict ourselves to symmetric matrices whose every row and every column, apart from a 0 in the main diagonal, consists of a sequence of 1's followed by a sequence of 0's. A typical matrix of this form is shown below.

$$\begin{matrix} 0 & 1 & 1 & 1 & 1 & 1 & 0 \\ 1 & 0 & 1 & 1 & 0 & 0 & 0 \\ 1 & 1 & 0 & 1 & 0 & 0 & 0 \\ 1 & 1 & 1 & 0 & 0 & 0 & 0 \\ 1 & 0 & 0 & 0 & 0 & 0 & 0 \\ 1 & 0 & 0 & 0 & 0 & 0 & 0 \\ 0 & 0 & 0 & 0 & 0 & 0 & 0 \end{matrix} .$$

For a fixed matrix $||f_{ij}||$ of this type, let m denote the greatest integer such that $f_{m,m+1} = 1$. Then $f_{ij} = 1$ if $i \leqq m+1$, $j \leqq m+1$ $(i \neq j)$, and $f_{ij} = 0$ if $i \geqq m+1$, $j \geqq m+1$. Hence

$$\begin{aligned} n^2\hat{p} &= \sum_{i=1}^{m}\sum_{j=1}^{m} f_{ij} + \sum_{i=1}^{m}\sum_{j=m+1}^{n} f_{ij} + \sum_{i=m+1}^{n}\sum_{j=1}^{m} f_{ij} \\ &= m(m-1) + 2\sum_{i=1}^{m}\sum_{j=m+1}^{n} f_{ij} , \end{aligned}$$

and

$$\begin{aligned} n^3\hat{r} &= \sum_{i=1}^{m}\left(\sum_{j=1}^{m} f_{ij} + \sum_{j=m+1}^{n} f_{ij}\right)^2 + \sum_{i=m+1}^{n}\left(\sum_{j=1}^{m} f_{ij}\right)^2 \\ &= m(m-1)^2 + 2(m-1)\sum_{i=1}^{m}\sum_{j=m+1}^{n} f_{ij} + \sum_{i=1}^{m}\left(\sum_{j=m+1}^{n} f_{ij}\right)^2 + \sum_{j=m+1}^{n}\left(\sum_{i=1}^{m} f_{ij}\right)^2 . \end{aligned}$$

If we put

$$a_i = \frac{1}{n-m}\sum_{j=m+1}^{n} f_{ij} , \qquad b_j = \frac{1}{m}\sum_{i=1}^{m} f_{ij} ,$$

$$d = \frac{1}{m(n-m)}\sum_{i=1}^{m}\sum_{j=m+1}^{n} f_{ij} = \frac{1}{m}\sum_{i=1}^{m} a_i = \frac{1}{n-m}\sum_{j=m+1}^{n} b_j ,$$

we obtain

$$n^2\hat{p} = m(m-1) + 2m(n-m)d\,, \tag{8}$$

and

$$\begin{aligned} n^3\hat{r} &= (m-1)n^2\hat{p} + (n-m)^2\sum_{i=1}^{m} a_i^2 + m^2\sum_{j=m+1}^{n} b_j^2 \\ &= (m-1)n^2\hat{p} + m(n-m)n\,d^2 \\ &\quad + m(n-m)\left[(n-m)\frac{1}{m}\sum_{i=1}^{m}(a_i-d)^2 + m\frac{1}{n-m}\sum_{j=m+1}^{n}(b_j-d)^2\right]. \end{aligned} \tag{9}$$

Now

$$\begin{aligned} &\frac{1}{m(n-m)}\sum_{i=1}^{m}\sum_{j=m+1}^{n}[(f_{ij}-d)-(a_i-d)-(b_j-d)]^2 \\ &\quad = \frac{1}{m(n-m)}\sum_{i=1}^{m}\sum_{j=m+1}^{n}(f_{ij}-d)^2 - \frac{1}{m}\sum_{i=1}^{m}(a_i-d)^2 - \frac{1}{n-m}\sum_{j=m+1}^{n}(b_j-d)^2 \\ &\quad = d - d^2 - \frac{1}{m}\sum_{i=1}^{m}(a_i-d)^2 - \frac{1}{n-m}\sum_{j=m+1}^{n}(b_j-d)^2\,. \end{aligned}$$

Since the left-hand side is non-negative, we obtain[2]

$$\frac{1}{m}\sum_{i=1}^{m}(a_i-d)^2 + \frac{1}{n-m}\sum_{j=m+1}^{n}(b_j-d)^2 \leqq d - d^2\,. \tag{10}$$

It now follows from (9) and (10) that

$$n^3\hat{r} \leqq (m-1)n^2\hat{p} + m(n-m)nd^2 + m(n-m)(d-d^2)\max(m,\ n-m)\,. \tag{11}$$

We thus have obtained an upper bound for $\hat{r}$ in terms of $\hat{p}$ for all matrices $||f_{ij}||$ with m fixed. We now put $c = m/n$.

From (11) we obtain $\hat{r} \leqq c\hat{p} + c(1-c)d^2 + c(1-c)(d-d^2)\max(c,\ 1-c)$. If we express d in terms of $\hat{p}$ and c by means of (8), we have $\hat{r} - \hat{p}^2 \leqq H(\hat{p},\ c) + \varepsilon_n$, where

$$H(p,c) = -p^2 + cp + \frac{p-c^2}{2}\left[\frac{p-c^2}{2c(1-c)} + \max(c, 1-c)\left(1 - \frac{p-c^2}{2c(1-c)}\right)\right],$$

and ε_n is independent of c and $\hat{p}$, and $\varepsilon_n \to 0$ as $n \to \infty$.

We now maximize $H(p,\ c)$ with respect to c. We can write

$$H(p,\ c) = \begin{cases} G(1-p,\ 1-c)\,, & c \leqq \frac{1}{2}\,, \\ G(p,\ c)\,, & c \geqq \frac{1}{2}\,, \end{cases}$$

where $G(p,\ c) = -p^2 + pc + \frac{1}{4}p^2c^{-1} - \frac{1}{4}c^3$.

[2] Inequalities (10) and (11) are closely related to the sharp upper bound for the variance of the Wilcoxon-Mann-Whitney two-sample statistic in terms of its mean; see van Dantzig ([8], footnote 4).

We calculate

$$G(p, p^{\frac{1}{2}}) - G(p, c) = \frac{1}{4c}(p^{\frac{1}{2}} - c)^2(c^2 + 2p^{\frac{1}{2}}c - p) .$$

If $c \geqq \frac{1}{2}$, the right side is positive, and hence $G(p, c) \leqq G(p, p^{\frac{1}{2}}) = p^{\frac{3}{2}} - p^2$. It follows that

$$H(p, c) \leqq \max [p^{\frac{3}{2}} - p^2, (1-p)^{\frac{3}{2}} - (1-p)^2] ,$$

and the function on the right is just $H(p)$ as defined in (1). This completes the proof for the case where $f(X_1, X_2)$ takes on the values 0 and 1 only.

Now let $f(X_1, X_2)$ be any function which satisfies our assumptions. Then $f(X_1, X_2)$ can be approximated by a function $f'(X_1, X_2)$ which also satisfies our assumptions and takes only finitely many values, in such a way that $p' = Ef'(X_1, X_2)$ and $r' = Ef'(X_1, X_2)f'(X_1, X_3)$ are as close as we please to p and r, respectively. It will therefore be sufficient to assume that $f(X_1, X_2)$ takes on the k values $f_1, f_2, \cdots, f_k$. Then it can be written in the form

$$f(X_1, X_2) = \sum_{i=1}^{k} f_i c_i(X_1, X_2) ,$$

where $0 \leqq f_i \leqq 1$,

$$c_i(X_1, X_2) = c_i(X_2, X_1) = 0 \text{ or } 1 ; \qquad c_i(X_1, X_2)c_j(X_1, X_2) = 0 , \qquad i \neq j .$$

Now let Y_1, Y_2, Y_3 be mutually independent random variables, independent of X_1, X_2, X_3, where each Y_i is uniformly distributed on the interval $0 \leqq Y_i \leqq 1$. The random vectors $Z_j = (X_j, Y_j)$, $j = 1, 2, 3$, are mutually independent and identically distributed. Now define

$$f^*(Z_1, Z_2) = \sum_{i=1}^{k} d_i(Y_1)\, d_i(Y_2)\, c_i(X_1, X_2) ,$$

where

$$d_i(Y) = \begin{cases} 1 , & 0 \leqq Y \leqq f_i^{\frac{1}{2}} , \\ 0 , & \text{otherwise} . \end{cases}$$

Then $f^*(Z_1, Z_2) = f^*(Z_2, Z_1) = 0$ or 1.

The conditional expected value of $f^*(Z_1, Z_2)$ for X_1, X_2 fixed is

$$E[f^*(Z_1, Z_2) \mid X_1, X_2] = \sum_{i=1}^{k} f_i c_i(X_1, X_2) = f(X_1, X_2) .$$

Hence

$$Ef^*(Z_1, Z_2) = Ef(X_1, X_2) = p . \tag{12}$$

Also

$$f^*(Z_1, Z_2)f^*(Z_1, Z_3) = \sum_{i=1}^{k} \sum_{j=1}^{k} d_i(Y_1)\, d_i(Y_2)\, c_i(X_1, X_2)\, d_j(Y_1)\, d_j(Y_3)\, c_j(X_1, X_3) ,$$

so that

$$
\begin{aligned}
E[f^*(Z_1, Z_2)f^*(Z_1, Z_3) \mid X_1, X_2, X_3] \\
&= \sum_{i=1}^{k} \sum_{j=1}^{k} \min (f_i^{\frac{1}{2}}, f_j^{\frac{1}{2}}) f_i^{\frac{1}{2}} f_j^{\frac{1}{2}} c_i(X_1, X_2) c_j(X_1, X_3) \\
&\geqq \sum_{i=1}^{k} \sum_{j=1}^{k} f_i f_j c_i(X_1, X_2) c_j(X_1, X_3) \\
&= f(X_1, X_2) f(X_1, X_3) .
\end{aligned}
$$

Thus,

$$E[f^*(Z_1, Z_2)f^*(Z_1, Z_3)] \geqq E[f(X_1, X_2)f(X_1, X_3)] . \tag{13}$$

Since, by the first part of the proof, Inequality (1) is true for f^*, it follows from (12) and (13) that it is also true for f. ||

References

[1] Hoeffding, W. A class of statistics with asymptotically normal distribution. *Ann. Math. Stat.*, 1948, **19**, 293–325.

[2] Kendall, M. G. *Rank Correlation Methods*, 2d ed. London: Griffin, New York: Hafner, 1955.

[3] Daniels, H. E., and Kendall, M. G. The significance of rank correlations where parental correlation exists. *Biometrika*, 1947, **34**, 197–208.

[4] Sundrum, R. M. Moments of the rank correlation coefficient τ in the general case. *Biometrika*, 1953, **40**, 409–20.

[5] Walsh, J. E. Some significance tests for the median which are valid under very general conditions, *Ann. Math. Stat.*, 1949, **20**, 64–81.

[6] [Tukey, J. W.] The simplest signed-rank tests. Statistical Research Group Memorandum No. 17, Princeton, N.J.: Princeton University [No date].

[7] Wilcoxon, F. Individual comparisons by ranking methods. *Biometrics Bull.*, 1945, **1**, 80–83.

[8] Dantzig, D. van. On the consistency and the power of Wilcoxon's two sample test. *Ned. Akad. Wetenschap. Proc.*, Ser. A, 1951, **54**, 1–8.

23 ON THE AMOUNT OF INFORMATION CONTAINED IN A σ-FIELD[1]

GOPINATH KALLIANPUR
Indian Statistical Institute and Michigan State University[2]

1. Introduction

Let $(\Omega, \mathscr{F})$ be an arbitrary measurable space and P_1 and P_2 two probability measures over $\mathscr{F}$. Let $\{A_i\}$ be a finite $\mathscr{F}$-measurable partition of Ω and $\mathscr{A}$ be the finite subfield of $\mathscr{F}$ given by the class of unions of the A_i. Then the quantity

$$I(\mathscr{A}) = \sum_i P_1(A_i) \log [P_1(A_i)/P_2(A_i)] \tag{1.1}$$

is commonly referred to as the information of P_1 with respect to P_2 ([1], p. 50). In the expression on the right-hand side of (1.1) we take $P_1(A_i) \log [P_1(A_i)/P_2(A_i)] = 0$ whenever $P_1(A_i) = 0$, even if $P_2(A_i) = 0$. $I(\mathscr{A})$ is always non-negative [2]; each of the terms in its definition exceeds $-\infty$, and $I(\mathscr{A}) = \infty$ if and only if $P_1(A_i) > P_2(A_i) = 0$ for some i. As our notation makes explicit, information as defined by (1.1) depends also on the finite field $\mathscr{A}$, so that it seems appropriate to call $I(\mathscr{A})$ the information of P_1 with respect to P_2 contained in $\mathscr{A}$, or briefly, the information contained in $\mathscr{A}$. The usefulness of this notion in statistics has been discussed by several authors ([1], [2], [3]), and it is pertinent to extend the definition (1.1) to cover the notion of the information contained in an arbitrary σ-field. An expression for the information in the general case has been assumed *ad hoc* by Kullback and Leibler in their paper [2], where, however, no attempt is made to relate it to the definition (1.1).

The present paper originated in an effort to prove certain results of I. M. Gelfand, A. N. Kolmogorov, and A. M. Yaglom on the general concept of information, announced without proof in [4]. The definition of information with which these authors are concerned is one peculiar to information theory (see Section 4), and it seemed advisable to derive the results for

[1] This work was supported by National Science Foundation Grant No. G-5011 and was carried out while the author was Visiting Professor of Statistics at Michigan State University, 1958-59.

[2] Now at Indiana University.

the more general situation in which the statistician is primarily interested. The principal result of Gelfand, Kolmogorov, and Yaglom is deduced in Section 4 as a special case of Theorem 3.

The motivation for extending (1.1) to the case of an arbitrary σ-field $\mathscr{F}$ rests on the following property. Let $\mathscr{A}_1$ be a field generated by a finer partition of Ω obtained by a subdivision of the partition $\{A_i\}$. Further, let A be any set of the A_i with $P_2(A) > 0$ and such that A is a finite disjoint union of sets B_j of $\mathscr{F}$. Then, in Jensen's inequality $\varphi(\sum_j \lambda_j x_j) \leqq \sum_j \lambda_j \varphi(x_j)$, ($\varphi$ convex, $\lambda_j \geqq 0, \sum_j \lambda_j = 1$), setting $\varphi(x) = x \log x$, $\lambda_j = [P_2(B_j)/P_2(A)]$, and $x_j = [P_1(B_j)/P_2(B_j)]$, we have $P_1(A) \log [P_1(A)/P_2(A)] \leqq \sum_j P_1(B_j) \log P_1(B_j)/P_2(B_j)]$. It immediately follows that if $\mathscr{A}_1$ is a finite field "finer" than $\mathscr{A}$ in the sense indicated above (observe also that $\mathscr{A} \subset \mathscr{A}_1$), then $I(\mathscr{A}_1) \geqq I(\mathscr{A})$. This entirely natural and desirable property of information leads to the following definition, the basis of which is to be found in [4]. The information $I(\mathscr{F})$ contained in the σ-field $\mathscr{F}$ is defined by

$$I(\mathscr{F}) = \sup_{\mathscr{A} \subset \mathscr{F}} I(\mathscr{A}) , \tag{1.2}$$

the supremum being taken over all finite fields $\mathscr{A}$ contained in $\mathscr{F}$. If $\mathscr{F}$ is a finite field, (1.2) coincides with (1.1).

In Sections 2 and 3 are derived necessary and sufficient conditions under which $I(\mathscr{F})$ is finite, as well as an explicit integral representation for $I(\mathscr{F})$. Properties of martingales and semi-martingales together with their applications to the theory of derivatives due to Doob [5] are the principal tools used in the proofs of these results. In the final section (Section 5), the property of the semi-martingale occurring in Theorem 1, on which the proof of that theorem rests, is extended to an important class of semi-martingales, viz., those which are convex functions of martingales. Since this result might be of use in applications, it is stated and proved separately as Theorem 5.

2. Information Contained in a Separable σ-Field

For each $n = 1, 2, \cdots$, let $\mathscr{M}_n = \{M_j^n\}$ be a finite $\mathscr{F}$-partition of Ω with the property that, for every n, M_j^{n+1} is a subset of some M_k^n. The sequence of partitions $\{\mathscr{M}_n\}$ will then be said to be *regular*.

Let $\mathscr{F}_n$ be the field consisting of finite unions of sets of $\mathscr{M}_n$, and $\mathscr{F}_\infty$ the σ-field generated by the field $\bigcup_{n=1}^\infty \mathscr{F}_n$. In general, $\mathscr{F}_\infty$ is a sub σ-field of $\mathscr{F}$ ([5], p. 343).

We shall begin by proving several lemmas. For this we first need the following definitions. For every positive integer n, let

$$x_n(\omega) = \begin{cases} P_1(M_j^n)/P_2(M_j^n) , & \text{if } \omega \in M_j^n \text{ and } P_2(M_j^n) > 0 , \\ 0 , & \text{if } \omega \in M_j^n \text{ and } P_2(M_j^n) = 0 , \end{cases} \tag{2.1}$$

and

$$y_n(\omega) = \begin{cases} x_n(\omega) \log x_n(\omega) , & \text{if } x_n(\omega) \neq 0 , \\ 0 , & \text{if } x_n(\omega) = 0 . \end{cases} \tag{2.2}$$

Then $\{x_n, n \geqq 1\}$ is a martingale on $(\Omega, \mathscr{F}_\infty, P_2)$ and hence $\{y_n, n \geqq 1\}$ is a semi-martingale on the same probability space ([5], pp. 343–45). It follows immediately that $I(\mathscr{F}_n) = \int y_n dP_2 = E_2(y_n)$ is a monotone non-decreasing function of n.

Employing the by-now-standard notation of Halmos and Savage [6], we shall write $P_1 \ll P_2, P_1 \equiv P_2, P_1 \perp P_2$, to denote P_1 absolutely continuous with respect to, equivalent to, and singular with respect to P_2, respectively.

LEMMA 1. *For any σ-field $\mathscr{F}$, a necessary condition for $I(\mathscr{F})$ to be finite is that $P_1 \ll P_2$ relative to $\mathscr{F}$.*

PROOF. If P_1 is not absolutely continuous with respect to P_2, we have $I(\mathscr{A}) = \infty$ upon taking $\mathscr{A} = \{S, \Omega - S\}$, where $S \in \mathscr{F}$ is the singular set of the singular component of P_1 relative to P_2. Lemma 1 then follows immediately. ||

LEMMA 2. *Let $P_1 \ll P_2$ relative to $\mathscr{F}_\infty$ and $\mathscr{A}$ be any finite subfield of $\mathscr{F}_\infty$. Then for every $\varepsilon > 0$, there is an integer m such that $I(\mathscr{F}_m) \geqq I(\mathscr{A}) - \varepsilon$.*

PROOF. Let $\mathscr{A} \subset \mathscr{F}_\infty$ be given by the partition $\{A_i\}$, $i = 1, \cdots, N$. For the purposes of this lemma there is no loss of generality in assuming $P_2(A_i) > 0$, $i = 1, \cdots, N$. For $\eta > 0$ and each $i = 1, \cdots, N$, take a set $F_i \in \bigcup_{n=1}^\infty \mathscr{F}_n$ such that $P_2(A_i \Delta F_i) < \eta$, where Δ denotes the symmetric difference. Let $G_i = F_i - \bigcup_{k \neq i} F_k$ for $i = 1, \cdots, N$ and let $G_{N+1} = \Omega - \bigcup_{i=1}^N G_i$. Then clearly the sets G_i, $i = 1, \cdots, N+1$, form a partition such that G_i belongs to $\bigcup_{n=1}^\infty \mathscr{F}_n$ for each i,

$$P_2(A_i \Delta G_i) \leqq \eta_1 = (N+1)\eta\,, \qquad \text{for } i = 1, \cdots, N\,, \tag{2.3}$$

and

$$P_2(G_{N+1}) \leqq \eta_1\,. \tag{2.4}$$

Furthermore, there exists an integer m depending on η such that

$$G_i \in \mathscr{F}_m\,, \qquad \text{for } i = 1, \cdots, N+1\,. \tag{2.5}$$

For any set A in $\mathscr{F}_\infty$ with $P_2(A) > 0$, let

$$h(A) = \begin{cases} [P_1(A)/P_2(A)] \log [P_1(A)/P_2(A)]\,, & \text{if } P_1(A) \neq 0\,, \\ 0\,, & \text{if } P_1(A) = 0\,. \end{cases}$$

If $\mathscr{F}_\infty$ is regarded as a metric space with the metric $\rho(A, B) = P_2(A \Delta B)$ ([7], pp. 168–69), on account of the absolute continuity of P_1 with respect to P_2, it is clear that $h(A)$ is continuous with respect to the metric ρ at $A = A_i$, $i = 1, \cdots, N$. If $\eta > 0$ and $d_i(\eta) = \sup_{\rho(A, A_i) \leq \eta} |h(A) - h(A_i)|$, the continuity of h at A_i implies that $d_i(\eta)$, $i = 1, \cdots, N$, and hence also $d(\eta) = \max_{1 \leq i \leq N} d_i(\eta)$, tends monotonely to zero as η approaches zero.

Let us now define the following step functions on Ω. Let

$$f = \sum_{i=1}^{N} h(A_i) I_{A_i}\,, \qquad f_\eta = \sum_{i=1}^{N} h(A_i) I_{G_i}\,, \quad \text{and} \quad g_\eta = \sum_{i=1}^{N} h(G_i) I_{G_i}\,.$$

The following relations are then easily verified:

$$\int |f - f_\eta|\, dP_2 \leqq K_N \eta_1\,, \tag{2.6}$$

$$\int |f_\eta - g_\eta|\, dP \leqq d(\eta_1)\,, \tag{2.7}$$

and

$$\int |f - g_\eta|\, dP_2 \leqq K_N \eta_1 + d(\eta_1)\,. \tag{2.8}$$

In these inequalities, K_N is a function only of $P_1(A_i)$ and $P_2(A_i)$, $i = 1, \cdots, N$. In other words, K_N is a constant depending only on $\mathscr{A}$ and on the probability measures P_1 and P_2, but not on η. Now we have $\int f\, dP_2 = I(\mathscr{A})$, and

$$\int g_\eta\, dP_2 = \sum_{i=1}^{N} P_1(G_i) \log [P_1(G_i)/P_2(G_i)] \leqq I(\mathscr{F}_m) - P_1(G_{N+1}) \log [P_1(G_{N+1})/P_2(G_{N+1})]$$

by Jensen's inequality, and (2.5). Finally, from (2.8) we have

$$I(\mathscr{F}_m) \geqq I(\mathscr{A}) - [K_N \eta_1 + d(\eta_1) - P_1(G_{N+1}) \log P_1(G_{N+1})]\,.$$

The quantity within the square brackets converges to zero as $\eta \to 0$ and the lemma is established. ||

LEMMA 3. *Let $P_1 \ll P_2$ relative to $\mathscr{F}_\infty$. Then* $\lim_{n\to\infty} I(\mathscr{F}_n) = I(\mathscr{F}_\infty)$.

PROOF. The limit on the left-hand side exists since $I(\mathscr{F}_n)$ is non-decreasing in n. First let $I(\mathscr{F}_\infty)$ be finite. By definition (1.2), if $\delta > 0$ there exists a finite subfield $\mathscr{A}$ of $\mathscr{F}_\infty$ such that $I(\mathscr{A}) > I(\mathscr{F}_\infty) - \delta$. By Lemma 2, for $\varepsilon > 0$ there exists an integer m with the property $I(\mathscr{F}_m) > I(\mathscr{A}) - \varepsilon$. Hence, for all $n \geqq m$, $I(\mathscr{F}_n) > I(\mathscr{F}_\infty) - \delta - \varepsilon$. The lemma follows from the fact that $I(\mathscr{F}_n) \leqq I(\mathscr{F}_\infty)$ for all n, and ε and δ are arbitrarily small. The truth of the lemma when $I(\mathscr{F}_\infty) = \infty$ follows in a similar manner. ||

With the help of the preceding lemmas we can now prove the following result concerning the representation of $I(\mathscr{F}_\infty)$.

THEOREM 1. *$I(\mathscr{F}_\infty)$ is finite if and only if the following two conditions are satisfied:*

(i) *$P_1 \ll P_2$ relative to $\mathscr{F}_\infty$,*

(ii) $E_2 |x_\infty \log x_\infty| < \infty$,

where x_∞ is the Radon-Nikodym derivative of P_1 with respect to P_2 relative to $\mathscr{F}_\infty$. Further, if (i) *holds, $I(\mathscr{F}_\infty) = E_2(x_\infty \log x_\infty) = E_1(\log x_\infty)$, whether $I(\mathscr{F}_\infty)$ is finite or infinite.*

In the proof of this theorem we make use of several properties of martingales and semi-martingales established by Doob in Theorems 1.1, 3.1, and 4.1s of [5] (pp. 295–325).

PROOF. Throughout this proof, whenever we say that a sequence of random variables is a martingale (semi-martingale), or is uniformly integrable, or converges with probability 1, these statements will be understood as referring to the probability space $(\Omega, \mathscr{F}_\infty, P_2)$. Suppose first that conditions (i) and (ii) are fulfilled. From Doob's example on the theory of derivatives ([5], Example 1, p. 344), it follows that the sequence $\{x_n, n \geqq 1\}$ defined at the beginning of this section forms a uniformly integrable martingale and that $\lim x_n = x_\infty$ with probability 1, x_∞ being the function defined in the statement of Theorem 1. Hence, from a well-known martingale property ([5], Theorem 4.1), the sequence $\{x_n, 1 \leqq n \leqq \infty\}$ is a martingale. Condition (ii) and the convexity of the function $t \log t \; (t \geqq 0)$ imply that $\{y_n, 1 \leqq n \leqq \infty\}$ is a semi-martingale. Since the function $t^+ \; (= t$ for $t \geqq 0, = 0$ for $t < 0)$ is monotone non-decreasing and convex, we can now deduce from Theorem 1.1 (i) and Theorem 3.1 (iii) of [5] that $\{y_n^+, 1 \leqq n \leqq \infty\}$ is a uniformly integrable semi-martingale.

Denoting the negative part of y_n by y_n^-, we have $y_n^- = -x_n \log x_n$ if $x_n \leqq 1$ and $y_n^- = 0$ otherwise, so that $0 \leqq y_n^-(\omega) \leqq 1$ for all n and ω. The random variables $\{y_n^-\}$ are obviously uniformly integrable. The uniform integrability of $\{y_n\}$ follows from that of $\{y_n^+\}$ and $\{y_n^-\}$. It has thus been shown that $\{y_n, 1 \leqq n \leqq \infty\}$ is a uniformly integrable semi-martingale such that $\lim_{n\to\infty} y_n = y_\infty = x_\infty \log x_\infty$ exists with probability 1. Theorem 4.1s(ii)(b) of [5] is now applicable, and we obtain the relation $\lim_{n\to\infty} E_2(y_n) = E_2(y_\infty)$. From Lemma 3 we conclude at once that $I(\mathscr{F}_\infty)$ is finite and equals $E_2(y_\infty)$.

Turning now to the necessity of conditions (i) and (ii), observe that the finiteness of $I(\mathscr{F}_\infty)$ and Lemma 1 imply (i). As in the proof of sufficiency, $\{x_n, 1 \leqq n \leqq \infty\}$ is a martingale, $\{y_n, n \geqq 1\}$ is a semi-martingale, $\lim y_n = y_\infty$ exists with probability 1, and the y_n^-'s are uniformly integrable. The boundedness of $E_2(y_n^+)$ follows from the relation

$$0 \leqq E_2(y_n^+) = E_2(y_n) + E_2(y_n^-) = I(\mathscr{F}_n) + E_2(y_n^-) \leqq I(\mathscr{F}_\infty) + 1 \, .$$

Hence

$$E_2|y_n| = E_2(y_n^+) + E(y_n^-) \leqq I(\mathscr{F}_\infty) + 2 \, .$$

From Fatou's lemma it follows that $E_2|y_\infty| < \infty$, which is condition (ii), since $y_\infty = x_\infty \log x_\infty$. The equality $I(\mathscr{F}_\infty) = E_2(x_\infty \log x_\infty)$ now follows as in the first part of the proof, and the theorem is established. ||

The σ-field $\mathscr{F}_\infty$ in Theorem 1 has the property that it is generated by a given regular sequence $\{\mathscr{M}_n\}$ of $\mathscr{F}$-measurable nets. Since the martingale $\{x_n\}$ in the proof is defined in terms of $\mathscr{M}_n$, it is relevant to inquire what σ-fields possess the above-mentioned property. It is clear from the definition of $\mathscr{F}_\infty$ that $\mathscr{F}_\infty$ is separable. We shall now show that every separable sub σ-field of $\mathscr{F}$ can be generated by a regular sequence of nets in the manner indicated at the beginning of this section. The importance of this fact is that, in the hypothesis of Theorem 1, we may replace $\mathscr{F}_\infty$ by an arbitrary separable sub σ-field of $\mathscr{F}$.

LEMMA 4. *Let $\mathscr{F}_0 (\subset \mathscr{F})$ be a separable σ-field. Then there exists a regular sequence $\{\mathscr{M}_n\}$ of finite $\mathscr{F}$-partitions of Ω such that $\mathscr{F}_0$ is generated by $\bigcup_{n=1}^{\infty} \mathscr{F}_n$, where $\mathscr{F}_n$ is the field generated by $\mathscr{M}_n$.*

PROOF. Let $E_n, n \geqq 1$ generate $\mathscr{F}_0$, and let $\mathscr{M}_n$ consist of the 2^n sets $\bigcap_{i=1}^{n} E_i^{\delta_i}$, where $\delta_i = 0$ or $1, E_i^0 = E_i$, and $E_i^1 = \Omega - E_i$. The conclusion of the lemma is then obvious. It also seems worth pointing out that in the limit as n tends to infinity, the meshes of $\mathscr{M}_n$ converge to the atoms of $\mathscr{F}_0$, the latter being sets of the form $\bigcap_{i=1}^{\infty} E_i^{\delta_i}$.

Theorem 1 and Lemma 4 lead immediately to the following result:

THEOREM 2. *Let $\mathscr{F}_0$ be any separable sub σ-field of $\mathscr{F}$. Then, the conclusions of Theorem 1 are valid for $\mathscr{F}_0$.*

3. Information Contained in an Arbitrary σ-Field

We are now in a position to discuss the explicit representation for the information $I(\mathscr{F})$ contained in an arbitrary σ-field $\mathscr{F}$ of subsets of Ω over which the two probability measures P_1 and P_2 are defined. With the aid of Theorems 1 and 2, this turns out to be surprisingly easy. From Lemma 1, for $I(\mathscr{F})$ to be finite it is necessary that $P_1 \ll P_2$ relative to $\mathscr{F}$. Denote by x the Radon-Nikodym (R-N) derivative of P_1 with respect to P_2 relative to $\mathscr{F}$. Let $\mathscr{F}_x$ be the class of sets of the form $x^{-1}(B)$, where B ranges over all linear Borel sets. $\mathscr{F}_x$ is then the minimal σ-field with respect to which x is measurable and is a separable sub σ-field of $\mathscr{F}$. In our next result it is shown that, if $P_1 \ll P_2$, $\mathscr{F}$ and $\mathscr{F}_x$ contain the same amount of information and Theorem 2 can be generalized by replacing $\mathscr{F}_0$ by $\mathscr{F}$.

THEOREM 3. *Let $I(\mathscr{F})$ be defined by* (1.2).

(i) *If $P_1 \not\ll P_2$ relative to $\mathscr{F}$, then $I(\mathscr{F}) = \infty$.*

(ii) *If $P_1 \ll P_2$ relative to $\mathscr{F}$, then*

$$I(\mathscr{F}) = I(\mathscr{F}_x) , \tag{3.1}$$

and

$$I(\mathscr{F}) = \int_\Omega x \log x dP_2 = \int_\Omega \log x dP_1 . \tag{3.2}$$

(iii) *$I(\mathscr{F})$ is finite if and only if* (a) *$P_1 \ll P_2$ relative to $\mathscr{F}$ and* (b) $E_2 | x \log x | < \infty$.

PROOF. Condition (i) is merely a restatement of Lemma 1. In order to prove (ii) and (iii), it suffices to prove (3.1) since the remaining conclusions of the theorem follow immediately from Theorem 2.

By (1.2), $I(\mathscr{F}) \geqq I(\mathscr{F}_x)$ so that, if $I(\mathscr{F}_x) = \infty$, the equality is trivially true. Let $I(\mathscr{F}_x) < \infty$. If $A \in \mathscr{F}$ is such that $P_1(A) > 0$, it can easily be shown by using Jensen's inequality that

$$\int_A x \log x \, dP_2 \geqq P_1(A) \log [P_1(A)/P_2(A)] \,.$$

(This is Lemma 3.2 of [2].) Now let $\mathscr{A}$ be any finite field of sets of $\mathscr{F}$ given by the partition $\{A_i\}$. From the inequality above (the equality holds trivially if $P_1(A) = 0$), it follows (by Theorem 2) that

$$I(\mathscr{A}) \leqq \sum_i \int_{A_i} x \log x \, dP_2 = \int_\Omega x \log x \, dP_2 = I(\mathscr{F}_x) \,.$$

Since this is true for every finite field $\mathscr{A} \subset \mathscr{F}$, we have the reverse inequality $I(\mathscr{F}) \leqq I(\mathscr{F}_x)$, thus proving (3.1). ||

Remark. It is known [2] that a σ-field $\mathscr{F}_1 \subset \mathscr{F}$ is sufficient for the set of probabilities $\{P_1, P_2\}$ (cf. [6] for the definition of sufficiency) if and only if $I(\mathscr{F}_1) = I(\mathscr{F})$. If $I(\mathscr{F})$ is finite, this result is an easy consequence of Jensen's inequality and the fact that if x and x_1 are the R-N derivatives of P_1 with respect to P_2 relative to $\mathscr{F}$ and $\mathscr{F}_1$ respectively, then $x_1 = E(x \mid \mathscr{F}_1)$, a.e. $[P_2]$.

From Theorem 3 we observe that $\mathscr{F}_x$ is a sufficient σ-field. $\mathscr{F}_x$, however, is further distinguished by the following additional property which is easy to prove. Let $\mathscr{F}_1 \subset \mathscr{F}$ be any σ-field sufficient for $\{P_1, P_2\}$. If $A \in \mathscr{F}_x$, then there exists a set $B \in \mathscr{F}_1$ such that $P_2(A \,\Delta\, B) = 0 = P_1(A \,\Delta\, B)$. This property has been called necessity in [8]. Hence $\mathscr{F}_x$ is both a necessary and a sufficient σ-field.

4. Generalization of Shannon's Information: Deduction of a Result of Gelfand, Kolmogorov, and Yaglom

Let $(\Omega, \mathscr{F}, P)$ be a probability space and let $\mathscr{A} = \{A_i\}$, $\mathscr{B} = \{B_j\}$ be finite $\mathscr{F}$-partitions of Ω. In information theory, following Shannon it is usual to call, as the information of $\mathscr{A}$ relative to $\mathscr{B}$, the quantity

(4.1) $$I(\mathscr{A}, \mathscr{B}) = \sum_{i,j} P(A_iB_j) \log [P(A_iB_j)/P(A_i)P(B_j)] \,.$$

Let $(X_i, \mathscr{S}_i)$, $i = 1, 2$, be arbitrary measurable spaces and x_i a measurable transformation of $(\Omega, \mathscr{F})$ into $(X_i, \mathscr{S}_i)$. For A_i in $\mathscr{S}_i$, the set

$$\{\omega\colon (x_1(\omega), x_2(\omega)) \in A_1 \times A_2\} = \{\omega\colon x_1(\omega) \in A_1\} \cap \{\omega\colon x_2(\omega) \in A_2\}$$

and hence belongs to $\mathscr{F}$. Writing $\boldsymbol{x} = (x_1, x_2)$, it then follows by the definition of the product σ-field $\mathscr{S}_1 \times \mathscr{S}_2$ that, for every set $E \in \mathscr{S}_1 \times \mathscr{S}_2$, the set $\boldsymbol{x}^{-1}(E) \in \mathscr{F}$. In other words, $\boldsymbol{x}$ is a measurable transformation of $(\Omega, \mathscr{F})$ into $(X_1 \times X_2, \mathscr{S}_1 \times \mathscr{S}_2)$.

Following the notation of Halmos ([7] p. 163), the transformation $\boldsymbol{x}$ induces a probability measure $P\boldsymbol{x}^{-1}$ over sets E of $\mathscr{S}_1 \times \mathscr{S}_2$ defined by $P\boldsymbol{x}^{-1}(E) = P\{\boldsymbol{x}^{-1}(E)\}$. Another probability measure over $\mathscr{S}_1 \times \mathscr{S}_2$ is the the product measure $\Pi = (Px_1^{-1}) \times (Px_2^{-1})$.

The generalization of (4.1) due to the above-mentioned authors [4] can now be given as follows: The amount of information contained in the measurable transformation x_1 relative to the measurable transformation x_2 and denoted by $I(x_1, x_2)$ is defined to be the information of Px^{-1} with respect to Π contained in the σ-field $\mathscr{S}_1 \times \mathscr{S}_2$. The latter quantity is given by the general definition (1.2). The results of the preceding sections are immediately applicable. In particular, from Theorem 3 we deduce the following theorem:

THEOREM 4 (Gelfand-Kolmogorov-Yaglom [4]). *$I(x_1, x_2)$ is finite if and only if both the following conditions are satisfied*:

(i) *$Px^{-1} \ll$ the product measure Π*;

(ii) $\int_{X_1 \times X_2} a(x_1, x_2) \log a(x_1, x_2)\, d\Pi < \infty$,
where $a(x_1, x_2)$ is the R-N derivative of Px^{-1} with respect to Π relative to $\mathscr{S}_1 \times \mathscr{S}_2$.

Finally,

$$I(x_1, x_2) = \int_{X_1 \times X_2} a(x_1, x_2) \log a(x_1, x_2)\, d\Pi(x_1, x_2)\,, \tag{4.1}$$

where both sides of (4.1) *are finite or both are infinite.*

5. A Property of Semi-martingales Which Are Convex Functions of a Martingale

In deriving the main results of this paper, we applied the martingale theorems of [5] to deduce certain properties of the semi-martingale $y_n = x_n \log x_n$. This part of the argument of Theorem 1 can be extended to yield a property of semi-martingales which are continuous convex functions of martingales. The following result is a specialization of Theorem 4.1s (ii) (b) of [5]:

THEOREM 5. *Let the following conditions hold*:

(i) *$\{x_n, n \geqq 1\}$ is a uniformly integrable martingale, so that $\lim_{n\to\infty} x_n = x_\infty$ exists with probability* 1;

(ii) *$\varphi(t)$ is a real, continuous, convex function of the real variable t, such that $E|\varphi(x_n)| < \infty$ for every finite n.*

Then,

$$\lim_{n\to\infty} E\,\varphi(x_n) = E\,\varphi(x_\infty)\,. \tag{5.1}$$

An equivalent conclusion of the theorem is to say that the sequence of random variables $\{\varphi(x_n), n \geqq 1\}$ is a uniformly integrable semi-martingale if and only if $E|\varphi(x_\infty)|$ is finite. Observe that both sides of (5.1) may be infinite.

PROOF. Write $y_n = \varphi(x_n)$, $1 \leqq n \leqq \infty$. The conditions of the theorem imply that x_∞ is integrable, $\lim_{n\to\infty} y_n = y_\infty$ with probability 1, and that $\{y_n, n \geqq 1\}$ is a semi-martingale. First suppose $E|\varphi(x_\infty)| < \infty$. Then, as in Theorem 1, it follows that the sequence $\{y_n^+\}$ is uniformly integrable. Let y_n^- denote the negative part of y_n, i.e., $y_n^- = -\varphi(x_n)$ if $\varphi(x_n) \leqq 0$, and $y_n^- = 0$ if $\varphi(x_n) > 0$. By the supporting line property of a convex function, for any t_0 and all t in the domain of φ, we have $\varphi(t) \geqq \varphi(t_0) + \lambda(t_0)(t - t_0)$, $\lambda(t_0)$ being a fixed real number. If we set $t = x_n$ and $t_0 = E(x_n)$, and observe that, x_n being a martingale, $t_0 = E(x_n) = E(x_1)$ independent of n, we have for $n \geqq 1$, $\varphi(x_n) \geqq \varphi(Ex_1) + \lambda(Ex_1)[x_n - Ex_1]$. From this inequality it follows that for all n and ω, $0 \leqq y_n^-(\omega) \leqq K_1 + K_2|x_n(\omega)|$, where K_1, K_2 are positive constants independent of n. The uniform integrability of $\{y_n^-\}$ follows from that of $\{x_n\}$ assumed in condition (i). The sequence $\{y_n\}$ is therefore uniformly integrable and (5.1) follows as an immediate consequence.

Now let $\lim_{n\to\infty} Ey_n < \infty$. This limit always exists since $\{y_n, n \geqq 1\}$ is a semi-martingale. As before, $\{y_n^-\}$ is uniformly integrable. Furthermore, since

$$0 \leqq y_n^+ = y_n + y_n^- , \qquad E(y_n^+) \leqq |E(y_n)| + E(y_n^-) ,$$

the right-hand side of this inequality is bounded. Hence $\sup_n E|y_n| < \infty$ and the integrability of y_∞ follows from Fatou's lemma. The conclusion (5.1) then follows as in the first part of the proof. It also follows that (5.1) is true always, since either both sides are infinite or both are finite and equal. The verification of the alternative conclusion of the theorem presents no difficulty. ||

References

[1] Savage, L. J. *The Foundations of Statistics*. New York: Wiley, 1954.

[2] Kullback, S., and Leibler, R. A. On information and sufficiency. *Ann. Math. Stat.*, 1951, **22**, 79–86.

[3] Schützenberger, M. P. On some measures of information used in statistics. In Colin Cherry (ed.), *Symposium on Information Theory*. New York: Academic Press, 1956.

[4] Gelfand, I. M., Kolmogorov, A. N., and Yaglom, A. M. *K* obshchemu opredeleniyu kolichestva informatsii. *Doklady Akad. Nauk, U.S.S.R.*, 1956, **111**(4), 745-48.

[5] Doob, J. L. *Stochastic Processes*. New York: Wiley, 1953.

[6] Halmos, P. R., and Savage, L. J. Application of the Radon-Nikodym theorem to the theory of sufficient statistics. *Ann. Math. Stat.*, 1949, **20**, 225-41.

[7] Halmos, P. R. *Measure Theory*, New York: Van Nostrand, 1950.

[8] Bahadur, R. R. Sufficiency and statistical decision functions. *Ann. Math. Stat.*, 1954, **25**, 423-62.

24 THE EVERGREEN CORRELATION COEFFICIENT

M. G. KENDALL

Research Technique Unit, London School of Economics

The derivation of the exact sampling distribution of the product-moment correlation coefficient in normal variation by R. A. Fisher in 1915 was one of the major early triumphs of mathematical statistics. The distribution was immediately subjected to intensive study by Karl Pearson and his co-workers at University College and was also discussed in later papers by Fisher. By 1921 the testing of a correlation was a routine matter, although it was not until 1938 that F. N. David published her tables of the distribution. Statistically, the subject appeared closed except for investigation of the distribution in non-normal variation.

However, there is a certain perennial freshness about the coefficient and its distribution. For a period of nearly twenty years Hotelling worked on the subject at intervals. His results were presented to the Royal Statistical Society in 1953 [1] and, extensive as they were, raised a number of new problems of general methodological interest, for example in the field of variate-transformations and of estimation.

The distribution of the correlation coefficient is a hypergeometric function and expressions for its moments are obtainable only as series of a hypergeometric type. In particular, it is known that for samples of size n the moments of r about ρ, $\lambda_k = E(r-\rho)^k$, are given by

$$\lambda_1 = (1-\rho^2)\left[-\frac{\rho}{2n} + \frac{\rho - 9\rho^3}{8n^2} + O(n^{-3})\right], \tag{1}$$

$$\lambda_2 = (1-\rho^2)^2\left[\frac{1}{n} + \frac{23\rho^2}{4n^2} + O(n^{-3})\right], \tag{2}$$

$$\lambda_3 = (1-\rho^2)^3\left[-\frac{15\rho}{2n^2} + O(n^{-3})\right], \tag{3}$$

$$\lambda_4 = (1-\rho^2)^4\left[\frac{3}{n^2} + O(n^{-3})\right]. \tag{4}$$

Thus the expectation of r is not equal to ρ, and the product-moment statistic r is a biassed estimator of the parameter ρ.

Hotelling was then led to consider the question: Does there exist a function of r, say $g(r)$, such that

$$Eg(r) = g(\rho)\,? \tag{5}$$

Imposing the light restriction that $g(r)$ can be expanded in powers of $r - \rho$ as a Taylor series up to some order at least, we have

$$g(r) = g(\rho) + (r - \rho)\, g'(\rho) + \tfrac{1}{2}(r - \rho)^2\, g''(\rho) + \text{etc.}\,, \tag{6}$$

where the primes denote differentiation. Taking expectations of both sides we find

$$g(\rho) = Eg(r) = g(\rho) + \lambda_1\, g'(\rho) + \tfrac{1}{2}\lambda_2\, g''(\rho) + \text{etc.}\,,$$

and on substituting for the moments from (1) to (4) and arranging terms in order of powers of $1/n$, we find

$$0 = \frac{1-\rho^2}{2n}\{-\rho g' + (1-\rho^2)g''\} + \frac{1-\rho^2}{8n^2}\{\rho(1-9\rho^2)g' + 23\rho^2(1-\rho^2)g'' - 10\rho(1-\rho^2)^2 g''' + 2(1-\rho^2)^3 g^{(\text{iv})}\} + \text{etc.} \tag{7}$$

If (5) is to be true for all n, the terms in n^{-1}, n^{-2}, etc., must separately vanish, and thus, from the coefficient of n^{-1} in (7), we have

$$-\rho g' + (1-\rho^2)g'' = 0\,, \tag{8}$$

which integrates at once to give

$$g = a \sin^{-1}\rho + b\,. \tag{9}$$

Thus, if any function at all (with a second derivative) is to exist satisfying (5), it must be the arc sine function.

At this point Homer nodded. Our required function must obey not only (8) but the other differential equations obtained by equating to zero coefficients of successive powers of n^{-1} in (7). Hotelling, misled by an algebraical slip (and no worker in this domain has ever been able to avoid one), came to the conclusion that the arc sine function did not obey the next equation and thus that no function $g(r)$ exists. This was an error and it is, in fact, remarkable that for all n and any ρ

$$E(\sin^{-1} r) = \sin^{-1}\rho\,. \tag{10}$$

The point was noted by Harley [2] who set out to establish Equation (10). The demonstration is by no means easy. By a partial integration she established the recurrence relation ·

$$E_{n+1}(\sin^{-1} r) = E_{n-1}(\sin^{-1} r)\,. \tag{11}$$

She was able to show that (10) holds for $n = 4$. It then follows that the equation holds for even n. In a later note [3] she established the result for $n = 3$, hence for odd n, and hence for all n.

The elegance and simplicity of (10) leads one to suspect that there is some direct way of looking at the problem which would make the relationship plain. Such an approach can, in fact, be made in the following way.

If we take a pair of values (x_1, y_1), (x_2, y_2) from a bivariate normal population, there is said to be a concordance if $x_1 - x_2$ and $y_1 - y_2$ have the same sign:

$$\operatorname{sgn}(x_1 - x_2)\operatorname{sgn}(y_1 - y_2) = 1 . \tag{12}$$

This is a very useful notion in the theory of rank correlation and it is known that the probability of a concordance is given by

$$P(C) = \frac{2}{\pi}\sin^{-1}\rho . \tag{13}$$

For our purposes, no generality is lost if the parent population is supposed to have zero means and unit variances. In such a case r is sufficient for ρ in the subclass of scale-invariant statistics and we may write

$$P(C \mid \rho) = \int P(C \mid r)P(r \mid \rho)\,dr . \tag{14}$$

The expression on the right is the expectation of $P(C \mid r)$. To establish Harley's theorem we then have only to show that

$$P(C \mid r) = \frac{2}{\pi}\sin^{-1} r. \tag{15}$$

This was done in a few lines of geometrical argument by Daniels and myself [4]. Effectively, the probability of concordance for fixed r depends on the angle $\cos^{-1} r$ divided by the total angle in a circle, 2π. The theorem thus derives, basically, from the geometrical interpretation of the correlation coefficient as the cosine of an angle between two vectors in the sample space.

It is interesting to look both backwards and forwards from this result. Harley's theorem rests on the formula of Equation (13) for a concordance. From one viewpoint, this formula is the same as Sheppard's well-known result [5] concerning the volume of a bivariate normal distribution in the positive quadrant; for the variates $\xi \equiv x_1 - x_2$ and $\eta \equiv y_1 - y_2$ are jointly distributed in the bivariate normal form with correlation ρ, and Sheppard's result gives the probability that they are both positive, which is one-half the probability that they have the same sign. We may trace the same result in a different context in some independent work by Greiner [6], who was concerned with the estimation of ρ by counting concordances in observed samples. Brokering backwards is an unprofitable occupation, but it is interesting to observe that Harley's theorem could have been derived fifty years earlier on the basis of Sheppard's theorem even before the explicit distribution of the correlation coefficient was known.

An interesting consequence of the proof of Harley's theorem given by Daniels and myself arises in connection with the rank correlation coefficient based on concordance, which I have called τ. If t is the observed coefficient in a sample of n from a bivariate population, it may easily be shown that

$$E(t \mid r) = \frac{2}{\pi}\sin^{-1} r . \tag{16}$$

This gives us the regression of t on r. The remarkable thing is that we can obtain the actual regression in this elementary form without being able to find simple expressions for the variance of r, the covariance of r and t, and hence for the *linear* regression of t on r.

Apart from the intrinsic interest of such results, they exhibit very well the extraordinary way in which independent strands of scientific thought sometimes come together. Many workers in many different countries have contributed to our knowledge of this particular branch of the subject, and an incidental result here or a chance remark there has proved the starting-point of a new wave of advance. Hotelling's work on the correlation coefficient is by no means the most important of his many distinguished contributions to theoretical statistics, but it has touched off research in several directions, notably on the theory of transformations and the theory of unbiassed estimating functions. As this note shows, even a rare mistake is fruitful. A man of his intellectual stature is to be measured, not merely by what he does himself, impressive though that is, but also by the powerful stimulus which his work provides to other people.

References

[1] Hotelling, H. New light on the correlation coefficient and its transforms. *J. Roy. Stat. Soc.*, Ser. B, 1953, **15**, 193-224, disc. 225-32.

[2] Harley, B. I. Some properties of an angular transformation for the correlation coefficient. *Biometrika*, 1956, **43**, 219-23.

[3] Harley, B. I. Further properties of an angular transformation of the correlation coefficient. *Biometrika*, 1957, **44**, 273-75.

[4] Daniels, H. E., and Kendall, M. G. Short proof of Miss Harley's theorem on the correlation coefficient. *Biometrika*, 1958, **45**, 571-72.

[5] Sheppard, W. F. On the application of the theory of error to cases of normal distributions and normal correlations. *Phil. Trans. Roy. Soc.* (*London*), Ser. A, 1899, **192**, 101-67.

[6] Greiner, R. Über das Fehlersystem der Kollektivmasslehre. *Z. Math. u. Phys.*, 1909, **57**, 121, 225, and 337.

25 ROBUST TESTS FOR EQUALITY OF VARIANCES[1]

HOWARD LEVENE
Columbia University

1. Introduction

The analysis of variance provides satisfactory and flexible methods for testing hypotheses about differences between means. Furthermore, it is well known that these methods are quite insensitive to departures from normality in the underlying populations being sampled. They are, in the words of Box, "robust tests." Comparably flexible and satisfactory methods for testing hypotheses about differences between variances have not, however, been developed. Under certain conditions the methods of the analysis of variance for the components of variance model do provide tests, but these are not nearly so satisfactory.

The classical method for testing for the difference between the variances of two samples is the ordinary F test, based on the ratio of the two sample variances, and Bartlett's test generalizes this to the case of k populations. It has been known for some time that these tests are quite sensitive to departures from normality, and this sensitivity has been emphasized by Box [1] who pointed out that the tests actually are quite satisfactory for use as tests of normality. Box and Andersen [2] have given modifications of the F test and Bartlett's test, based on permutation theory, which give approximate size α even under non-normality. These tests are based on the use of the fourth sample moment to obtain a correction factor by which the degrees of freedom in the conventional test are multiplied. Box and Andersen investigated the behavior of their test by empirical sampling and found that it gave nearly the correct α and good power for the normal distribution, as well as for the rectangular and double exponential distributions. This paper discusses a different approach to the problem, leading to a method of more generality, which can easily be extended to many of the problems dealt with by the ordinary analysis of variance.

2. The Test for the *p*-Sample Problem

Let the random variables $X_{i1}, \cdots, X_{iN_i}$ be independently identically distrib-

[1] This research was sponsored by the Office of Naval Research under Contract No. Nonr-266(33), Project No. NR 042-034. Reproduction in whole or in part is permitted for any purpose of the United States Government.

uted from the distribution $G_i(x)$ with mean μ_i and variance σ_i^2, for $i = 1, \cdots, p$. We will say that $X_{i1}, \cdots, X_{iN_i}$ are a random sample of observations on the random variable X_i. We wish to test the hypothesis that $\sigma_1^2 = \sigma_2^2 = \cdots = \sigma_p^2$. Now $(X_{ij} - \mu_i)^2$ has expected value σ_i^2, and so we can formally apply a one-way analysis of variance F test to the variables $(X_{ij} - \mu_i)^2$, $i = 1, \cdots, p$; $j = 1, \cdots, N_i$. Of course, these variables are never exactly normal and may be very non-normal; e.g., if X_{ij} is normally distributed, $(X_{ij} - \mu_i)^2$ is σ_i^2 times a χ_1^2 variable. Furthermore, even if the distributions of the X_{ij} differ only by a scale factor for different i, in general the $(X_{ij} - \mu_i)^2$ will have different variances as well as different means. Nevertheless, the robustness leads us to hope that the analysis of variance F statistic will have approximately the F distribution when $\sigma_1^2 = \sigma_2^2 = \cdots = \sigma_p^2$; and in any case this will be true asymptotically if $N_1 = N_2 = \cdots = N_p \to \infty$.

More generally, the F test can be applied to the variables $W_{ij} = g(|X_{ij} - \mu_i|)$ where $g(x)$ is any function monotonically increasing on $(0, \infty)$. In particular, $g(x) = x$ is suggested for its simplicity, and $g(x) = \log x$ because, if the distributions of the X_i differ only by a linear transformation, those of the W_i will differ only by a translation. The effect of non-normality of W on the validity of the F test can be measured approximately by $\gamma_1(W)$ and $\gamma_2(W)$, the normalized third and fourth cumulants. For X_i normally distributed we have $\gamma_1 = 2.83$, $\gamma_2 = 12$ for $(X_i - \mu_i)^2$; $\gamma_1 = 1.00$, $\gamma_2 = .39$ for $|X_i - \mu_i|$; $\gamma_1 = .08$, $\gamma_2 = -.52$ for $(|X_i - \mu_i|)^{\frac{1}{2}}$; and $\gamma_1 = -1.54$, $\gamma_2 = 4.00$ for $\log |X_i = \mu_i|$. Thus it might be suggested that F would more nearly have the F distribution for the first power and for the square root, while the square is even worse than the logarithm. However, the relative powers of the tests are even more important. The author's original feelings on this proved wrong, so no heuristic discussion will be attempted here.

Since in practice μ_i is unknown, the proposed method must be further modified, and the F test actually applied to the variables $B_{ij} = g(|X_{ij} - \bar{X}_i|)$, where $\bar{X}_i = \sum_{j=1}^{N_i} X_{ij}/N_i$. This adds the further difficulty that B_{ij} and B_{ik} are not independent. Temporarily dropping the subscript i for simplicity, for the case $B_j = (X_j - \bar{X})^2$, $j = 1, \cdots, N$, one easily finds quadratic moments of the B's (which are the ones primarily affecting the distribution of F). We have

$$\text{var}\,(B_j) = \sigma^4\,[2N(N-1)^2 + (N-1)(N^2 - 3N + 3)\,\gamma_2]\,/N^3\,,$$

$$\text{cov}\,(B_j, B_k) = \sigma^4[2N + (2N-3)\,\gamma_2]/N^3\,,$$

and

$$\text{correl}\,(B_j, B_k) = \frac{2N + (2N-3)\,\gamma_2}{2N(N-1)^2 + (N-1)(N^2 - 3N + 3)\gamma_2}\,,$$

where σ^2 and γ_2 are the variance and normalized fourth cumulant of X itself. It should be noted that the correlation of B_j and B_k is of order N^{-2}, except that where X takes on only two values, each with probability $\frac{1}{2}$, the correlation is $(-N+3)/(N-1)(2N-3) \sim -1/(2N)$. A correlation of order N^{-2} would probably not have any serious effect on the distribution of F, although a correlation of order N^{-1} (such as is present between $(X_{ij} - \bar{X}_i)$ and $(X_{ik} - \bar{X}_i)$) would. The correlation of $|X_{ij} - \bar{X}_i|$ and $|X_{ik} - \bar{X}_i|$ is known to be of order N^{-2} (Fisher [3]), and the correlations for the logarithm and square root of $|X_{ij} - \bar{X}_i|$ may be expected to be of the same order of magnitude.

3. The Sampling Experiment

Because of the well-known difficulty of obtaining the distribution of F from a non-normal population analytically, and the further complication of correlation between the B's of the previous section, the distributions were obtained by empirical sampling (Monte Carlo) methods. Box and Andersen [2] had done this for their proposed method, and the present study utilizes essentially the same case so that direct comparison is possible. The sampling procedure was programmed for an IBM 704 computer. The basic data were quasi-random numbers y_α, $\alpha = 1, \cdots, Nmp$, giving m values of F, each based on p samples of N observations each. Each y_α had 35 binary digits and was obtained by the recurrence relation

$$y_\alpha \equiv 5^{13}\, y_{\alpha-1} \pmod{2^{35}},$$

giving a sequence of y's with period 2^{33} if $y_0 \equiv 1 \pmod 4$. Short subsequences of this sequence behave as if the y_i, considered as binary fractions, were independently uniformly distributed on (0,1) (see Juncosa, [4]). The longest subsequence used contained 200,000 y's, which is short compared with the total period. The y's were used directly to give an essentially continuous uniform distribution on the interval (0,1). For other distributions y_α was transformed into a discrete random variable x_α by a table look-up program as follows. For the normal distribution, for

$$\begin{aligned} 0 \leqq y \leqq .0001,\ x &= 1.200; & .001 < y \leqq .004,\ x &= 2.160;\\ .0001 < y \leqq .0004,\ x &= 1.500; & .004 < y \leqq .01,\ x &= 2.530;\\ .0004 < y \leqq .001,\ x &= 1.800; \end{aligned}$$

and for $k/100 < y \leqq (k+1)/100$, x is the three-decimal probit corresponding to $(k+\frac{1}{2})/100$ for $k = 1\ (1)\ 98$. For $y > .99$, the table is symmetric about 5 with that for $y \leqq .01$.

The double exponential distribution was generated by the following table. For

$$\begin{aligned} 0 \leqq y \leqq .00005,\ x &= -.500; & .0005 < y \leqq .001,\ x &= 1.15;\\ .00005 < y \leqq .0001,\ x &= .150; & .001 < y \leqq .002,\ x &= 1.450;\\ .0001 < y \leqq .0002,\ x &= .450; & .002 < y \leqq .005,\ x &= 1.800;\\ .0002 < y \leqq .0005,\ x &= .800; & .005 < y \leqq .015,\ x &= 2.301; \end{aligned}$$

$$(k-\tfrac{1}{2})/100 < y \leqq (k+\tfrac{1}{2})/100,\ x = 2 + \log_{10}[2k] \text{ for } k = 2(1)\ 50;$$
$$(k-\tfrac{1}{2})/100 < y \leqq (k+\tfrac{1}{2})/100,\ x = 6 - \log_{10}[2(100-k)] \text{ for } k = 51(1)\ 99.$$

All logarithms are to three decimals, and for $y > .99$ the table is symmetric about 4 with that for $y \leqq .01$.

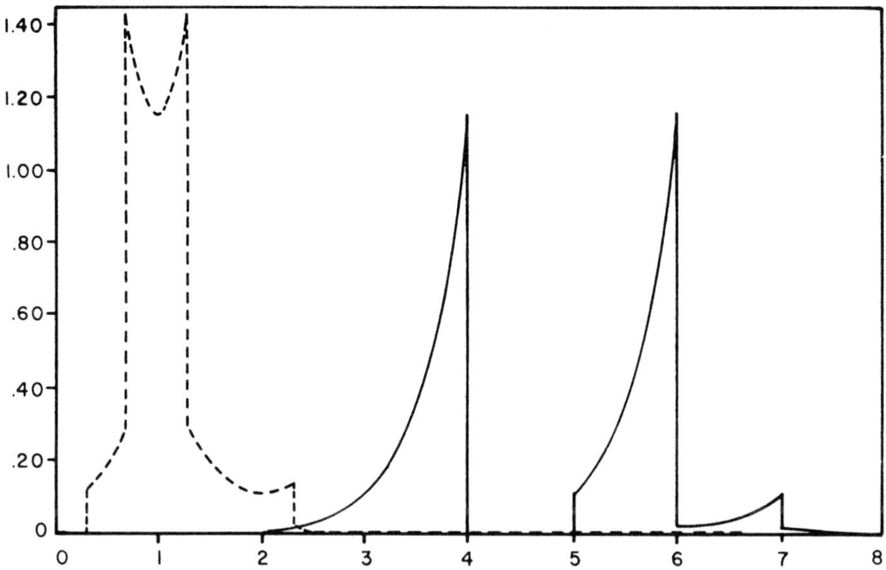

FIG. 1. Probability Density Functions Corresponding to the Discrete Distribution C for the Original Variable X (solid line) and $z = |X - E(X)|$ (dotted line).

The first coding for the double exponential distribution led to a bizarre distribution to be called distribution C. A number of results were obtained for this distribution before the error was discovered, and since distribution C is more severely non-normal than any likely to be met with in practice, these results are reported here. The table for distribution C is as follows. For

$$\begin{array}{llll}
.00005 < y \leqq .0001, & x = .301; & .0002 < y \leqq .0005, & x = 1.000; \\
.0001 < y \leqq .0002, & x = .602; & .001 < y \leqq .002, & x = 1.602; \\
.0005 < y \leqq .001, & x = 1.301; & .002 < y \leqq .005, & x = 2.000; \\
0 \leqq y \leqq .00005, & x = 0; & .005 < y \leqq .01, & x = 2.301;
\end{array}$$

$$k/100 < y \leqq (k+1)/100, \quad x = 2 + \log_{10}[2(k+1)] \text{ for } k = 1(1)\ 49;$$
$$k/100 < y \leqq (k+1)/100, \quad x = 4 + \log_{10}[200 - 2(k+1)] \text{ for } k = 50\,(1)\ 94;$$
$$k/100 < y \leqq (k+1)/100, \quad x = 6 + \log_{10}[200 - 2(k+1)] \text{ for } k = 95\,(1)\ 98;$$

$$\begin{array}{llll}
.99 < y \leqq .999, & x = 6.000; & .9995 < y \leqq .9998, & x = 7.398; \\
.995 < y \leqq .999, & x = 6.398; & .9998 < y \leqq .9999, & x = 7.6999; \\
.999 < y \leqq .9995, & x = 6.699; & .9999 < y \leqq 1, & x = 8.000.
\end{array}$$

Figure 1 gives the corresponding probability density functions for x and $|x - Ex|$.

For the discrete distributions used, the moments are, for the "normal," mean $= 5$, $\sigma^2 = .9995$, $\gamma_1 = 0$, $\gamma_2 = -.0121$. The tail values of the "double exponential" were chosen to give nearly the theoretical moments; the actual moments were mean $= 4$, $\sigma^2 = .3758$, $\gamma_1 = 0$, $\gamma_2 = 3.003$ compared with theoretical $\sigma^2 = 2(\log_{10} e)^2 = .3772$, $\gamma_2 = 3$. For distribution C the moments were mean $= 4.6719$, $\sigma^2 = 1.3322$, $\gamma_1 = -.0417$, $\gamma_2 = -1.303$.

Each machine run used one of these four distributions and one choice of p and N. The p and N used were (1) $p = 2$, $N = 10$; (2) $p = 2$, $N = 20$; (3) $p = 4$, $N = 20$; and (4) $p = 10$, $N = 20$. For each machine run the number of sets of p samples was $m = 1000$.

From the random variables x_{ij} there were obtained new variables $z_{ij} = |x_{ij} - \bar{x}_i|$, $s_{ij} = z_{ij}^2$ and either $L_{ij} = \log_{10} z_{ij}$ or $t_{ij} = z_{ij}^{\frac{1}{2}}$. The latter two functions were obtained by table look-up as follows. For $k_i < z \leqq k_{i+1}$, $L = 3 + \log_{10} k_{i+1}$, where k takes the values 0, .002, .0025, .005, .01(.01).1 (.02)1 (.2)10. For the square root, the same k were used, but $t = [(k_i + k_{i+1})/2]^{\frac{1}{2}}$. Logarithms were taken to four decimals and square roots to five decimals. The logarithm table was deliberately truncated near zero, with .002 the smallest argument, to eliminate the worst of the non-normality of the distribution of log z_{ij} caused by the very long left tail. For t this truncation is of no importance.

Computation then proceeded in parallel for the z_{ij}, s_{ij} and either L_{ij} or t_{ij}, giving in each case U, the mean square between the p samples; V, the mean square within samples; and $F = U/V$. For the runs on the two-sample problem, $p = 2$, the power was studied by setting $\sigma_2^2/\sigma_1^2 = 1, 2, 4$, and 6 successively. For $p = 4$, $H_1 : \sigma_i = a_i b$ (with $a_1 = 1$, $a_2 = 1.3$, $a_3 = 1.6$, $a_4 = 2$) was used in addition to $H_0 : a_i \equiv 1$. For $p = 10$, H_1 was $\sigma_i = a_i b$, with $a_1 = a_2 = a_3 = a_4 = 1$, $a_5 = a_6 = a_7 = 1.3$, $a_8 = a_9 = a_{10} = 1.6$, where b was the standard deviation of the basic distribution being used. To save machine time, the same samples were used for all alternatives. The basic calculations were done for H_0, and for the alternatives the mean square within samples for z was calculated as

$$V(z, H_1) = \frac{1}{p(N-1)} \sum_{i=1}^{p} a_1^2 Q_i ,$$

where

$$Q_i = \sum_{j}^{N} (z_{ij} - \bar{z}_i)^2 ,$$

and the mean square between samples was calculated as

$$U(z, H_1) = \frac{1}{p-1} \sum_{i=1}^{p} \left[a_i \bar{z}_i - \left(\frac{1}{p} \sum a_i \bar{z}_i \right) \right]^2 .$$

Similar formulas were used for s, L, and t.

The final result for a single set of p samples was 12 sets of values of U, V, and F when $p = 2$, and 6 sets of values when $p = 4$ and 10. Another set of p samples was then obtained and so on until $m = 1000$ sets had been obtained. During this process, for each variable z, s, and (L or t), and for each alternative a running total of U, V, U^2, V^2, and UV was maintained, while the F values were formed into a frequency distribution by adding a one to class h if $ah < F \leqq a(h+1)$. For $p=2$, $N=10$, $a=.2$, $h=0, 1, \cdots, 100$; for $p=2$, $N=20$, $a=.1$, $h=0, 1, \cdots, 150$; for $p=4$, $N=20$, $a=.06$, $h=0, 1, \cdots, 100$; for $p=10$, $N=20$, $a=.04$, $H=0,1, \ldots, 100$. In each case a final class gave the number of F larger than the largest ah. The output was the various frequency tables for F and the sums and sums of squares and products of U and V. It was originally proposed to include the correlation of U and V, but fixed-point arithmetic was used throughout, and because of scaling difficulties the correlations were not obtained. Even the sums of squares of U and V were not always reliable, and consequently no correlations are reported in this paper. Scaling difficulties did not affect the values of F, which were the primary results desired.

4. Results of the Sampling Experiments

The first sampling run used two normally distributed samples of size 20, the transformed variables z, s, and $L = \log z$, and used $y_0 = 1$ for the first (quasi)random number. The empirical distribution for 1000 values of F showed that all variables gave satisfactory values of α, but z and s had fairly good power, while L was greatly inferior. It also seemed that z was better than $s = z^2$. It was therefore decided that a less drastic transformation than the logarithm might be better, and the square root, t, was decided on. The first run was then repeated, with the final random number of the previous run, $y_{40,000} = 022\ 437\ 761\ 401)_8$, written as an integer to the base 8 (octal notation, used for all future random numbers) as the starting value. While $y_{40,000}$ can be obtained quite easily on a desk calculator, it was actually obtained by having the computer always print the last random number used. Because of a coding error, the run gave F values for t only for H_0, $\sigma_2^2/\sigma_1^2 = 1$. Accordingly, a third run was made, using $y_{80,000} = 205\ 422\ 143\ 001)_8$ as the starting random number. This time the power was also obtained for t, except for an error when $\sigma_2^2/\sigma_1^2 = 4$. It was not deemed worthwhile to repeat this run again, but the error was corrected for all the other runs except that with $p = 2$, $N = 10$ and a normal distribution. (For these runs, the power for $\sigma_2^2/\sigma_1^2 = 4$ in Table 2 is obtained by graphic interpolation.) There are thus 3000 independent values of F for z and s, 2000 for t under H_0, and 1000 for t under the alternatives, and 1000 for L, for the case of a normal distribution with $p = 2$ and $N = 20$. These results are summarized in Table 1.

In view of the evident superiority of t to L, t was used for all further runs. The runs for $p = 2$, $N = 20$, and the uniform and C distributions were done while "debugging" t, and do not include t. The starting random numbers used for these runs were y_0 for $p = 4$; $y_{40,000}$ for $p = 2$, $N = 10$, and the normal and C distributions; $y_{40,000}$ for $p = 2$, $N = 20$, and the uniform and C distributions; $y_{320,000} = 374\ 776\ 614\ 001)_8$ for $p = 10$, $N = 20$, and distribution C; $020\ 433\ 766\ 431)_8$ for $p = 2$, $N = 10$, and the uniform and double exponential distributions; $212\ 716\ 974\ 261)_8$ for $p = 10$, $N = 20$, and the uniform and double exponential distributions; and $250\ 350\ 276\ 125)_8$

for $p = 2$, $N = 20$, and the double exponential distribution. The last three starting values were obtained by choosing random digits between 0 and 7, except that the first was between 0 and 3 to give a 35-bit number, and the last was 1 or 5 to obtain a number $\equiv 1 \pmod 4$. The probability that two such numbers give sequences of 200,000 with any overlapping is $399{,}999/2^{33} \sim .000047$, so it can be presumed that each starting value used gave a distinct subsequence of y's.

Table 2 summarizes the results of all runs, giving 1000 times the empirical size and power of tests using z, s, and t for the tabular F values giving nominal size .01, .05, and .10. Considering first the normal distribution and α, we find that, if we suppose the number of sample F values greater than a given value has a binomial distribution, only one result (that for z with $\alpha = .10$, $p = 2$, $N = 20$, $m = 3000$) differs significantly from the theoretical. Observed values vary above and below the theoretical, with a tendency for the empirical α's for z and t to be too large and for s too small, and for all three to be large when p is large. Again, for the normal distribution, the power of tests based on z and s seems satisfactory while that for t is much poorer except when α and the power $1 - \beta$ are both small.

For the uniform distribution, the empirical α's tend to be too large, although none are significantly so. For this distribution, in general the test based on s is more powerful than that based on z, and much more powerful than that based on t; however, for $p = 2$, $N = 10$, $\alpha = .01$ these relations are reversed.

For the double exponential distribution, the empirical α's for z are too large, frequently significantly so, and this is even more pronounced for t. On the other hand, for s the empirical α is never significantly different from the theoretical value, but it is usually too small, although it is too large for $p = 10$, or $\alpha = .10$. Even so, the value for s is much smaller than that for z. The empirical powers more or less parallel the α-values, with t slightly more powerful than z, and z considerably more powerful than s.

Finally, for distribution C, the empirical α for z is usually too large, and for $p = 10$ importantly so. In this respect t is very much worse, and is unacceptable. The α for the test based on s is always too small, and frequently significantly so. As might be expected, the smaller size of the test based on s is accompanied by smaller power, but, surprisingly, the test based on t may have lower power than that based on z, even when the size of the test using t is much greater.

Although one would probably use the conventional critical values of F in applying these tests, it is of interest to see how their powers compare when the critical values of F are chosen to give each test the desired empirical α. Table 3 gives the necessary values of F and the resulting powers. In view of the unsatisfactory results for the test using t in Table 2, only z and s are used in Table 3. The critical empirical F value for $\alpha = .01$ is exceeded by only 10 sample F's out of 1000. Accordingly, this critical F value has a large sampling error, and this affects the observed powers. Hence, only the results for $\alpha = .05$ and .10 are reliable. We then find that for the normal distribution there is no consistent difference between z and s, for the double exponential distribution z has better power, and for the uniform distribution and distribution C, s has better power. It is perhaps surprising that distribution C should here resemble the uniform rather than the double exponential distribution. It is true that distribution C itself has a negative γ_2, but z has a large positive γ_2 for this distribution; furthermore, in Table 2 distribution C gives results more like those of the double exponential than of the uniform distribution.

Finally, it may be noted that, in general, F calculated using s has a smaller variance than it does for z and t. Thus for s the number of sample F's exceeding some given *large* value is usually relatively small, while s is relatively better when the power $1 - \beta$ is near 1.

5. The Efficiency of the Proposed Tests

For the normal distribution, the power of the conventional two-sample F test can be calculated for various sample sizes, N, while keeping the two samples equal in size. The efficiency of our tests can then be calculated as the sample size for which the conventional test has the power observed for our test, divided by the actual sample size (interpolating for fractional N). The results are given in Table 4. A few "efficiencies" of 1 or greater occur because the actual α exceeds the desired value. For the p-sample case, the power of our tests can be compared with the approximate power of Bartlett's test, giving efficiencies ranging from .83 to .90 for z and s.

For the double exponential distribution, the likelihood ratio statistic for the two samples is $(\sum_j|x_{1j}|)/(\sum_j|x_{2j}|)$. This statistic times σ_2/σ_1 has an F distribution with $2N$ and $2N$ degrees of freedom. Table 4 gives the efficiency of our test relative to this one for one special case.

For the uniform distribution, all tests based on means and moments are very inefficient compared with the most powerful tests using the special properties of this distribution; the same would be true of distribution C and for the same reason, namely, the presence in the probability density function of jumps from 0 to positive values at points depending on the unknown parameter. Nevertheless, these distributions do represent the behavior of our statistics with classes of distributions having similar moments.

The empirical behavior of these tests and those of Box and Andersen [2] can be compared. The tests here proposed, like all analysis of variance tests, are two-sided, as is Bartlett's test. On the other hand, the classical F test of σ_2^2/σ_1^2 for the two-sample problem is one-sided, as is the Box-Andersen test. Hence, the Box-Andersen powers for the two-sample case, all calculated for $\alpha = .05$, must be compared with the results of the present test at the 10% level. For the normal distribution, $p = 2$, $N = 20$, the test based on z gives almost the same power as the Box-Andersen test, but with α values estimated at .188 and .092 respectively. The difference of the estimated α's is not statistically significant. The test based on s, as noted above, has poorer power here. For the uniform distribution, the

Box-Andersen test seems to be slightly better (on the basis of their published graph) than that based on s, while for the double exponential the Box-Andersen test gives observed two-sided α about .11 and power comparable to s ($\alpha = .104$) and considerably below z ($\alpha = .117$) or even z adjusted to $\alpha = .10$. Finally, for $p = 10$ Box and Andersen considered the power only for the normal case, finding an efficiency of .81 compared with .83 for s and .87 for z, or .81 and .80 for s and z when the empirical α is .10.

Finally, the actual efficiencies of our proposed tests can be compared with their asymptotic efficiencies when the individual sample sizes $N \to \infty$. For the normal distribution and the two-sample case, the test based on s is equivalent to taking the difference of the two sample variances divided by its standard error, and the latter in turn is asymptotically equivalent to the standard F test and fully efficient. The test based on z is equivalent to using the difference of the two mean deviations and here has an asymptotic efficiency of $1/(\pi - 2) = .876$. The asymptotic efficiency of t and L will be even worse. Similar results hold for $p > 2$. On the other hand, for the double exponential distribution, the test based on z, being equivalent to using the mean deviation, will be asymptotically efficient, as is clear from the form of the likelihood ratio statistic or from the result of Fisher [3].

6. The Robustness of Ordinary Analysis of Variance

The present experiment can also be considered as a further test of the robustness of the ordinary analysis of variance, considering the distributions of z, s, and t as the basic distributions. Even for the normal case, z has approximately a χ_1 and s a χ_1^2 distribution, both extremely non-normal. The moments of $z = |x - \bar{x}|$ and its transforms may be approximated by those of $z' = |z - E(x)|$ and its transforms $s' = z'^2$, $t' = \sqrt{z'}$, $L' = \log(z')$. We then have for the normal distribution $\gamma_1(z') = .996$, $\gamma_2(z') = .386$, $\gamma_1(s') = 2.83$, $\gamma_2(s') = 12$, $\gamma_1(t') = .08$, $\gamma_2(t') = -.52$, $\gamma_1(L') = -1.54$, $\gamma_2(L') = 4$; for the uniform distribution $\gamma_1(z') = 0$, $\gamma_2(z') = -1.2$, $\gamma_1(s') = .64$, $\gamma_2(s') = -.86$; for the double exponential distribution $\gamma_1(z') = 2$ and 1.99, $\gamma_2(z') = 6$ and 6.01, $\gamma_1(s') = 6.62$ and 7.1, $\gamma_2(s') = 84.7$ and 88 for the exact distribution and for the discrete distribution used, respectively; and for distribution C, $\gamma_1(z') = 1.58$ and $\gamma_2(z') = 3.42$.

Many of these distributions are far more severely non-normal than any used in previous investigations of the behavior of the analysis of variance and Student's tests; yet we have seen that the empirical values of α are in satisfactory agreement with the theoretical for all of these. On the other hand, for the double exponential distribution, t gives rather large values of α. Yet for this case t' has $\gamma_1 = .63$, $\gamma_2 = .25$, rather closer to those of the normal distribution than most of those above, and the density function of t' is $2te^{-t^2}$, $t > 0$, which is closer to the normal than most of the other cases considered. The distribution of t itself should if anything be even more closely normal.

Another disturbing fact is that, for $N = 20$, the approximation of our F

statistic to the F distribution seems to get worse as p increases from 2 to 4 to 10. Nevertheless, for fixed N the distribution of F should tend to that under normal theory as $p \to \infty$. (See, e.g., Box and Andersen [2], who show that the goodness of the approximation for a given underlying distribution should depend essentially only on pN in our notation.)

There are thus some peculiarities remaining to be resolved. Nevertheless, the general picture emerging from this investigation is an almost unbelievable robustness of the analysis of variance for testing equality of the means of p equal-sized samples under H_0 when all variables have a common distribution.

7. A Generalization

The variables $z_{ij} = |x_{ij} - \bar{x}_i|$ may be looked upon as absolute values of residuals from a certain least-squares regression function. This suggests the possibility of using a similar analysis of variance on the absolute values of residuals from other regressions, in order to study the variances of the residuals. There seems no reason why this should not work just as well in general as in the special case here considered.

The present investigation was actually motivated by a problem in genetics [5] involving a two-way analysis of variance of the components of variance model with one observation per cell. The model was $x_{ij} = \mu + u_i + v_j + w_{ij}$, with μ a constant, the u_i having mean 0 and variance σ_u^2, the v_i having mean 0 and variance σ_v^2, and the w_{ij} having mean 0 and variances σ_{ij}^2. In addition to the usual tests, it seemed appropriate to test whether the value of σ_{ij}^2 depended on i (or on j). If the actual sample interactions are $g_{ij} = (x_{ij} - x_{i.} - x_{.j} + x_{..})$ in the usual notation, the problem could be attacked by doing an ordinary analysis of variance on the $|g_{ij}|$ or g_{ij}^2. In addition to the complications of the one-way analysis of variance described above, we have here two kinds of correlation. If all the w_{ij} have a common normal distribution with variance 1, and if $i = 1, \cdots, I$; $j = 1, \cdots, J$, we have $\operatorname{var} g_{ij}^2 = I^{-2}J^{-2}(2I^2J^2 - 4IJ^2 - 4I^2J + 2J^2 + 8IJ + 2I^2 - 4I - 4J + 2)$, $\operatorname{cov}(g_{ij}^2, g_{ik}^2) = 2I^{-2}J^{-2}(I-1)^2$, $\operatorname{cov}(g_{ij}^2, g_{hk}^2) = 2I^{-2}J^{-2}$. The correlation of two g_{ij}^2 in the same row is exactly the same as that of two z_j^2 in a single sample of size J and is of order J^{-2}, while the correlation between two g_{ij}^2 in different rows and columns is of order $I^{-2}J^{-2}$. Thus the relative contribution of the correlations to the various sums appearing in the analysis of variance is never of order of magnitude greater than $\max(I^{-1}, J^{-1})$. In particular, if $G_{i.} = \sum_j g_{ij}^2$, the correlation of $G_{i.}$ and $G_{h.}$ is again the same as that of z_i^2 and z_h^2 in a single sample of size I.

Thus, it seems quite certain that the analysis of variance can be applied to the $|g_{ij}|$ or g_{ij}^2 quite as safely as to the z_i and s_i of the present paper.

8. Conclusions

The tests based on z and $s = z^2$ are satisfactory, while those based on $t = \sqrt{z}$ and $L = \log z$ have poor power for nearly normal distributions and may have much too large an α for some non-normal distributions. The test based on z is likely to be more powerful in moderate samples for distributions with $\gamma_2 \geqq 0$, though at a possible cost of α being somewhat greater than it is supposed to be. Which statistic is chosen is a matter of taste; the author would tend to prefer z in most cases, particularly since it is easier to compute. The tests based on z have power comparable to the Box-Andersen test, but the latter seems to have α somewhat closer to the desired value. On the other hand, the present tests are more easily adapted to more complicated problems. These tests have in general an efficiency between .80 and .90 when the underlying distribution is actually normal; this loss in efficiency is not too high a price to pay for the robustness of the tests.

9. Acknowledgments

The author wishes to express his thanks to the Watson Scientific Computing Laboratory of Columbia University for providing technical assistance and the greater part of the time on the IBM 704 computer, and to Mrs. Joyce Alsop and particularly to Miss Frances Testa for programming and running the problem. He also wishes to thank the AEC Computing and Applied Mathematics Center of the Institute of Mathematical Sciences of New York University for providing additional computer time to permit completion of the work.

[*References are on page 292, following Table 4.*]

TABLE 1

CUMULATIVE AND EMPIRICAL DISTRIBUTIONS OF F

Cumulative distribution of F under $H_0: \sigma_2^2/\sigma_1^2 = 1$ for normal theory (α) and empirical distribution of F for various variables and values of σ_2^2/σ_1^2. Distribution for z and s based on 3000 values of F, those for t and L on 1000 values, except that t for $\sigma_2^2/\sigma_1^2 = 1$ is based on 2000 values.

Case $p = 2$, $N = 20$

Variable	F value / Theoret. α / σ_2^2/σ_1^2	15.0	12.7	10.0	7.4	6.0	4.1	2.8	1.4	1.0	.1
	Theoret. α	.0004	.001	.0032	.010	.019	.050	.100	.250	.324	.754
z	1	.000	.001	.005	.012	.019	.054	.118	.271	.346	.752
s	1	.000	.000	.001	.006	.015	.050	.109	.273	.359	.775
t	1	.001	.002	.004	.011	.021	.051	.100	.256	.342	.778
L	1	.001	.002	.004	.012	.022	.055	.115	.271	.347	.764
z	2	.010	.019	.047	.108	.168	.284	.411	.603	.683	.901
s	2	.001	.006	.021	.067	.135	.274	.431	.650	.717	.920
t	2	.002	.024	.055	.116	.162	.246	.355	.562	.629	.883
L	2	.009	.013	.027	.065	.096	.195	.301	.476	.545	.864
z	4	.113	.188	.306	.484	.589	.759	.860	.950	.967	.995
s	4	.024	.062	.166	.344	.490	.734	.878	.965	.980	.997
t	4	...	...	...	...	...	...	...	...	...	...
L	4	.065	.104	.186	.298	.382	.515	.651	.811	.854	.969
z	6	.265	.377	.545	.782	.826	.929	.970	.994	.996	1.000
s	6	.064	.141	.286	.534	.705	.901	.969	.996	.998	1.000
t	6	.283	.383	.513	.677	.761	.870	.937	.979	.985	.996
L	6	.166	.239	.355	.491	.582	.737	.832	.925	.937	.988

TABLE 2

OBSERVED NUMBER OF F VALUES

Observed number of F values per 1000 greater than the nominal α probability level for various numbers of populations (p) and sample sizes (N) for H_0 (boldface) and alternatives given by σ_2^2/σ_1^2 for $p = 2$ and in the text for $p > 2$.

Distribution	p	N	Alternative	$\alpha = .01$			$\alpha = .05$			$\alpha = .10$		
				z	s	t	z	s	t	z	s	t
Normal	2	10	H_0	**12**	**6.7**	**16**	**53**	**36**	**56**	**93**	**93**	**90**
	2	10	2	33	14.5	40	132	116	130	247	244	216
	2	10	4	151	65	165[a]	420	332	400[a]	582	554	540[a]
	2	10	6	276	114	293	607	495	572	778	717	717
	2	20	H_0	**11.7**	**5.7**	**10.5**	**53.7**	**50.3**	**50.5**	**118**	**109**	**100**
	2	20	2	108	66.7	116	284	274	246	411	431	355
	2	20	4	484	344	450[a]	759	734	695[a]	860	878	815[a]
	2	20	6	732	534	677	929	901	870[a]	970	969	937
	4	20	H_0	**8**	**10**	**8**	**48**	**42**	**49**	**97**	**85**	**95**
	4	20	H_1	368	298	310	622	589	541	795	790	705
	10	20	H_0	**11**	**12**	**9**	**63**	**57**	**59**	**108**	**114**	**108**
	10	20	H_1	432	443	328	680	656	577	746	730	671
Uniform	2	10	H_0	**13.5**	**13**	**13**	**61**	**56**	**58**	**118**	**117**	**119**
	2	10	2	73	62	60	215	228	183	312	351	283
	2	10	4	295	253	262	565	578	500	692	744	613
	2	10	6	468	377	552	753	739	684	860	881	787
	2	20	H_0	**7**	**7**	...	**47**	**47**	...	**107**	**99**	...
	2	20	2	205	267	...	422	524	...	546	673	...
	2	20	4	724	810	...	925	978	...	964	993	...
	2	20	6	931	936	...	967	996	...	998	1000	...
	10	20	H_0	**13**	**13**	**12**	**53**	**46**	**55**	**108**	**87**	**102**
	10	20	H_1	730	898	519	900	974	746	944	991	825
Double exponential	2	10	H_0	**12**	**4**	**18.5**	**72**	**35**	**84**	**142**	**110**	**154**
	2	10	2	26	7	39	140	82	153	228	164	240
	2	10	4	104	22	142	306	177	333	443	345	440
	2	10	6	168	45	202	449	260	464	598	468	572
	2	20	H_0	**10**	**4**	**16**	**62**	**46**	**63**	**117**	**104**	**125**
	2	20	2	57	20	64	191	141	183	317	258	297
	2	20	4	266	106	293	564	391	555	707	598	688
	2	20	6	464	184	510	771	555	760	869	770	856
	10	20	H_0	**28**	**13**	**25**	**85**	**59**	**90**	**154**	**110**	**162**
	10	20	H_1	254	124	246	457	305	465	592	443	582
C	2	10	H_0	**12.5**	**1.5**	**29**	**68**	**34**	**107**	**137**	**80**	**171**
	2	10	2	174	98	182	380	324	392	500	471	498
	2	10	4	585	440	596	780	743	762	843	851	812
	2	10	6	732	564	750	875	840	856	907	907	882
	2	20	H_0	**11**	**3**	...	**65**	**29**	...	**116**	**81**	...
	2	20	2	468	347	...	688	650	...	791	791	...
	2	20	4	935	899	...	974	987	...	986	997	...
	2	20	6	977	957	...	993	997	...	996	999	...
	4	20	H_0	**19**	**2**	**53**	**76**	**25**	**132**	**127**	**59**	**193**
	4	20	H_1	923	870	918	973	970	966	989	985	978
	10	20	H_0	**47**	**5.5**	**139**	**123**	**36**	**265**	**197**	**78**	**356**
	10	20	H_1	980	948	980	995	988	992	998	997	995

[a] Estimated.

TABLE 3

PROBABILITY POINTS FOR F AND OBSERVED NUMBER OF F VALUES

Empirical α upper probability points for F under H_0 (boldface) and observed number of F values per 1000 greater than these points for various p, N, for alternatives given, for $p = 2$, by σ_2^2/σ_1^2 (identification column), and in text for $p > 2$.

Distribution	p	N	Identi-fication	$\hat{\alpha} = .01$		$\hat{\alpha} = .05$		$\hat{\alpha} = .10$	
				z	s	z	s	z	s
Normal	2	10	F	**8.70**	**7.60**	**4.60**	**3.98**	**2.86**	**2.84**
	2	10	2	30	18	122	144	261	260
	2	10	4	138	90	400	402	606	584
	2	10	6	252	156	585	564	799	738
	2	20	F	**7.70**	**6.70**	**4.26**	**4.11**	**3.10**	**2.94**
	2	20	2	102	96	275	273	378	408
	2	20	4	465	412	744	733	838	863
	2	20	6	706	622	920	900	960	964
	4	20	F	**3.90**	**4.08**	**2.69**	**2.50**	**2.14**	**2.05**
	4	20	H_1	392	291	628	642	752	762
	10	20	F	**2.52**	**2.62**	**2.03**	**1.98**	**1.71**	**1.71**
	10	20	H_1	424	398	636	642	778	775
Uniform	2	10	F	**10.00**	**9.20**	**4.93**	**4.60**	**3.35**	**3.30**
	2	10	2	47	50	195	209	285	322
	2	10	4	225	197	524	553	663	708
	2	10	6	376	321	708	721	842	852
	2	20	F	**6.60**	**6.90**	**4.04**	**3.88**	**2.92**	**2.78**
	2	20	2	246	302	427	552	536	675
	2	20	4	782	843	926	984	960	993
	2	20	6	951	956	988	998	998	1000
	10	20	F	**2.56**	**2.64**	**1.95**	**1.91**	**1.69**	**1.60**
	10	20	H_1	712	865	897	976	939	992
Double exponential	2	10	F	**9.4**	**6.0**	**4.9**	**4.02**	**3.88**	**3.13**
	2	10	2	19	32	119	100	174	151
	2	10	4	70	77	261	205	355	323
	2	10	6	131	128	395	296	492	439
	2	20	F	**8.90**	**6.40**	**4.70**	**3.95**	**3.16**	**2.88**
	2	20	2	35	42	144	151	282	253
	2	20	4	163	157	509	417	659	582
	2	20	6	335	263	710	578	843	767
	10	20	F	**2.85**	**2.72**	**2.16**	**1.97**	**1.86**	**1.70**
	10	20	H_1	168	90	355	289	492	419
C	2	10	F	**8.9**	**6.28**	**5.32**	**3.73**	**3.67**	**2.75**
	2	10	2	151	193	312	384	441	500
	2	10	4	558	571	734	795	805	868
	2	10	6	710	713	850	876	886	926
	2	20	F	**7.45**	**5.60**	**4.70**	**3.51**	**3.20**	**2.48**
	2	20	2	465	510	642	711	759	820
	2	20	4	934	952	971	991	984	998
	2	20	6	978	987	993	999	996	1000
	4	20	F	**5.01**	**3.06**	**3.12**	**2.28**	**2.40**	**1.82**
	4	20	H_1	891	957	960	983	984	992
	10	20	F	**3.52**	**2.34**	**2.44**	**1.84**	**2.07**	**1.56**
	10	20	H_1	884	959	984	991	993	998

TABLE 4

EFFICIENCY OF TWO-SAMPLE TESTS

Empirical efficiency of two-sample tests using the theoretical critical values of F (lightface) and the empirical critical values of F (boldface).

Distribution	N	σ_2^2/σ_1^2	$\alpha = .01$		$\alpha = .05$		$\alpha = .10$	
			z	s	z	s	z	s
Normal	10	2	.80	.43	.82	.71	.96	.94
	10	2	**.74**	**.51**	**.76**	**.93**	**1.00**	**1.00**
	10	4	.78	.53	.85	.70	.90	.86
	10	4	**.74**	**.61**	**.83**	**.83**	**.95**	**.91**
	10	6	.75	.51	.83	.68	.90	.81
	10	6	**.72**	**.61**	**.79**	**.77**	**.95**	**.84**
	20	2	.88	.63	.91	.87	.94	1.05
	20	2	**.86**	**.83**	**.89**	**.88**	**.83**	**.92**
	20	4	.78	.63	.85	.79	.96	.92
	20	4	**.76**	**.65**	**.80**	**.78**	**.82**	**.88**
	20	6	.76	.62	.85	.75	.88	.88
	20	6	**.73**	**.64**	**.82**	**.75**	**.82**	**.85**
Exponential	10	4	.87	.22	.92	.50	>1.00	.70
	10	4	**.60**	**.65**	**.80**	**.58**	**.72**	**.67**

REFERENCES

[1] Box, G. E. P. Non-normality and tests on variances. *Biometrika*, 1953, **40**, 318-35.

[2] Box, G. E. P., and Andersen, S. L. Permutation theory in the derivation of robust criteria and the study of departures from assumption. *J. Roy. Stat. Soc.*, Ser. B, 1955, **17**, 1-34.

[3] Fisher, R. A. A mathematical examination of the methods of determining the accuracy of an observation by the mean error, and by the mean square error. *Monthly Not. Roy. Astron. Soc.*, 1920, **80**, 758-70.

[4] Juncosa, M. L. Random number generation on the BRL high speed computing machines. ABRL Report No. 855, Aberdeen Proving Ground, Md.: Army Ballistic Research Laboratories, 1954.

[5] Levene, H. Release of genetic variability through recombination, IV. Statistical theory. *Genetics*, 1959, **44**, 93–104.

26 INTRABLOCK AND INTERBLOCK ESTIMATES

HENRY B. MANN, Ohio State University

M. V. MENON, North Carolina State College

In a factorial design where the blocks are all of equal size and with $p+1$ treatment combinations, we consider p linear forms $S_1, \cdots, S_p$ of the treatment effects which are orthogonal to each other and to the mean. Corresponding to these we form, in each replication, linear forms $L_1, \cdots, L_p$ of the observations. We consider partially confounded designs and shall assume that the block-effects are normally and independently distributed. We compute L_α from the replications where S_α is not confounded and from the replications where S_α is confounded.

Let $L_\alpha^{(1)}, \cdots, L_\alpha^{(n_\alpha)}$ be the observed values of L_α in the replications where S_α is unconfounded, and $M_\alpha^{(1)}, \cdots, M_\alpha^{(m_\alpha)}$ the observed values in those replications where S_α is confounded. If $n_\alpha \neq 0$, let $(a_{ij}^{(\alpha)})$ denote an orthogonal n_α-rowed square matrix with first row $n_\alpha^{-\frac{1}{2}}, \cdots, n_\alpha^{-\frac{1}{2}}$. Similarly, if $m_\alpha \neq 0$, let $(b_{ij}^{(\alpha)})$ denote a similar m_α-rowed square matrix. For $n_\alpha \neq 0$, put $x_\alpha^{(j)} = \sum_s a_{js}^{(\alpha)} L_\alpha^{(s)}$, and for $m_\alpha \neq 0$, $y_\alpha^{(j)} = \sum_s b_{js}^{(\alpha)} M_\alpha^{(s)}$, where here and in the following all sums extend over the domain of definition of the quantities indexed with the summation index.

Certain of the S_α, say $S_{r+1}, \cdots, S_p$, may be known to be 0. We arrange the pairs (α, j), for $\alpha > r$ and for $j \geqq 2$, $\alpha \leqq r$, in some arbitrary order and according to this order put $x_\alpha^{(j)} = z_\beta$, $y_\alpha^{(j)} = t_\gamma$, respectively. For $\alpha \leqq r$ we put

$$x_\alpha = \begin{cases} x_\alpha^{(1)}, & \text{for } n_\alpha \neq 0, \\ 0, & \text{for } n_\alpha = 0, \end{cases} \qquad y_\alpha = \begin{cases} y_\alpha^{(1)}, & \text{for } m_\alpha \neq 0, \\ 0, & \text{for } m_\alpha = 0. \end{cases}$$

We thus have the following set of observations:

$$\begin{aligned} x_\alpha &= c_\alpha S_\alpha + \varepsilon_\alpha, & \alpha &= 1, \cdots, r, \\ y_\alpha &= d_\alpha S_\alpha + \eta_\alpha, & \alpha &= 1, \cdots, r, \\ z_\beta &= \varepsilon_\beta, & \beta &= r+1, \cdots, n, \\ t_\gamma &= \eta_\gamma, & \gamma &= r+1, \cdots, m, \end{aligned} \tag{1}$$

where $c_\alpha = \sqrt{n_\alpha}$, $d_\alpha = \sqrt{m_\alpha}$, $\varepsilon_\alpha = 0$ if $n_\alpha = 0$, $\eta_\alpha = 0$ if $m_\alpha = 0$, and the non-zero ε's and η's are jointly normally and independently distributed with mean 0, the ε's with variance σ_1^2 and the η's with variance σ_2^2. Moreover, we assume $c_\alpha^2 + d_\alpha^2 > 0$. The inequality $\sigma_1^2 \leqq \sigma_2^2$ holds but will not be used in this paper.

The situation described by Equations (1) carries over without change to symmetrical quasifactorial designs, where the S_α are the quasifactors and certain linear combinations of the quasifactors give the varietal effects. In other cases, as, for instance, in balanced incomplete block designs, the sum of squares of deviations from the mean can be decomposed as follows:

$$\sum_i (y_i - \bar{y})^2 = Q_a + c \sum_{\alpha=1}^{r+1} \hat{v}_\alpha^2 + d \sum_{\alpha=1}^{r+1} \bar{v}_\alpha^2 + \bar{Q}_a , \tag{2}$$

where the $\hat{v}_\alpha$ are the intrablock estimates and the $\bar{v}_\alpha$ the interblock estimates of the varietal effects v_α, and Q_a and $\bar{Q}_a$ are quadratic forms of the observations which are independent of the varietal effects. Moreover, for appropriate values of n and m we may write

$$Q_a = \sum_{\beta=r+1}^{n} z_\beta^2 , \qquad \bar{Q}_a = \sum_{\gamma=r+1}^{m} t_\gamma^2 ,$$

where z_β and t_γ satisfy the conditions of the system (1) and are orthogonal to $\hat{v}_\alpha$ and $\bar{v}_\alpha$. The $\hat{v}_\alpha$ and $\bar{v}_\alpha$ satisfy the relations $\sum_\alpha \hat{v}_\alpha = \sum_\alpha \bar{v}_\alpha = \sum_\alpha v_\alpha = 0$. An orthogonal transformation with first row $(r+1)^{-\frac{1}{2}}, \cdots, (r+1)^{-\frac{1}{2}}$ yields $2r$ orthogonal forms x_α, y_α, $\alpha = 1, \cdots, r$, satisfying (1), where now the S_α determine the varietal effects which we wish to estimate.

The argument above applies in fact to all incomplete block designs with equal block size u for which the intrablock equations have a unique solution. By Theorem 4.3 of [1], one obtains instead of (2) the decomposition

$$\sum_j (y_i - \bar{y})^2 = Q_a + \sum_{i=1}^{r+1} \sum_{j=1}^{r+1} c_{ij} \hat{v}_i \hat{v}_j + \frac{1}{u} \sum_j B_j^2 , \tag{2'}$$

where B_j denotes the sum of all observations in the jth block.

An orthogonal transformation with first row $(r+1)^{-\frac{1}{2}}, \cdots, (r+1)^{-\frac{1}{2}}$ carries (2′) into

$$\sum_i (y_i - \bar{y})^2 = Q_a + \sum_1^r \lambda_i \hat{K}_i^2 + \frac{1}{u} \sum_j B_j^2 \tag{2''}$$

and a further transformation gives

$$\sum_i \lambda_i \hat{K}_i^2 = \sum_i \hat{M}_i^2 ,$$

where

$$\hat{v}_i = \sum_j \alpha_{ij} \hat{M}_j , \qquad E(\hat{v}_i) = v_i = \sum_j \alpha_{ij} M_j , \qquad M_i = E(\hat{M}_i) ,$$

and the v_i are uniquely determined by the M_i and the restrictions $\sum_{i=1}^{r+1} v_i = 0$.

The $\hat{v}_i$ and therefore also the $\hat{M}_i$ are, moreover, orthogonal to the blocks B_j (Theorem 2.1 of [1]), and $E(B_i) = \sum_j a_{ij} M_j$ or, in matrix notation, $E(B) = AM$. By Cochran's theorem (Theorem 2.1 of [1]) the $\hat{M}_i$ are normally and independently distributed with variance σ_1^2.

Now, minimizing $(1/u) \sum_j [B_j - E(B_j)]^2$ with respect to the M_j leads to the interblock equations

$$A'B = A'AM . \tag{i}$$

Let T be an orthogonal matrix such that $T'A'AT = D = \operatorname{diag}(d_1, \cdots, d_s, 0, \cdots, 0)$. Then, writing $T'M = N$, we may write

$$T'A'B = DN . \tag{i'}$$

Since (i) and therefore (i′) must have a solution, the last $r - s$ rows of $T'A'B$ must vanish. From (i) we get the interblock estimates $\bar{N}_1, \cdots, \bar{N}_s$ of $N_i = E(\bar{N}_i)$.

Since T is an orthogonal matrix, we now have

$$\sum (y_i - \bar{y})^2 = Q_a + \sum_1^r \hat{N}_i^2 + \sum_1^s \sum_1^s p_{ij} \bar{N}_i \bar{N}_j + \bar{Q}_a . \tag{i''}$$

A further orthogonal transformation applied to $N_1, \cdots, N_s$ brings (i″) into the final form

$$\sum_i (y_i - \bar{y})^2 = Q_a + \sum_1^r \hat{H}_i^2 + \sum_1^s e_i \bar{H}_i^2 + \bar{Q}_a , \tag{2'''}$$

where, by Cochran's theorem applied to $(1/u) \sum B_j^2$, the quantities $\sqrt{e_i}\, \bar{H}_i$ are normally and independently distributed with variance σ_2^2.

We have with some values $c_{j\alpha}$,

$$v_j = \sum_{\alpha=1}^r c_{j\alpha} H_\alpha , \qquad j = 1, \cdots, r+1 .$$

If H_α^*, $\alpha = 1, \cdots, r$, are independent unbiased estimates of H_α, and $v_j^* = \sum_{\alpha=1}^r c_{j\alpha} H_\alpha^*$, then $\sigma_{v_j^*}^2 = \sum_\alpha c_{j\alpha}^2 \sigma_{H_\alpha^*}^2$.

We put $\sum_\alpha c_{j\alpha}^2 = \bar{c}_j^2$, $H_\alpha = c_\alpha S_\alpha$, and $d_\alpha = \sqrt{e_\alpha}\, c_\alpha$. If we substitute $\hat{H}_\alpha$ for x_α and $\bar{H}_\alpha$ for y_α in (1), the conditions of (1) are satisfied. We note for later use: If $S_1', \cdots, S_r'$ are estimates of the S_α, all with small variances, then the variance of $v_j' = \sum c_{j\alpha} S_\alpha'$ will be small so that precise estimates of all the S_α yield precise estimates of all the v_α. Moreover, if we have a design for which the variance of all the $\hat{v}_\alpha$ is small, then all the $c_{\alpha j}$ must be small and therefore c_α must be large.

Thus Equations (1) arise in partially confounded designs as well as in incomplete block designs with equal block size, where the intrablock equations have a unique solution.

If we put $\sigma_2^2/\sigma_1^2 = K$, $n + m = N$, and $\sigma_1 = \sigma$, we may write the likelihood function L of the joint distribution of the x_α and y_α as

$$(3)\qquad L = \log P = \text{const} - \frac{m}{2}\log K - N\log\sigma$$
$$- \frac{1}{2\sigma^2 K}\left[K\sum_\beta z_\beta^2 + K\sum_\alpha (x_\alpha - c_\alpha S_\alpha)^2 + \sum_\gamma t_\gamma^2 + \sum_\alpha (y_\alpha - d_\alpha S_\alpha)^2\right].$$

If K is known, then we obtain from this the maximum likelihood estimates

$$(4)\qquad \hat\sigma^2 = \frac{1}{NK}\left[K\sum_\beta z_\beta^2 + K\sum_\alpha (x_\alpha - c_\alpha \hat S_\alpha)^2 + \sum_\gamma t_\gamma^2 + \sum_\alpha (y_\alpha - d_\alpha \hat S_\alpha)^2\right];$$
$$\hat S_\alpha = \frac{Kc_\alpha x_\alpha + d_\alpha y_\alpha}{Kc_\alpha^2 + d_\alpha^2}.$$

Let $S_\alpha^{(1)} = x_\alpha / c_\alpha$, $S_\alpha^{(2)} = y_\alpha / d_\alpha$. The estimates $S_\alpha^{(1)}$ and $S_\alpha^{(2)}$ are the intra- and interblock estimates respectively, and $\hat S_\alpha = (Kc_\alpha^2 S_\alpha^{(1)} + d_\alpha^2 S_\alpha^{(2)})/(Kc_\alpha^2 + d_\alpha^2)$. Usually, however, K is not known and has to be estimated.

Applying the maximum likelihood method to (3) under the assumption that K is not known, one would get Equations (4) with K replaced by its maximum likelihood estimate $\hat K$, and the additional equation

$$(4')\qquad \hat K = \frac{1}{m\hat\sigma^2}\left[\sum_\gamma t_\gamma^2 + \sum_\alpha (y_\alpha - d_\alpha \hat S_\alpha)^2\right].$$

Together with (4), this gives

$$(4'')\qquad \hat\sigma^2 = \frac{1}{n}\left[\sum_\beta z_\beta^2 + \sum_\alpha (x_\alpha - c_\alpha \hat S_\alpha)^2\right],$$
$$\hat K = \frac{\left[\sum_\gamma t_\gamma^2 + \sum_\alpha (y_\alpha - d_\alpha \hat S_\alpha)^2\right]\Big/ m}{\left[\sum_\beta z_\beta^2 + \sum_\alpha (x_\alpha - c_\alpha \hat S_\alpha)^2\right]\Big/ n}.$$

The solution of (4″) together with (4) leads to a cubic equation in $\hat K$ and is too complicated to be useful. We therefore proceed in a different way. We take (4″) as a basis for an estimate of K and shall assume that $c_\alpha \neq 0$, $\alpha = 1, \cdots, r$. We replace $\hat S_\alpha$ in (4″) by the intrablock estimate $S_\alpha^{(1)}$. The numerator and denominator of (4″) then become biased estimates of σ_2^2 and $\sigma^2 = \sigma_1^2$, respectively. The bias in the denominator can be corrected by replacing n by $n - r$. To correct the bias in the numerator, we compute

$$(5)\qquad E\left[\sum_\alpha (y_\alpha - d_\alpha S_\alpha^{(1)})^2\right] = E\left[\sum_\alpha (y_\alpha - d_\alpha S_\alpha)^2\right] + E\left[\sum_\alpha d_\alpha^2 (S_\alpha - S_\alpha^{(1)})^2\right]$$
$$= s\sigma_2^2 + \sigma_1^2 \sum d_\alpha^2 / c_\alpha^2,$$

where s is the number of non-zero y_α.

In this way we are led to the estimate

$$(6)\qquad K^* = \frac{n-r}{m-r+s}\left[\sum_\gamma t_\gamma^2 + \sum_\alpha (y_\alpha - d_\alpha S_\alpha^{(1)})^2 - \sum_\alpha d_\alpha^2/c_\alpha^2 \sum_\beta z_\beta^2/(n-r)\right]\Big/ \sum_\beta z_\beta^2.$$

We now assume that we have a sequence of designs for estimating the

S_α for which $n \to \infty$, $m \to \infty$. Under these conditions, K^* is a consistent estimate of K. We obtain an estimate S_α^* of S_α by replacing in the second equation of (4) the parameter K by K^*. Thus

$$S_\alpha^* = \frac{K^* c_\alpha^2 S_\alpha^{(1)} + d_\alpha^2 S_\alpha^{(2)}}{K^* c_\alpha^2 + d_\alpha^2} . \tag{7}$$

The estimate (7) is the one customarily used. It seems to have been advanced on purely intuitive grounds. It is worthwhile to note its relation to the maximum likelihood estimates of S_α. If we replace K^* by K in (7), we obtain the maximum likelihood estimate $\hat{S}_\alpha$ for the case that K is known; if K^* is replaced by $\hat{K}$ of Equations (4″), one obtains the maximum likelihood estimate of S_α for the case that K is not known. The following equation (8) shows, moreover, that for large enough values of n and m the estimate S_α^* is, for all practical purposes, as good as the estimate $\hat{S}_\alpha$.

In the following, we shall use the notation proposed in [2] and shall repeatedly use the fact that all rules applicable to ordinary limit and order relations are also applicable to stochastic limit and order relations. (See [2], Corollary 1 to Theorem 1 ff.)

We shall prove

$$\operatorname{plim} (\hat{S}_\alpha - S_\alpha^*)/\sigma_{\hat{S}_\alpha} = 0 . \tag{8}$$

Proof. Putting $c_\alpha^2 = n_\alpha$, $d_\alpha^2 = m_\alpha$ we have

$$\hat{S}_\alpha - S_\alpha^* = \frac{n_\alpha m_\alpha (S_\alpha^{(1)} - S_\alpha^{(2)})(K - K^*)}{(K n_\alpha + m_\alpha)(K^* n_\alpha + m_\alpha)} .$$

Furthermore,

$$\sigma_{\hat{S}_\alpha}^2 = \frac{K}{K n_\alpha + m_\alpha} \sigma^2 , \qquad \sigma_{S_\alpha^{(1)} - S_\alpha^{(2)}}^2 = \frac{K n_\alpha + m_\alpha}{n_\alpha m_\alpha} \sigma^2 .$$

Hence

$$\sqrt{\frac{m_\alpha n_\alpha}{K n_\alpha + m_\alpha}} (S_\alpha^{(1)} - S_\alpha^{(2)}) = O_p(1) .$$

Furthermore,

$$\frac{\sqrt{m_\alpha n_\alpha}}{K^* n_\alpha + m_\alpha} \leqq \begin{cases} \min\left(\sqrt{\dfrac{m_\alpha}{n_\alpha}} \dfrac{1}{K^*} , \sqrt{\dfrac{n_\alpha}{m_\alpha}}\right) , \\ \max\left(1, \dfrac{1}{K^*}\right) = O_p(1) , \end{cases}$$

and since $K - K^* = o_p(1)$, we have

$$(\hat{S}_\alpha - S_\alpha^*)/\sigma_{\hat{S}_\alpha} = O_p(1)\, o_p(1) = o_p(1) ,$$

which is just another way of writing (8).

Let σ^* be any consistent estimate of σ (for instance, its intrablock estimate) and form

$$\bar{\sigma}^* = \sqrt{\frac{K^*}{K^*n_\alpha + m_\alpha}}\sigma^* .$$

Then

(9) $$\operatorname{plim} \frac{\sigma_{\hat{S}_\alpha}}{\bar{\sigma}^*} = \operatorname{plim} \sqrt{\frac{K}{K^*} \frac{K^*n_\alpha + m_\alpha}{Kn_\alpha + m_\alpha}} \frac{\sigma}{\sigma^*} = 1 .$$

We determine η so that

(10) $$P(| \hat{S}_\alpha - S_\alpha | \leqq \eta\sigma_{\hat{S}_\alpha}) = \beta .$$

Equations (8) and (9) show that the difference between the endpoints of the interval (10) and of the interval

(11) $$| S_\alpha^* - S_\alpha | \leqq \eta\bar{\sigma}^*$$

is, for sufficiently large values of n and m, small compared with the length of (10). Thus, if m and n are sufficiently large, we may for all practical purposes replace the exact confidence interval (10) by the interval (11).

Moreover, since $\hat{S}_\alpha$ is normally distributed, we have, for sufficiently large n and m and arbitrary $\varepsilon > 0$, $\varepsilon_1 > 0$,

$$\begin{aligned} P(| S_\alpha^* - S_\alpha | \leqq \eta\bar{\sigma}^*) &\geqq P(| \hat{S}_\alpha - S_\alpha | \leqq (\eta - \varepsilon)\bar{\sigma}^* \text{ and } | S_\alpha^* - \hat{S}_\alpha | \leqq \varepsilon\bar{\sigma}^*) \\ &\geqq P(| \hat{S}_\alpha - S_\alpha | \leqq (\eta - \varepsilon)\sigma_{\hat{S}_\alpha}) - 2\varepsilon_1 . \end{aligned}$$

By a proper choice of ε, we can get from this

(12) $$P(| S_\alpha^* - S_\alpha | \leqq \eta\bar{\sigma}^*) \geqq \beta - 3\varepsilon_1 .$$

Similarly,

$$\begin{aligned} P(| \hat{S}_\alpha - S_\alpha | \leqq (\eta + \varepsilon)\bar{\sigma}^*) &\geqq P(| S_\alpha^* - S_\alpha | \leqq \eta\bar{\sigma}^* \text{ and } | S_\alpha^* - \hat{S}_\alpha | \leqq \varepsilon) \\ &\geqq P(| S_\alpha^* - S_\alpha | \leqq \eta\bar{\sigma}^*) - \varepsilon_1 . \end{aligned}$$

By a proper choice of ε we can make the left-hand side of this inequality less than $\beta + \varepsilon_1$, so that for sufficiently large n and m

(13) $$\beta + 2\varepsilon_1 \geqq P(| S_\alpha^* - S_\alpha | \leqq \eta\bar{\sigma}^*) .$$

From (12) and (13) we have

(14) $$\lim P(| S_\alpha^* - S_\alpha | \leqq \eta\bar{\sigma}^*) = \beta . \quad ||$$

References

[1] Mann, H. B. *Analysis and Design of Experiments*. New York: Dover, 1949.

[2] Mann, H. B., and Wald, A. On stochastic limit and order relationships. *Ann. Math. Stat.*, 1943, **14**, 217-26.

27 A BIVARIATE CHEBYSHEV INEQUALITY FOR SYMMETRIC CONVEX POLYGONS[1]

ALBERT W. MARSHALL, Stanford University

INGRAM OLKIN, Michigan State University and Stanford University

1. Introduction

In this paper we obtain a lower bound for the probability that a two-dimensional random vector falls in a symmetric convex polygon. Berge [1] obtained such a bound when the polygon is a rectangle and (x_1, x_2) has a certain covariance matrix. This was extended by Lal [2] to an arbitrary covariance matrix, and to higher dimensions by Marshall [3], Olkin and Pratt [4], and Whittle [5]. In the current work, we lean heavily on the notation and results of [4].

One can view a rectangle as the intersection of two strips $|x_1| < w_1$ and $|x_2| < w_2$. Furthermore, the second strip coincides with the first strip after a rotation and dilation. This suggests consideration of a set which is the intersection of m strips, where the jth strip is obtained by rotating the strip $|x_1| < w_j$ through an angle α_j, $j = 1, \cdots, m$. The resulting set will be a convex polygon symmetric about the origin, and any such polygon can be represented as the intersection of such strips.

A general outline and some preliminaries are given in Section 2, and in Section 3 the bivariate inequality is obtained. Some additional results for the equal-width case $(w_1 = w_2 = \cdots = w_m)$ are given in Section 4, and sharpness is discussed in Section 5. A computing procedure and an example are given in Section 6. An application to distribution-free simultaneous confidence intervals is given in Section 7.

2. General Outline

Let $f(x) \equiv f(x_1, x_2)$ be a real-valued non-negative function such that

$$f(x) \geqq 1 , \qquad x \in \bar{T} , \tag{2.1}$$

where $\bar{T}$ denotes the complement of the set T. Then

$$Ef(x) \geqq \int_{\{x \in \bar{T}\}} f(x)\, dP \geqq \int_{\{x \in \bar{T}\}} dP = P(x \in \bar{T}) .$$

---

[1] This research was sponsored in part by the Office of Ordnance Research at Michigan State University and the Office of Naval Research at Stanford University.

We confine our attention to functions $f(x)$ which are quadratic forms, so that the bound will involve only the second moments. Let

$$f(x) = x\,A\,x' \equiv (x_1\,x_2)\begin{bmatrix} a_{11} & a_{12} \\ a_{12} & a_{22} \end{bmatrix} \begin{bmatrix} x_1 \\ x_2 \end{bmatrix}$$

satisfy (2.1), and let $T = \bigcap_1^m S_j$, where S_j is a strip obtained by rotating the strip $\{|x_1| < w_j\}$ through an angle α_j. Since $\bar{T} = \bigcup_1^m \bar{S}_j$, Condition (2.1) is satisfied if

$$x\,A\,x' \geqq 1\,, \qquad \text{whenever } x \in \bigcup_1^m \bar{S}_j\,. \tag{2.2}$$

A is positive definite if $m \geqq 2$, since T is bounded. We note that

$$E f(x) = E\,x\,A\,x' = \operatorname{tr} A\,\Sigma\,,$$

where $\Sigma = E\,x'x$ is the moment matrix of $(x_1\,x_2)$. The bound $\operatorname{tr} A\,\Sigma$ is to be minimized for positive definite A satisfying (2.2).

3. The Inequality

Let $\mathscr{A} = \{A \mid x\,A\,x' \geqq 0,\ x\,A\,x' \geqq 1 \text{ for } |x_1| \geqq w\}$. From Lemma 3.1 of [4], it follows that $A \in \mathscr{A}$ if and only if $B = A^{-1}$ is positive definite, and $b_{11} \leqq w^2$. Rotation of the axes through an angle $0 \leqq \alpha < \pi$ has the effect of transforming A into $\Gamma A \Gamma'$, where

$$\Gamma = \begin{bmatrix} \cos\alpha & \sin\alpha \\ -\sin\alpha & \cos\alpha \end{bmatrix}$$

is an orthogonal matrix. After multiplication in $\Gamma A \Gamma'$, the condition $b_{11} \leqq w^2$ is transformed to $b_{11}\cos^2\alpha + b_{22}\sin^2\alpha + b_{12}\sin 2\alpha \leqq w^2$. If we have m such rotations through angles $0 \leqq \alpha_1, \alpha_2, \cdots, \alpha_m < \pi$, $\alpha_i \neq \alpha_j$, with associated widths $w_1, \cdots, w_m$, we obtain the set of conditions

$$b_{11}\cos^2\alpha_j + b_{22}\sin^2\alpha_j + b_{12}\sin 2\alpha_j \leqq w_j^2\,, \qquad j = 1, \cdots, m\,. \tag{3.1}$$

The set of matrices B which are positive definite and which satisfy (3.1) will be denoted by $\mathscr{B}$.

As in [4], $\mathscr{B}$ is a convex set which is closed and bounded with respect to the usual norms. From a geometric point of view, the conditions (3.1) mean that the ellipse, $\{xB^{-1}x' < 1\}$, lies within T. Equality in the jth condition of (3.1) means that the boundary, $\{x\,B^{-1}x' = 1\}$, touches the boundary of the jth strip. Thus, to characterize the extreme points of $\mathscr{B}$, we investigate the kinds of contact that can occur.

The original problem of minimizing $\operatorname{tr} A\,\Sigma$ for positive definite A satisfying (2.2) is equivalent to the problem of minimizing $\operatorname{tr} B^{-1}\Sigma$ for $B \in \mathscr{B}$. We now investigate the nature of the extreme points of $\mathscr{B}$. We first note that A is extreme in $\mathscr{A}$ if and only if $B = A^{-1}$ is positive definite and $b_{11} = w^2$; this is proved as Theorem 3.3 in [4] for $i = 1$ only.

LEMMA 3.1. *If B is an extreme point of $\mathscr{B}$, then $B \in \mathscr{B}$ and in* (3.1) *equality holds for at least two values of j. Equality for three or more values of j determines B.*

PROOF. From Theorem 3.3 of [4], it follows that $B \in \mathscr{B}$. Suppose that $B \in \mathscr{B}$ is such that $(m-1)$ of the inequalities in (3.1) are strict, say all except the first. We then show that B is not extreme in $\mathscr{B}$. To accomplish this, we represent B as a convex combination of matrices B_1 and $B_2 \in \mathscr{B}$, namely $B = (B_1 + B_2)/2$, where

$$B = \begin{bmatrix} b_{11} & b_{12} \\ b_{12} & b_{22} \end{bmatrix}, \qquad B_1 = \begin{bmatrix} b_{11} - \varepsilon & b_{12} \\ b_{12} & b_{22} + \delta \end{bmatrix}, \qquad B_2 = \begin{bmatrix} b_{11} + \varepsilon & b_{12} \\ b_{12} & b_{22} - \delta \end{bmatrix}.$$

Furthermore, if ε, δ are chosen so that $\delta \sin^2 \alpha_1 = \varepsilon \cos^2 \alpha_1$, then the inequality (3.1) with $j = 1$ remains unchanged. In addition, ε and δ can be chosen so small that the remaining $(m-1)$ inequalities are undisturbed. Hence, equality must hold in at least two of the inequalities (3.1)

Suppose that in (3.1) equality is achieved for $j = 1, 2, 3$, say. These equations have a unique solution since the α's are distinct non-negative angles less than π, and

$$\text{(3.2)} \qquad \begin{vmatrix} \cos^2 \alpha_1 & \sin^2 \alpha_1 & \sin 2\alpha_1 \\ \cos^2 \alpha_2 & \sin^2 \alpha_2 & \sin 2\alpha_2 \\ \cos^2 \alpha_3 & \sin^2 \alpha_3 & \sin 2\alpha_3 \end{vmatrix}$$
$$= -\sin(\alpha_3 - \alpha_1)[\cos(2\alpha_2 - \alpha_1 - \alpha_3) - \cos(\alpha_3 - \alpha_1)] . \quad ||$$

LEMMA 3.2. *There is at least one extreme point B of $\mathscr{B}$ that is determined by three equalities in* (3.1).

PROOF. Choose any vertex of T with sides 1 and 2, say, and form any ellipse $\{x B^{-1} x' < 1\}$ touching these sides with sufficiently high eccentricity to lie within T. Then

$$\text{(3.3)} \qquad b_{11} \cos^2 \alpha_j + b_{22} \sin^2 \alpha_j + b_{12} \sin 2\alpha_j = w_j^2 , \qquad j = 1, 2 ,$$

$$\text{(3.4)} \qquad b_{11} \cos^2 \alpha_j + b_{22} \sin^2 \alpha_j + b_{12} \sin 2\alpha_j \leqq w_j^2 , \qquad j = 3, 4, \cdots, m .$$

In (3.3), solve for b_{11} and b_{22}, and substitute in (3.4). The left-hand side of (3.4) will still be linear in b_{12}, and it is clear that, by changing b_{12}, equality for some $j = 3, \cdots, m$ can be achieved in (3.4) without violating the remaining inequalities. ||

It is convenient to reparametrize by letting $u_j = (\cos \alpha_j)/w_j$, $v_j = (\sin \alpha_j/w_j$. This has the effect of changing (3.1) to

$$\text{(3.5)} \qquad b_{11} u_j^2 + b_{22} v_j^2 + 2b_{12} u_j v_j \leqq 1 , \qquad j = 1, 2, \cdots, m .$$

It also changes the boundary lines $x_1 \cos \alpha_j + x_2 \sin \alpha_j = w_j$ of S_j to $x_1 u_j + x_2 v_j = 1$, and T becomes $\{| u_j x_1 + v_j x_2 | < 1, \ j = 1, \cdots, m\}$.

One approach that can be used to find a best bound is to minimize $\operatorname{tr} B^{-1} \Sigma$ [considered as a function of b_{12} by using (3.3)] taking into account the constraints (3.4) and positive definiteness. Although the method is straightforward, the computations are involved and we give an alternative procedure.

It is convenient to now introduce some new terminology. Let S be any set in the plane and let $\mathscr{B}(S)$ be the set of all positive definite matrices B such that $xB^{-1}x' \geqq 1$ for all $x \in \bar{S}$, and let B_0 be such that

$$\min_{B \in \mathscr{B}(S)} \operatorname{tr} B^{-1}\Sigma = tr\, B_0^{-1}\Sigma .$$

Motivated by the fact that $xB^{-1}x' = 1$ is the equation of an ellipse for all B in $\mathscr{B}(S)$, we refer to the set $\{xB_0^{-1}x' < 1\}$ as the best ellipse for S, and call B_0 best for S. For the square $S = \{|x_1| < 1, |x_2| < 1\}$, a best ellipse exists and is unique ([4], Theorems 3.6 and 3.7).

The following lemma shows that it is reasonable to consider first the best ellipses for all parallelograms $S_i \cap S_j$, $i \neq j$.

LEMMA 3.3. *The best ellipse for the parallelogram* $S_1 \cap S_2$ *lies in* T *if and only if* (3.3) *and* (3.4) *hold, where*

$$b_{jj} = u_j^2 + 2u_jv_j\beta + v_j^2 , \qquad j = 1, 2 , \tag{3.6}$$

$$b_{12} = u_1u_2 + (u_1v_2 + u_2v_1)\beta + v_1v_2 , \tag{3.7}$$

and β *is given by* (3.10) *below. If the best ellipse for the parallelogrom* $S_1 \cap S_2$ *lies in* T, *then it is best for* T.

PROOF. It is known; e.g., [2], [3], [4, Equation (2.1)]; that the best B for the square $\{|y_1| < 1, |y_2| < 1\}$ is $B_s = \left(\begin{smallmatrix} 1 & \beta \\ \beta & 1 \end{smallmatrix}\right)$, where

$$\beta = \{\pi_{11} + \pi_{22} - \sqrt{[(\pi_{11} + \pi_{22})^2 - 4\pi_{12}^2]}\}/2\pi_{12} , \tag{3.8}$$

and $\pi_{ij} = Ey_iy_j$. The transformation mapping the square $\{|y_1| < 1, |y_2| < 1\}$ into the parallelogram $\{|u_1x_1 + v_1x_2| < 1, |u_2x_1 + v_2x_2| < 1\}$ has the matrix M given by

$$M = \begin{bmatrix} u_1 & u_2 \\ v_1 & v_2 \end{bmatrix}^{-1} . \tag{3.9}$$

Accordingly, we make the change of variable $x = yM$, and it follows that the best ellipse for the parallelogram has the matrix $B = M'B_sM$. This yields (3.6) and (3.7). Substituting for $\Pi = M'^{-1}\Sigma M^{-1}$ in (3.8), we obtain

$$\beta = \frac{c_1 + c_2 - \sqrt{(c_1 - c_2)^2 + 4(u_1v_2 - u_2v_1)^2(\sigma_{11}\sigma_{22} - \sigma_{12}^2)}}{2[u_1u_2\sigma_{11} + v_1v_2\sigma_{22} + (u_1v_2 + u_2v_1)\sigma_{12}]} , \tag{3.10}$$

where

$$c_j = u_j^2\sigma_{11} + v_j^2\sigma_{22} + 2u_jv_j\sigma_{12} , \qquad j = 1, 2 . \tag{3.11}$$

Clearly $\mathscr{B}(S_1 \cap S_2) \supset \mathscr{B}(T) \equiv \mathscr{B}$; hence, if $\min_{B \in \mathscr{B}(S_1 \cap S_2)} \operatorname{tr} B^{-1}\Pi = \operatorname{tr} B_0^{-1}\Pi$ and if $B_0 \in \mathscr{B}$, then $\min_{B \in \mathscr{B}} \operatorname{tr} B^{-1}\Pi = \operatorname{tr} B_0^{-1}\Pi$. But if the best ellipse $\{xB_0^{-1}x' < 1\}$ is in T, then $B_0 \in \mathscr{B}$. ||

LEMMA 3.4. *If* (x_1, x_2) *is a random vector with moment matrix* $\Sigma = (\sigma_{ij})$, *then*

$$P\{|u_1x_1 + v_1x_2| < 1, \quad |u_2x_1 + v_2x_2| < 1\} \geqq 1 - p\,, \tag{3.12}$$

where

$$p = \tfrac{1}{2}\{c_1 + c_2 + \sqrt{(c_1 - c_2)^2 + 4(\sigma_{11}\sigma_{22} - \sigma_{12}^2)(u_1v_2 - u_2v_1)^2}\}\,, \tag{3.13}$$

with c_j *given by* (3.11).

PROOF. The inequality for the square is

$$P\{|y_1| \geqq 1 \text{ or } |y_2| \geqq 1\} \leqq \tfrac{1}{2}[(\pi_{11} + \pi_{22}) + \sqrt{(\pi_{11} + \pi_{22})^2 - 4\pi_{12}^2}\,]\,, \tag{3.14}$$

where $y = (y_1, y_2)$ has moment matrix Π ([2], Equation B). As in the proof of Lemma 3.3, we make the change of variable $x = yM$, where M is given by (3.9), and obtain the relation $\Pi = M'^{-1}\,\Sigma\,M^{-1}$. Substitution for π_{ij} in (3.14) yields (3.13). ||

We note that, if $u_1 = 1, v_1 = 0, u_2 = 0, v_2 = 1$, the inequality (3.13) reduces to (3.14). If $u_1 = 1, v_1 = 0, u_2 = 0, v_2 = 0$, then the inequality reduces to Chebyshev's inequality. It is also of interest to note that, if $u_1 = -u_2$ and $v_1 = v_2$, (3.13) does not depend on σ_{12}.

We summarize the results of Lemmas 3.3 and 3.4 in

THEOREM 3.5. *If the best ellipse for some parallelogram, say* $S_1 \cap S_2$, *is best for* T, *i.e., if* (3.3) *and* (3.4) *hold with* b_{11}, b_{22}, b_{12} *as in* (3.6) *and* (3.7), *then* $P\{x \in T\} \geqq 1 - p$, *where* p *is given by* (3.13).

If for every parallelogram the corresponding best ellipse is not in T, then we look at the (unique) ellipse inscribed in each hexagon $S_i \cap S_j \cap S_k$. The motivation for this procedure stems from Lemma 3.1. By Lemma 3.2, we know that there exists one such ellipse in T.

For notational convenience, we consider the ellipse inscribed in $S_1 \cap S_2 \cap S_3$. This ellipse and the corresponding matrix B are determined by assuming equality in (3.5) for $j = 1, 2, 3$, namely,

$$(b_{11}, b_{22}, b_{12}) = (1, 1, 1)\begin{bmatrix} u_1^2 & u_2^2 & u_3^2 \\ v_1^2 & v_2^2 & v_3^2 \\ 2u_1v_1 & 2u_2v_2 & 2u_3v_3 \end{bmatrix}^{-1}. \tag{3.15}$$

The existence of the inverse is proved in Lemma 3.1.

For each such ellipse contained in T, i.e., satisfying

$$b_{11}u_j^2 + b_{22}v_j^2 + 2b_{12}u_jv_j \leqq 1\,, \qquad j = 4, \cdots, m\,,$$

$\operatorname{tr} B^{-1}\Sigma$ is computed, and the smallest value is the bound. In Section 5, we show that this procedure is sharp for $m = 3$, and conjecture that it is sharp in general.

4. Equal-width Case

If $w_1 = \cdots = w_m = w$, then we need look only at parallelograms $S_i \cap S_j$ whose associated angles are adjacent (Theorem 4.1). With no loss in generality we can take $w = 1$. If (3.3) and (3.4) hold with $w_j = 1$, and $(\alpha_1 + \alpha_2) \neq \pi$, then

$$b_{12} \sin(\alpha_1 - \alpha_j) \sin(\alpha_2 - \alpha_j)/\sin(\alpha_1 + \alpha_2) \geqq 0\,, \qquad j = 3, \cdots, m\,. \tag{4.1}$$

This inequality is obtained by solving (3.3) for b_{11} and b_{22} and inserting these values in (3.4).

If $\alpha_1 + \alpha_2 = \pi$, then (3.3) implies that $b_{12} = 0$, so that we get the circle.

We now show that the angles corresponding to the equalities (3.3) must be adjacent. To fix ideas, let $0 \leqq \theta_1 < \cdots < \theta_m < \pi$ be the ordered $\alpha_1, \cdots, \alpha_m$.

THEOREM 4.1. *If* (4.1) *holds, then* $(\alpha_1, \alpha_2) = (\theta_i, \theta_{i+1})$ *for some* $i = 1, \cdots, m-1$, *or* $(\alpha_1, \alpha_2) = (\theta_1, \theta_m)$.

PROOF. Suppose that (4.1) holds and that $b_{12}/\sin(\alpha_1 + \alpha_2) > 0$; then $\sin(\alpha_1 - \alpha_j)$ and $\sin(\alpha_2 - \alpha_j)$ are both negative or both positive, so that $\alpha_1 - \alpha_j$ and $\alpha_2 - \alpha_j$ have the same sign. Hence the relations $\alpha_1 < \alpha_j < \alpha_2$, $\alpha_2 < \alpha_j < \alpha_1$ are not possible; i.e., α_1 and α_2 are adjacent.

If $b_{12}/\sin(\alpha_1 + \alpha_2) < 0$, then, by a similar argument, one of the relations $\alpha_1 < \alpha_j < \alpha_2$, $\alpha_2 < \alpha_j < \alpha_1$ must hold for $j = 3, 4, \cdots, m$; hence α_1 and α_2 are extremes. ||

For any θ_j, θ_{j+1}, the best B for the associated parallelogram has b_{12} equal to

$$b_0(\theta_j, \theta_{j+1}) = \frac{(\sigma_{11} + \sigma_{22}) \sin(\theta_j + \theta_{j+1}) \sin(\theta_{j+1} - \theta_j)}{2\sin(\theta_{j+1} - \theta_j)\{\sigma_{22} \sin\theta_j \sin\theta_{j+1} + \sigma_{11} \cos\theta_j \cos\theta_{j+1} + \sigma_{12} \sin(\theta_j + \theta_{j+1})\}}$$
$$- \frac{\sin(\theta_j + \theta_{j+1})\sqrt{(\sigma_{11} + \sigma_{22})^2 - [(\sigma_{22} - \sigma_{11}) \cos(\theta_j + \theta_{j+1}) - 2\sigma_{12} \sin(\theta_j + \theta_{j+1})]^2}}{2\sin(\theta_{j+1} - \theta_j)\{\sigma_{22} \sin\theta_j \sin\theta_{j+1} + \sigma_{11} \cos\theta_j \cos\theta_{j+1} + \sigma_{12} \sin(\theta_j + \theta_{j+1})\}}\,. \tag{4.2}$$

This may be verified by computing $B = M'\begin{pmatrix}1 & \beta\\ \beta & 1\end{pmatrix}M$, where M and β are given by (3.9) and (3.10), respectively, or by minimizing $\operatorname{tr} B^{-1}\Sigma$ subject to the appropriate constraints.

THEOREM 4.2. *If* (4.1) *holds for* $b_0(\alpha_1, \alpha_2)$ *given by* (4.2) *with* $(\alpha_1, \alpha_2) = (\theta_j, \theta_{j+1})$ *for some* $j = 1, \cdots, m$, $(\theta_{m+1} = \theta_1 + \pi)$, *then*

$$P\{(x_1, x_2) \in \bar{T}\} \leqq \tfrac{1}{2}[(\sigma_{11} + \sigma_{22}) - z\cos(\theta_{j+1} - \theta_j) + \sin(\theta_{j+1} - \theta_j)\sqrt{(\sigma_{11} + \sigma_{22})^2 - z^2}\,]\,, \tag{4.3}$$

where $z = (\sigma_{22} - \sigma_{11}) \cos(\theta_j + \theta_{j+1}) - 2\sigma_{12} \sin(\theta_j + \theta_{j+1})$; *otherwise*

$$P\{(x_1, x_2) \in \bar{T}\} \leqq \sigma_{11} + \sigma_{22}\,. \tag{4.4}$$

PROOF. Inequality (4.3) follows directly from (3.12), noting that $u_i = \cos^2 \alpha_i$, $v_i = \sin^2 \alpha_i$, $i = 1, 2$. Since the unit circle (obtained with $B = I$) always lies in T, (4.4) always holds. ||

5. Sharpness

In the case where the best ellipse for some parallelogram $S_i \cap S_j$ is best for T, and p given by (3.13) is less than one, we exhibit a distribution which attains equality in (3.12) and thereby demonstrate sharpness. This distribution assigns probability to the origin and to points where the boundary of the ellipse is tangent to the boundary of $S_i \cap S_j$. The probabilities are chosen to yield the desired moment matrix. We can make use of results for the square, [3, 4], and obtain the distribution by the transformation $x = yM$, where M is given by (3.9). Equality is attained in (3.14) by the distribution

$$P\{y = (1, \beta)\} = P\{y = -(1, \beta)\} = p_1/2 ,$$
$$P\{y = (\beta, 1)\} = P\{y = -(\beta, 1)\} = p_2/2 ,$$
$$P\{y = 0\} = 1 - p_1 - p_2 ,$$

where β is given by (3.8), and

$$p_1 = (\pi_{11} - \beta^2\pi_{22})/(1 - \beta^4), \qquad p_2 = (\pi_{22} - \beta^2\pi_{11})/(1 - \beta^4) .$$

If we choose $\Pi = M'^{-1} \Sigma M^{-1}$, then we obtain $E x' x = M' \Pi M = \Sigma$.

If none of the best ellipses for the parallelograms lies within T, then we have to look at the ellipses inscribed in the hexagons. However, in this case the problem of sharpness is more involved. We prove sharpness only for $m = 3$.

THEOREM 5.1. *Let $m = 3$. If (4.1) does not hold for any (α_1, α_2), and if $\sigma_{11} + \sigma_{22} \leqq 1$, then there exists a distribution for which equality holds in (4.4).*

PROOF. First suppose the widths w_j are all equal. Since the inequality (4.4) emanated from a circle, we suspect that the equality is achieved by placing probabilities at the origin and at points of tangency of the circle and the boundary of T, namely at

$$(0, 0), \ \pm(\cos\theta_1, \sin\theta_1), \ \pm(\cos\theta_2, \sin\theta_2), \ \pm(\cos\theta_3, \sin\theta_3).$$

Equality is achieved if there exist probabilities, $1 - p_1 - p_2 - p_3$, $p_1/2$, $p_2/2$, and $p_3/2$, associated with these points which yield the given moment matrix Σ. If such probabilities exist, then p_1, p_2, and p_3 must be the solution of

$$pC \equiv (p_1\, p_2\, p_3)\begin{bmatrix} \cos^2\theta_1 & \sin^2\theta_1 & \sin 2\theta_1 \\ \cos^2\theta_2 & \sin^2\theta_2 & \sin 2\theta_2 \\ \cos^2\theta_3 & \sin^2\theta_3 & \sin 2\theta_3 \end{bmatrix} = (\sigma_{11}\ \sigma_{22}\ 2\sigma_{12}) \equiv \sigma . \tag{5.1}$$

We need to verify that, under the conditions of the theorem, p_1, p_2, p_3, and $1 - p_1 - p_2 - p_3$ are non-negative. From (5.1) we have

$$p_1 + p_2 + p_3 = \sigma_{11} + \sigma_{22} \equiv \delta , \tag{5.2}$$

(so that $\sigma_{11} + \sigma_{22} \leqq 1$ implies $1 - p_1 - p_2 - p_3 \geqq 0$) and

$$p_3 = -\frac{\sin(\theta_2 - \theta_1)}{\Delta}[\sigma_{11}\sin\theta_1\sin\theta_2 + \sigma_{22}\cos\theta_1\cos\theta_2 - \sigma_{12}\sin(\theta_1 + \theta_2)] , \tag{5.3}$$

where Δ is the determinant of C.

By hypothesis, $\delta \leqq 1$ and

$$(5.4)\quad 0 > \frac{2b_0(\theta_1, \theta_2)}{\sin(\theta_1+\theta_2)} = \frac{\delta\sin(\theta_2-\theta_1) - \sqrt{\delta^2 - \left[\dfrac{2p_3\Delta}{\sin(\theta_2-\theta_1)} + \delta\cos(\theta_2-\theta_1)\right]^2}}{p_3\Delta + \delta\sin(\theta_2-\theta_1)\cos(\theta_2-\theta_1)} .$$

We wish to show that (5.4) implies that $p_3 > 0$. Suppose the denominator is positive; then, after squaring and simplifying, we obtain from (5.4)

$$\frac{p_3\Delta}{\sin^2(\theta_2-\theta_1)}[p_3\Delta + \delta\sin(\theta_2-\theta_1)\cos(\theta_2-\theta_1)] < 0 ,$$

and hence $p_3\Delta < 0$. If we suppose that the denominator is negative, a similar argument yields the same result. But $2\theta_2 - \theta_1 - \theta_3 < \theta_3 - \theta_1 < \pi$, $\theta_1 < \theta_3 < \pi$, so that $\cos(\theta_3 - \theta_1) < \cos(2\theta_2 - \theta_1 - \theta_3)$ and $\sin(\theta_3 - \theta_1) > 0$. Hence, by (3.2), $\Delta < 0$ and $p_3 > 0$.

By a similar argument with the other angles, $p_1 > 0$, $p_2 > 0$.

We now show that, in the unequal-width case with $m = 3$, the procedure leads to a sharp inequality. For any ellipse $\{yB^{-1}y' < 1\}$, there exists a non-singular transformation $x = yG^{-1}$ with $yB^{-1}y' = xx'$. This maps lines tangent to the ellipse into lines tangent to the circle. We have exhibited a distribution for x such that $Ex'x = \Sigma$, which attains the bound. By the transformation $y = xG$, we obtain a distribution for y which attains the bound, such that $Ey'y = G\Sigma G' = \Sigma^*$. ||

The inequalities have been obtained with no assumptions concerning the expectations. With the additional assumption $Ex_1 = Ex_2 = 0$, the above example shows that the bound is still sharp. However, if non-zero means are specified, the bounds may not be sharp.

6. Computing Procedure and a Numerical Example

Let us first summarize the procedure given in Section 3 (and 4) for bounding $P\{\max_{1\leq i\leq m} |u_ix_1 + v_ix_2| > 1\} = P\{x \in \bar{T}\}$, where x has covariance matrix Σ. For each parallelogram $S_i \cap S_j = \{|u_ix_1 + v_ix_2| < 1\} \cap \{|u_jx_1 + v_jx_2| < 1\}$, compute the best B; if one of the corresponding ellipses $\{xB^{-1}x' < 1\}$ lies in T, then $\operatorname{tr} B^{-1}\Sigma$ is the desired bound. If none of the ellipses lie in T, then for each hexagon $S_i \cap S_j \cap S_k$, compute the matrix B corresponding to the inscribed ellipse and determine whether the ellipse lies in T. Amongst those that lie in T, compute $\operatorname{tr} B^{-1}\Sigma$; the minimum of these numbers is the desired bound. Thus the computing procedure is as follows:

1. For each pair $i \neq j$, compute $\Pi = (\pi_{rs}) = M'^{-1}\Sigma M^{-1}$, using (3.9) with the correspondence (i, j) for $(1, 2)$.
2. Determine β from Π and (3.8).
3. Compute $B = M'\begin{pmatrix} 1 & \beta \\ \beta & 1 \end{pmatrix}M$. This is the matrix of the best ellipse for the parallelogram $S_i \cap S_j$.
4. Determine whether

$$(6.1)\qquad b_{11}u_t^2 + b_{22}v_t^2 + 2b_{12}r_ts_t \leqq 1 \qquad \text{for all } t \neq i, j .$$

5. If (6.1) holds for some choice of i, j, compute the bound given by

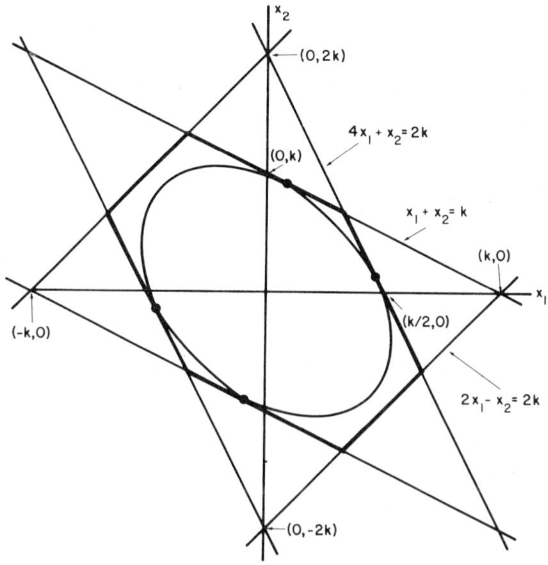

FIG. 1.

(3.14), where π_{rs} was obtained in step (1). This number is the desired bound.

5′. If (6.1) is violated for all choices of i, j, then consider hexagons. For all distinct i, j, k, compute (3.15) with the correspondence (i, j, k) for $(1, 2, 3)$.

6′. Determine whether

$$b_{11}u_t^2 + b_{22}v_t^2 + 2b_{12}u_t v_t \leqq 1 \qquad \text{for all } t \neq i, j, k . \tag{6.2}$$

7′. For all choices of i, j, k for which (6.2) holds, compute $\operatorname{tr} B^{-1}\Sigma$. The minimum of these numbers is the desired bound.

Using the above procedure, we have obtained the following example. If $E(x_1^2) = 1$, $E(x_2^2) = 2$, and $E(x_1 x_2) = \frac{1}{2}$, then

$$P\{|4x_1 + x_2| < 2c\ , \quad |x_1 + x_2| < c\ , \quad |2x_1 - x_2| < 2c\} \geqq 1 - 6.87/c^2 .$$

The bound is attained for

$$\begin{aligned} P\{(.4605c, .1580c)\} &= P\{-(.4605c, .1580c)\} = 4.65048/c^2 , \\ P\{(.0790c, .9210c)\} &= P\{-(.0790c, .9210c)\} = 2.22120/c^2 , \\ P\{(0, 0)\} &= 1 - 6.87/c^2 . \end{aligned}$$

The graph of the best ellipse is given in Fig. 1. The points of tangency are points of positive probability in the example of sharpness.

7. Simultaneous Confidence Intervals

Let the random vector (x_1, x_2) have a mean (μ_1, μ_2) and covariance matrix Σ, and suppose we wish to find a simultaneous (distribution-free) confidence interval for $u_j\mu_1 + v_j\mu_2$, $j = 1, \cdots, m$, with confidence coefficient $1 - \alpha$. Using the procedure outlined, we obtain an inequality

$$P\{|u_j(x_1 - \mu_1) + v_j(x_2 - \mu_2)| < k_j\ ,\quad j = 1, \cdots, m\} \geqq 1 - p(k_1, \cdots, k_m; \Sigma)\ .$$

and the desired confidence coefficient by setting

$$p(k_1, \cdots, k_m; \Sigma) = \alpha\ . \tag{7.1}$$

There will always be at least one solution of (7.1) for $k_1, \cdots, k_m$, namely when $k_1 = \cdots = k_m$. As an example, we obtain from Section 6

$$P\{|4x_1+x_2-(4\mu_1+\mu_2)| < 2k,\ |x_1 + x_2-(\mu_1 + \mu_2)| < k,\ |2x_1 - x_2 + (2\mu_1 - \mu_2)| < 2k\} \geqq 1 - 6.87/k^2\ ,$$

and may choose $k^2 = 6.87/\alpha$.

In order to find a "short interval," we choose $k_1, \cdots, k_m$ so as to minimize the volume of the convex polygon subject to (7.1).

This problem does not seem to be readily tractable. As a crude approximation, one might use the variances as weights, $k_j = k/\sigma_j$, where $\sigma_j^2 = \text{var}(u_jx_1 + v_jx_2)$. Since this ignores the covariances, it cannot be optimum. A more sensitive procedure is to determine the best bound for the parallelogram $S_1 \cap S_2$ with general k_1, k_2. Now minimize the area of the parallelogram subject to the constraint that the bound be equal to $1 - \alpha$. This will uniquely determine k_1 and k_2. The orientation of the third strip is prescribed by the problem, but now the width is chosen so as to make its boundary tangent to the ellipse. Thus a hexagon is determined with prescribed bound $1 - \alpha$. For each choice of indices 1 and 2 we get a hexagon, and the one with the minimum volume is chosen. It is still fairly complicated to determine the general solution, but results may be obtained numerically.

8. Acknowledgment

We are indebted to Miss Barbara Bonesteele for her aid in the computations.

References

[1] Berge, P. O. A note on a form of Tchebycheff's theorem for two variables. *Biometrika*, 1937, **29**, 405-6.

[2] Lal, D. N. A note on a form of Tchebycheff's inequality for two or more variables. *Sankhyā*, 1955, **15**, 317-20.

[3] Marshall, A. W. On the growth of stochastic processes. Technical Report No. 32, Department of Mathematics, Seattle: University of Washington, 1958.

[4] Olkin, I., and Pratt, J. W. A multivariate Tchebycheff inequality. *Ann. Math. Stat.*, 1958, **29**, 226-34.

[6] Whittle, P. A multivariate generalization of Tchebichev's inequality. *Quart. J. Math. (Oxford)*, Ser. 2, 1958, **9**, 232-40.

28 NOTES ON THE NUMERICAL CONVERGENCE OF ITERATIVE PROCESSES

SIGEITI MORIGUTI
University of Tokyo

1. Introduction

Iterative processes are extensively used in numerical analysis. For example, Newton's iteration for the square root is very famous. One sometimes finds that a theoretically convergent process fails to converge (i.e., oscillates) because of the rounding error. This phenomenon calls for special attention in programming for an automatic computer, because when this happens the control cannot get out of the loop by a "zero jump" order. In fact, we can point to some published routines for the square root, $\sqrt{a}$, which would fail for $a = 0.24999\ 99999$, provided that the computer is designed for ten-decimal-place fixed-point arithmetic and that the rounding is done by a simple truncation (see Section 2).

A general theory about this phenomenon was developed by Urabe [1]. On assumptions similar to those of Collatz [2], he concludes that "the numerical iteration process necessarily attains *the state of numerical convergence* after finite times of repetition of the process." His definition of the state of numerical convergence differs from Sibagaki's [3], in that it includes not only the stationary state but also the state of cyclic recurrence.

Let us define a *digital number* to be a multiple of a certain quantity ε (ε being, e.g., $= 10^{-10}$ for ten-decimal-place arithmetic) which may be called the *quantum* of the arithmetic. Only digital numbers can properly be stored and handled by a digital computer. (See, e.g., [4], Chap. 1.)

A numerical iterative process determines, for any digital number $\bar{x}_0$ as the starting value, a sequence of digital numbers. If the sequence does not include the starting value $\bar{x}_0$ again, then $\bar{x}_0$ is called *transient*. If the sequence does include $\bar{x}_0$ again, then it is called *recurrent*. In particular, if all values in the sequence are equal to $\bar{x}_0$, then $\bar{x}_0$ is *stationary*. A recurrent digital number which is not stationary belongs to a recurrent cycle of length $\geqq 2$. Some of this terminology comes from the theory of Markov chains (see, e.g., Feller [5], Chap. 15). Indeed, a numerical iterative process is a particular case of Markov chains, a very simple case where the transition probability from each point concentrates on a particular point.

In the present paper, we shall discuss some particular cases thoroughly. The author believes that this kind of result will be helpful when similar situations are encountered and may pave the way to a more powerful method of dealing with such situations.

2. Newton's Iteration for the kth Root

To extract the kth root of a given positive number a, we set

$$f(x) = x^k - a \tag{2.1}$$

and derive the Newtonian iteration formula

$$x_{n+1} = \varphi(x_n), \qquad n = 0, 1, 2, \cdots, \tag{2.2}$$

where

$$\varphi(x) = x - \frac{f(x)}{f'(x)} = x - \frac{x^k - a}{kx^{k-1}} = \frac{(k-1)x^k + a}{kx^{k-1}}. \tag{2.3}$$

Starting with a suitably chosen initial guess x_0, we get $x_1, x_2, \cdots$, successively from (2.2). This sequence would converge to the true value of $a^{1/k}$, if rounding were not applied in the process. Actually, what we get is the sequence of *digital* numbers $\bar{x}_0(= x_0), \bar{x}_1, \bar{x}_2, \cdots$ such that

$$\bar{x}_{n+1} \doteq \varphi(\bar{x}_n), \qquad n = 0, 1, 2, \cdots, \tag{2.4}$$

where $\doteq$ denotes that the two sides coincide after rounding.

Suppose we round to multiples of ε by a simple truncation. Then (2.4) can be interpreted as

$$\bar{x}_{n+1} \leqq \varphi(\bar{x}_n) < \bar{x}_{n+1} + \varepsilon, \qquad n = 0, 1, 2, \cdots. \tag{2.5}$$

Now, let

$$a^{1/k} \equiv b = \bar{b} + \beta, \tag{2.6}$$

where $\bar{b}$ is the largest digital number less than or equal to b, and hence

$$0 \leqq \beta < \varepsilon. \tag{2.7}$$

Let us assume that $k > 1$. We shall prove

THEOREM 1. *If*

$$(\bar{b}/k)[(1 + \beta/\bar{b})^k - 1] < \varepsilon, \tag{2.8}$$

then $\bar{b}$ is stationary and is the only recurrent point; if (2.8) *does not hold, then there is a recurrent cycle including $\bar{b}$ with length* $\geqq 2$.

First we have

LEMMA 2.1.

(i) *For any starting value $\bar{x}_0$, no succeeding approximate value is less than $\bar{b}$*, i.e.,

$$\bar{x}_n \geqq \bar{b}, \qquad n = 1, 2, \cdots. \tag{2.9}$$

(ii) *If $\bar{x}_n > \bar{b}$, then $\bar{x}_{n+1} < \bar{x}_n$.*

PROOF. (i) Differentiating (2.3), we get $\varphi'(x) = [(k-1)/k](1 - a/x^k)$, which shows that $\varphi'(x) \lesseqgtr 0$ according as $x \lesseqgtr a^{1/k} = b$. Hence we know that $\varphi(x)$ has its minimum at $x = b$, which is $\varphi(b) = b$. Consequently, for any $x > 0$, $\varphi(x) \geqq b$. Therefore, whatever $\bar{x}_n$ may be, we have

$$\bar{x}_{n+1} > \varphi(\bar{x}_n) - \varepsilon \geqq \bar{b} - \varepsilon \geqq b - \varepsilon\,, \qquad n = 0, 1, 2, \cdots . \tag{2.10}$$

Since both $\bar{x}_{n+1}$ and $\bar{b}$ are digital numbers, (2.10) implies (2.9).

(ii) The function

$$\varphi(x) - x = -x/k + a/(kx^{k-1}) \tag{2.11}$$

is strictly decreasing, for its derivative is always negative. Therefore, if $x > b$, then $\varphi(x) - x < \varphi(b) - b = 0$. On the other hand, if $\bar{x}_n > \bar{b}$, then $\bar{x}_n \geqq \bar{b} + \varepsilon > b$. Hence, if $\bar{x}_n > \bar{b}$, then we have

$$\bar{x}_{n+1} \leqq \varphi(\bar{x}_n) < \bar{x}_n\,, \tag{2.12}$$

which completes the proof. ||

PROOF OF THEOREM 1. Take any starting value $\bar{x}_0$. Then $\bar{x}_1 \geqq \bar{b}$, by Lemma 2.1 (i). If $\bar{x}_1 > \bar{b}$, then by Lemma 2.1 (ii) we have a strictly decreasing sequence $\bar{x}_1 > \bar{x}_2 > \cdots$, which will eventually terminate with some $\bar{x}_m = \bar{b}$. (This will occur after a finite number of steps, because there are only a finite number of digital values between $\bar{x}_1$ and $\bar{b}$.)

The value $\bar{b}$ is stationary if and only if

$$\varphi(\bar{b}) < \bar{b} + \varepsilon\,, \tag{2.13}$$

because the condition $\bar{b} \leqq \varphi(\bar{b})$ always holds. Inequality (2.13) is equivalent to (2.8). If this is not satisfied, then we can determine a positive integer, p such that

$$\bar{b} + p\varepsilon \leqq \varphi(\bar{b}) < \bar{b} + (p+1)\varepsilon\,. \tag{2.14}$$

In this case, the value $\bar{b} + p\varepsilon$ is recurrent. For, if we set $\bar{x}_m = \bar{b}$, then we get $\bar{x}_{m+1} = \bar{b} + p\varepsilon > \bar{b}$, whence we have a finite sequence

$$\bar{x}_{m+1} > \bar{x}_{m+2} > \cdots > \bar{x}_{m+N}\ (\text{say}) = \bar{b} = \bar{x}_m\,, \tag{2.15}$$

implying that every value contained in this sequence is recurrent. ||

Remark 1. For $k = 2$, Inequality (2.8) reduces to

$$\beta + \beta^2/(2\bar{b}) < \varepsilon\,. \tag{2.16}$$

It is important to note that the inequality fails to hold if β becomes sufficiently close to ε, however large $\bar{b}$ may be. For instance, when $\varepsilon = 10^{-5}$, say, (2.16) will not hold for $a = 0.24999$ and the state of convergence will be a cycle consisting of 0.49998 and 0.49999. In general, (2.16) will fail to hold for $a = 0.25 - \varepsilon$. Thus, oscillation does occur for some a, however small ε may be. This statement can be generalized into general k.

Remark 2. It usually happens that, when (2.8) is not satisfied, the integer

p defined by (2.14) is equal to 1, and consequently the recurrent cycle is of length 2 [i.e., in (2.15), $N = 2$]. This statement is supported by

LEMMA 2.2. *If* $\bar{b} = B\varepsilon$, *then* $\varphi(\bar{b}) < \bar{b} + 2\varepsilon$ *as long as* $k \leqq K$, *where* K *is the root of the equation*

$$K \log(1 + 1/B) = \log(1 + 2K/B) . \tag{2.17}$$

PROOF. $\varphi(b) \geqq \bar{b} + 2\varepsilon$ only if $(1 + \beta/\bar{b})^k \geqq 1 + 2k\varepsilon/\bar{b}$, whence only if $k \log(1 + \varepsilon/\bar{b}) > \log(1 + 2k\varepsilon/\bar{b})$ (notice that $\beta < \varepsilon$). This is equivalent to $k > K$, where K is as defined above. ||

TABLE 1

B	K	$(B+1)^{-K/(K-1)}$
1	2.660	0.329
2	3.939	0.229
3	5.205	0.180
4	6.467	0.149
5	7.726	0.128
6	8.985	0.112
7	10.243	0.100
8	11.501	0.090
9	12.758	0.082
10	14.015	0.076
20	26.583	0.042
50	64.279	0.018
100	127.101	0.010

Table 1 shows the values of K corresponding to a few values of B. For large B, it can be shown that

$$K \fallingdotseq 1.256\, B + 1.459 , \tag{2.18}$$

where the coefficient 1.256 is the root z_0 of the equation $\log_e(1 + 2z) = z$, and 1.459 is equal to $z_0(2z_0 + 1)/2(2z_0 - 1)$.

Now, if $k > K$, then, supposing that $B\varepsilon + \varepsilon < 1$,

$$a = (B\varepsilon + \beta)^k < (B\varepsilon + \varepsilon)^k < (B + 1)^K \varepsilon^K . \tag{2.19}$$

But the upper bound $(B + 1)^K\varepsilon^K$ is greater than ε only if $\varepsilon > (B + 1)^{-K/(K-1)}$. The right-hand member is also shown in Table 1. Usually, in practice, ε will be much smaller than those values, so that any a for which $\varphi(\bar{b}) \geqq \bar{b} + 2\varepsilon$ need not be considered.

3. Newton's Iteration for the Inverse of the kth Root

Let k be a positive integer, and let a be a positive number for which we wish to compute $a^{-1/k}$. For this purpose, we set

$$f(x) = C/x^k - a , \tag{3.1}$$

where C is a suitably chosen scale factor to hold the number to be dealt with well within the range of the computer. Equation (3.1) will lead to an iteration scheme of the same form as (2.2) with

$$\varphi(x) = x - \frac{C/x^k - a}{-kC/x^{k+1}} = \frac{k+1}{k}\, x - \frac{ax^{k+1}}{kC} \,. \tag{3.2}$$

The corresponding numerical process will generate, starting with some $\bar{x}_0$, a sequence of digital numbers $\bar{x}_1, \bar{x}_2, \cdots$, satisfying (2.5).

Now, let

$$(C/a)^{1/k} \equiv b = \bar{b} + \beta \,, \tag{3.3}$$

where $\bar{b}$ is the largest digital number less than or equal to b, and hence β satisfies (2.7).

THEOREM 2. *For the iterative process* (2.2) *with* (3.2), *any recurrent digital value is stationary and belongs to the closed interval* $[0, b]$. *In particular,* $\bar{b}$ *is always stationary, and* $\bar{b} - \varepsilon$ *is also stationary if and only if*

$$\bar{b} + \beta < (\bar{b} - \varepsilon)[1 - k\varepsilon/(\bar{b} - \varepsilon)]^{-1/k} \,. \tag{3.4}$$

PROOF. First we can prove that, for any $\bar{x}_0$, no succeeding value $\bar{x}_n$ is greater that $\bar{b}$, i.e. $\bar{x}_n \leqq \bar{b} \quad n = 1, 2, \cdots$. [The proof is similar to the proof of Lemma 2.1 (i).] This implies that all values greater than $\bar{b}$ are transient.

Now, $\varphi(x) - x$ takes a maximum value $(k+1)^{-(k+1)/k}b$ at $x = \{C/(k+1)a\}^{1/k}$. If the maximum is less than ε, then for any x in $[0, b]$, we have $0 \leqq \varphi(x) - x < \varepsilon$, whence all digital numbers in that interval are stationary. If, on the contrary, the maximum is $\geqq \varepsilon$, then there exist two values c_1 and c_1' such that

$$0 < c_1' \leqq [C/(k+1)a]^{1/k} \leqq c_1 < b \,, \tag{3.5}$$

and

$$\varphi(c_1') - c_1' = \varphi(c_1) - c_1 = \varepsilon \,. \tag{3.6}$$

In this case, for any x in the closed interval $[c_1', c_1]$, we have $\varepsilon \leqq \varphi(x) - x$, and for any x in the intervals $[0, c_1')$ and $(c_1, b]$, we have $0 \leqq \varphi(x) - x < \varepsilon$, whence all digital numbers in $[c_1', c_1]$ are transient, whereas all digital numbers in $[0, c_1')$ or $(c_1, b]$ are stationary. $\bar{b}$ is always stationary because $0 \leqq \varphi(\bar{b}) - \bar{b} \leqq b - \bar{b} = \beta < \varepsilon$; $\bar{b} - \varepsilon$ is also stationary if and only if $c_1 < \bar{b} - \varepsilon$; i.e., if and only if (3.4) holds. ||

Remark. Inequality (3.4) will hold when β is sufficiently close to ε, however large $\bar{b}$ may be. We must therefore always take into account the possibility of having two adjacent stationary points; but usually we do not have any wider range of indeterminacy than this, although we will not go into details here on this subject.

The stationary points which lie near 0 are to be interpreted as approximations to 0, another root of the equation $x = \varphi(x)$, thus they are of little interest here.

4. Rounding to the Nearest Digital Number

In the preceding two sections, we assumed that the rounding rule was to choose the greatest digital number less than or equal to the number to be rounded. If, instead, the rounding rule is to choose the nearest digital number (and to choose the larger one if there are two such numbers), then in place of (2.5) we have

$$\bar{x}_{n+1} - \varepsilon/2 \leqq \varphi(\bar{x}_n) < \bar{x}_{n+1} + \varepsilon/2\,, \qquad n = 0, 1, 2, \cdots . \tag{4.1}$$

Let us consider the case of Section 2 with the new rounding rule, and use the same notation as in (2.6). Lemma 2.1 (i) holds again, whence all values less than $\bar{b}$ are transient.

Then, $\bar{b}$ is stationary if and only if $\varphi(\bar{b}) < \bar{b} + \varepsilon/2$, i.e., if and only if

$$\bar{b} + \beta < \bar{b}(1 + k\varepsilon/2\bar{b})^{1/k}\,. \tag{4.2}$$

Also, $\bar{b} + \varepsilon$ is stationary if and only if $\varphi(\bar{b} + \varepsilon) \geqq \bar{b} + \varepsilon/2$, i.e., if and only if

$$\bar{b} + \beta \geqq (\bar{b} + \varepsilon)\{1 - k\varepsilon/2(\bar{b} + \varepsilon)\}^{1/k}\,. \tag{4.3}$$

(The right-hand member should be replaced by 0 if the expression in the braces becomes negative.) For the sake of simplicity, we shall consider the "usual" case where $\varphi(\bar{b}) < \bar{b} + 3\varepsilon/2$, so that any jump from $\bar{b}$ is at most to $\bar{b} + \varepsilon$.

The relation between (4.2) and (4.3) will become clear from

LEMMA 4.1. *Let*

$$\begin{aligned} h(u) &= \log(1+u) + \frac{1}{k}\log\left(1 - \frac{k}{2}\frac{u}{1+u}\right) - \frac{1}{k}\log\left(1 + \frac{k}{2}u\right) \\ &= \frac{k-1}{k}\log(1+u) + \frac{1}{k}\log\left\{1 + \left(1 - \frac{k}{2}\right)u\right\} - \frac{1}{k}\log\left(1 + \frac{k}{2}u\right). \end{aligned} \tag{4.4}$$

If $1 < k < 2$, *then* $h(u) > 0$ *for all* $u > 0$;
if $k = 2$, *then* $h(u) = 0$ *for all* u;
if $k > 2$, *then* $h(u) < 0$ *for all* u *in the interval* $(0, 2/(k-2))$.

PROOF. Differentiating (4.4), we get

$$\begin{aligned} h'(u) &= \frac{k-1}{k}\frac{1}{1+u} + \frac{1}{k}\frac{2-k}{2+(2-k)u} - \frac{1}{k}\frac{k}{2+ku} \\ &= \frac{(k-1)(2-k)u^2}{(1+u)\{2+(2-k)u\}(2+ku)}\,, \\ &\quad \begin{cases} > 0, & \text{if } 1 < k < 2 \text{ and } u > 0\,, \\ = 0, & \text{if } k = 2\,, \\ < 0, & \text{if } k > 2 \text{ and } 0 < u < 2/(k-2)\,. \end{cases} \end{aligned} \tag{4.5}$$

This result, together with the fact that $h(0) = 0$, proves the lemma. ||

If $1 < k < 2$, then Lemma 4.1 shows that the right-hand member of (4.3) is greater than that of (4.2), whence there can occur three cases:

(i) Inequality (4.2) is satisfied, implying (4.3) is not satisfied;
(ii) Inequality (4.2) is not satisfied and (4.3) is not satisfied either;
(iii) Inequality (4.3) is satisfied, implying that (4.2) is not.

In case (i), $\bar{b}$ is the unique stationary point; in case (ii), $\bar{b}$ and $\bar{b}+\varepsilon$ constitute a cycle of length two; in case (iii), $\bar{b}+\varepsilon$ is the unique stationary point.

If $k > 2$, then Lemma 4.1 shows that the right-hand member of (4.3) is less than that of (4.2), whence there can occur three cases:

(i) Inequality (4.3) is not satisfied, implying that (4.2) is satisfied;
(ii) Both (4.3) and (4.2) are satisfied;
(iii) Inequality (4.2) is not satisfied, implying that (4.3) is satisfied.

In case (i), $\bar{b}$ is the unique stationary point; in case (ii), both $\bar{b}$ and $\bar{b}+\varepsilon$ are stationary; in case (iii), $\bar{b}+\varepsilon$ is the unique stationary point.

If $k = 2$, then the right-hand member of (4.3) coincides with that of (4.2), whence the above-mentioned case (ii) disappears, so that there can occur only two cases, either $\bar{b}$ or $\bar{b}+\varepsilon$ being the unique stationary point in each case.

Remark. The case where $1 < k < 2$ may be of little practical interest. However, this case presents an interesting example of the possibility of final oscillation, whereas the other cases do not allow such an oscillation.

5. Linear Equation

Typical behavior of numerical processes will be illustrated by a very simple example, namely, the iterative solution of a linear algebraic equation for one unknown. Let the process be

$$x_{n+1} = a - bx_n , \qquad n = 0, 1, 2, \cdots , \tag{5.1}$$

where a and b are given constants ($|b| < 1$). The corresponding numerical process is

$$\bar{x}_{n+1} \doteq a - b\bar{x}_n , \qquad n = 0, 1, 2, \cdots , \tag{5.2}$$

where we assume that $\doteq$ denotes equality after a simple truncation.

Let us change the variable from x to u by the relation

$$x = (X + u)\varepsilon , \tag{5.3}$$

where X is a suitably chosen integer and ε is the quantum of the arithmetic. Then a digital value of x will correspond to an integral value of u. Now (5.2) becomes

$$\bar{u}_{n+1} \doteq a/\varepsilon - (1+b)X - b\bar{u}_n , \qquad n = 0, 1, 2, \cdots , \tag{5.4}$$

the bar upon u indicating that it is an integer. We can choose X so that

$$A = a/\varepsilon - (1+b)X \tag{5.5}$$

will satisfy the condition

$$0 \leqq A < 1 + b . \tag{5.6}$$

Then (5.4) can be interpreted as

$$\bar{u}_{n+1} \leqq A - b\bar{u}_n < \bar{u}_{n+1} + 1 , \qquad n = 0, 1, 2, \cdots . \tag{5.7}$$

The case where $b = 0$ is of no interest. For other cases, we have

LEMMA 5.1. (i) *Let* $b > 0$. *If* $\bar{u} \geqq 1$, *then*

$$-\bar{u}_n \leqq \bar{u}_{n+1} \leqq 0\,; \tag{5.8}$$

if $\bar{u}_n \leqq 0$, *then*

$$0 \leqq \bar{u}_{n+1} \leqq |\,\bar{u}_n\,| + 1\,. \tag{5.9}$$

(ii) *Let* $b < 0$. *If* $\bar{u}_n \geqq 1$, *then*

$$0 \leqq \bar{u}_{n+1} < \bar{u}_n\,; \tag{5.10}$$

if $\bar{u}_n \leqq 0$, *then*

$$\bar{u}_n \leqq \bar{u}_{n+1} \leqq 0\,. \tag{5.11}$$

PROOF. (i) From (5.7), (5.6), and $0 < b < 1$, we get, for $\bar{u}_n \geqq 1$,

$$\bar{u}_{n+1} > A - b\bar{u}_n - 1 \geqq 0 - b\bar{u}_n - 1 > -\bar{u}_n - 1\,, \tag{5.12}$$

and

$$\bar{u}_{n+1} \leqq A - b\bar{u}_n < 1 + b(1 - \bar{u}_n) \leqq 1\,. \tag{5.13}$$

But, since both $\bar{u}_{n+1}$ and $-\bar{u}_n - 1$ are integers, (5.12) and (5.13) imply (5.8).

Similarly, for $\bar{u}_n \leqq 0$, we get

$$\bar{u}_{n+1} \leqq A - b\bar{u}_n < 1 + b(1 - \bar{u}_n) < 1 + (1 - \bar{u}_n)\,, \tag{5.14}$$

and

$$\bar{u}_{n+1} > A - b\bar{u}_n - 1 \geqq -1\,, \tag{5.15}$$

which imply (5.9).

(ii) From (5.7), (5.6), and $-1 < b < 0$, we get, for $\bar{u}_n \geqq 1$,

$$\begin{aligned} \bar{u}_{n+1} &\leqq A - b\bar{u}_n < (1 + b) + \bar{u}_n - (1 + b)\bar{u}_n \\ &= \bar{u}_n - (1 + b)(\bar{u}_n - 1) \leqq \bar{u}_n\,, \end{aligned} \tag{5.16}$$

and

$$\bar{u}_{n+1} > A - b\bar{u}_n - 1 \geqq -1\,, \tag{5.17}$$

implying (5.10).

Similarly we get, for $\bar{u}_n \leqq 0$,

$$\bar{u}_{n+1} \leqq A - b\bar{u}_n < (1 + b) - b\bar{u}_n = 1 + b(1 - \bar{u}_n) < 1\,, \tag{5.18}$$

and

$$\bar{u}_{n+1} > A - b\bar{u}_n - 1 \geqq \bar{u}_n - (1 + b)\bar{u}_n - 1 \geqq \bar{u}_n - 1\,, \tag{5.19}$$

implying (5.11). ||

Let us first consider *the case where* $b > 0$. From Lemma 5.1(i), we can conclude that there are three limiting situations which may occur in this case:

(i) 0, as a stationary point;

(ii) k and $-k$ constituting a cycle, (k being a positive integer);

(iii) k and $-(k - 1)$ constituting a cycle, (k being a positive integer).

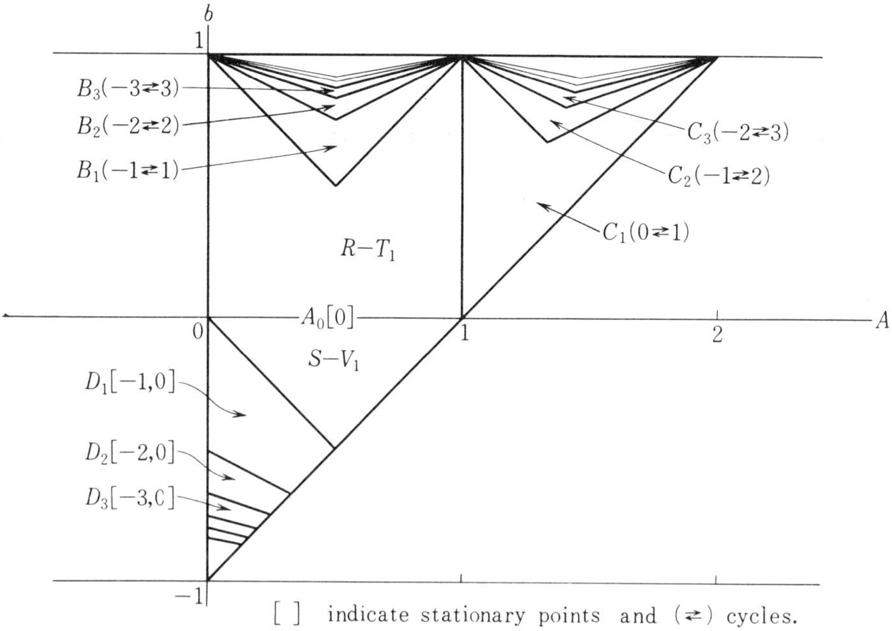

FIG. 1.

Type (i) will occur only if

$$0 \leqq A < 1 . \tag{5.20}$$

Type (ii) will occur only if

$$\begin{aligned} -k \leqq A - bk < -k + 1 , \\ k \leqq A + bk < k + 1 , \end{aligned} \tag{5.21}$$

i.e., only if

$$(1 - b)k \leqq A < 1 - (1 - b)k . \tag{5.22}$$

Type (iii) will occur only if

$$\begin{aligned} -k + 1 \leqq A - bk < -k + 2 , \\ k \leqq A + b(k - 1) < k + 1 , \end{aligned} \tag{5.23}$$

i.e., only if

$$1 + (1 - b)(k - 1) \leqq A < 2 - (1 - b)k . \tag{5.24}$$

On the (A, b) plane, in the region $0 \leqq b < 1$, (5.20) represents a rectangle R, (5.22) a triangle T_k, and (5.24) another kind of triangle U_k.

All points in the region $0 \leqq b < 1$, $0 \leqq A < 1 + b$ can be classified into three categories (see Fig. 1):

1. Those points which satisfy (5.20) but no condition of the type of (5.22), i.e., the points in $R - T_1$;

2. Those which satisfy (5.20) and conditions of the type of (5.22) up to some value K (say) of k, but no further, i.e., the points in $B_K = T_K - T_{K+1}$, $K = 1, 2, \cdots$;
3. Those which satisfy (5.24) up to some value K (say) of k, but no further, i.e., the points in $C_K = U_K - U_{K+1}$, $K = 1, 2, \cdots$.

For a point in $R - T_1$, the value 0 is the unique stationary state and no other recurrent state exists.

For a point in B_K, the state of convergence consists of a stationary state 0 and the cycles $(-k \rightleftarrows k)$ $k = 1, 2, \cdots, K$, i.e., the sequence either terminates in the value 0 or comes to alternate between the two values $-k$ and $k, 1 \leqq k \leqq K$. If $\bar{u}_0$ happens to be an integer such that $-K \leqq \bar{u}_0 \leqq K$, then of the above-mentioned alternatives that one which includes $\bar{u}_0$ will occur. If $\bar{u}_0$ is an integer either $\geqq K+1$ or $\leqq -(K+1)$, then the final state will always be the cycle $(-K \rightleftarrows K)$. Because, if $\bar{u}_n \geqq K+1$, then we have

$$\bar{u}_{n+1} \leqq A - b\bar{u}_n \leqq A - b(K+1) = A - bK - b < -K + 1 - b \,, \tag{5.25}$$

which implies that $\bar{u}_{n+1} \leqq -K$; and if $\bar{u}_n \leqq -(K+1)$, then we have

$$\bar{u}_{n+1} > A - b\bar{u}_n - 1 \geqq A + b(K+1) - 1 = A + bK + b - 1 \geqq K + b - 1 \,, \tag{5.26}$$

which implies that $\bar{u}_{n+1} \geqq K$. Thus there is no chance of getting into the range $[-K, K]$ from outside except into the cycle $(-K \rightleftarrows K)$.

For a point in C_K, the state of convergence consists of the cycles $(-k+1 \leftrightarrows k)$ $k = 1, 2, \cdots, K$. And here again we can show that the cycle $(-K+1 \rightleftarrows K)$ is what occurs when $\bar{u}_0$ lies outside of the range $[-K+1, K]$.

Next we shall consider *the case where* $b < 0$. From Lemma 5.1 (ii), we can conclude that the state of convergence will consist of a number of stationary values including 0. A positive number cannot be stationary, as seen from (5.10). In order that a negative integer $-k$ be stationary, it is necessary and sufficient that

$$A + bk < -k + 1 \,, \tag{5.27}$$

i.e., that

$$A < 1 - (1 + b)k \,. \tag{5.28}$$

On the (A, b) plane, in the region $S(-1 < b < 0, 0 \leqq A < 1 + b)$, (5.28) represents a triangle V_k. For a point in $S - V_1$, (5.28) holds for no k, and consequently 0 is the unique stationary value and no other recurrent value exists. For a point in $D_K = V_K - V_{K+1}$, (5.28) holds for $k = 1, \cdots, K$ but for no greater k, and we have $K+1$ stationary points $(0, -1, -2, \cdots, -K)$. The only points we can reach from outside are 0 and $-K$, as will be shown shortly. From the positive side, we can reach only 0 as seen from (5.10). On the other hand, if $\bar{u}_n \leqq -K-1$, then we have

$$\bar{u}_{n+1} \leqq A + b(K+1) < (-K+1) + b < -K + 1 \,, \tag{5.29}$$

implying that $\bar{u}_{n+1} \leqq -K$. Thus, from a point which lies to the left of $-K$, we can reach only the point $-K$.

To summarize the results, we have

THEOREM 3. *Let*

$$R = \{(A, b)\colon\ 0 \leqq A < 1, 0 \leqq b < 1\}\ ,$$
$$S = \{(A, b)\colon\ 0 \leqq A < 1 + b, -1 < b < 0\}\ ,$$
$$T_k = \{(A, b)\colon\ (1-b)k \leqq A < 1 - (1-b)k\}\ ,$$
$$U_k = \{(A, b)\colon\ 1 + (1-b)(k-1) \leqq A < 2 - (1-b)k\}\ ,$$
$$V_k = \{(A, b)\colon\ 0 \leqq A < 1 - (1+b)k, A < 1 + b, -1 < b < 0\}\ , \qquad k = 1, 2, \cdots,$$

and

$$A_0 = (R - T_1) + (S - V_1)\ ,$$
$$B_K = T_K - T_{K+1}\ ,$$
$$C_K = U_K - U_{K+1}\ ,$$
$$D_K = V_K - V_{K+1}\ , \qquad K = 1, 2, \cdots .$$

Then, for a point in A_0, 0 *is the unique stationary point; for a point in* B_K, *the final state will be the cycle* $(-K \leftrightarrows K)$ *if* $|\bar{u}_0| \geqq K+1$, *the cycle* $(-\bar{u}_0 \rightleftarrows \bar{u}_0)$ *if* $1 \leqq |\bar{u}_0| \leqq K$, *and* 0 *(stationary) if* $\bar{u}_0 = 0$; *for a point in* C_K, *the final state will be the cycle* $(-K+1 \leftrightarrows K)$ *if either* $\bar{u}_0 \geqq K+1$ *or* $\bar{u}_0 \leqq -K$, *the cycle* $(-\bar{u}_0 + 1 \rightleftarrows \bar{u}_0)$ *if* $1 \leqq \bar{u}_0 \leqq K$, *and the cycle* $(\bar{u}_0 \rightleftarrows -\bar{u}_0 + 1)$ *if* $-K+1 \leqq \bar{u}_0 \leqq 0$; *for a point in* D_K, *the final state will be the stationary point* 0 *if* $\bar{u}_0 \geqq 1$, *the stationary point* $-K$ *if* $\bar{u}_0 \leqq -K-1$, *and the stationary point* $\bar{u}_0$ *if* $-K \leqq \bar{u}_0 \leqq 0$.

Figure 1 shows some of the areas A_0, B_K, C_K, D_K, together with the "usual" terminal state which we obtain when the starting value does not happen to be recurrent.

Remark. Use of rounding to the nearest digital number will not change the situation materially. In fact, a shift of A, in (5.7) and thereafter, by one-half will take care of the change of the rounding rule perfectly.

6. Comments on More General Cases

The results in Section 5 can be generalized. For instance, Lemma 5.1 can be generalized into a scheme of the form (2.5) or (4.1) where $\varphi(x)$ is a function such that $\varphi'(x)$ exists and satisfies $0 > \varphi'(x) > -1$ in the region concerned. Likewise for Lemma 5.2. Moreover, we can prove that, *if* $0 > \varphi'(x) \geqq -\frac{1}{2}$, *then the state of convergence appears only either as a unique stationary point or as a pair of adjacent digital numbers constituting a cycle.* This statement has a considerable practical significance, because, when the condition given is satisfied, it will allow us to use the instruction sequence: "zero jump, raise, zero jump," in order to get out of the loop successfully when the state of convergence is reached. As an illustration, the author can cite his recent experience with an iteration for solving a transcendental equation. That is

$$x_{n+1} = \pi m - \text{arc tan}\,(C \tan Kx_n)\ , \qquad n = 0, 1, 2, \cdots . \tag{6.1}$$

Here, the condition is satisfied as long as $K \leqq C/2$, which was actually the case.

Some of the results in the last section also suggest an interpretation of a phenomenon which has been noticed by A. Okamoto,[1] a programmer for FACOM 128, a floating-point relay computer. He computed a root of an algebraic equation first by 8-decimal-place arithmetic and reached a stationary point which differed from the true value (which happened to be known) by an amount of order 10^{-4}. Next he used 16-decimal-place arithmetic and this time the error turned out to be of order 10^{-12}.

Now, let the equation be $f(x) = 0$. Then the iterative process may be represented by

$$x_{n+1} = x_n - f(x_n)/f'(x_n), \qquad n = 0, 1, 2, \cdots . \tag{6.2}$$

In the corresponding process, $f(x_n)$ will be computed and rounded to eight decimal places. If the rounded result turns out to be zero, then that value of x_n is a stationary point. In order that the rounded value of $f(x_n)$ be zero so that x_n be stationary, it is necessary and sufficient that

$$-\varepsilon/2 \leqq f(x_n) < \varepsilon/2 , \tag{6.3}$$

where ε is the quantum. Suppose $f'(x)$ is nearly equal to M in the neighborhood of the root α with which we are dealing. Then (6.3) will be approximately transformed into

$$-\varepsilon/2M \leqq x_n - \alpha < \varepsilon/2M . \tag{6.4}$$

In other words, any stationary value in this sense lies within about $\varepsilon/2M$ of α. This suggests that the rounding error will be magnified by the factor $1/M$. Thus if $M \approx 10^{-3}$, and $\varepsilon = 10^{-8}$, then the error in the solution will be of the order of 10^{-4}; and if we take $\varepsilon = 10^{-16}$ in the same situation, the error will become of the order 10^{-12}. This explains Okamoto's observation.

As a matter of fact, FACOM 128 which Okamoto used was a floating-point machine. But this fact does not much affect the preceding argument, because, if the computation of $f(x_n)$ is done by the nesting method as usual, then the rounding will usually take place for the last time just in the process of addition (or numerically, rather, subtraction) of the constant term, resulting in effect in rounding with quantum ε, the unit in the least significant position preserved of the constant term. Thus, even with a floating-point arithmetic, there can be a relatively wide range of indeterminacy when $f'(x)$ is moderately small in the neighborhood.

7. Final Remarks

We have shown some details of the interesting behavior of iterative processes under the influence of rounding errors. The author wishes to thank Messrs. Daisuke Miura and Hiroyuki Matsutani for their computational assistance.

[1] Computing Center, Yurin Denki Seiki Company.

References

[1] Urabe, M. Convergence of numerical iteration in solution of equations. *J. Sci. Hiroshima Univ.*, Ser. A, 1956, **19**, 479-89.

[2] Collatz, L. Einige Anwendungen funktionalanalytischer Methoden in der praktischen Analysis. *Z. angew. Math. u. Phys.*, 1953, **4**, 327–57.

[3] Sibagaki, W. On the idea of "numerical convergence" and its some applications. *Mem. Fac. Sci. Kyushu Univ.*, Ser. A, 1950, **5**, 89–97.

[4] Householder, A. S. *Principles of Numerical Analysis*. NewYork: McGraw-Hill, 1953.

[5] Feller, W. *An Introduction to Probability Theory and Its Applications*. Vol. 1. New York: Wiley 1950.

29 PREDICTION IN FUTURE SAMPLES

GEORGE E. NICHOLSON, JR.
University of North Carolina

1. Introduction

This paper treats a problem which arises in the field of psychology as well as in other fields. Suppose a test battery consisting of p aptitude tests (predictors) is applied to a group of students who are being considered for admission to a particular school. The battery is intended to furnish the admissions officer with information to assist him in predicting the success of the applicants in their first year. After the students have completed the first year, a criterion measure (average academic grade, say) is obtained for each student. The test battery is then validated; i.e., the multiple correlation coefficient between test score and criterion is computed. A prediction equation is obtained by using the regression coefficients, calculated on a least-squares basis, as weights to be applied to the test scores of succeeding students. This yields an estimate of the future criterion score of any student, and in this sense is useful for predicting the probable success or failure of members of succeeding classes.

When the criterion scores of a second group of students are available, these may be compared with the estimated criterion scores in order to determine how well the prediction equation is performing. The determination of the "goodness" of the prediction equation is usually made by comparing the second-sample multiple correlation coefficient with that obtained in the first (validation) sample. However, the objective of the procedure is to predict the criterion scores in the second sample and the correlation between the predicted scores and the observed scores in the second sample. The correlation between predicted and observed scores in the second sample can also be regarded as the actual measure of the "goodness" of the prediction equation. It has been observed by several writers that generally the second multiple correlation coefficient is smaller than the first and this effect has been called the "shrinkage" of the multiple correlation coefficient.

Examples of this effect have been given in [1] and [2]. In this paper we consider the nature of the problem, indicate the effect of sample size and number of predictors, and give an example.

2. Some Comments on Linear Regression

To fix ideas and establish a notation, let us consider a set of variates

$y, x_1, \cdots, x_p$, where y is the dependent or criterion variate and $x_1, \cdots, x_p$ are the independent or predictor variates. Let $y_{\nu_1}^{(1)}$ denote the νth observation on y, and $x_{i\nu_1}$ the ν_1th observation on x_i ($i = 1, 2, \cdots, p$; $\nu_1 = 1, 2, \cdots, N_1$) in a sample S_1. Let $y_{\nu_2}^{(2)}, x_{i\nu_2}^{(2)}$ ($i = 1, 2, \cdots, p$; $\nu_2 = 1, 2, \cdots, N_2$) be defined for an independent sample S_2. We assume that x_i are fixed constants. In particular, let $x_{1\nu_1}^{(1)} = x_{1\nu_2}^{(2)} = 1$ and $x_{j\nu_1}^{(1)}, x_{j\nu_2}^{(2)}$ ($j = 2, \cdots, p$) be deviations from sample means in S_1 and S_2, respectively. The observations y are random variables such that

1. $y_1^{(1)}, \cdots, y_{N_1}^{(1)}, y_1^{(2)}, \cdots, y_{N_2}^{(2)}$ are normally and independently distributed with a common unknown variance σ^2;
2. The expected value of $y = \beta_1 x_1 + \cdots + \beta_p x_p$ where β_i are unknown constants.

This implies that $y_\gamma = \eta_\gamma + \varepsilon_\gamma = \beta_1 x_{1\gamma} + \beta_2 x_{2\gamma} + \cdots + \beta_p x_{p\gamma} + \varepsilon_\gamma$ where ε_γ is a normal variate with zero mean, and such that $E(\varepsilon_i \varepsilon_j) = \sigma^2$ if $i = j$, and zero, otherwise.

Let $a_{ij} = S_\gamma x_{i\gamma} x_{j\gamma}$, $A = (a_{ij})$, and $Y_\alpha = b_1 x_{1\alpha} + \cdots + b_p x_{p\alpha}$, where the b_i are defined by the p-normal equations

$$S_\alpha (y_\alpha - Y_\alpha) x_{i\alpha} = 0 \,.$$

Denote by $A^{(1)}$ and $A^{(2)}$ respectively the two matrices of normal equations which define the two sets of regression coefficients with which we are concerned, i.e.,

$$A^{(1)} = (a_{ij}^{(1)}) = (S x_{i\alpha}^{(1)} x_{j\alpha}^{(1)}) \,,$$

and similarly for $A^{(2)}$. Also, $A^{(1)}/N_1$ is the sample covariance matrix for the x's in Sample 1 and a similar interpretation holds for $A^{(2)}/N_2$. Now consider the following 2×2 table.

	Sample 1	Sample 2
$Y^{(1)}$	R_{11}^2	R_{12}^2
$Y^{(2)}$	R_{21}^2	R_{22}^2

In the table above, $Y^{(1)}$ represents the least-squares predicting equation calculated from Sample 1, and $Y^{(2)}$ represents the least-squares predicting equation derived from Sample 2. R_{12}^2 is the result of correlating the estimated criterion variates in S_2 on the basis of the prediction equation derived from S_1.

It is clear that in this notation R_{11}^2 is simply the square of the multiple correlation coefficient calculated from the first sample, and R_{12}^2 is the square of the correlation between the observed criterion values in the second sample and those estimated from the predictors in the second sample and the regression coefficients calculated from the first sample. The problem which we consider arises from the observation of research workers that generally R_{12}^2 is lower than R_{11}^2.

If Sample 1 and Sample 2 are independently drawn from the same population, then R_{11}^2 and R_{22}^2 are independent estimates of the population parameter ρ^2. Under the assumption already made, the distribution of these sample estimates is well known. Accordingly, questions related to the behavior of the best linear prediction to be expected in any sample may be answered by using this distribution.

For example, the chance that R_{22}^2 will be less than R_{11}^2 is $\frac{1}{2}$ since both follow the same distribution. The chance that $R_{11}^2 - R_{22}^2 > d$ may be answered by working directly with the known distribution.

The distribution of R_{ii}^2 depends on N_i, the number of observations in the sample; p, the number of predictors; and ρ^2, the parameter. Previous considerations of this problem have resulted in proposals to correct the observed R_{ii}^2 for bias to obtain a better estimate of ρ^2. A first-order approximation to an unbiased estimate of ρ^2 (from [3]) is

$$\rho^2 = 1 - \frac{N_i - 2}{N_i - p}(1 - R_{ii}^2) .$$

Exact expressions are available for obtaining unbiased estimates of ρ^2 ([4], [5]), but the problem of inferring how well a predicting equation will perform in a new sample is not answered by this kind of analysis.

3. Formulation of the Shrinkage Problem and Definition of Measure of Efficiency

The regression coefficients estimated from the first sample are known to be unbiased estimates of the population regression coefficients. For example, let $\beta = (\beta_1, \beta_2, \cdots, \beta_p)$ be the vector of population coefficients and let $b_1 = (b_{11}, b_{21}, \cdots, b_{p1})$ be the vector of regression coefficients estimated on the basis of Sample 1. Then $E(b_1)$ is equal to β. If the first sample is large, then it may be expected that $\max_i (b_i - \beta_i) = \delta$ is very small, and it may be considered that this problem reduces itself to considering the sampling fluctuations of the simple correlation coefficient, since for a fixed set of β the multiple correlation coefficient reduces to the correlation between the criterion and a fixed linear function of normally distributed variables which is a single normal variable. Then the problem becomes that of studying the distribution of the ordinary correlation coefficient. The question, however, remains: Why there should be any shrinkage, since it would be expected that greater values of the second coefficient would occur as frequently as smaller ones? The intuitive answer seems to run as follows. If the first sample is large and has ample degrees of freedom for estimating β, then it is reasonable to assume that the regression weights are close to the population weights. Since b is the unique set of coefficients which maximizes the multiple correlation coefficient in the sample, the observed R_{11}^2 is an overestimate of ρ^2. However, a second sample, of approximately the same size, (the b used being close to the β of the population) will tend to make R_{12}^2 an unbiased estimate of ρ^2 and hence less than R_{11}^2, which is biased upward.

Accordingly, the phenomenon of shrinkage should be observed whenever two fairly large samples are used and when the degrees of freedom available for estimation of β are large. Furthermore, in this case the phenomenon ought to be adequately explained by appealing to the distribution of the simple correlation coefficient. The problem, however, will depend on the number of degrees of freedom available for estimating β in Sample 1 and the size of the second sample relative to the first.

Let us consider the following formulation of the problem. The best prediction in a particular sample is obtained when the least-squares prediction equation derived from that sample is used. A measure of the adequacy of this prediction equation is given by R^2_{11}. The actual measure of prediction efficiency is, however, R^2_{12}, since it is the second sample in which the prediction is required and not the first. Now, among the various interpretations which can be made of the correlation coefficient, one is that the square of the correlation coefficient is the proportion of variance of one of the variables which is explained by a linear least-squares regression on the other. It is this interpretation which we adopt. We propose that the accuracy of a prediction equation in a sample be measured by this means. If then the y_α's in a sample of N_i observations are to be estimated by an estimation procedure $Y(\alpha)$ which gives rise to N_i estimates $Y^{(k)}_\alpha$, then we shall define the correlation between y and $Y^{(k)}$ to be

$$R^2(I) = 1 - \frac{S_\alpha(y_\alpha - Y^{(k)}_\alpha)^2}{S_\alpha(y_\alpha - \bar{y})^2} .$$

It should be noted that the relation,

$$\frac{[S_\alpha(y_\alpha - \bar{y})(Y^{(k)}_\alpha - \bar{Y}^{(k)})]^2}{S_\alpha(y_\alpha - \bar{y})^2\, S_\alpha(Y^{(k)}_\alpha - \bar{Y}^{(k)})^2} = 1 - \frac{S_\alpha(y_\alpha - Y^{(k)}_\alpha)^2}{S_\alpha(y_\alpha - \bar{y})} ,$$

is not true unless $Y^{(k)}_\alpha$ is the estimate based on the least-squares regression calculated from the same sample. We therefore select as our definition

$$R^* = 1 - \frac{S_\alpha(y^{(2)}_\alpha - Y^{(1)}_\alpha)^2}{S_\alpha(y^{(2)}_\alpha - \bar{y}^{(2)})^2} . \tag{1}$$

The failure of $Y^{(1)}_\alpha$ to be equal to $y^{(2)}_\alpha$ is measured by

$$S_\alpha(y^{(2)}_\alpha - Y^{(1)}_\alpha)^2 = S_\alpha(y^{(2)}_\alpha - Y^{(2)}_\alpha)^2 + S_\alpha(Y^{(2)}_\alpha - Y^{(1)}_\alpha)^2 ,$$

which may be partitioned into the two components shown. The first term represents the failure of the observations to follow the best least-squares prediction equation, and the second part represents the failure of the prediction equation derived from Sample 1 to be equal to that from Sample 2. It therefore seems natural to use the ratio of $S_\alpha(y^{(2)}_\alpha - Y^{(2)}_\alpha)^2$ to $S_\alpha(y^{(2)}_\alpha - Y^{(1)}_\alpha)^2$ as a measure of the efficiency of the predictor $Y^{(1)}_\alpha$ in Sample 2. Let

$$E = \frac{S_\alpha(y^{(2)}_\alpha - Y^{(2)}_\alpha)^2}{S_\alpha(y^{(2)}_\alpha - Y^{(1)}_\alpha)^2}$$

measure the efficiency of the predicting equation. It is clear that $E \leqq 1$ and assumes its maximum value for $Y_\alpha^{(1)} = Y_\alpha^{(2)}$. Also from (1),

$$E = \frac{(1 - R_{12}^2)\, S_\alpha(y_\alpha^{(2)} - \bar{y}^{(2)})^2 - S_\alpha(Y_\alpha^{(2)} - Y_\alpha^{(1)})^2}{S_\alpha(y_\alpha^{(2)} - Y_\alpha^{(1)})^2} .$$

If $Y_\alpha^{(1)}$ is taken as $\bar{y}^{(2)}$, we see that $E < 1 - R^2$, showing that (for a large R^2) a simple predicting equation like the mean of the y's has very low efficiency.

4. Distribution of E

Let

$$Q_1 = S(y_\alpha^{(2)} - Y_\alpha^{(2)})^2$$

$$Q_2 = \sum_j \sum_i (b_i^{(2)} - b_i^{(1)})(b_j^{(2)} - b_j^{(1)}) a_{ij}^{(2)} .$$

Then

$$S(y_\alpha^{(2)} - Y_\alpha^{(1)})^2 = S(y_\alpha^{(2)} - Y_\alpha^{(2)})^2 + \sum_j \sum_i (b_i^{(2)} - b_i^{(1)})(b_j^{(2)} - b_j^{(1)}) a_{ij}^{(2)} = Q_1 + Q_2 .$$

It follows from all the assumptions that Q_1 is distributed like $\sigma^2 \chi^2_{N_2 - p}$ and is independent of Q_2. If we let $b_i^{(2)} - b_i^{(1)} = w_i$, it follows that the set w_i has a normal multivariate distribution with zero means and covariance matrix $\Sigma = [(A^{(1)}) + (A^{(2)})]^{-1}$. This covariance matrix Σ and the matrix of the form $Q_2 = A^{(2)}$ are both positive definite. Therefore, Q_2 may (by a familiar transformation) be transformed into the form, $Q_2 = \lambda_1 z_1^2 + \cdots + \lambda_p z_p^2$, where the real positive λ_i are the roots of the determinantal equation

(2) $$| A^{(2)}(A^{(1)})^{-1} - (\lambda - 1)I | = 0 ,$$

and the z_i are normally distributed random variables with zero means and unit variance. The final result then is

$$Q_2/Q_1 = 1 + \left(\sum_i \lambda_i z_i^2\right) \Big/ (\chi^2_{N_2 - p}) .$$

We now consider the reciprocal of the ratio $S(y_\alpha^{(2)} - Y_\alpha^{(2)})^2 / S(Y_\alpha^{(2)} - Y_\alpha^{(1)})^2 = E$, which is

$$\frac{1}{E} = \frac{Q_1 + Q_2}{Q_1} = 1 + \frac{Q_2}{Q_1} = 1 + \frac{Q_2}{\chi^2_{N_2 - p}}$$

and

$$p = \frac{1}{E} - 1 = \frac{Q_2}{\chi^2_{N_2 - p}} .$$

Special Cases. Assume that $b_i^{(1)} = \beta_i$; i.e., the first sample is so large that the estimates of the β_i are equal to the β_i. In this case Q_2 is distributed exactly as χ_p^2 and

$$\Pr\left\{\frac{1}{E} \leqq x\right\} = \Pr\left\{1 + \frac{\chi_p^2}{\chi_{N_2-p}^2} \leqq x\right\} = \Pr\left\{F_{p,N_2-p} \leqq \frac{N_2 - p}{p}(x-1)\right\},$$

where F_{p,N_2-p} is Snedecor's F with p and $N_2 - p$ degrees of freedom.

Assume that $A^{(2)} = kA^{(1)}$ and consider the determinantal equation (2). Under this assumption all of the roots are equal, $\lambda_i = k + 1$, and $Q_2 = (1 + k)\chi_p^2$. Then ,

$$\Pr\left\{\frac{1}{E} \leqq x\right\} = \Pr\left\{F_{p,N_2-p} \leqq \frac{(x-1)(N_2-p)}{p(1+k)}\right\}.$$

In order to obtain an upper confidence limit for P at the ε level of significance, choose x such that it satisfies the equation $F_\varepsilon = (x-1)(N_2-p)/[p(1+k)]$, that is, $x = 1 + F_\varepsilon[p(1+k)/(N_2 - p)]$, where

$$\Pr\{F_{p,N_2-p} \leqq F_\varepsilon\} = 1 - \varepsilon .$$

The confidence interval is $1 \leqq 1/E \leqq x$, with confidence coefficient $1 - \varepsilon$.

Since E is the relative efficiency of the prediction equation $Y^{(1)}$ derived from S_1, the interpretation of the confidence interval gives an estimate of how well $Y^{(1)}$ will predict in S_2. The practical significance of this result is obvious; for large samples, $A^{(2)}/N_2$ should be approximately equal to $A^{(1)}/N_1$.

For $k = N_2/N_1$,

$$\frac{1}{E} = 1 + \frac{[1 + (N_2/N_1)]\chi_p^2}{\chi_{N_2-p}^2}.$$

The expected value of $1/E$ is

$$\mathscr{E}\left(\frac{1}{E}\right) = \mathscr{E}\left[1 + \left(1 + \frac{N_2}{N_1}\right)F_{p,N_2-p}\frac{p}{N_2 - p - 2}\right] = \frac{N_1(N_2 - 2) + N_2 p}{N_1(N_2 - p - 2)},$$

which for large N_2 is approximately

$$\mathscr{E}\left(\frac{1}{E}\right) = \frac{1 + (p/N_1)}{1 - (p/N_2)}.$$

We note from this that, if p is fixed and N_1 and N_2 are both large, then $\mathscr{E}(1/E) = 1$, which indicates that the regression equation calculated from a very large sample will predict equally well in another very large sample provided the number of predictors is relatively small. Let $N_2 = rN_1$ and $p = sN_1$, where $0 \leqq s \leqq 1$ and $r > 0$. Then, $\mathscr{E}(1/E) = (1 + s)r/(r - s)$, and

$$\frac{1}{\mathscr{E}(1/E)} = \frac{r - s}{r(1 + s)}.$$

For this expression to have its maximum value, s must be a minimum. Hence, we have the conclusion above that the ratio of the number of predictors to the sample size N_1 should be small. However, if r is very large for small s, we have an efficient arrangement. For $A^{(2)} = (N_2/N_2)A^{(1)}$, then, Table 1 gives the expected relative efficiency of the predictor $Y^{(1)}$ in S_2.

TABLE 1

N_2/N_1 \ p/N_1	.01	.05	.1	.5	1
.1	.89	.47			
.2	.94	.71	.45		
.5	.97	.86	.73		
1	.98	.90	.82	.33	
2	.99	.93	.86	.50	.25
10	.99	.95	.90	.63	.45
∞	.99	.95	.91	.66	.50

5. An Alternative Measure of Efficiency

Another possible measure of efficiency is

$$\frac{1}{E'} = \frac{N_1}{N_2} \frac{S(\bar{y}_\alpha^{(2)} - Y_\alpha^{(1)})^2}{S(y_\gamma^{(1)} - Y_\gamma^{(1)})^2} .$$

This is the ratio of the sample variance of the residuals about the predictor from S_1 applied in S_2 to the predictor of S_1 applied in S_1. By the same arguments as used before, $1/E'$ may be expressed as

$$\frac{1}{E'} = \frac{N_1}{N_2} \frac{\chi^2_{N_2-p} + \sum_{i=1}^p \lambda_i z_i^2}{\chi^2_{N_1-p}} ,$$

where $\chi^2_{N_1-p}$ is independent of the numerator and the quantities in the numerator have already been defined. If we assume $A^{(1)} = A^{(2)}/k$ by the previous argument, we have

$$\frac{1}{E'} = \frac{N_1}{N_2} \frac{\chi^2_{N_2-p} + (1+k)\chi^2_p}{\chi^2_{N_1-p}}$$

and, therefore,

$$\mathscr{E}\left(\frac{1}{E'}\right) = \frac{N_1}{N_2} \frac{N_2 + kp}{N_1 - p - 2} ,$$

from which it follows that if $k = N_2/N_1$,

$$\mathscr{E}\left(\frac{1}{E'}\right) = \frac{1 + (p/N_1)}{1 - (p+2)/N_1} .$$

It is of interest to note this expression is independent of N_2.

6. An Illustration

In order to present an empirical example based on the analysis described above, two sets of data were used,[1] from the Law School Admission Test

[1] Data were obtained through the courtesy of Dr. Ledyard Tucker and Miss Marjorie Olsen, of the Educational Testing Service, Princeton, N.J.

as given in Table 2. The notation

$$R'_{12} = \frac{S(y_\alpha^{(1)} - \bar{y}^{(1)})(Y_\alpha^{(2)} - \bar{Y}^{(2)})}{\sqrt{S(y_\alpha^{(1)} - \bar{y}^{(1)})^2 S(Y_\alpha^{(2)} - \bar{Y}^{(2)})^2}} \quad \text{and} \quad R_{12}^2 = 1 - \frac{S(y_\gamma^{(2)} - Y_\gamma^{(1)})^2}{S(y_\gamma^{(2)} - \bar{y}^{(2)})^2}$$

is used.

TABLE 2

RECORDS FOR 1,000 FRESHMAN STUDENTS, FROM LAW SCHOOL ADMISSION TEST

School	Sample I	Sample II	Total	School	Sample I	Sample II	Total
1. Buffalo	44	44	88	6. Rutgers	55	55	110
2. Columbia	64	64	128	7. So. Calif.	62	62	124
3. Cornell	47	47	94	8. Virginia	27	27	54
4. N.Y.U.	35	35	70	9. Yale	69	69	138
5. Penn.	97	97	194	Total	500	500	1000

Raw scores for six of the ten subtests of the Law School Admission Test and the total scores over all ten subtests were available. The criterion was the first-year average grades converted to a common scale for all schools.

Since Pennsylvania had the greatest number of students (194), this sample was used. Sample I was used for determining the multiple correlation. The regression weights were determined and used in Sample II, to predict the criterion. Sample II was divided into two parts. The first 50 constituted the first group for which the criterion was predicted and then the entire 97 were used and the criterion predicted.

This was done for three predictors and then for six predictors. The subtests used as predictors are identified as follows:

Part 2. Sentence Completion	Part 6. Figure Classification
Part 3. Paragraphs	Part 7. Debates
Part 5. Reading Comprehension	Part 10. Reading Comprehension.

Parts 2, 3, and 10 were used as the three predictors, while all six were used in the second case.

TABLE 3

RESULTS

N_1	N_2	p	R_{12}^2	$R_{12}^{\prime 2}$	R_{12}^2	$\frac{1}{E^*} = \frac{S(y - Y^{(1)})^2}{S(y - Y^{(2)})^2}$	$\mathscr{E}\left(\frac{1}{E^*}\right)$	Confid. inter.	%
97	50	3	.2436	.1256	.1139	1.17	1.0967	1-1.27 1-1.40	95 99
97	97	3	.2436	.1229	.0697	1.23	1.0638	1-1.17 1-1.26	95 99
97	50	6	.3345	.2004	.1939	1.21	1.13	1-1.46 1-1.65	95 99
97	97	6	.3345	.1652	.0771	1.38	1.13	1-1.29 1-1.39	95 99

In the calculations above, the estimate of P is taken to be $1/E^* = (1-R_{12}^2)/(1-R_{11}^2)$ instead of $1/E = (1-R_{12}^2)/(1-R_{22}^2)$.

Tables 4 and 5 give the calculations upon which this example is based.

TABLE 4

INTERCORRELATIONS FOR SAMPLE I
(Pennsylvania: Two samples of 97 each)

Part 2	.....					
Part 3	.7387	.....				
Part 5	.7162	.6484	.....			
Part 6	.2759	.3995	.3286	.....		
Part 7	.5495	.5192	.5542	.2772	.....	
Part 10	.8230	.7570	.7329	.3321	.5932	.....
Criterion	.4576	.4319	.4651	.4017	.4357	.4672

TABLE 5

RAW-SCORE REGRESSION WEIGHTS

Three predictors[a] (Part scores 2, 3, & 10)	Six predictors[b] (Part scores 2, 3, 5, 6, 7, & 10)
$b_2 = .1097$	$b_2 = .0878$
$b_3 = .0961$	$b_3 = .0005$
$b_{10} = .1126$	$b_5 = .1055$
$b_0 = 6.4245$	$b_6 = .1642$
Multiple $R_{11} = .4919$	$b_7 = .1204$
	$b_{10} = .0320$
	$b_0 = .9254$
	Multiple $R_{11} = .5784$

[a] Predicting criterion in Sample II: $N = 50$, $R'_{12} = .3544$; $N = 97$, $R'_{12} = .2497$.

[b] Predicting criterion in Sample II: $N = 50$, $R'_{12} = .4477$; $N = 97$, $R'_{12} = .4064$.

REFERENCES

[1] Horst, P. The prediction of personal adjustment. *Social Sci. Res. Council Bull.*, 1941, **48**, 105.

[2] Larson, S. C. The shrinkage of the multiple correlation coefficient. *J. Educ. Psychol.*, 1931, **22**, 45-55.

[3] Wherry, R. J. A new formula for predicting the shrinkage of the multiple correlation coefficient. *Ann. Math. Stat.*, 1931, **2**, 440-57.

[4] Fisher, R. A. The influence of rainfall on the yield of wheat at Rothamsted. *Phil. Trans. Roy. Soc. (London)*, Ser. B., 1924, **213**, 89-142.

[5] Olkin, I., and Pratt, J. W. Unbiased estimation of certain correlation coefficients. *Ann. Math. Stat.*, 1958, **29**, 201-11.

30 RANKING IN TRIPLE COMPARISONS[1]

R. N. PENDERGRASS, Radford College

R. A. BRADLEY,[2] Virginia Polytechnic Institute

1. Summary

Paired comparisons are widely used in subjective testing. While small block-size designs are usually necessary, it is not always desirable to use paired comparisons. This paper deals with methods of analysis for the ranking of t treatments in sets or subsets of the possible $\binom{t}{3}$ triple comparisons formed from t treatments.

A mathematical model for triple comparisons has been devised as an extension of the model for paired comparisons. Treatment parameters are postulated and estimated by the method of maximum likelihood and the asymptotic variances and convariances of these estimators are found. In general, we consider n_{ijk} rankings on the comparison of treatments i, j, and k, and in some cases take each $n_{ijk} = n$, in which event it is sometimes possible to group the triplets into n complete repetitions of the basic set of comparisons. Most of the discussion is for the cases where $n_{ijk} = n$, but the extensions are shown for the more general situation in which some n_{ijk} may in fact be zero.

Test procedures are also derived. The nature of small-sample distributions of test statistics is indicated, but main results are asymptotic ones for large values of n or of the n_{ijk}. Tests are provided for hypotheses on (i) treatment effects, (ii) treatment by group interactions, and (iii) the appropriateness of the model used. The properties and asymptotic power function of the main test of treatment effects are investigated. The asymptotic relative efficiencies of triple comparisons relative to paired comparisons and the analysis of variance are respectively 3/2 and $9/4\pi$.

The paper concludes with a brief numerical example and discussion.

[1] Research sponsored by the Office of Naval Research under the contract with the Virginia Polytechnic Institute on statistical methods in quality control and surveillance testing. Reproduction in whole or in part is permitted for any purpose of the United States Government.

[2] Present address: Department of Statistics, The Florida State University, Tallahassee, Florida.

2. Introduction

The analysis of experiments based on ranking items within the experiment, or within subdivisions of it, has received considerable attention in statistical and psychological methodology. In the latter discipline, much effort has been devoted to problems of scaling, while, in the former, more emphasis has been placed on testing and in particular on tests of random rankings. Ranking methods are helpful particularly where quantitative observations cannot be obtained easily but where estimation and test procedures are desired nevertheless.

Various methods have been used to analyze ranks. Use is made of Table XX of Fisher and Yates [1] by Bliss *et al.* [2], Galinat [3], Greenwood [4], and many others. Methods of rank analysis have been devised by Friedman [5], Kendall and Babington Smith [6], Durbin [7], Kruskal and Wallis [8], and others following them. Paired comparisons have had particular attention from many, including Thurstone [9], Morrissey [10], Gulliksen [11], Kendall and Babington Smith [12], Guttman [13], Scheffé [14], and Bradley and Terry [15]. The comprehensive bibliography of non-parametric methods prepared by I. R. Savage [16], and now under revision, contains many additional references.

Here we shall develop a method for ranking in triple comparisons as an extension of the Bradley-Terry method for paired comparisons. The method involves postulating a mathematical model involving treatment parameters, the estimation of these parameters together with their large-sample variances and covariances (thus solving the scaling problem), the development of test procedures, and investigation of the properties of the model and procedures developed. The methods employed parallel those of Bradley [17], [18] very closely, and in some arguments proofs will not be given in detail because of this parallelism.

Experiments involving ranking within small groups of treatments or items seem particularly appropriate in sensory-judgment investigations. Paired comparisons have been widely used in this context, and preliminary use of triple comparisons indicates the utility of this method also.

3. Models for Triple Comparisons

We shall state two models for triple comparisons and indicate our choice. Review of the simple model for paired comparisons is first necessary.

In paired comparisons, the existence of non-negative parameters $\pi_1, \cdots, \pi_t$ associated with t treatments $T_1, \cdots, T_t$ is postulated such that

$$\sum_i \pi_i = 1 , \tag{3.1}$$

and the probability that T_i is "preferred" to T_j is

$$P(T_i > T_j) = \pi_i/(\pi_i + \pi_j) , \qquad i \neq j, \quad i, j = 1, \cdots, t . \tag{3.2}$$

Probabilities associated with pairs (or triplets) of treatments are taken to be independent.

We shall also use $P(r_i < r_j) = P(T_i > T_j)$ where r_i is the rank of T_i, rank 1 being assigned to the more (most) preferred treatment and so on. The notation $\sum_i$ indicates summation with respect to i for values $1, \cdots, t$. These and additional *notational conventions*[3] will be used throughout with departures from these conventions specified as required.

Several extensions of the model for paired comparisons are possible to triple comparisons. We retain the concept of non-negative parameters $\pi_1, \cdots, \pi_t$ associated with $T_1, \cdots, T_t$ and the restriction (1), which is for convenience only, in each of the two models (i) and (ii) discussed.

(i) A model for triple comparisons was proposed by Bradley and Terry [19], and it supposes that in a comparison of T_i, T_j, and T_k,

$$(3.3) \qquad P(T_i > T_j > T_k) = P(r_i < r_j < r_k) = P(r_i = 1, r_j = 2, r_k = 3) = \pi_i\pi_j/[(\pi_i + \pi_j + \pi_k)(\pi_j + \pi_k)] .$$

(ii) The model now proposed and fully discussed in this paper is that

$$(3.4) \qquad P(T_i > T_j > T_k) = \pi_i^2\pi_j/\Delta_{ijk} ,$$

where

$$(3.5) \qquad \Delta_{ijk} = \pi_i^2(\pi_j + \pi_k) + \pi_j^2(\pi_i + \pi_k) + \pi_k^2(\pi_i + \pi_j) .$$

Both models have desirable features. Model (i) in (3.3) has the properties that

$$P(r_i < r_j) = P(r_i < r_j \mid r_k = 1) = \pi_i/(\pi_i + \pi_j) .$$

Model (ii) in (3.4) has the properties that

$$P(r_i < r_j \mid r_k = 3) = P(r_i < r_j \mid r_k = 1) = \pi_i/(\pi_i + \pi_j) .$$

It can be shown that it is impossible to develop a model of the class desired for which more than two of $P(r_i < r_j \mid r_k = 1)$, $P(r_i < r_j \mid r_k = 2)$, $P(r_i < r_j \mid r_k = 3)$, and $P(r_i < r_j)$ have the value $\pi_i/(\pi_i + \pi_j)$ similar to paired comparisons. Details on such a demonstration were given by Pendergrass [20].

Models based on (3.3) and (3.4) yield very similar results in applications and indeed there is little to choose between them. Both also give results similar to those obtained with the same treatments in paired comparisons in applications. However, the model based on (3.4) has an element of symmetry lacking in (3.3) since inverse ranking (ranking by "dislike" rather than "preference") permits replacement of π_i by $1/\pi_i^*$, $i = 1, \cdots, t$, and subsequent use of $\pi_1^*, \cdots, \pi_t^*$ in (3.4) instead of $\pi_1, \cdots, \pi_t$ where π_i^* is a measure of dislike and π_i a measure of preference.

The model based on (3.4) with the restriction (3.1) will be used for triple comparisons. We shall develop our main results for experiments with n rankings on all possible triplets formed by $T_1, \cdots, T_t$, but indicate special

[3] We use Π_i to mean $\Pi_{i=1}^t$, $\sum_{i<j<k}$ and $\Pi_{i<j<k}$ to represent sums and products with $i<j<k$, $i, j, k = 1, \cdots, t$ and $\sum_{j<k}$ and $\sum_{j\neq k}$ to indicate summations with $j<k$ and $j \neq k$ with $j, k = 1, \cdots, t$. $\sum'_{j<k}$ is a sum with $j \neq i$, $k \neq i$, $j < k$, $i, j, k = 1, \cdots, t$, where i is the additional subscript in the argument of the summation.

results in the more general case where n_{ijk} rankings are made on the triplet T_i, T_j and T_k, $i, j, k = 1, \cdots, t$, $i \neq j \neq k$. A test of the appropriateness of the model chosen is developed in Section 9.

The models chosen for paired and triple comparisons lead to procedures with considerable flexibility, and these procedures may be preferred to methods of concordance (see, for example, Durbin [7]) for incomplete block designs in that the nature of departures from random orderings within incomplete blocks is more clearly discerned. The models may be extended to ranking in larger incomplete sets, but procedures become increasingly more complex.

4. Estimation

The method of maximum likelihood will be used to obtain estimators $p_1, \cdots, p_t$ of $\pi_1, \cdots, \pi_t$. The rank of T_i in the αth comparison of T_i, T_j, and T_k will be written $r_{i\alpha,jk}$, $\alpha = 1, \cdots, n$. Tied ranks are not permitted in the following theoretical work, but the average of tied ranks is used in practice if only a few ties occur.

Consistent with the model (3.4), the probability of a specified ranking in the αth triple comparison of T_i, T_j, and T_k is

$$P(r_{i\alpha,jk}, r_{j\alpha,ik}, r_{k\alpha,ij}) = \pi_i^{3-r_{i\alpha,jk}} \pi_j^{3-r_{j\alpha,ik}} \pi_k^{3-r_{k\alpha,ij}} / \Delta_{ijk} . \tag{4.1}$$

Often, judgments in triplets may be logically grouped into repetitions consisting of one set of judgments on each possible triplet, although grouping into repetitions is not necessary for our development. If repetitions have been formed, the likelihood function for the αth repetition is

$$f(a_\alpha, \pi) = \prod_i \pi_i^{a_{i\alpha}} / \prod_{i<j<k} \Delta_{ijk} , \tag{4.2}$$

where a_α and π represent the vectors $(a_{1\alpha}, \cdots, a_{t\alpha})$ and $(\pi_1, \cdots, \pi_t)$ respectively, while

$$a_{i\alpha} = {\sum_{j<k}}' (3 - r_{i\alpha,jk}) . \tag{4.3}$$

Note that $a_{i\alpha}$ is twice the number of first-place ranks plus the number of second-place ranks for T_i in all comparisons involving T_i in the αth repetition. To obtain (4.2) we have assumed independence in probability of judgments on different triplets, and, now assuming independence between repetitions, we may write the complete likelihood function as

$$L(a, \pi) = \prod_{\alpha=1}^{n} f(a_\alpha, \pi) = \prod_i \pi_i^{a_i} / \prod_{i<j<k} \Delta_{ijk}^n , \tag{4.4}$$

in which

$$a_i = \sum_{\alpha=1}^{n} a_{i\alpha} = \sum_\alpha {\sum_{j<k}}' (3 - r_{i\alpha,jk}) = \tfrac{3}{2} n(t-1)(t-2) - \sum_\alpha {\sum_{j<k}}' r_{i\alpha,jk} . \tag{4.5}$$

The final term in (4.5) is the total sum of ranks for T_i.

The maximum likelihood estimators $p_1, \cdots, p_t$ are obtained by maximizing

$\ln L(a, \pi)$ with respect to $\pi_1, \cdots, \pi_t$ subject to (3.1) and represent the values of these parameters for which the maximum occurs. The resulting normal equations after minor simplifications are

$$a_i/p_i = n \sum_{j<k}{}' \{[2p_i(p_j + p_k) + p_j^2 + p_k^2]/D_{ijk}\}, \qquad i = 1, \cdots, t, \tag{4.6}$$

which together with (3.1) yield solutions $p_1, \cdots, p_t$. In (4.6) we use

$$D_{ijk} = p_i^2(p_j + p_k) + p_j^2(p_i + p_k) + p_k^2(p_i + p_j), \tag{4.7}$$

similar to (3.5).

A closed-form solution of Equations (4.6) and (3.1) is not tractable, but an iterative solution is available and leads to a unique solution. Convergence is proved in a manner similar to the proof by Ford [21] of convergence for the paired-comparisons problem. Iteration proceeds as follows: Let $p_1^{(0)}, \cdots, p_t^{(0)}$ be first trial values for $p_1, \cdots, p_t$. Second trial values are

$$Cp_i^{(1)} = a_i\{n \sum_{j<k}{}' [2p_i^{(0)}(p_j^{(0)} + p_k^{(0)}) + p_j^{(0)2} + p_k^{(0)2}]/D_{ijk}^{(0)}\}^{-1}, \tag{4.8}$$

where C is eliminated through use of (3.1) and $D_{ijk}^{(0)}$ is D_{ijk} evaluated by using $p_1^{(0)}, \cdots, p_t^{(0)}$. The procedure indicated is continued through repeated use of (4.8) and (3.1) until the process converges to the accuracy required. This method is readily adaptable to machine use, and the rapidity of convergence is good if the initial trial values are good.

Dykstra [22] suggested means of obtaining good values of $p_1^{(0)}, \cdots, p_t^{(0)}$ for paired-comparisons, and the same approach may be used here. The procedure is to set $p_j = (1 - p_i)/(t - 1)$ in the right-hand member of (4.6), $j \neq i, j = 1, \cdots, t$. Such a substitution yields a quadratic equation in p_i,

$$\begin{aligned} p_i^2[a_i(2t^2 - 6t + 6) - n(2t^4 - 11t^3 + 22t^2 - 19t + 6)] \\ + p_i[a_i(2t - 6) - n(t^3 - 4t^2 + 5t - 2)] + 2a_i = 0, \end{aligned} \tag{4.9}$$

$i = 1, \cdots, t$. The positive roots of these equations yield values $Cp_i^{(0)}$, and (3.1) is used to determine C and hence the $p_i^{(0)}$. These values are good first approximations in most cases; the exceptional cases are those in which there are great differences among the a_i and consequently the p_i.

Sometimes extreme sets of values of $a_1, \cdots, a_t$ occur as in paired comparisons (Bradley [23]). This happens when $T_1, \cdots, T_t$ have a subset that always outranks the complementary subset. Then values of p_i for the dominated subset will be zero and indeterminacies occur. (Examples can be found in Table 1.) As in the reference, the procedure is to replace those p_i for the dominated subset (if it contains more than one treatment) by $b_i\delta$ and let $\delta \to 0$ in $L(a, \pi)$ of (4.4). Values of p_i for the dominant subset and of b_i for the dominated subset are now obtained by special maximization of $L(a, \pi)$ after the limit as $\delta \to 0$. The b_i are regarded as secondary estimators and provide information about the relative merits of the dominated subset. Sometimes further subdivision is necessary. Similar use of δ and the b_i is sometimes necessary in evaluating statistics based on the p_i in the following work. Additional details are given by Pendergrass [20].

5. The Distribution of Estimators

Under general conditions (such as those given by Chanda [8]) on $f(a_\alpha, \pi)$ that can be verified for triple comparisons, p_i is a consistent estimator of π_i. If we regard $\pi_t = 1 - \sum_{i=1}^{t-1}\pi_i$ as a function of the first $(t-1)$ parameters and p_t similarly as a function of the first $(t-1)$ estimators, $\sqrt{n}(p_1 - \pi_1), \cdots, \sqrt{n}(p_{t-1} - \pi_{t-1})$ have a joint $(t-1)$-variate normal distribution asymptotically with n, with zero means and dispersion matrix given by the inverse of the matrix (λ'_{ij}) where

$$\lambda'_{ij} = E\left[\frac{\partial \ln f(a_\alpha, \pi)}{\partial \pi_i} \cdot \frac{\partial \ln f(a_\alpha, \pi)}{\partial \pi_j}\right], \qquad i, j = 1, \cdots, (t-1) . \tag{5.1}$$

It follows directly from (4.2) that

$$\lambda'_{ij} = \lambda_{ij} - \lambda_{it} - \lambda_{jt} + \lambda_{tt} , \tag{5.2}$$

where

$$\lambda_{ij} = \text{cov}\,(a_{i\alpha}, a_{j\alpha})/\pi_i\pi_j , \qquad i, j = 1, \cdots, t . \tag{5.3}$$

The covariances (or variances if $i = j$) in (5.3) are found by noting the definition (4.3) and that terms in that sum are independent in probability. We may write $a_{i\alpha,jk} = 3 - r_{i\alpha,jk}$ and $a_{i\alpha} = \sum'_{j<k} a_{i\alpha,jk}$. The variances and covariances needed and basic to evaluating λ_{ij} are

$$\text{var}\,(a_{i\alpha,jk}) = \pi_i(\pi_j + \pi_k)[4\pi_i\pi_j\pi_k(\pi_j + \pi_k) + (\pi_j^2 + \pi_k^2)(\pi_i^2 + \pi_j\pi_k)]/\Delta_{ijk}^2 , \tag{5.4}$$

and

$$\text{cov}\,(a_{i\alpha,jk}, a_{j\alpha,ik}) = -\pi_i\pi_j[2\pi_i\pi_j\pi_k(\pi_i + 2\pi_k + \pi_j) + 2\pi_k^3(\pi_i + \pi_j) - \pi_k^2(\pi_i^2 + \pi_j^2) + \pi_i^2\pi_j^2 + \pi_k^4]/\Delta_{ijk}^2 . \tag{5.5}$$

From (5.4), (5.5), and (4.3), it follows that

$$\lambda_{ii} = \frac{1}{\pi_i}\sum_{j<k}{}'\{[4\pi_i\pi_j\pi_k(\pi_j + \pi_k)^2 + (\pi_j + \pi_k)(\pi_j^2 + \pi_k^2)(\pi_i^2 + \pi_j\pi_k)]/\Delta_{ijk}^2\} , \qquad i = 1, \cdots, t , \tag{5.6}$$

and

$$\lambda_{ij} = -\sum_{k\neq i,j}\{[2\pi_i\pi_j\pi_k(\pi_i + 2\pi_k + \pi_j) + 2\pi_k^3(\pi_i + \pi_j) - \pi_k^2(\pi_i^2 + \pi_j^2) + \pi_i^2\pi_j^2 + \pi_k^4]/\Delta_{ijk}^2\} , \qquad i \neq j, \quad i, j = 1, \cdots, t . \tag{5.7}$$

The matrix elements λ'_{ij} in (5.1) may now be evaluated and the dispersion matrix of $\sqrt{n}(p_1 - \pi_1), \cdots, \sqrt{n}(p_{t-1} - \pi_{t-1})$ obtained. In practice, estimates are used by replacing $\pi_1, \cdots, \pi_t$ with $p_1, \cdots, p_t$ in the calculations.

Although we have now provided means of obtaining the dispersion matrix, a more symmetric form is desirable. We notice that $\pi_i\lambda_{ii} + \sum'_j \pi_j\lambda_{ij} = 0$ and may resort to the same algebraic manipulations used by Bradley [18] for paired comparisons. If elements σ_{ij} of $(\lambda'_{ij})^{-1}$ are defined, the result is that

$$\sigma_{ij} = \frac{\text{cofactor of } \lambda_{ij} \text{ in} \begin{bmatrix} (\lambda_{ij}) & e' \\ e & 0 \end{bmatrix}}{\begin{vmatrix} (\lambda_{ij}) & e' \\ e & 0 \end{vmatrix}}, \qquad i, j = 1, \cdots, (t-1), \tag{5.8}$$

where (λ_{ij}) denotes the t-square matrix with elements in (5.6) and (5.7), e denotes the t-element row vector of unit elements, and e' is the transpose of e. But it follows from symmetry that (5.8) holds for $i, j = 1, \cdots, t$, and we may define a t-square matrix $\Sigma = (\sigma_{ij})$ that is singular with rank $(t-1)$. It follows that $\sqrt{n}\,(p_1 - \pi_1), \cdots, \sqrt{n}\,(p_t - \pi_t)$ have, asymptotically with n, the singular t-variate normal distribution with zero means and dispersion matrix Σ.

If $\pi_1, \cdots, \pi_t$ are near equality, good approximations to the values of the σ_{ij} may be obtained by taking $\pi_1 = \cdots = \pi_t = 1/t$ and then

$$\lambda_{ij} = -t^2(t-2)/3\,, \qquad i \neq j, \quad i, j = 1, \cdots, t\,, \tag{5.9}$$

and

$$\lambda_{ii} = t^2(t-1)(t-2)/3\,, \qquad i = 1, \cdots, t\,. \tag{5.10}$$

The elements of Σ can be evaluated directly from (5.8), and we obtain

$$\sigma_{ij} = -3/t^4(t-2)\,, \qquad i \neq j, \quad i, j = 1, \cdots, t\,, \tag{5.11}$$

and

$$\sigma_{ii} = 3(t-1)/t^4(t-2)\,, \qquad i = 1, \cdots, t\,. \tag{5.12}$$

The p_i, as estimators of true treatment parameters, provide a means of quantifying the results of a triple-comparisons experiment. The results concerning their distribution are useful in large-sample comparisons and contrasts involving the p_i or functions of them. The analogy with results for paired comparisons suggests the use of $\ln p_i$ as a scale point for T_i, and contrasts among these values may be considered and confidence limits obtained at least approximately.

6. Some Generalizations on Estimation

Bradley and Terry [15] considered the somewhat more general model in which the treatments might be grouped so that $\pi_i = \pi(b)$, $b = 1, \cdots, m$, $i = s_{b-1} + 1, \cdots, s_b$ where $s_0 = 1$, $s_m = t$ and $\sum_{b=1}^{m}(s_b - s_{b-1})\pi(b) = 1$ for paired comparisons. This may also be done for triple comparisons and simply involves substitution of the $\pi(b)$ in (4.4) for the π_i in appropriate places, and maximization subject to the new restraint of this paragraph instead of (3.1) formerly used. Then maximum likelihood estimators $p(b)$ for $\pi(b)$ may be obtained. We shall not give details here but refer to [20] where some additional details are given.

The work discussed in preceding sections represents a special use of the more general model of this section where $m = t$. Another special case of some interest is that for which $m = 2$. When $m = 2$, suppose $\pi_i = \pi$,

$i = 1, \cdots, s$ and $\pi_i = (1 - s\pi)/(t - s)$, $i = s + 1, \cdots, t$. In this situation, p the estimator of π is the permissible root of the quadratic

$$(6.1)\quad [b'(t-s)(2t-3s) + a's(t-3s)]p^2 + [b'(t-s) - a'(t-5s)]p - 2a' = 0 .$$

Definitions of a' and b' follow: Let a be twice the number of times treatments among the first group of s treatments receive rank 1 plus the number of times they receive rank 2. A parallel definition of b is made for the second group of the remaining $(t - s)$ treatments. More mathematically,

$$(6.2)\qquad a = \sum_{i=1}^{s} a_i \quad \text{and} \quad b = \sum_{i=s+1}^{t} a_i .$$

Then

$$(6.3)\qquad a' = a - 3n\binom{s}{3} - (t-s)n\binom{s}{2},$$

and

$$(6.4)\qquad b' = b - 3n\binom{t-s}{3} - sn\binom{t-s}{2}.$$

In (6.3) and (6.4) delete those terms containing combination symbols undefined if s or $(t - s)$ is too small. Note also that

$$a' + b' = 2sn\binom{t-s}{2} + 2(t-s)n\binom{s}{2}.$$

Definitions in (6.2) are based on the assumption that all $\binom{t}{3}$ triple comparisons are used—usually this will have been the case—but sometimes the $\binom{s}{3}$ and $\binom{t-s}{3}$ triplets of treatments formed from within the two groups need not be used, for they yield no information on π. If these triplets are omitted, delete the terms $3n\binom{s}{3}$ and $3n\binom{t-s}{3}$ in (6.3) and (6.4) respectively and proceed. From (6.1), if $a' = 0$, then $p = 0$, and this occurs when treatments of the second group are always preferred to treatments of the first group; if $b' = 0$, the reverse situation holds and $p = 1/s$ is a root of (6.1).

The asymptotic variance of $\sqrt{n}(p - \pi)$ is obtained by developing the required special form of $f(a_\alpha, \pi)$ in (4.2) for this case and evaluating the reciprocal of $E(\partial \ln f/\partial\pi)^2$. The result is

$$(6.5)\qquad \operatorname{var}[\sqrt{n}(p-\pi)] = \frac{2\pi(1-s\pi)[(t^2-3st+3s^2)\pi^2 + (t-3s)\pi + 1]}{s(t-2)[(t^2-6st+6s^2)\pi^2 + 2(2t-3s)\pi + 1]} .$$

If each $\pi = 1/t$, $\operatorname{var}[\sqrt{n}(p - \pi)]$ becomes $(t - s)/[st^2(t - 2)]$. Again, for large samples, (6.5) is evaluated by substitution of p for π.

Another generalization is possible and was suggested in Section 3. Now we again consider t distinct treatments but use n_{ijk} observations on the triplet T_i, T_j, T_k, $i \neq j \neq k$, $i, j, k = 1, \cdots, t$. Results already obtained may be modified. The normal equations (4.6) become simply

$$(6.6)\qquad \frac{a_i}{p_i} = \sum_{j<k}{}' n_{ijk}\{[2p_i(p_j + p_k) + p_j^2 + p_k^2]/D_{ijk}\}, \qquad i = 1, \cdots, t ,$$

and (3.1) is unchanged. In (4.9), for trial values of the p_i, n is replaced by $\sum'_{j<k} n_{ijk}/\binom{t-1}{2}$. Results on the limiting distribution are also changed. We now define

$$N = \sum_{i<j<k} n_{ijk}, \tag{6.7}$$

and

$$\gamma_{ijk} = n_{ijk}/N. \tag{6.8}$$

Then $\sqrt{N}(p_1 - \pi_1), \cdots, \sqrt{N}(p_t - \pi_t)$ have, asymptotically with N for fixed ratios γ_{ijk}, the singular t-variate normal distribution previously defined but with modified definitions,

$$\lambda_{ii} = \frac{1}{\pi_i} \sum_{j<k}' \{\gamma_{ijk}[4\pi_i\pi_j\pi_k(\pi_j + \pi_k)^2 + (\pi_j + \pi_k)(\pi_j^2 + \pi_k^2)(\pi_i^2 + \pi_j + \pi_k)]/\Delta_{ijk}^2\} \qquad i = 1, \cdots, t, \tag{6.9}$$

and

$$\lambda_{ij} = -\sum_{k \neq i,j}' \{\gamma_{ijk}[2\pi_i\pi_j\pi_k(\pi_i + 2\pi_k + \pi_j) + 2\pi_k^3(\pi_i + \pi_j) - \pi_k^2(\pi_i^2 + \pi_j^2) + \pi_i^2\pi_j^2 + \pi_k^4]/\Delta_{ijk}^2\}, \qquad i \neq j,\ i, j = 1, \cdots, t. \tag{6.10}$$

Modification of the usual limit theorems is required and has been discussed by Bradley and Gart [25].

7. Tests of Main Effects

Asymptotic tests of significance on the equality of treatment effects may be made. Consider

$$H_0 : \pi_1 = \cdots = \pi_t = 1/t \tag{7.1}$$

and the alternative

$$H_{a_1} : \pi_i \neq \pi_j, \qquad \text{for some } i \neq j,\ i, j = 1, \cdots, t. \tag{7.2}$$

The π_i are completely specified under H_0 and have values $p_1, \cdots, p_t$ already obtained under H_{a_1}. $L(a, \pi)$ in (4.4) is used to obtain the likelihood ratio λ_1 and

$$Z = -2 \ln \lambda_1 = 2n\binom{t}{3} \ln 6 + 2 \sum_i a_i \ln p_i - 2n \sum_{i<j<k} \ln D_{ijk}. \tag{7.3}$$

For large n, Z may be taken to have the chi-square distribution with $(t-1)$ degrees of freedom (Wilks [26]) under the null hypothesis H_0.

Small-sample tables for the distribution of Z given H_0 may be developed but are laborious to construct and would be extremely voluminous if sufficiently extensive for practical use. The procedure for developing such tables is similar to that used by Bradley and Terry [15] and Bradley [23]. An example of such tables is given in Table 1, wherein Z_0 indicates the values of Z for specified sets of sums of ranks in the table.

TABLE 1

DISTRIBUTION OF $Z = -2 \ln \lambda_1$ FOR $t = 4$, $n = 1$

Rank sums				Estimates of π_i				Distribution	
$\sum r_1$	$\sum r_2$	$\sum r_3$	$\sum r_4$	p_1	p_2	p_3	p_4	Z_0	$P(Z \geq Z_0)$
3	5	7	9	1.00	....	....	....	14.33	.02
3	6	6	9	1.00	....	....	....	11.56	.06
4	4	7	9	0.50	0.50	....	....	11.56	
4	4	8	8	0.50	0.50	....	....	8.79	.07
4	5	6	9	0.53	0.31	0.16	....	7.66	.32
3	6	7	8	1.00	....	....	....	7.66	
3	5	8	8	1.00	....	....	....	7.02	.34
3	7	7	7	1.00	....	....	....	6.59	.38
5	5	5	9	0.33	0.33	0.33	....	6.59	
4	5	7	8	0.52	0.31	0.11	0.06	4.04	.49
4	6	6	8	0.53	0.20	0.20	0.07	3.49	.56
5	5	6	8	0.35	0.35	0.22	0.08	2.46	.69
4	6	7	7	0.54	0.20	0.13	0.13	2.46	
5	5	7	7	0.35	0.35	0.15	0.15	1.59	.78
5	6	6	7	0.36	0.24	0.24	0.16	0.78	.98
6	6	6	6	0.25	0.25	0.25	0.25	0.00	1.00

Some indication of the rapidity of convergence of Z to the chi-square distribution is given in Tables 2 and 3. The means and variances of the exact distribution of Z in Table 2 indicate a fairly rapid convergence to the limiting values of the mean and variance even for relatively small values of n. The convergence is probably even more rapid for larger values of t since this is true in paired comparisons. Table 3 compares the exact

TABLE 2

MEANS AND VARIANCES FOR Z FOR $t = 3$

n	$E(Z)$	$V(Z)$
2	3.02	6.28
3	2.66	6.97
4	2.41	6.30
5	2.28	5.67
6	2.24	5.41
8	2.16	4.73
∞	2.00	4.00

TABLE 3

DISTRIBUTION OF Z AND OF CHI-SQUARE FOR $t = 3$ FOR SELECTED VALUES Z_0 OF Z

$n = 6$			$n = 8$		
Z_0	$P(Z \leqq Z_0)$	$P(\chi^2 \leqq Z_0)$	Z_0	$P(Z \leqq Z_0)$	$P(\chi^2 \leqq Z_0)$
2.48	.75	.71	4.34	.880	.880
3.26	.82	.80	5.13	.921	.919
4.55	.86	.90	5.76	.953	.942
4.85	.93	.91	7.01	.963	.970
6.38	.95	.96	7.70	.970	.977
7.58	.97	.98	8.07	.982	.981
8.60	.99	.98	9.07	.990	.989
10.87	.992	.995	10.85	.992	.995
13.18	.995	.998	11.42	.995	.996
13.87	.998	.999	13.91	.999	.999

cumulative distribution of Z with the limiting chi-square distribution having two degrees of freedom. There is very little difference between the significant values of Z for eight repetitions and those of the chi-square distribution.

The test of significance may still be developed if all n_{ijk} are not equal to n. Then the estimators $p_1, \cdots, p_t$ are obtained through use of (6.6), and

$$Z = -2 \ln \lambda_1 = N \ln 6 + 2 \sum a_i \ln p_i - 2 \sum_{i<j<k} n_{ijk} \ln D_{ijk} . \tag{7.4}$$

Demonstration that Z still has the chi-square distribution with $(t - 1)$ degrees of freedom as N becomes large with γ_{ijk} fixed proportions follows from [25].

When H_0 of (7.1) is compared with

$$H_{a_2} : \pi_i = \pi, \quad i = 1, \cdots, s, \qquad \pi_i = (1 - s\pi)/(t - s), \quad i = s + 1, \cdots, t, \tag{7.5}$$

the resulting statistic is

$$\begin{aligned} 2 \ln \lambda_2 = {} & 2n\left\{\binom{t}{3} - \left[\binom{s}{3} + \binom{t-s}{3}\right]\right\} \ln 6 \\ & - 2n\left[(t-s)\binom{s}{2} + s\binom{t-s}{2}\right] \ln 2 - 2n\left[(t-s)\binom{s}{2}\right. \\ & \left. + s\binom{t-s}{2}\right] \ln \{[1 + (t - 3s)p + (t^2 - 3st + 3s^2)p^2]/(t-s)^2\} \\ & + 2a' \ln p + 2b' \ln [(1 - sp)/(t - s)] , \end{aligned} \tag{7.6}$$

where p results from (6.1) and a' and b' are defined in (6.3) and (6.4). Asymptotically with n, $-2 \ln \lambda_2$ has the chi-square distribution with one degree of freedom.

8. Tests of Interaction

In certain situations, repetitions of triple comparisons may be completed in groups of repetitions by different judges at different times, or under different circumstances. Then the over-all experiment may be considered as one with g groups of repetitions, the uth of which has n_u repetitions, $n = \sum_{u=1}^{g} n_u$. Distinct treatment parameters $\pi_{1u}, \cdots, \pi_{tu}$ may exist for each group, $u = 1, \cdots, g$. The failure of the treatment parameters to be the same for each group represents a group-by-treatment interaction or lack of agreement. We now develop a test to detect such interactions.

The required test of interaction is a test of the null hypothesis

$$H_0 : \pi_{iu} = \pi_i , \qquad i = 1, \cdots, t, \quad u = 1, \cdots, g , \tag{8.1}$$

against the alternative

$$H_{a_3} : \pi_{iu} \neq \pi_i , \qquad \text{for some } i \text{ and } u. \tag{8.2}$$

We again use large-sample likelihood ratio tests and here this is most easily done by considering an allied test. Consider

$$H_0 : \pi_{iu} = 1/t , \qquad \text{for all } i \text{ and } u, \tag{8.3}$$

and

$$H_{a_4} : \pi_{iu} \neq 1/t , \qquad \text{for some } i \text{ and } u, \tag{8.4}$$

but with no assumption that $\pi_{iu} = \pi_{iv}$, $u \neq v$, $u, v = 1, \cdots, g$, $i = 1, \cdots, t$. It turns out that if λ_C is the likelihood ratio in this latter situation,

$$Z_C = -2 \ln \lambda_C = \sum_{u=1}^{g} Z_u , \tag{8.5}$$

where Z_u is the Z of (7.3) computed for the uth group. Asymptotically with the n_u, $-2 \ln \lambda_C$ has the chi-square distribution with $g(t-1)$ degrees of freedom under H_0 of (8.3). This test is designated as the "combined test" of treatment equality or for main effects. It may sometimes be useful, but its usefulness is limited. The likelihood ratio test of interaction, H_0 of (8.1) versus H_{a_3} of (8.2), depends on

$$Z_C - Z \tag{8.6}$$

and asymptotically with the n_u has the chi-square distribution with $(g-1)(t-1)$ degrees of freedom. In (8.6), Z_C is as defined in (8.5), and Z as in (7.3), based on pooling the totality of repetitions.

The computing procedures are clear. To obtain Z_u, $p_{1u}, \cdots, p_{tu}$ are obtained as estimates of $\pi_{1u}, \cdots, \pi_{tu}$ through consideration of only the uth group and use of the methods of Sections 4 and 5. To obtain Z, $p_1, \cdots, p_t$ are obtained as estimates of $\pi_1, \cdots, \pi_t$ on the assumption that all groups of repetitions may be pooled into a single group.

The large-sample test procedures are summarized in Table 4. If the n_{ijk} are not equal to n_u in each group, then values of Z in Table 4 may all be modified through use of (7.4) instead of the form in (7.3).

9. A Test of the Model

In any statistical method, it is desirable that means be available to test

TABLE 4

LARGE-SAMPLE TESTS OF SIGNIFICANCE

Test	Hypothesis	Statistic	References	Limiting distribution
Main effects (No interaction)	$H_0: \pi_i = 1/t$ $H_{a_1}: \pi_i$	Z	(7.1) (7.2) (7.3)	χ^2_{t-1}
Interaction	$H_0: \pi_{iu} = \pi_i$ $H_{a_3}: \pi_{iu} \neq \pi_i$	$Z_c - Z$	(8.1) (8.2) (8.6)	$\chi^2_{(g-1)(t-1)}$
Main effect (Interaction)	$H_0: \pi_{iu} = 1/t$ $H_{a_4}: \pi_{iu}$	Z_c	(8.3) (8.4) (8.5)	$\chi^2_{g(t-1)}$

the appropriateness of the model on which the method is based.

The most general model for triple comparisons, and one that fits data exactly, is formed by postulating for each triplet the existence of positive parameters $\pi_{ijk}, \cdots, \pi_{kji}$, six in number, corresponding to the probabilities of occurrence of the six possible rankings of T_i, T_j, and T_k. Thus, π_{ijk} is the probability that T_i, T_j, T_k receive ranks 1, 2, and 3 respectively in a triplet. The six parameters sum to unity for each triplet, and their maximum likelihood estimators are $f_{ijk}/n, \cdots, f_{kji}/n$ for the n comparisons on this triplet and where f_{ijk} is the number of times the ranking 1, 2, and 3 for T_i, T_j, and T_k respectively occurs in the n triplets.

The basic model for triple comparisons implies that

$$H_0: \pi_{ijk} = \pi_i^2\pi_j/\Delta_{ijk}\,, \qquad i \neq j \neq k, \quad i, j, k = 1, \cdots, t\,, \tag{9.1}$$

and the most general model leads to the alternative

$$H_{a_5}: \pi_{ijk} \neq \pi_i^2\pi_j/\Delta_{ijk}\,, \qquad \text{for some } i,\ j,\ k\,. \tag{9.2}$$

Let us write $\Pi_{(i,j,k)}$ and $\Sigma_{(i,j,k)}$ to indicate respectively products and sums over the six permutations of i, j, k. Then the general likelihood function for triple comparisons is

$$L(\pi_{ijk}) = \prod_{i<j<k} \prod_{(i,j,k)} \pi_{ijk}^{f_{ijk}} \tag{9.3}$$

and this reduces to $L(a, \pi)$ in (4.4) if the relations of (9.1) apply since

$$a_i = \sum_{j\neq k}{}' (2f_{ijk} + f_{ijk})\,, \qquad i = 1, \cdots, t\,. \tag{9.4}$$

The likelihood ratio statistic for testing H_0 of (9.1) against H_{a_5} of (9.2) is

$$-2\ln\lambda_5 = 2\left[\sum_{i<j<k}\sum_{(i,j,k)} f_{ijk}\ln f_{ijk} - n\binom{t}{3}\ln n + n\sum_{i<j<k}\ln D_{ijk} - \sum_i a_i \ln p_i\right]. \tag{9.5}$$

This test constitutes a test of the model for triple comparisons and, for large n, $-2\ln\lambda_5$ of (9.5) has the chi-square distribution with

$$\left[5\binom{t}{3}-(t-1)\right]$$

degrees of freedom.

The usual procedure for tests of goodness of fit is to form chi-square by taking sums of terms of the form $(f_{ijk}-f'_{ijk})^2/f'_{ijk}$, where f_{ijk} is the observed frequency as defined above and f'_{ijk} is the estimate of the expected frequency corresponding to f_{ijk}. Under H_0 of (9.1),

$$f'_{ijk} = np_i^2 p_j/D_{ijk} \ . \tag{9.6}$$

In terms of frequencies, $-2\ln\lambda_5$ of (9.5) may be written

$$-2\ln\lambda_5 = 2\sum_{i<j<k}\ \sum_{(i,j,k)} f_{ijk}\ln(f_{ijk}/f'_{ijk}) \ . \tag{9.7}$$

An approximation procedure, similar to that used by Bradley [17] for paired comparisons, shows that

$$-2\ln\lambda_5 \approx \sum_{i<j<k}\ \sum_{(i,j,k)} (f_{ijk}-f'_{ijk})^2/f'_{ijk} \ . \tag{9.8}$$

The forms for $-2\ln\lambda_5$ in (9.5) and (9.8) have very nearly equal values and differ very little in computational convenience, although it may be desirable actually to compute values f'_{ijk} in applied work.

If the n_{ijk} are not all equal, forms (9.7) and (9.8) may be used directly for the test of the model. They require the new definitions

$$f'_{ijk} = n_{ijk}p_i^2 p_j/D_{ijk} \tag{9.9}$$

and computation of $p_1, \cdots, p_t$ based on (6.6). When group-by-treatment interactions exist, the basic model cannot be expected to apply over all, but tests on the model should be made within groups of repetitions in just the way described here but through treatment of each group of repetitions separately. If an over-all test is then required, values of chi-square should be added over groups of repetitions.

The Bradley-Terry model for paired comparisons has been used extensively and has appeared to fit many practical applications. An experimental methodological study for paired comparisons by Hopkins [27] led to the conclusion that the proposed model was appropriate. Sufficient experimental evidence is not yet available to assess the appropriateness of the model for triple comparisons. No doubt there will be some departures from the model, but it is conjectured that they will be rare and that assignable causes may usually be found to explain them when they do occur.

10. The Limiting Power Function of the Test of Main Effects

The likelihood ratio test based on $Z = -2\ln\lambda_1$ of (7.3) is a consistent test of H_0 of (7.1) given non-equal values of the parameters $\pi_1, \cdots, \pi_t$. Consequently, to obtain the limiting power function of the test, we adopt the frequently used procedure of considering alternative values of $\pi_1, \cdots, \pi_t$ increasingly close to the null value $1/t$ of these parameters with large n.

We set

$$\pi_i = \frac{1}{t} + \frac{\delta_{in}}{\sqrt{n}}\,, \qquad i = 1, \cdots, t\,, \tag{10.1}$$

where δ_{in} represents a sequence of constants converging to δ_i as $n \to \infty$.

The development of the limiting distribution of Z follows that of [18] very closely. The first step involves showing that

$$Z = \frac{nt(t-2)}{3} \sum y_i^2 + R_1(y)\,, \tag{10.2}$$

where

$$y_i = t\left(p_i - \frac{1}{t}\right), \qquad i = 1, \cdots, t\,, \tag{10.3}$$

and $R_1(y)$ converges to zero in probability, given that the $\sqrt{n}\, y_i$ have limiting distributions. The form (10.2) results from the expansion of the right-hand member of (7.3) after substitution based on (10.3). It will follow that Z has the same limiting distribution as $[nt(t-2)/3)] \sum_i y_i^2$.

The limiting distribution of the $\sqrt{n}\, y_i$ is considered from an examination of the normal equations (4.6). Substitution of the y_i in (4.6) and reduction of the resulting form leads to

$$\frac{a_i}{n} = \binom{t-1}{2} + \frac{t(t-2)}{3} y_i + R_2(y)\,. \tag{10.4}$$

In (10.4) $R_2(y)$ is such that $t(t-2)y_i/3 \to (a_i/n) - \binom{t-1}{2}$ for fixed a_i/n and as $n \to \infty$. We shall show that $3\sqrt{n}\,[(a_i/n) - \binom{t-1}{2}]/t(t-2)$ have a joint limiting distribution, and consequently it follows that the $\sqrt{n}\, y_i$ have the same joint limiting distribution.

Let

$$z_i = \sqrt{n}\left[\frac{a_i}{n} - \binom{t-1}{2}\right], \qquad i = 1, \cdots, t\,, \tag{10.5}$$

and then

$$z_i = \frac{1}{\sqrt{n}} \sum_{\alpha=1}^{n} \sum_{j<k}{}' (a_{i\alpha,jk} - 1)\,, \tag{10.6}$$

from (4.5) and the definitions of $a_{i\alpha,jk}$ preceding (5.4). The characteristic functions of $z_1, \cdots, z_t$ may be written as

$$\begin{aligned} \phi_n(\tau) &= E\left\{\exp\left[\frac{\sqrt{-1}}{\sqrt{n}} \sum_{\alpha=1}^{n} \sum_i \sum_{j<k}{}' (a_{i\alpha,jk} - 1)\tau_i\right]\right\} \\ &= \left\{E \exp\left[\frac{\sqrt{-1}}{\sqrt{n}} \sum_i \sum_{j<k}{}' (a_{i,jk} - 1)\tau_i\right]\right\}^n, \end{aligned} \tag{10.7}$$

the second form depending on the assumed independence of the $a_{i\alpha,jk}$ for different α, and $a_{i,jk}$ representing a representative $a_{i\alpha,jk}$. The exponential of the second form is expanded to yield

$$\phi_n(\tau) = \left\{E\left[1 - \frac{\sqrt{-1}}{\sqrt{n}} \sum_i \sum_{j<k}{}' (a_{i,jk} - 1)\tau_i - \frac{1}{2n}\left(\sum_i \sum_{j<k}{}' (a_{i,jk} - 1)\tau_i\right)^2 + \frac{1}{n^{3/2}} R_1'\right]\right\}^n . \tag{10.8}$$

The $a_{i,jk}$ are functions of multinomial variates with the multinomial parameters dependent on π_i, π_j, and π_k in the triplet comparing T_i, T_j, and T_k. It is easy, but tedious, to show that with the π_i of (10.1), from whence $\sum_i \delta_{in} = 0$,

$$E\left[\sum_{j<k}{}' (a_{i,jk} - 1)\right] = \sum_{j<k} \frac{t^2(t-2)\delta_{in}}{3\sqrt{n}} + \frac{R_2'}{n} , \tag{10.9}$$

and

$$E\left[\sum_i \sum_{j<k}{}' (a_{i,jk} - 1)\tau_i\right]^2 = \sum_i \frac{(t-1)(t-2)}{3}\tau_i^2 - \sum_{i \neq j} \frac{(t-2)}{3}\tau_i\tau_j + \frac{R_3'}{\sqrt{n}} . \tag{10.10}$$

Then,

$$\phi_n(\tau) = \left\{1 + \frac{\sqrt{-1}}{3n} \sum_i t^2(t-2)\delta_{in}\tau_i - \frac{1}{2n}\left[\sum_i \frac{(t-1)(t-2)\tau_i^2}{3} - \sum_{i \neq j} \frac{(t-2)\tau_i\tau_j}{3}\right] + \frac{R}{n^{3/2}}\right\}^n , \tag{10.11}$$

where R_1', R_2', and R_3' of (10.8), (10.9), and (10.10) combine into R of (10.11) so that R is bounded as $n \to \infty$. We let $\phi(\tau) = \lim_{n\to\infty} \phi_n(\tau)$ and obtain

$$\phi(\tau) = \exp\left\{\frac{\sqrt{-1}}{3} \sum_i t^2(t-2)\delta_i\tau_i - \tfrac{1}{2}\left[\frac{(t-1)(t-2)}{3} \sum_i \tau_i^2 - \frac{(t-2)}{3} \sum_{i \neq j} \tau_i\tau_j\right]\right\} . \tag{10.12}$$

But (10.12) is the characteristic function of the singular multivariate normal distribution in t variates and rank $(t-1)$ with means $t^2(t-2)\delta_i/3$, $i = 1, \cdots, t$, and variances and covariances $(t-1)(t-2)/3$ and $-(t-2)/3$ respectively. Then the $\sqrt{n}\, y_i$ have a similar limiting distribution with means $t\delta_i$ and variances and covariances $3(t-1)/t^2(t-2)$ and $-3/t^2(t-2)$.

The limiting distribution of Z now follows. Let

$$u_i = \sqrt{nt(t-2)/3}\; y_i \qquad i = 1, \cdots, t , \tag{10.13}$$

and Z has the same limiting distribution as $\sum_i u_i^2$. But the u_i have the t-variate, singular, normal distribution of rank $(t-1)$ with means $t^{3/2}\sqrt{t-2}\,\delta_i/\sqrt{3}$ and variances and covariances $(t-1)/t$ and $-1/t$ respectively. The dispersion matrix for the u_i, say Σ_u, is such that $\Sigma_u = \Sigma_u^2$ and $\operatorname{tr} \Sigma_u = (t-1)$. This is sufficient to state that $\sum_i u_i^2$ and Z have the limiting, non-central chi-square distribution with $(t-1)$ degrees of freedom and parameter of non-centrality

(10.14) $$\lambda = \frac{t^3(t-2)}{3} \sum_i \delta_i^2 .$$

This limiting distribution may be written

(10.15) $$f(Z) = \frac{\exp[-\frac{1}{2}(Z+\lambda)]}{2^{(t-1)/2}} \sum_{h=0}^{\infty} \frac{Z^{\frac{1}{2}(t-1)+h-1}\lambda^h}{\Gamma[\frac{1}{2}(t-1)+h]2^{2h}h!} .$$

The limiting distribution $f(Z)$ may be used for large-sample comparisons of test procedures, as is done in the following section, and for large-sample approximations to powers of the test. In the latter situation, it seems appropriate to take $\delta_{in}/\sqrt{n} = \pi_i - (1/t)$ and then for given $\pi_1, \cdots, \pi_t$ and n to use

(10.16) $$\lambda = nt^3(t-2) \sum_i \left(\pi_i - \frac{1}{t}\right)^2 \Big/ 3 .$$

Tables by Tang [28] or Fix [29] or the charts of Pearson and Hartley [30] may be used to evaluate integrals of $f(Z)$.

Note that when each $\pi_i = 1/t$, $\lambda = 0$, and Z has the central chi-square distribution with $(t-1)$ degrees of freedom as noted earlier.

Additional details on the developments of this section have been given by Pendergrass [20].

11. Asymptotic Relative Efficiencies of the Test of Main Effects

The method of triple comparisons will now be compared with the method of paired comparisons and the analysis of variance.

For paired comparisons, Bradley [18] defined $\pi_i = (1/t) + (\delta'_{in}/\sqrt{n'})$, $i = 1, \cdots, t$ in much the same way as we proceeded in Section 10. He then showed that the test procedure has the limiting non-central chi-square distribution with $(t-1)$ degrees of freedom and parameter of non-centrality,

(11.1) $$\lambda' = t^3 \sum_i \delta_i'^2/4 ,$$

where $\delta'_{in'} \to \delta'_i$ as $n' \to \infty$. The δ'_i of paired comparisons may be associated with the δ_i of triple comparisons as follows. For paired comparisons.

(11.2) $$P(r_i < r_j) = \pi_i/(\pi_i + \pi_j) ,$$

and for triple comparisons

(11.3) $$P(r_i < r_j) = (\pi_i^2\pi_j + \pi_i^2\pi_k + \pi_k^2\pi_i)/\Delta_{ijk} .$$

Replacement of the π_i by their expressions in terms of $\delta'_{in'}$ and δ_{in} in (11.2) and (11.3), respectively, yields

(11.4) $$P(r_i < r_j) = \frac{1}{2} + \frac{t}{4}\left(\frac{\delta'_{in'} - \delta'_{jn'}}{\sqrt{n'}}\right) + R_2 ,$$

for (11.2) and

(11.5) $$P(r_i < r_j) = \frac{1}{2} + \frac{t}{3}\left(\frac{\delta_{in} - \delta_{jn}}{\sqrt{n}}\right) + R_3 ,$$

for (11.3) where R_2 and R_3 are $O(n'^{-1})$ and $O(n^{-1})$. Then the alternatives are essentially equal for large n' and n if

$$\frac{t}{3}\delta_{in}/\sqrt{n} = \frac{t}{4}\delta'_{in'}/\sqrt{n'} \,. \tag{11.6}$$

To put paired comparisons and triple comparisons on an equal basis, we should consider treatment replications and not numbers of repetitions. These replication numbers are respectively $n_2 = n'(t-1)$ and $n_3 = n(t-1)(t-2)/2$. Then (11.6) yields

$$\sqrt{t-2}\,\delta_{in}/(3\sqrt{2n_3}) = \delta'_{in'}/(4\sqrt{n_2}) \,. \tag{11.7}$$

For the power functions of paired and triple comparisons to be identical, we require $\lambda = \lambda'$, λ in (10.13) and λ' in (11.1). The form (11.7) provides means of doing this, and it follows that the result is

$$\lim_{n,n' \to \infty} n_2/n_3 = 3/2 \,. \tag{11.8}$$

But (11.8) yields the asymptotic relative efficiency of triple comparisons to paired comparisons and the demonstration is complete. In term of repetitions, this means that $3(t-2)/4$ complete repetitions of paired comparisons are needed for each repetition of triple comparisons as an asymptotic ratio.

To compare triple comparisons with the analysis of variance for the similar balanced incomplete block design with blocks of size 3, we can obtain an association through results already obtained for paired comparisons. Bradley [18] showed that the asymptotic relative efficiency of paired comparisons compared with analysis of variance in complete blocks is $t/\pi(t-1)$. (This leads to the asymptotic relative efficiency of $2/\pi$ for paired comparisons in comparison with analysis of variance for balanced incomplete blocks of size 2.) From our examination of triple comparisons in contrast with paired comparisons, it follows that the asymptotic relative efficiency of triple comparisons to analysis of variance in complete blocks is $\frac{3}{2}t/[\pi(t-1)] = 3t/2\pi(t-1)$. The efficiency of analysis of variance in incomplete blocks of size 3 in comparison with complete blocks is $2t/3(t-1)$ and hence the asymptotic relative efficiency of triple comparisons to analysis of variance in blocks of size 3 is $[3t/2\pi(t-1)]/[2t/3(t-1)] = 9/4\pi$. The efficiencies of $2/\pi$ and $9/4\pi$ for paired and triple comparisons respectively in comparison with analysis of variance for similar designs have been confirmed by G. Noether in private correspondence.

The contrast between triple comparisons and the analysis of variance is dependent on conditions for the valid use of analysis of variance. The triple-comparisons test was formulated for use with subjective evaluations as an extension of paired comparisons and should be useful in many cases where the analysis of variance is not applicable.

12. An Example

To illustrate some of the procedures developed, we include a numerical example with $t = 4$ and $n = 40$. The data are given in Table 5, and f_{ijk} and f'_{ijk} are as defined in Section 9.

TABLE 5

FREQUENCIES AND EXPECTED FREQUENCIES OF RANKINGS IN AN EXPERIMENT WITH $t = 4$ AND $n = 40$

(Expected frequencies, f'_{ijk}, are in parentheses.)

$f_{123} = 10$ (9)	$f_{124} = 12$ (11)	$f_{134} = 10$ (11)	$f_{234} = 8$ (9)
$f_{132} = 8$ (8)	$f_{142} = 8$ (8)	$f_{143} = 8$ (8)	$f_{243} = 6$ (7)
$f_{213} = 8$ (7)	$f_{214} = 8$ (9)	$f_{341} = 8$ (4)	$f_{342} = 6$ (6)
$f_{231} = 6$ (5)	$f_{241} = 6$ (5)	$f_{314} = 8$ (8)	$f_{324} = 8$ (8)
$f_{312} = 4$ (6)	$f_{412} = 4$ (4)	$f_{413} = 4$ (5)	$f_{423} = 6$ (5)
$f_{321} = 4$ (5)	$f_{421} = 2$ (3)	$f_{431} = 2$ (4)	$f_{432} = 6$ (5)

We first obtain p_1, p_2, p_3, and p_4. From (9.4) and Table 5, $a_1 = 148$, $a_2 = 126$, $a_3 = 116$, and $a_4 = 90$. First approximations to the p's follow from (4.9) which has the form in this example,

$$(7a_i - 1800)p_i^2 + (a_i - 360)p_i + a_i = 0\,, \qquad i = 1, \cdots, 4\,.$$

The positive roots of these equations, after normalization so the $\sum_i p_i = 1$, are given in the first row of Table 6. Successive approximations in Table 6 show the convergence of the estimates and of Z from (7.3) and the excellence of the first approximations. The final Z taken as χ^2 with three degrees of freedom indicates highly significant treatment main effects.

TABLE 6

SUCCESSIVE APPROXIMATIONS TO $p_1, \cdots, p_4$ AND CORRESPONDING VALUES OF Z

Approx.	p_1	p_2	p_3	p_4	Z
0	.319	.261	.238	.183	16.53
1	.3207	.2598	.2362	.1834	16.54
2	.3211	.2596	.2360	.1833	16.53
3	.3212	.2595	.2360	.1833	16.53

Estimated variances and covariances of $\sqrt{40}(p_1 - \pi_1), \cdots, \sqrt{40}(p_4 - \pi_4)$ may be calculated by inserting values of $p_1, \cdots, p_4$ for $\pi_1, \cdots, \pi_4$ in (5.11) and (5.7) and thence in (5.8). From these substitutions we obtain

$$\hat{\Sigma} = [\hat{\sigma}_{ij}] = \begin{bmatrix} .0256 & -.0105 & -.0089 & -.0063 \\ -.0105 & .0192 & -.0053 & -.0034 \\ -.0089 & -.0053 & .0165 & -.0023 \\ -.0063 & -.0034 & -.0023 & .0119 \end{bmatrix}.$$

The approximation or null Σ based on (5.11) and (5.12) would have diagonal elements .0177 and non-diagonal elements $-.0059$. The actual diagonal elements of $\hat{\Sigma}$ above average .0183, and the non-diagonal elements average

$-.0061$. The matrix $\hat{\Sigma}$ may be used to develop confidence intervals as desired for functions of $\sqrt{n}(p_i - \pi_i)$ based on the large-sample normality of the $\sqrt{n}(p_i - \pi_i)$.

The goodness-of-fit test may be applied after appropriate substitution in (9.5) or (9.8). The f'_{ijk} are obtained from (9.6) and have been shown in parentheses in Table 5. Using (9.5), we find that $-2 \ln \lambda_5 = 7.94$, and this is taken to be a value from a chi-square distribution with 17 degrees of freedom. Use of the form (9.8) yields the comparable value of 8.00. The proposed model appears to be quite satisfactory for these data.

The data provided were not grouped and consequently we do not illustrate the test for interaction.

13. Discussion

A second derivation of the model (3.4) may be found by noticing that $P(r_i < r_j < r_k)$ involves three paired comparisons (that must be consistent) in which $r_i < r_j$, $r_i < r_k$, and $r_j < r_k$. Hence, we could have considered

$$P(r_i < r_j < r_k) = CP(r_i < r_j)\, P(r_i < r_k)\, P(r_j < r_k)$$

where $P(r_i < r_j) = \pi_i/(\pi_i + \pi_j)$ and so on for the three probabilities. The result is that

$$P(r_i < r_j < r_k) = C\pi_i^2\pi_j/(\pi_i + \pi_j)(\pi_i + \pi_k)(\pi_j + \pi_k)\,,$$

with C of necessity being such as to lead again to (3.4) for the expression for $P(r_i < r_j < r_k)$.

The approach here is capable of obvious generalizations to ranking in blocks of size greater than 3. Pendergrass [20] did consider models for larger blocks but algebraic details become increasingly complex. Larger block sizes were not further explored because it is thought that, in subjective testing involving tastes or odors particularly, paired or triple ranking will satisfy most needs.

References

[1] Fisher, R. A., and Yates, F. *Statistical Tables for Biological, Agricultural and Medical Research*, 4th ed. Edinburgh: Oliver and Boyd, 1953.

[2] Bliss, C. I., Anderson, E. O., and Marland, R. E. A technique for testing consumer preferences with special references to the constituents of ice cream. *Storrs Agr. Exp. Stat., Bull.*, No. 251, 1943, 1–20.

[3] Galinat, W. C., and Everett, H. L. A technique for testing flavor in sweet corn. *Agron. J.*, 1949, **41**, 443–45.

[4] Greenwood, M. L., and Salerno, R. Palatability of kale in relation to cooking. *Food Research*, 1949, **14**, 314–19.

[5] Friedman, M. The use of ranks to avoid the assumption of normality implicit in the analysis of variance. *J. Amer. Stat. Assoc.*, 1937, **32**, 675–701.

[6] Kendall, M. G., and Babington Smith, B. The problem of *m* rankings. *Ann. Math. Stat.*, 1939, **10**, 275–93.

[7] Durbin, J. Incomplete blocks in ranking experiments. *Brit. J. Psychol.*, 1951, **4**, 85–90.

[8] Kruskal, W. H., and Wallis, W. A. Use of ranks in one-criterion variance analysis. *J. Amer. Stat. Assoc.*, 1952, **47**, 583–621.

[9] Thurstone, L. L. Psychophysical analysis. *Amer. J. Psychol.*, 1927, **38**, 368–89.

[10] Morrissey, J. H. New method for assignment of psychometric scale values from incomplete paired comparisons. *J. Opt. Soc. Amer.*, 1955, **45**, 373–78.

[11] Gulliksen, H. A least squares solution for paired comparisons with incomplete data. *Psychometrika*, 1956, **21**, 125–34.

[12] Kendall, M. G., and Babington Smith, B. On the method of paired comparisons. *Biometrika*, 1940, **31**, 324–45.

[13] Guttman, L. An approach for quantifying paired comparisons and rank order. *Ann. Math. Stat.*, 1946, **17**, 144–63.

[14] Scheffé, H. An analysis of variance for paired comparisons. *J. Amer. Stat. Assoc.*, 1952, **47**, 381–400.

[15] Bradley, R. A., and Terry, M. E. The rank analysis of incomplete block designs. I. The method of paired comparisons. *Biometrika*, 1952, **39**, 324–45.

[16] Savage, I. R. Bibliography of nonparametric statistics and related topics. *J. Amer. Stat. Assoc.*, 1953, **48**, 844–906.

[17] Bradley, R. A. Incomplete block rank analysis: On the appropriateness of the model for a method of paired comparisons. *Biometrics*. 1954, **10**, 375–90.

[18] Bradley, R. A. Rank analysis of incomplete block designs. III. Some large-sample results on estimation and power for a method of paired comparisons. *Biometrika*, 1955, **42**, 450–70.

[19] Bradley, R. A., and Terry, M. E. The rank analysis of incomplete block designs. II. The method for blocks of size three. Appendix A, Bi-annual Report No. 4, Blacksburg, Va: Va. Agr. Exp. Stat., 1952.

[20] Pendergrass, R. N. *Ranking in Triple Comparisons* (Ph.D. dissertation). Library, Virginia Polytechnic Institute, Blacksburg, Va., 1958.

[21] Ford, L. R., Jr. Solution of a ranking problem from binary comparison. *Amer. Math. Month.*, 1957, **64** (Pt. II), 28.

[22] Dykstra, O. A note on the rank analysis of incomplete block designs—applications beyond the scope of existing tables. *Biometrics*, 1956, **12**, 301–6.

[23] Bradley, R. A. Rank analysis of incomplete block designs. II. Additional tables for the method of paired comparisons. *Biometrika*, 1954, **41**, 502–37.

[24] Chanda, K. C. A note on the consistency and maxima of the roots of likelihood equations. *Biometrika*, 1959, **46**, 55–60.

[25] Bradley, R. A., and Gart, J. J. The asymptotic properties of M. L. estimators when sampling from associated populations. Submitted for publication (1959).

[26] Wilks, S. S. *Mathematical Statistics*. Princeton, N.J.: Princeton Univ. Press, 1946.

[27] Hopkins, J. W. Incomplete block rank analysis: Some taste test results. *Biometrics*, 1954, **10**, 391–98.

[28] Tang, P. C. The power function of analysis of variance tests with tables and illustrations of their use. *Stat. Res. Mem.*, 1938, **2**, 126–57.

[29] Fix, E. Tables of non-central χ^2. *Univ. Calif. Publ. Stat.*, 1949, **1**, 15–21.

[30] Pearson, E. S., and Hartley, H. O. Charts for the power function of analysis of variance tests, revised from the non-central F-distribution. *Biometrika*, 1951, **38**, 112–16.

31 A STATISTICAL SCREENING PROBLEM[1]

HERBERT ROBBINS

Columbia University

We consider the following decision problem: Θ is an unknown quantity which when measured yields a random variable $X = \Theta + Y$, where Y, the error in measurement, is a random variable with known distribution function $G(y) = \Pr\{Y \leqq y\}$. The two possible actions are to "take" or to "leave" Θ. If we take Θ, we gain the amount Θ, which may be negative. If we leave Θ, we gain the amount 0. Denote our gain by Θ^*. The decision whether to take or leave Θ is assumed to be based on a rule of the following form:

$$R_a:\ \begin{array}{ll} \text{take } \Theta, & \text{if } X > a\,, \\ \text{leave } \Theta, & \text{if } X \leqq a\,, \end{array} \tag{1}$$

where a is a constant at our disposal. When using such a rule, our expected gain for fixed $\Theta = \theta$ is

$$\begin{aligned} E_a(\Theta^* \mid \theta) &= \theta \Pr\{X > a\} + 0 \cdot \Pr\{X \leqq a\} \\ &= \theta \Pr\{Y > a - \theta\} = \theta[1 - G(a - \theta)]\,. \end{aligned} \tag{2}$$

Even though G is a known function, the value of a which maximizes (2) depends on θ, which is unknown. Accordingly, we shall assume that Θ is itself a random variable, independent of Y, with a distribution function $F(\theta) = \Pr\{\Theta \leqq \theta\}$. In this case the over-all expected gain when using (1) is

$$\begin{aligned} h(a) = E_a(\Theta^*) &= \int_{-\infty}^{\infty} E_a(\Theta^* \mid \theta)\, dF(\theta) \\ &= \int_{-\infty}^{\infty} \theta\,[1 - G(a - \theta)]\, dF(\theta) \\ &= E(\Theta) - \int_{-\infty}^{\infty} \theta\, G(a - \theta)\, dF(\theta)\,. \end{aligned} \tag{3}$$

Assuming for the moment that F, as well as G, is known, we can then choose a so as to maximize (3).

We shall carry out this problem in detail for the case in which both Y and Θ are normal:

[1] This research was supported by the Office of Naval Research under Contract No. Nonr-266(59), Project No. NR-042-205. Reproduction in whole or in part is permitted for any purpose of the United States Government.

(4) $$Y \sim N(0, c^2)\,, \qquad \Theta \sim N(\mu, \sigma^2)\,.$$

Thus

$$F'(\theta) = \frac{1}{\sqrt{2\pi}\,\sigma} \exp\left[-\frac{(\theta-\mu)^2}{2\sigma^2}\right],$$

$$G'(y) = \frac{1}{\sqrt{2\pi}\,c} \exp\left(-\frac{y^2}{2c^2}\right),$$

and hence from (3),

$$\begin{aligned} h'(a) &= -\int_{-\infty}^{\infty} \theta\, G'(a-\theta)F'(\theta)\,d\theta \\ &= -\frac{1}{2\pi c\sigma}\int_{-\infty}^{\infty} \theta \exp\left\{-\left[\frac{\sigma^2(a-\theta)^2 + c^2(\theta-\mu)^2}{2c^2\sigma^2}\right]\right\} d\theta\,. \end{aligned}$$

After some algebraic reduction, we can write this in the form

$$h'(a) = -\frac{1}{2\pi c\sigma} \exp\left[-\frac{(\mu-a)^2}{2(\sigma^2+c^2)}\right]\int_{-\infty}^{\infty} \theta \exp\left\{-\frac{(\sigma^2+c^2)\left[\theta - \dfrac{\sigma^2 a + c^2\mu}{\sigma^2+c^2}\right]^2}{2c^2\sigma^2}\right\} d\theta\,,$$

and making the substitution $t = \theta - (\sigma^2 a + c^2\mu)/(\sigma^2+c^2)$, we find without difficulty that

(5) $$h'(a) = -\frac{1}{\sqrt{2\pi}}\,\frac{\sigma^2 a + c^2\mu}{(\sigma^2+c^2)^{3/2}} \exp\left[-\frac{(\mu-a)^2}{2(\sigma^2+c^2)}\right].$$

Thus $\operatorname{sgn} h'(a) = -\operatorname{sgn}(\sigma^2 a + c^2\mu)$, and therefore $h(a)$ is maximized for a equal to

(6) $$\bar{a} = (-c^2/\sigma^2)\mu\,.$$

Since $h(-\infty) = E(\Theta) = \mu$, we have from (5)

$$\begin{aligned} h(a) &= \mu + \int_{-\infty}^{a} h'(x)\,dx \\ &= \mu - \frac{1}{\sqrt{2\pi}}\int_{-\infty}^{a} \frac{\sigma^2 x + c^2\mu}{(\sigma^2+c^2)^{3/2}} \exp\left[-\frac{(x-\mu)^2}{2(\sigma^2+c^2)}\right] dx\,. \end{aligned}$$

The substitution $t = (x-\mu)/(\sigma^2+c^2)^{\frac{1}{2}}$ yields

$$\begin{aligned} h(a) = \mu - \frac{1}{\sqrt{2\pi}\,(\sigma^2+c^2)}\Big\{&\mu(\sigma^2+c^2)\int_{-\infty}^{(a-\mu)/(\sigma^2+c^2)^{\frac{1}{2}}} \exp(-t^2/2)\,dt \\ &+ \sigma^2(\sigma^2+c^2)^{\frac{1}{2}}\int_{-\infty}^{(a-\mu)/(\sigma^2+c^2)^{\frac{1}{2}}} t\exp(-t^2/2)\,dt\Big\}\,. \end{aligned}$$

Let $\Phi(x) = \dfrac{1}{\sqrt{2\pi}}\displaystyle\int_{-\infty}^{x} \exp(-t^2/2)\,dt$, $\phi(x) = \Phi'(x)$; then we can write

$$\begin{aligned} h(a) &= \mu\Phi\left(\frac{\mu-a}{(\sigma^2+c^2)^{\frac{1}{2}}}\right) + \frac{\sigma^2}{(\sigma^2+c^2)^{\frac{1}{2}}}\,\phi\left(\frac{\mu-a}{(\sigma^2+c^2)^{\frac{1}{2}}}\right) \\ &= k(b)\,, \end{aligned}$$

where

$$b = \frac{\mu - a}{(\sigma^2 + c^2)^{\frac{1}{2}}}\,, \qquad k(b) = \mu\,\Phi(b) + \frac{\sigma^2}{(\sigma^2 + c^2)^{\frac{1}{2}}}\,\phi(b)\,.$$

The maximum of $h(a)$ was found in (6) to be at $\bar{a}$, which corresponds to the quantity

$$\bar{b} = \frac{\mu(\sigma^2 + c^2)^{\frac{1}{2}}}{\sigma^2}$$

which maximizes $k(b)$.

Example. Let $\mu = \sigma = 1$, $c = \sqrt{3}$. Here $b = \frac{1}{2}(1 - a)$, $a = 1 - 2b$, $\bar{a} = -3$, $\bar{b} = 2$, and $k(b) = \Phi(b) + \frac{1}{2}\phi(b)$. Corresponding values are given in Table 1.

TABLE 1

a	b	$\Phi(b)$	$\phi(b)$	$h(a) = \Phi(b) + \frac{1}{2}\phi(b)$
$-\infty$	∞	1	0	1
⋮	⋮	⋮	⋮	⋮
−5	3	.9987	.0044	1.0009
−3	2	.9773	.0540	1.0043
−1	1	.8413	.2420	.9623
0	.5	.6915	.3521	.8675
⋮	⋮	⋮	⋮	⋮
2.85	−.93	.1762	.2589	.3057
⋮	⋮	⋮	⋮	⋮
∞	$-\infty$	0	0	0

We see that little improvement over the choice $a = -\infty$ is possible (always take Θ) even for $a = \bar{a} = -3$ (take Θ when $X > -3$).

To explain why the entry $a = 2.85$ appears in Table 1 and at the same time to give some point to the whole discussion, suppose we were to handle the problem by the methods of hypothesis testing (or confidence intervals) at the 5% level of significance. Let

$$H_0\colon\ \Theta = 0\,, \qquad H_1\colon\ \Theta > 0\,.$$

Reject H_0 (i.e., take Θ) when $X > x_0$, where x_0 is chosen so that $\Pr\{\text{rej } H_0 \mid H_0\} = .05$, i.e.,

$$\Pr\{Y > x_0 \mid H_0\} = \Pr\left\{\frac{Y}{\sqrt{3}} > \frac{x_0}{\sqrt{3}} \mid H_0\right\} = \frac{1}{\sqrt{2\pi}}\int_{x_0/\sqrt{3}}^{\infty} \exp\left[-t^2/2\right] dt = .05\,,$$

$$\frac{x_0}{\sqrt{3}} = 1.65\,, \qquad x_0 = 2.85\,.$$

Thus we would take Θ (reject H_0) when $X > 2.85$, and from the foregoing we would have $E(\Theta^*) = .3057$, as compared with the optimal value $\bar{a} = -3$; by taking Θ when $X > -3$ we would have $E(\Theta^*) = 1.0043$.

This example shows how a statistical screening procedure operating at a conventional significance level may be worse than useless, since by *always* taking Θ we would have $E(\Theta^*) = 1$.

The catch in the foregoing discussion is, of course, that to compute $\bar{a}$, the maximizing argument in h, we needed to know the exact form of the distribution F of Θ. If we are willing to assume that F is normal, but with unknown μ and σ, we can proceed as follows. If observations $x_1, \cdots, x_n$ on X are available and if we set

$$\bar{x} = \frac{1}{n} \sum_1^n x_i \,, \qquad s^2 = \frac{1}{n-1} \sum_1^n (x_i - \bar{x})^2 \,,$$

then, since $X = \Theta + Y$,

$$E(X) = E(\Theta) + E(Y) = E(\Theta) = \mu \,,$$
$$\operatorname{var}(X) = \operatorname{var}(\Theta) + \operatorname{var}(Y) = \sigma^2 + c^2 \,,$$

so that $\bar{x}$ and $s^2 - c^2$ are unbiased and consistent estimates of μ and σ^2, respectively. Hence, it would be reasonable to take as an estimate of $\bar{a}$ given by (6) the value $A_n = -c^2\bar{x}/(s^2 - c^2)$, either for use with the next observation x_{n+1} or with the set $x_1, \cdots, x_n$ if these decisions have not yet been made. It then becomes necessary to evaluate the expectations of Θ^*_{n+1} or of $(1/n) \sum_1^n \Theta_i^*$, using A_n as the cut-off points; we shall not consider this problem here. And if the assumption that F is normal is dropped, then even the use of A_n becomes questionable.

Up to now we have been considering a somewhat special decision problem: whether to "take" or "leave" an unknown value Θ on the basis of the observed value of $X = \Theta + Y$. Let us now pose the more usual problem of estimation: What is the best point estimate of Θ of the form $t(X)$? Assume as before that (4) holds and that Θ and Y are independent; then the conditional probability density function of Θ for given $X = x$ is

$$f(\theta \mid x) = \frac{f(\theta) f(x \mid \theta)}{f(x)}$$
$$= \frac{\dfrac{1}{\sqrt{2\pi}\,\sigma} \exp\left[-\dfrac{(\theta-\mu)^2}{2\sigma^2}\right] \dfrac{1}{\sqrt{2\pi}\,c} \exp\left[-\dfrac{(x-\theta)^2}{2c^2}\right]}{\dfrac{1}{\sqrt{2\pi}\sqrt{\sigma^2+c^2}} \exp\left[-\dfrac{(x-\mu)^2}{2(\sigma^2+c^2)}\right]} \,,$$

which reduces to

$$\frac{\sqrt{\sigma^2+c^2}}{\sqrt{2\pi}\,\sigma c} \exp\left[-\frac{c^2+\sigma^2}{2\sigma^2 c^2}\left(\theta - \frac{c^2\mu + \sigma^2 x}{c^2+\sigma^2}\right)^2\right] .$$

Thus *the conditional distribution of Θ given $X = x$ is*

$$N\left(\frac{c^2\mu + \sigma^2 x}{c^2+\sigma^2} \,, \frac{\sigma^2 c^2}{\sigma^2 + c^2}\right) . \tag{7}$$

It follows that the best estimator of Θ of the form $t(X)$, in the sense of minimizing, e.g., $E[t(X) - \Theta]^2$, is given by choosing

$$t(x) = \frac{c^2\mu + \sigma^2 x}{c^2 + \sigma^2}, \tag{8}$$

and that

$$E[t(X) - \Theta]^2 = \frac{\sigma^2 c^2}{\sigma^2 + c^2}. \tag{9}$$

The usual estimator of Θ, based on the assumption that Θ is not a random variable but an unknown constant, is, of course, X itself, for which

$$E[X - \Theta]^2 = c^2, \tag{10}$$

which is greater than (9). Of course, for fixed $\Theta = \theta$ the estimator $t(X)$ is biased, with

$$E_\theta[t(X)] = \frac{c^2\mu + \sigma^2\theta}{c^2 + \sigma^2}, \tag{11}$$

$$E_\theta[t(X) - \theta]^2 = c^2 \frac{\sigma^4 + c^2(\mu - \theta)^2}{(c^2 + \sigma^2)^2},$$

as contrasted with

$$E_\theta(X) = \theta, \tag{12}$$

$$E_\theta[X - \theta]^2 = c^2.$$

The values of θ for which $E_\theta[t(X) - \theta]^2 \leqq E_\theta(X - \theta)^2$ are those for which

$$\frac{\sigma^4 + c^2(\mu - \theta)^2}{(c^2 + \sigma^2)^2} \leqq 1,$$

i.e.,

$$|\theta - \mu| \leqq \sqrt{c^2 + 2\sigma^2}. \tag{13}$$

We are now in a position to pose an important question. Suppose we are in the business of screening applicants for admission to college by administering a standard test. If Θ denotes ability and X the test score, then $X = \Theta + Y$ where Θ and Y are independent and $Y \sim N(0, c^2)$, c known. Suppose further it is known that the applicants are drawn at random from a population in which $\Theta \sim N(\mu, \sigma^2)$, μ and σ known. Then the best estimate of an applicant's unknown Θ in terms of his observed X is

$$t(X) = \frac{c^2\mu + \sigma^2 X}{c^2 + \sigma^2}, \tag{14}$$

in the sense of minimizing the average squared error of estimate in repeated drawings from the population. Now suppose there are *two* populations from which applicants are drawn, such that in population i ($i = 1, 2$), $\Theta \sim N(\mu_i, \sigma_i^2)$. Then the "best" estimate of an applicant's Θ in terms of his X is

$$t_i(X) = \frac{c^2\mu_i + \sigma_i^2 X}{c^2 + \sigma_i^2} \tag{15}$$

if the applicant comes from population i. To fix the ideas, suppose $\sigma_1^2 = \sigma_2^2 = \sigma^2$, say, and $\mu_1 < \mu_2$. Then

$$t_1(x) = \frac{c^2\mu_1 + \sigma^2 x}{c^2 + \sigma^2}, \qquad t_2(y) = \frac{c^2\mu_2 + \sigma^2 y}{c^2 + \sigma^2}, \tag{16}$$

and

$$t_1(x) > t_2(y) \quad \text{if and only if} \quad x - y > \frac{c^2}{\sigma^2}(\mu_2 - \mu_1). \tag{17}$$

Thus an applicant from population 1 will be preferred to an applicant from population 2 only if the test score x of the former exceeds the test score y of the latter by $(c^2/\sigma^2)(\mu_2 - \mu_1)$. But this criterion is clearly unfair to the applicants from population 1. To what extent can the statistician condone the use of the two functions (15) which weight the X-values by two different quantities which are in no way influenced or controlled by the applicant's own qualities or performance?

32 ON THE POWER OF SOME RANK-ORDER TWO-SAMPLE TESTS[1]

JOAN RAUP ROSENBLATT
National Bureau of Standards

1. Introduction

The objective of this paper is to report a collection of small-sample results concerning properties of some rank-order two-sample tests (and related tests) with respect to various types of decision problems. Some of these results are "negative" in the sense that the intuitively appealing Wilcoxon-Mann-Whitney test is shown to have some "undesirable" properties (Section 2). It is noted later, however, that the relevant theorems fail if the class of distributions is mildly restricted (Section 5).

The methods used involve chiefly the manipulation of integrals arising in expressions for the probabilities of orderings of observations from two populations. A certain pair of distribution functions, defined by (2.4), (2.5), is found to have "least favorable" properties (Theorem 1), and more generally to be useful in exhibiting the attainment of extreme values for certain functionals arising in the expressions for probabilities of orderings (e.g., Lemma 1).

In Sections 3 and 4, an approach introduced by I. R. Savage [1] is attempted and is found to suggest the need for restrictions on the class of alternatives to be admitted against the null hypothesis $F = G$. In the absence of restrictions, the criterion of admissibility excludes relatively few rank-order tests. A test may have an "optimum" property for some but not all levels of significance, and (uniformly) for only a subset of a set of alternatives (Table 1).

We shall be concerned chiefly with a class of pairs (F, G) of distributions for which the functional

$$\theta(F, G) = \Pr(X < Y) = \iint \phi(x, y)\, dF(x)\, dG(y) = E_{FG}\,\phi(X, Y) \tag{1.1}$$

is defined, where

[1] Most of this work was done at the University of North Carolina, with the support of the Air Force Office of Scientific Research.

(1.2)
$$\phi(x, y) = \begin{cases} 1, & \text{if } x < y, \\ 0, & \text{if } x \geqq y. \end{cases}$$

Two types of null hypotheses are of particular interest. The most usual null hypothesis for the two-sample problem is $F = G$. The Wilcoxon-Mann-Whitney statistic was proposed for tests of this hypothesis. When considering that statistic, however, one is led naturally to a partition of alternatives according to the value of $\theta(F, G)$. The next step is to consider a null hypothesis stated in terms of the value of θ.

We shall suppose that we are given the $(n + m)$ mutually independent random variables $X_1, \cdots, X_n$ and $Y_1, \cdots, Y_m$, distributed according to $F(x)$, $G(y)$, respectively. Suppose that (F, G) is a pair of distributions from a class $\mathscr{D}$ of pairs of distributions for which $0 < \theta(F, G) < 1$. Let $\mathscr{D}(\theta)$ be the subset of $\mathscr{D}$ in which $\theta(F, G) = \theta$.

The decision problem is to select one or the other of two hypotheses, each of which is given by a subset of the class $\mathscr{D}$, on the basis of information given by two observed samples. Thus, we have in general H: $(F, G) \in \mathscr{D}_0$ and H': $(F, G) \in \mathscr{D}_1$, where $\mathscr{D}_0$ and $\mathscr{D}_1$ are disjoint subsets of $\mathscr{D}$.

Let z_{mn} represent the vector $(x_1, \cdots, x_n; y_1, \cdots, y_m)$. Given $\alpha > 0$ and a statistic $t_{mn}(z_{mn})$, we will mean by a "test," a function $\varphi(z_{mn}; t_{mn})$ which gives the probability of selecting H' (rejecting H) when z_{mn} has been observed, such that

(1.3)
$$\varphi = \begin{cases} 1, & \text{if } t_{mn} > c_{mn}, \\ a_{mn}, & \text{if } t_{mn} = c_{mn}, \\ 0, & \text{if } t_{mn} < c_{mn}, \end{cases}$$

where c_{mn} and a_{mn} $(0 \leqq a_{mn} < 1)$ are chosen so that

(1.4)
$$\sup_{(F, G) \in \mathscr{D}_0} E_{FG}\, \varphi(Z_{mn}; t_{mn}) = \alpha .$$

Thus we consider a test which is randomized only to the extent necessary in order that the level of significance be exactly α. For simplicity, a_{mn} and c_{mn} will be written without their subscripts when it is not desired to emphasize their dependence on m, n.

A family $\mathscr{T}$ of tests will consist of tests of the form (1.3) based on the sequence $\{t_{mn}\}$, $m, n = 1, 2, \cdots$.

We will in particular be interested in the families of tests $\mathscr{U}$ and $\mathscr{W}$ based on the statistics (1.5) and (1.6), respectively. Suppose $n \leqq m$ and let

(1.5)
$$mnU_{mn} = \sum_{i=1}^{n} \sum_{j=1}^{m} \phi(x_i, y_j) ,$$

(1.6)
$$n W_{mn} = \sum_{i=1}^{n} \phi(x_i, y_i) ,$$

where $\phi(x, y)$ is given by (1.2). The symbols $\mathscr{U}'$ and $\mathscr{W}'$ will denote the families of two-sided tests based on $|U_{mn} - \frac{1}{2}|$, $|W_{mn} - \frac{1}{2}|$, respectively.

2. Tests of $\theta \leqq \theta_1$ against $\theta \geqq \theta_2$

Let us consider $H_1: \theta \leqq \theta_1$ and $H_2: \theta \geqq \theta_2$, $\theta_1 < \theta_2$. We show first that the family $\mathscr{W}$ provides a test which maximizes the minimum power with respect to H_2, among tests of size α for testing H_1. Second, by comparison with this test from $\mathscr{W}$, a test from $\mathscr{U}$ is shown to be inadmissible for testing H_1 against H_2 (for continuous F and G). These results are "weakened," however, by observations included below in Section 5.

For testing H_1 against H_2, we have a test $\varphi(z_{mn}; W_{mn})$ of the form (1.3) from the family $\mathscr{W}$. nW_{mn} has binomial distribution $B(n, \theta)$. Thus, for $(F, G) \in \mathscr{D}(\theta)$,

$$E_{FG}\,\varphi(Z_{mn}; W_{mn}) = \sum_{j=c+1}^{n} \binom{n}{j} \theta^j (1-\theta)^{n-j} + a\binom{n}{c}\theta^c(1-\theta)^{n-c} = H(\theta; c, a), \tag{2.1}$$

which is an increasing function of θ, and we have a and c determined by

$$\alpha = H(\theta_1; c, a). \tag{2.2}$$

Let μ be a measure on the real line R_1 having the following property. There exist disjoint intervals I_1, I_2, I_3 of R_1 such that

$$\begin{aligned} &\text{(i)} \quad u_j \in I_j \quad (j = 1, 2, 3) \quad \text{implies } u_1 < u_2 < u_3, \\ &\text{(ii)} \quad 0 < \mu(I_j) < \infty, \qquad j = 1, 2, 3. \end{aligned} \tag{2.3}$$

THEOREM 1. *Let $\mathscr{D}_\mu$ be a class of pairs (F, G) of distributions having generalized density with respect to some fixed measure μ satisfying* (2.3). *Then a test from the family $\mathscr{W}$ of size α for testing H_1 maximizes the minimum power with respect to H_2, provided $\mathscr{D}_\mu$ contains the pairs $(K(x, \theta), H(y))$, $0 < \theta < 1$, defined by* (2.4), (2.5).

PROOF. By Theorem 2 of [2], it will be sufficient to exhibit $(F_1, G_1) \in \mathscr{D}_\mu(\theta_1)$ and $(F_2, G_2) \in \mathscr{D}_\mu(\theta_2)$ such that $\varphi(z_{mn}; W_{mn})$ is a most powerful test of size α for testing the simple hypothesis $\bar{H}_1: (F, G) = (F_1, G_1)$ against the simple alternative $\bar{H}_2: (F, G) = (F_2, G_2)$.

Let $H(y)$ be a distribution having the probability density

$$h(y) = \begin{cases} 1/\mu(I_2), & \text{if } y \in I_2, \\ 0, & \text{otherwise}. \end{cases} \tag{2.4}$$

Let $K(x, \theta)$ be a distribution with density

$$k(x, \theta) = \begin{cases} \theta/\mu(I_1), & \text{if } x \in I_1, \\ (1-\theta)/\mu(I_3), & \text{if } x \in I_3, \\ 0, & \text{otherwise}. \end{cases} \tag{2.5}$$

Let $G_1(y) = G_2(y) = H(y)$ and $F_i(x) = K(x, \theta_i)$, $i = 1, 2$. The desired result follows from a familiar application of the Neyman-Pearson lemma. ||

We might broaden the statement of the theorem slightly, for the test from $\mathscr{W}$ actually maximizes the minimum power uniformly with respect to a partition according to values of θ of the subset of $\mathscr{D}_\mu$ with $\theta \geqq \theta_2$.

It is also possible, for example, to prove Theorem 1 for a class $\mathscr{D}$ restricted to contain only distributions F, G which are continuous and everywhere increasing (unless equal to zero or one). The distributions K, H are replaced by sequences $\{K_{it}\}, \{H_{it}\}$ with $(K_{it}, H_{it}) \in \mathscr{D}(\theta_i)$, $i = 1, 2$; $t = 1, 2, \cdots$, such that the power of a most powerful test of level α for testing (K_{1t}, H_{1t}) against (K_{2t}, H_{2t}), tends, as $t \to \infty$, to the power of $\varphi(z_{mn}; W_{mn})$ for testing $(F, G) \in \mathscr{D}(\theta_1)$ at level α, against $(F, G) \in \mathscr{D}(\theta_2)$.

A similar result may be obtained for two-sided tests from the family $\mathscr{W}'$ for testing H_1': $\theta = \frac{1}{2}$ against H_2': $|\theta - \frac{1}{2}| \geqq \delta > 0$. The proof follows as above. A MP test is obtained for testing the simple hypothesis that the joint distribution of $(X_1, \cdots, X_n)$ is $\Pi_{i=1}^n K(x_i, \frac{1}{2})$ against the simple alternative that it is

$$\left(\frac{1}{2}\right)\prod_{i=1}^{n} K(x_i, \tfrac{1}{2} - \delta) + \left(\frac{1}{2}\right)\prod_{i=1}^{n} K(x_i, \tfrac{1}{2} + \delta) .$$

Examination of the proof of Theorem 2 shows that a test from the family $\mathscr{U}$ does not in general maximize the minimum power. First, we establish the following lemma.

LEMMA 1. *Let $\mathscr{D}_c$ be the class of pairs of continuous distributions. Then $\theta(F, G) = \int F\,dG$ and*

$$\max_{(F,G)\in\mathscr{D}_c(\theta)} \int (1 - G)^m\, dF^n = \theta^s , \tag{2.6}$$

$$\max_{(F,G)\in\mathscr{D}_c(\theta)} \left\{\int (1 - G)^m\, dF^n + \int (1 - F)^n\, dG^m\right\} = \theta^s + (1 - \theta)^s , \tag{2.7}$$

where $s = \min(m, n)$.

PROOF. We have three steps in the proof of (2.6).

(i) $\int (1 - G)^m\, dF^n \leqq \int (1 - G)^s\, dF^s$. If $m > n$, then $(1 - G)^m \leqq (1 - G)^n$. If $m < n$, integrate by parts to obtain

$$\int (1 - G)^m\, dF^n = m\int (1 - G)^{m-1} F^n\, dG \leqq m\int (1 - G)^{m-1} F^m\, dG = \int (1 - G)^m\, dF^m .$$

(ii) $\int (1 - G)^s\, dF^s \leqq \theta^s$. The integral is equal to the probability of the event A: $\max(X_1, \cdots, X_s) < \min(Y_1, \cdots, Y_s)$, while θ^s is the probability of the event B: $X_1 < Y_1, \cdots, X_s < Y_s$; and $\Pr(A) \leqq \Pr(B)$.

(iii) There exist $(F, G) \in \mathscr{D}_c(\theta)$, $0 < \theta < 1$, such that the upper bound θ^s is attained. Consider the distributions $K(x, \theta)$, $H(y)$ defined in (2.4), (2.5). Let μ be Lebesgue measure. If $n \leqq m$, let $F(x) = K(x, \theta)$, $G(y) = H(y)$.

Then $\int(1-G)^m dF^n = \Pr(A) = \Pr(B) = \theta^s$. If $n > m$, let $F(x) = H(x)$, $G(y) = K(y, 1-\theta)$. Again, $\Pr(A) = \Pr(B)$.

To prove (2.7), we see that steps (i) and (ii) may be repeated with F and G interchanged and θ replaced by $1-\theta = \int G\,dF$. That the upper bound $\theta^s + (1-\theta)^s$ may be attained for every θ is verified with the same distributions used in step (iii). ||

THEOREM 2. *If $(F, G) \in \mathscr{D}_c$, a test from the family $\mathscr{U}$ is inadmissible for testing H_1 against H_2 whenever $\alpha < \theta_1^s$, where $s = \min(m, n) > 1$.*

PROOF. We show that a uniformly better test is available, namely a test from the family $\mathscr{W}$. For $m = n = 1$, of course, U_{mn} and W_{mn} are identical.

If $\alpha < \theta_1^s$, then the following tests are of size α with respect to $H_1\colon \theta \leqq \theta_1$.

$$\varphi(z_{mn}; W_{mn}) = \begin{cases} \theta_1^{-s}\alpha, & \text{if } W_{mn} = 1, \\ 0, & \text{otherwise;} \end{cases}$$

$$\varphi(z_{mn}; U_{mn}) = \begin{cases} \theta_1^{-s}\alpha, & \text{if } U_{mn} = 1, \\ 0, & \text{otherwise}, \end{cases}$$

in the latter case by Lemma 1. If $(F, G) \in \mathscr{D}_c(\theta)$,

$$E_{FG}\,\varphi(Z_{mn}; W_{mn}) = \theta_1^{-s}\alpha\theta^s,$$

$$E_{FG}\,\varphi(z_{mn}; U_{mn}) = \theta_1^{-s}\alpha\int(1-G)^m\,dF^n.$$

Thus, by Lemma 1, $E_{FG}\varphi(Z_{mn}; W_{mn}) \geqq E_{FG}\varphi(Z_{mn}; U_{mn})$ and it is easily seen that strict inequality holds for some (F, G). ||

Let $\mathscr{U}'$ be the family of (two-sided) tests of the form (1.3) based on the statistic $|U_{mn} - \frac{1}{2}|$, for testing $H_1'\colon \theta = \frac{1}{2}$ against $H_2'\colon |\theta - \frac{1}{2}| \geqq \delta > 0$. The second part of Lemma 1 may be used to show that a test from $\mathscr{U}'$ is inadmissible for testing H_1' against H_2' wherever $\alpha < 2^{-(s-1)}$, $s > 1$.

The curious properties exhibited by tests based on U_{mn} in small samples are strikingly illustrated by the following example. Suppose $(F, G) \in \mathscr{D}_c$. Let $m = n = 2$, $\alpha \leqq \theta_1^2$. It is shown that for any π, $0 \leqq 2\pi \leqq \alpha\theta_1^{-2}$, the following test is of size α:

$$\bar{\varphi}_\pi = \begin{cases} \alpha\theta_1^{-2}, & \text{if } U_{22} = 1, \\ \pi, & \text{if } U_{22} = \frac{3}{4}, \\ 0, & \text{otherwise}. \end{cases}$$

This test is not in general of the form (1.3). We have

$$E_{FG}\bar{\varphi}_\pi = 2\pi\theta^2 + (\theta_1^{-2}\alpha - 2\pi)\int(1-G)^2\,dF^2,$$

since $\Pr(U_{22} = \frac{3}{4}) = 2\theta^2 - 2\int(1-G)^2\,dF^2$. Now

$$\max_{(F,G)\in\mathscr{D}_c(\theta)} E_{FG}\bar{\varphi}_\pi = \theta_1^{-2}\alpha\theta^2$$

by Lemma 1.

1. The test $\bar{\varphi}^*$ corresponding to $\pi = \frac{1}{2}\alpha\theta_1^{-2}$ is uniformly better than $\bar{\varphi}_\pi$ for smaller π.

2. The test $\bar{\varphi}^*$ has the same power function as the test $\varphi(z_{22}; W_{22})$ from the family $\mathscr{W}$. Hence $\bar{\varphi}^*$ is not necessarily inadmissible.

3. For $\alpha < \theta_1^2$, a test $\bar{\varphi}_\pi$ is not of the form (1.3). Thus, by permitting greater randomization, we may obtain a test based on U_{22} which is no worse than $\varphi(z_{22}; W_{22})$. This weakens somewhat the effect of Theorem 2.

4. For $\alpha = \theta_1^2$, $\bar{\varphi}_\pi$ is a test of the form (1.3), and we see that it is possible that a test of the form (1.3) may not be uniquely determined by the fixing of the level of significance. Also, Theorem 2 cannot hold for $m = n = 2, \alpha = \theta_1^2$.

3. Rank-order Tests

In this and the following sections, we turn to rank-order tests for testing the null hypothesis $H_0: F = G$. It is assumed that $(F, G) \in \mathscr{D}_c$; that is, with probability one, the $m + n$ observed values are distinct. In this section, some notation is introduced for rank-order tests, and some of their properties are noted. At the end of the section there are some remarks on the role of rank-order tests in testing $H_1: \theta \leqq \theta_1$ against $H_2: \theta \geqq \theta_2$.

Let $u_j = u_j(m, n)$, $j = 1, \cdots, M$, denote an arrangement of n x's and m y's; let $\mathscr{Y}_{mn}$ denote the class of $M = \binom{m+n}{m}$ possible arrangements. For example,

$$\mathscr{Y}_{12} = \{(xxy), (xyx), (yxx)\} = \{u_1, u_2, u_3\} .$$

A rank-order test (r.o. test) assigns to each u_j a number π_j $(0 \leqq \pi_j \leqq 1)$ which is the probability of rejecting the null hypothesis (usually $H_0: F = G$) if the (ordered) x's and y's of the sample z_{mn} have arrangement u_j, this correspondence being denoted by $z_{mn} \sim u_j$. We will use the notation $\rho(z_{mn}; \pi)$, where $\rho(z_{mn}; \pi) = \pi_j$, if $z_{mn} \sim u_j$, $j = 1, \cdots, M$, $\pi = (\pi_1, \cdots, \pi_M)$, for r.o. tests. For a test of size α, for testing $H_0: F = G$,

$$\sum_{j=1}^{M} \pi_j = \alpha M . \tag{3.1}$$

Observe that a r.o. test, as defined, is in general a randomized test and not necessarily of the form (1.3).

We will find it useful also to define a *symmetrical rank-order* (s.r.o.) *test*, for testing H_0 against two-sided alternatives. Let u_j' denote the arrangement obtained from u_j by reversing the order; e.g., $u_1(1, 2) = (xxy)$, $u_1'(1, 2) = (yxx)$. The set $\mathscr{Y}_{mn}$ may be divided into subsets, $\mathscr{Y}_{mn}^0$ in which $u_j' = u_j$, and $\mathscr{Y}_{mn}'$ in which $u_j' \neq u_j$. For a s.r.o. test, $\pi_k = \pi_j$ if $u_k = u_j'$. We will write $\rho_s(z_{mn}; \pi)$, where $\rho_s(z_{mn}; \pi) = \pi_j$, if $z_{mn} \sim u_j$, $j = 1, \cdots, M$, for s.r.o. tests.

Thus, the arrangements $u_j \in \mathscr{Y}_{mn}'$ are considered in pairs in a s.r.o. test. We have for all $u_j \in \mathscr{Y}_{mn}$,

$$U_{mn}(z_{mn}) = 1 - U_{mn}(z_{mn}') \text{ if } z_{mn} \sim u_j,\ z_{mn}' \sim u_j' . \tag{3.2}$$

Rank-order tests and s.r.o. tests have as special cases tests based on the statistics U_{mn} and $|U_{mn} - \frac{1}{2}|$, respectively, which assign the π_j according to the value of $U_{mn}(z_{mn})$ for $z_{mn} \sim u_j$. Since more than one arrangement may be associated with a given value of U_{mn}, a test based on U_{mn} assigns the π_j to groups of arrangements. By (3.2) a test based on $|U_{mn} - \frac{1}{2}|$ assigns the same probability to u_j and u'_j.

The power function of a r.o. test is given by

$$E_{FG}\,\rho(Z_{mn}\,;\pi) = \sum_{j=1}^{M} \pi_j \Pr(Z_{mn} \sim u_j)\,. \tag{3.3}$$

We say that a test $\rho(z_{mn}\,;\pi')$ maximizes the minimum power among r.o. tests of size α for testing H_0 against H': $(F, G) \in \mathscr{D}_1$ if for all π such that (3.1) holds,

$$\inf_{(F,G)\in\mathscr{D}_1} E_{FG}\,\rho(Z_{mn}\,;\pi') \geqq \inf_{(F,G)\in\mathscr{D}_1} E_{FG}\,\rho(Z_{mn}\,;\pi)\,. \tag{3.4}$$

And, of course, $\rho(z_{mn}\,;\pi')$ is UMP among r.o. tests if for all π satisfying (3.1),

$$E_{FG}\,\rho(Z_{mn}\,;\pi') \geqq E_{FG}\,\rho(Z_{mn}\,;\pi) \qquad \text{for all } (F, G) \in \mathscr{D}_1\,. \tag{3.5}$$

For s.r.o. tests, replace ρ by ρ_s in (3.3) to (3.5).

Consider now the use of r.o. tests for testing $H_1\colon \theta \leqq \theta_1$ against $H_2\colon \theta \geqq \theta_2$. In the example given at the end of Section 2, it is found that a test based on U_{mn} maximizes the minimum power. This does not happen in general. It is, however, always possible to construct a *rank-order test* which maximizes the minimum power among tests of size α for testing $H_1\colon \theta \leqq \theta_1$ against $H_2\colon \theta \geqq \theta_2$. This is done by constructing a test whose power function is always equal to the power function of the test based on W_{mn}, as follows:

1. A test of the form (1.3) based on W_{mn} may be written

$$\varphi(z_{mn}\,; W_{mn}) = p_t \quad \text{if} \quad s\,W_{mn}(z_{mn}) = t\,,$$
$$t = 0, 1, \cdots, s = \min(m, n)\,,$$

where $0 \leqq p_t \leqq 1$. We construct the symmetrized test $\varphi^*(z_{mn}\,; W_{mn})$ given by

$$m!\,n!\;\varphi^*(z_{mn}\,; W_{mn}) = \Sigma'\varphi(z'_{mn}\,; W_{mn})\,,$$

where Σ' denotes two-fold summation over the permutations $z'_{mn} = (x_{i_1}, \cdots, x_{i_n}\,; y_{j_1}, \cdots, y_{j_m})$ of the x's and y's of z_{mn}. Now

$$m!\,n!\;\varphi^*(z_{mn}\,; W_{mn}) = \sum_{t=0}^{s} p_t\,[\text{number of } z'_{mn} \text{ with } s\,W_{mn}(z'_{mn}) = t]\,,$$

or

$$\varphi^*(z_{mn}\,; W_{mn}) = \sum_{t=0}^{s} p_t \Pr[s\,W_{mn}(Z_{mn}) = t \mid Z_{mn} \sim u(m, n)]\,,$$

so that φ^* is clearly a r.o. test.

2. Evidently, the power function of φ^* is the same as the power function of φ.

In the same way, we can show that the symmetrized form of $\varphi(z_{mn}\,; |W_{mn} - \frac{1}{2}|)$ is a s.r.o. test.

We may remark that the test $\bar{\varphi}^*$ based on U_{22}, which maximizes the minimum power, in the example at the end of Section 2 happens to be identical with $\varphi^*(z_{22}; W_{22})$.

4. Tests of $F = G$

Various results concerning the properties of small-sample rank-order tests of the hypothesis $H_0: F = G$ against certain classes of alternatives have been obtained by Savage [1]. Savage's approach is to examine the ordering induced on the set of possible arrangements of (ordered) x's and y's by the distributions (F, G) of the alternative hypothesis (all arrangements being equally likely under H_0). He considers alternatives which are all special cases of the slippage alternatives, H_s: $F(x) > G(x)$ for all x.

Savage shows that it is impossible, for most sample sizes, to construct uniformly most powerful r.o. tests for testing H_0 against H_s. His remarks apply *a fortiori* to the problem of constructing UMP rank-order tests against H_1: $\theta(F, G) \leqq \theta_1$ or H_2: $\theta(F, G) \geqq \theta_2 (\theta_1 < \frac{1}{2} < \theta_2)$.

The general conclusion reached by Savage is that the class of admissible tests for testing $F = G$ against H_s is very large. The same remark may be made about the classes of admissible tests for testing $F = G$ against H_1, H_2, or H_2': $|\theta(F, G) - \frac{1}{2}| \geqq \delta > 0$.

We may, however, suggest that tests based on W_{mn} cannot be expected in general to have even the weak optimum properties that were shown for the problem of testing $\theta \leqq \theta_1$ against $\theta \geqq \theta_2 (\theta_1 < \theta_2)$.

THEOREM 3. *If* $m = n = 2$, $\alpha < \frac{1}{2}$, $(F, G) \in \mathscr{D}_c$, *the two-sided test of the form* (1.3) *based on* $|W_{22} - \frac{1}{2}|$ *is inadmissible for testing* H_0 *against* H_2'.

PROOF. We show that a test based on $|U_{22} - \frac{1}{2}|$ is uniformly better. Let

$$\varphi(z_{22}; |W_{22} - \tfrac{1}{2}|) = \begin{cases} 2\alpha, & \text{if } |W_{22} - \frac{1}{2}| = \frac{1}{2}, \\ 0, & \text{otherwise}. \end{cases}$$

This test is of size α and $E_\theta \varphi(Z_{22}; |W_{22} - \frac{1}{2}|) = 2\alpha[\theta^2 + (1-\theta)^2]$. Let

$$\varphi_1(z_{22}; |U_{22} - \tfrac{1}{2}|) = \begin{cases} 3\alpha, & \text{if } |U_{22} - \frac{1}{2}| = \frac{1}{2}, \\ 0, & \text{otherwise}, \end{cases}$$

$$\varphi_2(z_{22}; |U_{22} - \tfrac{1}{2}|) = \begin{cases} 1, & \text{if } |U_{22} - \frac{1}{2}| = \frac{1}{2}, \\ 3\alpha - 1, & \text{if } |U_{22} - \frac{1}{2}| = \frac{1}{4}, \\ 0, & \text{otherwise}. \end{cases}$$

These tests are of size α for $0 < \alpha \leqq \frac{1}{3}$, $\frac{1}{3} < \alpha < \frac{1}{2}$, respectively. We have

$$E_{FG}\varphi_1(Z_{22}; |U_{22} - \tfrac{1}{2}|) = 3\alpha(\tfrac{1}{3} + 2\Delta^2),$$

$$E_{FG}\varphi_2(Z_{22}; |U_{22} - \tfrac{1}{2}|) = (\tfrac{1}{3} + 2\Delta^2) + 4(3\alpha - 1)[\tfrac{1}{3} - \Delta^2 - \theta(1-\theta)],$$

where $\Delta^2 = \int (F - G)^2 dG$. Now,

$$\Delta^2 \geqq (\theta - \tfrac{1}{2})^2 . \tag{4.1}$$

With (4.1), we establish for $0 < \alpha \leqq \frac{1}{3}$ (with an obvious abbreviation of notation):

$$\begin{aligned} &E_{FG}\varphi_1(Z_{22};U_{22}) - E_\theta\varphi(Z_{22};W_{22}) \\ &\qquad = 6\alpha\Delta^2 - 4\alpha(\theta-\tfrac{1}{2})^2 \geqq 2\alpha(\theta-\tfrac{1}{2})^2 > 0, \quad \text{if } |\theta - \tfrac{1}{2}| \geqq \delta , \end{aligned}$$

and for $\frac{1}{3} < \alpha < \frac{1}{2}$:

$$\begin{aligned} E_{FG}\varphi_2(Z_{22};U_{22}) - E_\theta\varphi(Z_{22};W_{22}) &= 2(1-2\alpha)[3\Delta^2 - 2(\theta-\tfrac{1}{2})^2] \\ &\geqq 2(1-2\alpha)(\theta-\tfrac{1}{2})^2 > 0 , \quad \text{if } |\theta-\tfrac{1}{2}| \geqq \delta . \quad \| \end{aligned}$$

A corollary to Theorem 3 is that $\varphi(z_{22};|W_{22}-\frac{1}{2}|)$ cannot maximize the minimum power for testing H_0 against H_2'.

We may ask whether, in the absence of "better" optimum properties, it is possible to find tests which maximize the minimum power *among rank-order tests* of size α for testing $H_0: F = G$.

Table 1, at the end of this section, summarizes the results that have been obtained. It is clear that, unless the alternative hypothesis is restricted, the class of admissible rank-order tests is too large to permit satisfactory results from this approach.

We consider one statistic not defined above, namely, the two-sample statistic proposed by Lehmann [3] for testing H_0 against alternatives $H_3: F \neq G$. Let

$$K(x_1, x_2, y_1, y_2) = \begin{cases} 1 , & \text{if } \max(x_1, x_2) < \min(y_1, y_2) , \\ & \text{or } \max(y_1, y_2) < \min(x_1, x_2) , \\ 0 , & \text{otherwise} , \end{cases}$$

and

$$\binom{m}{2}\binom{n}{2} L_{mn}(z_{mn}) = \sum K(x_i, x_j, y_k, y_l) ,$$

where the summation is extended over sets of integers (i, j, k, l) with $1 \leqq i < j \leqq n$, $1 \leqq k < l \leqq m$. Then we have

$$E L_{mn}(Z_{mn}) = \tfrac{1}{3} + 2\Delta^2 , \tag{4.2}$$

where $\Delta^2 = \int (F-G)^2 \, dG$.

The remainder of this section is devoted to the derivation of Table 1.

4.1. Conditions for Table 1. Let $\mathscr{D}$ be the class of pairs of continuous distributions (F, G). Let the one-sided alternatives $\mathscr{D}_2 = \{(F,G): \theta(F,G) > \frac{1}{2}\}$ be partitioned according to values of $\theta(F,G)$. Let the two-sided alternatives $\mathscr{D}_2' = \{(F,G): |\theta(F,G) - \frac{1}{2}| > 0\}$ be partitioned according to values of $d = |\theta(F,G) - \frac{1}{2}|$. Let the alternatives $\mathscr{D}_3 = \{(F,G): \Delta^2 > 0\}$ be partitioned according to values of Δ^2. The null hypothesis is H_0: $F = G$.

Tests denoted by $\varphi(z_{mn}; t_{mn})$ are of the form (1.3); there is no ambiguity of definition. Tests denoted by $\bar{\varphi}(z_{mn}; t_{mn})$ are tests based on t_{mn} but employing more extensive randomization. The following are the ones which appear in the table:

$$\bar{\varphi}(z_{12}; U_{12}) = \begin{cases} 2\alpha , \\ \alpha , \\ 0 , \end{cases} \quad \text{if } U_{12} = \begin{cases} 1 , \\ \frac{1}{2} , \\ 0 , \end{cases}$$

$$\bar{\varphi}(z_{12}; | U_{12} - \tfrac{1}{2} |) = \alpha .$$

4.2. Derivation of Table 1. Theorem 3 gives an example of a proof of inadmissibility.

The set of r.o. tests (or of s.r.o. tests in the two-sided case) is reduced by use of necessary conditions for admissibility. Let $\mathscr{C}_1$ be a class of alternatives.

(i) If $\Pr(Z_{mn} \sim u_j) \geqq \Pr(Z_{mn} \sim u_k)$ for all $(F, G) \in \mathscr{C}_1$, then $\rho(z_{mn}; \pi)$ will be admissible only if $\pi_k > 0$ implies $\pi_j = 1$ (Savage [1]).

(ii) If $\Pr(Z_{mn} \sim u_j) + \Pr(Z_{mn} \sim u_j') \geqq \Pr(Z_{mn} \sim u_k) + \Pr(Z_{mn} \sim u_k')$ for all $(F, G) \in \mathscr{C}_1$, where $u_j, u_k \in \mathscr{Y}'_{mn}$, then $\rho_s(z_{mn}; \pi)$ will be admissible only if $\pi_k > 0$ implies $\pi_j = 1$.

(iii) If $\Pr(Z_{mn} \sim u_j) + \Pr(Z_{mn} \sim u_j') \geqq 2\Pr(Z_{mn} \sim u_k)$ for all $(F, G) \in \mathscr{C}_1$, where $u_j \in \mathscr{Y}'_{mn}$, $u_k \in \mathscr{Y}^0_{mn}$, then $\rho_s(z_{mn}; \pi)$ will be admissible only if $\pi_k > 0$ implies $\pi_j = 1$.

When there is only one r.o. (s.r.o.) test which is not inadmissible, then that is most powerful among r.o. tests.

It will be seen that the results are "complete" for $m = 1$, $n = 2$. For $m \geqq 2, n \geqq 2$, there are only certain significance levels and certain subsets of the alternatives for which results may be obtained without restrictions on the class of distributions considered. For larger sample sizes, it does not seem to be possible to obtain the power functions of the r.o. (s.r.o.) tests in manageable form.

5. Restriction of the Class of Alternatives

It is of interest, finally, to consider the behavior of tests from the two families $\mathscr{U}$ and $\mathscr{W}$ with respect to a slightly modified problem, for which tests from $\mathscr{W}$ do not maximize the minimum power for testing $H_1: \theta \leqq \theta_1$ against $H_2: \theta \geqq \theta_2, \theta_1 < \theta_2$. In this section, for simplicity, we consider only families of tests based on samples of equal size. The assumption $(F, G) \in \mathscr{D}_c$ is also made.

It has been shown [4] that the statistics U_{nn} and W_{nn} are asymptotically equally "good" for testing H_1 against H_2, and that the reason for this is essentially that the variance of $n^2 U_{nn}$ can attain its maximum value $n\theta(1-\theta)$ for $(F, G) \in \mathscr{D}_c(\theta)$, where $n\theta(1-\theta)$ is the value of the variance of $n W_{nn}$ for $(F, G) \in \mathscr{D}_c(\theta)$. Now, for $(F, G) \in \mathscr{D}_c(\theta)$,

TABLE 1

Sample sizes	Level of significance	Alternatives for which properties hold (uniformly)	Tests which maximize minimum power among r.o. (s.r.o.) tests	Inadmissible tests
$m=1$ $n=2$	$0<\alpha<\frac{1}{2}$	$\mathscr{D}_2(\theta)$ $\frac{1}{2}<\theta<\frac{2}{3}$	$\bar{\varphi}(z_{12};\ U_{12})$ $\varphi(z_{12};\ W_{12})$	
		$\mathscr{D}_2(\theta)$ $\frac{2}{3}\leqq\theta<1$	$\varphi(z_{12};\ U_{12})^{a}$	$\varphi(z_{12};\ W_{12})$ $\bar{\varphi}(z_{12};\ U_{12})$
		$\mathscr{D}'_2(d)$ $0<d^2<\frac{1}{12}$	$\bar{\varphi}(z_{12};\ \lvert U_{12}-\frac{1}{2}\rvert)$ $\varphi(z_{12};\ \lvert W_{12}-\frac{1}{2}\rvert)$	$\varphi(z_{12};\ \lvert U_{12}-\frac{1}{2}\rvert)$
		$\mathscr{D}'_2(d)$ $\frac{1}{12}\leqq d^2<\frac{1}{4}$	$\varphi(z_{12};\ \lvert U_{12}-\frac{1}{2}\rvert)^{a}$	$\varphi(z_{12};\ \lvert W_{12}-\frac{1}{2}\rvert)$
$m=2$ $n=2$	$0<\alpha\leqq\frac{1}{6}$	$\mathscr{D}_2(\theta)$ $\frac{3}{4}\leqq\theta^2<1^{b}$	$\varphi(z_{22};\ U_{22})^{a}$	$\varphi(z_{22};\ W_{22})$
	$0<\alpha<\frac{1}{4}$	$\mathscr{D}_2(\theta)$ $\frac{1}{2}\leqq\theta^2<1^{b}$		$\varphi(z_{22};\ W_{22})$
	$0<\alpha\leqq\frac{1}{3}$	$\mathscr{D}'_2(d)$ $\frac{1}{12}\leqq d^2<\frac{1}{4}$	$\varphi(z_{22};\ \lvert U_{22}-\frac{1}{2}\rvert)^{a}$ $\varphi(z_{22};\ L_{22})^{a}$	$\varphi(z_{22};\ \lvert W_{22}-\frac{1}{2}\rvert)$
	$0<\alpha<\frac{1}{2}$	$\mathscr{D}'_2(d)$ $0<d^2<\frac{1}{4}$		$\varphi(z_{22};\ \lvert W_{22}-\frac{1}{2}\rvert)$
	$0<\alpha\leqq\frac{1}{3}$	$\mathscr{D}_3(\Delta^2)$ $\frac{1}{6}\leqq\Delta^2<\frac{1}{3}$	$\varphi(z_{22};\ L_{22})^{a}$ $\varphi(z_{22};\ \lvert U_{22}-\frac{1}{2}\rvert)^{a}$	$\varphi(z_{22};\ \lvert W_{22}-\frac{1}{2}\rvert)$
	$0<\alpha\leqq\frac{3}{7}$	$\mathscr{D}_3(\Delta^2)$ $0<\Delta^2<\frac{1}{3}$		$\varphi(z_{22};\ \lvert W_{22}-\frac{1}{2}\rvert)$
$m=2$ $n=3$	$0<\alpha\leqq\frac{1}{5}$	$\mathscr{D}'_2(d^2)$ $\frac{11}{52}\leqq d^2<\frac{1}{4}^{b}$	$\varphi(z_{23};\ L_{23})^{a}$ $\varphi(z_{23};\ \lvert U_{23}-\frac{1}{2}\rvert)^{a}$	
	$0<\alpha\leqq\frac{3}{7}^{b}$	$\mathscr{D}'_2(d^2)$ $0<d^2<\frac{1}{4}$		$\varphi(z_{23};\ \lvert W_{23}-\frac{1}{2}\rvert)$
	$0<\alpha\leqq\frac{1}{5}$	$\mathscr{D}_3(\Delta^2)$ $\frac{1}{6}\leqq\Delta^2<\frac{1}{3}$	$\varphi(z_{23};\ L_{23})^{a}$ $\varphi(z_{23};\ \lvert U_{23}-\frac{1}{2}\rvert)^{a}$	
	$0<\alpha\leqq\frac{3}{7}^{b}$	$\mathscr{D}_3(\Delta^2)$ $0<\Delta^2<\frac{1}{3}$		$\varphi(z_{23};\ \lvert W_{23}-\frac{1}{2}\rvert)$

[a] This test is most powerful among r.o. (s.r.o.) tests.
[b] This condition is sufficient but undoubtedly not necessary.

(5.1) $$\sigma^2(n^2 U_{nn}) = n\theta(1-\theta) - (n-1)[\tfrac{1}{3} - \theta(1-\theta) - \Delta^2]\,,$$

where $\Delta^2 = \Delta^2(F, G) = \int (F - G)^2\, dF$ (see, e.g., [5]). The maximum is attained when $\Delta^2(F, G) = \frac{1}{3} - \theta(1-\theta)$, by (4.2) and Lemma 1. But the pairs of distributions for which this maximum is attained are rather unusual: In a typical case, the random variable distributed according to F must take on values in two intervals separated by a third interval in which must be found the entire range of the random variable distributed according to G (see Section 2).

Suppose we impose the condition that, for all $(F, G) \in \mathscr{D}(\theta)$,

(5.2) $$\Delta^2(F, G) \leqq \tfrac{1}{3} - \tfrac{4}{3}\theta(1-\theta)\,.$$

First, under this condition, the proof of Theorem 1 fails, as do also the proofs of the modifications and extensions of that theorem mentioned in Section 2. Second, under this condition, we retain in $\mathscr{D}(\theta)$ pairs of distributions of several common types, among which are the following:

1. *Normal.* Let $N(\mu, \sigma)$ denote the normal distribution with mean μ and variance σ^2. Let $F = N(0, \sigma)$, $G = N(\mu\sigma, \sigma)$. Then θ, Δ^2 depend only on μ (see Sundrum [6]), and (5.2) holds.

2. *Rectangular.* Let $R(a, b)$ denote the uniform distribution on the interval (a, b). Condition (5.2) holds if we let $F = R(0, 1)$, $G = R(\delta, 1 + \delta)$, $\delta \geqq 0$.

3. *Exponential.* Let $E(a, b)$ denote the distribution whose c.d.f. is

$$E(x; a, b) = \begin{cases} 0\,, & x < a\,, \\ 1 - \exp\left[\dfrac{-(x-a)}{b}\right], & x \geqq a\,. \end{cases}$$

Then (5.2) holds if we let $F = E(0, b), G = E(ab, b)$; or if we let $F = E(a, 1)$, $G = E(ab, b)$.

4. *"Lehmann's Alternatives"* (see [7]). Condition (5.2) holds if $(F, G) = (F, F^a)$, $a > 0$.

6. Acknowledgment

The author wishes to acknowledge the guidance of Professor Wassily Hoeffding, who proposed the problems considered here and essentially conjectured the result embodied in Theorem 1.

References

[1] Savage, I. R. Contributions to the theory of rank order statistics: Two-sample case. *Ann. Math. Stat.*, 1956, **27**, 590–615.

[2] Hoeffding, W. "Optimum" nonparametric tests. In J. Neyman (ed.), *Proceedings of the Second Berkeley Symposium on Mathematical Statistics and Probability.* Berkeley, Calif.: Univ. California Press, 1951.

[3] Lehmann, E. L. Consistency and unbiasedness of certain non-parametric tests. *Ann. Math. Stat.*, 1951, **22**, 165–79.

[4] Rosenblatt, J. R. On a class of nonparametric tests. Institute of Statistics Mimeograph Series Report No. 138, Chapel Hill, N.C.: University of North Carolina, August 1955.

[5] Sundrum, R. M. The power of Wilcoxon's two-sample test. *J. Roy. Stat. Soc.*, Ser. B, 1953, **15**, 246–52.

[6] Sundrum, R. M. On Lehmann's two-sample test. *Ann. Math. Stat.*, 1954, **25**, 139–45.

[7] Lehmann, E. L. The power of rank tests. *Ann. Math. Stat.*, 1953, **24**, 23–43.

33 SOME NON-PARAMETRIC ANALOGS OF "NORMAL" ANOVA, MANOVA, AND OF STUDIES IN "NORMAL" ASSOCIATION[1]

S. N. ROY and V. P. BHAPKAR
University of North Carolina

1. Summary

We imagine the experimental data to be given in the form of frequencies in cells determined by a (finitely) multiway cross-classification, with predefined categories, finite in number, along each "way of classification." According as the marginal frequencies along any way or dimension are held fixed or left free, that dimension or "way" will be said to be associated with a "factor" or a "response." The general probability model is that of a product of several multinomial distributions. To start with, the probabilities of the different cells are assumed unknown, except for certain weak restrictions. According as a set of (real) values is or is not associated with the categories along any way of classification (with or without fixed marginal frequencies) that way of classification will be said to be a structured or unstructured response or factor. It is the purpose of this paper (which is a follow-up of previous papers, [1], [2], [3], [4], [5]) to pose hypotheses that the authors consider to be generalizations of the usual hypotheses in (1) classical, "normal," univariate, "fixed-effects" analysis of variance or ANOVA, (2) "normal," multivariate, "fixed-effects" analysis of variance or MANOVA, and (3) analysis of various kinds of "normal" independence; and to offer large-sample tests for such hypotheses. Point estimation and confidence bounds of meaningful parametric functions would follow on the lines discussed in the previous papers, [1], [2], [4], [5], and are not discussed here. Nor are more general decision procedures (e.g., ranking, choosing the best, etc.), which are no doubt physically more meaningful than the usual testing of hypotheses or even estimation, discussed in this paper. Another severe restriction

[1] This research was supported partly by the Office of Naval Research under Contract No. Nonr-855(06) for research in probability and statistics at Chapel Hill, North Carolina, and partly by the United States Air Force through the Office of Scientific Research, USAF Air Research and Development Command, under Contract No. AF 49(638)-213. Reproduction in whole or in part is permitted for any purpose of the United States Government.

is this: For each (kind of) response or factor, we make a distinction between the structured and the unstructured case; but even for structured factors we give the pre-Box treatment that avoids the realistic problem of exploring the nature of the dependence of the response on the structured factor or factors. Work is now in progress on this realistic problem. Part of the material presented here on responses that are categorical without a structure is a recapitulation of earlier discussions ([1], [2], [4], [5]), while the material on responses that are categorical with a structure is entirely new.

2. Introduction and Preliminaries

Let

$$\Phi = \prod_j \left[\frac{n_{0j}!}{\prod_j n_{ij}!} \prod_i p_{ij}^{n_{ij}} \right] \tag{2.1}$$

denote a product-multinomial distribution of a certain number of experimental units such that $\sum_i p_{ij} \equiv p_{0j} = 1$ and $\sum_i n_{ij} \equiv n_{0j}$ is fixed with $i = 1, 2, \cdots, r$ and $j = 1, 2, \cdots, s$. In other words, the model might be regarded as one of stratified sampling in which "j" refers to a stratum and "ij" to the ith category in the jth stratum, n_{0j} the preassigned sample size from the jth stratum, n_{ij} the observed frequency of experimental units in the ith category of the jth stratum, and p_{ij} the probability of having an experimental unit from the jth stratum in the ith category. It should be further noticed that i may be a multiple subscript, say, $i_1 i_2 \cdots i_k$ with $i_1 = 1, 2, \cdots, r_1$; $i_2 = 1, 2, \cdots, r_2$; $\cdots$; $i_k = 1, 2, \cdots, r_k$; with all combinations allowed, $r = r_1 r_2 \cdots r_k$. Likewise, j also might be a multiple subscript, say, $j_1 j_2 \cdots j_l$ with $j_1 = 1, 2, \cdots, s_1$; $j_2 = 1, 2, \cdots, s_2$; $\cdots$; $j_l = 1, 2, \cdots, s_l$, but with the important distinction that all combinations may not be allowed. In other words, the classification with respect to the j's might be based on a multidimensional incomplete design such that, for a given set $(j_1, \cdots, j_{l-1})$, $j_l \in (s_l)_{j_1 j_2 \cdots j_{l-1}}$ (j_l takes a set of values which is a subset of the integers $1, 2, \cdots, s_l$, depending upon $j_1 j_2 \cdots j_{l-1}$), and similarly, $j_{l-1} \in (s_{l-1})_{j_1 j_2 \cdots j_{l-2}}$, $\cdots$; $j_2 = (s_2)_{j_1}$, and $j_1 = 1, 2, \cdots, s_1$. This means that, basically, an incomplete design is just a rule by which we can make this kind of choice in successive stages. Thus s is not necessarily $s_1 s_2 \cdots s_l$, but can be computed from this characterization of the incomplete design. This will be called a k-response (or k-variate) and l-factor problem, "i_1", "i_2", $\cdots$, "i_k" being called responses (or variates) and "j_1", "j_2", $\cdots$, "j_l" being called factors. Summation over a subscript will be indicated by replacing that subscript by 0; on n, such a summation will, of course, indicate a marginal frequency. On p, however, such a summation will indicate a marginal probability only when the subscript in question is associated with a response, not when it is associated with a factor. It may be noticed that n_{0j} in (2.1) and p_{0j} a few lines below really stand for $n_{0 \cdots 0 j_1 \cdots j_l}$ and $p_{0 \cdots 0 j_1 \cdots j_l}$. To fix our ideas, consider, for example, three subscripts, say $i_1 i_2 i_3$. Then a typical quantity (other than n or p) depending upon $i_1 i_2 i_3$ will be denoted by, say $q_{i_1 i_2 i_3}$. If it depends on i_1 and i_3 but not i_2 it will be denoted by $q_{i_1*i_3}$, and so on. Thus a star in place of a subscript will indicate that the quantity in question is independent of that subscript.

Next, as to the hypothesis, it may be noticed that (2.1) assumes constraints on the p_{ij}'s, namely, that $\sum_i p_{ij} = p_{0j} = 1$, aside from inequalities of the type discussed in [1], [2], [4], [5]. If, with (2.1) as the model, a hypothesis H_1 is given in the form of certain constraints on p_{ij}'s (in addition, of course, to $\sum_i p_{ij} = 1$), then the large-sample test of H_1 under (2.1) as model is in terms of a statistic given by

$$\sum_{i,j} (n_{ij} - n_{0j}\hat{p}_{ij})^2/n_{0j}\hat{p}_{ij} , \tag{2.2}$$

in which the $\hat{p}_{ij}$'s are estimates (in terms of n_{ij}'s) of the p_{ij}'s which maximize (2.1) subject to $\sum_i p_{ij} = 1$ and to the further constraints on p_{ij}'s that define the hypothesis. This statistic, in the limit as $n \to \infty$ subject to the (n_{0j}/n)'s being held fixed, is distributed as a chi-square variable with degrees of freedom equal to the number of *independent* constraints on the p_{ij}'s (beyond $\sum_i p_{ij} = 1$) that define the hypothesis. It may be observed that, in place of the maximum likelihood estimates of the p_{ij}'s, we may consider the minimum chi-square estimates or any set of estimates belonging to the broader well-known class of BAN estimates. Instead of the statistic (2.2), one could use the slightly different one known as χ_1^2 defined by (2.2) as modified by the substitution of n_{ij} for $n_{0j}\hat{p}_{ij}$ in the denominator of the typical term in the summation (with the same asymptotic chi-square distribution). At this stage we can go further, introduce some additional constraints on the p_{ij}'s and thus define a new hypothesis $H_2 \subset H_1 \subset (2.1)$. The test of H_2, under (2.1) as model, is in terms of a statistic given by

$$\sum_{i,j} (n_{ij} - n_{0j}\hat{\hat{p}}_{ij})^2/n_{0j}\hat{\hat{p}}_{ij} , \tag{2.3}$$

which, in the limit as $n \to \infty$ subject to the (n_{0j}/n)'s being held fixed, has the chi-square distribution with degrees of freedom equal to the number of independent constraints on the p_{ij}'s (beyond $\sum_i p_{ij} = 1$) that define H_1 plus the additional ones that define H_2. However, if we want to test H_2 with H_1 as the model, then the test will be given in terms of a statistic ([1], [2]),

$$\sum_{i,j} \frac{(n_{ij} - n_{0j}\hat{\hat{p}}_{ij})^2}{n_{0j}\hat{\hat{p}}_{ij}} - \sum_{i,j} \frac{(n_{ij} - n_{0j}\hat{p}_{ij})^2}{n_{0j}\hat{p}_{ij}} , \tag{2.4}$$

which, in the limit as $n \to \infty$ subject to the (n_{0j}/n)'s being held fixed, has the chi-square distribution with degrees of freedom equal to the number of free parameters under the model minus the number of free parameters under the hypothesis, which is equal to the number of additional independent constraints on the p_{ij}'s that define H_2 with H_1 as model ([1], [2], [4], [5]). The asymptotic power of such tests for H_1 or H_2 with (2.1) as the model, and H_2 with H_1 as the model has been defined and obtained (in an abstract form) in [1] and [2]. As far as these large-sample tests (as also their asymptotic power) are concerned, the treatment given in an abstract form ([1], [2], [4], [5]) is perfectly general and can take care of both situations—namely when the responses (or variates) and/or the factors are structured or unstructured.

Now let us consider the case of a structured response. A response, say "i_m", will be said to be structured if there is a set of (real) values $a_{i_m}^{(m)}$, $i_m = 1, 2, \cdots, r_m$, associated as weights with the set of categories "i_m" as follows.

1. If the categories "i_m" are class intervals (numbered in the proper order) for a continuous stochastic variate, then the distances from an arbitrary origin of, say, the mid-points of such intervals form a natural set of $a_{i_m}^{(m)}$'s.

2. Likewise, if there is a discrete (real)-valued stochastic variate taking on values $a_1^{(m)} \leqq a_2^{(m)} \leqq \cdots \leqq a_{r_m}^{(m)}$, then these might be the natural weights to go with the categories in that direction, which also might be defined and numbered in a natural manner.

3. If the response is not even a discrete (real)-valued variate but is categorical with an implied ranking (like good, fair, poor, etc.), then the categories can be defined and numbered in a natural way and scores can be assigned to these categories.

4. Finally, if the response is terms of categories without even an implied ranking (for example, like black, yellow, red, etc.), then the numbering of these categories is purely nominal and there is no natural system of weights to go *prima facie* with the categories. However, because of economic or other considerations, we might assign a system of (real-valued) weights to these categories which allow initially unstructured responses to become structured.

But even in the case of (1), (2), or (3), such considerations might induce a system of weights different from the ones that are "natural" for (1), (2), or (3). In most realistic problems, especially of a utilitarian (rather than purely scientific) nature, each kind of response eventually becomes structured although it may be unstructured to start with. On the other hand, so far as the factors are concerned, any factor could well be either continuous or discrete, or categorical with an implied ranking, or purely categorical. Thus, in general, in a complex realistic situation some of the responses are of type (1), some of type (2), some of type (3), and the rest of type (4), with a similar division into four types for the factors. In the present paper, we shall discuss three types of problems, namely, (i) where all responses are unstructured and so also are all factors; (ii) some responses are structured, and all factors are unstructured; (iii) some responses are structured, and some factors are structured. As far as the problems of interest are concerned, the main distinction between the unstructured and the structured case seems to be this. It is feasible and also useful, in the structured case, to go ahead and define certain over-all (or, to borrow a phrase from differential geometry, *im grossen*) aspects of the distribution which are not possibly defined for the unstructured case. Then, while in the structured case it is still possible to study the same problems as in the unstructured case, such problems (interesting for the unstructured case) are hardly of the same interest, for the structured case, and the problems involving the newly-defined over-all aspects are the ones that become more meaningful. Problems involving other possibilities will be discussed in later papers.

To make everything concrete, we shall consider in this paper only some three-way, four-way, and five-way tables. For the three-way table (ijk) we shall consider the following cases: (a) "i", "j", "k" are all responses, where one or more are unstructured; (b) "i" is an unstructured response and "j" and "k" unstructured factors; (c) "i" is a structured response and "j" and "k" are factors, where one or both are unstructured. For four-way $(ijkl)$ tables we consider problems similar to (a) above; and for five-way $(ijklm)$ tables we consider the cases where (a) "i" and "j" are unstructured responses, "k", "l", and "m" are unstructured factors; (b) "i" and "j" are structured responses, "k", "l", and "m" are all unstructured factors, or "k" is structured, and "l" and "m" are unstructured factors. This will serve as an illustration, and will indicate what happens as we increase the dimensionality of the table (but otherwise stay within the limitations indicated earlier).

3. A Three-way Table (ijk) in Which "i", "j", and "k" All Are Responses or Variates

Let us consider a single-multinomial distribution

$$\Phi = \frac{n!}{\prod_{i,j,k} n_{ijk}!} \prod_{i,j,k} p_{ijk}^{n_{ijk}} , \tag{3.1}$$

where $\sum_{i,j,k} p_{ijk} = 1$ and $\sum_{i,j,k} n_{ijk} = n$ (fixed).

3.1 The Case Where "i", "j", and "k" Are All Unstructured. As an illustration, we consider some important specific hypotheses, as follows:

$$H_{01} : p_{ijk} = p_{i0k}\, p_{0jk}/p_{00k} , \tag{3.2}$$

which can be described as the hypothesis that "i" and "j" are independent in the conditional distribution of "i" and "j", given "k", or, in other words, that there is no partial association between "i" and "j", given "k". There are two equivalent ways to get at (3.2). One is to notice that the conditional joint distribution of "i" and "j", given "k", is p_{ijk}/p_{00k}, and the conditional marginal distribution of "i", given "k", is p_{i0k}/p_{00k}, that of "j", given "k", is p_{0jk}/p_{00k}, which leads to (3.2) as the condition that there is no partial association between "i" and "j". Another way is to start with the condition

$$p_{ijk}/p_{0jk} = q_{i*k} , \tag{3.3}$$

which means that the conditional distribution of "i", given "j" and "k", is independent of "j", and then rewrite (3.3) in the form

$$p_{ijk} = p_{0jk}\, q_{i*k} . \tag{3.4}$$

Summing over "j", we find that (3.3) is equivalent to (3.2). This hypothesis and the appropriate chi-square test have already been discussed ([2], [4]), except that the approach (3.3) was not considered.

We next proceed to the hypothesis that "j" and "k" are independent and so also are "i" and "k", i.e.,

$$H_{02}: p_{0jk} = p_{0j0}p_{00k}\,, \quad \text{and} \quad p_{i0k} = p_{i00}p_{00k}\,. \tag{3.5}$$

We note that $H_{01} \cap H_{02} \Rightarrow$

$$H_{03}: p_{ijk} = p_{i00}p_{0j0}p_{00k}\,, \tag{3.6}$$

which is the hypothesis of over-all independence of "i", "j" and "k". The hypothesis that "k" is multi-independent of "ij", or, in other words, that there is no multiple association between "ij" and "k" is given by

$$H_{04}: p_{ijk} = p_{ij0}p_{00k}\,. \tag{3.7}$$

It is easy to check by summing the two sides of (3.7) over "i" and over "j" separately that $H_{04} \Rightarrow H_{02}$, but not conversely (unlike the multivariate normal case). It is also easy to check that one could also get to (3.7) by starting either from the assumption

$$p_{ijk}/p_{ij0} = q_{**k}\,, \tag{3.8}$$

or from the assumption

$$p_{ijk}/p_{00k} = q_{ij*}\,. \tag{3.9}$$

Another hypothesis which might be of interest is that of two × two independence of "i", "j", and "k", or in symbols,

$$H_{05}: p_{ij0} = p_{i00}p_{0j0}, \quad p_{i0k} = p_{i00}p_{00k}, \quad \text{and} \quad p_{0jk} = p_{0j0}p_{00k}\,. \tag{3.10}$$

It is easy to check that $H_{03} \Rightarrow H_{05}$ but, unlike the multivariate normal case, not conversely, which is well known to probabilists. We observe that the experimenter might very well be interested in H_{02} but not necessarily in H_{04}, and likewise in H_{05} but not necessarily in H_{03}.

3.2 The Case Where "i" Is Structured but Not Necessarily "j" or "k". Let us start by considering a hypothesis which does not have any analog under the previous model, namely,

$$H_{00}: \sum_i a_i p_{ijk}/p_{0jk} = q_{*j*}q_{**k}\,, \tag{3.11}$$

or

$$H_{00}^{(1)}: \sum_i a_i p_{ijk}/p_{0jk} = q_{*j*}^{(1)} + q_{**k}^{(1)}\,, \tag{3.12}$$

where the q's are unknown functions, and the a_i's are given. We next consider a hypothesis analogous to H_{01}, using the version of H_{01} given by (3.2):

$$H_{01}^{(1)}: \sum a_i p_{ijk}/p_{0jk} = q_{**k}\,. \tag{3.13}$$

This is, of course, a special case of (3.11). By rewriting (3.13) as $\sum_i a_i p_{ijk} =$

$p_{0jk}q_{**k}$ and summing both sides over "j", it is easy to check that (3.13) could be rewritten in the equivalent form

$$H_{01}^{(1)}: \sum_i a_i \left[p_{ijk} - \frac{p_{i0k}p_{0jk}}{p_{00k}} \right] = 0 . \tag{3.14}$$

It should be noticed that $H_{01} \Rightarrow H_{01}^{(1)}$ but not conversely, for a given set of a_i's. However, if (3.14) is to hold for all sets of a_i's, then (3.14) is equivalent to H_{01}.

Also analogous to H_{02} (but with an interchange of subscripts), we now have

$$H_{02}^{(1)}: \sum_i a_i p_{ij0}/p_{0j0} = \lambda , \qquad \sum_i a_i p_{i0k}/p_{00k} = \mu , \tag{3.15}$$

which is equivalent to

$$H_{02}^{(1)}: \sum_i a_i(p_{ij0} - p_{i00}p_{0j0}) = 0 , \qquad \sum_i a_i(p_{i0k} - p_{i00}p_{00k}) = 0 . \tag{3.16}$$

The same remarks concerning the relation between H_{01} and $H_{01}^{(1)}$ are applicable here. Likewise, analogous to H_{04} (but with an interchange of subscripts), we have

$$H_{04}^{(1)}: \sum_i a_i p_{ijk}/p_{0jk} = \nu , \tag{3.17}$$

which is equivalent to

$$H_{04}^{(1)}: \sum_i a_i (p_{ijk} - p_{i00}p_{0jk}) = 0 . \tag{3.18}$$

The same remarks are applicable. The analogs of H_{03} and H_{05} are obvious and are left to the reader. The more interesting case is one where more than one of "i", "j", and "k" are structured.

3.3 The Case Where "i" and "j" Are Structured but Not Necessarily "k". For concreteness, we single out "i"; then the hypotheses of independence of "i" with respect to "j" and "k" of the types (3.13), (3.15), and (3.17) (as well as the tests) remain as before. If these are tested and rejected, then (in subsequent experiments with the same type of material) we have other interesting and useful possibilities. Assuming a set of weights b_j's to go with "j", we can write down, for example, a hypothesis related to $H_{01}^{(1)}$ in the form

$$H_{01}^{(2)}: \sum_i a_i p_{ijk}/p_{0jk} = (\lambda + \mu b_j)q_{**k} , \quad \text{or} \quad = (\lambda_k + \mu_k b_j) , \tag{3.19}$$

(the latter being more general), where q_{**k} and λ_k and μ_k are unknown functions of "k", and λ and μ are unknown constants. It is easy to see that (3.19) is a special case of (3.11) when "j" is structured. We can test this in the spirit of testing for linearity of regression; or, assuming this as a model, we can proceed to test the hypothesis that $\mu = 0$ or $\mu_k = 0$.

A hypothesis naturally related to $H_{02}^{(1)}$ is

(3.20) $$H_{02}^{(2)} : \sum_i a_i p_{ij0}/p_{0j0} = \lambda + \mu b_j \, , \qquad \sum_i a_i p_{i0k}/p_{00k} = \nu \, .$$

In this case again, treating λ and μ as unknown, we can test this hypothesis (in the spirit of testing for linearity of regression for the "normal" case) or, assuming this as a model, we can proceed to test for, say, $\mu = 0$.

In this model a hypothesis naturally related to $H_{04}^{(1)}$ and more general than (3.19) has not so far suggested itself to us. It may be observed that (3.19) itself might, in a sense, be regarded as being related to $H_{04}^{(1)}$.

3.4 The Case Where "i", "j", and "k" Are All Structured. As in Section 3.3, for concreteness, we single out "i"; then the hypotheses of independence of "i" with respect to "j" and "k" of the types (3.13), (3.15), and (3.17) (as well as the tests) remain as before. If these are tested and rejected, then (in subsequent experiments with the same type of material) we have other important and interesting possibilities. Assuming a set of weights b_j's to go with "j" and c_k's with "k", we can write down, for example, a hypothesis related to $H_{04}^{(1)}$ in the form

(3.21) $$H_{04}^{(3)} : \sum_i a_i p_{ijk}/p_{0jk} = \lambda + \mu b_j + \nu c_k \, .$$

We can test this hypothesis (in the spirit of testing for linearity of regression in "normal" analysis of variance), in this case, treating λ, μ, and ν as unknown and using a chi-square test. Also assuming (3.21) as a model in addition to (3.1) we can also test that $\mu = 0$ or that $\nu = 0$. Likewise, a hypothesis related to (3.15) is

(3.22) $$H_{02}^{(3)} : \sum_i a_i p_{ij0}/p_{0j0} = \lambda_1 + \mu b_j \, , \qquad \sum_i a_i p_{i0k}/p_{00k} = \lambda_2 + \nu c_k \, .$$

In this case also, treating λ_1, λ_2, μ, and ν as unknown, we can test this hypothesis (in the spirit of testing for linearity of regression in the "normal" case) or, assuming this as a model, we can proceed to test for $\mu = 0$ or $\nu = 0$, or for both.

Similarly, a hypothesis related to (3.13), although not too natural in this case, is

(3.23) $$H_{01}^{(3)} : \sum_i a_i p_{ijk}/p_{0jk} = (\lambda + \mu b_j) q_{**k} \quad \text{or} \quad = \lambda_k + \mu_k b_j \, ,$$

where q_{**k}, λ_k, and μ_k are unknown functions of "k", and λ and μ are unknown constants. We can test this in the spirit of testing for linearity of regression, or, assuming this as a model, we can proceed to test for $\mu = 0$ or $\mu_k = 0$. We can also set up a similar hypothesis by interchanging "j" and "k" and replacing b_j's by c_k's. It is easy to see that, if both "j" and "k" are structured, then (3.21) and (3.22) are among the natural hypotheses to test. Hypothesis (3.23) is not too natural but is permissible. On the other hand, (3.23) would be quite natural if "k" were unstructured while "j" is structured. Likewise, a hypothesis similar to (3.23) in which the roles of "j" and "k" were interchanged and b_j's were replaced by c_k's would be natural if "j" were unstructured while "k" were structured. Nevertheless, even if "j"

and "k" were both structured, the hypothesis (3.23), one similar to this with "j" and "k" interchanged with c_k in place of b_j's, and also one in which the right side is just replaced by $q_{**k} \times q_{*j*}$, would all be permissible, although none of them would be too natural.

The reader must have perceived that in the study of association in this non-parametric situation or model we have been working in the spirit of "regression" (i.e., "prediction") rather than in the spirit of "correlation." In other words, we have not been trying to use a single measure for any of the various types of association. We feel that such a single measure has, in general, only a limited use and not too much should be made of it in the non-parametric case. The asymptotic power functions of the respective tests for independence of various kinds, already obtained in the references [1], [2], [4], [5], involved, in each case, a non-centrality parameter which, in these publications, was also incidentally interpreted as a single measure of association for which some limited uses were mentioned.

4. A Three-way Table (ijk) in Which "i" Is a Variate or Response and "j" and "k" Are Factors

Let us consider a product-multinomial distribution

$$\Phi = \prod_{j,k}\left[\frac{n_{0jk}!}{\prod_i n_{ijk}!}\prod_i p_{ijk}^{n_{ijk}}\right], \tag{4.1}$$

where $\sum_i p_{ijk} = p_{0jk} = 1$ and $\sum_i n_{ijk} = n_{0jk}$ (fixed).

4.1 The Case Where "i", "j", and "k" Are All Unstructured. The hypothesis of no interaction between "j" and "k" means, essentially, that for a given "i", there is a lesser number of unknown parameters than would be given by all allowable (jk) combinations. Two specializations are in the same spirit as in ordinary analysis of variance. From here on we must keep in mind that the unknown parameters or parametric functions introduced on the right sides must be such that (a) $0 < p_{ijk} < 1$ for all allowable (ijk) combinations and (b) $\sum_i p_{ijk} = 1$, again for all allowable (jk) combinations. The same kind of constraint also carries over into tables of dimensions higher than three, whenever we have factorial situations to deal with. The specializations are as follows.

$$H_{06} : p_{ijk} = q_{i*k} q_{ij*} \tag{4.2}$$

and

$$H_{06}^{(1)} : p_{ijk} = q_{i*k}^{(1)} + q_{ij*}^{(1)} . \tag{4.3}$$

The physical interpretations are as follows. For (4.2), $p_{ijk}/p_{ij'k}$ is independent of "k", and $p_{ijk}/p_{ijk'}$ is independent of "j"; and, for (4.3), $(p_{ijk} - p_{ij'k})$ is independent of "k", and $(p_{ijk} - p_{ijk'})$ is independent of "j".

Now, if the design is complete, then summing both sides of (4.2) over "j" and "k" separately and then jointly, it is easy to check that (4.2) is equivalent to

(4.4) $$H_{06}^{(cb)}: p_{ijk} = p_{i0k}p_{ij0}/p_{i00}\ ,$$

(where cb stands for complete block). It must be remembered that none of p_{i0k}, p_{0jk}, or p_{i00} is a probability, as each is based on summation over the levels of a factor, i.e., a summation over subscripts belonging to different multinomial distributions. Thus (4.4) is only formally similar to (3.2), but it is easy to check that (4.4) would be completely identical with the condition that there is no partial association between "j" and "k", for given "i", if "j" and "k" were variates and not factors as in the present case. Thus the no-interaction hypothesis in the form (4.4) (but only in the case of a complete design) is related to the hypothesis of no partial association in the case where "j" and "k" are variates.

Next consider (4.3). Let $s^{(k)}$ be the number of "j"'s that go with a given "k", $t^{(j)}$ be the number of "k"'s that go with a given "j", and let $u = \sum_{k=1}^{t} s^{(k)} = \sum_{j=1}^{s} t^{(j)}$ equal the number of "jk" combinations. If the design is complete, then $s^{(k)} = s$, $t^{(j)} = t$, and $u = st$. In this case, summing (4.3) over "j" and "k" separately, and then over both "j" and "k", it is easy to check that (4.3) is equivalent to

(4.5) $$H_{06}^{(cb)(1)}: p_{ijk} = \frac{1}{s}p_{i0k} + \frac{1}{t}p_{ij0} - \frac{1}{st}p_{i00}\ .$$

Going back to H_{06} [expressed in the form (4.2)] we can test the narrower hypothesis that q_{ij*} is independent of "j", or, in other words, that p_{ijk} is independent of "j", which, without any loss of generality, is equivalent to

(4.6) $$H_{07}: p_{ijk} = q_{i*k}\ ,$$

or

(4.7) $$H_{07}: p_{ijk} = \frac{1}{s^{(k)}}p_{i0k}\ .$$

It is easy to check that we shall eventually get the same hypothesis expressed in either of the equivalent forms (4.6) or (4.7), if we start from $H_{06}^{(1)}$ [expressed in the form of (4.3)] and try to state the narrower hypothesis that $q_{ij*}^{(1)}$ is independent of "j", or, in other words, that p_{ijk} is independent of "j". Now, assuming for concretness that "j" stands for treatments and "k" for blocks in a (possibly) incomplete block design, (i) H_{06} [expressed in the form (4.2)] and (ii) $H_{06}^{(1)}$ [again expressed in the form (4.3)], we may define respectively as the hypothesis or model of no interaction (i) in the multiplicative sense and (ii) in the additive sense. The hypothesis H_{07}, given by (4.7), defines the hypothesis of no-treatment effect. It is open to us (depending upon past knowledge) (a) to start from (3.1) as the model and test as a hypothesis either H_{06} or $H_{06}^{(1)}$ or directly even H_{07}, or (b) to start from a model which is (3.1) together with either H_{06} or $H_{06}^{(1)}$, and then (in either case) to test H_{07} as a hypothesis.

4.2 The Case Where "i" Is Structured but Not Necessarily "j" or "k". In this case, some natural analogs of H_{06} and $H_{06}^{(1)}$ (with a given set of a_i's) are

(4.8) $$H_{08} : \sum_i a_i p_{ijk} = q_{**k} q_{*j*} ,$$

and

(4.9) $$H_{08}^{(1)} : \sum_i a_i p_{ijk} = q_{**k}^{(1)} + q_{*j*}^{(1)} ,$$

Remembering that $q_{**k}^{(1)}$ and $q_{*j*}^{(1)}$ are completely unknown functions of "j" and "k" respectively, we can rewrite (4.9) in the equivalent forms

(4.10) $$H_{08}^{(1)} : \sum_i a_i p_{ijk} = \sum_i a_i q_{i*k}^{(1)} + \sum_i a_i q_{ij*}^{(1)} , \quad \sum_i a_i [p_{ijk} - (q_{i*k}^{(1)} + q_{ij*}^{(1)})] = 0 .$$

It is now easy to see that $H_{06}^{(1)}$ of (4.3) $\Rightarrow H_{08}^{(1)}$ but not conversely, and that (4.10) (for all sets of a_i's) $\Rightarrow H_{06}^{(1)}$. On the other hand, neither $H_{06} \Rightarrow H_{08}$, nor does the converse hold (even for all sets of a_i's). Suppose we ask what happens if (4.8) is to hold for all sets of a_i's? Rewriting the right side as, say, $q_{*j*} \sum_i a_i q_{i*k}$, we can rewrite (4.8) as

(4.11) $$H_{08} : \sum_i a_i [p_{ijk} - q_{i*k} q_{*j*}] = 0 .$$

This means that if we set up a hypothesis

(4.12) $$H_{081} : p_{ijk} = q_{i*k} q_{*j*} ,$$

then $H_{081} \Rightarrow H_{08}$ but not conversely. However, H_{08} (for all a_i's) $\Rightarrow H_{081}$ with a counterpart formed by interchanging "j" and "k" on the right-hand side of (4.12). Thus it turns out that H_{08} does not have a natural connection with H_{06} in the sense in which $H_{08}^{(1)}$ has a natural connection with $H_{06}^{(1)}$. The connection of H_{08} (in this sense) is with H_{081} or its counterpart. However, if we sum up both sides of (4.12) over i, then, since $p_{0jk} = 1$, we should have $1 = q_{*j*} \sum_i q_{i*k}$, whence $q_{*j*} = 1/\sum_i q_{i*k}$, which really means that both are pure constants (not depending on "j" or "k" or "i"). Thus H_{081} is essentially H_{07}, and its counterpart is essentially the counterpart of H_{07}, and (4.8) (for all sets of a_i's) is the same as either (4.13) (for all sets of a_i's) or a statement (for all sets of a_i's) similar to (4.13), with q_{*j*} replacing q_{**k} on the right-hand side.

Going back to H_{07}, we write down its analog in the form

(4.13) $$H_{09} : \sum_i a_i p_{ijk} = q_{**k} ,$$

which, as in the case just preceding, can also be written in the equivalent form

(4.14) $$H_{09} : \sum_i a_i \left[p_{ijk} - \frac{1}{s^{(k)}} p_{i0k} \right] = 0 .$$

Again, as before, $H_{07} \Rightarrow H_{09}$ but not conversely. However, H_{09} (for all sets of a_i's) $\Longleftrightarrow H_{07}$. The other remarks made after (4.17) would also carry over to this case, covering (4.8), (4.9), and (4.13).

Going back to (4.8), we have another hypothesis which has the same relation to H_{06} as $H_{08}^{(1)}$ has to H_{06}:

(4.15) $$H_{082}: \prod_i p_{ijk}^{a_i} = q_{*j*}q_{**k} .$$

Rewriting the right-hand side as $\Pi_i\, q_{ij*}^{a_i} q_{i*k}^{a_i}$, we can then rewrite (4.15) as

(4.16) $$H_{082}: \prod_i (p_{ijk}/q_{ij*}q_{i*k})^{a_i} = 1 .$$

It is easy to check that $H_{06} \Rightarrow H_{082}$ but not conversely, and also that H_{082} (for all a_i's) $\Rightarrow H_{06}$. It remains now to interpret the left side of (4.15). We notice that $[\Pi_i p_{ijk}^{a_i}]^{1/a}$, where $a = \sum_i a_i$, is a kind of weighted geometric mean of p_{ijk} (over "i") with a_i's for weights and that, since $\sum_i a_i$ is independent of "i", the left side of (4.15) is a certain power of this weighted geometric mean. We still do not know how meaningful or useful this interpretation of the model or hypothesis (4.15) would be in the end. We offer it quite tentatively at this stage.

4.3 The Case Where "*i*" and "*j*" Are Both Structured but Not Necessarily "*k*". In this case (assuming a given set of weights b_j's to go with "j") the natural analogs of H_{08}, $H_{08}^{(1)}$, and H_{09} are

(4.17) $$H_{010}: \sum_i a_i p_{ijk} = q_{**k}(\lambda + \mu b_j) \quad \text{or} \quad = \lambda_k + \mu_k b_j ,$$

(4.18) $$H_{010}^{(1)}: \sum_i a_i p_{ijk} = q_{**k}^{(1)} + (\lambda + \mu b_j) ,$$

and

(4.19) $$H_{011}: \sum_i a_i p_{ijk} = q_{**k} ,$$

where λ_k and μ_k are supposed to be unknown functions of k, and λ and μ just unknown constants. The remarks on interpretation and inference procedures made after (4.7) would also be applicable here. However, on the question of interpretation, this case involves some additional possibilities (in terms of the conceptual universe or framework). While treating "k" as an unstructured factor we may treat "j" as a structured factor in a two-dimensional (jk) design or a concomitant variate in a one-way (k) classification. But for several reasons we shall, in this paper, omit this aspect and interpretation.

No matter which interpretation we make, if "k" is a factor and not a set of blocks (i.e., not a factor that is irrelevant and yet enters into the experiment in an unavoidable manner), then another meaningful hypothesis parallel to H_{011} is

(4.20) $$H_{012}: \sum_i a_i p_{ijk} = \lambda + \mu b_j ,$$

(where λ and μ are, of course, assumed to be unknown). Going back to (4.17), we observe that instead we may consider the same kind of possibility that we had in the previous case in terms of H_{082} of (4.15). But we will not pursue that here. The same remark applies to later cases whenever we consider the multiplicative model or hypothesis of no interaction with responses that are structured and factors that are either all unstructured or some structured and the rest unstructured—for example, responses considered in (6.6), (6.8), (6.10), and (6.12).

4.4 The Case Where "i", "j", and "k" Are All Structured. In this case, assuming furthermore a given set of weights c_k's to go with "k", it is possible to set up the same H_{010}, $H_{010}^{(1)}$, and H_{012} and also similar ones in which the roles of "j" and "k" are interchanged and b_j's are replaced by c_k's. However, in this case, the more natural hypothesis would seem to be

$$H_{013}: \sum_i a_i p_{ijk} = \lambda + \mu b_j + \nu c_k , \tag{4.21}$$

(say, for the sake of concretness and simplicity, where λ, μ and ν are, of course, supposed to be unknown). (i) Taking (4.1) as a model we can either test this hypothesis or directly test one in which $\mu = 0$ and/or $\nu = 0$; or (ii) taking (4.1) together with (4.21) as a model we can test the hypothesis that $\mu = 0$ and/or $\nu = 0$.

5. A Four-way ($ijkl$) Table in Which "i", "j", "k", and "l" Are All Responses or Variates

The distribution is

$$\Phi = \frac{n!}{\prod_{i,j,k,l} n_{ijkl}!} \prod_{i,j,k,l} p_{ijkl}^{n_{ijkl}} , \tag{5.1}$$

where $\sum_{i,j,k,l} p_{ijkl} = 1$ and $\sum_{i,j,k,l} n_{ijkl} = n$ (fixed). Whether these variates are all unstructured, or all structured, or whether some are unstructured and the rest structured, there is much in common between this four-variate case and the three-variate table or case discussed in Section 3. However, the four-variate table or case presents certain new features and we shall discuss a few of these, for purposes of illustration. These are by no means exhaustive.

5.1 The Case Where "i", "j", "k", and "l" Are All Unstructured. We have the hypothesis

$$H_{014}: p_{ijk0} = p_{i0k0} p_{0jk0}/p_{00k0} , \qquad p_{ij0l} = p_{i00l} p_{0j0l}/p_{000l} . \tag{5.2}$$

5.2 The Case Where "l" Is Unstructured and "i", "j", and "k" Are Structured with Given Sets of Weights a_i's, b_j's, and c_k's To Go with Them. In this case,

$$H_{015}: \text{(i)} \sum_i a_i p_{ijk0}/p_{0jk0} = \lambda + \mu b_j + \nu c_k , \tag{5.3}$$

or

$$\text{(ii)} \sum_j b_j p_{ijk0}/p_{i0k0} = \lambda^{(1)} + \mu^{(1)} a_i + \nu^{(1)} c_k ,$$

together with

$$\text{(iii)} \sum_i a_i p_{ij0l}/p_{0j0l} = q_{***l} \times (\lambda^{(2)} + \mu^{(2)} b_j) ,$$

or

$$\text{(iv)} \sum_j b_j p_{ij0l}/p_{i00l} = q_{***l}^{(1)} \times (\lambda^{(3)} + \mu^{(3)} a_i) ,$$

where (i) is taken with (iii), or (ii) is taken with (iv). A more general possibility with $\lambda_l^{(2)}$ and $\mu_l^{(2)}$ or with $\lambda_l^{(3)}$ and $\mu_l^{(3)}$ could also be considered here, as in similar situations in previous cases.

6. A Five-way (*ijklm*) Table in Which "*i*" and "*j*" Are Responses or Variates and "*k*", "*l*", and "*m*" Are Factors

The distribution here is

$$\Phi = \prod_{k,l,m} \left[\frac{n_{ooklm}!}{\prod_{i,j} n_{ijklm}!} \prod_{i,j} p_{ijklm}^{n_{ijklm}} \right], \tag{6.1}$$

where $\sum_{i,j} p_{ijklm} = p_{ooklm} = 1$ and $\sum_{i,j} n_{ijklm} = n_{ooklm}$ (fixed).

Again, whether all these are structured or some of them are structured and the rest unstructured, there is much in common between this model and the corresponding three-way model already discussed in Section 4. However, there are some new features and we shall discuss a few of these as follows.

6.1 The Case Where All Are Unstructured. We shall consider the hypothesis

$$H_{016}: p_{ioklm} = q_{i**lm}\, q_{i*k*m}\, q_{i*kl*}\,, \qquad p_{ojklm} = q_{*j*lm}\, q_{*jk*m}\, q_{*jkl*}\,, \tag{6.2}$$

or

$$H_{016}^{(1)}: p_{ioklm} = q_{i**lm}^{(1)} + q_{i*k*m}^{(1)} + q_{i*kl*}^{(1)}\,, \qquad p_{ojklm} = q_{*j*lm}^{(1)} + q_{*jk*m}^{(1)} + q_{*jkl*}^{(1)}\,, \tag{6.3}$$

and next, the hypothesis

$$H_{017}: p_{ioklm} = q_{i*k**}\, q_{i**l*}\, q_{i***m}\,, \qquad p_{ojklm} = q_{*jk**}\, q_{*j*l*}\, q_{*j**m}\,, \tag{6.4}$$

or

$$H_{017}^{(1)}: p_{ioklm} = q_{i*k**}^{(1)} + q_{i**l*}^{(1)} + q_{i***m}^{(1)}\,, \qquad p_{ojklm} = q_{*jk**}^{(1)} + q_{*j*l*}^{(1)} + q_{*j**m}^{(1)}\,, \tag{6.5}$$

and finally H_{018} and $H_{018}^{(1)}$, which state that p_{ioklm} and p_{ojklm} are independent of one or more of the factors "k", "l", and "m", in the multiplicative and additive models respectively. It is easy to check that $H_{016} \supset H_{017} \supset H_{018}$ and $H_{016}^{(1)} \supset H_{017}^{(1)} \supset H_{018}^{(1)}$, and it seems to be natural to start from (6.1) and test H_{016} or $H_{016}^{(1)}$ as a hypothesis, or from H_{016} or $H_{016}^{(1)}$ as a model and test H_{017} or $H_{017}^{(1)}$ as a hypothesis, or from H_{017} or $H_{017}^{(1)}$ as a model and test H_{018} or $H_{018}^{(1)}$ as a hypothesis. There are various interesting intermediate cases which we omit and which the reader can supply for himself.

6.2 The Case Where "*i*" and "*j*" Are Structured and "*k*", "*l*", and "*m*" Are Unstructured. The analogs of the hypotheses in the previous case are (in the same order):

(6.6) $$H_{019}:\sum_i a_i p_{i0klm}=q^{(1)}_{***lm}\,q^{(1)}_{**k*m}\,q^{(1)}_{**kl*}\,,$$

$$\sum_j b_j p_{0jklm}=q^{(2)}_{***lm}\,q^{(2)}_{**k*m}\,q^{(2)}_{**kl*}\,,$$

or

(6.7) $$H^{(1)}_{019}:\sum_i a_i p_{i0klm}=q^{(3)}_{***lm}+q^{(3)}_{**k*m}+q^{(3)}_{**kl*}\,,$$

$$\sum_j b_j p_{0jklm}=q^{(4)}_{***lm}+q^{(4)}_{**k*m}+q^{(4)}_{**kl*}\,,$$

(6.8) $$H_{020}:\sum_i a_i p_{i0klm}=q^{(1)}_{**k**}\,q^{(1)}_{***l*}\,q^{(1)}_{****m}\,,$$

$$\sum_j b_j p_{0jklm}=q^{(2)}_{**k**}\,q^{(2)}_{***l*}\,q^{(2)}_{****m}\,,$$

or

(6.9) $$H^{(1)}_{020}:\sum_i a_i p_{i0klm}=q^{(3)}_{**k**}+q^{(3)}_{***l*}+q^{(3)}_{****m}\,,$$

$$\sum_j b_j p_{0jklm}=q^{(4)}_{**k**}+q^{(4)}_{***l*}+q^{(4)}_{****m}\,,$$

and finally H_{021} and $H^{(1)}_{021}$, which state respectively that $\sum_i a_i p_{i0klm}$ and $\sum_j b_j p_{0jlm}$ are independent of one or more of the factors "k", "l", and "m", in the multiplicative and additive models respectively. The remarks made about $H^{(1)}_{018}$ after (6.5) would also apply here.

6.3 The Case Where "i", "j", and "k" Are Structured While "l" and "m" Are Unstructured. The natural analogs of the hypotheses of the previous case are (in the same order) as follows:

(6.10) $$H_{022}:\sum_i a_i p_{i0klm}=(\lambda^{(1)}_{lm}+\mu^{(1)}_{lm}c_k)\,,$$

$$\sum_j b_j p_{0jklm}=(\lambda^{(2)}_{lm}+\mu^{(2)}_{lm}c_k)\,,$$

or

(6.11) $$H^{(1)}_{022}:\sum_i a_i p_{i0klm}=q^{(1)}_{***lm}+\lambda^{(1)}+\mu^{(1)}c_k\,,$$

$$\sum_j b_j p_{0jklm}=q^{(2)}_{***lm}+\lambda^{(2)}+\mu^{(2)}c_k\,,$$

(6.12) $$H_{023}:\sum_i a_i p_{i0klm}=(\lambda^{(1)}_{l*}\lambda^{(1)}_{*m}+\mu^{(1)}_{l*}\mu^{(1)}_{*m}c_k)\,,$$

$$\sum_j b_j p_{0jklm}=(\lambda^{(2)}_{l*}\lambda^{(2)}_{*m}+\mu^{(2)}_{l*}\mu^{(2)}_{*m}c_k)\,,$$

or

(6.13) $$H^{(1)}_{023}:\sum_i a_i p_{i0klm}=q^{(1)}_{***l*}+q^{(1)}_{****m}+\lambda^{(1)}+\mu^{(1)}c_k\,,$$

$$\sum_j b_j p_{0jklm}=q^{(2)}_{***l*}+q^{(2)}_{****m}+\lambda^{(2)}+\mu^{(2)}c_k\,,$$

and finally H_{024} and $H^{(1)}_{024}$ similar to H_{021} and $H^{(1)}_{021}$ of Section 6.4. The remarks made after (6.5) apply here as in the previous case. It may be observed that (6.11) might also be regarded as a rather remote analog of (6.6′) given by

$$(6.6') \qquad H_{0191}: \sum_i a_i p_{i0klm} = q^{(1)}_{***lm}\, q^{(1)}_{**k**}\,,$$

$$\sum_j b_j p_{0jklm} = q^{(2)}_{***lm}\, q^{(2)}_{**k**}\,,$$

[which is a special case of (6.6)] and likewise, (6.11) might also be regarded as a rather remote analog of (6.6″) given by

$$(6.6'') \qquad H^{(1)}_{0191}: \sum_i a_i p_{i0klm} = q^{(3)}_{***lm} + q^{(3)}_{**k**}\,,$$

$$\sum_j b_j p_{0jklm} = q^{(4)}_{***lm} + q^{(4)}_{**k**}\,,$$

[which again is a special case of (6.7)]. There is a question whether in this case there is anything more general than (6.10) or (6.11) but along the same lines. So far we have failed to hit on any such thing.

7. Concluding Remarks

The foregoing discussion suggests the proper treatment for a complex experimental situation in which some of the responses or variates are structured, say v_1 in number, the rest unstructured, say v_2 in number, some of the factors are structured, say f_1 in number, while the rest are unstructured, say f_2 in number. In theory or principle the tests in terms of the appropriate chi-square statistic (or difference of two chi-square statistics) can be set up for all the specific hypotheses mentioned, as well as for the corresponding hypotheses in the larger framework indicated in this section. The asymptotic power functions can also be calculated ([1], [2], [4], [5]). The actual form of the chi-square has been given for only some of the specific hypotheses discussed in Sections 3 to 6, and the asymptotic power functions for even fewer of such tests. Work is now in progress to close these gaps.

The treatment given in this paper is all in terms of single or product-multinomial distributions which, again, are based on sampling with replacement. This is no doubt somewhat unrealistic, the more realistic operation being sampling without replacement, which would generate a hypergeometric or product-hypergeometric distribution. However, even under sampling without replacement, the multinomial or product-multinomial distribution would be a good enough approximation if the sampling fraction were small. If the sampling fraction were not small, a development in terms of hypergeometric or product-hypergeometric distributions parallel to the first operation would be called for, and this is under consideration.

References

[1] Diamond, E. L. Asymptotic power and independence of certain classes of tests on categorical data. Institute of Statistics Mimeograph Series No. 196, Chapel Hill, N.C.: Univ. North Carolina, 1958.

[2] Mitra, S. K. Statistical analysis of categorical data. Institute of Statistics Mimeograph Series No. 142, Chapel Hill, N.C.: Univ. North Carolina, 1956.

[3] Roy, S. N., and Kastenbaum, M. A. On the hypothesis of "no-interaction" in a multiway contingency table. *Ann. Math. Stat.*, 1956, **27**, 749–57.

[4] Roy, S. N., and Mitra, S. K. An introduction to some nonparametric generalizations of analysis of variance and multivariate analysis. *Biometrika*, 1956, **43**, 361–76.

[5] Roy, S. N. *Some Aspects of Multivariate Analysis*. New York: Wiley, 1958.

34 RELATIONS BETWEEN CERTAIN INCOMPLETE BLOCK DESIGNS[1]

S. S. SHRIKHANDE

University of North Carolina

1. Summary

Relations between balanced incomplete block designs derived from the symmetrical balanced incomplete block designs with $\lambda = 1$ and 2 and certain partially balanced incomplete block designs are investigated in this paper.

2. Introduction

Let v, b, r, k, λ be the parameters of a balanced incomplete block (b.i.b.) design [1]. Then we have

$$vr = bk \,, \tag{2.1}$$

$$\lambda(v-1) = r(k-1) \,. \tag{2.2}$$

For a symmetrical b.i.b. design it is known that any two blocks have exactly λ treatments in common. Hence, from the design with parameters

$$v = b \,, \quad r = k \,, \quad \lambda \,, \tag{2.3}$$

we get by the method of block section [1] another b.i.b. design with parameters

$$v' = v - k \,, \quad b' = v - 1 \,, \quad k' = k - \lambda \,, \quad r' = k \,, \quad \lambda' = \lambda \,. \tag{2.4}$$

Thus the existence of a b.i.b. design with parameters (2.3) implies the existence of another b.i.b. design with parameters (2.4). The converse is not true in general [2]. However, when $\lambda = 1$, conditions (2.4) correspond to a finite Euclidean plane and it is well known that such a plane can be embedded into a finite projective plane which is the symmetrical design with parameters (2.3) for $\lambda = 1$. A similar result has been proved by Hall and Connor [3] for the case $\lambda = 2$. We thus have the following result.

Let C, C_1, D, D_1 be b.i.b. designs with parameters as follows.

[1] This research was supported by the United States Air Force through the Office of Scientific Research, USAF Air Research and Development Command, under Contract No. AF 49(638)-213. Reproduction in whole or in part is permitted for any purpose of the United States Government.

(2.5) C: $v = b = s^2 + s + 1\,, \quad r = k = s + 1\,, \quad \lambda = 1\,,$

(2.6) C_1: $v = s^2\,, \quad b = s^2 + s\,, \quad r = s + 1\,, \quad k = s\,, \quad \lambda = 1\,,$

(2.7) D: $v = b = (n^2 + n + 2)/2\,, \quad r = k = n + 1\,, \quad \lambda = 2\,,$

(2.8) D_1: $v = n(n-1)/2\,, \quad b = n(n+1)/2\,, \quad r = n + 1\,, \quad k = n - 1\,, \quad \lambda = 2\,.$

Then either C and C_1 (D and D_1) are both existent or both nonexistent, and any solution of one yields a solution of the other.

From [4], [5] it follows that the dual of C_1, is the group-divisible design C_1^*.

$$C_1^*:\ v = s^2 + s\,,\quad b = s^2\,,\quad r = s\,,\quad k = s + 1\,, \qquad (2.9)$$
$$\lambda_1 = 1\,,\quad \lambda_2 = 0\,,\quad p = s + 1,\quad q = s\,;$$

where p is number of groups of treatments, q is size of each group, λ_1 is number of times any two treatments from, different groups occur together, λ_2 is number of times any two treatments from any group occur together.

Again from [4] it follows that the dual D_1^* of D_1 is a partially balanced incomplete block (p.b.i.b.) design with two associate classes with parameters

$$D_1^*:\ v = n(n+1)/2\,,\quad b = n(n-1)/2\,,\quad r = n - 1\,,\quad k = n + 1\,, \qquad (2.10)$$
$$n_1 = 2n - 2\,,\quad n_2 = (n-1)(n-2)/2\,,\quad \lambda_1 = 1\,,$$
$$\lambda_2 = 2\,,\quad p_{11}^1 = n - 1\,,\quad p_{11}^2 = 4\,.$$

Conversely, if we use the results of J. Roy and Laha [6], it can be seen that the designs with parameters given by (2.9) and (2.10) are linked block designs, and hence the dual of a design with parameters (2.9) is the b.i.b. design with parameters (2.6). Similarly, the dual of a design with parameters (2.10) is the b.i.b. design with parameters (2.8). Hence, the designs C_1 and C_1^* (D_1 and D_1^*) are either both existent or both nonexistent.

A p.b.i.b. design with two associate classes is said to be triangular [7] if the number of treatments is $n(n-1)/2$ and the association scheme is an array of n rows and n columns with the following properties:

(a) The positions in the principal diagonal are blank.

(b) The $n(n-1)/2$ positions above the principal diagonal are filled by the numbers $1, 2, \cdots, n(n-1)/2$ corresponding to the treatments.

(c) The positions below the principal diagonal are filled so that the array is symmetrical about the principal diagonal.

(d) For any treatment i, the first associates are exactly those treatments which lie in the same row and the same column as treatment i.

The following relations obviously hold:

1. The number of first associates of any treatment is $n_1 = 2n - 4$.
2. With respect to any two treatments θ_1 and θ_2 which are first associates, the number of treatments which are first associates of both θ_1 and θ_2 is $p_{11}^1 = n - 2$.
3. With respect to any two treatments θ_3 and θ_4 which are second associates, the number of treatments which are first associates of both θ_3 and θ_4 is $p_{11}^2 = 4$.

Conversely, Connor [8] has shown that for a p.b.i.b. design with two associate classes if $v = n(n-1)/2, n \geqq 9$, relations (1), (2), and (3) above imply that the association scheme is triangular, denoted by (T_n). In a recent paper [9], the author has shown that the same result is true for $n = 5$ and 6. The cases $n = 7$ and 8 are as yet undecided.

3. Results for b.i.b. Design with $\lambda = 1$

In this section we derive some results for designs obtained from the symmetrical b.i.b. design with $\lambda = 1$. They provide the motivation for the results obtained in Sections 4 and 5.

We have already seen that a design C_1 with parameters (2.6) is the dual of a design C_1^* with parameters (2.9) and conversely. Now, since C_1^* is a group-divisible design, the blocks of C_1 can be divided into $(s+1)$ sets of s, each such that any two blocks of the same set have no treatment in common, whereas any two blocks from different sets have exactly one treatment in common. Hence, if we omit one set of blocks, we get a group-divisible design C_{11} given by parameters

$$C_{11}: \quad v = b = s^2\,, \quad r = k = s\,, \quad \lambda_1 = 1\,, \quad \lambda_2 = 0\,. \tag{3.1}$$

Further, the omitted set of blocks gives rise to another disconnected group-divisible design C_{12} with parameters

$$C_{12}: \quad v = s^2\,, \quad b = s\,, \quad r = 1\,, \quad k = s\,, \quad \lambda_1 = 0\,, \quad \lambda_2 = 1\,. \tag{3.2}$$

In C_{11} and C_{12}, the s groups of s treatments are given by the s blocks of C_{12}. We may indicate this result symbolically by $C_1 = C_{11} + C_{12}$.

Now consider the dual C_{11}^* of C_{11}. By virtue of the relation $C_1 = C_{11} + C_{12}$, C_{11}^* can be obtained from C_1^*, the dual of C_1, by removing from the blocks of C_1^* the treatments corresponding to the blocks of C_{12}. We note that each block of C_1^* contains just one of the s treatments corresponding to the s blocks of C_{12} and that any two of these s treatments are second associates in C_1^*. Now C_1^* is a group-divisible design with $(s+1)$ groups of s treatments each, where any two treatments from the same group are second associates. Hence, obviously the s treatments corresponding to the blocks of C_{12} must all belong to the same group. Omitting these treatments from C_1^*, we are left with C_{11}^*. The association scheme of C_{11}^* is obviously group-divisible with the remaining s groups of size s each. Thus C_{11}^* is again a group-divisible design with exactly the same parameters as C_{11}.

4. Results Connected with b.i.b. Design with $\lambda = 2$

In this section we prove four theorems for designs connected with b.i.b. designs with $\lambda = 2$.

THEOREM 1. *If, for any value of n, a design D_1^* with parameters* (2.10) *exists, then its association scheme is of the type (T_{n+1}).*

PROOF. We note that the design D_1 given by parameters (2.8) does not exist for $n = 6$ and 7 [10]. Hence D_1^* does not exist for these values of n.

For $n \geqq 8$, the theorem follows from Connor's result [8], and for $n = 4$ and 5, it follows from the result of the author [9] referred to previously. ||

THEOREM 2. *For any value of n, the existence of the design D_1 with parameters (2.8) implies the existence of two designs D_{11} and D_{12} given respectively by the following parameters.*

$$(4.1)\quad D_{11}:\ v = b = n(n-1)/2\,,\quad r = k = n-1\,,\quad n_1 = 2n-4\,,$$
$$n_2 = (n-2)(n-3)/2\,,\quad \lambda_1 = 1\,,\quad \lambda_2 = 2\,,\quad p^1_{11} = n-2\,,\quad p^2_{11} = 4\,.$$
$$(4.2)\quad D_{12}:\ v = n(n-1)/2, b = n, k = n-1, r = 2, n_1 = 2n-4\,,$$
$$n_2 = (n-2)(n-3)/2\,,\quad \lambda_1 = 1\,,\quad \lambda_2 = 0\,,\quad p^1_{11} = n-2\,,\quad p^2_{11} = 4\,,$$

such that $D_1 = D_{11} + D_{12}$.

PROOF. Instead of D_1 we consider its dual D_1^* given by (2.10). Let D_{11}^* and D_{12}^* denote the duals of D_{11} and D_{12} respectively. It is sufficient to prove that the treatments of D_1^* can be divided into two distinct sets T_1 and T_2, such that by retaining treatments of the set T_1 in the blocks of D_1^* we get the design D_{11}^*, whereas by omitting these treatments (and hence retaining treatments of the set T_2) in the blocks of D_1^* we get the design D_{12}^*. We note that D_{12}^* is nothing but the unreduced b.i.b. design with $v = n, k = 2$, $\lambda = 1$. From Theorem 1, we know that the association scheme of D_1^* is triangular (T_{n+1}), which may be written without loss of generality as follows:

$$(4.3)\quad \begin{bmatrix} x & 1 & 2 & 3 & \cdots & \cdot & n \\ 1 & x & (n+1) & (n+2) & \cdots & \cdot & (2n-1) \\ 2 & (n+1) & x & \cdot & \cdots & \cdot & \cdot \\ 3 & (n+2) & \cdot & \cdot & \cdots & \cdot & \cdot \\ & \cdots & & & & \cdots & \\ \cdot & \cdot & \cdot & \cdot & \cdots & x & (n^2+n)/2 \\ n & (2n-1) & \cdot & \cdot & \cdots & (n^2+n)/2 & x \end{bmatrix}.$$

We show that any block of D_1^* contains exactly two treatments from any row of (4.3). Consider $(n+1)$ blocks S_i of size $(n+1)$, each containing a new treatment α and all the treatments of row i of (4.3), $i = 1, 2, \cdots, (n+1)$. Adding these $(n+1)$ blocks S_i to the blocks of D_1^*, we obviously get a design with $v = (n^2+n+2)/2$, $r = k = n+1$. The treatment α occurs with any treatment of D_1^* in exactly two of these blocks. Any two treatments of D_1^*, say l and m, occurring in the same row or column of (4.3) occurred once in a block of D_1^*. They also occur once in a block S_i. Again, if l and m do not lie in the same row or same column of (4.3), they occur together in two blocks of D_1^*, but they do not occur together at all in the new blocks S_i. Thus, any two treatments l and m of D_1^* occur together twice in the new design, giving $\lambda = 2$. Hence, the design so obtained is nothing but the b.i.b. design with parameters (2.7), and hence any two blocks of this design have two treatments in common. Let θ_1 and θ_2 be two different treatments of D_1^* belonging to the same row of (4.3). Then θ_1 and θ_2 occur together in exactly one block of D_1^*.

Further, this block of D_1^* cannot contain any other treatment θ_3 of the same row of (4.3). For otherwise the design (2.7) obtained previously will have two blocks containing three treatments in common, which is impossible. Consider in particular the treatments $1, 2, \cdots, n$ occurring in the first row of (4.3). Then each of the the $n(n-1)/2$ pairs of these treatments uniquely determines a block of D_1^*. Noticing that each pair determines only one block and that the number of blocks in D_1^* is exactly $n(n-1)/2$, we can set up this correspondence in a one-to-one manner. This proves the statement made at the beginning of this paragraph. If we now retain only the treatments of the set $T_2 = (1, 2, \cdots, n)$ in the blocks of D_1^*, we get the design D_{12}^*, which is the unreduced b.i.b. design for $v = n$ and $k = 2$. We now write the n_{c_2} combination of the n treatments of T_2 in the following array of n rows and n columns,

$$\tag{4.4} \begin{bmatrix} x & (1,2) & (1,3) & (1,4) & \cdots & \cdot & (1,n) \\ (2,1) & x & (2,3) & (2,4) & \cdots & \cdot & (2,n) \\ (3,1) & (3,2) & x & (3,4) & \cdots & \cdot & (3,n) \\ (4,1) & (4,2) & (4,3) & x & \cdots & \cdot & \cdot \\ & \cdots & & & & \cdots & \\ \cdot & \cdot & \cdot & \cdot & \cdots & \cdot & (n-1,n) \\ (n,1) & (n,2) & (n,3) & \cdot & \cdots & (n,n-1) & x \end{bmatrix}.$$

As mentioned before, each combination determines a unique block of D_1^* to which the combination belongs. By removing these treatments from the corresponding blocks, we are left with a design with $v = b = n(n-1)/2$, $r = k = n-1$, the $n(n-1)/2$ treatments being the treatments of the set $T_1 = \{n+1, n+2, \cdots, (n^2+n)/2\}$. Now in D_1^* any two blocks have two treatments in common. Hence, in this design for treatments of the set T_1, any two blocks corresponding to two combinations which are either θ_3 in the same row or in the same column have exactly one treatment in common, since the corresponding combinations have one treatment in common; but these are now omitted. Similarly, since any two combinations not in the same row and not in the same column have no treatment in common, the corresponding blocks for the above-described design for the treatment of T_1 have two treatments in common. This implies that the blocks of the above design can be written in a triangular array like (4.4) such that any two blocks which are either in the same row or the same column have exactly one treatment in common, whereas any other pair of blocks have two treatments in common. We may say that the block structure for this design is triangular (T_n). It is now obvious that this design is nothing but D_{11}^*, the dual of D_{11}.

Now, considering the duals of D_1^*, D_{11}^*, D_{12}^*, we see that the $n(n+1)/2$ blocks of D_1 can be divided into sets of $n(n-1)/2$ and n respectively, such that the first set of blocks gives the design D_{11} and the second set gives the design D_{12}. We thus have $D_1 = D_{11} + D_{12}$. ||

We now prove a partial converse of Theorem 2.

THEOREM 3. *The existence of the design D_{11} with parameters* (4.1), *having triangular association scheme* (T_n) *for any value of n, implies the existence of the corresponding design D_1 with parameters* (2.8).

PROOF. Without loss of generality we can represent the association scheme (T_n) of D_{11} as follows:

$$\begin{bmatrix} x & 1 & 2 & \cdots & \cdot & (n-1) \\ 1 & x & n & \cdots & \cdot & (2n-3) \\ 2 & n & x & \cdots & \cdot & \cdot \\ & \cdots & & & & \cdots \\ \cdot & \cdot & \cdot & \cdots & x & (n^2-n)/2 \\ (n-1) & (2n-3) & \cdot & \cdots & (n^2-n)/2 & x \end{bmatrix}. \tag{4.5}$$

Consider the n blocks of size $(n-1)$ given by the rows of (4.5). They constitute the design D_{12} given by (4.2). Adding the blocks of D_{12} to the blocks of D_{11}, we get a design with $v = n(n-1)/2$, $b = n(n+1)/2$, $r = n+1$, $k = n-1$. Further, any two treatments of D_{11} occurring in the same row of the same column of (4.5) occur together once more in these new blocks. Also, two treatments which are neither in the same row nor in the same column of (4.5) already occur together twice in blocks of D_{11}, but they do not occur together at all in these new blocks. Hence, in the new design we have $\lambda = 2$, for all pairs of treatments. Thus the new design is exactly the design D_1. Further, we have $D_1 = D_{11} + D_{12}$. ||

The design D_{11} is known to be impossible [11] for $n = 6$, as is the design D_1 for the same value of n [10]. The design D_1 for $n = 7$ is also known to be impossible [10], but the impossibility of D_{11} for this value of n has not been proved. If, however, D_{11} exists for $n = 7$, the association scheme cannot be triangular, as from Theorem 3 this would imply the existence of D_1 for $n = 7$. From [8], [9] we know that the association scheme for D_{11} is triangular (T_n) for $n \geqq 9$ and $n = 5$. We thus have the following corollary.

COROLLARY. *If $n = 5$ or $n \geqq 9$, the existence of D_{11} with parameters* (4.1) *implies the existence of D_1 with parameters* (2.8).

We now state and prove another theorem for the design D_{11}.

THEOREM 4. *If $n = 5$ or $n \geqq 9$, the dual of the design D_{11} given by* (4.1) *is another p.b.i.b. design with the same parameters as* (4.1).

PROOF. From Theorem 3 and its corollary we have $D_{11} + D_{12} = D_1$. Consider the dual D_1^* of D_1. Then D_{11}^* is obtained by omitting from D_1^* the treatments corresponding to the blocks of D_{12}. We note that in each block of D_1^* there are two treatments corresponding to the blocks of D_{12} and $(n-1)$ treatments corresponding to the blocks of D_{11}. Further, since the dual of D_{12} is D_{12}^*, which is a b.i.b. design with parameters $v=n$, $b=n(n-1)/2$, $k = 2$, $r = n-1$, $\lambda = 1$, any two of the n treatments $\alpha_1, \alpha_2, \cdots, \alpha_n$ of D_{12}^* are first associates in D_1^*. Now the association scheme of D_1^* is triangular (T_{n+1}), from [8], [9] since $n+1 = 6$ or $n+1 \geqq 10$.

We now show that $\alpha_1, \alpha_2, \cdots, \alpha_n$ must all be in the same row of the association scheme (T_{n+1}). For, consider any two treatments α_1 and α_2, say. Since they are first associates, they must lie either in the same row or the same column of the association scheme. Suppose they lie in the same row; then, since the association scheme is unchanged if we interchange any two rows and the same two columns of the array of (T_{n+1}), we can, without loss of generality, assume that α_1 and α_2 are in the first row in positions (1, 2) and (1, 3) respectively. If $\alpha_3, \alpha_4, \cdots, \alpha_n$ all lie in the first row, we are finished. Otherwise, there is at least one treatment, say α_3, which does not lie in this row. Since the first associates of α_1 must lie in the first and the second rows, α_3 lies in the second row in position $(2, j)$, say. Similarly, since α_3 is first associate of α_2, it must be in the third row in position $(3, j_1)$, say. Since these two positions of α_3 must be symmetrical with respect to the main diagonal, we have $j = 3$ and $j_1 = 2$; i.e., α_3 occurs in the array of (T_{n+1}) in positions (2, 3) and (3, 2).

Now consider α_4. Since it is first associate of α_1, it must occur either in the first row or in the second row in column position $l \geqq 4$. Suppose it occurs in position $(1, l)$, $l \geqq 4$; then α_4 does not occur in the same row or same column as α_3 and hence is second associate of α_3, which is a contradiction. Suppose α_4 occurs in position $(2, l)$, $l \geqq 4$; then it is not in the same row or the same column as α_2 and hence is second associate of α_2, which is again a contradiction. Thus if α_1 and α_2 lie in a row, all other treatments $\alpha_3, \cdots, \alpha_n$ must lie in the same row. We get a similar result if α_1 and α_2 lie in the same column. This proves the result that all the treatments $\alpha_1, \alpha_2, \cdots, \alpha_n$ must lie in the same row and hence the same column. We now omit the treatments $\alpha_1, \alpha_2, \cdots, \alpha_n$ from the blocks of D_1^*. This leaves the designs D_{11}^*.

Further, the association scheme of D_{11}^* is obviously obtained by omitting the row and column in which the treatments $\alpha_1, \alpha_2, \cdots, \alpha_n$ lie. The scheme is therefore triangular (T_n). It is now easy to see that the parameters of D_{11}^* are exactly those of D_{11}. ||

5. A Constructive Proof for Embedding the Design D_1 into the Design D

As mentioned earlier, Hall and Connor [3] proved that the existence of the design D_1 given by (2.8) implies the existence of the design D given by (2.7). Their proof is not constructive. We give in this section a constructive proof of their result.

Suppose the design D_1 exists, then D_1^* also exists, and by Theorem 1 the association scheme of D_1^* is triangular (T_{n+1}). Hence, the blocks $B_1, B_2, \cdots, B_{n(n+1)/2}$ of D_1 can be exhibited in the scheme (4.3) when i stands for B_i, such that any two blocks of D_1 in the same row or in the same column of (4.3) have exactly one treatment in common, whereas any two blocks neither in the same row nor in the same column of (4.3) have two treatments in common. Take a new block B' consisting of new treatments $u_1, u_2, \cdots, u_{n+1}$. Let the $n(n+1)/2$ combinations of these treatments be listed as follows:

$$(5.1)\qquad \begin{bmatrix} x & (u_1, u_2) & (u_1, u_3) & \cdots & \cdot & (u_1, u_{n+1}) \\ (u_2, u_1) & x & (u_2, u_3) & \cdots & \cdot & (u_2, u_{n+1}) \\ (u_3, u_1) & (u_3, u_2) & x & \cdots & \cdot & (u_3, u_{n+1}) \\ \cdots & & & & \cdots & \\ \cdot & \cdot & \cdot & \cdots & x & (u_n, u_{n+1}) \\ (u_{n+1}, u_1) & (u_{n+1}, u_2) & (u_{n+1}, u_3) & \cdots & (u_{n+1}, u_n) & x \end{bmatrix}.$$

We now assign these combinations to the respective blocks B_i in the corresponding positions of (4.3) to get new blocks B_i'. Thus B_1' consists of treatments of B_1 besides the treatments u_1 and u_2. It is now easy to verify that B' and B_i', $i = 1, 2, \cdots, (n^2+n)/2$ give a design with $v = b = (n^2+n+2)/2$, $r = k = n + 1$. Further, if i and j belong to the same row or same column of (4.3), then B_i and B_j have one treatment in common and the corresponding combinations added to these blocks to give B_i' and B_j' have again one treatment in common. Thus B_i' and B_j' have two treatments in common. If on the other hand i and j neither belong to the same row nor to the same column of (4.3), then B_i and B_j have two treatments in common; but now the corresponding combinations to be added to them have no treatment in common. Thus again B_i' and B_j' have two treatments in common. Again B' has obviously two treatments in common with any B_i'. Thus we get a symmetric design in which every two blocks have two treatments in common. This implies that $\lambda = 2$. The design thus obtained is exactly the design D. ||

References

[1] Bose, R. C. On the construction of balanced incomplete block designs. *Ann. Eugen.*, 1939, **9**, 353-99.

[2] Bhattacharya, K. N. A new balanced incomplete block design. *Science and Culture*, 1944, **9**, 508.

[3] Hall, M., Jr., and Connor, W. S. An embedding theorem for balanced incomplete block designs. *Canad. J. Math.*, 1953, **6**, 35-41.

[4] Shrikhande, S. S. On the dual of some balanced incomplete block designs. *Biometrics*, 1952, **8**, 66-72.

[5] Bose, R. C., and Connor, W. S. Combinatorial properties of group divisible incomplete block designs. *Ann. Math. Stat.*, 1952, **23**, 367-83.

[6] Roy, J., and Laha, R. G. Classification and analysis of linked block designs. *Sankhyā* 1956, **17**, 115-32.

[7] Bose, R. C., and Shimamoto, T. Classification and analysis of partially balanced designs with two associate classes. *J. Amer. Stat. Assoc.*, 1952, **47**, 151-90.

[8] Connor, W. S. The uniqueness of the triangular association scheme. *Ann. Math. Stat.*, 1958, **29**, 262-66.

[9] Shrikhande, S. S. On a characterization of the triangular association scheme. *Ann. Math. Stat.*, 1959, **30**, 39-47.

[10] Connor, W. S. On the structure of balanced incomplete block designs. *Ann. Math. Stat.*, 1952, **23**, 57-71.

[11] Connor, W. S., and Clatworthy, W. H. Some theorems for partially balanced designs. *Ann. Math. Stat.*, 1954, **25**, 100-113.

35 INFINITESIMAL RENEWAL PROCESSES[1]

WALTER L. SMITH
University of North Carolina

1. Introduction

Renewal theory (Smith [1]) is concerned primarily with the *renewal function* $H(x)$, which gives the expected number of partial sums $S_n = X_1 + X_2 + \cdots + X_n$ of the independent and identically distributed random variables $\{X_i\}$ which satisfy the inequality $S_n \leqq x$. Generally the X_i are non-negative, and we then call the sequence $\{X_i\}$ a *renewal process*; when we wish to drop the non-negativity restriction, i.e., consider unrestricted X_i, we can draw attention to this extra generality by speaking of an *extended renewal process*.

The renewal function $H(x)$ can be related to the following physical model concerned with the motion of a particle P on the real axis. At zero time P is assumed to be at the origin, and P remains there until time $t = 1$, at which instant P moves instantaneously to the point $S_1 = X_1$ where it is to remain for a further time interval of unit duration. The motion of P then continues in an obvious way so that, for instance, P is at S_n during the time-interval $[n, n + 1)$. In this model, the renewal function $H(x)$ measures the expected amount of time which P spends in the interval $(-\infty, x]$. It is assumed here, as is usual in renewal theory, that $\mathscr{E} X_i > 0$, and it is known that in this case $H(x)$ is necessarily finite for all finite x.

One can now imagine the following development of our model of a renewal process. Suppose we change our time and distance scales so that the individual shifts in the position of P become, in general, very small, but at the same time occur extremely frequently. We are led to consider a process $x(t)$ with stationary independent increments for which $x(0) = 0$ and $\mu_1 = \mathscr{E} x(1) > 0$, and to consider the expected amount of time that this process satisfies the condition $x(t) \leqq x$. We shall call this expected duration $H(x)$ and show that $H(x)$ has all the properties of a conventional renewal function. In particular we shall show that, *under suitable conditions*, the following three theorems hold.

Blackwell's Theorem. For any $\alpha > 0$, $H(x + \alpha) - H(x) \to \alpha\mu_1^{-1}$ as $x \to +\infty$; the limit being zero if $\mu_1 = +\infty$.

[1] This research was supported by the Office of Naval Research under Contract No. Nonr-855(06), for research in probability and statistics, at Chapel Hill, N.C. Reproduction in whole or in part for any purpose of the United States Government is permitted.

Second Renewal Theorem. For a certain constant μ_2 (which is actually the second moment, assumed finite, of the renewal "lifetimes" in a certain ordinary renewal process which is associated with the process $x(t)$; the details of this ordinary renewal process are given below),

$$H(x) = \frac{x}{\mu_1} + \frac{\mu_2}{2\mu_1^2} - 1 + o(1)$$

as $x \to +\infty$.

Renewal Density Theorem. If $h(x) = H'(x)$ exists then

$$h(x) \to \mu_1^{-1}, \qquad \text{as } x \to +\infty,$$
$$h(x) \to 0, \qquad \text{as } x \to -\infty.$$

The name for the second renewal theorem is coined here, for the first time, as a convenience. It has previously been proved only for renewal processes satisfying the non-negativity assumption (Smith [2]); we give a proof of the second renewal theorem which is valid for extended renewal processes in Section 3 below.

Because of the physical model which we have described, and which gives rise to the analogy between our present study and standard renewal theory, we shall speak of *infinitesimal renewal processes*.

In Theorems 1, 2, and 3, we, essentially, complete the program outlined above. In particular, when $x(t)$ has a frequency function $f(x; t)$, we shall see that $h(x) = \int_0^\infty f(x; t)\,dt$. Integrals of this kind are of particular interest in the theory of queues and dams, as we hope to show elsewhere, and it was in considering their asymptotic behavior that we were led to the present investigation. Unfortunately, the most general sufficient conditions available for the renewal density theorem (those of Smith [3]) are, in the present application, bound up with the behavior of the density function $f(x) = \int_0^\infty e^{-t} f(x; t)\,dt$. One obstacle is that we need $f(x)$ to belong to the class $L_{1+\delta}$ for some $\delta > 0$. Theorems 4 and 5 are concerned with providing sufficient restraints under which this condition will be automatically satisfied. We should point out, incidentally, that the renewal density theorem for the infinitesimal renewal process can apply to cases where $x(t)$ is not absolutely continuous.

Finally, we consider the particular process associated with the density

$$f(x; t) = \frac{e^{-x} x^{t-1}}{\Gamma(t)}, \qquad t > 0.$$

We shall see that none of our conditions cover this simple case even though the renewal density theorem does hold. In fact, we shall deduce from this process a conventional renewal process for which the renewal density theorem holds, but for which $f(x)$ does not belong to any class $L_{1+\delta}$ for $\delta > 0$. This raises the intriguing question of how the class of functions considered in [3], for which the renewal density theorem holds, can be widened.*

* *Note added in proof*: We have recently obtained necessary and sufficient conditions for the renewal density theorem.

A heuristic discussion of the present ideas has already been given in [1]. It is proposed to consider elsewhere the theory of infinitesimal renewal processes in relation to the particular homogeneous process which arises in the theory of dams. It will there be shown that the infinitesimal renewal density function $h(x)$ gives the expected number of up-crosses of the real number x by the process.

2. The Infinitesimal Renewal Function

Let $\{X_i\}$ be a sequence of independent, identically distributed random variables such that $0 < \mathscr{E}X_i \leqq +\infty$. Let $S_n = X_1 + X_2 + \cdots + X_n$, and write $F_n(x)$ for the distribution function of S_n, and $F(x)$ for $F_1(x)$. Then the renewal function corresponding to $\{X_i\}$ is determined by

$$H(x) = \sum_{k=1}^{\infty} F_k(x)\,, \tag{2.1}$$

(see, e.g., [1]).

We write $x(t)$ for a process with stationary independent increments such that $x(0) = 0$ and $0 < \mathscr{E}x(1) \leqq \infty$ and suppose, wherever necessary, that $x(t)$ is both separable and measurable. (For a discussion of such processes we refer to Doob [4], Gnedenko and Kolmogorov [5], and Lévy [6]). We shall also write $F(x; t) = \Pr\{x(t) \leqq x\}$, defined for all $t \geqq 0$.

The function $U(x)$ is defined by

$$U(x) = \begin{cases} 1, & \text{for } x \geqq 0\,, \\ 0, & \text{for } x < 0\,. \end{cases}$$

Since $U(\cdot)$ is Borel-measurable, $U(x - x(t))$ will be a measurable process for any fixed x (Halmos [7], p. 81). Thus, by Fubini's theorem, we can introduce the random variable

$$T(x) = \int_0^{\infty} U(x - x(t))\,dt\,. \tag{2.2}$$

Evidently $T(x)$ is the total amount of time for which $x(t) \leqq x$. Further, since $\mathscr{E}U(x - x(t)) = F(x; t)$, we can deduce from (2.2) by a second appeal to Fubini's theorem that

$$\mathscr{E}T(x) = \int_0^{\infty} F(x; t)\,dt\,, \tag{2.3}$$

although the finiteness of $\mathscr{E}T(x)$ is not yet established.

THEOREM 1. *If we define the distribution function*

$$F(x) = \int_0^{\infty} e^{-t} F(x; t)\,dt\,, \tag{2.4}$$

and the renewal function $H(x)$ by (2.1), in which of course $F_k(x)$ will be the k-fold convolution of $F(x)$ with itself, then $H(x) = \mathscr{E}T(x)$. Consequently, $\mathscr{E}T(x)$ is a finite and monotone non-decreasing function of x.

PROOF. Let $\{Y_n\}$ be a sequence of independent, non-negative, and identically distributed random variables with the distribution: $\Pr\{Y_n \leqq y\} = U(y)[1 - e^{-y}]$; i.e., the Y_n are exponential variables. It is assumed that the variables $\{Y_n\}$ are independent of $x(t)$. Write $Z_n = Y_1 + Y_2 + \cdots + Y_n$, for the nth partial sum of the Y's. Then

$$\Pr\{x(Y_1) \leqq x\} = \int_0^\infty e^{-y} \Pr\{x(y) \leqq x\}\, dy = F(x) ,$$

by (2.4). Consider the equation

$$x(Z_n) = \sum_{r=1}^{n} [x(Z_r) - x(Z_{r-1})] . \tag{2.5}$$

Since $x(t)$ is a process with stationary independent increments, the n brackets in the summation in (2.5) enclose n independent and identically distributed random variables. For $[x(Z_r) - x(Z_{r-1})]$ is the change in the value of $x(t)$ during a period of time of length Y_r; thus all terms in the summation will have the same distribution function as $x(Y_1)$, namely $F(x)$. Hence $x(Z_n)$ will have $F_n(x)$ as its distribution function.

By a well-known property of the exponential distribution, the random variable Z_n has the probability density function

$$\frac{e^{-x}x^{n-1}}{(n-1)!} , \qquad \text{for } x \geqq 0 .$$

Consequently, we can infer that

$$\Pr\{x(Z_n) \leqq x\} = \int_0^\infty \frac{e^{-y}y^{n-1}}{(n-1)!} F(x; y)\, dy .$$

Thus, by equating the distribution functions we have obtained for the left- and right-hand sides of (2.5), we conclude that

$$F_n(x) = \int_0^\infty \frac{e^{-y}y^{n-1}}{(n-1)!} F(x; y)\, dy .$$

Hence, by a standard theorem from integration theory ([7], p. 112, Theorem B), we find that

$$\begin{aligned} H(x) &= \sum_{n=1}^{\infty} F_n(x) \\ &= \int_0^\infty \sum_{n=1}^{\infty} \frac{e^{-y}y^{n-1}}{(n-1)!} F(x; y)\, dy \\ &= \int_0^\infty F(x; y)\, dy . \end{aligned}$$

This, by (2.3), proves $H(x) = \mathscr{E}\, T(x)$. The fact that $H(x)$ is finite and non-decreasing is well known, since $H(x)$ is a conventional renewal function (see, e.g., [8]).

We can call $F(x)$ the *associated distribution function*, and the random variables $\{X_n\}$ defined by $X_n = x(Z_n) - x(Z_{n-1})$ the *associated renewal process*.

The argument in which we have just indulged to prove Theorem 1 has glossed over a non-trivial point of rigor, namely the question of whether $x(Y_1)$, say, is a random variable. It can be shown fairly easily that this question has an affirmative answer, but as the necessary measure-theoretic elaborations would seem a trifle out of place in the present setting, we shall not discuss this point further here; moreover, Theorem 1 can equally well be proved without mentioning random variables, either by an argument using characteristic functions or by a somewhat tiresome manipulation of multiple integrals and Jacobians. However, we prefer the present demonstration as being more vivid.

Write

$$\mu_r = \int_{-\infty}^{+\infty} x^r \, dF(x) \,, \quad \text{and} \quad m_r(t) = \int_{-\infty}^{+\infty} x^r \, d_x F(x; t),$$

whenever these moments are definable. Also write $m_r = m_r(1)$. Since, as is well known, the cumulant generating function of $x(t)$ [given $x(0) = 0$] is a direct multiple of t, the relations between the moments $\{m_r(t)\}$ and the moments $\{m_r\}$ are easy to determine. In particular, one finds that when the appropriate moments are finite

$$m_1(t) = m_1 t \,, \qquad m_2(t) = m_2 t + m_1^2 t(1 - t) \,. \tag{2.6}$$

Thus, by (2.4), we can find that

$$\mu_1 = m_1 \int_0^\infty e^{-t} t \, dt = m_1 \,, \qquad \mu_2 = \int_0^\infty e^{-t}[m_2 t + m_1^2 t(1 - t)] \, dt = m_2 + m_1^2 \,. \tag{2.7}$$

Renewal processes may be *continuous* or *discrete* (see, e.g., [1]). For every theorem which is true for a continuous renewal process there is a closely analogous result for a discrete renewal process. *For the remainder of this section and for Section* 3 *we shall assume that the associated renewal process is continuous.* Corresponding results for the case in which the associated renewal process is discrete may be proved on lines similar to those adopted here (by appealing to appropriate standard theorems for discrete, instead of continuous, renewal processes and by dealing with sums instead of integrals, etc.). We shall briefly discuss in Section 4 the characterization of those stationary processes with independent increments which lead to discrete associated renewal processes.

THEOREM 2. (i) *For any $\alpha > 0$, as $x \to \infty$,*

$$\mathscr{E}[T(x + \alpha) - T(x)] \to \frac{\alpha}{m_1} \,,$$

where the limit is to be zero if $m_1 = +\infty$.

(ii) *If $m_2 < \infty$, then as $x \to +\infty$,*

$$\mathscr{E}\, T(x) = \frac{x}{m_1} + \frac{\sigma^2}{2m_1^2} + o(1) \,,$$

where $\sigma^2 = m_2 - m_1^2$.

PROOF. Part (i) is simply a restatement of Blackwell's theorem for the extended renewal process [8]. We have merely substituted $\mathscr{E}T(x)$ for $H(x)$ and, by (2.7), m_1 for μ_1.

Part (ii) follows immediately from the second renewal theorem quoted in Section 1 above. We have used (2.7) to express the result in terms of the moments $\{m_r\}$ instead of the moments $\{\mu_r\}$. However, the second renewal theorem has nowhere been proved for the extended renewal process, so we must now prove this theorem.

3. Proof of the Second Renewal Theorem

Consider the function $Q(x)$ defined by

$$Q(x) = \mu_1 U(x) - \int_{-\infty}^{x} [U(y) - F(y)]\,dy\,. \tag{3.1}$$

Since $\int_0^\infty [1 - F(x)]\,dx = \int_0^\infty x\,dF(x)$, and $\int_{-\infty}^0 F(x)\,dx = -\int_{-\infty}^0 x\,dF(x)$, we can also write (3.1) as

$$Q(x) = \begin{cases} \displaystyle\int_x^\infty [1 - F(y)]\,dy, & \text{for } x \geqq 0\,, \\ \displaystyle -\int_{-\infty}^x F(y)\,dy, & \text{for } x < 0\,. \end{cases} \tag{3.2}$$

Notice that, since we are supposing μ_2 to be finite, the first absolute moment of F must also be finite.

Let us write $F_+(x) = U(x)F(x)$, and μ_1^+ for the first moment of $F_+(x)$. Then for $x \geqq 0$ we have, by (3.2), that $Q(x) = \mu_1^+ - \int_0^x [1 - F_+(y)]\,dy$. Then, by the results of Smith [9] concerning "derived distributions," or by direct computation, it follows that

$$\int_0^\infty Q(y)\,dy = \frac{1}{2}\int_0^\infty y^2\,dF_+(y)\,.$$

A similar argument, coupled with this result, thus proves

LEMMA 1. *$\int_{-\infty}^{+\infty} Q(y)\,dy = \frac{1}{2}\mu_2$, the integral converging absolutely.*

Given any $\varepsilon > 0$ we can, by Lemma 1, find a large Δ so that

$$\int_{|y| \geqq \Delta - 1} |\,Q(y)\,|\,dy < \varepsilon\,.$$

Consider

$$\begin{aligned} I_+ &= \int_{y - z \geqq \Delta} Q(y - z)\,dH(z) \\ &= \sum_{n=1}^{\infty} \int_{\Delta + n - 1 \leqq y - z < \Delta + n} Q(y - z)\,dH(z)\,. \end{aligned}$$

Because $Q(\cdot)$ is non-increasing for positive values of its argument, we therefore have $I_+ \leqq \sum_{n=1}^{\infty} Q(\Delta + n - 1)[H(y - \Delta - n + 1) - H(y - \Delta - n)]$. But there is a finite β such that $H(x+1) - H(x) < \beta$ for all x, for obviously $H(x+1) - H(x) < 1 + H(1) - H(-1)$. Thus

$$I_+ \leqq \beta \sum_{n=1}^{\infty} Q(\Delta + n - 1) \leqq \beta \int_{\Delta-1}^{\infty} Q(y)\, dy ,$$

by the monotonicity of Q again. Hence I_+ can be made arbitrarily small for all y by choosing Δ large enough.

A similar argument can be applied to

$$I_- = \int_{y-z \leqq -\Delta} | Q(y-z) |\, dH(z) ,$$

which, coupled with the result just obtained, shows that if

(3.3)
$$I = \int_{|y-z| \geqq \Delta} Q(y-z)\, dH(z) ,$$

then I can be made arbitrarily small, uniformly in y, by choosing Δ large enough.

Let $\mathscr{G}$ be the class of bounded step-functions each of which vanishes outside some bounded interval (which may depend on the function) and has finitely many discontinuities in that interval. Then it is an easy deduction from Blackwell's theorem for the extended renewal process that, as $x \to +\infty$,

(3.4)
$$\int_{-\infty}^{+\infty} G(x-z)\, dH(z) \to \frac{1}{\mu_1} \int_{-\infty}^{+\infty} G(z)\, dz ,$$

whenever $G(\cdot) \in \mathscr{G}$. Since $Q(x)$ is monotone in $(-\infty, 0)$ and in $[0, +\infty)$, it is not difficult to see that for any $\varepsilon > 0$ we can find $G^-(\cdot) \in \mathscr{G}$ and $G^+(\cdot) \in \mathscr{G}$ such that for all $|x| \leqq \Delta$ we have

(3.5)
$$G^-(x) \leqq Q(x) \leqq G^+(x) ,$$

and

(3.6)
$$0 \leqq \int_{|x| \leqq \Delta} [G^+(x) - G^-(x)]\, dx \leqq \varepsilon .$$

Plainly,

$$\int_{|x-z| \leqq \Delta} G^-(x-z)\, dH(z) \leqq \int_{|x-z| \leqq \Delta} Q(x-z)\, dH(z) \leqq \int_{|x-z| \leqq \Delta} G^+(x-z)\, dH(z) ,$$

and, since (3.4) holds for all functions in $\mathscr{G}$, we discover that

$$\begin{aligned} \int_{|y| \leqq \Delta} G^-(y)\, dy &\leqq \liminf_{x=\infty} \int_{|x-z| \leqq \Delta} Q(x-z)\, dH(z) \\ &\leqq \limsup_{x=\infty} \int_{|x-z| \leqq \Delta} Q(x-z)\, dH(z) \leqq \int_{|y| \leqq \Delta} G^+(y)\, dy . \end{aligned}$$

This chain of inequalities in conjunction with (3.5), (3.6), and the arbitrariness of ε prove that as $x \to \infty$,

$$\int_{|x-z| \leq \Delta} Q(x-z)\, dH(z) \to \int_{|y| \leq \Delta} Q(y)\, dy .$$

Because of Lemma 1 and of what we have already proved concerning the integral I of (3.3), we have thus established

LEMMA 2. *As* $x \to +\infty$, $\int_{-\infty}^{+\infty} Q(x-z)\, dH(z) = \frac{1}{2}\mu_2 + o(1)$.

We shall presently make use of the *elementary renewal theorem*, which states that, as $x \to +\infty$, $H(x)/x \to \mu_1^{-1}$. It will be found discussed in many places, in particular in [1]. The proof given in [1] can be adapted easily to cover the *extended* renewal process.

Since $\mu_2 < \infty$, it is an easy deduction that $z^2 F(z) \to 0$ as $z \to -\infty$. Thus we can write

$$\int_{-\infty}^{x-z} F(y)\, dy = \frac{\rho(x-z)}{x-z} ,$$

where $\rho(x-z) \to 0$ as $z \to +\infty$, x fixed. Hence

$$\begin{aligned} \lim_{z=+\infty} H(z) \int_{-\infty}^{x-z} [U(y) - F(y)]\, dy &= \lim_{z=+\infty} -H(z) \int_{-\infty}^{x-z} F(y)\, dy \\ &= \lim_{z=+\infty} -\frac{H(z)\rho(x-z)}{x-z} . \end{aligned}$$

In view of the elementary renewal theorem, we can thus conclude

$$\lim_{z=+\infty} H(z) \int_{-\infty}^{x-z} [U(y) - F(y)]\, dy = 0 . \tag{3.7}$$

The function $H(z)$ must decrease to 0 as $z \to -\infty$, or the strong law of large numbers would be contradicted. Thus

$$\lim_{z=-\infty} H(z) \int_{-\infty}^{x-z} [U(y) - F(y)]\, dy = \lim_{z=-\infty} \mu_1 H(z) = 0 , \tag{3.8}$$

where we have used the fact that $\mu_1 = \int_{-\infty}^{+\infty} [U(y) - F(y)]\, dy$ in the intermediate step.

If we now write $K = \int_{-\infty}^{+\infty} dH(z) \int_{-\infty}^{x-z} [U(y) - F(y)]\, dy$ and appeal to (3.7) and (3.8), then integration by parts yields the absolutely convergent

$$\begin{aligned} K &= \int_{-\infty}^{+\infty} [U(x-z) - F(x-z)] H(z)\, dz \\ &= \int_{-\infty}^{+\infty} dz \int_{-\infty}^{+\infty} U(z-y)[U(x-z) - F(x-z)]\, dH(y) . \end{aligned}$$

By Fubini's theorem, therefore,

$$K = \int_{-\infty}^{+\infty} dH(y) \int_{-\infty}^{+\infty} U(z-y)[U(x-z) - F(x-z)]\, dz .$$

We change the variable of integration in the inner integral from z to u by putting $z = x + y - u$; this yields

$$K = \int_{-\infty}^{+\infty} dH(y) \int_{-\infty}^{+\infty} U(x-u)[U(u-y) - F(u-y)]\, du .$$

A further appeal to Fubini's theorem then shows

$$K = \int_{-\infty}^{x} du \int_{-\infty}^{+\infty} [U(u-y) - F(u-y)]\, dH(y) .$$

However, the integral equation of renewal theory (see, e.g., [1]) shows that the inner integral is simply $F(u)$. Thus

$$K = \int_{-\infty}^{x} F(u)\, du . \tag{3.9}$$

If we refer back to equation (3.1), which defines $Q(\cdot)$, we realize that (3.9) implies

LEMMA 3. $\int_{-\infty}^{+\infty} Q(x-z)\, dH(z) = \mu_1 H(x) - \int_{-\infty}^{x} F(u)\, du$.

It is now possible to finish the proof, for $\int_{-\infty}^{x} F(u) du = xU(x) - \mu_1 + o(1)$, as $x \to \infty$. Thus Lemma 2 and Lemma 3 jointly imply

$$H(x) = \frac{x}{\mu_1} + \frac{\mu_2}{2\mu_1^2} - 1 + o(1) ,$$

as $x \to \infty$.

4. The Renewal Density Theorem for Infinitesimal Processes

We prove first

THEOREM 3. (i) *If,* (D1): $F(x) = pU(x) + q\hat{F}(x)$, *where* p *and* q *are non-negative,* $p + q = 1$, *and* $\hat{F}(x)$ *is an absolutely continuous distribution function;* (D2): $\hat{f}(x)$ *is a frequency function corresponding to* $\hat{F}(x)$ *and belongs to* $L_{1+\delta}$ *for some* $\delta > 0$; *then* $h(x) = H'(x)$ *exists for all* $x \neq 0$ *and*

(a) *if* $\hat{f}(x) \to 0$ *as* $x \to +\infty$, $h(x) \to \mu_1^{-1}$ *as* $x \to +\infty$;

(b) *if* $\hat{f}(x) \to 0$ *as* $x \to -\infty$, $h(x) \to 0$ *as* $x \to -\infty$.

(ii) *A sufficient condition for* $F(x)$ *to satisfy* (D1) *is that*

$$F(x, t) = p(t)U(x) + q(t)F(x, t) , \tag{4.1}$$

where $p(\cdot)$ *and* $q(\cdot)$ *are non-negative measurable functions such that* $p(t)+q(t)=1$, *and* $\hat{F}(x, t)$ *is an absolutely continuous distribution function. In this case* $p(t) = e^{-\gamma t}$ *for some* $\gamma > 0$.

(iii) *If* $F(x, t)$ *is of the form* (4.1), *then a sufficient condition for* $F(x)$ *to satisfy* (D2) *is that if we define*

$$I_{1+\delta}(t) = \int_{-\infty}^{+\infty} [\hat{f}(x, t)]^{1+\delta}\, dx ,$$

then, for some $\delta > 0$, $e^{-t} I_{1+\delta}(t)$ *shall be in* $L_1(0, \infty)$.

PROOF. (i) If $F(x) = pU(x) + q\hat{F}(x)$ then, in an obvious notation, (2.1) shows that

$$\begin{aligned} H &= \sum_{k=1}^{\infty}\sum_{r=1}^{k}\binom{k}{r}p^{k-r}q^{r}\hat{F}_r + \sum_{k=1}^{\infty}p^kU \\ &= pq^{-1}U + \sum_{r=1}^{\infty}\left[q^r\hat{F}_r\sum_{s=0}^{\infty}\binom{r+s}{r}p^s\right] \\ &= pq^{-1}U + \sum_{r=1}^{\infty}q^{-1}\hat{F}_r\ ; \end{aligned}$$

i.e, $H(x) = pq^{-1}U(x) + q^{-1}\hat{H}(x)$, so that for all $x \neq 0$, $H'(x) = q^{-1}\hat{H}'(x)$. The renewal function $\hat{H}(x)$ is one to which the available renewal density theorem will apply [3], and so 3(i) is proved. Notice that $p < 1$, since $F(x)$ is to have a strictly positive first moment.

(ii) If (4.1) holds then, by (2.4),

$$F(x) = U(x)\int_0^{\infty}e^{-t}p(t)\,dt + \int_0^{\infty}e^{-t}q(t)\hat{F}(x,t)\,dt\ . \tag{4.2}$$

But there will be a frequency function $\hat{f}(x, t)$ such that $\hat{F}(x,t) = \int_{-\infty}^{x}\hat{f}(z,t)\,dz$, and so (4.2) leads, via Fubini's theorem, to

$$F(x) = U(x)\int_0^{\infty}e^{-t}p(t)\,dt + \int_{-\infty}^{x}dz\int_0^{\infty}e^{-t}q(t)f(z,t)\,dt\ .$$

This shows $F(x)$ must satisfy (D1).

If (4.1) holds, moreover, then

$$\Pr\{x(t_1+t_2)=0\} = \Pr\{x(t_1)=0\}\Pr\{x(t_1+t_2)-x(t_1)=0\}\ ,$$

for any positive t_1, t_2. Thus $p(t_1+t_2) = p(t_1)p(t_2)$, by the stationarity of the $x(t)$ process. Since $p(t)$ must be monotone non-increasing, we must have $p(t) = e^{-\gamma t}$, as announced. The constant γ may not vanish, for this would make $F(x) = U(x)$ and then $F(x)$ would have a zero mean value.

(iii) When (4.1) holds, $q\hat{f}(x) = \int_0^{\infty}e^{-t}(1-e^{-\gamma t})\hat{f}(x,t)\,dt$. Hölder's inequality then gives

$$q^{1+\delta}[\hat{f}(x)]^{1+\delta} \leqq \int_0^{\infty}e^{-t}(1-e^{-\gamma t})[\hat{f}(x,t)]^{1+\delta}\,dt\ ,$$

so that, by Fubini's theorem,

$$q^{1+\delta}\int_{-\infty}^{+\infty}[\hat{f}(x)]^{1+\delta}\,dx \leqq \int_0^{\infty}e^{-t}(1-e^{-\gamma t})I_{1+\delta}(t)\,dt\ .$$

This proves (iii) of Theorem 3.

COROLLARY 3A. *When $F(x;t)$ has the form* (4.1) *then, for all $x \neq 0$,*

$$h(x) = \int_0^{\infty}e^{-t}(1-e^{-\gamma t})f(x,t)\,dt\ .$$

Notice that, if $F(x, t)$ is absolutely continuous for all $t > 0$, with a density function $f(x, t)$, then (4.1) automatically holds [with $p(t) = 0$]. We can then take $\hat{f} = f$. The more "general" form (4.1) with $p(t) \neq 0$ can arise, however, when we have

$$F(x, t) = \sum_{n=0}^{\infty} \frac{e^{-\gamma t}(\gamma t)^n}{n!} A_n(x) ,$$

where $A(\cdot)$ is some distribution function and $A_n(\cdot)$ is its n-fold convolution.

It is desirable to obtain, if possible, conditions, under which (D1) and (D2) are satisfied, which relate to more intimate properties of the $x(t)$ process. We shall now introduce the cumulant-generating function $\psi(\theta)$ of $x(1)$, i.e., we suppose

$$\mathscr{E} e^{i\theta x(t)} = e^{t\psi(\theta)} . \tag{4.3}$$

[Recall that $x(0) = 0$]. Then, since (4.3) is to be a characteristic function for all t, we must have

$$\mathscr{R}\psi(\theta) \leqq 0 , \qquad \text{for all } \theta , \tag{4.4}$$

for otherwise the modulus of this characteristic function would exceed unity for large values of t.

If $\varphi(\theta)$ is the characteristic function of $F(x)$, then it follows from (2.4) and (4.3) that

$$\varphi(\theta) = \int_0^{\infty} e^{-t+t\psi(\theta)} dt = \frac{1}{1 - \psi(\theta)} . \tag{4.5}$$

THEOREM 4. (i) *A necessary condition for F to satisfy* (D1) *is that there exist*

$$\lim_{|\theta| = \infty} \psi(\theta) = \lambda \geqq -\infty .$$

(ii) *If λ is finite, then $p = (1 - \lambda)^{-1} > 0$, and in this case a sufficient condition for F to satisfy* (D1) *and* (D2) *is that* $[\psi(\theta) - \lambda]$ *be the Fourier transform of a function in L_1 and $L_{1+\delta}$, for some $\delta > 0$.*

(iii) *If λ is infinite, then $p = 0$, and in this case a sufficient condition for F to satisfy* (D1) *and* (D2) *is that for some $\varepsilon > 0$*

$$\liminf_{|\theta| = \infty} \frac{|\psi(\theta)|}{|\theta|^{\frac{1}{2}+\varepsilon}} > 0 . \tag{4.6}$$

PROOF. (i) If (D1) holds, then $\varphi(\theta) = p + q\hat{\varphi}(\theta)$, where $\hat{\varphi}(\theta)$ is the Fourier transform of the density function $\hat{f}(x)$. Thus, by the Riemann-Lebesgue lemma, $\varphi(\theta) \to p$ as $|\theta| \to \infty$. Reference to (4.5) then proves 4(i).

(ii) From the previous remarks and (4.5) it is clear that if $\lambda > -\infty$ then $p = (1 - \lambda)^{-1} > 0$. From (4.5) it transpires that

$$q\hat{\varphi}(\theta) = p\varphi(\theta)[\psi(\theta) - \lambda] . \tag{4.7}$$

Suppose $g(x)$ is the function in L_1 and $L_{1+\delta}$ whose Fourier transform is $[\psi(\theta) - \lambda]$. Then (4.7) implies that for almost all x, $q\hat{f}(x) = p \int_{-\infty}^{+\infty} g(x - z)\, dF(z)$, where we have appealed to the fact that $g \in L_1$ in writing down the right-hand side. A familiar application of Hölder's inequality then gives

$$q^{1+\delta}\int_{-\infty}^{+\infty}[\hat{f}(x)]^{1+\delta}\,dx \leqq p^{1+\delta}\int_{-\infty}^{+\infty}[g(x)]^{1+\delta}\,dx < \infty\ .$$

(iii) If (4.6) holds, then for some $\eta > 0$ and all sufficiently large $|\,\theta\,|$

$$|\,\psi(\theta)\,| > \eta\,|\,\theta\,|^{\frac{1}{2}+\varepsilon}\ .$$

Thus, by (4.5), $\varphi(\theta) = O(\,|\,\theta\,|^{-\frac{1}{2}-\varepsilon})$. Hence $\varphi(\theta) \in L_2$ and so, by Titchmarsh ([10], p. 69) there must exist a density $f(x)$ which belongs to L_2.

It follows from the discussion of Lévy [6] that for the most general possible process with stationary independent increments one can write $\psi(\theta)$ in the form

$$\text{(4.8)}\qquad i\theta\alpha - \frac{1}{2}\sigma^2\theta^2 + \lambda_1\int_{-1-0}^{+1+0}\frac{e^{i\theta u}-1-i\theta u}{u^2}\,dN(u) + \lambda_2\int_{|u|>1}(e^{i\theta u}-1)\,dN(u)\ .$$

In this expression $\sigma^2, \lambda_1, \lambda_2$ are arbitrary non-negative real numbers, α is an arbitrary real number, and $N(\cdot)$ is an arbitrary distribution function *which is continuous at* 0. The form (4.8) is not the most compact one given in the literature (see, e.g., [5]) but is more convenient for our present purposes. We shall say there is a *drift* α if $\alpha \neq 0$, and a *Brownian component* if $\sigma^2 > 0$.

If it turns out that the distribution function N in (4.8) is such that

$$\int_{-1}^{+1}\frac{1}{|\,u\,|}\,dN(u) < \infty\ ,$$

then $\psi(\theta)$ can be given the simpler and more convenient form*

$$\text{(4.9)}\qquad i\theta\alpha - \frac{1}{2}\sigma^2\theta^2 + \lambda\int_{-\infty}^{+\infty}\frac{e^{i\theta u}-1}{|\,u\,|}\,dM(u)\ ,$$

where, here, σ^2 and λ are arbitrary non-negative numbers, α is an arbitrary real number, and M is an arbitrary distribution function which is continuous at 0. We might call the third term in (4.9) the *pure jump* term.

Notice that, if N is such that (4.8) can be thrown into the form (4.9), then the constant α will be modified in the conversion. This observation is relevant to Theorem 5(iii) below. Before we proceed to Theorem 5, however, the introduction of the canonical form for $\psi(\theta)$ provides us with an opportunity to give the brief discussion, promised in Section 1, of conditions under which the associated renewal process will be discrete. For such a discrete process to arise there must be a real $\tilde{\omega} > 0$ such that, with probability 1, $x(Y_1)$ is divisible by $\tilde{\omega}$. Plainly this requires that $\varphi(2\pi/\tilde{\omega}) = 1$, and hence, by (4.5), that $\psi(2\pi/\tilde{\omega}) = 0$. The real part of (4.8) is, in general, strictly negative and can only vanish at $2\pi/\tilde{\omega}(\neq 0)$ if $\sigma^2 = 0$ and if the sole points of increase of $N(u)$ are where u is a positive or negative (but not zero) multiple of $\tilde{\omega}$. Thus $\psi(\theta)$ may be given the simpler form (4.9) with $\sigma^2 = 0$ and $M(u)$ the distribution function of a lattice variable. But Theorem

* *Note added in proof*: Provided m_1 is finite; otherwise replace $|u|$ in (4.9) by $\min(1, |u|)$. Similarly for later appearances of this formula.

5(iii) below shows that we must have $\alpha = 0$, or else $\psi(\theta)$ would be the cumulant-generating function of an absolutely continuous distribution. We may therefore conclude that discrete renewal processes arise only if, for some $\tilde{\omega} > 0$ and some sequence $\{p_n\}$ such that $p_0 = 0$ and $\Sigma_{-\infty}^{+\infty} p_n = 1$, we can write

$$\psi(\theta) = \lambda \sum_{n=-\infty}^{n=+\infty} \frac{e^{n\tilde{\omega}i\theta} - 1}{n\tilde{\omega}} p_n .$$

THEOREM 5. *Sufficient conditions for* $F(\cdot)$ *to satisfy* (D1) *and* (D2) *are that, in the expression* (4.8) *for* $\psi(\theta)$, *either*

(i) $\sigma^2 > 0$. i.e. $x(t)$ *has a Brownian component*;

or

(ii) $\sigma^2 = 0$, *but for some* $\varepsilon < \frac{1}{2}$

$$\liminf_{h=0+} \frac{\lambda_1}{h^{1+\varepsilon}} \int_{-h}^{+h} dN(u) > 0 ;$$

or

(iii) $N(\cdot)$ *is such that* $\int_{-1}^{+1} (1/|u|) dN(u) < \infty$, *and when* $\psi(\theta)$ *is cast in the form* (4.9), *then* $\alpha \neq 0$, i.e. *there is a drift.*

Notice that sufficient conditions 5(ii) and 5(iii) might apply when $x(t)$ is not absolutely continuous; in fact, 5(iii) might apply when $x(t)$ has a discrete distribution. Thus it is not necessary, apparently, for $x(t)$ to be absolutely continuous before the infinitesimal renewal density theorem holds.

PROOF. (i) Let $G(x; t)$ be the distribution function of the process that would have the cumulant-generating function (4.8), but with $\sigma^2 = 0$. Then, if $F(x; t)$ corresponds to (4.8) as it stands, it is clear that $F(x; t)$ is obtained by convoluting $G(x; t)$ and a normal distribution function with zero mean and variance $\sigma^2 t$. Thus $F(x; t)$ must be absolutely continuous, with a frequency function

$$f(x, t) = \int_{-\infty}^{+\infty} \frac{\exp[-(x-z)^2/2\sigma^2 t]}{\sigma\sqrt{2\pi t}} dG(z, t) .$$

Thus, again by Hölder's inequality,

$$[f(x, t)]^2 \leqq \int_{-\infty}^{+\infty} \frac{\exp[-(x-z)^2/\sigma^2 t]}{2\sigma^2 \pi t} dG(z, t) .$$

A use of Fubini's theorem yields, in the notation of Theorem 3(iii),

$$I_2(t) \leqq \frac{1}{2\sigma\sqrt{\pi t}} .$$

Thus $e^{-t} I_2(t)$ is in L_1 and 5(i) follows from Theorem 3(iii).

(ii) If $\psi(\theta)$ has the form (4.9), but with $\sigma^2 = 0$, then

$$\begin{aligned}\mathscr{R}\psi(\theta) &= \lambda_1 \int_{-1-0}^{+1+0} \frac{\cos(\theta u) - 1}{u^2} dN(u) + \lambda_2 \int_{|u|>1} [\cos(\theta u) - 1] \, dN(u) \\ &= J_1(\theta) + J_2(\theta) , \quad \text{say} .\end{aligned}$$

Trivially we have $|J_2(\theta)| \leqq 2\lambda_2$. But

$$|J_1(\theta)| = 2\lambda_1 \int_{-1-0}^{+1+0} \frac{\sin^2(\frac{1}{2}u\theta)}{u^2}\, dN(u)\,.$$

Thus, for fixed $\eta > 0$, and $|\theta|$ sufficiently large

$$|J_1(\theta)| > 2\lambda_1 \int_{-\eta/|\theta|}^{+\eta/|\theta|} \frac{\sin^2(\frac{1}{2}u\theta)}{u^2}\, dN(u) > \frac{\lambda_1\theta^2}{2} \int_{-\eta/|\theta|}^{+\eta/|\theta|} \frac{\sin^2(\frac{1}{2}u\theta)}{(\frac{1}{2}u\theta)^2}\, dN(u)\,.$$

If η is small enough, $\sin^2 x/x^2 > \frac{1}{2}$ for all $|x| < \eta$. Thus

$$|J_1(\theta)| > \frac{\lambda_1\theta^2}{4} \int_{-\eta/|\theta|}^{+\eta/|\theta|} dN(u)\,,$$

and so

$$\liminf_{|\theta|=\infty} \frac{|J_1(\theta)|}{|\theta|^{1-\varepsilon}} \geqq \liminf_{|\theta|=\infty} \frac{\lambda_1 |\theta|^{1+\varepsilon}}{4} \int_{-\eta/|\theta|}^{+\eta/|\theta|} dN(u)\,,$$

which is greater than 0, by hypothesis.

Since $\varepsilon < \frac{1}{2}$ and $J_2(\theta)$ is bounded, we can conclude that for some $\nu > 0$,

$$\liminf_{|\theta|=\infty} \frac{|\psi(\theta)|}{|\theta|^{\frac{1}{2}+\nu}} > 0\,,$$

for $|\psi(\theta)| \geqq |\mathscr{R}\psi(\theta)|$. An appeal to Theorem 4(iii) finishes the proof of 5(ii).

(iii) If $\psi(\theta)$ has the form (4.9) with $\alpha \neq 0$, then

$$\frac{\mathscr{I}\psi(\theta)}{\theta} = \alpha + \lambda \int_{-\infty}^{+\infty} \frac{\sin(\theta u)}{\theta|u|}\, dM(u)\,. \tag{4.10}$$

If we can show that the integral on the right of (4.10) tends to zero as $|\theta| \to \infty$, then it will appear that $|\mathscr{I}\psi(\theta)| \sim |\alpha|\,|\theta|$ for large values of $|\theta|$. Since $|\psi(\theta)| \geqq |\mathscr{I}\psi(\theta)|$, we will then have

$$\liminf_{|\theta|=\infty} \frac{|\psi(\theta)|}{|\theta|} > 0\,,$$

and 5(iii) follows from Theorem 4(iii).

For any $\varepsilon > 0$, by the Riemann-Lebesgue lemma,

$$\theta J_1 = \int_{|u|>\varepsilon} \frac{\sin(\theta u)}{|u|}\, dM(u) \to 0$$

as $|\theta| \to \infty$. Thus J_1 approaches zero even more rapidly, and we need only consider

$$J_2 = \int_{|u|\leqq\varepsilon} \frac{\sin\theta u}{\theta\,|u|}\, dM(u)\,.$$

For all x, $|\sin x| \leqq |x|$, so $|J_2| \leqq \int_{|u|\leqq\varepsilon} dM(u)$.
But $M(u)$ is continuous at 0, so by choosing ε sufficiently small $|J_2|$ can be made arbitrarily small. This shows that the integral on the right of (4.10) tends to zero as $|\theta| \to \infty$, and thereby completes the proof of 5(iii).

In connection with Theorem 5(ii), it is discouraging to note the following.

THEOREM 6. *If $w(\cdot)$ is any increasing function, then there is a distribution function $N(\cdot)$ such that $N(0+) = 0$,*

$$\int_0^1 \frac{1}{u}\, dN(u) = +\infty ,$$

but

$$\liminf_{h=0+} w\left(\frac{1}{h}\right) \int_0^h dN(u) = 0 .$$

PROOF. We have to show there is a strictly positive random variable X, say, such that $\mathscr{E} X^{-1} = +\infty$, but such that

$$\liminf_{h=0+} w\left(\frac{1}{h}\right) \Pr\{X < h\} = 0 .$$

Our task is made easier if rephrased in terms of $Y = X^{-1}$. We want to find a positive random variable Y with infinite expectation such that

$$\liminf_{l=+\infty} w(l) \Pr\{Y > l\} = 0 .$$

With no loss of generality we may suppose $w(x) > 1$ for $x > 1$. Let $\{\varepsilon_n\}$ be any strictly decreasing sequence of strictly positive real numbers such that $\varepsilon_1 < 1$, and $\varepsilon_n \to 0$ as $n \to \infty$. We define $\xi_1, \xi_2, \cdots$, and $\zeta_n = \xi_1 + \xi_2 + \cdots + \xi_n$, as follows:

$$\xi_1 = 1, \qquad \xi_{n+1} = w(\zeta_n)/\varepsilon_n , \qquad \text{for } n > 1 .$$

Thus

$$\xi_2 = \frac{w(\xi_1)}{\varepsilon_1} = \frac{1}{\varepsilon_1} > 1 = \xi_1 ,$$

and for $n > 1$, $\xi_{n+1} > w(\zeta_{n-1})/\varepsilon_{n-1} = \xi_n$. Therefore, $\{\xi_n\}$ is a strictly increasing sequence and $\zeta_n \to \infty$ as $n \to \infty$.

Define the function $G(x)$ as follows. In the interval (ζ_n , ζ_{n+1}), $n = 1, 2, \cdots$, $1 - G(x) = \varepsilon_n/w(\zeta_n)$. Then $G(x)$ increases to unity as x tends to infinity, and hence, if we define $G(x)$ to be zero for all $x < 1$, it will be a distribution function. However,

$$\int_{\zeta_n}^{\zeta_{n+1}} [1 - G(x)]\, dx = \frac{\varepsilon_n}{w(\zeta_n)} \xi_{n+1} = 1 ,$$

so that

$$\int_0^\infty [1 - G(x)]\, dx = \infty ,$$

and therefore $G(x)$ has an infinite mean value.

But

$$w(\zeta_n)[1 - G(\zeta_n)] = w(\zeta_n) \frac{\varepsilon_n}{w(\zeta_n)} = \varepsilon_n .$$

Since $\varepsilon_n \to 0$ as $n \to \infty$, this implies that

$$\liminf_{x=\infty} w(x)[1 - G(x)] = 0 ,$$

and the theorem is complete.

We close this section by drawing attention to a serious omission. We have been able to obtain some reasonably general conditions, in Theorems 4 and 5, which ensure that $F(\cdot)$ satisfies (D1) and (D2). But to apply the density theorem (Theorem 3), we must also have that $\hat{f}(x)$ is vanishingly small at infinity. Indeed, this requirement is a *necessary* one for the validity of the density theorem. So far we have obtained no worthwhile results in this direction.

5. A Special Infinitesimal Renewal Process

Consider the process given by the density function

$$f(x; t) = \frac{e^{-x}x^{t-1}}{\Gamma(t)} , \qquad t > 0 .$$

The associated cumulant-generating function is $\psi(\theta) = -\log(1 - i\theta)$ so that, by (4.5), $\varphi(\theta) = 1/[1 + \log(1 - i\theta)]$. Plainly, this characteristic function belongs to no class L_p, so there is no hope that the various conditions derived in the last section will be of any use in applying the density theorem to this special case.

However, by Corollary 3A,

$$h(x) = \frac{e^{-x}}{x}\int_0^\infty \frac{x^t}{\Gamma(t)}\,dt . \tag{5.1}$$

The integral on the right of (5.1) may be evaluated by a method similar to the one of steepest descents (see, e.g., Jeffreys and Jeffreys [11], p. 501 *et seq.*). However, the present case is not quite routine, so we shall give some details of the calculation.

Let A be a large positive constant, and write

$$h(x) = \int_0^{x-A\sqrt{x}} + \int_{x-A\sqrt{x}}^{x+A\sqrt{x}} + \int_{x+A\sqrt{x}}^\infty \frac{e^{-x}x^{t-1}}{\Gamma(t)}\,dt = I_1 + I_2 + I_3, \text{ say} .$$

We first show that I_1 and I_3 are asymptotically negligible for large x, if A is large enough. If we use the inequality $\Gamma(t) > t^{t-\frac{1}{2}}e^{-t}\sqrt{(2\pi)}$ (see, e.g., Whittaker and Watson [12], p. 253) then we easily obtain

$$I_1 < \frac{1}{\sqrt{(2\pi x)}}\int_0^{x-A\sqrt{x}} e^{-(x-t)}\left(\frac{x}{t}\right)^t dt .$$

If we put $u = x - t$ and use the further inequality $(x/t)^t < \exp(u - u^2/2x)$, which is valid for $t < x$, then we find

$$I_1 < \frac{1}{\sqrt{(2\pi x)}}\int_{A\sqrt{x}}^x \exp(-u^2/2x)\,du .$$

Evidently I_1 can be made arbitrarily small by choosing A large enough.

Next we find, in a similar way, that

$$I_3 < \frac{1}{\sqrt{(2\pi x)}} \int_{x+A\sqrt{x}}^{\infty} e^{-(x-t)} \left(\frac{x}{t-\frac{1}{2}} \right)^{t-\frac{1}{2}} dt .$$

On putting $t - \frac{1}{2} = \tau$ and $B = A - 1$ we have, more tidily,

$$I_3 < \frac{e^{\frac{1}{2}}}{\sqrt{(2\pi x)}} \int_{x+B\sqrt{x}}^{\infty} e^{-(x-\tau)} \left(\frac{x}{\tau} \right)^{\tau} d\tau .$$

We now observe that, for $\tau > x$, one has

$$\left(\frac{x}{\tau} \right)^{\tau} < \exp\left[-(\tau - x) - (\tau - x)^2/2\tau\right] .$$

Thus

$$I_3 < \frac{e^{\frac{1}{2}}}{\sqrt{(2\pi x)}} \int_{x+B\sqrt{x}}^{\infty} \exp\left[-(\tau - x)^2/2\tau\right] d\tau .$$

But in the range of integration

$$\frac{(\tau - x)^2}{2\tau} > \frac{1}{4} B^2 + \frac{B}{2\sqrt{x}} (\tau - x - B\sqrt{x}) .$$

Thus $I_3 < (\sqrt{2}/B\sqrt{\pi}) \exp(\frac{1}{2} - \frac{1}{4}B^2)$, and so I_3, also, will be arbitrarily small if A is sufficiently large.

Finally we consider I_2. For large x, in view of Stirling's asymptotic approximation to the gamma function, we can evidently write

$$I_2 \sim \int_{x-A\sqrt{x}}^{x+A\sqrt{x}} \frac{e^{-(x-t)} x^{t-1}}{t^{t-\frac{1}{2}} \sqrt{2\pi}} dt \sim \frac{1}{\sqrt{(2\pi)}} \int_{x-A\sqrt{x}}^{x+A\sqrt{x}} e^{g(x,t)} dt ,$$

where $g(x, t) = t - x + (t - 1)(\log x) - (t - \frac{1}{2})(\log t)$.

One finds that

$$\frac{\partial g}{\partial t} = \frac{1}{2t} + \log\left(\frac{x}{t}\right), \quad \frac{\partial^2 g}{\partial t^2} = -\frac{1}{t} - \frac{1}{2t^2} , \quad \frac{\partial^3 g}{\partial t^3} = \frac{1}{t^2} + \frac{1}{t^3} .$$

It is then simple to show that g has a unique maximum in the range of integration for I_2; it occurs at $t_0 = x + \frac{1}{2} + O(1/x)$. Routine analysis will then show that

$$g(x, t_0) = -\frac{1}{2} \log x + O\left(\frac{\log x}{x}\right),$$

$$\left.\frac{\partial^2 g}{\partial t^2}\right|_{t_0} = -\frac{1}{x} + O\left(\frac{1}{x^2}\right),$$

and for all t in the range under consideration

$$\frac{\partial^3 g}{\partial t^3} = O\left(\frac{1}{x^2}\right).$$

Thus, for all t in this range, we have, by Taylor's theorem,

$$g(x, t) = -\frac{1}{2}\log x - \frac{(t-x)^2}{2x} + O\left(\frac{1}{\sqrt{x}}\right);$$

in this expansion the final correction term is uniform in t. It follows from this last expansion that as x tends to infinity we have

$$I_2 \sim \int_{x-A\sqrt{x}}^{x+A\sqrt{x}} \frac{\exp[-(t-x)^2/2x]}{\sqrt{(2\pi x)}}\, dt\,.$$

This final integral can be made arbitrarily near unity by choosing A large enough. Thus, as $x \to \infty$, $h(x) \to 1$, and the renewal density theorem holds.*

The density function

$$f(x) = \int_0^\infty \frac{e^{-t-x}x^{t-1}}{\Gamma(t)}\, dt$$

is therefore one which does not satisfy the conditions required by [3], but it is yet one for which the density theorem is valid.

References

[1] Smith, W. L. Renewal theory and its ramifications. *J. Roy. Stat. Soc.*, Ser. B, 1958, **20**, 243–302.

[2] Smith, W. L. Asymptotic renewal theorems. *Proc. Roy. Soc. (Edinburgh)*, Ser. A, 1954, **64**, 9–48.

[3] Smith, W. L. Extensions of a renewal theorem. *Proc. Camb. Phil. Soc.*, 1955, **51**, 629–38.

[4] Doob, J. L. *Stochastic Processes*. New York: Wiley, 1953.

[5] Gnedenko, B. V., and Kolmogorov, A. N. *Limit Distributions for Sums of Independent Random Variables*. (Translated from the Russian and annotated by K. L. Chung.) Cambridge, Mass.: Addison-Wesley, 1954.

[6] Lévy, P. *Théorie de l'Addition des Variables Aléatoires*. Paris: Gauthier-Villars, 1937.

[7] Halmos, P. R. *Measure Theory*. New York: Van Nostrand, 1950.

[8] Blackwell, D. Extension of a renewal theorem. *Pacific J. Math.*, 1953, **3**, 315–20.

[9] Smith, W. L. On the cumulants of renewal processes. *Biometrika*, 1959, **46**, 1–29.

[10] Titchmarsh, E. C. *Introduction to the Theory of Fourier Integrals*, 2d ed. Oxford: Oxford Univ. Press, 1948.

[11] Jeffreys, H., and Jeffreys, B. S. *Methods of Mathematical Physics*, 2d ed. Cambridge: Cambridge Univ. Press, 1950.

[12] Whittaker, E. T., and Watson, G. N. *A Course of Modern Analysis*, 4th ed. Cambridge: Cambridge Univ. Press, 1927.

* *Note added in proof*: At the time of writing this section, the author was unaware of a result of G. H. Hardy (*Ramanujan*, Cambridge; Cambridge Univ. Press, 1940, p. 196) from which the asymptotic behavior of $h(x)$ can easily be determined.

36 CLASSIFICATION PROCEDURES BASED ON DICHOTOMOUS RESPONSE VECTORS[1]

HERBERT SOLOMON
Stanford University

1. Introduction

In many fields, measurement is still rather primitive, and at best only qualitative information can be secured. This is especially true in the social and behavioral sciences, where attempts at making more than a dichotomization frequently represent superfluous sophistication. In addition, a number of variables, sex and handedness, for example, are naturally dichotomous. At other times, dichotomization is not necessary but provides conveniences in scoring procedures. Despite the fact that these dichotomous situations present deviations from the usual normality assumptions of classical prediction theory and the classical theory of discriminatory analysis, one often finds that a linear function of a set of dichotomous responses is used in classifying an element or an individual into one of two populations. For example, in achievement tests, the linear function is usually an unweighted sum of binomial variables which represents total score; in other situations the linear function is weighted in a special way as, for example, R. A. Fisher's discriminant function. This latter function, of course, possesses optimal properties if the measurements are continuous and are normally distributed with the same covariance matrix in each population.

In psychological evaluation, several measurement devices are common and used routinely. Among these are (1) tests of mental ability (achievement tests, IQ tests), (2) tests of personal behavior (projective tests, personality inventories), (3) attitude scales, and (4) biographical inventories. Because of long-standing assumptions in mental measurement, based mainly on Spearman's single-factor theory, classification by tests of mental ability is based on a single dimension, namely the test score, which is typically a sum of responses to test items. However, measurement theory for personality behavior does not yet have a tradition or a long development, and thus care

[1] This study was supported in part under Contract No. AF 41(657)-214 for the USAF School of Aviation Medicine, Randolph Air Force Base, Texas, while the author was at Teachers College, Columbia University.

should be exercised to prevent the indiscriminate use of mental measurement techniques, i.e., linear functions of test-item responses, for this purpose. Attitude measurement has a more substantial development and tradition. Beginning with Thurstone's pioneering work through the present developments in Lazarsfeld's latent-structure analysis, a body of theory has grown in this area.

Some of the techniques explored in this paper bear some resemblance to Lazarsfeld's work, in the main because both latent-structure analysis and the analysis in this paper study dichotomous systems. From the assumptions made in attitude measurement, the use of linear functions of responses to attitudinal items for classification purposes may be relevant in some situations and may be quite poor at other times. The use of biographical information for measurement and classification is relatively unexploited, and the use of linear functions in that situation is not supported by experience.

This paper will explore the worth of several classification procedures based on test-item response vectors in a situation where an external, dichotomous criterion is available and where for each response vector we know the number of individuals giving this response in each of the two classes. Each classification procedure is a rule applied either to the vector of responses given by an individual or to a function of this vector (e.g., sum of the binomial variables), the rule which places the individual into one of the two classes. In addition, we will explore the representation of the joint frequency distribution of response vectors. Classification procedures based on various approximations to the joint distribution of frequencies of response vectors will be evaluated in order to assess the loss of information incurred by the approximations. The measure of worth is two-dimensional and is given by the proportions of misclassification possible in the assignment of an individual into each of the two groups by the use of the classification procedure, and these two values can be obtained directly in this situation.

2. Population and Data

Before discussing the representation of the joint distribution of response vectors and the analysis and evaluation of various classification procedures, we shall describe the population and the data available for our study. Even though we are discussing the simplest kind of scoring, namely dichotomous responses, the number of response vectors quickly rises with the addition of test items. While for four items we have 16 response vectors, there are 1024 vectors for ten items. The latter number of items would be considered rather small for many psychological tests, yet analysis would be required which would seem inordinately extensive. Naturally we are not going to find complete and exhaustive populations where the total picture of frequencies for each response vector is known. However, to insure reliable values for the frequency of each response vector, so that a population is adequately reproduced, a very large number of individuals is essential and, more important, their actual response vectors must be available. This required extensive searching. Two studies, suitable for our purposes, were found, one of which is reported here. The second study has been partially analyzed and will be reported subsequently.

The individuals and data in the study under analysis in this paper were obtained in the following way.[2] A total of 2982 New Jersey high-school seniors were given an attitude questionnaire in 1957 entitled, "Attitudes Toward Science and Scientific Careers." This questionnaire was planned, for purposes quite different from that of the present study, by the Science Manpower Project, Teachers College, Columbia University. In addition to the responses given by the students, a score was obtained for each on a brief IQ vocabulary test, hereafter referred to as the IQ test. For the purposes of this study, the IQ variable is considered an external criterion. Two classes of IQ values were set up, to which an individual could be assigned on the basis of his attitude toward science. These two classes are the upper half (high IQ) and lower half (low IQ), 1491 individuals in each class, obtained by score on the IQ test. This arrangement may be a bit artificial but it was quite useful for our purposes.

A total of 95 items concerned with the attitudes of high-school seniors toward science comprised the questionnaire. Since we require reliable values for the frequencies of the response vectors and there are 1491 observations available in each of the two IQ groups, it was felt that at most six items would be appropriate for the analysis. This leads to 64 response vectors, or an average of approximately 23 per pattern, so that zero frequencies, if they occur, are probably meaningful and not due to scanty data. The frequency distributions for the 64 response vectors in each IQ group were tabulated. For some of our subsequent analyses the use of even six items proved to be too cumbersome for the computing resources available, so that only four of the six items were used in several studies. The six items are as follows: (When only four items are analyzed, the first four items listed are employed.)

- x_1 The development of new ideas is the scientist's greatest source of satisfaction.
- x_2 Scientists and engineers should be eliminated from the military draft.
- x_3 The scientist will make his maximum contribution to society when he has freedom to work on problems which interest him.
- x_4 The monetary compensation of a Nobel Prize winner in physics should be at least equal to that given popular entertainers.
- x_5 The free flow of scientific information among scientists is essential to scientific progress.
- x_6 The neglect of basic scientific research would be the equivalent of "killing the goose that laid the golden eggs."

Their selection as items which, taken singly, had good discrimination

[2] I would like to thank Mr. Hugh Allen and Professor Hubert Evans, Science Manpower Project, Teachers College, Columbia University, for the basic data used in this study.

value between the high- and low-IQ groups, was based on both the data on hand and correlations with the IQ variable. While the individuals originally were allowed five choices along an Agree–Disagree scale, their responses

TABLE 1

$x_1x_2x_3x_4$	Low IQ		High IQ	
	Frequency	Relative frequency	Frequency	Relative frequency
1111	62	.042	122	.082
1110	70	.047	68	.046
1101	31	.020	33	.022
1100	41	.027	25	.017
1011	283	.190	329	.221
1010	253	.170	247	.166
1001	200	.134	172	.115
1000	305	.205	217	.146
0111	14	.009	20	.013
0110	11	.007	10	.007
0101	11	.007	11	.007
0100	14	.009	9	.006
0011	31	.020	56	.038
0010	46	.031	55	.037
0001	37	.025	64	.043
0000	82	.055	53	.036
Totals	1491	.998	1491	1.002

for our study were collapsed and recorded as 1 or 0 for "agree" or "disagree." The frequency distributions for the four items are given in Table 1.

One of the purposes of the study is to contrast the efficiency of linear functions of item responses with the total-response vector. Suppose, therefore, that we use as our index of attitude toward science the sum of the scores over the four items. This provides a uni-dimensional variable rang-

TABLE 2

Score	Low IQ		High IQ	
	Frequency	Relative frequency	Frequency	Relative frequency
4	62	.042	122	.082
3	398	.267	450	.302
2	547	.367	521	.349
1	402	.270	345	.231
0	82	.055	53	.036
Totals	1491	1.001	1491	1.000

ing from 0 to 4 with frequency distribution for low-IQ groups and high-IQ groups as given in Table 2.

3. Analysis

Before we examine the worth of classification procedures, it is interesting to assess the information contained in the response vector and compare it with a more simple measure of four responses made by an individual. We know that a classification based on a score of three must be less informative than a classification procedure based on knowledge of how the three "agree" responses were arranged over the four items. One natural way of assessing how much more informative the total-response vector can be is to compute the errors made in misclassification through the use of the total-response vector and other procedures. We will also discuss departures from total information intrinsic to the system and without regard to classification. Both Lazarsfeld [1] and Bahadur [2] have demonstrated how the joint distribution of responses for n dichotomous items can be represented in a more instructive form. We will use the Bahadur representation here. Let X denote the set of all points $x = (x_1, x_2, \cdots, x_n)$ with each $x_i = 0$ or 1. Let $p(x)$ be a given probability distribution on X, that is, $p(x) \geqq 0$ for each x and $\sum p(x) = 1$ for all x belonging to X. For each $i = 1, 2, \cdots, n$, let $\alpha_i = \Pr\{x_i = 1\} = E_p(x_i)$, where E_p denotes expected value with respect to $p(x)$. In other words, the α_i, $i = 1, 2, \cdots, n$, represent the marginal frequencies of the x_i's. Now set

$$z_i = (x_i - \alpha_i)/\sqrt{\alpha_i(1 - \alpha_i)}$$

and define

$$r_{ij} = E_p(z_i z_j) , \qquad i < j ,$$

$$r_{ijk} = E_p(z_i z_j z_k) , \qquad i < j < k ,$$

$$\cdots \quad \cdots \quad \cdots \quad \cdots \quad \cdots$$

$$r_{123 \cdots n} = E_p(z_1 z_2 z_3 \cdots z_n) .$$

The second-order correlations, r_{ij}, the third-order correlations, r_{ijk}, and the nth order correlations, $r_{123 \cdots n}$, together comprise

$$n_{C_2} + n_{C_3} + \cdots + n_{C_n} = 2^n - n - 1$$

parameters. These together with the n marginal frequencies α_i comprise the $2^n - 1$ parameters which determine the probability distribution $p(x)$. Another way, but much less instructive, of viewing the $2^n - 1$ parameters is to recall there are 2^n response vectors and the 2^n frequencies must sum to one.

Let $p_{[1]}(x_1, x_2, \cdots, x_n)$ denote the joint probability distribution of the x_i when (a) the x_i's are independently distributed, and (b) they have the same marginal distributions as under the given distribution $p(x_1, x_2, \cdots, x_n)$. Then we have

$$p_{[1]}(x_1, x_2, \cdots, x_n) = \sum_{i=1}^{n} \alpha_i^{x_i}(1-\alpha_i)^{1-x_i} ,$$

and

$$p(x) = p_{[1]}(x) f(x) ,$$

where

$$f(x) = 1 + \sum_{i<j} r_{ij} z_i z_j + \sum_{i<j<k} r_{ijk} z_i z_j z_k + \cdots + r_{12\cdots n}\, z_1 z_2 \cdots {}_n .$$

It now becomes possible to approximate the actual frequencies $p(x)$ by dropping out terms of the series, especially where the deleted terms have some heuristic or instructive meaning. For example, one could assume, as in those instances where joint normal distributions prevail, that all correlations higher than second order are zero. On the other hand, in some very non-linear regression situations we could just as easily assume that only fourth-order correlations do not vanish. Notice that any approximation to the joint distribution we are discussing is obtained by deleting terms. This provides less information; therefore the classification procedure using the approximation must have larger proportions of errors in assignments than the same classification procedure using the total-response vector.

The parameters were evaluated for the four items in the scientific-attitude study. It was too cumbersome to evaluate the parameters for the six items. The results are as follows:

	r_{12}	r_{13}	r_{14}	r_{23}	r_{24}	r_{34}	r_{123}	r_{124}	r_{134}	r_{234}	r_{1234}
High IQ	.003	.143	.111	.180	.020	.140	−.010	−.064	−.041	−.033	.003
Low IQ	.049	.144	.043	.155	.096	.125	−.006	−.012	.002	.002	.002

and

	α_1	α_2	α_3	α_4
High IQ	.821	.159	.505	.436
Low IQ	.801	.189	.599	.530 .

Whichever representation of $p(x)$ is used, we are now faced with the problem of an appropriate classification procedure. Let $g = g(x)$ and $f = f(x)$ be the probability distributions of x in the high- (H) and low- (L) IQ groups, respectively. Then, as is well known, a complete theoretical solution of the classification problem is as follows: Take the likelihood ratio $L(x)$ where

$$L(x) = g(x)/f(x)$$

and adopt the procedure:

> "classify an individual with response vector x as a member of the high-IQ group (H) if and only if $L(x) > c$, otherwise classify the individual in the low-IQ group (L),"

where c is a fixed constant, equal to some possible value of $L(x)$.

The worth of this classification procedure can be evaluated in the following way. Consider all choices of the cut-off point c, and for each c compute the probabilities of misclassification. Let α_c denote the probability of misclassifying an individual from the high-IQ group (H) into the low-IQ group (L), when c is the cut-off point, that is

$$\alpha_c = \sum_{x:L(x)\leqq c} g(x) .$$

Likewise,

$$\beta_c = \sum_{x:L(x)>c} f(x)$$

is the probability of misclassifying an individual from low-IQ group L into high-IQ group H. Let $c_1, c_2, \cdots$, be the ordered values of c. Then the set of points $(\alpha_{c_i}, \beta_{c_i})$ together with the line segments joining the pairs $(\alpha_{c_i}, \beta_{c_i})$, $(\alpha_{c_{i+1}}, \beta_{c_{i+1}})$ is an error curve representing the risks in employing the classification procedure $L(x)$. If we have additional information about the *a priori* probabilities of the two groups L and H, only part of the error curve might be necessary for an evaluation.

The best risk curve for four items, that is, the curve with maximum curvature, will occur when we use the joint distribution of responses to the four items. Obviously, the curve is always bounded by the straight line in the α, β plane connecting (1, 0) and (0, 1). The construction of the curve can be easily routinized. Take $g(x)/f(x)$ for all 16 response vectors, then realign them in descending order according to the values of the likelihood ratios. These values provide the cut-off points c_i which lead to 17 decision rules, the ith rule given by $g(x)/f(x) > c_i$. The first rule is classify as low-IQ, L, for every response vector; the second rule is classify in L unless the newly selected first response vector is observed; the third rule is classify in L unless one of the first two newly selected response vectors is observed, etc.; the seventeenth rule is classify as high-IQ, H, for every response vector. The risks in using each of these 17 rules are now easily obtained by summing over the appropriate values of $f(x)$ and $g(x)$, which can be easily obtained from the new relisting.

The following assessments are now possible. We can obtain the risk curve associated with the likelihood ratio classification procedure when

(a) The joint distribution for the total-response vector is employed;

(b) The joint distribution for the total-response vector is approximated by the joint distribution obtained through the assumption of independence of responses over the four items, that is, only the marginal frequencies are used in the series representation for the joint frequencies;

(c) The joint distribution is approximated by ignoring all correlations higher than second order, the situation that would prevail if the joint distribution is given by a multivariate normal distribution;

(d) The joint distribution is approximated by any other function obtained by operating on the series representation, for example, in the four-item case, by considering only the fourth-order correlation term and of course the marginals, a situation that could very well arise in the use of personality and biographical inventories.

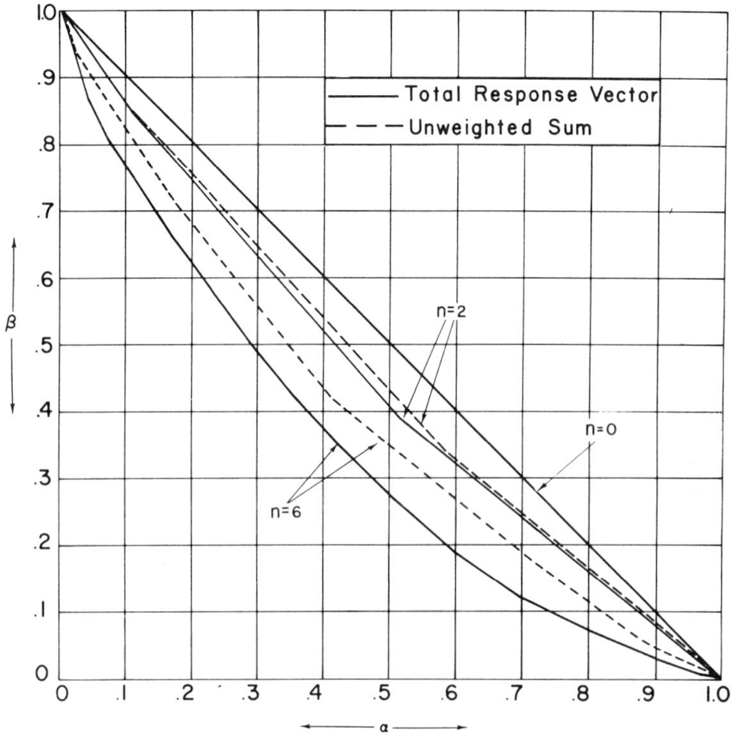

FIG. 1. The α, β Risk Curves for Two Items and Six Items

In addition, the joint distribution for the total-response vector could be collapsed to a univariate distribution by using a function of the four-item responses to obtain a single-valued response; for example, two linear functions which have had much use in this regard are an unweighted sum of the four binomial variables or a sum weighted in the sense of Fisher's discriminant function.

Actually, for each of these situations, a risk curve was constructed for two items and for three items in addition to the four-item situation. Also, risk curves were constructed for five items and for six items, both when the response vector was reduced to the unweighted sum of binomial variables and for the total-response vector situation. For the two-item case, Items 2 and 3 were employed; for the three-item case, Items 2, 3, and 4 were employed; for the five-item case, Items 1, 2, 3, 4, and 5 were employed.

4. Conclusions

Some risk curves are reproduced here to illustrate what has been done and to suggest some findings. Figure 1 depicts risk curves for two and

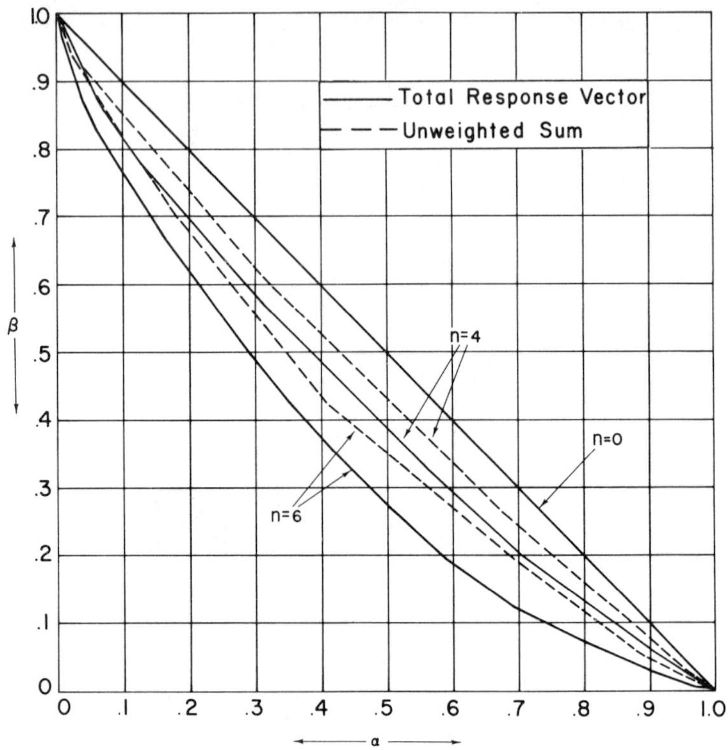

FIG. 2. The α, β Risk Curves for Four Items and Six Items

six items when the total-response vector is employed and when the unweighted sum of responses ("test score") is employed. Figure 2 presents comparisons of risk curves for response vector versus test score for four items and six items. Figure 3 demonstrates risk curves for four items where the use of the total-response vector is contrasted with several approximations. In this figure there are also several points plotted which were obtained to assess the use of Fisher's discriminant statistic classification procedure.

It is true that the number of items employed in this study is quite small, and therefore it is possible that not much difference in risk can be demonstrated by the α, β curves either for different number of items or for different approximations using the same number of items. However, the amount of work is so formidable even for four to six items that studies employing a larger number of items require more resources and a much larger staff than the usual investigator and an assistant or two. What is surprising is that, for the small number of items employed, one can detect differences in risks and thus gain a little feeling for the added information made possible by an increase in number of items (Figs. 1 and 2) and the information lost in

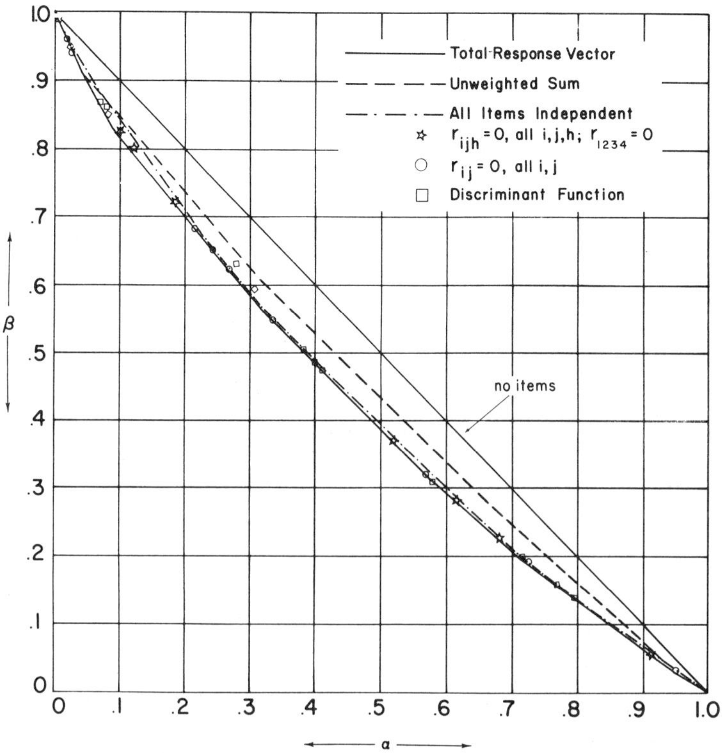

FIG. 3. The α, β Risk Curves for Four Items

using an approximation for the total-response vector (Fig. 3). In addition, it becomes possible to make decisions in test design—namely, how many items and what types should be used in a "test." From our study, for example, it is possible to state that the minimax risk in using the sum of the responses to the six items is equivalent for all practical purposes to the use of the total-response vector for three items (Items 2, 3, 4). To attain the ability to make statements of this type is essentially the major objective for this kind of study.

References

[1] Lazarsfeld, P. F. Some observations on dichotomous systems. Sociology Department, Columbia University, New York. Duplicated report (second draft), October 1956.

[2] Bahadur, R. R. A representation of the joint distribution of responses to n dichotomous items. USAF SAM Series in Statistics, Report No. 59-42, Randolph AFB, Texas: USAF School of Aviation Medicine, January 1959.

37 MULTIPLE REGRESSION

CHARLES STEIN
Stanford University

1. Introduction

The problem treated here is that of estimating regression coefficients when the predictors and the predictand have a joint normal distribution. We take as the loss function the suitably normalized mean square error of prediction when the estimated regression coefficients are used to make a prediction on the basis of a random observation on the predictors. In Section 2 we formulate the problem in detail. In Section 3 we discuss the transformations leaving the problem invariant and prove the minimax property of the maximum likelihood estimator. The admissibility of the maximum likelihood (m.l.) estimator in certain cases, where the number of predictors is small, is proved in Section 4. The inadmissibility of the m.l. estimator, when the mean is known and there are three or more predictors, is proved in Section 5. The results of Sections 4 and 5, which are still incomplete, are similar to the results in [1] for estimating the mean of a multivariate normal distribution with known covariance matrix, but some new complications occur in the proofs. In Section 6 we indicate (very roughly) the practical consequences of the results of Section 5 and mention a number of unsolved problems.

2. Formulation of the Problem

Let $X_1, \cdots, X_n$ be independently, normally distributed $(p+1)$-dimensional random vectors with mean ξ and non-singular covariance matrix Σ. Unless the contrary is explicitly stated, we shall consider Σ completely unknown. Detailed results are obtained only in the case where ξ is known, but for the present we shall also consider the case where ξ is unknown. We partition the X_i, ξ, and Σ in the following way

$$X_i = \begin{pmatrix} Y_i \\ Z_i \end{pmatrix}, \tag{2.1}$$

$$\xi = \begin{pmatrix} \eta \\ \zeta \end{pmatrix}, \tag{2.2}$$

$$\Sigma = \begin{pmatrix} A & B' \\ B & \Gamma \end{pmatrix}, \tag{2.3}$$

where the Y_i and η are 1×1, the Z_i and ζ are $p\times 1$, A is 1×1, B is $p\times 1$ and Γ is $p\times p$. Then

$$E(Y_i \mid Z_i) = \beta' Z_i + \alpha\,, \tag{2.4}$$

where

$$\beta = \Gamma^{-1}B\,, \tag{2.5}$$

and

$$\alpha = \eta - \beta'\zeta\,. \tag{2.6}$$

We are interested in estimating α and β, say by $\hat\alpha$ and $\hat\beta$, and take as the measure of the error (when ξ is unknown)

$$L((\xi, \Sigma), (\alpha, \beta)) = \frac{[(\hat\alpha - \alpha) + (\hat\beta - \beta)'\zeta]^2 + (\hat\beta - \beta)'\,\Gamma\,(\hat\beta - \beta)}{A - B'\Gamma^{-1}B}\,. \tag{2.7}$$

This choice is determined by the following considerations. Let us imagine that there is another random vector, $X_0 = \binom{Y_0}{Z_0}$, distributed in the same way as the other X_i and independent of them, but we observe only Z_0. As a prediction of Y_0, we take

$$\hat Y_0 = \hat\alpha + \hat\beta' Z_0\,. \tag{2.8}$$

Considering $\hat\alpha$ and $\hat\beta$ constant, we find for the mean square error

$$\begin{aligned} E_{\xi,\Sigma}(\hat Y_0 - Y_0)^2 &= E_{\xi,\Sigma}[(\hat\alpha + \hat\beta' Z_0) - Y_0]^2 \\ &= E_{\xi,\Sigma}\{[(\alpha + \beta' Z_0) - Y_0] + [(\hat\alpha - \alpha) + (\hat\beta - \beta)' Z_0]\}^2 \\ &= E_{\xi,\Sigma}[(\alpha + \beta' Z_0) - Y_0]^2 + E_{\xi,\Sigma}[(\hat\alpha - \alpha) + (\hat\beta - \beta)' Z_0]^2 \\ &= A - B'\Gamma^{-1}B + (\hat\alpha - \alpha)^2 + 2(\hat\alpha - \alpha)(\hat\beta - \beta)'\zeta \\ &\qquad + E_{\xi,\Sigma}(\beta - \beta)' Z_0 Z_0'(\beta - \beta) \\ &= A - B'\Gamma^{-1}B + [(\hat\alpha - \alpha) + (\hat\beta - \beta)'\zeta]^2 \\ &\qquad + (\hat\beta - \beta)'\Gamma(\hat\beta - \beta)\,. \end{aligned} \tag{2.9}$$

Dividing by $A - B'\Gamma^{-1}B$ and subtracting 1, we obtain (2.7). These operations merely reduce the risk of the maximum likelihood estimator to a constant and the loss of a perfect estimate to 0. When $\xi = \binom{\eta}{\zeta}$ is known, we assume the prediction of Y_0 to have the form

$$\hat Y_0 = \eta + \hat\beta'(Z_0 - \zeta)\,.$$

A computation similar to (2.9) yields

$$E_\Sigma(\hat Y_0 - Y_0)^2 = A - B'\Gamma^{-1}B + (\hat\beta - \beta)'\Gamma(\hat\beta - \beta)\,, \tag{2.10}$$

and we take for the loss

$$L(\Sigma, \hat\beta) = \frac{(\hat\beta - \beta)'\Gamma(\hat\beta - \beta)}{A - B'\Gamma^{-1}B}\,, \tag{2.11}$$

related to (2.10) as (2.7) is to (2.9).

The maximum likelihood estimator $(\hat{\alpha}_0, \hat{\beta}_0)$ of (α, β) when ξ is unknown is given by

$$\hat{\beta}_0 = V^{-1}U \,, \tag{2.12}$$

$$\hat{\alpha}_0 = \bar{Y} - \hat{\beta}_0'\bar{Z} \,, \tag{2.13}$$

where

$$\bar{Y} = \frac{1}{n}\sum_1^n Y_i \,, \quad \bar{Z} = \frac{1}{n}\sum_1^n Z_i \,, \tag{2.14}$$

and

$$S = \begin{pmatrix} T & U' \\ U & V \end{pmatrix} = \sum_1^n X_i X_i' - \frac{1}{n}\left(\sum_1^n X_i\right)\left(\sum_1^n X_i\right)' \,, \tag{2.15}$$

where T is 1×1. The risk, that is, the expected value of the loss (2.7), of the maximum likelihood estimator is obtained as follows:

$$\begin{aligned}
(2.16)\quad & E_{\xi,\Sigma}[(\hat{\alpha}_0 - \alpha) + (\hat{\beta}_0 - \beta)'\zeta]^2 \\
&= E_{\xi,\Sigma}[(\bar{Y} - \hat{\beta}_0'\bar{Z}) - (\eta - \beta'\zeta) + (\hat{\beta}_0 - \beta)'\zeta]^2 \\
&= E_{\xi,\Sigma}[(\bar{Y} - \eta) - \hat{\beta}_0'(\bar{Z} - \zeta)]^2 \\
&= E_{\xi,\Sigma}\{[(\bar{Y} - \eta) - \beta'(\bar{Z} - \zeta)] - (\hat{\beta}_0 - \beta)'(\bar{Z} - \zeta)\}^2 \\
&= E_{\xi,\Sigma}[(\bar{Y} - \eta) - \beta'(\bar{Z} - \zeta)]^2 + E_{\xi,\Sigma}(\hat{\beta}_0 - \beta)'(\bar{Z} - \zeta)(\bar{Z} - \zeta)'(\hat{\beta}_0 - \beta) \\
&= [A - B'\Gamma^{-1}B + E_{\xi,\Sigma}(\hat{\beta}_0 - \beta)'\Gamma(\hat{\beta}_0 - \beta)]/n \,.
\end{aligned}$$

Also

$$\begin{aligned}
(2.17)\quad E_{\xi,\Sigma}(\hat{\beta}_0 - \beta)'\Gamma(\hat{\beta}_0 - \beta) &= E_{\xi,\Sigma}(V^{-1}U - \beta)'\Gamma(V^{-1}U - \beta) \\
&= E_{\xi,\Sigma}\operatorname{tr}\Gamma(V^{-1}U - \beta)(V^{-1}U - \beta)' \,.
\end{aligned}$$

But

$$E_{\xi,\Sigma}(U \mid V) = E_{\xi,\Sigma}\left(\sum_1^n (Z_i - \bar{Z})(Y_i - \bar{Y})' \,\middle|\, \sum_1^n (Z_i - \bar{Z})(Z_i - \bar{Z})'\right) = V\beta \tag{2.18}$$

and

$$E_{\xi,\Sigma}(UU' \mid V) = (V\beta)(V\beta)' + (A - B'\Gamma^{-1}B)V \,. \tag{2.19}$$

It follows that

$$\begin{aligned}
(2.20)\quad & E_{i,\Sigma}(\hat{\beta}_0 - \beta)'\Gamma\hat{\beta}(_0 - \beta) \\
&= E_{\xi,\Sigma}\operatorname{tr}\Gamma\{V^{-1}[V\beta\beta'V + (A - B'\Gamma^{-1}B)V]V^{-1} - 2V^{-1}V\beta\beta' + \beta\beta'\} \\
&= (A - B'\Gamma^{-1}B)\,E_{\xi,\Sigma}\operatorname{tr}\Gamma V^{-1} \,.
\end{aligned}$$

Since $E_{\xi,\Sigma}\operatorname{tr}\Gamma V^{-1}$ does not depend on the parameters, we may assume $\Gamma = I$ (the $p \times p$ identity matrix). But the diagonal elements of V^{-1} are then distributed as $\chi^{-2}_{n-1-(p-1)} = \chi^{-2}_{n-p}$ (see, for example, [2], p. 85), so that

$$E_{\xi,\Sigma}\operatorname{tr}\Gamma V^{-1} = pE\chi^{-2}_{n-p} = \frac{p}{(n-p-2)} \,. \tag{2.21}$$

Thus by (2.16), (2.20), and (2.21)

$$(2.22)\quad E_{\xi,\Sigma}\{[(\hat{\alpha}_0 - \alpha) + (\hat{\beta}_0 - \beta)'\zeta]^2 + (\hat{\beta}_0 - \beta)'\Gamma(\hat{\beta}_0 - \beta)\}$$

$$= (A - B\Gamma^{-1}B)\left[\frac{1}{n}\left(1 + \frac{p}{n-p-2}\right) + \frac{p}{n-p-2}\right]$$

$$= (A - B'\Gamma^{-1}B)\frac{n(p+1)-2}{n(n-p-2)} .$$

We should remark that this formula is correct only when $n \geqq p + 3$, for otherwise $E_{\xi,\Sigma}\operatorname{tr}\Gamma V^{-1} = \infty$. Thus the risk of the maximum likelihood estimator $(\hat{\alpha}_0, \hat{\beta}_0)$ is given by

$$(2.23)\qquad E_{\xi,\Sigma}L((\xi, \Sigma), \hat{\alpha}(X), \hat{\beta}_0(X)) = \begin{cases} \dfrac{n(p+1)-2}{n(n-p-2)}, & \text{if } n \geqq p+3 , \\ \infty, & \text{if } n \leqq p+2 . \end{cases}$$

The formula for the m.l. estimator makes sense only when $n \geqq p + 1$. The preceding computation could have been shortened somewhat by the use of the notions of Section 3, but it seems desirable to give it here. In the case where ξ is known, the risk of the m.l. estimator $\hat{\beta}_0$ is similarly computed to be

$$(2.24)\qquad E_{\xi,\Sigma}L(\Sigma, \beta_0(X)) = \begin{cases} \dfrac{p}{n-p-1}, & \text{if } n \geqq p+2 , \\ \infty, & \text{if } n \leqq p+1 . \end{cases}$$

In this case, the formula for the m.l. estimator makes sense only when $n \geqq p$.

Three problems which are partially solved in this paper are the following:

(i) Is the m.l. estimator minimax, that is, is it true that for any $\varepsilon < 0$ any other estimator must have a risk which for some parameter point falls short of the constant risk of the m.l. estimator by at most ε? The answer given in Section 3 is "yes."

(ii) Is the m.l. estimator admissible, that is, must any essentially different estimator have a risk which at some parameter point is greater than that of the m.l. estimator? When ξ is unknown, $p = 1$, and $n \geqq 6$, the answer is "yes," and the remaining cases, other than trivial ones, have not been settled. When ξ is known, for $p = 1$, $n \geqq 4$ and for $p = 2$, $n \geqq 6$, the answer is "yes." For $p \geqq 3$ the answer is "no." Of course even for $p = 1$ or 2 but $n \leqq p + 1$, the answer is also "no," since the risk of the m.l. estimator is infinite while it is easy to write down estimators with everywhere finite but unbounded risk, e.g., $\hat{\beta} = 0$. For $p = 1, n = 3$ and for $p = 2, n = 4$ or 5, the problem is unsolved.

(iii) Since the m.l. estimator is not admissible for $p \geqq 3$, what are practical methods of improving on this estimator? This problem is discussed (very inadequately) in Section 6, again only in the artificial case where ξ is known.

3. Transformations Leaving the Problem Invariant

In order to prove the minimax property of the maximum likelihood estimator and to improve our understanding of the problem, we shall look at the group of affine transformations leaving the problem invariant. We consider first a subgroup of the group of all such transformations since only this subgroup seems to be useful. We consider the case where the mean is unknown and the loss function is given by (2.7). The modifications required, when the mean is known and the loss function is given by (2.11), are not difficult. The group of transformations we consider is

$$\binom{Y_i}{Z_i} \to \begin{pmatrix} a & b' \\ 0 & I \end{pmatrix}\binom{Y_i}{Z_i} + \binom{d}{0} = \binom{aY_i + b'Z_i + d}{Z_i}, \tag{3.1}$$

$$\xi = \binom{\eta}{\zeta} \to \begin{pmatrix} a & b' \\ 0 & I \end{pmatrix}\binom{\eta}{\zeta} + \binom{d}{0} = \binom{a\eta + b'\zeta + d}{\zeta}, \tag{3.2}$$

$$\begin{aligned} \Sigma = \begin{pmatrix} A & B' \\ B & \Gamma \end{pmatrix} &\to \begin{pmatrix} a & b' \\ 0 & I \end{pmatrix}\begin{pmatrix} A & B' \\ B & \Gamma \end{pmatrix}\begin{pmatrix} a & 0 \\ b & I \end{pmatrix}, \\ &= \begin{pmatrix} a^2A + 2ab'B + b'\Gamma b, & aB' + b'\Gamma \\ \Gamma b + aB & \Gamma \end{pmatrix}, \end{aligned} \tag{3.3}$$

$$\binom{\hat{\alpha}}{\hat{\beta}} \to \binom{a\hat{\alpha} + d}{a\hat{\beta} + b}. \tag{3.4}$$

Here a is a non-zero real number, b is a p-dimensional coordinate vector, d is a real number, and I is the $p \times p$ identity matrix. The problem is invariant under this group; that is, if the $\binom{Y_i}{Z_i}$ are independently normally distributed with mean ξ and covariance matrix Σ, then the $\binom{aY_i+b'Z_i+d}{Z_i}$ are independently, normally distributed with mean $\binom{a\eta+b'\zeta+d}{\zeta}$ and covariance matrix $\begin{pmatrix} a^2A+2ab'B+b'\Gamma b & aB'+b'\Gamma \\ \Gamma b+aB & \Gamma \end{pmatrix}$, and

$$\begin{aligned} &L((\xi, \Sigma), (\hat{\alpha}, \hat{\beta})) \\ &= L\left(\left(\binom{a\eta + b'\zeta + d}{\zeta}, \begin{pmatrix} a^2A + 2ab'B + b'\Gamma b & aB' + b'\Gamma \\ \Gamma b + aB & \Gamma \end{pmatrix}\right), (a\hat{\alpha} + d, a\hat{\beta} + b)\right), \end{aligned} \tag{3.5}$$

since

$$\binom{\alpha}{\beta} = \binom{\eta - B'\Gamma^{-1}\zeta}{\Gamma^{-1}B} \tag{3.6}$$

also transforms according to (3.4). This group consists of the transformations which would be relevant even if the predictors were treated as constants rather than as random variables with a partially known distribution. If the reader keeps this in mind, he ought to have no difficulty in understanding intuitively the proof of the minimax property and (in broad outline) the proof of admissibility when the m.l. estimator is admissible.

It will be convenient to use, instead of the sample point, the sufficient statistic

$$(\hat{\alpha}_0, \hat{\beta}_0, \bar{Z}, V, T - U'V^{-1}U), \tag{3.7}$$

where

$$\hat{\alpha}_0 = \bar{Y} - \beta_0'\bar{Z}\,, \tag{3.8}$$

$$\hat{\beta}_0 = V^{-1}U\,, \tag{3.9}$$

$$\begin{pmatrix} T & U' \\ U & V \end{pmatrix} = \sum_{i=1}^{n} \begin{pmatrix} Y_i \\ Z_i \end{pmatrix}(Y_i', Z_i') - n\begin{pmatrix} \bar{Y} \\ \bar{Z} \end{pmatrix}(\bar{Y}', \bar{Z}')\,, \tag{3.10}$$

$$\begin{pmatrix} \bar{Y} \\ \bar{Z} \end{pmatrix} = \frac{1}{n}\sum_{i=1}^{n}\begin{pmatrix} Y_i \\ Z_i \end{pmatrix}.$$

Then $(\hat{\alpha}_0, \hat{\beta}_0)$ also transform according to (3.4), and the condition that an estimator $(\hat{\alpha}, \hat{\beta})$ be invariant (see, for example, Kiefer [3]) is

$$\hat{\alpha}(a\hat{\alpha}_0 + d, a\hat{\beta}_0 + b, \bar{Z}, V, a^2(T - U'V^{-1}U)) = a\hat{\alpha}(\hat{\alpha}_0, \hat{\beta}_0, \bar{Z}, V, T - U'V^{-1}U) + d\,, \tag{3.11}$$

$$\hat{\beta}(a_0\hat{\alpha} + d, a\hat{\beta}_0 + b, \bar{Z}, V, a^2(T - U'V^{-1}U)) = a\hat{\beta}(\hat{\alpha}_0, \hat{\beta}_0, \bar{Z}, V, T - U'V^{-1}U) + b\,. \tag{3.12}$$

From (3.11) and (3.12), even if we restrict ourselves to $a = \pm 1$, it follows that

$$\hat{\alpha}(\hat{\alpha}_0, \hat{\beta}_0, \bar{Z}, V, T - U'V^{-1}U) = \hat{\alpha}_0\,, \tag{3.13}$$

$$\hat{\beta}(\hat{\alpha}_0, \hat{\beta}_0, \bar{Z}, V, T - U'V^{-1}U) = \hat{\beta}_0\,. \tag{3.14}$$

For if we write $\hat{\gamma} = \binom{\hat{\beta}}{\hat{\alpha}}$, $\hat{\gamma}_0 = \binom{\hat{\alpha}_0}{\hat{\beta}_0}$, $g = \binom{a}{b}$, $W = (\bar{Z}, V, T - U'V^{-1}U)$, then (3.11) and (3.12) for $a = \pm 1$ become

$$\hat{\gamma}(\pm\hat{\gamma}_0 + g, W) = \pm\hat{\gamma}(\hat{\gamma}_0, W) + g\,. \tag{3.15}$$

Putting $\hat{\gamma}_0 = 0$ in (3.15), we get

$$\hat{\gamma}(g, W) = \pm\hat{\gamma}(0, W) + g\,. \tag{3.16}$$

Since the $\pm$ does not occur on the left-hand side, we must have $\hat{\gamma}(0, W) = 0$ so that

$$\hat{\gamma}(g, W) = g\,, \tag{3.17}$$

i.e.,

$$\hat{\gamma}(\hat{\gamma}_0, W) = \hat{\gamma}_0\,, \tag{3.18}$$

which is the same as (3.13) and (3.14).

The minimax property of the m.l. estimator follows from these considerations and from the theorem of Section 3 of Kiefer [3]. We shall prove that the m.l. estimator is minimax even when ζ, Γ, and $A - B'\Gamma^{-1}B$ are known. In the discussion below of the group of all affine transformations leaving the problem invariant, we shall see that there is no essential loss of generality in supposing $\zeta = 0, \Gamma = I$, and $A - B'\Gamma^{-1}B = 1$, and we shall do so. The only unknown parameters are then (α, β), and the original parameters are given by

$$\binom{\eta}{\zeta} = \binom{\alpha}{0}, \tag{3.19}$$

$$\begin{pmatrix} A & B' \\ B & \Gamma \end{pmatrix} = \begin{pmatrix} 1+\beta'\beta & \beta' \\ \beta & I \end{pmatrix}. \tag{3.20}$$

The loss function (2.7) reduces to

$$L((\alpha, \beta), (\hat{\alpha}, \hat{\beta})) = (\hat{\alpha} - \alpha)^2 + (\hat{\beta} - \beta)'(\hat{\beta} - \beta). \tag{3.21}$$

We consider first the subgroups of the group of transformations (3.1) through (3.4), determined by the condition $a = 1$. This is the additive group of a $(p+1)$-dimensional real linear space. The conditions of Kiefer's theorem are easy to verify in the present case. Assumptions 1 and 2 follow immediately if we take

$$g(\hat{\alpha}_0, \hat{\beta}_0, \bar{Z}, V, T - U'V^{-1}U) = (\hat{\alpha}_0, \hat{\beta}_0), \tag{3.22}$$

considered as a translation. Assumption 4 follows from Kiefer's condition 4b. Assumption 5 follows from his condition 5b(1). The group of transformations (3.1)–(3.4) with $a = 1$ is a normal subgroup of the group (3.1)–(3.4) with $a = \pm 1$, and the quotient group has two elements. Assumption 3, the minimax theorem for the estimators invariant under (3.1)–(3.4) with $a = 1$, follows by averaging over the two-element quotient group. Thus Kiefer's theorem applies to the group (3.1)–(3.4), with $a = 1$ and the m.l. estimator, which was shown in (3.13) and (3.14) to be the only estimator invariant under the group (1)–(4) with $a = \pm 1$, must be minimax.

As we mentioned earlier, the transformations (3.1)–(3.4) are not the only ones leaving the problem invariant. A more general transformation leaving the problem invariant is

$$\binom{Y_i}{Z_i} \to \begin{pmatrix} a & b' \\ 0 & c \end{pmatrix}\binom{Y_i}{Z_i} + \binom{d}{e} = \binom{aY_i + b'Z_i + d}{cZ_i + e}, \tag{3.23}$$

$$\xi = \binom{\eta}{\zeta} \to \begin{pmatrix} a & b' \\ 0 & c \end{pmatrix}\binom{\eta}{\zeta} + \binom{d}{e} = \binom{a\eta + b'\zeta + d}{c\zeta + e}, \tag{3.24}$$

$$\begin{aligned} \Sigma = \begin{pmatrix} A & B' \\ B & \Gamma \end{pmatrix} &\to \begin{pmatrix} a & b' \\ 0 & c \end{pmatrix}\begin{pmatrix} A & B' \\ B & \Gamma \end{pmatrix}\begin{pmatrix} a & b' \\ 0 & c \end{pmatrix} \\ &= \begin{pmatrix} a^2A + 2ab'B + b'\Gamma b, & (aB' + b'\Gamma)c' \\ c(\Gamma b + Ba) \quad , & c\Gamma c' \end{pmatrix}, \end{aligned} \tag{3.25}$$

$$\binom{\hat{\alpha}}{\hat{\beta}} \to \binom{a\hat{\alpha} - ae'c'^{-1}\beta + d - b'c^{-1}e}{ac'^{-1}\hat{\beta} + c'^{-1}b}. \tag{3.26}$$

Here a is a non-zero real number, b is a p-dimensional coordinate vector, c is a non-singular $p \times p$ matrix, d is a real number, e is a p-dimensional coordinate vector, and 0 is the p-dimensional coordinate vector all of whose coordinates are 0. The 0 is needed in the lower left corner of (3.23) because it would otherwise be impossible to define a transformation of the estimate $(\hat{\alpha}, \hat{\beta})$ which would leave the loss function (2.7) invariant. Intuitively the 0 avoids mixing the predictand into the predictors. The somewhat complicated formula for the transformation of $(\hat{\alpha}, \hat{\beta})$ is the same as the rule of transformation of (α, β).

This larger group of transformations seems to be useful mainly for reducing problems to canonical form, but in view of the fact that, when many predictors are present, the m.l. estimator, which was derived as the only estimator invariant under the group (3.1)–(3.4) with $a = \pm 1$, is not admissible; estimators invariant under other subgroups may later be found to be of some interest. Earlier in this section we mentioned that, if we are considering ζ, Γ, and $A - B'\Gamma^{-1}B$ known, there is no essential loss of generality in assuming $\zeta = 0, \Gamma = I$, and $A - B'\Gamma^{-1}B = 1$. To see this, suppose ζ, Γ, and $A - B'\Gamma^{-1}B$ known, and consider the transformations (3.23) through (3.26) with

$$c'c = \Gamma^{-1}\,, \tag{3.27}$$

$$e = -c\zeta\,, \tag{3.28}$$

$$a = (A - B'\Gamma^{-1}B)^{-\frac{1}{2}}\,. \tag{3.29}$$

It is easy to see that the transformed variates have ζ replaced by $0, \Gamma$ by I, and $A - B'\Gamma^{-1}B$ by 1.

It seems likely that the transformations (3.23) through (3.26) are the only transformations leaving the problem invariant, but we have not attempted to carry through the program of [4] in this case.

4. Admissibility of the m.l. Estimator in Certain Low-dimensional Cases

The results on admissibility (or inadmissibility) will be largely restricted to the case where the mean ξ is known. The only result dealing with the case of unknown mean is Theorem 2 which was obtained very recently and is very incomplete. For the present, we shall suppose the mean $\xi = \binom{\eta}{\zeta}$ is known, and without essential loss of generality we take $\xi = 0$. This assumption that the mean is known is unrealistic in nearly all applications. The part of the first theorem below concerned with admissibility will be proved in this section, and the part concerned with inadmissibility in Section 5.

THEOREM 1. *When the mean is known, the m.l. estimator $\hat{\beta}_0$ is admissible for estimating β with the loss function L given by* (2.11), *when $p = 1, n \geqq 4$ or $p = 2, n \geqq 6$, and is inadmissible when $p \geqq 3$.*

The cases $p = 1, n = 3$ and $p = 2, n = 4$ or 5 have not yet been settled. For $n = p$ or $n = p + 1$, $\hat{\beta}_0$ is inadmissible in a trivial way since its constant risk is infinite, and for $n < p$ the m.l. estimator is not defined. In this section we prove the result for $p = 1$ and 2, and in the next section the result for $p \geqq 3$. The results of this section are proved by using the following lemmas.

LEMMA 1. *Let $\mathscr{B}$ be the σ-algebra of all Borel subsets of the real line $\mathscr{X}$ and $\mathscr{C}$ a σ-algebra of subsets of a set $\mathscr{Y}$. Let μ be Lebesgue measure on $\mathscr{B}$ and ν a probability measure on $\mathscr{C}$. Let p be a non-negative-valued $\mathscr{B}\mathscr{C}$ measurable function on $\mathscr{X} \times \mathscr{Y}$ such that*

$$\int p(x, y)\, dx = 1\,, \qquad \text{for all } y\,, \tag{4.1}$$

$$\int xp(x, y)\, dx = 0\,, \qquad \text{for all } y\,, \tag{4.2}$$

$$\int d\nu(y) \left[\int x^2 p(x, y)\, dx\right]^{\frac{3}{2}} < \infty\,, \tag{4.3}$$

where we write dx instead of $d\mu(x)$. Then, if we observe (X, Y) distributed so that for some unknown ξ, $(X - \xi, Y)$ has probability density p with respect to $\mu\nu$, the estimator X is an admissible estimator of ξ with squared error as loss.

LEMMA 2. *Let everything be as in Lemma 1 except that $\mathscr{X}$ is the two-dimensional real coordinate space, x^2 denotes the sum of squares of the coordinates of x, and (4.3) is replaced by*

$$\int d\nu(y) \left[\int x^{2+\delta} p(x, y)\, dx\right]^{2} < \infty\,, \tag{4.3$'$}$$

for some $\delta > 0$. Then again X is an admissible estimator of ξ with loss function L given by

$$L(\xi, \hat{\xi}) = (\hat{\xi} - \xi)' A (\hat{\xi} - \xi)\,, \tag{4.4}$$

for any constant positive semidefinite matrix A.

Lemmas 1 and 2 are the principal results of the author's papers [5] and [6], respectively.

We are now supposing the $X_i = \binom{Y_i}{Z_i}$ for $i = 1, \cdots, n$ are independently, normally distributed with mean 0 and common unknown non-singular covariance matrix $\Sigma = \begin{pmatrix} A & B' \\ B & \Gamma \end{pmatrix}$. As in the beginning of Section 3, it is convenient to consider, instead of the sample point $(X_1, \cdots, X_n)$ the sufficient statistic $(V, \hat{\beta}_0, T - U'V^{-1}U)$. Let $Y_i^* = Y_i - \beta' Z_i$. It is easy to show that the Y_i^* are independently normally distributed with mean 0 and variance $A - B'\Gamma^{-1}B$. Let

$$U^* = \sum Z_i Y_i^* = U - V\beta\,, \tag{4.5}$$

$$T^* = \sum Y_i^{*2}\,. \tag{4.6}$$

Then

$$T^* - U^{*\prime} V^{-1} U^* = T - U' V^{-1} U\,, \tag{4.7}$$

and

$$\hat{\beta}_0 = V^{-1}U = V^{-1}(U^* + V\beta) = V^{-1}U^* + \beta\,. \tag{4.8}$$

Thus we can express our sufficient statistic

$$(4.9)\qquad (V, \hat{\beta}_0, T - U'V^{-1}U) = (V, V^{-1}U^* + \beta, T^* - U^{*\prime}V^{-1}U^*) .$$

The third component $T - U'V^{-1}U$ of this sufficient statistic is distributed as $(A - B'\Gamma^{-1}B)\chi^2_{n-p}$ independent of the remaining components; V has a Wishart distribution with n degrees of freedom; and the conditional distribution of $\hat{\beta}_0 = V^{-1}U^* + \beta$ given V is normal with mean 0 and covariance matrix V^{-1}. In order to apply Lemma 1 if $p = 1$ or Lemma 2 if $p = 2$, we identify $(V, T - U'V^{-1}U)$ with Y, $\hat{\beta}_0$ with X, and β with ξ. We first suppose Γ and $A - B'\Gamma^{-1}B$ known. Then $p(\hat{\beta}_0, (v, t - u'v^{-1}u))$ is the normal density in $\hat{\beta}_0$ with mean β and covariance matrix v^{-1}. The distribution of $Y = (V, T - U'V^{-1}U)$ is known, and the conditions (4.1) and (4.2) are satisfied. In the case $p = 1$, condition (4.3) becomes

$$(4.10)\qquad E(V^{-1})^{\frac{3}{2}} < \infty .$$

But, for $n \geqq 4$,

$$(4.11)\qquad E(V^{-1})^{\frac{3}{2}} = E\chi_n^{-\frac{3}{2}} = \frac{1}{2^{\frac{n}{2}}\Gamma\left(\frac{n}{2}\right)}\int_0^\infty x^{\frac{n}{2}-1-\frac{3}{2}}e^{-\frac{x}{2}}\,dx < \infty .$$

In the case $p = 2$, the condition (4.3) becomes

$$(4.12)\qquad Ef^2(V) < \infty ,$$

where

$$(4.13)\qquad f(V) = E\{(U^{*\prime}V^{-2}U^*)^{1+\delta/2} \mid V\} .$$

Without essential loss of generality, we may suppose Γ to be the identity, since the value of Γ does not influence the finiteness of (4.12). Since the conditional distribution of U^* given V is normal with mean 0 and covariance matrix V, the unconditional distribution of $U^*V^{-2}U^*$ is that of Hotelling's T^2, that is

$$(4.14)\qquad U^{*\prime}V^{-2}U^* = \frac{\chi_2^2}{\chi_{n-1}^2} ,$$

where the numerator is independent of the denominator. Thus

$$(4.15)\qquad \begin{aligned} Ef^2(V) &\leqq E(U^*V^{-2}U^*)^{2+\delta} \\ &= E\left(\frac{\chi_2^2}{\chi_{n-1}^2}\right)^{2+\delta} \leqq cEX_{n-1}^{-4-2\delta} \\ &= \frac{c}{2^{\frac{n-1}{2}}\Gamma\left(\frac{n-1}{2}\right)}\int x^{\frac{n-1}{2}-1-2-\delta}e^{-\frac{x}{2}}\,dx < \infty , \end{aligned}$$

for $n \geqq 6$, provided we choose $\delta < \frac{1}{2}$. Thus, in the cases indicated, $\hat{\beta}_0$ would be admissible even if Γ and $A - B'\Gamma^{-1}B$ were known, and it remains admissible when Γ and $A - B'\Gamma^{-1}B$ are not known.

THEOREM 2. *When $p = 1$ and the mean is unknown, the m.l. estimator* $(\hat{\alpha}_0, \hat{\beta}_0)$ *given by* (2.12) *and* (2.13) *is admissible for estimating* $(\hat{\alpha}, \hat{\beta})$ *with the loss function* (2.7) *when* $n \geqq 1$.

The theorem is proved by applying Lemma 2, above, together with some of the considerations in the first part of Section 3. We shall prove that $(\hat{\alpha}_0, \hat{\beta}_0)$ would be an admissible estimator of (α, β) even if $\Gamma, A - B'\Gamma^{-1}B$ and ζ were known. As in Section 3, there is no essential loss of generality in supposing $\Gamma = I, A - B'\Gamma^{-1}B = 1$ and $\zeta = 0$, and we shall do so. To apply Lemma 2, we take

$$X = \begin{pmatrix} \hat{\alpha}_0 \\ \hat{\beta}_0 \end{pmatrix}, \qquad Y = (\bar{Z}, V, T - U'V^{-1}U), \tag{4.16}$$

and

$$\xi = \begin{pmatrix} \alpha \\ \beta \end{pmatrix}. \tag{4.17}$$

The distribution ν given in Lemma 2 is then the distribution of $Y = (\bar{Z}, V, T - U'V^{-1}U)$, and the density $p(\cdot \mid Y)$ is the conditional density of $(\hat{\alpha}_0, \hat{\beta}_0)$ given Y when $\alpha = \beta = 0$. The distribution of $Y = (\bar{Z}, V, T - U'V^{-1}U)$ may be described as follows: $T - U'V^{-1}U$ has a χ^2_{n-p-1} distribution independent of everything else, $\bar{Z}$ is normally distributed with mean 0 and covariance matrix $(1/n)I$, and V is independent of $\bar{Z}$ and has a p-variate Wishart distribution with $n - 1$ degrees of freedom and $EV = (n-1)I$. The conditional density $p(\cdot \mid Y)$ occurring in Lemma 2 may be described as follows: The conditional distribution of $\hat{\beta}_0$ given $\bar{Z}$ and V is normal with mean 0 and covariance matrix V^{-1}. The conditional distribution of $\hat{\alpha}_0 = \bar{Y} - \beta_0'\bar{Z}$ given $\bar{Z}$, V, and $\hat{\beta}_0$ is normal with mean $-\hat{\beta}_0'\bar{Z}$ and variance $1/n$. Thus the density $p(\cdot \mid Y)$ is

$$\sqrt{\frac{V}{2\pi}}\, e^{-\frac{1}{2}\hat{\beta}_0^2 V} \sqrt{\frac{n}{2\pi}}\, e^{-\frac{n}{2}(\hat{\alpha}_0 + \hat{\beta}_0 \bar{Z})^2}. \tag{4.18}$$

This is the density of a normal distribution with mean 0 and covariance matrix

$$\begin{pmatrix} \dfrac{1}{n} + \dfrac{\bar{Z}^2}{V} & -\dfrac{\bar{Z}}{V} \\ -\dfrac{\bar{Z}}{V} & \dfrac{1}{V} \end{pmatrix} \tag{4.19}$$

The condition (4.3′) is then equivalent to

$$EV^{-2-\delta} < \infty. \tag{4.20}$$

Since V is distributed as χ^2_{n-1}, (4.20) holds if $n \geqq 6$.

The problem with unknown mean and $p = 2$ seems fairly difficult, and we make no attempt to guess the answer. It follows from the above analysis and the principal result of [1] that, if $\Gamma, A - B'\Gamma^{-1}B$, and ζ are known, then $(\hat{\alpha}_0, \hat{\beta}_0)$ is inadmissible, but it is not easy to see what happens when these are not known. It seems likely that it will not be difficult to prove that, with unknown mean and $p \geqq 3$, $(\hat{\alpha}_0, \hat{\beta}_0)$ is inadmissible, but we have not yet succeeded in carrying out the details.

5. Inadmissibility of the m.l. Estimator $\hat{\beta}_0$ when $p \geqq 3$

We shall show that for $p \geqq 3$ the m.l. estimator $\hat{\beta}_0$ is not an admissible estimator of β. We need consider only the case $n \geqq p+2$, since for $n < p+2$ the constant risk of $\hat{\beta}_0$ is infinite so that any estimator (for example, 0) with an everywhere finite (but unbounded) risk is better. For $p \geqq 3, n \geqq p+2$, we shall show that for sufficiently small b and large a (with both $a, b > 0$) the estimator

$$\hat{\beta} = \left(1 - \frac{b(1-R^2)}{a(1-R^2)+R^2}\right)\hat{\beta}_0 \tag{5.1}$$

has everywhere smaller risk than $\hat{\beta}_0$. Here R^2 is the squared sample multiple-correlation coefficient

$$R^2 = \frac{U'V^{-1}U}{T}, \tag{5.2}$$

and $\hat{\beta}_0$ is the m.l. estimator

$$\hat{\beta}_0 = V^{-1}U . \tag{5.3}$$

Since the risk of $\hat{\beta}$ obviously depends only on the population multiple-correlation coefficient, ρ^2, there is no essential loss of generality in supposing

$$\Gamma = I, \tag{5.4}$$

$$A - B'\Gamma^{-1}B = 1 . \tag{5.5}$$

We use the fact that $T - U'V^{-1}U$ is distributed as χ^2_{n-p} independent of U, V, and as in Section 4 we let

$$U^* = U - V\beta , \tag{5.6}$$

and use the fact that the conditional distribution of U^* given V is normal with mean 0 and covariance matrix V. Then

$$\frac{b(1-R^2)}{a(1-R^2)+R^2} = \frac{b\chi^2_{n-p}}{a\chi^2_{n-p} + U^{*\prime}V^{-1}U^* + \beta'V\beta + 2U^{*\prime}\beta}, \tag{5.7}$$

and

$$\hat{\beta}_0 = V^{-1}(U^* + V\beta) = V^{-1}U^* + \beta . \tag{5.8}$$

We shall drop the star on U^*.

The risk of $\hat{\beta}$ is given by

$$(5.9)\quad E(\hat\beta-\beta)'(\hat\beta-\beta)$$

$$=E\left[\left(1-\frac{b(1-R^2)}{a(1-R^2)+R^2}\right)\hat\beta_0-\beta\right]'\left[\left(1-\frac{b(1-R^2)}{a(1-R^2)+R^2}\right)\hat\beta_0-\beta\right]$$

$$=E(\hat\beta_0-\beta)'(\hat\beta_0-\beta)-2bE\frac{1-R^2}{a(1-R^2)+R^2}\hat\beta_0'(\hat\beta_0-\beta)$$

$$+b^2E\left(\frac{1-R^2}{a(1-R^2)+R^2}\right)^2\hat\beta_0'\hat\beta_0 .$$

Thus our aim is to prove that for all β

$$(5.10)\qquad E\frac{1-R^2}{a(1-R^2)+R^2}\hat\beta_0'(\hat\beta_0-\beta)-\frac{b}{2}E\left(\frac{1-R^2}{a(1-R^2)+R^2}\right)^2\hat\beta_0'\hat\beta_0>0 .$$

But

$$(5.11)\qquad E\frac{1-R^2}{a(1-R^2)+R^2}\hat\beta_0'(\hat\beta_0-\beta)=E\frac{\chi^2_{n-p}(\beta+V^{-1}U)'V^{-1}U}{a\chi^2_{n-p}+U'V^{-1}U+\beta'V\beta+2U'\beta} .$$

We use the identity

$$(5.12)\qquad \frac{1}{1+x}=1-x+\frac{x^2}{1+x}$$

and, except in the remainder, take conditional expectations given χ^2_{n-p}, V, and UU', using the fact that the conditional distribution of U given these is symmetrical about 0 so that

$$E(\beta'V^{-1}U|\chi^2_{n-p},V,UU')=0 \quad\text{and}\quad E(U'\beta|\chi^2_{n-p},V,UU')=0.$$

Thus,

$$(5.13)\quad E\frac{1-R^2}{a(1-R^2)+R^2}\hat\beta_0'(\hat\beta_0-\beta)=E\frac{\chi^2_{n-p}(\beta'V^{-1}U+U'V^{-2}U)}{a\chi^2_{n-p}+U'V^{-1}U+\beta'V\beta}$$

$$\left\{1-\frac{2U'\beta}{a\chi^2_{n-p}+U'V^{-1}U+\beta'V\beta}\right.$$

$$\left.+\frac{4(U'\beta)^2}{(a\chi^2_{n-p}+U'V^{-1}U+\beta'V\beta)(a\chi^2_{n-p}+U'V^{-1}U+\beta'V\beta+2U'\beta)}\right\}$$

$$=E\frac{\chi^2_{n-p}}{a\chi^2_{n-p}+U'V^{-1}U+\beta'V\beta}\left\{U'V^{-2}U-\frac{2U'\beta\beta'V^{-1}U}{a\chi^2_{n-p}+U'V^{-1}U+\beta'V\beta}\right.$$

$$\left.+\frac{4(U'\beta)^2[\beta'V^{-1}U+U'V^{-2}U]}{[a\chi^2_{n-p}+U'V^{-1}U+\beta'V\beta][a\chi^2_{n-p}+U'V^{-1}U+\beta'V\beta+2U'\beta]}\right\}$$

$$=E\frac{\chi^2_{n-p}}{a\chi^2_{n-p}+U'V^{-1}U+\beta'V\beta}\left\{U'V^{-2}U-\frac{2U'\beta\beta'V^{-1}U}{a\chi^2_{n-p}+U'V^{-1}U+\beta'V\beta}\right\}$$

$$+o\left(\frac{1}{a+\lambda}\right),$$

where $\lambda=\rho^2/(1-\rho^2)=\beta'\beta$. The remainder term is $o(1/a+\lambda)$ as $a\to\infty$, uniformly in λ. This and similar remainder terms occurring subsequently are collected at the end of this section. Similarly,

$$(5.14)\quad E\left(\frac{1-R^2}{a(1-R^2)+R^2}\right)^2\hat{\beta}_0'\hat{\beta} = E\frac{\chi_{n-p}^4(\beta+V^{-1}U)'(\beta+V^{-1}U)}{(a\chi_{n-p}^2+U'V^{-1}U+\beta'V\beta+2U'\beta)^2}$$

$$= E\frac{\chi_{n-p}^4\lambda}{(a\chi_{n-p}^2+U'V^{-1}U+\beta'V\beta)^2}+o\left(\frac{1}{a+\lambda}\right).$$

Thus our aim is to prove that there exist constants $a, b, c > 0$ with a arbitrarily large such that for all β,

$$(5.15)\quad E\frac{\chi_{n-p}^2}{a\chi_{n-p}^2+U'V^{-1}U+\beta'V\beta}\left\{U'V^{-2}U-\frac{2U'\beta\beta'V^{-1}U}{a\chi_{n-p}^2+U'V^{-1}U+\beta'V\beta}\right\}$$

$$-\frac{b}{2}E\frac{\chi_{n-p}^4\lambda}{(a\chi_{n-p}^2+U'V^{-1}U+\beta'V\beta)^2} > \frac{c}{a+\lambda}.$$

But

$$(5.16)\quad E\frac{\chi_{n-p}^2}{a\chi_{n-p}^2+U'V^{-1}U+\beta'V\beta}\left\{U'V^{-2}U-\frac{2U'\beta\beta'V^{-1}U}{a\chi_{n-p}^2+U'V^{-1}U+\beta'V\beta}\right\}$$

$$-\frac{b}{2}E\frac{\chi_{n-p}^4\lambda}{(a\chi_{n-p}^2+U'V^{-1}U+\beta'V\beta)^2}$$

$$> E\left\{\frac{\chi_{n-p}^2U'V^{-2}U}{a\chi_{n-p}^2+\beta'V\beta}\left(1-\frac{U'V^{-1}U}{a\chi_{n-p}^2+U'V^{-1}U+\beta'V\beta}\right)\right.$$

$$\left.-\frac{2\chi_{n-p}^2U'\beta\beta'V^{-1}U}{(a\chi_{n-p}^2+\beta'V\beta)^2}-\frac{b\lambda}{2}E\frac{\chi_{n-p}^2}{(a\chi_{n-p}^2+\beta'V\beta)^2}\right\}$$

$$= E\left\{\frac{\chi_{n-p}^2\operatorname{tr}V^{-1}}{a\chi_{n-p}^2+\beta'V\beta}-\frac{2\lambda\chi_{n-p}^4}{(a\chi_{n-p}^2+\beta'V\beta)^2}\right.$$

$$\left.-\frac{b}{2}\frac{\lambda\chi_{n-p}^4}{(a\chi_{n-p}^2+\beta'V\beta)^2}\right\}+o\left(\frac{1}{a+\lambda}\right),$$

since

$$(5.17)\qquad E(U'V^{-2}U\mid V,\chi_{n-p}^2) = E(\operatorname{tr}V^{-2}UU'\mid V) = \operatorname{tr}V^{-1},$$

and

$$(5.18)\qquad E(U'\beta\beta'V^{-1}U\mid V,\chi_{n-p}^2) = E(\operatorname{tr}\beta\beta'V^{-1}UU'\mid V) = \operatorname{tr}\beta\beta' = \beta'\beta = \lambda.$$

In order to evaluate the right-hand side of (5.16) we may suppose without essential loss of generality that

$$(5.19)\qquad \beta = \begin{bmatrix}\sqrt{\lambda}\\ 0\\ \cdot\\ \cdot\\ \cdot\\ 0\end{bmatrix}.$$

Then

$$(5.20)\qquad \beta'V\beta = \lambda V_{11},$$

and $\operatorname{tr}V^{-1}$ is a sum of p reciprocals of χ^2's with $n-p+1$ degrees of freedom, of which $(p-1)$ are independent of V_{11} (and of χ_{n-p}^2, of course), and the remaining one is a component of V_{11}. That is, given V_{11}, it is distributed as WV_{11} where W is a B-variate independent of V_{11} with $n-p+1$ and $p-1$ degrees of freedom. Thus

$$\frac{\chi^2_{n-p}\operatorname{tr} V^{-1}}{a\chi^2_{n-p}+\beta' V\beta} = E\frac{\chi^2_{n-p}\left(\frac{p-1}{n-p-1}+\frac{n-2}{(n-p-1)V_{11}}\right)}{a\chi^2_{n-p}+\lambda V_{11}} . \tag{5.21}$$

But V_{11} is a χ^2_n independent of χ^2_{n-p} so that if we write

$$\chi^2_{2n-p} = \chi^2_n + \chi^2_{n-p} , \tag{5.22}$$

$$Q = \frac{\chi^2_{n-p}}{\chi^2_n + \chi^2_{n-p}} , \tag{5.23}$$

then Q is independent of χ^2_{2n-p} and has a B-distribution with $n-p$ and n degrees of freedom. It follows that

$$\begin{aligned} E\frac{\chi^2_{n-p}\operatorname{tr} V^{-1}}{a\chi^2_{n-p}+\beta' V\beta} &= \frac{p-1}{n-p-1}E\frac{Q}{aQ+\lambda(1-Q)} \\ &\quad + \frac{n-2}{n-p-1}\left(E\frac{1}{\chi^2_{n-p}}\right)E\frac{Q}{(1-Q)[aQ+\lambda(1-Q)]} \\ &= \frac{p-1}{n-p-1}E\frac{Q}{aQ+\lambda(1-Q)} \\ &\quad + \frac{n-2}{(n-p-1)(2n-p-2)}E\frac{Q}{(1-Q)[aQ+\lambda(1-Q)]} \end{aligned} \tag{5.24}$$

and

$$E\frac{2\lambda\chi^2_{n-p}}{(a\chi^2_{n-p}+\lambda V_{11})^2} = 2\lambda\left(E\frac{1}{\chi^2_{2n-p}}\right)E\frac{Q}{[aQ+\lambda(1-Q)]^2} , \tag{5.25}$$

and

$$E\frac{2\lambda\chi^2_{n-p}}{(a\chi^2_{n-p}+\beta' V\beta)^2} = \lambda E\left(\frac{Q}{aQ+\lambda(1-Q)}\right)^2 . \tag{5.26}$$

From (5.15), (5.16), (5.24), (5.25), and (5.26), we see that it will suffice to prove that there exist $a, b, c < 0$ with a arbitrarily large such that for all $\lambda > 0$

$$\begin{aligned} &\frac{p-1}{n-p-1}E\frac{Q}{aQ+\lambda(1-Q)} \\ &\quad + \frac{n-2}{(n-p-1)(2n-p-2)}E\frac{Q}{(1-Q)[aQ+\lambda(1-Q)]} \\ &\quad - \frac{2\lambda}{2n-p-2}E\frac{Q}{[aQ+\lambda(1-Q)]^2} - \frac{b\lambda}{2}E\left[\frac{Q}{aQ+\lambda(1-Q)}\right]^2 > \frac{c}{a+\lambda} . \end{aligned} \tag{5.27}$$

The left-hand side of (5.27) is larger than

$$(5.28)\quad \frac{p-1}{n-p-1}E\frac{Q}{aQ+\lambda(1-Q)}$$
$$+\left[\frac{n-2}{(n-p-1)(2n-p-2)}-\left(\frac{2}{2n-p-2}+\frac{b}{2}\right)\right]E\frac{Q}{(1-Q)[aQ+\lambda(1-Q)]}\,.$$

If $(n-2)/(n-p-1)>2$, i.e., $2p>n$, (5.27) follows from the trivial fact that

$$(5.29)\quad E\frac{Q}{aQ+\lambda(1-Q)}>\frac{c}{a+\lambda}\,.$$

To evaluate (5.28) in the case $n \geqq 2p$ we shall prove below that

$$(5.30)\quad \frac{E\dfrac{Q}{(1-Q)[aQ+\lambda(1-Q)]}}{\dfrac{Q}{aQ+\lambda(1-Q)}}<\frac{E\dfrac{Q}{(1-Q)^2}}{E\dfrac{Q}{1-Q}}$$
$$=\frac{B\left(\dfrac{n-p+2}{2},\dfrac{n-4}{2}\right)}{B\left(\dfrac{n-p+2}{2},\dfrac{n-2}{2}\right)}=\frac{2n-p-2}{n-4}\,.$$

Thus the left-hand side of (5.28) is larger than

$$(5.31)\quad \left\{\frac{p-1}{n-p-1}+\frac{2n-p-2}{n-4}\left[\frac{n-2}{(n-p-2)(2n-p-2)}\right.\right.$$
$$\left.\left.-\left(\frac{2}{2n-p-2}+\frac{b}{2}\right)\right]\right\}E\frac{Q}{aQ+\lambda(1-Q)}$$
$$>\left[\frac{p}{n-p-1}-\frac{2}{n-4}-\frac{b(2n-p-2)}{2(n-4)}\right]E\frac{Q}{aQ+\lambda(1-Q)}>\frac{c}{a+\lambda}\,.$$

for b and c sufficiently small.

To prove (5.30), we observe that, with Q_1, Q_2 independently distributed the same as Q and $\alpha = a/\lambda$,

$$(5.32)\quad \frac{Q}{(1-Q)[\alpha Q+(1-Q)]}E\frac{Q}{1-Q}-E\frac{Q}{\alpha Q+(1-Q)}E\frac{Q}{(1-Q)^2}$$
$$=E\left\{\frac{Q_1}{(1-Q_1)(\alpha Q_1+1-Q_1)}\frac{Q_2}{1-Q_2}-\frac{Q_2}{\alpha Q_2+1-Q_2}\frac{Q_1}{(1-Q_1)^2}\right\}$$
$$=E\left\{\frac{1}{\alpha+\dfrac{1-Q_1}{Q_1}}\frac{Q_2}{(1-Q_1)(1-Q_2)}-\frac{1}{\alpha+\dfrac{1-Q_2}{Q_2}}\frac{Q_1}{(1-Q_1)^2}\right\}$$
$$=E\frac{1}{1-Q_1}\frac{\alpha\left(\dfrac{Q_2}{1-Q_2}-\dfrac{Q_1}{1-Q_1}\right)}{\left(\alpha+\dfrac{1-Q_1}{Q_1}\right)\left(\alpha+\dfrac{1-Q_2}{Q_2}\right)}$$

$$= \frac{\alpha}{2} E \frac{\left(\frac{1}{1-Q_1} - \frac{1}{1-Q_2}\right)\left(\frac{Q_2}{1-Q_2} - \frac{Q_1}{1-Q_1}\right)}{\left(\alpha + \frac{1-Q_1}{Q_1}\right)\left(\alpha + \frac{1-Q_2}{Q_2}\right)} < 0 .$$

We have only to list the remainder terms dropped in the course of the preceding proof. These are

$$(5.33)\quad R_1 = E \frac{4\chi^2_{n-p}(U'\beta)^2[\beta' V^{-1}U + U'V^{-2}U]}{[a\chi^2_{n-p} + U'V^{-1}U + \beta' V\beta]^2 \cdot [a\chi^2_{n-p} + U'V^{-1}U + \beta' V\beta + 2U'\beta]} ,$$

$$(5.34)\quad R_2 = E \frac{\chi^4_{n-p}}{(a\chi^2_{n-p} + U'V^{-1}U + \beta' V\beta)^2} \Big\{ 2\beta' V^{-1}U + U'V^{-2}U$$
$$- 4 \frac{(\beta + V^{-1}U)'(\beta + V^{-1}U)[U'\beta(a\chi^2_{n-p} + U'V^{-1}U + \beta' V\beta) + (U'\beta)^2)]}{(a\chi^2_{n-p} + U'V^{-1}U + \beta' V\beta + 2U'\beta)^2} \Big\} ,$$

$$(5.35)\quad R_3 = E \frac{\chi^2_{n-p}(U'V^{-2}U)(U'V^{-1}U)}{(a\chi^2_{n-p} + \beta' V\beta)(a\chi^2_{n-p} + U'V^{-1}U + \beta' V\beta)} .$$

All of these are easily proved to be $o[1/(a + \lambda)]$ as $a \to \infty$, uniformly in λ, by considering separately the two cases $\lambda \geqq a$, $\lambda < a$ as in formulas (18) and (19) of [1].

6. Miscellaneous Remarks

First let us try to get some notion of the possible magnitude of the improvement over the usual estimator. In the case where the mean is known (and taken to be 0), let us look at the estimator (5.1) with $a = 0$, i.e.,

$$(6.1)\quad \hat{\beta} = \left(1 - \frac{b(1 - R^2)}{R^2}\right)\hat{\beta}_0 .$$

We shall compute its risk, first, at $\beta = 0$, and then for large $\beta'\beta$. When $\beta = 0$, the risk is given by

$$(6.2)\quad E\hat{\beta}'\hat{\beta} = E\left(1 - \frac{b(1 - R^2)}{R^2}\right)^2 \hat{\beta}_0'\hat{\beta}_0 = E\left(1 - \frac{b\chi^2_{n-p}}{U'V^{-1}U}\right)^2 U'V^{-2}U$$
$$= EU'V^{-2}U - 2b(n-p)\frac{U'V^{-2}U}{U'V^{-1}U} + b^2[(n-p)^2 + 2(n-p)]\frac{U'V^{-2}U}{(U'V^{-1}U)^2}$$
$$= EW'V^{-1}W - 2b(n-p)\frac{W'V^{-1}W}{W'W} + b^2[(n-p)^2 + 2(n-p)]\frac{W'V^{-1}W}{(W'W)^2} ,$$

where W' is normally distributed with mean 0 and the p-dimensional identity matrix as covariance matrix, independent of V. We have

$$(6.3)\quad E(W'V^{-1}W \mid V, W'W) = E(\mathrm{tr}\, V^{-1}WW' \mid V, W'W) = \frac{W'W}{p}\,\mathrm{tr}\, V^{-1},$$

so that

$$(6.4)\quad E(W'V^{-1}W \mid W'W) = \frac{W'W}{p}\,\frac{p}{n-p-1} = \frac{W'W}{n-p-1} .$$

From (6.2) and (6.4) we get

$$(6.5)\qquad E\hat{\beta}'\hat{\beta} = \frac{EW'W}{n-p-1} - 2b\frac{n-p}{n-p-1} + b^2\frac{[(n-p)^2 + 2(n-p)]}{n-p-1}E\frac{1}{W'W}$$
$$= \frac{p}{n-p-1} - 2b\frac{n-p}{n-p-1} + b^2\frac{(n-p)^2 + 2(n-p)}{(p-2)(n-p-1)} .$$

This attains its minimum of

$$(6.6)\qquad \frac{p}{n-p-1} - \frac{(n-p)(p-2)}{(n-p-1)(n-p+2)} = \frac{2n}{(n-p-1)(n-p+2)} ,$$

when

$$(6.7)\qquad b = \frac{p-2}{n-p+2} .$$

Now let us look at the case of large $\beta'\beta$. In this case we use (5.9), (5.13), and (5.14) with $a = 0$, the remainders in the latter two being $o(1/\beta'\beta)$ as $\beta'\beta \to \infty$. As in Section 5, we put $\lambda = \beta'\beta$. Thus we have

$$(6.8)\quad E_\beta(\hat{\beta} - \beta)'(\hat{\beta} - \beta)$$
$$= \frac{p}{n-p-1} - 2bE\frac{\chi^2_{n-p}}{U'V^{-1}U + \beta'V\beta}\left[U'V^{-2}U - \frac{2U'\beta\beta'V^{-1}U}{U'V^{-1}U + \beta'V\beta}\right]$$
$$+ b^2E\frac{\chi^4_{n-p}\lambda}{(U'V^{-1}U + \beta'V\beta)^2} + o\left(\frac{1}{\lambda}\right)$$
$$= \frac{p}{n-p-1} - 2bE\frac{\chi^2_{n-p}}{\beta'V\beta}\left[U'V^{-2}U - 2\frac{U'\beta\beta'V^{-1}U}{\beta'V\beta}\right]$$
$$+ b^2E\frac{\chi^4_{n-p}\lambda}{(\beta'V\beta)^2} + o\left(\frac{1}{\lambda}\right)$$
$$= \frac{p}{n-p-1} - 2b(n-p)E\left[\frac{\operatorname{tr} V^{-1}}{\beta'V\beta} - 2\frac{\lambda}{(\beta'V\beta)^2}\right]$$
$$+ b^2\lambda[(n-p)^2 + 2(n-p)]E\frac{1}{(\beta'V\beta)^2} + o\left(\frac{1}{\lambda}\right)$$
$$= \frac{p}{n-p-1} - 2b(n-p)E\left[\frac{\dfrac{p-1}{n-p-1} + \dfrac{n-2}{(n-p-1)V_{11}}}{\lambda V_{11}} - \frac{2}{\lambda V_{11}^2}\right]$$
$$+ b^2[(n-p)^2 + 2(n-p)]E\frac{1}{\lambda V_{11}^2} + o\left(\frac{1}{\lambda}\right)$$
$$= \frac{p}{n-p-1} - 2\frac{b}{\lambda}(n-p)\left[\frac{p-1}{(n-p-1)(n-2)}\right.$$
$$\left. + \left(\frac{n-2}{n-p-1} - 2\right)\frac{1}{(n-2)(n-4)}\right]$$
$$+ \frac{b^2}{\lambda}[(n-p)^2 + 2(n-p)]\frac{1}{(n-2)(n-4)} + o\left(\frac{1}{\lambda}\right)$$

$$= \frac{p}{n-p-1} - 2\frac{b}{\lambda}\frac{(n-p)(p-2)}{(n-p-1)(n-4)} + \frac{b^2}{\lambda}\frac{(n-p)(n-p+2)}{(n-2)(n-4)} + o\left(\frac{1}{\lambda}\right).$$

To the approximation we are using, this attains a minimum of

$$\frac{p}{n-p-1} - \frac{(n-2)(p-2)^2}{\lambda(n-p+2)(n-p-1)^2} + o\left(\frac{1}{\lambda}\right) \tag{6.9}$$

at

$$b = \frac{(n-p)(p-2)}{(n-p-1)(n-p+2)}\,. \tag{6.10}$$

We observe that this value of b is larger than the value (6.7) which minimizes the risk at $p = 0$. Until better calculations have been made, it would seem wise to use (6.7) rather than (6.10), because the behavior for small λ is more important than that for large λ and because the use of a smaller b runs less danger of increasing the risk over that of the usual estimator in the unexplored range of moderate values of λ. Thus we would tentatively recommend the estimator

$$\hat{\beta} = \max\left[\left(1 - \frac{p-2}{n-p+2}\frac{1-R^2}{R^2}\right), 0\right]\hat{\beta}_0\,. \tag{6.11}$$

It is clear that the estimator given by (6.11) has everywhere a smaller risk than that given by (6.1) and (6.7).

It will often be desirable to apply an appropriate transformation before using the estimator (6.11) or a similar one. The risk given above has its minimum at $\rho^2 = 0$, but if we suspect that ρ is close to 0, we shall not usually be interested in estimating a multiple regression. However, it may happen that, before the experiment is carried out, we can guess β to be, say, β^* with some hope that we are close. We can then apply the estimator to the Z_i, $Y_i - \beta^* Z_i$ instead of the Z_i, Y_i and add β^* to obtain an estimate of the original β. The risk of the resulting estimator of β is then the same function of the multiple correlation coefficient between Z_i and $Y_i - \beta^* Z_i$ as the risk of the estimator (6.11) is of that between Z_i and Y_i. If β^* is a good guess, the multiple correlation between Z_i and $Y_i - \beta^* Z_i$ may be much smaller than that between Z_i and Y_i. Of course, the guess β^* must not be influenced by the observations.

In many cases it will be hard to make a good guess at β, but it may still be possible to improve on the direct use of the estimator (6.11) or any estimator of the form $\varphi(R^2)\hat{\beta}_0$. For example, there may be a small number of predictors which are expected to yield a multiple correlation which is a large proportion of the total multiple correlation. Then we might estimate the multiple regression for the preferred predictors in the usual way and use a procedure analogous to (6.11) on the residuals. We have not attempted to work out the details. As another example, if we believe a Spearman single-factor model to be a good approximation, it should be possible to find an estimator everywhere better than the usual one which is especially good when our belief is correct.

We mention two problems in addition to those which have been discussed or hinted at above:

(i) In the case of three or more predictors, are there any admissible minimax estimators of the form $\varphi(R^2)\hat{\beta}_0$? What are some admissible minimax estimators?

(ii) In the notation of the beginning of Section 2, suppose η and Σ are unknown, but ζ is known to be 0. Are there estimators of η which have everywhere smaller mean square error than $\bar{Y} - S_{12}S_{22}^{-1}\bar{Z}$?

References

[1] Stein, C. Inadmissibility of the usual estimator for the mean of a multivariate normal distribution. In Vol. 1, J. Neyman, ed., *Proceedings of the Third Berkeley Symposium on Mathematical Statistics and Probability*, Berkeley, Calif.: Univ. California Press, 1956.

[2] Anderson, T. W. *Introduction to Multivariate Statistical Analysis*. New York: Wiley, 1958.

[3] Kiefer, J. Invariance, minimax sequential estimation, and continuous time processes. *Ann. Math. Stat.*, 1957, **28**, 573–601.

[4] Lehmann, E. L., and Stein, C. The totality of transformations leaving a family of normal distributions invariant (abst.). *Ann. Math. Stat.*, 1953, **24**, 142.

[5] Stein, C. Admissibility of Pitman's estimator for a single location parameter. *Ann. Math. Stat.*, 1959, **30**, 970–79.

[6] Stein, C. Admissibility of Pitman's estimator for two location parameters. (Submitted to *The Annals of Mathematical Statistics*.)

38 AN OPTIMUM REPLICATED TWO-SAMPLE TEST USING RANKS[1]

MILTON E. TERRY

Bell Telephone Laboratories

1. Summary

The hypothesis that k pairs of samples, the ith pair containing m_i, n_i observations, are independent and come from a common continuous population is tested by a rank-order test which is most powerful against a normal alternative. The pertinent statistic is shown to be asymptotically normal under certain restrictions. A laboratory example of the use of the test together with tables is presented.

2. Introduction

The rank-order test, most powerful against the normal alternative that the two samples come from two normal populations with common variance σ^2 and different means $\mu_{1.1}$ and $\mu_{1.2}$, with $\delta = (\mu_{1.2} - \mu_{1.1})/\sigma$ sufficiently small, was presented as a special case of a more general test in a previous paper by the author [1]. This paper presents a detailed study of a special case of Hoeffding's general most powerful rank-order test [2].

3. The Case of k Pairs of Samples

Consider k pairs of samples, each pair having m_i, n_i observations respectively; $m_i + n_i = N_i$, $i = 1, \cdots, k$. Let H_0 be the hypothesis that all variables are independent, and that each pair came from a common continuous population. Denote by $z_{i1}, \cdots, z_{iN_i}$ the random variables in the ith pair. Under H_0, the joint probability function $P(S)$ of $z_{i1}, \cdots, z_{iN_i}$ is invariant under all permutations within $(z_{i1}, \cdots, z_{iN_i})$, for $i = 1, \cdots, k$. The probability of the set W, where the coordinates of each pair of samples are all different, is one.

Let r_{ij} be the rank of z_{ij} within $(z_{i1}, \cdots, z_{iN_i})$, and write R_i for $(r_{i1}, \cdots, r_{iN_i})$ and R^k for $(R_1, \cdots, R_k)$. Under the null hypothesis, the probability of each vector R^k is

[1] These results were included in the thesis submitted to the Department of Mathematical Statistics, University of North Carolina, in partial fulfillment of the requirements for the degree of Doctor of Philosophy, June 1951.

(1) $$P(R^k \mid H_0) = \left(\prod_{i=1}^{k} N_i!\right)^{-1}.$$

Consider the alternative where all variables are independent and normally distributed with $E(z_{ij}) = \mu_{i1}$, $j = 1, \cdots, m_i$; $E(z_{ij}) = \mu_{i2}$, $j = m_i + 1, \cdots, m_i + n_i$; and $\text{var}\,(z_{ij}) = \sigma_i^2$ for all j, such that $(\mu_{i2} - \mu_{i1})/\sigma_i = \delta$, $i = 1, \cdots, k$. It has been shown in [1] that the probability $P_i(R_i, \delta)$ of R_i is a function of the parameter δ only and that

(2) $$P_i(R_i, \delta) = (N_i!)^{-1}[1 + c_{i1}(R_i)\,\delta + 0(\delta^2)]\,,$$

where

$$c_{i1}(R_i) = \sum_{j=m_i+1}^{N_i} E(Z_{N_{i1}, r_{ij}})\,,$$

if $Z_{N,r}$ denotes the rth order statistic in a random sample of size N from a standard normal population. Since $R_1, \cdots, R_k$ are mutually independent, the probability of the vector R^k under the alternative is

(3) $$P(R^k, \delta) = \prod_{i=1}^{k} P_i(R_i, \delta)\,,$$

and we have

(4) $$P(R^k, \delta) = \left[\prod_{i=1}^{k} N_i!]^{-1}[1 + c_1(R^k)\delta + 0(\delta^2)\right],$$

where

(5) $$c_1(R^k) = \sum_{i=1}^{k} c_{i1}(R_i) = \sum_{i=1}^{k} \sum_{j=m_i+1}^{N_i} E(z_{N_i, r_{ij}})\,.$$

Therefore, for δ sufficiently small and positive, the rank-order test of size $M/\Pi_{i=1}^{k} N_i!$ which is most powerful against $P(S, \delta)$ is determined by the M permutations R^k for which $c_1(R^k)$ takes its largest values.

Table 1 gives the critical values of $c_1(R^k)$ for probability levels less than or equal to .10 for some small values of N_i, k. $c_1(R)$ has been tabulated for all values of m and n, $(m + n) \leqq 10$, in [1].

4. Asymptotic Distribution of $c_1(R^k)$

From (5) we have $c_1(R^k) = \sum_{i=1}^{k} c_{i1}(R_i)$, and we note that the random variables $c_1(R_i), \cdots, c_k(R_k)$ are independent. It then follows from the theorem of Section 6 of the previous paper [1] that, under H_0, the variance of $c_{i1}(R_i)$ is

(6) $$V_0(c_{i1}(R_i)) = \frac{m_i n_i}{N_i(N_i - 1)} \sum_{j=1}^{N_i} (EZ_{N_i, j})^2\,,$$

and if, as $N_i \to \infty$,

(7) $$0 < k' < \frac{n_i}{N_i} < k'' < 1, \qquad \text{for all } i,$$

TABLE 1

VALUES OF $c_1(R^k) = c_1^*(R^k)$ FOR SMALL VALUES OF $N, n, m,$ AND k

Also included is the probability p of $c_1(R^k)$ exceeding these values under H_0, where N is the total sample size; m, n are the subsample sizes; and k is the number of replications of samples of size N.

Sample size	$c_1^*(R^k)$	p
$N=4$, $m=2$, $n=2$; $k=2$	2.66	.028
	2.06	.083
$N=4$, $m=2$, $n=2$; $k=3$	3.39	.018
	2.79	.032
	2.66	.050
	2.06	.120
$N=5$, $m=2$, $n=3$; $k=2$	3.32	.010
	2.82	.030
	2.32	.060
	2.16	.080
	1.82	.100
$N=5$, $m=2$, $n=3$; $k=3$	3.98	.010
	3.87	.019
	3.32	.033
	2.82	.062
	2.32	.092
$N=6$, $m=3$, $n=3$; $k=2$	3.42	.010
	2.98	.030
	2.75	.045
	2.54	.065
	2.35	.075
	2.31	.085
	2.10	.095
$N=6$, $m=3$, $n=3$; $k=3$	4.42	.010
	3.81	.024
	3.58	.032
	3.22	.039
	3.18	.053
	2.97	.060
	2.78	.072
	2.74	.085
	2.55	.096

Sample size	$c_1^*(R^k)$	p
$N_1=4$, $N_2=5$, $m_1=2$, $n_1=2$, $m_2=2$, $n_2=3$; $k=2$	2.49	.033
	2.39	.050
	1.99	.067
	1.89	.100
$N_1=4$, $N_2=6$, $m_1=2$, $n_1=2$, $m_2=3$, $n_2=3$; $k=2$	2.84	.025
	2.60	.042
	2.44	.050
	2.11	.083
	2.00	.092
$N_1=5$, $N_2=6$, $m_1=2$, $n_1=3$, $m_2=3$, $n_2=3$; $k=2$	3.37	.010
	2.93	.025
	2.49	.045
	2.43	.055
	2.37	.060
	2.11	.080
	1.93	.100

then $[c_{i1}(R_i)]/\sqrt{V_0(c_{i1}(R_i))}$ has a standard normal limiting distribution. Hence the theorem below follows.

THEOREM. *If H_0 is true, $V_0(c_1(R^k)) = \sum_{i=1}^{k} V_0(c_{i1}(R_i))$, and if as $N \to \infty$, $i=1, \cdots, k$, condition (7) is satisfied for $i=1, \cdots, k$, then $[c_1(R^k)]/\sqrt{V_0(c_1(R^k))}$ has a standard normal limiting distribution.*

5. Example of $c_1(R^k)$ in Experimental Use

For some time, L. L. Davis of the Department of Horticulture, Virginia Polytechnic Institute, has been conducting research on the effect of anti-oxidants on the odor and on the preservation of flavor of frozen broilers. The following pilot experiment was conducted. Three treated birds and three controls were thawed, halved, and broiled. The six treated halves were randomly paired, as were the six control halves, and placed in six dishes, three dishes with treated birds and three with controls, with integers $1,\cdots,6$ randomly assigned to the dishes. Three judges of "known and proven" ability then ranked the six dishes in order of preference for flavor of dark meat, white meat, and over-all flavor. Since it is presumed that the anti-oxidant acts by inhibiting the breakdown of fats, it does not add flavor, but merely preserves; and since it is presumed that there is a gradual breakdown of fats in untreated birds, we should expect the treated birds to taste better. It is also reasonable to consider the distribution of fleshing quality, location and amount of fat as normal. Hence the underlying assumptions for the $c_1(R^3)$ test are satisfied, and the test was used.

The following results were obtained:

Dark Meat:
$$R^3 = yyxyxx;\ yyxyxx;\ yyxxyx\ ,$$
$$c_1(R^3) = 4.69, \quad p \sim 0.005\ .$$

White Meat:
$$R^3 = yyyxxx;\ yyyxxx;\ yxyyxx\ ,$$
$$c_1(R^3) = 5.49, \quad p \sim 0.002$$

Over-all Flavor:
$$R^3 = yyxyxx;\ yyxyxx;\ yxyyxx\ ,$$
$$c_1(R^3) = 4.69, \quad p \sim 0.005\ ,$$

where y is a treated sample and x is a control.

In each case the results are highly significant, and we reject the hypothesis that these samples come from identical populations.

On the basis of limited experimentation, this test would appear to be practical and useful in research of this type. Further use of this test by research groups will show its true merit.

References

[1] Terry, M. E. Some rank order tests which are more powerful against specific parametric alternatives. *Ann. Math. Stat.*, 1952, **23**, 346-66.

[2] Hoeffding, W. "Optimum" nonparametric tests. In J. Neyman (ed.), *Proceedings of the Second Berkeley Symposium on Mathematical Statistics and Probability.* Berkeley, Calif.: Univ. California Press, 1951.

39 A SURVEY OF SAMPLING FROM CONTAMINATED DISTRIBUTIONS[1]

JOHN W. TUKEY
Princeton University

1. Historical Introduction

It is particularly fitting that the results produced to date on sampling from contaminated distributions should be summarized in a volume dedicated to Harold Hotelling. Toward the end of World War II, several groups under contract to the Applied Mathematics Panel, National Defense Research Council, had common concern with problems of the effectiveness of bomber machine-gun fire against attacking fighters. Two of these were the Princeton University Fire Control Research Group, of which the writer was a member, and SRG–C (Statistical Research Group–Columbia) for which Harold Hotelling was official investigator. The work summarized in this paper had its origin in those particular problems of accuracy of fire, and, in particular, in the statistical problems of inferring effectiveness from an observed pattern of shots fired with fixed aim or fixed intended aim.

Two contributions to the effectiveness of fire are easily identified: (i) Individual projectiles do not go in the direction in which they are are aimed, or intended to be aimed; (ii) the aim is neither perfect nor constant, and the vulnerability of the target is neither uniform nor concentrated at a point. As a consequence, the probable effectiveness of a projectile traveling at a given angle to the direction of aim, or intended aim, averaged over all situations, decreases smoothly as the angle between the directions of travel and aim increases. Thus, the smaller the dispersion of directions of motion around the direction of aim, the greater probable effectiveness of fire. (Aim-wander, and the possibility of overhitting, cf., for example, Cunningham and Hynd [1], Fraser [2], reverse this trend for sufficiently small dispersions, but the dispersions in question did not fall in this range.)

In assessing records of dispersion, then, the purpose was to assess and compare the average values of such smooth functions. (Because the angular

[1] The basic results surveyed here were obtained in the course of research sponsored by the Office of Naval Research. Their integration and exposition was carried out under Contract No. DA-36-034-ORD-2297 sponsored by the Office of Ordnance Research, U. S. Army.

deviations are small, deviations in a plane perpendicular to the direction of aim or intended aim can replace angles. This is particularly convenient when holes in a flat target produce such deviations as raw data.)

The problem can be further simplified without doing violence to its essentials by treating as known the location of the distribution, as it might be expressed by the coordinate means. Two quite distinct arguments support

> **THE FIRST QUESTION**
>
> Given two normal populations with the same mean, one having three times the standard deviation of the other, it is proposed to prepare a sequence of mixed populations by adding varying small amounts of the wider normal population to the narrower one. It is well known that, in large samples, the relative efficiency as a measure of scale of the mean deviation compared with the standard deviation is 88% when the underlying population is normal. As specific small amounts of the wider normal population are added to the narrower one, thus defining new classes of distributions of fixed shape, will the relative efficiency for scaling of the mean deviation compared to the standard deviation increase or decrease?
>
> NOTES: 1. The possible answers are "increase," "stay the same," and "decrease."
> 2. The problem is a large-sample problem.
> 3. To find the answer, after giving the question long and careful thought, turn over two pages.

this simplification. On the one hand, experience with large-sample location-and-scaling problems shows that they behave in very much the same way as the corresponding pure scaling problems. In a sense the location part of the problem is either separable from, or simpler than, the scaling part. On the other hand, it is quite appropriate to believe that, in due course, the adjustment of sighting devices will, in each instance, become accurate enough to place the center of the distribution at the direction of aim, or intended aim. While either argument would probably suffice, together they make the simplification very reasonable.

Thus the assessment-of-dispersion-for-the-purpose-of-assessing-or-comparing-effectiveness problem reduces to a two-dimensional analog of a familiar problem: Given a sample, to assess the scale of the distribution from which it was drawn. If the distribution is normal, then the classical statistical results tell us that we do best to use the three second-degree sample moments. But what if the distribution is not normal?

Some years of close contact with the late C. P. Winsor had taught the writer to beware of extreme deviates, and in particular to beware of using them with high weights. Using second moments to assess variability means giving very high weights to extremely deviant observations. Thus the use of second moments, unquestionably optimum for normal distributions, comes into serious question. When this point was raised in conversation, real differences of opinion between some of the statisticians concerned showed themselves. The earliest work on sampling from contaminated distributions was carried out in an attempt to develop facts which would help to quiet this clash of opinion. (See references [3] through [10] for some of the work done on contaminated distributions.)

2. The Broad Nature of the Problem

Most of the standard statistical techniques are optimum, or recommended, or perhaps merely correctly calibrated, given certain standard assumptions. How do they behave under alternative assumptions? The work to be summarized here can be recognized as another step toward the solution of this broad problem.

It is interesting to speculate why more examples of this broad problem have not been treated. Each specific example can be formulated, and, with the expenditure of sufficient effort, solved within the scope of classical mathematical statistics. But statisticians seem to have been reluctant to approach problems of this type. Is it because the statement of such problems implies that a standard statistical technique might be used when its standard assumptions are not fulfilled? Perhaps a desire to "sweep the dirt under the rug" is responsible for much of the seeming reluctance to tackle such problems. Yet this cannot be the whole explanation, for there have been a fair number of inquiries into the behavior of tests of significance (and, consequently, of techniques of confidence estimation) when something other than the standard assumption holds (e.g., Pearson [11]; Eden and Yates [12]; Pitman [13], [14], [15]; Welch [16]; Box and Andersen [17]). But there has not been any comparable amount of work on questions of the relative precision or accuracy of different point estimates, including questions of relative efficiency. Is this (i) because statisticians do not mean their statements of relative efficiency to be taken seriously, (ii) because statisticians believe that, in contrast to statements of significance, statements of relative efficiency will not be compared with experience, or (iii) because of some reason or reasons not yet suggested? No definitive answer to this question will be presented here, but statisticians owe the problem some consideration.

It would seem that questions of robustness of efficiency are intrinsically at least as important as questions of robustness of significance level. And as soon as it develops, as it shortly will, that failures of robustness of efficiency may be very substantial, the need for more work on the robustness of efficiency will be clear and pressing.

3. Sharpening and Reduction of the Problem

With the problem reduced to a distribution in a plane, a mathematical statistician's first instinct is to consider the special case where this distribution is bivariate normal. This special case is simple, since the first and second moments of the deviations become sufficient statistics for the five parameters of the general case, and, with location assumed known, only second moments need be considered. Any average over the distribution, including the average effectiveness whose assessment is the purpose of the study, is a function of these five parameters and can be efficiently estimated on the basis of (i.e., as a function of) these first and second moments. All this results if the assumption of normality is either to be taken seriously, or to be utilized as a fully effective approximation.

The smooth function of deviation of shot from aim, whose average we seek to assess, needs now to be specified in more detail. The choice

$$\phi = e^{-Q} ,$$

where Q is a quadratic function of the deviations, is likely to represent the true situation with sufficient accuracy for practical purposes, in part because of our limited knowledge, and in part because great precision is not required.

It is very natural to ask, perhaps in a loud and bitter tone, why this assumed form for the measure of effectiveness should be adequate, when the same assumed form for the probability density (the assumption of normality) is thought to be quite inadequate. There is a simple answer. In dealing with the mathematical statistics of the assumed distribution, there is a temptation to give a very high weight to a very aberrant deviation. If this occurs, very fine detail in the tails of the distribution can matter greatly. But a very aberrant deviation will never be assigned a large probability or a large effectiveness, so that details of the tails of the expression to be averaged will never matter greatly. The artificialities of "efficient" estimation emphasize the tails in a way in which the realities of probability assessment never do.

There is nothing in the problem, as now formulated, which suggests that the two-dimensionality of the situation is playing an essential role. In order to separate essentials more clearly from details, it is thus natural to begin by considering the one-dimensional analog of the two-dimensional problem. The basic problem then becomes: Given a sample $z_1, z_2, \cdots, z_n$ from an unknown, moderately normal, univariate distribution of known location, and given a function $\phi(z)$, which may be taken to be of the form e^{-cz^2}, how it is reasonable to estimate the average value of the function over the distribution? It is relevant, natural, and important to specify the distribution as moderately normal—relevant and natural since we would not have given much, if any, consideration to guidance from normal theory if the sample made it clear that the population was markedly non-normal; and important since only those procedures of estimation will be considered which would be acceptable if the population were in fact normal.

Note: References [18] through [27] report some of the approaches to the problem of compound distributions, while Refs. [28] through [35] relate to compound normal distribution. The literature on contagious (and negative binomial) distributions has been omitted; many references can be traced from the lists given on pages 67–68, 105–6, 194, and 248–49 of Volume 14 of *Biometrics*.

4. Location and Scaling: Standard and Alternative Pencils

A pure location problem is one in which the specified cumulatives (cumulative distributions) are transformations carrying any distribution of the specification into any other. That is, the cumulatives are $G(z - \mu)$ where $G(z)$ is the prototype cumulative, and μ is a location parameter. A pure scaling problem is one in which the specified cumulatives are $G(z/\lambda)$, where $G(z)$ is again the prototype cumulative, and λ is a scaling parameter. The transformation $z = e^w$, $\lambda = e^\tau$ reduces any pure scaling problem to a pure location problem. Both cases provide estimation problems of the simplest form involving

1. A random sample from a one-dimensional distribution,
2. A one-dimensional parameter,
3. A group of transformations carrying each distribution into all the distributions concerned.

If a general situation can be persuaded to show itself in such an estimation problem, one may expect its essential nature to be most clearly revealed there.

We face *a* (location or) scaling parameter, changes in whose values sweep us back and forth along a one-dimensional family of distributions, in loosely geometric terms, along a *pencil* of distributions. In our general problem, which has been reduced and analogized to certain pure scaling problems, we must consider not one pencil, but many. We face a *nearly* normal distribution, but we do not know its precise shape in detail. Yet, with location fixed, each shape corresponds to a unique pencil, different pencils corresponding to different shapes. One of these pencils, that consisting of normal distributions, plays a unique and leading role. All methods of estimation to be considered must produce satisfactory results for the normal pencil.

But in reality we face not the normal pencil but an alternative pencil. The behavior of each method of estimation which is seriously considered must be studied on each of many pencils. This may be done one pencil at a time, but it will eventually be necessary to combine results across pencils.

The basic problem is to study the behavior *on the alternative pencil* of methods of estimation which are *correctly calibrated on the standard reference pencil* (in our instances the normal one).

5. Choice of Alternatives

Our aim is to study the behavior of estimators of location and scale, especially those optimized for normality, for a wider range of specifications. It would be possible to try to study their behavior in very varied circumstances, perhaps, even, for translations or scalings of an arbitrary distribution with a continuous density function. For the present this does not seem analytically manageable. Moreover, since gross non-normality will be detected, practical concern must be focussed on the effects of indetectable or

THE FIRST ANSWER

The relative efficiency of the mean deviation will increase.

THE SECOND QUESTION

Will the relative efficiency ever reach 100%; in other words, will the mean deviation be as good a measure of scale as the standard deviation for any of the contaminated populations obtained by mixing two normal populations which have the same mean but whose standard deviations are in the ratio 3: 1?

NOTES: 1. The possible answers are "never reach," "just reach," "reach and go beyond."
2. The problem is a large-sample problem.
3. To find the answer, after giving the question long and careful thought, turn over two pages.

barely detectable non-normality. We shall learn that such effects may be large.

Thus it is natural to consider as simple a family of pencils as possible, selected to possess non-normality with large consequences and poor detectability. Convenience of computation is rather well served by confining attention to some family of compound normal distributions with two constituents, that is, to certain mixtures of two different normal distributions, whose two constituents can differ in scale, or in location, or in both.

Various special choices of family can be made. Different choices by different statisticians helped avoid agreement about the original problem. The writer, basing his choice partly upon empirical experience with largish "samples" of range-finder fluctuations and partly upon the great sensitivity of many procedures to "dirt in the tails," favored reference to "contaminated distributions" where the two constituents have equal means and different variances. R. D. Bennett tended to choose the case where the constituents had equal variances and unequal means. Since the qualitative nature of the results is opposite for these two cases, this choice is of considerable importance. Sections 7 and 15 will discuss this point further.

For the present, the prototype distributions will be *contaminated distributions* of contamination γ, at scale ratio h, where a fraction, γ, of the wider normal distribution, whose scale is h times that of the narrower normal

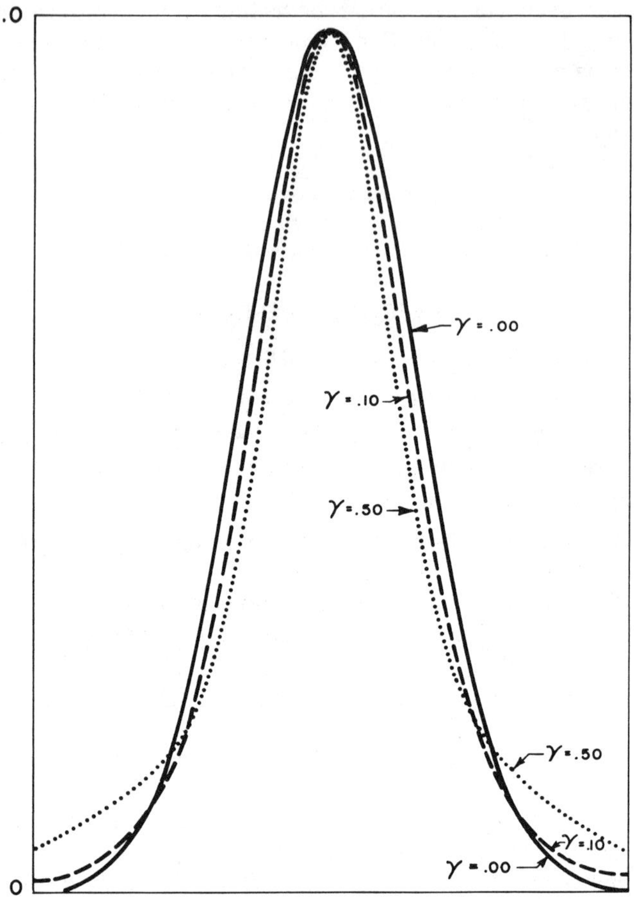

FIG. 1. Probability Densities of Normal and Contaminated Distributions (Central Densities Equated)

distribution, is combined with a fraction, $1-\gamma$, of the narrow one. The corresponding prototype cumulative is

$$N_{\gamma,h}(z) = (1-\gamma)\int_{-\infty}^{z} \frac{1}{\sqrt{2\pi}} e^{-u^2/2}\, du + \gamma \int_{-\infty}^{z} \frac{1}{h\sqrt{2\pi}} e^{-u^2/2h^2}\, du ,$$

corresponding to a probability density

$$n_{\gamma,h}(z)\, dz = (1-\gamma)\frac{1}{\sqrt{2\pi}} e^{-z^2/2}\, dz + \gamma \frac{1}{h\sqrt{2\pi}} e^{-z^2/2h^2}\, dz .$$

If $\gamma = 0$ or $\gamma = 1$, the corresponding pencil consists of normal distributions. Mainly because of experience with distributions of range-finder fluctuations, the value $h = 3$ was used throughout the investigations to be summarized

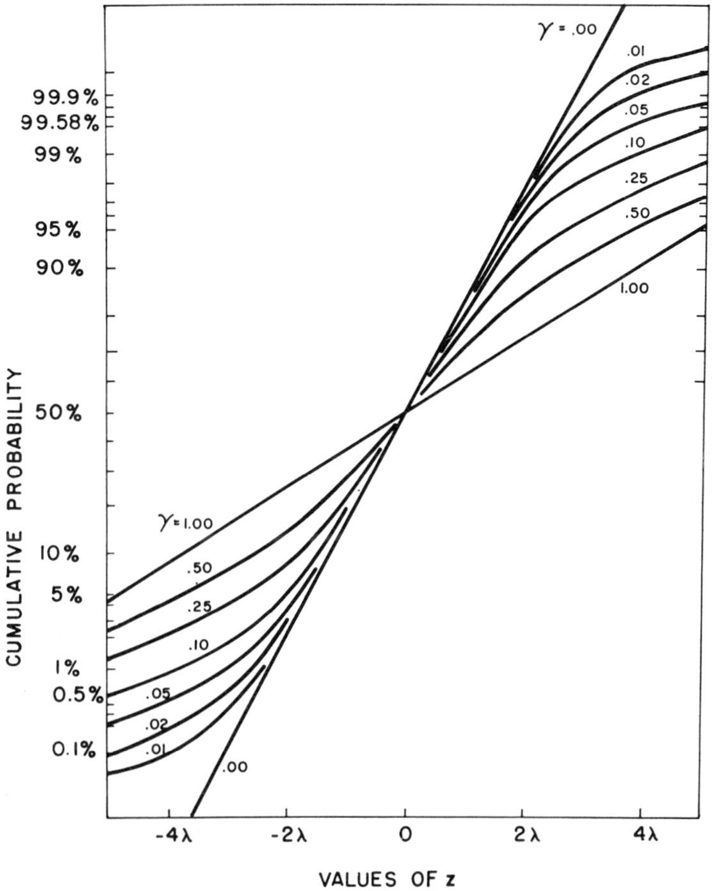

FIG. 2. Cumulatives of Normal and Contaminated Distributions

here. Thus the contaminated pencil will take the form $N_{\gamma,3}(z - \mu)$ in location problems and $N_{\mu,3}(ze^{-\tau})$ in scaling problems.

6. The Natural History of Contaminated Distributions

Before going on to consider the behavior of estimates based on samples from such contaminated distributions, it is well to learn something of the nature of the distributions themselves. This is probably best done graphically. Figures 1 and 2 show probability density functions (on a linear scale) and cumulatives (on probability paper), respectively, for a selected set of values of γ. It is quite clear from Fig. 1 that simple frequency distributions, even based on extremely large samples, are very unlikely to reveal the difference between $\gamma = .00$ and $\gamma = .10$ to the eye.

Throughout the discussion, just as in these figures, there will be continued emphasis on values of γ between .00 and .10. There are two reasons for

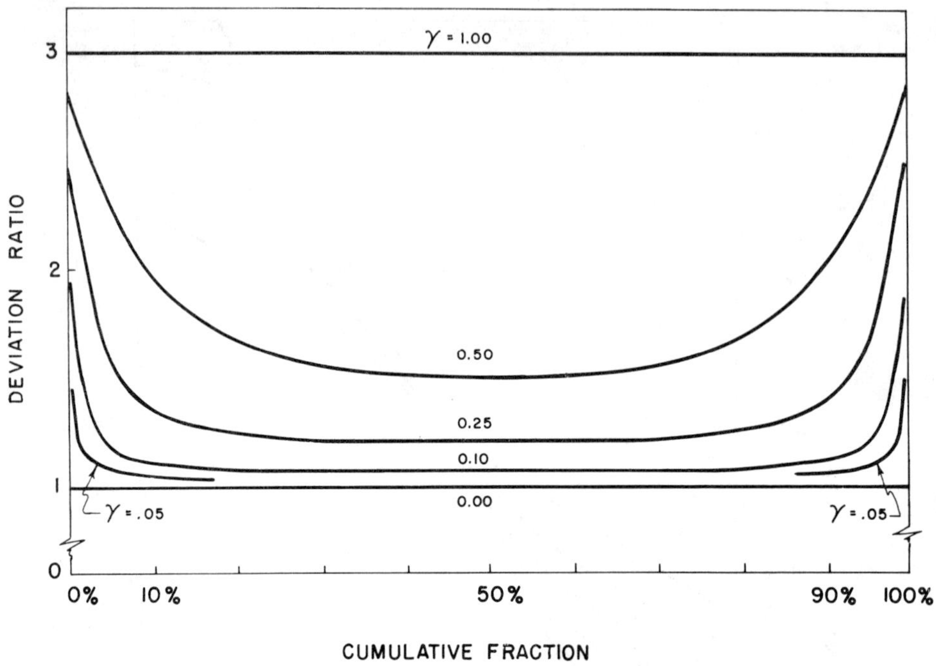

FIG. 3. Deviation Ratios for Normal and Contaminated Distributions

this. On the one hand, we are concerned with deviations from normality which are not likely to be striking. And, on the other hand, many of the effects of contamination have their more striking effects near $\gamma = .10$. R. A. Fisher (now Sir Ronald), who pointed out that the two constituents contribute *equal* amounts to the variance of the contaminated distribution when $\gamma = .10$, was the only statistician queried by the author to anticipate the relatively large effects of small γ. (In those figures where γ's larger than .10 appear, the scale of γ is linear in the proportions of the total variance contributed by the two constituents.)

Given a sample of several thousand from some distribution, how is it reasonable to inquire what value of γ might provide a contaminated distribution with "similar" behavior so far as tails are concerned? If such sample sizes seem unrealistic, consider a sample of 1000 from a population with $\gamma = .01$, which we shall learn is already large enough to produce behavior quite different from that for $\gamma = .00$ (i.e., that for normality). The average number of observations which come from the broader, rarer constituent will be only ten. Reference to Fig. 2 shows that the cumulatives do not deviate from one another by relevant amounts between -2.5λ and $+2.5\lambda$. Only some 40 percent of the observations from the wider component will fall outside these bounds, and we may thus expect only two observations from the upper tail of the broader, rarer constituent, and another two from the lower tail.

Beyond the same limits we will expect about six (in each tail) from the narrower constituent. Unless one or both of the two are very extreme, the indication of non-normality will be very slight. And if one or both are very extreme, we are likely to think of them as "strays," "wild shots," "errors in technique," or "errors in processing" and to focus our attention

THE SECOND ANSWER

For some contaminated populations, the mean deviation will be a better estimate of scale, in large samples, than the standard deviation.

THE THIRD QUESTION

What fraction of the wider normal population must be added to the narrower one in order for the mean deviation to be as good a large-sample measure of scale as the standard deviation?

NOTES: 1. The possible answers are fractions between .000 and 1.000.
2. To find the answer, after giving the question long and careful thought, turn over two pages.

on how normally distributed the rest of the distribution appears to be. (One who does this commits two oversights, forgetting Winsor's principle that "all distributions are normal in the middle," and forgetting that the distribution relevant to statistical practice is that of the values actually provided and not that of the values which ought to have been provided.) A sample of *one* thousand is likely to be of little help.

A convenient and effective way to display the behavior of the sample of several thousand (or of a theoretical distribution) is to plot the deviation ratio (fractile = % point):

$$\frac{\text{(deviation of fractile from mean)}}{\text{(deviation of same fractile from mean for normality)}}$$

against the cumulative fraction. For normality, the plot is a horizontal line. For contaminated distributions, the ends are raised. Figure 3 shows the results for several values of γ. Given a sufficiently large sample, this type of plot is quite revealing.

Finally, how can we conceive of contaminated distributions, or close approximation to contaminated distributions, arising in practice? Surveyors have long carefully distinguished between "errors," which afflict all their measurement, and "blunders," which affect only a small fraction and are usually so large as to be findable by checking procedures. Although surveyors' blunders are (i) not normally distributed, and (ii) more than three times as large as surveyors' errors, this example suggests a model in which some source of fluctuation affects only a fraction, γ, of the observations. If this intermittent source supplies eight times the total variance supplied by all other sources, and all sources provide normally distributed contributions, the resulting distribution will be contaminated by fraction γ at scale ratio 3.

Other more general situations, where the scale of variability is heterogeneous, but where variation at each separate scale is normal, will produce distributions too long in the extreme tails for normality, distributions which clearly resemble contaminated distributions.

7. Long Tails or Short?

It is easy to think of many situations which produce distributions with longer tails than a normal distribution. (Heterogeneity of variance and causes acting only a small part of the time are, of course, the main possibilities.) But it is much more difficult to think of realistic situations which lead to tails shorter than a normal distribution. Sets of observations which have been de-tailed by over-vigorous use of a rule for rejecting outliers are inappropriate, since they are not samples. Quantitative characteristics of manufactured articles that have been sorted by gaging or measurement will tend to have short-tailed distributions, but, when the sorting is at fixed cut-offs, the mean of the sorted populations is not a location parameter, and their standard deviation is not a scale parameter.

A contribution taking two values, each with substantial probability, can shorten tails somewhat, as when it creates a mixture of two normal distributions with the same variance and different means. But, to be effective in tail-shortening, it must meet many requirements. It must be a large contribution, and it must be almost alone of its kind, for the sum of many such will converge to normality. Its values can only rarely be widely different from the two particular values, since intermediate values with substantial probability will reduce its effect greatly. (And a few unusually extreme contributions will erase its effect entirely.) The most tail-shortening it can contribute corresponds to $\gamma_2 = -2$ or $\beta_2 = 1$ (Fisher's and Pearson's measures of normality, respectively), while tail-lengthening contributions can provide much larger positive γ_2's.

It is not suprising that, in practice, long tails seem more frequent than short. As Box and Andersen say ([17], p. 2), "Published data are comparatively meagre but the frequency distributions given in the older issues of *Biometrika* give little ground for supposing that distributions usually follow the normal law." They go on to cite certain examples where large samples reveal the long-tailedness of populations. To their list should certainly be added Student's paper on the errors of routine analyses ([36], also in [37]), which provides much evidence, based on very many small samples, on the long-tailedness of experimental measurement. All in all, the experimental evidence seems to lead us to beware long tails as strongly as does our theoretical insight.

8. Asymptotic Techniques

By definition, large-sample results are asymptotic results. Asymptotic results have always tended to be either simple and heuristic or complex and rigorous.

The first step in attacking the problem of sampling from contaminated distributions was to set up a system of asymptotic techniques whose use could be both simple and rigorous [38], [39], [40]. Definitions and results are set forth in Appendix A. Four points deserve remark here:

(i) Both the asymptotic average value of $y\{Z_n\} = \vec{\text{ave}}\, g(Z_n)$ and the asymptotic variance of $g(Z_n) = \vec{\text{var}}\, y(Z_n)$ will exist for reasonable $g(z)$ whenever the distributions of the Z_n's converge in a reasonable way to a single point, as is almost always the case when the Z_n's are reasonable bases of estimate.

(ii) $\vec{\text{Ave}}\, Z_n$ and $\vec{\text{var}}\, Z_n$ are deliberately not defined exactly, in the sense that terms which go to zero faster than $\vec{\text{var}}\, Z_n$ itself may be freely added and subtracted.

(iii) Only the distributions of the individual Z_n appear in the definitions.

(iv) The so-called "Δ-process," or "method of propagation of error," is usually valid in terms of these asymptotic techniques.

In this connection it is interesting to note that Harold Hotelling once wrote ([41], p. 36): "Another example of nonrigorous mathematics used extensively in statistics is the whole business of asymptotic standard errors found by the differential method. It is desirable that good mathematics replace bad in such connections."

In the sections that follow, the word "asymptotic" will have the meaning that is defined in Appendix A.

9. The Location Problem

Since contaminated distributions studied here are symmetric, most estimates of location are unbiased for all amounts of contamination. Thus only variability need be considered, and the asymptotic variances of the various estimates will be used to indicate preferable choices.

If both the amount of contamination, γ, and the scale, λ, of the underlying distribution were known, one could be both simple and "efficient" (asymptotically) by using maximum likelihood ideas, either directly or in terms of Fisher's score (the logarithmic likelihood derivative). It is natural to ask how well the score appropriate for one fraction of contamination, γ', (and the correct value of λ) will do if it is used on observations drawn from a distribution with another fraction, γ, of contamination. Calculation shows ([6], Table 6) that, if we assume $\gamma' = .00$, we may suffer rather low efficiency (90% for $\gamma = .02^+$, 80% for $\gamma = .05$, 70% for $\gamma = .10$), but if we assume γ' between .01 and .10, we shall obtain at least 96% efficiency over the whole range $.00 \leqq \gamma \leqq .10$. Thus there is more to gain than to lose by assuming a moderate degree of contamination.

This result is not yet a practical answer for at least two reasons. The

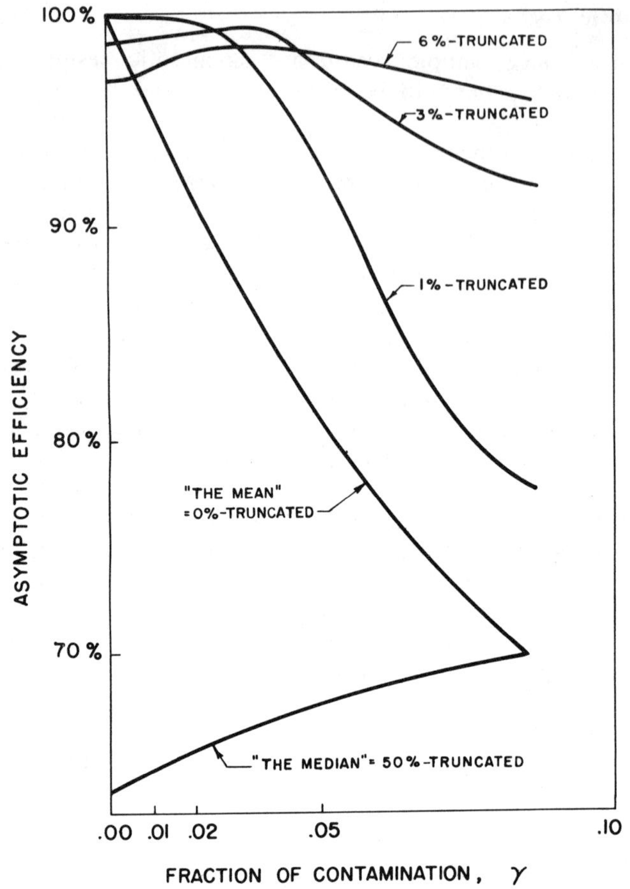

FIG. 4. Asymptotic Efficiency, for Location, of Truncated Mean

estimator studied was specifically constructed in terms of a specific contaminated distribution, so that we may rightfully distrust its efficiency for other types of deviation from non-normality. And it makes use of knowledge (except when $\gamma = .00$) of the value of λ which we would not have in practice.

Since our concern is with large samples and asymptotic results, the latter difficulty can be easily gotten over. Maximum likelihood technique can be replaced by the use of a weighted linear combination of order statistics, where the weights are chosen to approximate the derivative of the score, evaluated at the corresponding order-statistic expectations. Asymptotically, the corresponding estimating techniques will be equivalent (cf. [6] and Jung [42]).

When the resulting weighting functions were examined, it appeared that they could be rather well approximated by appropriate truncated means. A *p%-truncated mean* (sometimes called a *broadened median*) is the mean of the observations remaining when the *p%* highest and the *p%* lowest have been set aside. The truncated means form a natural bridge between two classical estimates of location, since the 0%- and 50%-truncated means are the mean and the median, respectively.

THE THIRD ANSWER

When just less than .008 of the mixed population comes from the wider normal population, the mean deviation has the same large-sample precision as the measure of scale that the standard deviation has.

NOTE: Do not judge an answer which deviates widely from .008 harshly. Many distinguished and experienced statisticians have given answers between .150 and .250.

THE FOURTH AND FIFTH QUESTIONS

Can I expect to know whether or not the populations from which I draw large samples deviate from normality as much as, say, a contaminated population containing .008, .02, or .05 of the wider normal?

If I cannot, what should I do about estimating the scale of an actual population from a large sample?

NOTES: 1. To obtain insight into whether or not you may know, examine Fig. 1 and read Section 5.
2. To obtain insight into what to do, read Sections 11 to 13.

Figure 4 displays the asymptotic efficiencies of the 0%-, 1%-, 3%-, 6%-, and 50%-truncated means. Clearly, a small amount of truncation brings great reward for small loss. If 1% of the sample is deleted from each end, for example, the greatest loss is 0.4% and the gain may exceed 12%. If there is serious fear of contamination, the use of the 6%-truncated mean, with a maximum loss of 3% and a maximum gain of 26%, seems entirely reasonable.

Although the changes in efficiency here are small in comparison with those which we will face in considering scaling, they are not negligible, as they rise for substantial contamination to the level of the difference in efficiency, on normal theory, between the mean and the median. [See Jeffreys ([43], Sec. 5.7, pp. 287 ff.) for another approach to the location problem for non-normal samples.]

Investigation of the terms in $1/n^2$ indicates that, in samples as small as 100, the asymptotic formulas probably overstate variances by only a few percentage points [7].

10. Problems of Scaling Type: Bases of Scaling

In Section 4, λ was called *a* scaling parameter. Why the caution? Because once alternative pencils must be dealt with, there are many distinct numerical functions of distributions which can rightly be called scaling parameters. Consider, for example, the scaling of pencils consisting of distributions centered at the origin with finite second moments, and, in particular, consider the pencil of centered normal distributions and the pencil of centered logistic distributions.

The parameters α and β, defined by

$$\alpha^2 \doteq \text{ave}\, z^2 \,,$$

$$\Pr\{z \geqq 3.090\,\beta\} = 0.001 \,,$$

are both scaling parameters. Along any one pencil they enlarge and contract in strict proportion to the enlargement and contraction of the distribution. For the normal pencil their values agree. For the logistic pencil their values are in a constant ratio of 1.58. For any one pencil for which they are both finite, their values are in a constant ratio.

So long as any one pencil is considered, the estimation of α and the estimation of β are equivalent problems. Once two or more pencils are considered, the problems are no longer equivalent. For more than one pencil, there is no unique scaling problem; instead there are very many scaling problems.

The well-known fact that location seems to be simpler than scaling is reflected in the simplicity gained by adopting a logarithmic measure of scale. We shall write $\lambda = e^{\tau}$, and consider our contaminated pencils to be specified in the form $N_{3,\gamma}(ze^{-\tau})$. It is now simple to construct a variety of estimators of τ whose bias is independent of the actual value of τ. For instance:

(i) From the *pth degree absolute* (sample) *moment* $(\sum |z_i|^p)/n$ we can form

$$\frac{1}{p}\log\frac{1}{n}\sum|z_i^p| \,.$$

(ii) The *p%-interfractile distance* is the distance from the observation $p\%$ from the bottom of the sample to the observation $p\%$ from the top of the sample. Its logarithm has constant bias.

(iii) The *p%-truncated variance*, whose square root is the *p%-truncated standard deviation*, is the variance of those observations which remain when the $p\%$ highest and $p\%$ lowest have been set aside. The logarithm of the truncated standard deviation has constant bias.

In particular, we shall also wish to estimate the population average of e^{-cz^2}, which will be called a Gaussian average. The only *completely unbiased* (that is, unbiased for every possible distribution) estimate of such an average is the sample mean of the same expression. Hence, we consider the Gaussian means $(\sum e^{-cz_i^2})/n$ as a base of estimate. We can find a function of a given Gaussian mean which is asymptotically unbiased along the normal pencil. However, its asymptotic bias along a contaminated pencil will not be constant (although an asymptotically equivalent base of estimate does have constant bias; see Section B.3 of Appendix B). Thus, even after changing to the logarithmic scaling parameter it may be sometimes necessary or desirable to deal with varying bias.

It will be helpful to notice (i) that it is often useful to write $e^{-b(z/\lambda)^2}$ for e^{-cz^2}, thus replacing c by the non-dimensional $b = c\lambda^2$, (ii) that as $b \to 0$, Gaussian means, considered as bases of estimates, behave more and more like the second moment (like the variance), (iii) that as $b \to \infty$, Gaussian means, considered as bases of estimate, behave more and more like $p\%$-interfractile distances for $p\% \to 50\%$. (In the last instance, asymptotic variances approach ∞.)

11. Size of Sample

A statistical problem is usually called a large-sample problem when $1/n$ is so small that everything else can be neglected. If this is not the case, some will automatically call it a small-sample problem. Others will reserve the latter term for problems which (a) can be solved explicitly, or (b) are treated by experimental sampling. A more refined classification is needed in the present situation, where, by either definition, there is a small-sample problem. Many derived distributions can be written down in closed, but apparently impractically complicated, forms. Many specific problems can be effectively attacked for specific small-sample sizes by experimental sampling. To date, little if anything has been done in this direction. All the work reported here concerns large sample sizes.

But it will often be essential to recognize different "largenesses" of samples, since the relative importance of variance and bias will vary. Typically, asymptotic variability falls off with $1/n$, while asymptotic bias tends to a non-zero limit. For *very* large samples, then, bias outweighs variance, and the choice of estimator is primarily made to match bias, and only secondarily to reduce variance.

But for slightly large samples, it may well be that bias is unimportant in comparison with variability *and yet* asymptotic results are precise enough to be useful. A similar reliance on variability alone arises when the aim is to *compare* the scales of two or more distributions where, perhaps because of the similarity of their sources, it is reasonable to assume that all the distributions come from a single pencil (e.g., have the same value of γ). No matter how large the samples may be, they then act as if they were slightly large.

So long as estimators are considered whose bias (referred, say, to λ) is constant along a pencil, to consider "variability alone" means to consider variance. But, if bias varies along the pencil, it is necessary to make allowance for this fact, just as Mandel and Stiehler [44] have done in defining "sensitivity" and just as is exemplified by the more flexible forms of the Cramér-Rao inequality (e.g., [45], Sec. 32.3). The "effective variance" takes the form

$$\text{effective variance} = \frac{(\text{variance})}{\left[1 + \dfrac{d}{d\tau}(\text{bias})\right]^2},$$

so that the *asymptotic effective variance*, which is the appropriate measure, for this problem in our situation, becomes

$$\text{asymptotic effective variance} = \frac{\text{asymptotic variance}}{\left[1 + \frac{d}{d\tau}(\text{asymptotic bias})\right]^2} .$$

And, finally, for *moderately* large samples, it may be best to accept a carefully chosen amount of bias, in order to decrease variance enough to make a net reduction in some combined measure, such as the asymptotic average square deviation. (This latter case will be discussed only in Appendix B.)

12. Variability of Scaling

The bases of estimate considered in [8] were the following:

(i) pth-degree absolute moments for $0.2 \leqq p \leqq 5$,

(ii) Gaussian means for $0 \leqq c \leqq \infty$,

(iii) $p\%$-interfractile distances for $0.1\% \leqq p\% < 50\%$,

(iv) Selected truncated variances (2% and 5%).

Figure 5 shows the behavior of the product of asymptotic effective variance and sample size for $.00 \leqq \gamma \leqq .50$ and selected examples of each type. Clearly the second moment, which corresponds to the standard deviation, is the least safe of all. Its use can be recommended only when γ is far less than .01, and we are rarely sure of this. More detail is provided by Fig. 6, which shows, for the range $.00 \leqq \gamma \leqq .10$ and some of the better-behaved estimates, the percentage of excess of the effective variance of estimate over the Cramér-Rao lower bound ([45], Sec. 32.5). Probably the most promising of the alternatives shown, if substantial γ is to be feared, are (i) the mean of $e^{-z^2/4}$, and, if γ is quite likely to be $\leqq .07$, say, (ii) the 2%-truncated variance. In general, it would appear that a suitable truncated variance would do better over a narrow range of γ, while a suitable Gaussian mean would do better over a broader region. (It is almost certain that a procedure of varying truncation, which sets aside varying percentage of the observations from different samples, would do even better. Reasonable procedures do not seem easy to write down. The study of their asymptotic variability seems likely to be much more difficult.)

It is hard to imagine a situation where contamination would appear, and yet appear in such small amounts as to make the standard deviation either as good as the mean deviation or nearly as good as the 2%-truncated standard deviation, at least insofar as variability of scaling goes.

Because of practical questions of computing, we *may* find averaging e^{-cz^2} uncomfortable. If so, then the most reasonable solutions for this problem are:

1. The truncated variance (and its square root, the truncated standard deviation) with 2% to 5% of the observations deleted from each tail,
2. The mean deviation.

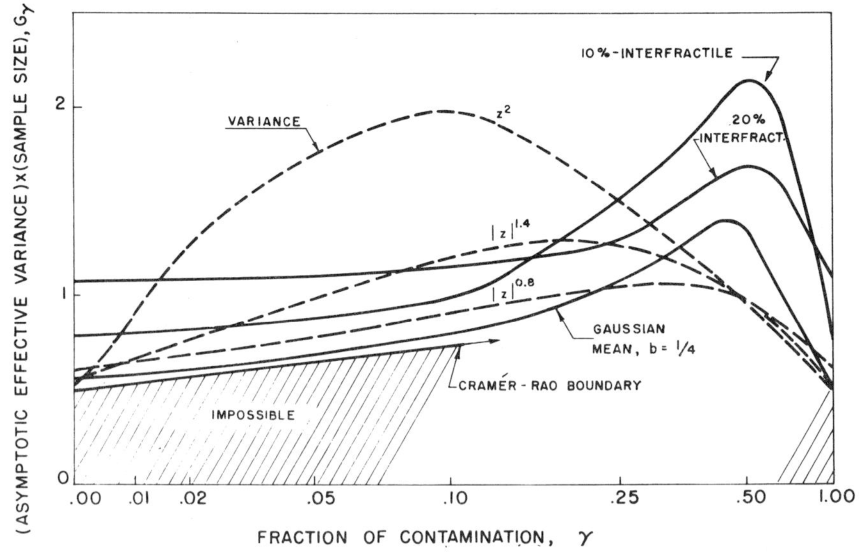

FIG. 5. Variability When Using Various Estimators in Scaling Slightly Large Samples

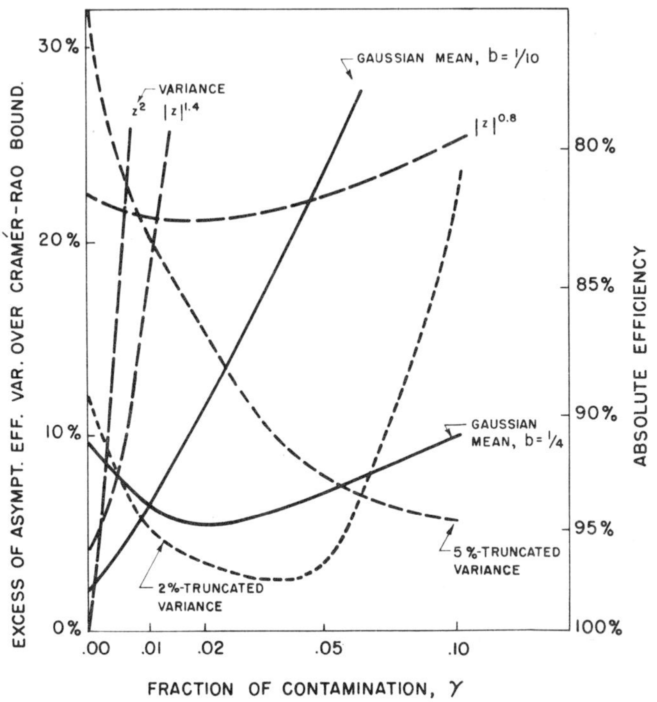

FIG. 6. Details of Variability When Using Various Estimators in Scaling Slightly Large Samples

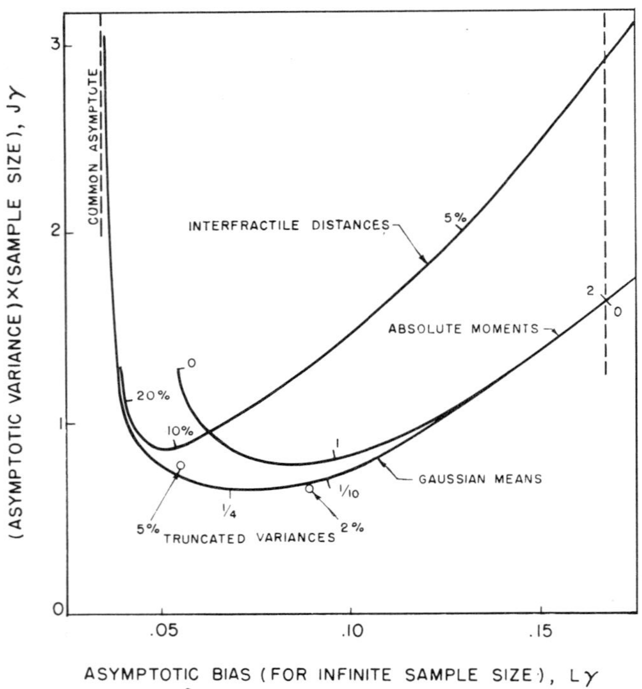

FIG. 7. Relation of Asymptotic Variance to Asymptotic Bias When Scaling Contaminated Distributions with $\gamma = .05$

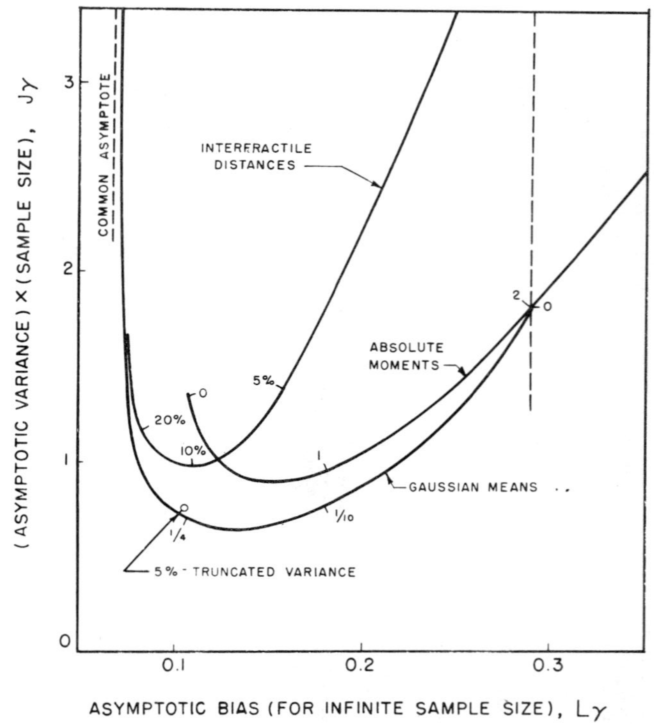

FIG. 8. Relation of Asymptotic Variance to Asymptotic Bias When Scaling Contaminated Distributions with $\gamma = .10$

The latter will surely be inferior to the former unless the distribution is quite heavily contaminated.

Since the variance (and its square root, the standard deviation) cannot be more than 11% better than the 2%-truncated variance, while it can be 140% worse (for $\gamma = .05$), the variance will never be a safe choice.

13. Very Large Samples

In very large samples, the matching of bias of estimate to bias of estimand, as judged from any standard of "zero bias," is of prime importance, and the reduction of variability is secondary. This thought is familiar in classical estimation, but not of great importance, since biases can usually be matched by adding a constant to the estimate, which reduces the problem to one of minimizing variance. The situation here is different, because the biases that concern us are due to the degree of contamination which we face in a particular instance. This amount is unknown, and even a very large sample may not suffice to estimate it closely enough to make adequate correction possible. Thus we may really have to match biases by choice of estimator and take what variances we can have for given biases.

We look, then, to plots of asymptotic variance, J_γ, against asymptotic bias, L_γ, to see what type of estimator gives the desired biases with least variance. Figures 7 and 8 do this for $\gamma = 5\%$ and $\gamma = 10\%$. In each case, the bias corresponding to the use of the variance, which is represented by the point where the curve of Gaussian means joins the curve of absolute moments, is a crucial dividing value.

To the left of this line, Gaussian means do better than both absolute-moment and interfractile distances. In this same region, truncated variances do almost as well. Interfractile distances for $p\% > 7\%$ do better than the corresponding moments, if any, while for $p\% < 7\%$ they do worse.

To the right of this line, a region occupied only by absolute moments of degree $\geqq 2$, and by extreme interfractile distances, the moments do better for fixed bias. This is, of course, a high-variability region, one to be avoided if possible.

Optimum procedure would seem to call for:

1. Use of a Gaussian mean whenever one exists with matching bias, or possibly the use of a truncated variance,
2. In other circumstances, use of an absolute moment with matching.

Whether or not these recommendations are realistic depends on whether or not one can match biases over a range of values of γ. Figures 9, 10, and 11 give the b-value of the Gaussian mean, or the degree of the absolute moment, required for matching at $\gamma = .05$ and .10. The match from Gaussian means to absolute moments (Fig. 9) shows remarkably close agreement. The matches from Gaussian means (Fig. 10) or absolute moments (Fig. 11) to interfractile distances are not so precise, but are quite close enough to make compromise quite effective.

14. The Reduced Form of the Original Problem

Location and scaling have been discussed in some generality, but what of

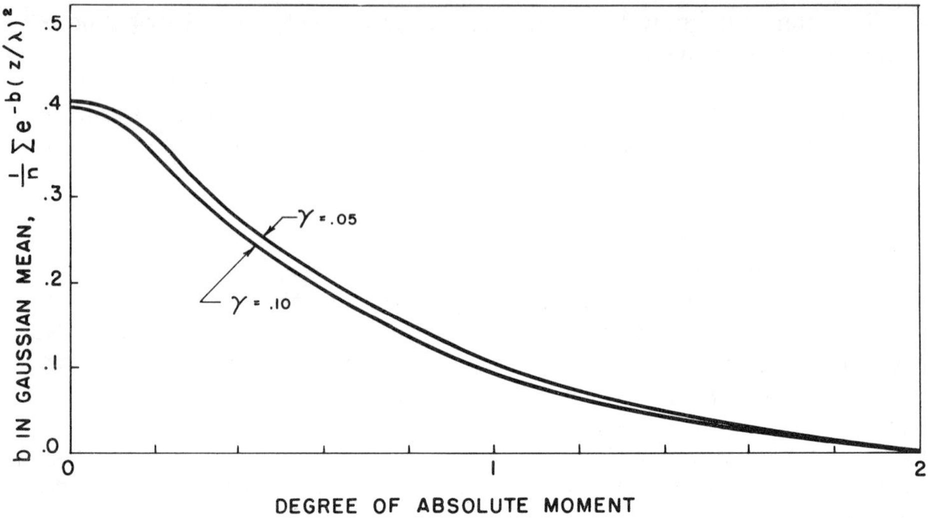

FIG. 9. Gaussian Means Matched in Bias to Moderate Absolute Moments

the special case which generated the study? Suppose that $b = 1$, so that the average value of $e^{-(z/\lambda)^2}$ is to be assessed from a sample which was drawn from a population known to be centered at zero.

The principal competitors to be considered are the mean deviation and the sample mean of $e^{-(z/\lambda)^2}$ itself. The resulting variances and biases are shown in Figs. 12 and 13. In slightly large samples, where variance dominates bias, the mean deviation is slightly preferred to the sample mean of $e^{-(z/\lambda)^2}$. Both are far better, if contamination is at all likely, than the sample variance (or standard deviation). In larger samples, where bias plays a controlling role, the sample mean of $e^{-(z/\lambda)^2}$ far outdoes its competitors.

The specific example treated was the average of the $e^{-b(z/\lambda)^2}$ for $b = 1$. In the original application, larger values of b would be more likely to occur than smaller values. If the same three competitors were considered for the estimation of such an average with $b > 1$, the relative asymptotic variances would be the same, and the relative asymptotic biases would increase in the same rank order as the worst asymptotic variances. Clearly, no one should dare to use the variance as a base of estimate, even though it is a sufficient statistic for precise normality.

15. How Far Have We Advanced?

How much better off are we after considering contaminated distributions than when we considered only the normal case? For a unique normal pencil, we have substituted a one-parameter family of contaminated pencils. This is still a very thin representation of all reasonable pencils, of all reasonable

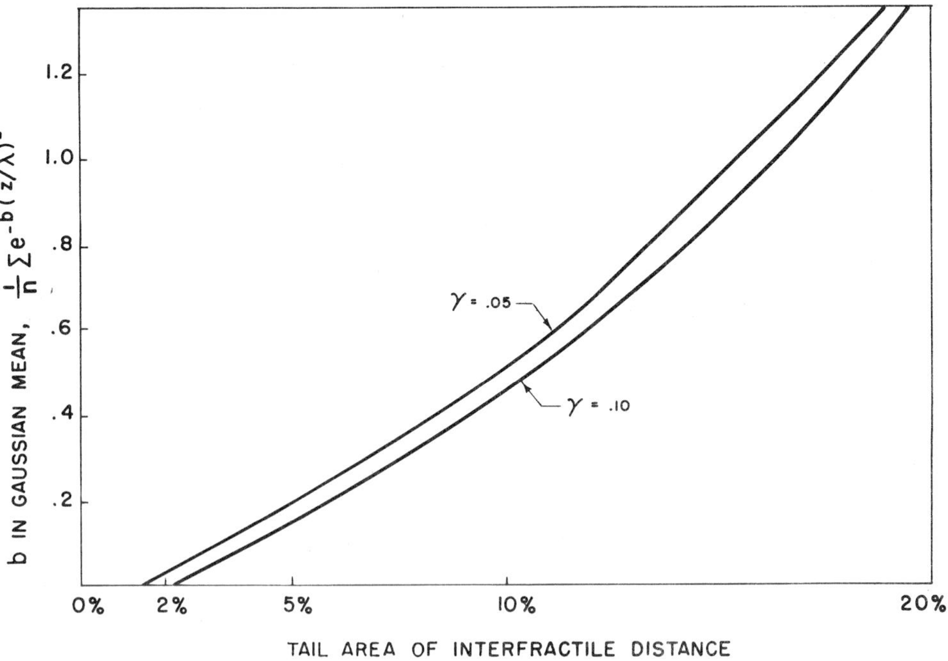

FIG. 10. Gaussian Means Matched in Bias to Moderate Interfractile Distances

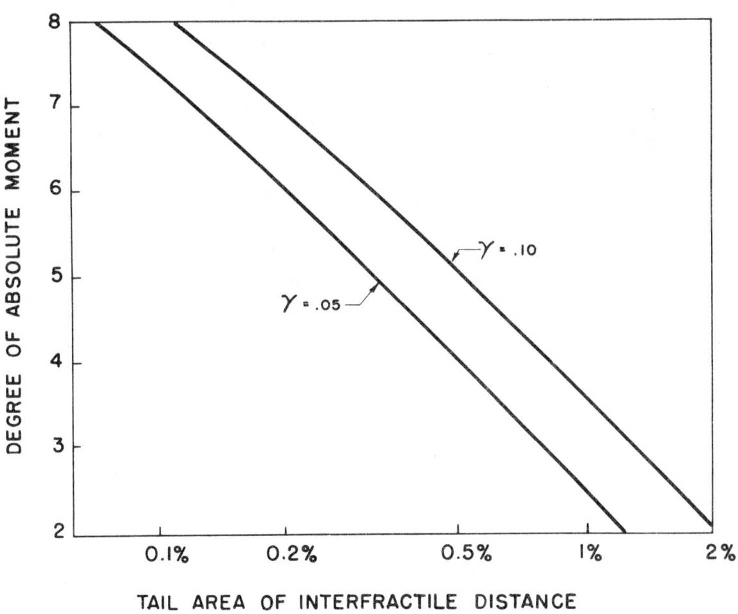

FIG. 11. Absolute Moments Matched in Bias to Extreme Interfractile Distances

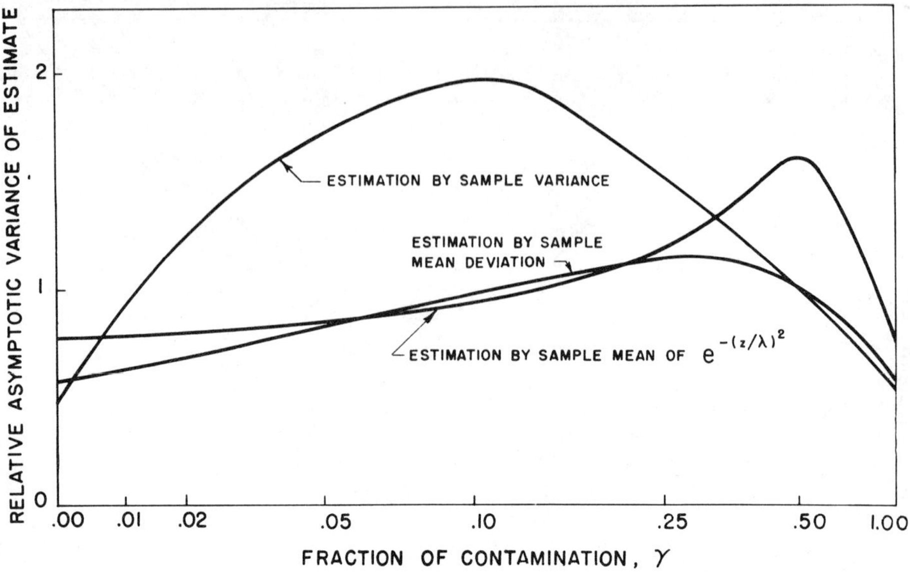

FIG. 12. Relative Variance of Estimation, in Large Samples, of the Population Average of $e^{-(z/\lambda)^2}$

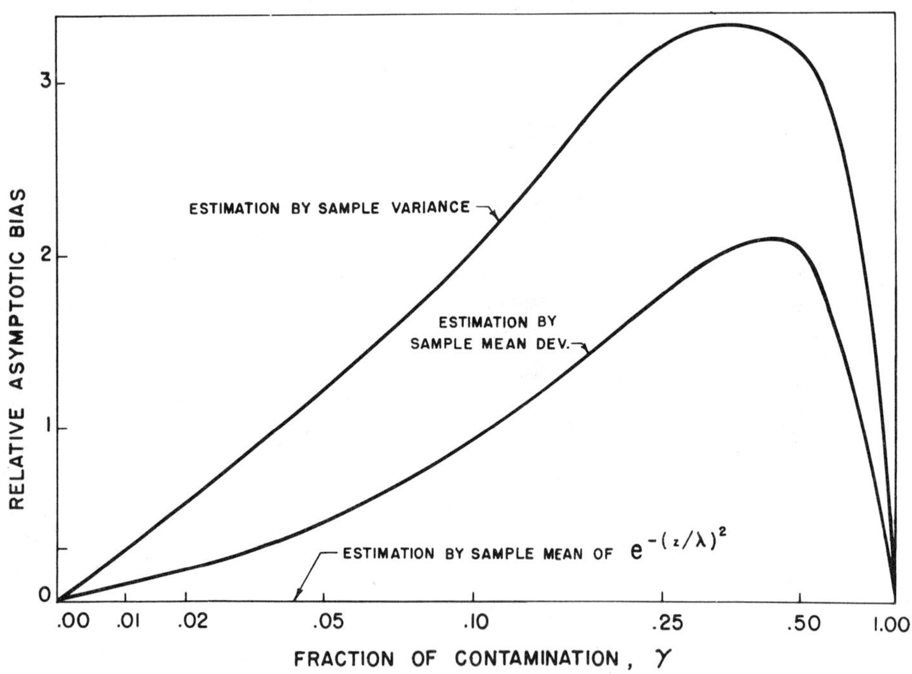

FIG. 13. Relative Bias of Estimation, in Large Samples, of the Population Average of $e^{-(z/\lambda)^2}$

distributional shapes. In particular, the mixtures of two normal distributions with equal variances and unequal means proposed by R. D. Bennett have not been considered.

Solid answers to such questions must be founded on a general philosophy. In balancing behavior for one value of γ against that for another, we have already used a rather clear, though implicit, philosophy. Its major foundation seems to be a belief that it is *not* wise to seek a small percentage reduction in variability under one set of circumstances at the price of incurring a large percentage increase in variability under another set of circumstances. If we are to be reasonable, and not necessarily formal, we must admit the desirability of using judgment to determine which sets of circumstances should be considered, and with what weight they should be considered. But we still feel that it is unwise to recommend a procedure which sometimes gains a little and sometimes loses a lot.

Within the framework of such an approach, the study of non-normality ought to begin by examining situations which display great sensitivity, where deviations from normality which are difficult to detect have large effects on efficiency. Examining contaminated distributions does just this. If there be other types of deviation from normality which produce effects of a different nature, but of a comparable sensitivity, they deserve urgent attention. But there is reason to doubt their existence, at least so long as we deal with smooth distributions.

Under slippage and expansion-contraction—in problems of location and scale—the tails of a distribution play a peculiarly important role. For normal pencils, the tails, as conventionally defined by the two-sided 5% points, contribute 28% to the information integral for location and 60% to the information integral for scaling. No other 5% contributes nearly as much. Shortening the tails of the distribution, as by considering mixtures of two normal distributions with the same variance and different means, will also have a noticeable effect on efficiencies. Placing high weight on extreme deviations will be rewarded even more than for normality. Exploration of this situation is clearly desirable and worthwhile; but it would be easy to overvalue the practical importance of its results, not for reasons of general theory or philosophy, but for reasons related to the real world, to the populations and distributions which actually do arise, reasons which were briefly summarized in Section 7. All available evidence and discussion leads the writer to believe that, although contaminated distributions are a thin representation of non-normality, they are an exploration of the most important area of non-normality. Consequently, he judges the step from normality alone to contaminated distributions to be a large part of the step from normality alone to all smooth, nearly normal distributions.

16. The Mean Deviation and the Standard Deviation

Fisher's second major paper in pure statistics was "A Mathematical Examination of the Methods of Determining the Accuracy of an Observation by the Mean Error and by the Mean Square Error" [46] (also in [47]). This paper, as Fisher says ([47] p. 2.757a), arose from an examination of a statement by Eddington. In a footnote to page 762 of Fisher's paper, Eddington acknowledges the need for Fisher's correction adding, however, "I think it accords with the general experiences of astronomers that, for the errors commonly occurring in practice, the mean error is a safer criterion of accuracy than the mean square error ...." The present investigation seems to support Eddington's conclusions, rather than Fisher's. We shall shortly need to ask why.

In his paper Fisher showed clearly, for the case of *exact* normality of distribution, that the standard deviation gave a much more precise estimate of scale than did the mean deviation. Moreover, he treated their joint distribution and pointed out ([46], p. 769) that the standard deviation was, in modern language, sufficient. (This is probably the first appearance of this concept.) He went on to point out that, if the distribution resembles a double exponential distribution rather than a normal distribution, it did not seem unreasonable to employ the mean deviation as an estimator of scale. The problem of bias did not evade him, since he pointed out that the factor connecting the average sample mean deviation with the population standard deviation was 12% greater for double exponential distributions than for normal distributions.

What did Fisher not do that we have done? Why did he reach a conclusion almost opposite to ours? First, he assumed that the purpose of scaling was the estimation of the population standard deviation, and this alone. (The assumption may well be far more reasonable in the astronomical applications he contemplated than in the situation which started the investigation surveyed here.) Second, he did not investigate in what sense, and to what degree, distributions would have to resemble the double exponential in order to make the use of the mean deviation preferable.[2] It is, I believe, fair to say that his analysis was clearly not carried far enough but that it contained the essential bases for a mathematical justification of Eddington's statement.

In his 1922 paper, "On the Mathematical Foundations of Theoretical Statistics," ([49], p. 314), Fisher said: "As regards problems of specification, these are entirely a matter for the practical statistician, for those cases where the qualitative nature of the hypothetical population is known do not involve any problems of this type. In other cases we may know by experience what forms are likely to be suitable, and the adequacy of our choice may be tested *a posteriori*. We must confine ourselves to those forms which

[2] He did state that β_2 seemed well fitted to provide an indication of this, remarking that $\beta_2 = 3$ for the normal, but $\beta_2 = 6$ for the double exponential. For contaminated distributions with $\gamma = .02$ at scale ratio 3, the value of β_2 is 5.75. Thus Fisher's criterion would suggest using the mean deviation for γ greater than about .022, a value we realize that we would be unable to discriminate from $\gamma = .00$ with the aid of a random sample of one thousand observations. See Burrau [48] for the use of the mean deviation.

we know how to handle, or for which any tables which may have been necessary have been constructed. More or less elaborate forms will be suitable according to the volume of the data. Evidently these are considerations the nature of which may change greatly during the work of a single generation . . . ." How should these words be translated, now that more than a generation has passed? The *a posteriori* testing of the "goodness of our choice" is now seen to require much more than a simple test of goodness of fit. There is a responsibility to explore, empirically or mathematically, all specifications which would be found to fit adequately, if tested, and to determine to what extent our conclusions, both about the situation underlying the data *and* about the statistical techniques most likely to illuminate such a situation, depend on which of the adequately fitted specifications we select. We are not yet able to shoulder the whole of this responsibility, but we badly need to acknowledge its existence.

In 1922, the theoretical specification of a population was almost always made in terms of a Pearson curve. Today, the Pearson curves serve us frequently as useful approximations to theoretical sampling distributions, and only rarely (except for the normal curve) as specifications. A wide variety of non-parametric techniques which use very broadly formulated specifications have been invented, investigated, and exploited. The time has come to study problems where the specification is neither so precise as a Pearson curve nor so diffuse as all distributions with a continuous density function. The precise formulation of such specifications as "the underlying distribution is nearly normal" in such a way as to be precise, manageable, and relevant to practice is now both important and overdue. It is time to return to the problem of specification.

17. Conclusions

What then are reasonable conclusions to draw from these investigations? Certainly these:

1. Problems of robustness of efficiency are probably as important as problems of robustness of validity, and, because of their relatively undeveloped state, deserve even more attention from statisticians.

2. Unless there is good reason to believe in its robustness, there is little if any sense in paying attention to an efficiency figure to greater precision than $\pm 10\%$ of itself.

3. In assessing the relative merits of competitive estimators in very large, or moderately large, samples, the ultimate purpose of the estimate is important, because biases due to apparently small mis-specifications of the problem (in our instances, changes in shape of distribution) cannot be neglected.

4. In large samples the sample mean is not nearly so safe an indicator of location as is the mean of the observations which remain after a small percentage of the highest, and an equal percentage of the lowest, have been set aside (use of a lightly truncated mean).

5. In slightly large samples, there is ground for doubt that the use of the variance (or the standard deviation) as a basis for estimates of scaling type is ever truly safe.

6. In moderately or very large samples, where matching of bias is important, the variance or standard deviation is safely used only to estimate one of the following (among all the estimands we have considered):

a. The variability (i) of a component, (ii) from a source, or (iii) of a value entering, as a datum, into a least-squares analysis,

b. Fractiles (% points) for tail areas between 1.5% and 2.5%.

Note: The tables of cell values entering an analysis of variance are rarely in any sense a large sample. When they are, the robustness of the associated significance and confidence procedures (cf. Pitman [15], Welch [16], and later developments) provides a good excuse for the use of means and variances in such analyses of cell values. If individual cell values themselves summarize large samples, then means or variances of these samples may well be replaced, as such summaries, by appropriately modified estimators.

7. Nearly imperceptible non-normalities may make conventional relative efficiencies of estimates of scale and location entirely useless.

8. If contamination is a real possibility (and when is it not?), neither mean nor variance is likely to be a wisely chosen basis for making estimates from a large sample.

9. As an interim measure, the use of truncated variances is likely to be quite satisfactory.

10. In smaller samples, the use of the mean deviation may be a frequently useful compromise,

11. More work is needed on the specific problem considered here, especially (i) in smaller samples, (ii) on lower bounds for J_γ when L_γ is given, and (iii) on a wider variety of estimators, including weighted order-statistic analogs of Gaussian means, and schemes of variable truncation.

Appendix A: ASYMPTOTIC TECHNIQUES

A.1 Possibilities

What can be expected from an organized asymptotic technique? At most, that it provides us with all terms through a given order, say $1/n$. The typical result might be of the form

$$\text{criterion} = A + \frac{B}{n} + o\left(\frac{1}{n}\right)$$

where $o(1/n)$ symbolizes additional terms of which we know only that they tend to zero faster than $1/n$. Or it might be of a more complex form, with an error term $o(1/n^2)$, and so on. But, if the algebraic manipulations are to be reasonable, it is almost certain that we must be left facing an error term of which we know only that it vanishes faster than some given power of $1/n$. This is the necessary nature of a simple asymptotic result.

If the technique is to be generally applicable, reflection shows that it cannot draw a sharp line between terms of order precisely $1/n$ and terms of higher order. Since, in particular,

$$7 + \frac{13}{n} + 97e^{-n}, \qquad 7 + \frac{13}{n} - \frac{1400}{n^2}, \qquad \text{and} \qquad 7 + \frac{13}{n}$$

all differ by $o(1/n)$, an asymptotic technique which goes only to $o(1/n)$ cannot appropriately distinguish among these three. It is thus convenient to take as asymptotic values not single numerical sequences but whole equivalence-classes of sequences.

The results so far described have been obtained and stated in terms of asymptotic average values and variances which are only *defined* up to $o(1/n)$. Consequently, their errors for any fixed value of n are unknown. (This need not prevent us from substituting finite values of n to obtain numerical illustrations.) But these errors will surely become negligible in comparison with $1/n$ for sufficiently large n, whatever definition and whatever evaluation of the expressions we take.

A.2 Adequate Intervals and Asymptotic Moments

Consider any sequence of chance quantities $\{Z_n\}$ and any finite open interval, I. Consider the *restrictions* $\{Z_n \| I\}$ of the Z_n to I. $Z_n \| I$ has as its distribution that part of the distribution of Z_n which is on I, inflated by the factor required to make the total probability of $Z_n \| I$ in I equal to unity.

The bounded open interval I is *adequate* for $\{Z_n\}$ if the key condition,

$$\Pr\{Z_n \text{ not in } I\} = o(\text{var}\{Z_n \| I\}),$$

holds. If I' and I'' are both adequate for $\{Z_n\}$, (i) the ratio of var$\{Z_n \| I\}$ to var$\{Z_n \| I'\}$ tends to unity, and (ii) the common part of I' and I'' is also adequate for $\{Z_n\}$.

If s_n^2 is *any* sequence of positive numbers such that

$$\frac{s_n^2}{\text{var}\{Z_n \| I\}} \to 1$$

for any one adequate I, and hence for all adequate I's, the *asymptotic variance* of $\{Z_n\}$, written $\vec{\text{var}}\, Z_n$, is defined to be s_n^2. (More precisely, the asymptotic variance is the equivalence class of such sequences, which consists of all t_n^2 with $t_n^2/s_n^2 \to 1$, but the looser language will be simpler, and safe enough to use.)

A notion of *asymptotic average value* can be defined by analogy with this definition. Its values are again equivalence classes of sequences, although again we shall speak of them as if they were sequences. We shall write $\vec{\text{ave}}\{g(Z_n)\} = g_n$, if there is an adequate interval I on which $g(z)$ is bounded, and for which $\vec{\text{ave}}\{g(Z_n \| I)\} = g_n + o(s_n^2)$, where $o(s_n^2)$ means a quantity which tends to zero faster than s_n^2. If this relation holds for one such I, then it holds for all such I [i.e., all those (i) which are adequate for $\{Z_n\}$, and (ii) on which $g(z)$ is bounded].

Given a sequence $\{Z_n\}$ of chance quantities, what can be said about the structure of the family of bounded open intervals which are adequate for $\{Z_n\}$? Because the common part of two adequate I's is itself adequate, there are only eight possibilities, which fall in three groups:

1. There are no adequate I's.
2. There exist numbers s and t, with $s < t$, such that an I is adequate if, and only if, it includes (a) the closed interval $[s, t]$, (b) the half-open interval $[s, t)$, (c) the half-open interval $(s, t]$, or (d) the open interval (s, t).
3. There exists a number z_0 such that an I is adequate if, and only if, it (a) includes z_0, (b) includes z_0 or has z_0 as left-hand end point, or (c) includes z_0 or has z_0 as right-hand end point.

In case (1) our machinery is not applicable at all. In cases (2) the sequence $\{Z_n\}$ does not behave the way a sequence of estimates or (normalized) bases of estimate should behave in large samples, though some of our machinery is applicable. The cases that really concern us are cases (3), where we say that $\{Z_n\}$ has asymptotic variance s_n^2 *near* z_0, written $\vec{\text{var}}\,\{Z_n\} = s_n^2$ near z_0.

There is a further simple characterization of cases (3). These are the cases where there are arbitrarily short adequate intervals. Thus "$\{Z_n\}$ has some asymptotic variance near some point" and "there exist arbitrarily short intervals which are adequate for $\{Z_n\}$" are equivalent statements, either of which characterizes a kind of behavior which is, in particular, shown (in large samples) by respectable estimates.

If $\{Z_n\}$ has asymptotic variance near z_0, then $\vec{\text{ave}}\,\{Z_n\} = m_n$ always exists, and converges to z_0. Clearly we should call m_n [again defined up to $o(s_n^2)$] the asymptotic average value of $\{Z_n\}$. It is natural to ask about higher asymptotic moments. The curious result is that the asymptotic moments of $\{Z_n\}$ about $\{m_n\}$ of order greater than two all "vanish" [i.e., are all $o(s_n^2)$, all belong to the zero equivalence class],

$$\vec{\text{ave}}\,\{(Z_n - m_n)^p\} = 0\,, \qquad p > 2\,,$$

and, as a corollary

$$\vec{\text{ave}}\,\{(Z_n - z_0)^p\} = (m_n - z_0)^p\,, \qquad p \neq 2\,.$$

It might seem that, if asymptotic moments (about the asymptotic mean) of order higher than two vanish, leaving only the asymptotic mean and the asymptotic variance, at least in the present sense of "asymptotically," all trace of shape of distribution is lost. This proves not to be the case.

A.3 Shapes, and Their Justification

A distribution is standardized if it has zero mean and unit variance. The distribution of $(Z - \text{ave}\,\{Z\})/(\text{var}\,\{Z\})^{\frac{1}{2}}$ is standardized, provided ave $\{Z_n\}$

and var $\{Z_n\}$ exist. Thus, if Pr $\{Z_n \text{ in } I\}$ is never zero, the distributions of the chance quantities

$$\frac{(Z_n \| I) - \text{ave}\,\{(Z_n \| I)\}}{(\text{var}\,\{Z_n \| I\})^{\frac{1}{2}}}$$

are all standardized. This sequence of standardized distributions may converge, or it may not; it may converge to a distribution, or it may not. If it does converge to a distribution, the limiting distribution is naturally regarded as expressing the *limiting shape* of $\{Z_n \| I\}$. (Note that it is completely determined.)

If I and I' are both adequate for $\{Z_n\}$, and $\{Z_n \| I\}$ has a limiting shape, then $\{Z_n \| I'\}$ has the same limiting shape. Thus it is appropriate to call this common limit the asymptotic limiting shape of $\{Z_n\}$.

Limiting shapes, ordinary or asymptotic, need not be expressed by standardized distributions. Any distribution with ave $\{z^2\}$ *less than or equal to* 1 and ave $\{z\}$ *equal to* 0 can be a limit of standardized distributions. Only those with ave $\{z^2\}$ *equal to* 1 are standardized or, as we shall now say, represent *faithful shapes*. The others represent *shrunken*, or *degenerate shapes* (they have ave $\{z\} = 0$ and ave $\{z^2\} < 1$). Standardized distributions can converge either to a faithful shape or to a shrunken shape.

In the other direction, a sequence of standardized distributions cannot be wholly divergent, for any such sequence must contain a convergent subsequence. Thus, if a sequence of standardized distributions does not converge, it contains subsequences converging to two or more different limiting shapes. The limiting *shape behavior* of a sequence of distributions is thus determined by specifying (i) those limiting distributions toward which at least one (standardized) subsequence converges, and (ii) what subsequences converge to which of these limiting distributions. If all these limiting distributions represent faithful shapes, then the limiting behavior is *non-shrinking*. A special case of *non-shrinking limiting behavior* occurs when the standardized distributions converge to a single distribution of unit variance.

The result about the limiting shapes of $\{Z_n \| I\}$ and $\{Z_n \| I'\}$ can now be extended to assert that these two sequences have identical limiting behavior when I and I' are both adequate. The *asymptotic limiting shape behavior* of $\{Z_n\}$ can thus be meaningfully defined whenever some interval I is adequate for $\{Z_n\}$.

We can now state results which show to what extent the notions "asymptotic average value" and "asymptotic variance" are extensions of the notions "ordinary average value" and "ordinary variance." If $\mu_n = \text{ave}\,\{Z_n\} \to z_0$ and $\sigma_n^2 = \text{var}\,\{Z_n\} = 0$, *and if* the limiting shape behavior $\{Z_n\}$ is non-shrinking, then $\vec{\text{ave}}\,\{Z_n\} = \mu_n$ and $\vec{\text{var}}\,\{Z_n\} = \sigma_n^2$ near z_0. (Examples show that the hypothesis on the limiting behavior cannot be eliminated.) As a simple consequence, it follows that, if some I is adequate for $\{Z_n\}$ whose limiting behavior is non-shrinking, and for which var $\{Z_n \| I\} \to 0$ and ave $\{Z_n \| I\} \to z_0$, then $\{Z_n\}$ has asymptotic variance $\vec{\text{var}}\,\{Z_n \| I\}$ near z_0. In particular, if var $\{Z_n\}$ exists and converges to zero, then $\vec{\text{var}}\,\{Z_n\} = \text{var}\,\{Z_n\}$.

A.4 Propagation Results

The next group of results relate the behavior of $\{g(Z_n)\}$ to that of $\{Z_n\}$. It is useful to do this under as weak hypotheses on $g(z)$ as possible. Thus parallel results with differing hypotheses may be desirable.

The *typical conclusion* is that

$$\text{ave}\{g(Z_n) \,||\, J\} = g(z_n) + o(s_n)\,, \qquad \text{var}\{g(Z_n) \,||\, J\} = A^2 s_n^2 + o(s_n^2)\,,$$

where $A = g'(z_0)$, and where, moreover, if $A \neq 0$, $\vec{\text{var}}\{g(Z_n)\} = A^2 s_n^2$ near z_0. This conclusion follows from the hypotheses:

(i) $g'(z)$ exists and is continuous (and not infinite) in an interval I continuing z_0, and

(ii) $\vec{\text{var}}\{Z_n\} = s_n^2$ near z_0,

or from the hypotheses, weaker for $g(z)$, but stronger for $\{Z_n\}$;

(i′) $g'(z_0)$ exists and is not infinite (only at the point z_0!),

(ii′) $\vec{\text{var}}\{Z_n\} = s_n^2$ near z_0, and

(iii) $m_n = \text{ave}\{Z_n \,||\, I\} = z_0 + o(s_n)$.

These latter hypotheses are also sufficient to establish that the asymptotic limiting shape behavior of $\{g(Z_n)\}$ is the same as that of $\{Z_n\}$.

The latter two results can be put in the following form: Provided that

(i) $A = g'(z_0)$ exists finite and unequal to 0,

then

(ii) $\vec{\text{var}}\{Z_n\} = s_n^2$ near z_0,

(iii) $\vec{\text{ave}}\{Z_n\} = z_0 + o(s_n)$, and

(iv) the asymptotic limiting shape behavior of $\{Z_n\}$ being so-and-so; together imply that

(ii_g) $\vec{\text{var}}\{g(Z_n)\} = A^2 s_n^2$ near $g(z_0)$,

(iii_g) $\vec{\text{ave}}\{g(Z_n)\} = z_0 + o(s_n)$, and

(iv_g) the asymptotic limiting shape behavior of $\{Z_n\}$ being so-and-so; and, indeed, (ii) and (iii) together imply (ii_g) and (iii_g).

Thus the class of sequences of chance quantities which satisfy analogs of (ii) and (iii) is closed under those changes of $\{Z_n\}$ to $\{g(Z_n)\}$ for which $g'(z_0)$ exists finite and non-zero. And the sáme is true for the subclasses defined by particular asymptotic limiting behavior.

Generalizations of these results to the case of $\{g_n(Z_n)\}$ are given in [40], which also contains details and proofs of the material summarized here.

Appendix B: THE EXPRESSION AND BALANCING OF BIAS AND VARIANCE

B.1 The General Formulation

The general formulation will be described in notation that (i) applies immediately to the specific instances of sampling from contaminated distributions and (ii) can also be interpreted as applying to general instances of estimation in the alternative pencil. Asymptotic operators without subscripts, like $\vec{\text{ave}}$ and $\vec{\text{var}}$, refer to evaluations for the reference pencil, in our special instances a pencil of normal distributions. Asymptotic operators with subscript γ, like $\vec{\text{ave}}_\gamma$ and $\vec{\text{var}}_\gamma$, refer to evaluations for the alternative pencil, in our special instances a pencil of distributions each contaminated by a fraction γ at scale ratio 3.

If $u = u_n = u(z_1, \cdots, z_n)$ is a function of observations suitable as a base of estimate for the parameter θ, then, for a suitable $d(\theta)$, $\vec{\text{ave}}\, u_n - d(\theta)$ and $\vec{\text{var}}\, u_n$ both tend to zero like $1/n$. It is easy to describe ([8], Sec. 9) how to form a function t of u such that

$$\vec{\text{ave}}\, t = \theta \,, \qquad \vec{\text{var}}\, t = \frac{1}{n} i(\theta) \,.$$

This new function of observations estimates θ in the reference pencil of distributions. In the alternative pencil, we have

$$\vec{\text{ave}}_\gamma\, t = H_\gamma + \frac{1}{n} K_\gamma \,, \qquad \vec{\text{var}}_\gamma\, t = \frac{1}{A} J_\gamma \,,$$

where all of H_γ, J_γ, K_γ may depend on θ, and where manageable general expressions can be given for these three functions of θ ([8], Sec. 10).

If bias is not a concern, then the appropriate measure of variability of t as an indicator of θ in the alternative pencil is related to the sensitivity of t as defined by Mandel and Stiehler [44]. The appropriate asymptotic measure of variability is, explicitly,

$$G_\gamma = \frac{J_\gamma}{(H'_\gamma)^2} \,,$$

where $H'_\gamma = (d/d\theta) H_\gamma$.

If both bias and variability are involved, it is convenient to put $H_\gamma = \theta + L_\gamma$ so that

$$\vec{\text{ave}}\, t = \theta + L_\gamma + \frac{1}{n} K_\gamma \,, \qquad \vec{\text{var}}\, t = \frac{1}{n} J_\gamma \,.$$

Here L_γ indicates the asymptotically important part of the bias, and J_γ is now the more appropriate measure of variability, since, referred to θ,

$$\text{average square } \vec{\text{deviation}} = \vec{\text{var}}_\gamma\, t + (\vec{\text{ave}}_\gamma\, t - \theta)^2 = \frac{1}{n} J_\gamma + L_\gamma^2 + \frac{2}{n} L_\gamma K_\gamma \,,$$

while if it is referred to $\theta + L_{e\gamma}$, where $L_{e\gamma}$ is the asymptotic bias of the quantity [e.g., the average of $\phi(z)$] which we wish to assess, then

$$\begin{aligned} \text{average square } \vec{\text{deviation}} &= \vec{\text{var}}_\gamma\, t + \vec{\text{ave}}_\gamma\, (t - \theta - L_{e\gamma})^2 \\ &= (L_\gamma - L_{e\gamma})^2 + \frac{1}{n}[J_\gamma + 2(L_\gamma - L_{e\gamma})K_\gamma] \,. \end{aligned}$$

Any serious attempt to reduce the average square error by choosing a basis of estimate from a one-parameter family of possibilities will have to make $L_\gamma - L_{e\gamma} \to 0$ as $n \to \infty$, so that $(1/n)(L_\gamma - L_{e\gamma})K_\gamma$ will be $o(1/n)$, and

$$\text{average square } \overrightarrow{\text{deviation}} = (L_\gamma - L_{e\gamma})^2 + \frac{1}{n} J_\gamma$$

since the left-hand side is defined only up to $o(1/n)$.

B.2 Balancing Bias and Variance

In order to discuss the balancing of bias and variance by the choice of a base of estimate out of a one-parameter family of bases of estimate, we introduce a somewhat peculiar assumption. It is that some rule has been given so that $\overrightarrow{\text{ave}}_\gamma\, t$ and $\overrightarrow{\text{var}}_\gamma\, t$ are determined to $o(1/n^2)$ and not just to $o(1/n)$. The peculiarity is that we shall find that the nature of this rule, so long as there is one, will not affect the results.

The basic expressions now lengthen to (we shall drop the γ subscripts throughout this discussion for clarity of formulas)

$$\overrightarrow{\text{ave}}\, t = \theta + L + \frac{1}{n}K + \frac{1}{n^2}M\,, \qquad \overrightarrow{\text{var}}\, t = \frac{1}{n}J + \frac{1}{n^2}N\,.$$

If our estimand has bias L_e, then the minimum of the asymptotic average square deviation, when a base of estimate is to be selected from a one-parameter family, will occur when

$$L = L_e - \left(K + \frac{1}{2}\frac{dJ}{dL}\right)\frac{1}{n}\,,$$

and the corresponding values of the asymptotic average square deviation will be, when $L = L_e$,

$$\frac{1}{n}J + \frac{1}{n^2}(K^2 + N)\,,$$

but will be less by

$$\frac{1}{n^2}\left(K + \frac{1}{2}\frac{dJ}{dL}\right)^2,$$

when L is offset as above. Thus the improvement due to offset is by a factor asymptotic to

$$\frac{\left(K + \frac{1}{2}\frac{dJ}{dL}\right)^2}{nJ}\,.$$

Notice that neither the formula for $L - L_e$, nor that for the fractional improvement in variance due to its use involves the coefficients M and N of $1/n^2$. Whatever modification of the definitions of $\overrightarrow{\text{ave}}$ and $\overrightarrow{\text{var}}$ is adopted, the leading terms in the amount of, and the profit from, optimum offset are the same.

B.3 Choice of Expression

If θ is replaced by θ^2, $\sqrt{\theta}$, or $\log \theta$, the discussion of the last section can be repeated. What effect will there be on asymptotic variances and biases?

Investigation shows ([8], Sec. 14) that the main changes are

1. An over-all scale change by a factor expressing the derivative of one mode of expressing the parameter with respect to the other,
2. The addition of a term in K_γ,
3. Non-linear effects for large biases.

These effects do not have an important influence on any of the uses to which we expect to put H_γ, J_γ, K_γ, L_γ, or Δ_γ. Consequently, we may take the particular mode of expression of the parameter as not being of major importance.

B.4 Estimates of Scaling Type

When the parameter θ is to be a function of a scaling parameter, λ, there are certain conveniences in the choice of a certain mode of expression. (And, indeed, it can be argued that, insofar as this choice has minor effects on the comparison of biases and variabilities, this convenient choice is also the most proper choice, so far as these minor changes go.) This choice is a logarithmic expression of scale, say in the form $\lambda = e^\tau$, where τ is now to play the role played by θ above.

What are some of these conveniences?

1. If $G_\gamma((z - z_0)/\lambda)$ is a scale pencil of distributions, and if we write $w = \log(z - z_0)$ and let $H(w, \tau)$ be the corresponding family of distributions of w parametrized by τ, then τ is a location parameter.
2. Most, and by a device all, of the most natural bases of scaling have equivalent forms in which the bias of estimating λ is independent of λ.

These equivalent forms include

$$\frac{1}{p} \log \frac{1}{n} \sum |z_i|^p \quad \log(p\%\text{-interfractile distance})\,,$$

and (here is a device)

$$\log \lambda_A\,, \qquad \text{where } \frac{1}{n} \sum e^{-b(z/\lambda_A)^2} = A\,,$$

the last being asymptotically equivalent, as a base of estimate, to the Gaussian mean $(1/n) \sum e^{-cz^2}$ where $c\lambda^2$, which is the asymptotic value of $c\lambda_A^2$, equals b.

This choice is recommended for all problems of scaling type, and was used in the treatment of scaling problems in contaminated distributions reported in Section 10.

B.5 Balancing Bias and Variance in the Example

Formulas for balancing bias and variance were attained in Section B.2, but no indication has so far been given that they may be practically usable. For a specific alternative pencil can never be identified in practice, where the alternative is always an unknown pencil, one of many possibilities. If the formula yields widely different choices of bases of estimate for

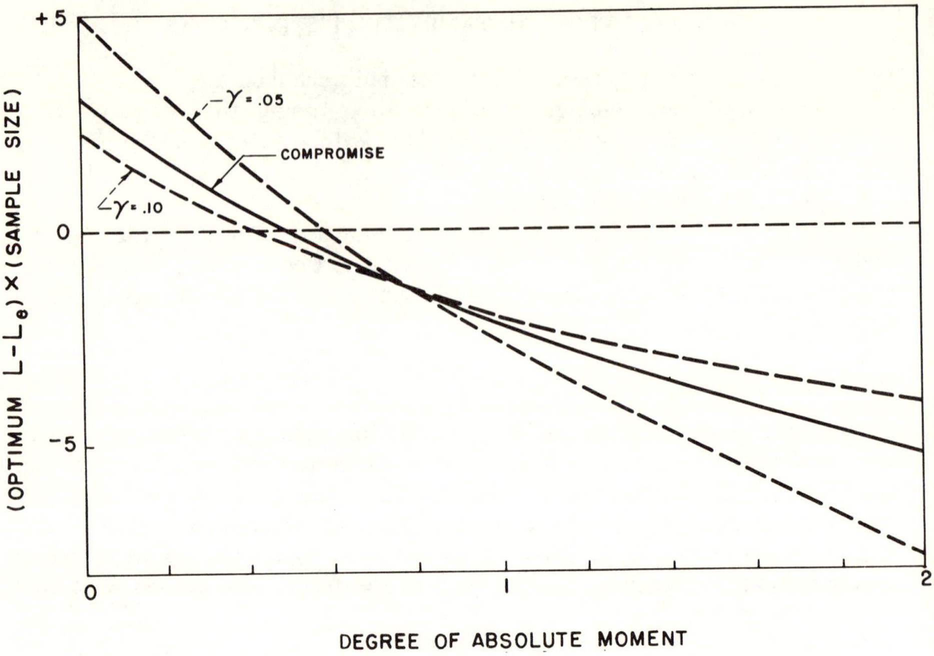

FIG. B-1. Optimum (for Specific γ) and Compromise Offsets in Bias

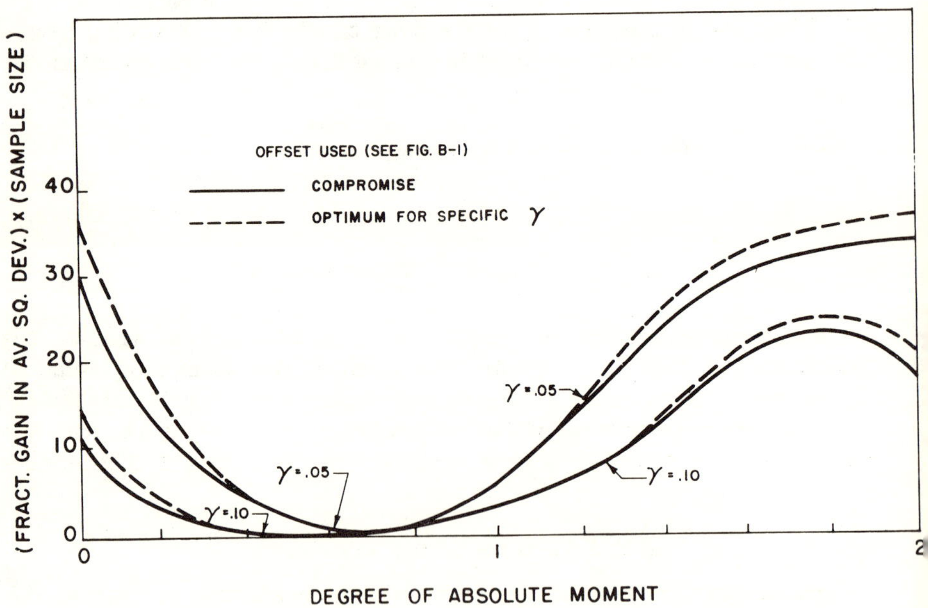

FIG. B-2. Fractional Gain in Average Square Deviation from Offset in Bias

possible alternative pencils, in our instances for different γ, it will be of little practical use.

A brief look at the scaling of contaminated distributions thus seems in order. The case of replacing a moderate absolute moment (one with $p < 2$) by a Gaussian mean of similar bias is a convenient example. Figure B-1 shows values of the offset in L as "(optimum $L - L_e$) × (sample size)" for $\gamma = .05$ and $\gamma = .10$ for $0 \leqq p \leqq 2$. The curve marked "compromise" is, in most places, the geometric mean of the curves for individual γ. Figure B-2 shows curves for "(fractional gain in average square deviation) × (sample size)" for the same situations. Because the loss in adopting a non-optional offset is a quadratic function, the compromise gives quite satisfactory results.

The general order of *fractional* gain is from $0/n$ to $30/n$. This would be negligible for samples of size 1000, so that these should presumably be treated as very large. Depending upon the degree of the moment, the gain for samples of size 100 would frequently be important. Samples of size 100 might thus frequently be wisely treated as moderately large.

This whole discussion is incomplete, because it has not taken into account the fact (see Fig. 9 in Section 12) that slightly different b values are required to match bias with a single absolute moment for $\gamma = .05$ and $\gamma = .10$. However, the indications (i) that useful offsets for several alternative families are possible, and (ii) that "moderately large" may mean approximately 100, are still useful and illuminating.

References

[1] Cunningham, L. B. C., and Hynd, W. R. B. Random processes in problems of air warfare. *J. Roy. Stat. Soc.*, Ser. B, 1946, **8**, 62–85, disc. 85–97.

[2] Fraser, D. A. S. Generalized hit probabilities with a Gaussian target. *Ann. Math. Stat.*, 1951, **22**, 248–55; 1955, **24**, 288–94.

[3] Tukey, J. W. Sampling from contaminated distributions, preliminary report (abst.). *Ann. Math. Stat.*, 1946, **17**, 501.

[4] Tukey, J. W. Sampling from contaminated distributions, preliminary report (abst.). *Bull. Amer. Math. Soc.*, 1946, **52**, 828–29.

[5] Tukey, J. W. Sampling from contaminated distributions, 2. Scaling by and for power-means—the asymptotic case. Statistical Techniques Group, Memorandum Report No. 25, Princeton, N.J.: Princeton University, (1949 draft, unpublished).

[6] Harris, T. E., and Tukey, J. W. Sampling from contaminated distributions, 3. Measures of location and scale which are relatively insensitive to contamination. Statistical Research Group, Memorandum Report No. 31, Princeton, N.J.: Princeton University, July 1949, (duplicated).

[7] Tukey, J. W. Sampling from contaminated distributions, 4. The truncated mean in moderately large samples. Statistical Research Group, Memorandum Report No. 32, Princeton, N.J.: Princeton University, July 1949 (duplicated).

[8] Tukey, J. W. Sampling from contaminated distributions, 5. Scaling by and for percentiles and exponential averages. Statistical Research Group, Memorandum Report No. 33, Princeton, N.J.: Princeton University (1949 draft, unpublished).

[9] Tukey, J. W. Sampling from contaminated distributions, 6. Skeleton tables related to contaminated distributions at scale 3. Statistical Research Group, Memorandum Report No. 34, Princeton, N.J.: Princeton University, July 1949 (duplicated).

[10] Tukey, J. W. Estimation in the alternative family of distributions (abst.). *Proc. Int. Cong. Math.*, 1950, **1**, 586, 1952.

[11] Pearson, E. S. The analysis of variation in cases of non-normal variation. *Biometrika*, 1931, **23**, 114–33.

[12] Eden, T., and Yates, F. On the validity of Fisher's z-test when applied to an actual example of non-normal data. *J. Agric. Sci.*, 1933, **23**, 6–16.

[13] Pitman, E. J. G. Significance tests which may be applied to any population. *J. Roy. Stat. Soc.*, 1937 suppl., **4**, 119–30.

[14] Pitman, E. J. G. Significance tests which may be applied to samples from any population, II. The correlation coefficient test. *J. Roy. Stat. Soc.*, 1937 suppl., **4**, 225–32.

[15] Pitman, E. J. G. Significance tests which may be applied to samples from any population, III. The analysis of variance test. *Biometrika*, 1937, **29**, 322–35.

[16] Welch, B. L. On the z-test in randomized blocks and Latin squares. *Biometrika*, 1937, **29**, 21–52.

[17] Box, G. E. P., and Andersen, S. L. Permutation theory in the derivation of robust criteria and the study of departures from assumptions, *J. Roy. Stat. Soc.*, Ser. B, 1955, **17**, 1–26, disc. 26–34.

[18] Baker, G. A. Empiric investigation of a test for homogeneity for populations composed of normal distributions, *J. Amer. Stat. Assoc.*, 1958, **53**, 551–57.

[19] de Helguero, F. Per la risoluzione delle curve dimorphiche. *Biometrika*, 1905(-06), **4**, 230–31.

[20] de Helguero, F. Per la risoluzione delle curve dimorphiche, *Mem. R. Accad. dei Lincei* (*Rome*), Ser. 5, 1906, **6**, 163–203.

[21] Edgeworth, F. Y. On the representation of statistics by mathematical formulae, Pt. 2. *J. Roy. Stat. Soc.*, 1899, **62**, 125–40.

[22] Feller, W. On a general class of contagious distributions. *Ann. Math. Stat.*, 1943, **14**, 389–400.

[23] Gurland, J. Some interrelations among compound and generalized distributions. *Biometrika*, 1957, **44**, 265–68.

]24] Pearson, K. Contributions to the mathematical theory of evolution, 7. *Phil. Trans. Roy. Soc.* (*London*), Ser. A, 1894, **185**, 71–110.

[25] Pearson, K. On some applications of the theory of chance to racial differentiation. *Phil. Mag.*, 1901, **6**(1), 110–24.

[26] Pollard, H. On the relative stability of the median and arithmetic mean, with particular reference to certain frequency distributions which can be dissected into normal distributions. *Ann. Math. Stat.*, 1934, **5**, 227–62.

[27] Preston, E. J. A graphical method for the analysis of statistical distributions into two normal components. *Biometrika*, 1954, **40**, 460–64.

[28] Jeffreys, H. An alternative to the rejection of observations. *Proc. Roy. Soc.* (*London*), Ser. A, 1932, **137**, 78–87.

[29] Levi, B. On the form of composite frequency curves (Spanish). *Math. Notae*, 1951, **11**, 87–109.

[30] Medgyessy, P. Anwendungsmöglichkeiten der Analyse der Wahrscheinlichkeitsdichtefunktionen bei der Auswertung von Messungsergebnissen. *Z. angew. Math. u. Mech.*. 1957, **37**, 128–39.

[31] Navratil, J. Determination of the parameters of a compound normal distribution. *Pokroky Mat. Fys. Astr.*, 1958, **3**, 41–45.

[32] Putnam, C. R., and Wintner, A. On the addition of symmetric normal frequency curves. *Math. Notae*, 1951, **11**, 79–86.

[33] Teichroew, D. The mixture of normal distributions with different variances. *Ann. Math. Stat.* 1957, **28**, 510–12.

[34] Watanake, Y., Wajiki, I., and Kawashiro, T. On the compound normal distributions. *J. Gakugei Tokushima Univ. Nat. Sci. Math.*, 1956, **7**, 53-65.

[35] Wintner, A. Infinite divisible symmetric laws and normal stratifications. *Publ. Inst. Stat. Univ. Paris*, 1957, **6**, 326-36.

[36] Student. Errors of routine analysis. *Biometrika*, 1927, **19**, 151–69.

[37] Student (William Sealy Gosset). *"Student's" Collected Papers*. London: Biometrika Office, 1942.

[38] Tukey, J. W. Asymptotic moments and expectations (abst.). *Bull. Amer. Math. Soc.*, 1948, **54**, 644–45.

[39] Tukey, J. W. Asymptotic moments and expectations. Statistical Techniques Group, Memorandum Report No. 4, Princeton, N.J.: Princeton University, May 1949 (duplicated).

[40] Tukey, J. W. Asymptotic variances and average values. (Being revised for the *Annals of Mathematical Statistics*.)

[41] Hotelling, H. The place of statistics in the university. In *Proceedings of the Berkeley Symposium on Mathematical Statistics and Probability*. Berkeley, Calif.: Univ. California Press, 1949.

[42] Jung, J. On linear estimates defined by a continuous weight function. *Arkiv för Matematik*, 1955(1954–58), **3**(15), 199–209.

[43] Jeffreys, H. *Theory of Probability*, 2d ed. Oxford: Clarendon Press, 1948.

[44] Mandel, J., and Stiehler, R. O. Sensitivity—a criterion for the comparison of methods of test, *J. Research N.B.S.*, 1954, **53**, 155–59.

[45] Cramér, H. *Mathematical Methods of Statistics*, Princeton, N.J.: Princeton Univ. Press, 1946.

[46] Fisher, R. A. A mathematical examination of the methods of determining the accuracy of an observation by the mean error, and by the mean square error. *Monthly Not. Roy. Astron. Soc.*, **80**, 758–70.

[47] Fisher, R. A. *Contributions to Mathematical Statistics*, New York: Wiley, 1950.

[48] Burrau, Ø. The mean error as a measure of uncertainty (Danish). *Mat. Tidskr.*, Ser. B, 1943, 9-16 (cf. *Math. Rev.*, 1946, **7**, 130).

[49] Fisher, R. A. On the mathematical foundations of theoretical statistics. *Phil. Trans. Roy. Soc. (London)*, Ser. A, 1922, **222**, 309–68.

40 MULTIDIMENSIONAL STATISTICAL SCATTER[1]

S. S. WILKS

Princeton University

1. Introduction

The purpose of this paper is to present a simple and perhaps novel expository approach to some of the basic ideas of multivariate statistical analysis for the non-specialist in this field. Since Professor Hotelling, to whom this volume is dedicated, has made a number of fundamental contributions to multivariate statistical analysis, the author believes it is particularly fitting to attempt an expository article of this kind.

It turns out that a considerable amount of multivariate statistical machinery, including statistical tests, can be effectively described in terms of quantities which we shall call *multidimensional statistical scatters*. We shall be concerned with these scatters and their use in the description of certain multivariate statistical tools and tests. No attempt has been made to cover all of the important multivariate methods in this expository treatment. The author's aim is to consider primarily those situations where the notions of multidimensional statistical scatter are most useful. Except for Sections 3 and 4, which deal with basic properties of what we call *statistical scatter*, we shall not give proofs and derivations, since these can be found in various original research articles referred to in the paper. Proofs of most of the results in Sections 5 to 9 can also be found in Anderson's book [1].

At the outset we shall need some notation and definitions. It will be sufficient to consider a k-dimensional continuous random variable $(x_1, \cdots, x_k)$ with distribution function $F(x_1, \cdots, x_k)$ and density function $f(x_1, \cdots, x_k)$ given by

$$f(x_1, \cdots, x_k) = \frac{\partial^k F(x_1, \cdots, x_k)}{\partial x_1 \cdots \partial x_k}, \tag{1.1}$$

where $f(x_1, \cdots, x_k) \geqq 0$ and

$$\int_{\mathscr{R}_k} f(x_1, \cdots, x_k)\, dx_1, \cdots, dx_k = 1, \tag{1.2}$$

$\mathscr{R}_k$ being the entire k-dimensional Euclidean space of $(x_1, \cdots, x_k)$.

[1] Preparation of this paper was partially supported by a contract of the Office of Naval Research.

The *vector of means* $(\mu_1, \cdots, \mu_k)$ of the distribution is defined by

$$\mu_i = \int_{\mathscr{R}_k} x_i \, dF(x_1, \cdots, x_k) , \qquad i = 1, \cdots, k , \tag{1.3}$$

and $f(x_1, \cdots, x_k)\, dx_1, \cdots, dx_k$ is denoted by $dF(x_1, \cdots, x_k)$. It is sometimes convenient to call $(\mu_1, \cdots, \mu_k)$, as a point in k-space, the *center of gravity* of the distribution.

The *covariance matrix* $\|\sigma_{ij}\|$ is defined by

$$\sigma_{ij} = \int_{\mathscr{R}_k} (x_i - \mu_i)(x_j - \mu_j)\, dF(x_1, \cdots, x_k) . \tag{1.4}$$

This matrix is symmetric and is assumed to be positive definite. The determinant $|\sigma_{ij}|$ is sometimes called the *generalized variance* of the distribution $F(x_1, \cdots, x_k)$.

A sample O_n of size n from $F(x_1, \cdots, x_k)$ is an nk-dimensional random variable, $(x_{1\alpha}, \cdots, x_{k\alpha}, \alpha = 1, \cdots, n)$, having distribution function

$$\prod_{\alpha=1}^{n} F(x_{1\alpha}, \cdots, x_{k\alpha}) , \tag{1.5}$$

or density function $\Pi_{\alpha=1}^{n} f(x_{1\alpha}, \cdots, x_{k\alpha})$.

There will be no ambiguity if we use the same notation for the sample and a realized (or observed) value of the sample, since it will be evident from the context which sense is intended. This dual usage will provide considerable simplification of notation.

2. One-dimensional Scatter

We shall begin with the completely familiar case of one-dimensional scatter defined for a sample $(x_1, \cdots, x_n)$ from a one-dimensional distribution $F(x)$. Since x is a continuous random variable, the probability is 1 that all components of a sample $(x_1, \cdots, x_n)$ will be different, in which case an observed value of the sample may be graphically represented by a *cluster* of n points $x_1, \cdots, x_n$ on the x-axis. Let x_0 be an arbitrary point on the x-axis. We define the (one-dimensional) *scatter* $S_{1,x_0,n}$ of the sample about the *pivotal point* x_0 as the following sum of squares

$$S_{1,x_0,n} = \sum_{\alpha=1}^{n} (x_\alpha - x_0)^2 . \tag{2.1}$$

Note that

$$S_{1,x_0,n} = S_{1,\bar{x},n} + n(\bar{x} - x_0)^2 , \tag{2.2}$$

where $\bar{x} = \sum_{\alpha=1}^{n} x_\alpha / n$, the sample mean. From (2.2) we observe at once the familiar result that $S_{1,x_0,n}$ has a minimum if x_0 is chosen as $\bar{x}$, the minimum value of $S_{1,x_0,n}$ being $S_{1,\bar{x},n}$.

If the mean and variance of the distribution $F(x)$ are μ and σ^2, we observe that

$$\mathscr{E}(S_{1,x_0,n}) = n\sigma^2 + n(\mu - x_0)^2 , \tag{2.3}$$

and that $\mathscr{E}(S_{1,x_0,n})$ is a minimum if x_0 is chosen as μ, the value of the minimum being $n\sigma^2$.

3. Two-dimensional Scatter

Now, let us proceed to the case of a sample $(x_{1\alpha}, x_{2\alpha}, \alpha = 1, \cdots, n), n > 2$, from a two-dimensional distribution function $F(x_1, x_2)$. A realized value of this sample can be graphically represented as a cluster of n points in the x_1x_2-plane. Let (x_{10}, x_{20}) be an arbitrary point in the plane. If we take (x_{10}, x_{20}) together with any two points $(x_{1\alpha}, x_{2\alpha})$ and $(x_{1\beta}, x_{2\beta})$ from the cluster, these three points determine a triangle whose squared area will be denoted by $\Delta^2_{\alpha\beta,0}$. If we let

$$V^2_{\alpha\beta,0} = (2\Delta_{\alpha\beta,0})^2\,, \tag{3.1}$$

then $V^2_{\alpha\beta,0}$ is the squared area (content) of any of the three parallelograms which can be formed having $(x_{1\alpha}, x_{2\alpha})$, $(x_{1\beta}, x_{2\beta})$ and (x_{10}, x_{20}) as three of its vertices.

Since (α, β) can be any pair of the integers $1, \cdots, n$, and since $V^2_{\alpha\beta,0} = V^2_{\beta\alpha,0}$, we can take $\alpha < \beta$. Thus, there are $\binom{n}{2}$ different quantities $V^2_{\alpha\beta,0}$. The (two-dimensional) *scatter* $S_{2,x_0,n}$ of the sample $(x_{1\alpha}, x_{2\alpha}, \alpha = 1, \cdots, n)$, about the *pivotal point* (x_{10}, x_{20}), is defined as the sum of $V^2_{\alpha\beta,0}$ over all possible choices of (α, β), $\alpha < \beta$, that is,

$$S_{2,x_0\,n} = \sum_{1\leqq\alpha<\beta\leqq n} V^2_{\alpha\beta,0}\,. \tag{3.2}$$

We know from elementary coordinate geometry that $V^2_{\alpha\beta,0}$ can be expressed in terms of the coordinates of these three points as the following squared determinant:

$$V^2_{\alpha\beta,0} = \begin{vmatrix} x_{1\alpha} - x_{10} & x_{2\alpha} - x_{20} \\ x_{1\beta} - x_{10} & x_{2\beta} - x_{20} \end{vmatrix}^2 . \tag{3.3}$$

Squaring the determinant, we find it may be expressed as a determinant which can be written as the sum of four determinants, namely

$$V^2_{\alpha\beta,0} = D_{\alpha\alpha} + D_{\alpha\beta} + D_{\beta\alpha} + D_{\beta\beta}\,, \tag{3.4}$$

where

$$D_{\alpha\beta} = \begin{vmatrix} (x_{1\alpha} - x_{10})^2 & (x_{1\beta} - x_{10})(x_{2\beta} - x_{20}) \\ (x_{2\alpha} - x_{20})(x_{1\alpha} - x_{10}) & (x_{2\beta} - x_{20})^2 \end{vmatrix}, \tag{3.5}$$

with similar expressions for $D_{\alpha\alpha}$, $D_{\beta\alpha}$, and $D_{\beta\beta}$; $D_{\alpha\alpha}$ and $D_{\beta\beta}$ are both 0, of course.

But it is evident from (3.4) and the fact that $D_{\alpha\alpha} = 0$ that the sum on the right-hand side of (3.2) is exactly the same as $\sum_{\beta=1}^n \sum_{\alpha=1}^n D_{\alpha\beta}$, that is,

$$S_{2,x_0,n} = \sum_{\beta=1}^{n} \sum_{\alpha=1}^{n} D_{\alpha\beta}\,. \tag{3.6}$$

But $D_{\alpha\beta}$ is of form $\begin{vmatrix} a_\alpha & b_\beta \\ c_\alpha & d_\beta \end{vmatrix}$, from which we see that

$$\sum_{\beta=1}^{n}\sum_{\alpha=1}^{n}\begin{vmatrix} a_\alpha & b_\beta \\ c_\alpha & d_\beta \end{vmatrix} = \begin{vmatrix} \sum_\alpha a_\alpha & \sum_\beta b_\beta \\ \sum_\alpha c_\alpha & \sum_\beta d_\beta \end{vmatrix}. \tag{3.7}$$

Hence

$$S_{2,x_0,n} = \begin{vmatrix} \sum_\alpha(x_{1\alpha}-x_{10})^2 & \sum_\beta(x_{1\beta}-x_{10})(x_{2\beta}-x_{20}) \\ \sum_\alpha(x_{2\alpha}-x_{20})(x_{1\alpha}-x_{10}) & \sum_\beta(x_{2\beta}-x_{20})^2 \end{vmatrix}. \tag{3.8}$$

Now $S_{2,x_0,n}$ will vanish if and only if all points of the sample and (x_{10}, x_{20}) are collinear. But it can be shown that if the covariance matrix $||\sigma_{ij}||$, $i, j = 1, 2$, is positive definite, the probability is 0 that the sample points are collinear. The matrix whose determinant is shown on the right of (3.8) will be called the *scatter matrix* of the sample about (x_{10}, x_{20}). It is of form $X'X$ where X is the matrix

$$\begin{Vmatrix} x_{11}-x_{10} & x_{21}-x_{20} \\ x_{12}-x_{10} & x_{22}-x_{20} \\ \vdots & \vdots \\ x_{1n}-x_{10} & x_{2n}-x_{20} \end{Vmatrix}.$$

If we let $\bar{x}_i = \sum_\alpha x_{i\alpha}/n$, $i = 1, 2$, and

$$u_{ij} = \sum_\alpha (x_{i\alpha}-\bar{x}_i)(x_{j\alpha}-\bar{x}_j) = u_{ji}, \tag{3.9}$$

then $\sum_\alpha(x_{i\alpha}-x_{i0})(x_{j\alpha}-x_{j0}) = u_{ij} + n(\bar{x}_i - x_{i0})(\bar{x}_j - x_{j0})$, and we have

$$S_{2,x_0,n} = \begin{vmatrix} u_{11}+n(\bar{x}_1-x_{10})^2 & u_{12}+n(\bar{x}_1-x_{20})(\bar{x}_2-x_{20}) \\ u_{21}+n(\bar{x}_2-x_{20})(\bar{x}_1-x_{10}) & u_{22}+n(\bar{x}_2-x_{20})^2 \end{vmatrix}, \tag{3.10}$$

which is the two-dimensional analog of $S_{1,x_0,n}$ as given by (2.2).

It will sometimes be convenient to call $(\bar{x}_1, \bar{x}_2)$ the *center of gravity* of the sample cluster, and the matrix $||u_{ij}||$ the *internal scatter matrix*, and $|u_{ij}|$ the *internal scatter* of the sample since this determinant is the scatter $S_{2,\bar{x},n}$ of the sample about its own center of gravity.

The matrix $||u_{ij}||$, $i, j = 1, 2$, is positive definite with probability 1 and hence will have inverse $||u^{ij}||$. It then follows that we can write (3.10) as follows

$$S_{2,x_0,n} = |u_{ij}| \cdot [1 + n\sum_{i,j=1}^{2} u^{ij}(\bar{x}_i - x_{i0})(\bar{x}_j - x_{j0})]. \tag{3.11}$$

Since $||u_{ij}||$ is positive definite, so is $||u^{ij}||$, and the sum in brackets on the right-hand side of (3.11) is a positive definite quadratic form in $(\bar{x}_1 - x_{10})$ and $(\bar{x}_2 - x_{20})$, vanishing if and only if $x_{10} = \bar{x}_1$ and $x_{20} = \bar{x}_2$. We therefore have the result that:

The choice of the pivotal point (x_{10}, x_{20}) *which minimizes the scatter* $S_{2,x_0,n}$ *is the center of gravity* $(\bar{x}_1, \bar{x}_2)$ *of the sample cluster, the minimum scatter being the internal scatter of the sample, that is*

$$\min_{x_0} (S_{2,x_0,n}) = S_{2,\bar{x},n} = |u_{ij}| . \tag{3.12}$$

Now consider the problem of finding the value of $\mathscr{E}(S_{2,x_0,n})$. Since $S_{2,x_0,n}$ is the sum of $V^2_{\alpha\beta,0}$ for all $\binom{n}{2}$ possible choices of (α, β), we concentrate first on $\mathscr{E}(V^2_{\alpha\beta,0})$. We have from (3.4), since $D_{\alpha\alpha} = D_{\beta\beta} = 0$, that

$$\mathscr{E}(V^2_{\alpha\beta,0}) = \mathscr{E}(D_{\alpha\beta}) + \mathscr{E}(D_{\beta\alpha}) = 2\mathscr{E}(D_{\alpha\beta}) . \tag{3.13}$$

But

$$\mathscr{E}(D_{\alpha\beta}) = \int_{\mathscr{R}_2}\int_{\mathscr{R}_2} D_{\alpha\beta}\, dF(x_{1\alpha}, x_{2\alpha})\, dF(x_{1\beta}, x_{2\beta}) . \tag{3.14}$$

Since relation (3.7) holds for integrals as well as sums, we thus find

$$\mathscr{E}(D_{\alpha\beta}) = \begin{vmatrix} \tau_{11} & \tau_{12} \\ \tau_{21} & \tau_{22} \end{vmatrix} , \tag{3.15}$$

where

$$\tau_{ij} = \mathscr{E}(x_i - x_{i0})(x_j - x_{j0}) = \sigma_{ij} + (\mu_i - x_{i0})(\mu_j - x_{j0}) . \tag{3.16}$$

Therefore we may write

$$\mathscr{E}(D_{\alpha\beta}) = |\sigma_{ij}| \cdot \left[1 + \sum_{i,j=1}^{2} \sigma^{ij}(\mu_i - x_{i0})(\mu_j - x_{j0})\right] , \tag{3.17}$$

where $||\sigma^{ij}||$ is the inverse of $||\sigma_{ij}||$. Since the scatter $S_{2,x_0,n}$ is given by

$$S_{2,x_0,n} = \sum_{1 \leq \alpha < \beta \leq n} A^2_{\alpha\beta,0} = 2 \sum_{1 \leq \alpha < \beta \leq n} D_{\alpha\beta} ,$$

and since $\mathscr{E}(D_{\alpha\beta})$ does not depend on (α, β), we have

$$\begin{aligned} \mathscr{E}(S_{2,x_0,n}) &= 2 \sum_{1 \leq \alpha < \beta \leq n} \mathscr{E}(D_{\alpha\beta}) \\ &= 2\binom{n}{2} \cdot |\sigma_{ij}| \cdot \left[1 + \sum_{i,j=1}^{n} \sigma^{ij}(\mu_i - x_{i0})(\mu_j - x_{j0})\right] . \end{aligned} \tag{3.18}$$

Since $||\sigma_{ij}||$ is positive definite, $||\sigma^{ij}||$ is also positive definite, and hence we have the result that:

The mean value of the scatter $S_{2,x_0,n}$ of the sample about the point (x_{10}, x_{20}) attains its minimum if (x_{10}, x_{20}) is chosen as the vector of means (μ_1, μ_2) of the distribution from which the sample is drawn. The minimum attained by $\mathscr{E}(S_{2,x_0,n})$ is $\mathscr{E}(S_{2,\mu,n})$ where

$$\mathscr{E}(S_{2,\mu,n}) = 2\binom{n}{2}|\sigma_{ij}| , \qquad i, j = 1, 2 . \tag{3.19}$$

4. *k*-Dimensional Scatter

The results of Section 3 extend readily to the case of a sample of size n, i.e., $(x_{1\alpha}, \cdots, x_{k\alpha}, \alpha = 1, \cdots, n)$, from a k-dimensional distribution having means $(\mu_1, \cdots, \mu_k)$ and covariance matrix $||\sigma_{ij}||$. A realized value of such a sample can be represented as a cluster of n points in Euclidean k-space. If $(x_{10}, \cdots, x_{k0})$ is an arbitrary point in the k-space, this point together with any k points $(x_{1\alpha_i}, \cdots x_{k\alpha_i})$, $i = 1, \cdots, k$, of the sample cluster determine a k-dimensional simplex whose squared volume (k-dimensional content) would be denoted by $\Delta^2_{\alpha_1,\cdots,\alpha_k,0}$. If $V^2_{\alpha_1,\cdots,\alpha_k,0}$ denotes the squared content of any one of the $k+1$ k-dimensional parallelotopes which can be formed having $(x_{10}, \cdots, x_{k0})$ and the k selected points of the sample as $k+1$ of its $k+2$ vertices, then

$$V^2_{\alpha_1,\cdots,\alpha_k,0} = (k!\, \Delta_{\alpha_1,\ldots,\alpha_k,0})^2 . \tag{4.1}$$

Note that the value of $V^2_{\alpha_1,\cdots,\alpha_k,0}$ remains unchanged under permutations of $\alpha_1, \cdots, \alpha_k$.

We define the scatter $S_{k,x_0,n}$ of the sample $(x_{1\alpha}, \cdots, x_{k\alpha}, \alpha = 1, \cdots, n)$ about $(x_{10}, \cdots, x_{k0})$ as the sum of $V^2_{\alpha_1,\cdots,\alpha_k,0}$ for all $\binom{n}{k}$ possible selections of $(\alpha_1, \cdots, \alpha_k)$ from $(1, \cdots, n)$ which we may write as

$$S_{k,x_0,n} = \sum_{1 \leq \alpha_1 < \cdots < \alpha_k \leq n} V^2_{\alpha_1,\cdots,\alpha_k,0} . \tag{4.2}$$

The value of $V^2_{\alpha_1,\cdots,\alpha_k,0}$ is given by the squared determinant

$$V^2_{\alpha_1,\cdots,\alpha_k,0} = \begin{vmatrix} x_{1\alpha_1} - x_{10} & \cdots & x_{k\alpha_1} - x_{k0} \\ \vdots & & \vdots \\ x_{1\alpha_k} - x_{10} & \cdots & x_{k\alpha_k} - x_{k0} \end{vmatrix}^2 . \tag{4.3}$$

Generalizing (3.4), one can see that, if the determinant on the right in (4.3) is squared, it can be expressed as a determinant which can be expanded into the sum of k^k determinants $D_{\beta_1,\ldots,\beta_k}$, where each β can take on any one of the integers $\alpha_1, \cdots, \alpha_k$, all such determinants having the value 0, except when $\beta_1, \cdots, \beta_k$ is a permutation of $\alpha_1, \cdots, \alpha_k$. Hence, there are only $k!$ determinants having values different from 0 (with probability 1). If $\beta_1, \cdots, \beta_k$ is a permutation of $\alpha_1, \cdots, \alpha_k$ then

$$D_{\beta_1,\ldots,\beta_k} = \begin{vmatrix} (x_{1\beta_1} - x_{10})^2 & \cdots & (x_{1\beta_k} - x_{10})(x_{k\beta_k} - x_{k0}) \\ \vdots & & \vdots \\ (x_{k\beta_1} - x_{k0})(x_{1\beta_1} - x_{10}) & \cdots & (x_{k\beta_k} - x_{k0})^2 \end{vmatrix} . \tag{4.4}$$

Hence

$$\sum_{1 \leq \alpha_1 < \cdots < \alpha_k \leq n} V^2_{\alpha_1,\cdots,\alpha_k,0} = \sum_{\beta_k=1}^{n} \cdots \sum_{\beta_1=1}^{n} D_{\beta_1,\ldots,\beta_k} . \tag{4.5}$$

But, using the extension of (3.7) for kth-order determinants, we see that the right-hand side of (4.5) may be written as the kth-order extension of the right-hand side of (3.8). Hence, adopting the notation of (3.9) for $i, j = 1, \cdots, k$, we have

$$S_{k,x_0,n} = |u_{ij} + n(\bar{x}_i - x_{i0})(\bar{x}_j - x_{j0})| , \tag{4.6}$$

which may be written in the alternative form

$$(4.7) \qquad S_{k,x_0,n} = |u_{ij}| \cdot \left[1 + n \sum_{i,j=1}^{k} u^{ij}(\bar{x}_i - x_{i0})(\bar{x}_j - x_{j0})\right],$$

where $||u^{ij}||$ is the inverse of the *internal-scatter matrix* $||u_{ij}||$ of the sample, which will be positive definite with probability 1 if the covariance matrix $||\sigma_{ij}||$ of the distribution from which the sample is drawn is positive definite. Hence:

The k-dimensional scatter $S_{k,x_0,n}$ is minimized by taking the pivotal point $(x_{10}, \cdots, x_{k0})$ as $(\bar{x}_1, \cdots, \bar{x}_k)$, the minimum of $S_{k,x_0,n}$ being the internal scatter of the sample; that is,

$$(4.8) \qquad \min_{x_0} (S_{k,x_0,n}) = S_{k,\bar{x},n} = |u_{ij}| .$$

Suppose we consider any non-singular linear transformation

$$(4.9) \qquad x_i = \sum_{j=1}^{k} c_{ij} y_j , \qquad i = 1, \cdots, n ,$$

having inverse

$$y_i = \sum_{j=1}^{k} c^{ij} x_j , \qquad i = 1, \cdots, n ,$$

and let the *image-sample* points in the y-space corresponding to $(x_{1\alpha}, \cdots, x_{k\alpha}, \alpha = 1, \cdots, n)$ be $(y_{1\alpha}, \cdots, y_{k\alpha}, \alpha = 1, \cdots, n)$, with $(y_{10}, \cdots, y_{k0})$ being the point corresponding to $(x_{10}, \cdots, x_{k0})$. Then the relation between $S_{k,x_0,n}$, the scatter of the sample cluster in the x-space, about $(x_{10}, \cdots, x_{k0})$ and $S'_{k,y_0,n}$, the scatter of the corresponding image-sample cluster in the y-space about $(y_{10}, \cdots, y_{k0})$, is given by

$$(4.10) \qquad S_{k,x_0,n} = |c_{ij}|^2 \cdot S'_{k,y_0,n} .$$

The problem of finding the value of $\mathscr{E}(S_{k,x_0,n})$ is a fairly straightforward extension of that in two dimensions. We have

$$(4.11) \qquad \mathscr{E}(V^2_{\alpha_1,\cdots,\alpha_k,0}) = \sum{}^{*}\mathscr{E}(D_{\beta_1,\cdots,\beta_k}) ,$$

where $\beta_1, \cdots, \beta_k$ is any permutation of $\alpha_1, \cdots, \alpha_k$ and $\sum^*$ denotes summation over all $k!$ such permutations. But for any permutation $\beta_1, \cdots, \beta_k$ we have, as an extension of (3.15), $\mathscr{E}(D_{\beta_1,\cdots,\beta_k}) = |\sigma_{ij} + (\mu_i - x_{i0})(\mu_j - x_{j0})|$. Hence

$$(4.12) \qquad \mathscr{E}(V^2_{\alpha_1,\cdots,\alpha_k,0}) = k! \; |\sigma_{ij} + (\mu_i - x_{i0})(\mu_j - x_{j0})| ,$$

which does not depend on $\alpha_1, \cdots, \alpha_k$. Since $S_{k,x_0,n}$ consists of the sum of $V^2_{\alpha_1,\cdots,\alpha_k,0}$ for all possible $\binom{n}{k}$ selections of k points from the cluster of n sample points, we have

$$(4.13) \qquad \mathscr{E}(S_{k,x_0,n}) = k! \binom{n}{k} |\sigma_{ij} + (\mu_i - x_{i0})(\mu_j - x_{j0})| .$$

This can be written alternatively as

$$(4.14)\qquad \mathscr{E}(S_{k,x_0,n}) = k!\binom{n}{k}\cdot|\sigma_{ij}|\cdot\left[1+\sum_{i,j=1}^{k}\sigma^{ij}(\mu_i - x_{i0})(\mu_j - x_{j0})\right],$$

from which we see that:

The mean value of the k-dimensional scatter $S_{k,x_0,n}$ about the pivotal point $(x_{10}, \cdots, x_{k0})$ attains its minimum if $(x_{10}, \cdots, x_{k0})$ is chosen as the center of gravity $(\mu_1, \cdots, \mu_k)$ of the distribution from which the sample is drawn. The minimum of $\mathscr{E}(S_{k,x_0,n})$ is $\mathscr{E}(S_{k,\mu,n})$; that is,

$$(4.15)\qquad \min_{x_0}\,[\mathscr{E}(S_{k,x_0,n})] = \mathscr{E}(S_{k,\mu,n}) = k!\binom{n}{k}|\sigma_{ij}|\,.$$

Since $\mathscr{E}(S_{1,\bar{x},n}) = (n-1)\sigma^2$ for the case of one-dimensional scatter, the question arises whether a corresponding result holds for the case of k-dimensional scatter, namely

$$(4.16)\qquad \mathscr{E}(S_{k,\bar{x},n}) = k!\binom{n-1}{k}|\sigma_{ij}|\,.$$

This result is true if the distribution from which the sample comes is a k-dimensional normal distribution with covariance matrix $||\sigma_{ij}||$. If $k=2$, it can be shown that (4.16) holds if the sample comes from any distribution in which $\{u_{11}, u_{22}, u_{12}\}$ and $\{\bar{x}_1, \bar{x}_2\}$ are independent sets of random variables. The most general conditions under which (4.16) holds for $k>2$ is an open question.

At this point it should be mentioned that, if the sample O_n comes from a k-dimensional normal distribution with covariance matrix $||\sigma_{ij}||$, we know much more about the distribution function of $S_{k,\bar{x},n}$ than is indicated in (4.16). As a matter of fact, the exact distribution function of the elements of the internal-scatter matrix $||u_{ij}||$ under this assumption is known. The probability density function $f(\{u_{ij}\})$ of $\{u_{ij}, i, j = 1, \cdots, k\}$ is given by

$$(4.17)\qquad f(\{u_{ij}\}) = \frac{|\sigma_{ij}|^{\frac{1}{2}(n-1)}\,|u_{ij}|^{\frac{1}{2}(n-k-2)}\exp\left(-\frac{1}{2}\sum_{i,j=1}^{k}\sigma^{ij}u_{ij}\right)}{2^{\frac{1}{2}k(n-1)}\pi^{\frac{1}{4}k(k-1)}\Gamma\left(\frac{n-1}{2}\right)\cdots\Gamma\left(\frac{n-k}{2}\right)}$$

over the region in the $\frac{1}{2}k(k+1)$-dimensional space of the $\{u_{ij}\}$ for which $||u_{ij}||$ is positive definite. This is the most fundamental distribution function in normal multivariate statistical theory. It was established by Wishart in 1928 [2]. The hth moment of $S_{k,\bar{x},n}$ is readily found from the Wishart distribution to be given by

$$(4.18)\qquad \mathscr{E}(S_{k,\bar{x},n})^h = |2\sigma_{ij}|^h\prod_{i=1}^{k}\frac{\Gamma\left(\frac{n-i}{2}+h\right)}{\Gamma\left(\frac{n-i}{2}\right)},$$

from which (4.16) follows, putting $h=1$. We also point out that the elements of the scatter matrix $||v_{ij}||$, where $v_{ij} = u_{ij} + n(\bar{x}_i - \mu_i)(_j\bar{x} - \mu_j)$, are identical with the distribution of the $\{u_{ij}\}$ with $n-1$ replaced by n.

5. Hotelling's T^2 and k-Dimensional Scatters

We have seen in Section 4 that the k-dimensional scatter of a sample is a minimum if the pivotal point is taken as the center of gravity $(\bar{x}_1, \cdots, \bar{x}_k)$ of the sample cluster, the minimum scatter being $S_{k,\bar{x},n}$. A simple comparison of this minimum scatter with the scatter $S_{k,\mu,n}$, about the mean $(\mu_1, \cdots, \mu_k)$ of the distribution from which the sample is drawn, by means of the ratio

(5.1) $$R_1 = S_{k,\bar{x},n}/S_{k,\mu,n} ,$$

yields [by utilizing (4.7) and (4.8)]

(5.2) $$R_1 = \frac{1}{[1 + T^2/(n-1)]} ,$$

where T^2 is given by

(5.3) $$T^2 = n(n-1) \sum_{i,j=1}^{k} u^{ij}(\bar{x}_i - \mu_i)(\bar{x}_j - \mu_j) .$$

If the sample comes from any multivariate normal distribution with means $(\mu_1, \cdots, \mu_k)$, *the probability element of* T^2 *is*

(5.4) $$\frac{\Gamma(n/2)}{(n-1)^{k/2}\Gamma(k/2)\Gamma((n-k)/2)}\left(1 + \frac{T^2}{n-1}\right)^{-n/2}(T^2)^{(k/2)-1}\, d(T^2) .$$

This distribution was established by Hotelling [3] and is one of the classical results of multivariate statistical theory.

It has been shown by Wilks [4] that R_1, and hence T^2, are equivalent to the Neyman-Pearson likelihood ratio test for the statistical hypothesis that a sample comes from a normal distribution with specified means $(\mu_1, \cdots, \mu_k)$, given that the sample is from *some* normal multivariate distribution.

It should be noted from (4.10) that under any non-singular linear transformation (4.9)

(5.5) $$R_1 = S_{k,\bar{x},n}/S_{k,\mu,n} = S_{k,\bar{y},n}/S_{k,\mu',n} ,$$

where $(\mu_1', \cdots, \mu_k')$ is the mean of distribution of $(y_1, \cdots, y_k)$. Thus R_1, and hence T^2, are invariant under (non-singular) linear transformations.

Now let us consider two independent samples O_{n_1} and O_{n_2} from k-dimensional distributions $F_1(x_1, \cdots, x_k)$ and $F_2(x_1, \cdots, x_k)$. Let $(\bar{x}_1^{(1)}, \cdots, \bar{x}_k^{(1)})$ and $(\bar{x}_1^{(2)}, \cdots, \bar{x}_k^{(2)})$ be the centers of gravity of the two samples, and let $||u_{ij}^{(1)}||$ and $||u_{ij}^{(2)}||$ be the two internal-scatter matrices of the samples, respectively. Let $||u_{ij}^{T}||$ be the internal-scatter matrix of the *grand sample* of size $n_1 + n_2$ obtained by pooling together the two samples O_{n_1} and O_{n_2}. Then we have the following relationship:

(5.6) $$||u_{ij}^{T}|| = ||u_{ij}^{W} + b_i b_j|| ,$$

where

$$b_i = \sqrt{\frac{n_1 n_2}{n_1 + n_2}} (\bar{x}_i^{(1)} - \bar{x}_i^{(2)}) , \tag{5.7}$$

$$u_{ij}^W = u_{ij}^{(1)} + u_{ij}^{(2)} .$$

The determinant $|u_{ij}^T|$ is the internal scatter of the grand sample and $|u_{ij}^W|$ is the *within-sample scatter* of the two samples O_{n_1} and O_{n_2}. Geometrically interpreted, $|u_{ij}^W|$ is the scatter one obtains as follows. Suppose one of the sample clusters is rigidly translated so that its center of gravity coincides with the center of gravity of the other sample cluster. Then $|u_{ij}^W|$ is the internal scatter of the resulting grand cluster of $n_1 + n_2$ points.

We observe from (5.6) that we may write

$$|u_{ij}^T| = |u_{ij}^W| \cdot \left[1 + \sum_{i,j=1}^{k} u_W^{ij} b_i b_j\right] , \tag{5.8}$$

where $||u_W^{ij}||$ is the inverse of $||u_{ij}^W||$. Now if we define R_2 as the following ratio of scatters

$$R_2 = |u_{ij}^W|/|u_{ij}^T| , \tag{5.9}$$

it will be seen from (5.8) that R_2 reduces to

$$R_2 = 1/[1 + T_2^2/(n_1 + n_2 - 1)] , \tag{5.10}$$

where

$$T_2^2 = \frac{n_1 n_2 (n_1 + n_2 - 2)}{n_1 + n_2} \sum_{i,j=1}^{k} u_W^{ij} (\bar{x}_i^{(1)} - \bar{x}_i^{(2)})(\bar{x}_j^{(1)} - \bar{x}_j^{(2)}) \tag{5.11}$$

is Hotelling's T^2-statistic for two samples. He showed [3] that, if both samples come from identical k-dimensional normal distributions, T_2^2 has the distribution (5.4) with n replaced by $n_1 + n_2 - 1$.

If two k-dimensional distributions (populations) have centers of gravity $(\mu_1, \cdots, \mu_k)$ and $(\mu_1', \cdots, \mu_k')$ and identical covariance matrices, namely $||\sigma_{ij}||$, Mahalanobis [5] has called the quantity

$$\Delta^2 = \sum_{i,j=1}^{k} \sigma^{ij} (\mu_i - \mu_i')(\mu_j - \mu_j') \tag{5.12}$$

the squared *statistical distance* between the two distributions.

The analog of (5.12) for two samples is

$$D^2 = T^2(n_1 + n_2)/(n_1 n_2) , \tag{5.13}$$

which may be similarly regarded as the squared *statistical distance* between the two samples. It should be noted that T^2/n, where T^2 is defined in (5.3), is, in a sense, a squared statistical distance between the sample and population.

Note that R_2, and hence T_2^2, and D_2, are invariant under non-singular linear transformations (the same transformation being applied to both samples).

It can be shown that R_2, and hence T_2^2 or D_2^2, are equivalent to the Neyman-Pearson likelihood ratio test for the statistical hypothesis that two samples come from identical k-dimensional normal distributions, given that they come from normal distributions having identical covariance matrices.

6. Statistical Scatter in Several Samples

The scatter ratios R_1 and R_2 can be readily extended to the case of r independent samples $O_{n_1}, \cdots, O_{n_r}$. Let $(\bar{x}_1^{(t)}, \cdots, \bar{x}_k^{(t)})$ be the center of gravity and $||u_{ij}^{(t)}||$ the internal-scatter matrix of sample O_{n_t}, $t = 1, \cdots, r$. Let $||u_{ij}^T||$ be the internal-scatter matrix of the grand sample of size $n = n_1 + \cdots + n_r$ obtained by pooling samples $O_{n_1}, \cdots, O_{n_r}$. Let $||u_{ij}^W|| = ||u_{ij}^{(1)} + \cdots + u_{ij}^{(r)}||$ and $||u_{ij}^B|| = ||u_{ij}^T - u_{ij}^W||$, and let R_r be the ratio of the within-sample scatter $|u_{ij}^W|$ to the internal scatter $|u_{ij}^T|$ of the grand sample, that is

$$R_r = |u_{ij}^W|/|u_{ij}^T| \, . \tag{6.1}$$

As in the two-sample case, R_r is equivalent to the Neyman-Pearson likelihood ratio test for the statistical hypothesis that the r samples come from identical k-dimensional normal distributions, given that they come from k-dimensional normal distributions with identical covariance matrices. If this hypothesis is true, that is, if the samples come from identical k-dimensional normal distributions, Wilks [4] has determined the moments of R_r from which it follows that if (a) $r - 1 < k$, the distribution of R_r is identical with the distribution of the product of $r - 1$ independent beta variables $u_1, \cdots, u_{r-1}$ where u_i has probability element on the interval $(0, 1)$ given by

$$\frac{\Gamma(\frac{1}{2}(n-i))}{(\Gamma\frac{1}{2}k)\Gamma(\frac{1}{2}(n+k-i))} u_i^{\frac{1}{2}(n+k-i-2)}(1-u_i)^{\frac{1}{2}(k-2)}\,du_i\,, \qquad i = 1, \cdots, r-1\,; \tag{6.2}$$

and if (b) $r - 1 \geqq k$, the distribution of R_r is identical with the distribution of the product of k independent beta variables $v_1, \cdots, v_k$ where v_j has probability element on $(0, 1)$

$$\frac{\Gamma(\frac{1}{2}(n-j))}{\Gamma(\frac{1}{2}(r-1)\Gamma(\frac{1}{2}(n-r+1-j))} v_i^{\frac{1}{2}(n-r-1-j)}(1-v_j)^{\frac{1}{2}(r-3)}\,dv_j\,, \qquad i = 1, \cdots, k\,. \tag{6.3}$$

More generally, if $||u_{ij}||$ is any symmetric matrix whose elements have the Wishart distribution with density function given by (4.17), and if $(y_{1\gamma}, \cdots, y_{k\gamma})$, $\gamma = 1, \cdots, r-1$, are $r - 1$ independent sets of random variables, each set having a k-dimensional normal distribution with 0 means and covariance matrix $||\sigma_{ij}||$, then the ratio

$$R_r^* = \frac{|u_{ij}|}{\left|u_{ij} + \sum_{\gamma=1}^{r-1} y_{i\gamma}y_{j\gamma}\right|} \tag{6.4}$$

has identically the same distribution as R_r defined in (6.1). There are various applications of R_r^* in multivariate analysis of variance and multivariate regression analysis, but space does not permit a discussion of these applications here.

7. Principal Components and Statistical Scatter

Suppose that we have a sample of size n from a k-dimensional distribution and that the internal-scatter matrix of the sample is $||u_{ij}||$. We wish to linearly map (project) the cluster of sample points in our k-dimensional Euclidean space $\mathscr{R}_k$ orthogonally onto an s-dimensional Euclidean space $\mathscr{R}_s$ so that, for any $s \leqq k$, the s-dimensional scatter of the resulting n points in $\mathscr{R}_s$ will be as large as possible. The problem is to find the linear mapping which will accomplish this objective. This amounts to finding s orthogonal unit vectors

$$(c_{1p}, \cdots, c_{kp}), \qquad p = 1, \cdots, s, \tag{7.1}$$

that is, vectors satisfying

$$\sum_{i=1}^{k} c_{ip} c_{iq} = \delta_{pq}, \tag{7.2}$$

where $\delta_{pq} = 1$ if $p = q$, and $\delta_{pq} = 0$ if $p \neq q$, such that the internal s-dimensional scatter $||v_{pq}||$ of the *image sample* $(y_{1\alpha}, \cdots, y_{s\alpha}, \quad \alpha = 1, \cdots, n)$ in $\mathscr{R}_s$ is a maximum where

$$y_{p\alpha} = \sum_{i=1}^{k} c_{ip} x_{i\alpha}, \qquad p = 1, \cdots, k. \tag{7.3}$$

The required pth vector $(c_{1p}, \cdots, c_{kp})$ is the (unit vector) solution of the set of equations

$$\sum_{j=1}^{k} (u_{ij} - l_p \delta_{ij}) c_{jp} = 0, \qquad i = 1, \cdots, k, \tag{7.4}$$

where l_p is the pth largest root of the equation

$$|u_{ij} - l\delta_{ij}| = 0. \tag{7.5}$$

The probability is 1 that all s roots of (7.5) will be different. The maximum value of the scatter $|v_{pq}|$, which occurs for the above choice of the unit vectors, is the product of the s largest roots of (7.5), i.e.,

$$\max |v_{pq}| = l_1 l_2 \cdots l_s. \tag{7.6}$$

Thus, by ordering the vectors to correspond to $l_1, \cdots, l_s$, where $l_1 > \cdots > l_s$, we obtain vectors such that, no matter what value of s we choose among the integers $1, \cdots, k$, the set of s vectors thus chosen will yield an orthogonally mapped image sample $(y_{1\alpha}, \cdots, y_{k\alpha}, \alpha = 1, \cdots, n)$ in $\mathscr{R}_s$ which will maximize the scatter $|v_{pq}|$. Thus if $s = 1$, the components of the unit vector $(c_{11}, \cdots, c_{k1})$, corresponding to the largest root of (7.5), provide direction cosines such that if the points of the sample cluster in $\mathscr{R}_k$ are projected onto any line in $\mathscr{R}_k$ with these direction cosines by means of $y_{1\alpha} = \sum_{i=1}^{k} c_{i1} x_{i\alpha}$, the resulting one-dimensional cluster of image-sample points $(y_{11}, \cdots, y_{1n})$ along the line have maximum scatter, namely l_1. If $s = 2$, the unit vectors $(c_{11}, \cdots, c_{k1})$ and $(c_{12}, \cdots, c_{k2})$, which are orthogonal, determine an orientation for an ordinary (two-dimensional) plane in $\mathscr{R}_k$ such that the orthogonal projection of the cluster in $\mathscr{R}_k$ onto any plane parallel to such a plane by means of

$$y_{1\alpha} = \sum_{i=1}^{k} c_{i1} x_{i\alpha} ,$$

$$y_{2\alpha} = \sum_{i=1}^{k} c_{i2} x_{i\alpha} ,$$

yields a two-dimensional cluster of image-sample points $(y_{1\alpha}, y_{2\alpha}, \alpha = 1, \cdots, n)$ having maximum two-dimensional scatter, namely $l_1 l_2$. And so on for larger values of s.

It will be noted from (7.5) that

$$l_1 + \cdots + l_k = u_{11} + \cdots + u_{kk} . \tag{7.7}$$

If we divide (7.7) by $n - 1$ and put $u_{ii}/(n-1) = s_i^2$, then s_i^2 is an unbiased estimator for the variance σ_{ii} and we have

$$\frac{l_1}{n-1} + \cdots + \frac{l_k}{n-1} = s_1^2 + \cdots + s_k^2 . \tag{7.8}$$

The sum on the right-hand side of (7.8) is the total of *sample variances* and is an unbiased estimator for the total of the population variances $\sigma_{11} + \cdots + \sigma_{kk}$. The set of quantities $l_1/(n-1), \cdots, l_k/(n-1)$, which decrease in value, are called ***principal components*** of the total of sample variances by Hotelling [6], who developed the theory of principal components in mathematical statistics.

The ordered roots $l_1 > \cdots > l_k$ of Equation (7.5) are, of course, random variables. The distribution theory of $l_1, \cdots, l_k$ is known only in the case where the distribution from which the sample is drawn is the k-dimensional normal distribution with covariance matrix $||\sigma_{ij}||$ equal to $||\sigma^2\delta_{ij}||$ where $||\delta_{ij}||$ is the unit matrix and $\sigma^2 > 0$. More generally, if the sample is from a k-dimensional normal distribution with covariance matrix $||\sigma_{ij}||$, the distribution of the roots $l_1 > \cdots > l_k$ of the equation,

$$|u_{ij} - l\sigma_{ij}| = 0 , \tag{7.9}$$

has the probability density function

$$\frac{\pi^{k/2} (\prod_{i=1}^{k} l_i)^{\frac{1}{2}(n-k-2)} \prod_{1 \le i < j \le k} (l_i - l_j) \exp(-[\frac{1}{2}\sum_{i=1}^{k} l_i])}{2^{\frac{1}{2}k(n-1)} \prod_{i=1}^{k} \Gamma((k+1-i)/2)\Gamma((n-i)/2)} , \tag{7.10}$$

over the region $0 < l_k < \cdots < l_1 < +\infty$, a result originally obtained by Fisher [7] and Hsu [8], and more recently by Mood [9], James [10], and Olkin [11], using different methods.

8. Discriminant Analysis and Statistical Scatter

8.1 The Case of Two Samples. Suppose we have two independent samples $(x_{1\alpha_1}^{(1)}, \cdots, x_{k\alpha_1}^{(2)}, \alpha_1 = 1, \cdots, n)$ and $(x_{1\alpha_2}^{(2)}, \cdots, x_{k\alpha_2}^{(2)}, \alpha_2 = 1, \cdots, n_2)$, from k-dimensional distributions. These samples may, of course, be represented as two clusters of points in Euclidean k-space. We want to project these two clusters of points orthogonally onto a line so that the means of the two (one-dimensional) image samples are as far apart as possible relative to within-sample scatter of the two (one-dimensional) image samples. The idea is to separate the two samples as much as possible by using a suitably chosen linear function of k measurements on each element in each sample.

We shall use the notation of Section 5 and let $||u_{ij}^{(1)}||$ and $||u_{ij}^{(2)}||$ be the internal-scatter matrices of the two samples, while $||u_{ij}^{W}|| = ||u_{ij}^{(1)} + u_{ij}^{(2)}||$. Let $||u_{ij}^{T}||$ be the internal-scatter matrix of the grand sample of size $n_1 + n_2$ obtained by pooling the two samples.

For an arbitrary vector $(c_1, \cdots, c_k)$ consider the linear mapping of O_{n_1} and O_{n_2} onto a line (the y-axis, say) where

$$y_{\alpha}^{(1)} = \sum_{i=1}^{k} c_i x_{i\alpha}^{(1)} , \tag{8.1}$$

$$y_{\alpha}^{(2)} = \sum_{i=1}^{k} c_i x_{i\alpha}^{(2)} .$$

Let $v^{(1)}$ and $v^{(2)}$ be the internal scatters of the image samples $(y_1^{(1)}, \cdots, y_n^{(1)})$ and $(y_1^{(2)}, \cdots, y_n^{(2)})$, respectively. Let v^T be the internal one-dimensional scatter of the grand sample of size $n_1 + n_2$ obtained by pooling the two image samples. Let v^W be the within-sample scatter of the two samples, that is, $v^W = v^{(1)} + v^{(2)}$. Our problem, which we can now state more precisely than in the first paragraph, is to find the particular vector $(c_1', \cdots, c_k')$ which minimizes the ratio v^W/v^T for a fixed value K of v^T. The required vector is the solution $(c_1', \cdots, c_k')$ of the equations

$$\sum_{j=1}^{k} (u_{ij}^{B} - l_1 u_{ij}^{W})c_j = 0 , \qquad i = 1, \cdots, k \tag{8.2}$$

subject to $v^W = K$, where $||u_{ij}^{B}|| = ||u_{ij}^{T} - u_{ij}^{W}||$, and l_1 is the non-zero root of

$$|u_{ji}^{B} - l u_{ij}^{W}| = 0 . \tag{8.3}$$

For the value of l_1 we have

$$l_1 = T_2^2/(n_1 + n_2 - 2) , \tag{8.4}$$

where T_2^2 is the Hotelling T^2 for two samples, given by (5.11). The minimum value of the ratio v^W/v^T, which is provided by the vector $(c_1', \cdots, c_k')$ is $1/(1 + l_1)$.

The procedure by which the vector $(c_1', \cdots, c_k')$ is determined so as to minimize v^W/v^T for a fixed value K of v^W is called (two-sample) *discriminant analysis*.

8.2 The Case of Several Samples. In the preceding paragraphs we have discussed the problem of finding the linear function of the k observations on each member of each sample which separates the two samples as much as possible. In the case of three k-dimensional samples, we would want to linearly map the three samples onto a two-dimensional space (ordinary plane) so that, for a fixed internal scatter of the grand image sample obtained by pooling the three two-dimensional image samples, the within-sample scatter of the three two-dimensional image samples is minimized. In general, if we have $s + 1$ k-dimensional samples, $s < k$, we want to linearly map these $s + 1$ samples onto an s-dimensional space so that, for a fixed value of the internal scatter of the grand image sample obtained by pooling all $s+1$ s-dimensional image samples, the within-sample scatter of the $s+1$ s-dimensional samples is made as small as possible.

More precisely, suppose

$$(x_{1\alpha_\gamma}^{(\gamma)}, \cdots, x_{k\alpha_\gamma}^{(\gamma)}, \quad \alpha_\gamma = 1, \cdots, n_\gamma, \gamma = 1, \cdots, s+1) \tag{8.5}$$

are $s+1$ k-dimensional samples, where $s < k$. Let $||u_{ij}^{(\gamma)}||$ be the internal scatter matrix and $(\bar{x}_1^{(\gamma)}, \cdots, \bar{x}_k^{(\gamma)})$ the center of gravity of the γth sample. Let $||u_{ij}^T||$ be the internal-scatter matrix of the grand sample of size $n = n_1 + \cdots + n_{s+1}$ obtained by pooling the $s+1$ samples. Let

$$||u_{ij}^W|| = ||u_{ij}^{(1)} + \cdots + u_{ij}^{(s+1)}|| \text{ and } ||u_{ij}^B|| = ||u_{ij}^T - u_{ij}^W|| .$$

Let $(c_{1p}, \cdots, c_{kp})$, $p = 1, \cdots, s$, be s linearly independent vectors and consider the mapping of the γth k-dimensional sample cluster into the γth s-dimensional image-sample cluster provided by

$$y_{p\alpha_\gamma}^{(\gamma)} = \sum_{i=1}^{k} c_{ip} x_{i\alpha_\gamma}^{(\gamma)}, \quad p = 1, \cdots, s, \alpha_\gamma = 1, \cdots, n_\gamma . \tag{8.6}$$

Let $||v_{pq}^{(\gamma)}||$ be the internal scatter of the γth s-dimensional image sample. Let $||v_{pq}^T||$ be the internal scatter of the grand s-dimensional image sample obtained by pooling the $s+1$ s-dimensional image samples. Let

$$||v_{pq}^W|| = ||v_{pq}^{(1)} + \cdots + v_{pq}^{(s+1)}|| \text{ and } ||v_{pq}^B|| = ||v_{pq}^T - v_{pq}^W|| .$$

Our problem is to determine the s vectors $(c_{1p}, \cdots, c_{kp})$, $p = 1, \cdots, s$, so as to minimize the ratio of scatters

$$|v_{pq}^W|/|v_{pq}^T| , \tag{8.7}$$

for a fixed value K of $|v_{pq}^T|$. The required vector $(c_{1p}, \cdots, c_{kp})$, $p = 1, \cdots, s$, is the solution of the equations

$$\sum_{j=1}^{k} (u_{ij}^B - l_p u_{ij}^W) c_{jp} = 0 , \qquad i = 1, \cdots, k . \tag{8.8}$$

where $l_1, \cdots, l_s$ are the s non-zero roots of

$$|u_{ij}^B - l u_{ij}^W| = 0 \tag{8.9}$$

and where for convenience we order them as follows: $l_1 > \cdots > l_s$. The vectors $(c_{1p}, \cdots, c_{kp})$, $p = 1, \cdots, s$, are scaled so that $|v_{pq}^W| = K$. The value of $|v_{pq}^W|/|v_{pq}^T|$ subject to fixed $|v_{pq}^W|$, which is produced by the choice of the vectors given above, is a minimum, which, as a matter of fact, is exactly the same as the value of $|u_{ij}^W|/|u_{ij}^T|$, which is $1/[(1+l_1) \cdots (1+l_s)]$, that is,

$$\min \frac{|v_{pq}^W|}{|v_{pq}^T|} = \frac{|u_{ij}^W|}{|u_{ij}^T|} = \frac{1}{(1+l_1) \cdots (1+l_s)} . \tag{8.10}$$

Finally, we should remark that we have considered only the case where $s < k$. The results extend with no particular difficulty to the less interesting case $s \geqq k$. In this situation, Equation (8.9) will have exactly k roots $l_1 > \cdots > l_k$, and k vectors $(c_{1p}, \cdots, c_{kp})$, $p = 1, \cdots, k$, are required.

The roots $l_1 > \cdots > l_s$ for $s < k$ or $l_1 > \cdots > l_k$ for $s \geqq k$ are, of course, random variables. Their distribution is known only in the case where all samples come from identical k-dimensional normal distributions. In the case $s \geqq k$ the probability function of $l_1, \cdots, l_s$ is

$$K\left[\prod_{i=1}^{k}(1+l_i)\right]^{-\frac{1}{2}(n+s)}\left[\prod_{i=1}^{k} l_i\right]^{\frac{1}{2}(s-k-1)}\prod_{1\leq i<j\leq k}(l_i - l_j)\,, \tag{8.11}$$

over the region for which $0 < l_k < \cdots < l_1 < +\infty$, where

$$K = \pi^{k/2}\prod_{i=1}^{k}\frac{\Gamma\left(\frac{n+s+1-i}{2}\right)}{\Gamma\left(\frac{n+1-i}{2}\right)\Gamma\left(\frac{s+1-i}{2}\right)\Gamma\left(\frac{k+1-i}{2}\right)}\,.$$

and $n = n_1 + \cdots + n_{s+1} - s - 1$.

For the case $s < k$, the distribution of $l_1, \cdots, l_s$ is given by (8.11) with k, n, s replaced by $s, n + s - k, k$, respectively. These distributions were originally obtained by Fisher [7] and Hsu [8], and more recently by Mood [9], James [10], Olkin [11], and Olkin and Roy [12], using other methods. Roy [13] determined the distributions except for the value of K. His results appeared almost simultaneously with those of Fisher and Hsu.

9. Canonical Correlation and Statistical Scatter

Suppose we have a sample O_n from a k-dimensional distribution, and that we wish to find a linear function of the first s variables of the sample and a linear function of the last t variables of the sample $(s + t = k)$ so that these two linear functions will have the highest possible correlation coefficient. In a practical situation the two linear functions may be regarded as indices constructed from the first s and last t variables, respectively, and one of these indices may be used to predict or estimate the other.

More precisely, suppose $||u_{ij}||$ is the internal-scatter matrix of the sample O_n and let $||u_{pq}||$, $p, q = 1, \cdots, s$ be the internal-scatter matrix of the first s variables and $||u_{vw}||$, $v, w = s+1, \cdots, k$ that of the last t variables, and for convenience we take $t \geqq s > 0$. We may write

$$||u_{ij}|| = \left\|\begin{array}{c:c} u_{pq} & u_{pw} \\ \hdashline u_{vq} & u_{vw}\end{array}\right\|. \tag{9.1}$$

Now consider two vectors, $(c_{11}, \cdots, c_{1s})$ and $(c_{2,s+1}, \cdots, c_{2k})$, and put

$$y_{1\alpha} = \sum_{p=1}^{s} c_{1p}x_{p\alpha}\,, \tag{9.2}$$

$$y_{2\alpha} = \sum_{v=s+1}^{k} c_{2v}x_{v\alpha}\,.$$

Let $\left\|\begin{matrix} U_{11} & U_{12} \\ U_{21} & U_{22} \end{matrix}\right\|$ be the internal-scatter matrix of the two-dimensional image sample $(y_{1\alpha}, y_{\alpha 2}, \alpha = 1, \cdots, n)$. Our problem is to determine the two vectors so as to minimize the scatter $\left|\begin{matrix} U_{11} & U_{12} \\ U_{21} & U_{22} \end{matrix}\right|$ for fixed values of U_{11} and U_{22} which we may take as unity. This requirement is equivalent, of course, to determining the vectors so that the correlation coefficient

$$R = \frac{U_{12}}{\sqrt{U_{11}U_{22}}} \tag{9.3}$$

is a maximum subject to fixed values of U_{11} and U_{22}, which we may take as unity.

The required vectors are the solutions of the equations

$$\begin{aligned} -\sqrt{g_1}\sum_{q=1}^{s} u_{pq}c_{1q} + \sum_{w=s+1}^{k} u_{pw}c_{2w} &= 0\,, \\ \sum_{q=1}^{s} u_{vq}c_{1q} - \sqrt{g_1}\sum_{w=s+1}^{k} u_{vw}c_{2w} &= 0\,, \end{aligned} \tag{9.4}$$

where g_1 is the largest root of the equation

$$\left|\begin{array}{c:c} gu_{pq} & -u_{pw} \\ \hdashline -u_{vq} & u_{vw} \end{array}\right| = 0\,. \tag{9.5}$$

The maximum value of R which is produced by the vectors satisfying (9.4) subject to $U_{11} = U_{22} = 1$ is $\sqrt{g_1}$.

Since $t \geqq s > 0$, it can be shown that (9.5) has s roots $g_1, \cdots, g_s$, $0 < g_s < \cdots < g_1 < 1$, and s corresponding pairs of vectors obtained from (9.4) by substituting the various roots for g_1. The values of R corresponding to these roots are $\sqrt{g_1}, \cdots, \sqrt{g_s}$ and are called the *canonical correlation coefficients* between the first s variables and the last t variables in the sample. Canonical correlation theory was developed by Hotelling [14].

The roots $g_1, \cdots, g_s$ are, of course, random variables, and their distribution is known only in the case where the first s variables and the last t variables in the k-dimensional distribution from which O_n comes are independent, with the distribution of the first s variables being k-dimensional normal, and the last t variables having an arbitrary t-dimensional distribution. The distribution of $g_1, \cdots, g_s$, under these conditions, has the density function

$$K_0\left[\prod_{p=1}^{s}(1-g_p)\right]^{\frac{1}{2}(n-t-s-2)}\left[\prod_{p=1}^{s} g_p\right]^{\frac{1}{2}(t-s-1)}\prod_{1\leqq p<q\leqq s}(g_p - g_q)\,, \tag{9.6}$$

where

$$K_0 = \pi^{s/2}\prod_{p=1}^{s}\frac{\Gamma\left(\frac{n-p}{2}\right)}{\Gamma\left(\frac{n-t-p}{2}\right)\Gamma\left(\frac{t+1-p}{2}\right)\Gamma\left(\frac{s+1-p}{2}\right)}\,, \tag{9.7}$$

over the region $0 < g_s < \cdots < g_1 < 1$. Except for the exponents and constants K and K_0, the distribution theory of $g_1, \cdots, g_s$ is exactly the same as the distribution theory of the ratios $l_1/(1 + l_1), \cdots, l_s/(1 + l_s)$, where $l_1, \cdots, l_s$ are the roots of (8.9).

References

[1] Anderson, T. W. *Introduction to Multivariate Statistical Analysis*. New York: Wiley, 1957.

[2] Wishart, J. The generalized product moment distribution in samples from a normal multivariate population. *Biometrika*, 1928, **20A**, 32-52.

[3] Hotelling, H. The generalization of Student's ratio. *Ann. Math. Stat.*, 1931, **2**, 360-78.

[4] Wilks, S. S. Certain generalizations in the analysis of variance. *Biometrika*, 1932, **24**, 471-94.

[5] Mahalanobis, P. C. On the generalized distance in statistics. *Proc. Nat. Inst. Sci. India*, 1936, **12**, 49-55.

[6] Hotelling, H. Analysis of a complex of statistical variables into principal components. *J. Educ. Psychol.*, 1933, **24**, 417-41, 498-520.

[7] Fisher, R. A. The sampling distribution of some statistics obtained from non-linear equations. *Ann. Eugen.*, 1939, **9**, 238-49.

[8] Hsu, P. C. On the distribution of the roots of certain determinantal equations. *Ann. Eugen.*, 1939, **9**, 250-58.

[9] Mood, A. M. On the distribution of normal second-moment matrices. *Ann. Math. Stat.*, 1951, **22**, 266-73.

[10] James, A. T. Normal multivariate analysis and the orthogonal group. *Ann. Math. Stat.*, 1954, **25**, 40-75.

[11] Olkin, I. On distribution problems in multivariate analysis. Institute of Statistics Mimeograph Series Report No. 43, Chapel Hill, N.C.: University of North Carolina, 1951.

[12] Olkin, I., and Roy, S. N. On multivariate distribution theory. *Ann. Math. Stat.*, 1954, **25**, 329-39.

[13] Roy, S. N. *p*-statistics or some generalizations in analysis of variance appropriate to multivariate problems. *Sankhyā*, 1939, **4**, 381-96.

[14] Hotelling, H. Relations between two sets of variates. *Biometrika*, 1936, **28**, 321-77.

41 CONVERGENCE OF THE EMPIRIC DISTRIBUTION FUNCTION ON HALF-SPACES[1]

J. WOLFOWITZ
Cornell University

Let (X, Y) be a chance variable, and let $(X_1, Y_1), (X_2, Y_2), \cdots$ be a sequence of independent chance variables with the same distribution function (d.f.) as (X, Y). Define the function

$$\psi_x(a) = \begin{cases} 0, & x \leqq a, \\ 1, & x > a, \end{cases}$$

and the empiric d.f. V_n of any n chance variables $Z_1, \cdots, Z_n$ by

$$V_n(x) = \frac{1}{n} \sum_{i=1}^{n} \psi_x(Z_i), \qquad -\infty < x < \infty.$$

Let $F(x \mid p)$ denote the d.f. of $pX + Y$, where p is a real parameter. Let $F_n(x \mid p)$ denote the empiric d.f. of $pX_i + Y_i$, $i = 1, \cdots, n$. Finally, if V and W are two (one-dimensional) d.f.'s, define $\delta(V, W) = \sup_{-\infty<x<\infty} | V(x) - W(x)|$.

The theorem below was proved by Fortet and Mourier [1] (see also Blum [2]) under the condition that the d.f. of (X, Y) be absolutely continuous (with respect to Borel measure), and by the present author [3] under the condition that X and Y be independent. In the present paper we impose no restrictions whatever on the d.f. of (X, Y). The present theorem can also be extended, as indicated in [1] and [3], so as to weaken the requirement that $(X_1, Y_1), (X_2, Y_2), \cdots$ be independent. The extension to a chance variable $(X, Y, \cdots, Z)$ with any finite number of components is trivial. It is also possible to extend the result to much more complicated regions than half-spaces.

We shall prove the following:

Theorem. *We have*

$$\Pr\left\{\lim_{n\to\infty} \sup_{-\infty<p<\infty} \delta(F(x \mid p), F_n(x \mid p)) = 0\right\} = 1.$$

---

[1] Research supported by the U. S. Air Force under Contract No. AF18(600)-685, monitored by the Office of Scientific Research, USAF Air Research and Development Command.

The proof of the theorem will begin with the proof of several lemmas.

LEMMA 1. *Suppose that the d.f. of (X, Y) is such that no line*[2] *in the plane of (X, Y) has positive probability. Suppose that $0 \leqq Q \leqq 1$, and α and ε are positive. There exist positive functions $\beta(\alpha, Q)$ and $N(\alpha, Q, \varepsilon)$ such that*

$$\Pr\{\delta(F(x \mid p), F_n(x \mid p)) < \alpha, \mid p - Q \mid \leqq \beta \text{ and } p \geqq 0, \quad n = N, N+1, \cdots\} > 1 - \varepsilon. \tag{1}$$

PROOF. We take $\alpha < \frac{1}{2}$, as indeed we may. Let H be so large that

$$\Pr\{\mid X \mid > H\} < \frac{\alpha}{64}, \tag{2}$$

so that

$$\sup_x \mid F(x \mid Q) - \Pr\{QX + Y < x, \mid X \mid \leqq H\} \mid < \frac{\alpha}{64}. \tag{3}$$

Under the hypotheses of the lemma, the d.f. $F(x \mid p)$ is continuous and hence uniformly continuous, for every p. For suitable h, therefore,

$$\sup_x \mid F(x + h \mid Q) - F(x \mid Q) \mid < \frac{\alpha}{64}. \tag{4}$$

From (2), (3), and (4) we easily obtain

$$\delta(F(x \mid Q), F(x \mid p)) < \frac{\alpha}{8}, \tag{5}$$

for all positive p in a sufficiently small neighborhood of Q. By the Glivenko-Cantelli theorem, for N sufficiently large,

$$\Pr\left\{(\text{number of } i,\ 1 \leqq i \leqq n, \text{ such that } \mid X_i \mid > H) < \frac{n\alpha}{32}, n = N, N+1, \cdots\right\} > 1 - \frac{\varepsilon}{3}, \tag{6}$$

and

$$\Pr\left\{\sup_x \mid (\text{number of } i,\ 1 \leqq i \leqq n, \text{ such that } \mid X_i \mid \leqq H \text{ and } QX_i + Y_i < x) - nF(x \mid Q) \mid < \frac{n\alpha}{16}, n = N, N+1, \cdots\right\} > 1 - \frac{\varepsilon}{3}. \tag{7}$$

From (4), (6), and (7), therefore, for β sufficiently small,

$$\Pr\left\{\delta(F(x \mid Q), F_n(x \mid p)) < \frac{\alpha}{2}, \mid p - Q \mid \leqq \beta \text{ and } p \geqq 0, n = N, N+1, \cdots\right\} > 1 - \varepsilon. \tag{8}$$

The lemma follows from (5) and (8).

[2] By "line" we mean straight line.

LEMMA 2. *Suppose that the d.f. of (X, Y) is such that no line in the plane of (X, Y) has positive probability. Let α and ε be positive. For N sufficiently large we have*

$$\Pr\{\delta(F(x \mid p), F_n(x \mid p)) < \alpha,\ 0 \leqq p \leqq 1,\ n = N, N+1, \cdots\} > 1 - \varepsilon . \tag{9}$$

PROOF: To each p, $0 \leqq p \leqq 1$, assign the neighborhood of half-width $\beta(\alpha, p)$. A finite number, say K, of these neighborhoods suffice to cover the interval $[0, 1]$. Let these neighborhoods be centered at $t_1, \cdots, t_K$. Let

$$N = \max_{1 \leqq i \leqq K} \{N(\alpha, t_i, \varepsilon/K)\} .$$

Lemma 2 follows at once.

LEMMA 3. *The theorem holds if the d.f. of (X, Y) is such that no line in the plane of (X, Y) has positive probability.*

PROOF. Let α and ε be *any* positive numbers. We apply Lemma 2 to the chance variable $pY + X$ for $0 \leqq p \leqq 1$. This gives, for suitable N,

$$P\{\delta(F(x \mid p), F_n(x \mid p)) < \alpha,\ 1 \leqq p < \infty,\ n = N,\ N+1, \cdots\} > 1 - \varepsilon . \tag{10}$$

Combining (9) and (10), noting that corresponding results hold for negative p, taking the largest of the four N, and replacing ε by $\varepsilon/4$, we obtain Lemma 3, since α and ε are arbitrary.

PROOF OF THE THEOREM. Let $d_1, d_2, \cdots$ be the (necessarily denumerable) points of the plane of (X, Y) to which the d.f. of (X, Y) assigns positive probability. Let $L_1, L_2, \cdots$ be the (necessarily denumerable) lines of the plane of (X, Y) to which the d.f. of (X, Y) would assign positive probability even if the points $d_1, d_2, \cdots$ were deleted from the plane. Write the d.f. G of (X, Y) as

$$G(x, y) \equiv a G_1(x, y) + (1 - a) G_2(x, y) , \tag{11}$$

where the d.f. G_1 assigns probability 1 to the plane with $\{d_1, d_2, \cdots\}$ and $\{L_1, L_2, \cdots\}$ deleted, and the d.f. G_2 assigns probability 1 to the union U of $\{d_1, d_2, \cdots\}$ and $\{L_1, L_2, \cdots\}$. If $a = 1$, the theorem is already proved (Lemma 3). If $a = 0$, the proof will be obvious. We therefore assume $0 < a < 1$.

Let M_2 be the number of i, $1 \leqq i \leqq n$, for which (X_i, Y_i) is in U. Let $M_1 = n - M_2$. Define $M_2G_{2n}(x \mid p)$ as the number of i, $1 \leqq i \leqq n$, such that $pX_i + Y_i < x$ and (X_i, Y_i) is in U. Define $M_1G_{1n}(x \mid p)$ as the number of i, $1 \leqq i \leqq n$, such that $pX_i + Y_i < x$ and (X_i, Y_i) is not in U. Obviously

$$nF_n(x \mid p) = M_1G_{1n}(x \mid p) + M_2G_{2n}(x \mid p) , \tag{12}$$

and

$$F(x \mid p) = a\, G_1(x \mid p) + (1 - a)\, G_2(x \mid p) , \tag{13}$$

where $G_1(x \mid p)$ and $G_2(x \mid p)$ are defined analogously to $F(x \mid p)$.

Let α and ε be any positive numbers. Let s be sufficiently large so that G_2 assigns probability greater than $(1 - \alpha/16)$ to the set $S = \{d_1, \cdots, d_s, L_1, \cdots, L_s\}$. Let M_2' be the number of i, $1 \leqq i \leqq n$, such that (X_i, Y_i) is in S, and let $M_2'' = M_2 - M_2'$. We define $G_2'(x \mid p)$, $G_{2n}'(x \mid p)$, $G_2''(x \mid p)$, $G_{2n}''(x \mid p)$ analogously to the unaccented functions. Obviously

$$nF_n(x \mid p) = M_1 G_{1n}(x \mid p) + M_2' G_{2n}'(x \mid p) + M_2'' G_{2n}''(x \mid p) \ . \tag{14}$$

For N sufficiently large we have

$$\Pr\left\{\delta(G_{2n}'(x \mid p), G_2'(x \mid p)) < \frac{\alpha}{16}, \ -\infty < p < \infty, \ n = N, N+1, \cdots\right\} > 1 - \frac{\varepsilon}{4} \tag{15}$$

by the law of large numbers and the Glivenko-Cantelli theorem,

$$\Pr\left\{\delta(G_{1n}(x \mid p), G_1(x \mid p)) < \frac{\alpha}{16}, \ -\infty < p < \infty, \ n = N, N+1, \cdots\right\} > 1 - \frac{\varepsilon}{4} \tag{16}$$

by Lemma 3,

$$\Pr\left\{\left|\frac{M_1}{n} - a\right| < \frac{\alpha}{16}, \ n = N, \ N+1, \cdots\right\} > 1 - \frac{\varepsilon}{4} \tag{17}$$

and

$$\Pr\left\{\left|\frac{M_2''}{n}\right| < \frac{\alpha}{8}, \ n = N, \ N+1, \cdots\right\} > 1 - \frac{\varepsilon}{4} \tag{18}$$

by the law of large numbers. The theorem follows from (13), (14), (15), (16), (17), and (18), since α and ε were arbitrary.

References

[1] Fortet, R., and Mourier, E. Convergence de la répartition empirique vers la répartition théorique. *Ann. Ecole Normale Sup.*, 1953, **60**, 267.

[2] Blum, J. R. On the convergence of empiric distribution functions. *Ann. Math. Stat.*, 1955, **26**, 527.

[3] Wolfowitz, J. Generalization of the theorem of Glivenko-Cantelli. *Ann. Math. Stat.*, 1954, **25**, 131.

42 ANALYSIS OF TWO-FACTOR CLASSIFICATIONS WITH RESPECT TO LIFE TESTS[1]

M. ZELEN

National Bureau of Standards

1. Summary

Recently, much attention has been given to a study of the exponential distribution with particular emphasis on applications in life testing, e.g., Epstein and Sobel [1], [2]; Gupta and Sobel [3]. These papers have mainly considered techniques and procedures associated with observations from a single exponential distribution. This paper outlines procedures to be used for analyzing the results of a two-way classification with respect to the scale factor of the exponential distribution. The test procedures are based on likelihood ratio tests and have their analogs in the usual analysis-of-variance tests for main effects and interactions. Approximations are given for tests based on small samples. These same procedures carry over directly to a two-way analysis of variance, if the quantities to be analyzed are variance estimates following chi-square distributions.

2. A Remark on the Generality of the Exponential Distribution

This paper will deal mainly with life-test observations following the exponential distribution with probability density function (p.d.f.)

$$p(t;\theta,A)=\begin{cases}\dfrac{1}{\theta}\exp\left[-\dfrac{t-A}{\theta}\right], & \text{for } t\geqq A\,,\\ 0\,, & \text{for } t<A\,.\end{cases}\tag{2.1}$$

However, all results immediately apply to populations having the p.d.f.

$$f(t)=\begin{cases}h(t)\exp-\displaystyle\int_0^t h(x)\,dx\,, & \text{for } t\geqq 0\,,\\ 0\,, & \text{for } t<0\,,\end{cases}\tag{2.2}$$

where $h(t)$ is a non-negative function. The quantity $h(t)$ is sometimes referred to as the "hazard" or "conditional rate of failure." Actually, it is the

[1] Part of this work was carried out by the author when he was at the University of California (Berkeley) during the spring, 1957.

conditional probability density function of the failure time, given that the particular component or mechanism has not failed up to time t, i.e.,

$$h(t) = \frac{f(t)}{1 - F(t)}, \tag{2.3}$$

where $F(t)$ is the cumulative distribution function.

The conditional p.d.f. (2.3) has long been used by actuaries and is often referred to as the force of mortality. It forms the basis of the famous Gompertz and Makeham curves of mortality. It is apparent that these same concepts can be immediately adapted to life-testing situations.

A recent paper on this subject is one by Birnbaum and Saunders [4]. Thus, if $h(t) = t^{p-1}/\theta$ for $p > 1$, we have the Weibull distribution with p.d.f.

$$f(t) = \begin{cases} \dfrac{t^{p-1}}{\theta} \exp - \dfrac{t^p}{\theta p}, & \text{for } t \geqq 0, \\ 0, & \text{for } t < 0, \end{cases} \tag{2.4}$$

which has been used to describe ball-bearing failure, e.g., Weibull [5], Lieblein and Zelen [6].

If $h(t) = e^t/\theta$, we have

$$f(t) = \begin{cases} \dfrac{1}{\theta} \exp\left[-\dfrac{e^t - 1}{\theta} + t\right], & \text{for } t \geqq 0, \\ 0, & \text{for } t < 0, \end{cases} \tag{2.5}$$

which is similar to the extreme-value distribution except that the range is for $t \geqq 0$.

In the "language of equipment failures," if $h(t)$ is independent of t, we have "catastrophic failures"; when $h(t)$ is an increasing function of t, the failure situation is characterized as "wear-out failures"; when $h(t)$ decreases for small t, we have a situation termed as "early failures." Thus the use of $h(t)$ gives rise to a large class of distributions which have important interpretations.

For all $h(t)$ it is easy to see that if

$$\int_0^t h(x)\,dx = \frac{H(t)}{\theta}, \tag{2.6}$$

then $H(t)$ will follow an exponential distribution having the p.d.f. $p(t; \theta, 0)$. Thus, with distributions having p.d.f.'s of the form (2.2), it is always possible to find a transformation that will make the random variable exponentially distributed.

3. Two-factor Experiment

3.1 Preliminaries. Consider a factorial experiment involving the factors A and B having levels a and b respectively. Let the experiment be planned so that n components or items are placed on test for each of the ab factórial combinations. The test for a particular combination is terminated when exactly r ($r \leqq n$) of the test items have failed. We shall assume that the underlying p.d.f. for the (i, j) factorial combination, $i = 1, 2, \cdots, a$; $j = 1, 2, \cdots, b$, is $p(t; \theta_{ij}, A_{ij})$.

Then it is well known that the p.d.f. of the r ordered failures for a particular factorial combination is

$$\left\{\frac{n}{\theta_{ij}} \exp\left[-\frac{n(t_{ij1} - A_{ij})}{\theta_{ij}}\right]\right\}\left\{\frac{(n-1)!}{(n-r)!}\frac{1}{\theta_{ij}^{r-1}} \exp\left[-\frac{V_{ij}}{\theta_{ij}}\right]\right\}, \tag{3.1}$$

where $t_{ij1} \leqq t_{ij2} \leqq \cdots \leqq t_{ijr}$ are the failure times in ascending order and

$$V_{ij} = \sum_{k=1}^{r} (t_{ijk} - t_{ij1}) + (n - r)(t_{ijr} - t_{ij1}) . \tag{3.2}$$

It is also well known (see Epstein and Sobel [2]) that (t_{ij1}, V_{ij}) are sufficient statistics for (A_{ij}, θ_{ij}) and that $n(t_{ij1} - A_{ij})$ and V_{ij} are independently distributed as $\theta_{ij}\chi^2/2$ with 2 and $2(r-1)$ degrees of freedom, respectively.

In this paper we shall be concerned with testing hypotheses on θ_{ij} regardless of the values of A_{ij}. We shall take as a model for θ_{ij} that

$$\theta_{ij} = m a_i b_j c_{ij} \qquad i = 1, 2, \cdots, a;\ j = 1, 2, \cdots, b , \tag{3.3}$$

where the parameters a_i, b_j, and c_{ij} are subject to the $a + b + 1$ independent restrictions

$$\prod_{i=1}^{a} a_i = \prod_{j=1}^{b} b_j = \prod_{i=1}^{a} c_{ij} = \prod_{j=1}^{b} c_{ij} = 1 . \tag{3.4}$$

These parameters are the analogs of the main-effects and interaction parameters used in the usual fixed-effects model in the analysis of variance.

Since the values of a_i, b_j, and c_{ij} are invariant with respect to a change in location of the failure times, we can restrict our considerations to the joint distribution of the V_{ij}'s, i.e.,

$$P(V) = [\Gamma(r-1)]^{-ab} \frac{\prod_{i,j} V_{ij}^{r-2}}{m^{ab(r-1)}} \exp\left[-\sum_{i,j} \frac{V_{ij}}{m a_i b_j c_{ij}}\right] \tag{3.5}$$

having the log likelihood function[2]

$$\begin{aligned} L = {} & -ab \log \Gamma(r-1) + (r-2) \sum_{i,j} \log V_{ij} \\ & - ab(r-1) \log m - \sum_{i,j} \frac{V_{ij}}{m a_i b_j c_{ij}} . \end{aligned} \tag{3.6}$$

The maximum likelihood equations for estimating m, a_i, b_j, and c_{ij} can readily be found by setting equal to zero the partial derivatives with respect to the parameters of

$$Q = L - \lambda_1 \sum_i \log a_i - \lambda_2 \sum_j \log b_j - \sum_i \lambda_{3i} \sum_j \log c_{ij} - \sum_j \lambda_{4j} \sum_i \log c_{ij} ,$$

where the λ's are Lagrange multipliers. This leads to the solutions

[2] All logarithms in this paper are to the base e.

(3.7)
$$\hat{m} = \frac{1}{(r-1)}\Big(\prod_{i,j} V_{ij}\Big)^{1/ab},$$
$$\hat{a}_i = \frac{1}{(r-1)\hat{m}}\Big(\prod_{j} V_{ij}\Big)^{1/b},$$
$$\hat{b}_j = \frac{1}{(r-1)\hat{m}}\Big(\prod_{i} V_{ij}\Big)^{1/a},$$
$$\hat{c}_{ij} = \frac{V_{ij}}{(r-1)\hat{m}\hat{a}_i\hat{b}_j}.$$

3.2 Moments of the Estimates. Since V_{ij} is distributed as $(ma_ib_jc_{ij})\chi^2/2$ with $\nu = 2(r-1)$ degrees of freedom, we have

(3.8)
$$E(V_{ij}^p) = (ma_ib_jc_{ij})^p \frac{\Gamma(\frac{1}{2}\nu + p)}{\Gamma(\frac{1}{2}\nu)}, \qquad p \geqq -\tfrac{1}{2}\nu.$$

All the estimates involve products of ratios of the V_{ij}. Hence it is an easy matter to find the moments. Equations (3.10) summarize the results for the moments where

(3.9)
$$g(p) = \Gamma\Big(\frac{\nu}{2} + \frac{p}{ab}\Big)\Big/\Gamma\Big(\frac{\nu}{2}\Big).$$

(3.10)
$$E(\hat{a}_i^p) = a_i^p\{g[p(a-1)]^b g[-p]^{b(a-1)}\},$$
$$E(\hat{b}_j^p) = b_j^p\{g[p(b-1)]^a g[-p]^{a(b-1)}\},$$
$$E(\hat{c}_{ij}^p) = c_{ij}^p\{g[p(a-1)(b-1)]\, g[-p(a-1)]^{b-1} \cdot g[-p(b-1)]^{a-1} g[p]^{(a-1)(b-1)}\},$$
$$E(\hat{a}_i^p\hat{a}_{i'}^q) = a_i^p a_{i'}^q\{g[p(a-1)-q]^b g[q(a-1)-p]^b g[-(p+q)]^{b(a-2)}\}, \qquad i \neq i',$$
$$E(\hat{b}_j^p\hat{b}_{j'}^q) = b_j^p b_{j'}^q\{g[p(b-1)-q]^a g[q(b-1)-p]^a g[-(p+q)]^{a(b-2)}\}, \qquad j \neq j',$$
$$E(\hat{c}_{ij}^p\hat{c}_{i'j'}^q) = c_{ij}^p c_{i'j'}^q\{g[p(a-1)(b-1)+q]\, g[q(a-1)(b-1)+p] \cdot g[q-p(a-1)]^{b-2} g[p-q(a-1)]^{b-2} \cdot g[q-p(b-1)]^{a-2} g[p-q(b-1)]^{a-2} \cdot g[-p(a-1)-q(b-1)]\, g[-p(b-1)-q(a-1)] \cdot g[p+q]^{(a-2)(b-2)}\}, \qquad i \neq i',\ j \neq j',$$
$$E(\hat{c}_{ij}^p\hat{c}_{ij'}^q) = c_{ij}^p c_{ij'}^q\{g[(a-1)(bp-p-q)]\, g[(a-1)(bq-p-q)] \cdot g[p+q-pb]^{a-1} g[p+q-qb]^{a-1} g[-(a-1)(p+q)]^{b-2} \cdot g(p+q)^{(a-1)(b-2)}\}, \qquad j \neq j'.$$

3.3 Likelihood Ratio Tests. This section considers the likelihood ratio tests for testing the null hypotheses, $H_1 : a_i = 1$ for all i; $H_2 : b_j = 1$ for all j; $H_3 : c_{ij} = 1$ for all i and j; $H_4 : b_jc_{ij} = 1$ for all i and j; $H_5 : a_ic_{ij} = 1$ for all i and j; $H_6 : a_ib_jc_{ij} = 1$ for all i and j. These are the analogs of the tests for main effects and interactions one usually associates with the analysis of variance.

The hypotheses H_1 and H_2 do not seem to be of much practical interest, as they state that θ_{ij} is of the form $\theta_{ij} = b_j c_{ij}$ and $\theta_{ij} = a_i c_{ij}$, respectively. However, these hypotheses are included for completeness. For practical situations H_3, the hypothesis of absence of interaction, should be of utmost importance. This would be especially true if one were interested in accelerating life tests by testing at some extreme environmental condition and then extrapolating the results to "normal" conditions. In order to make meaningful extrapolations to normal operating conditions, it would seem that one requirement should be the absence of interaction, i.e., $c_{ij} = 1$ for all i and j.

The likelihood ratio test for H_i reduces to

$$\mu_i = \left(\frac{\hat{m}}{\tilde{m}_i}\right)^{ab(r-1)}, \tag{3.11}$$

where $\tilde{m}_i$ is the maximum likelihood estimate of m under H_i. The values of $\tilde{m}_i$ are

$$\begin{aligned}
\tilde{m}_1 &= \frac{1}{a(r-1)} \sum_i \left(\prod_j V_{ij}\right)^{1/b}, \\
\tilde{m}_2 &= \frac{1}{b(r-1)} \sum_j \left(\prod_i V_{ij}\right)^{1/a}, \\
\tilde{m}_3 &= \frac{1}{ab(r-1)} \sum_{i,j} \frac{V_{ij}\left(\prod_{i,j} V_{ij}\right)^{2/ab}}{\left(\prod_i V_{ij}\right)^{1/a}\left(\prod_j V_{ij}\right)^{1/b}}, \\
\tilde{m}_4 &= \frac{1}{b(r-1)} \prod_i \left(\sum_j V_{ij}\right)^{1/a}, \\
\tilde{m}_5 &= \frac{1}{a(r-1)} \prod_j \left(\sum_i V_{ij}\right)^{1/b}, \\
\tilde{m}_6 &= \frac{1}{ab(r-1)} \sum_{i,j} V_{ij}.
\end{aligned} \tag{3.12}$$

Since, under H_i, $M_i = -2 \log \mu_i$ is asymptotically distributed as χ^2 with (say) f_i degrees of freedom, Table 1 summarizes the expressions for M_i in convenient form.

3.4 Small-sample Approximations to the Distribution of the Likelihood Ratio. For large numbers of observations, the asymptotic chi-square property of the likelihood ratio will probably be sufficient for most practical work. However, for small samples one would like to modify the likelihood ratio tests in order to obtain closer approximations to the actual distributions.

The test for the hypothesis $H_6 : a_i b_j c_{ij} = 1$ is exactly the test first given by Neyman and Pearson and commonly referred to as "Bartlett's test," for testing the homogeneity of ab variances each with $\nu = 2(r-1)$ degrees of freedom arising from normal populations. Also, it can be shown that the tests for the hypotheses $H_4 : b_j c_{ij} = 1$ and $H_5 : a_i c_{ij} = 1$ are composed of sums of independent Bartlett test statistics.

Work has been done by Bartlett [7], Box [8], and Hartley [9] on finding

TABLE 1

SUMMARY OF $M_i = -2 \log \mu_i$

Null hypothesis	Degrees of freedom	$M_i = -2 \log \mu_i$
$H_1 : a_i = 1$	$(a - 1)$	$2ab(r-1) \log\left(\frac{1}{a} \sum_i \hat{a}_i\right)$
$H_2 : b_j = 1$	$(b - 1)$	$2ab(r-1) \log\left(\frac{1}{b} \sum_j \hat{b}_j\right)$
$H_3 : c_{ij} = 1$	$(a - 1)(b - 1)$	$2ab(r-1) \log\left(\frac{1}{ab} \sum_{i,j} \hat{c}_{ij}\right)$
$H_4 : b_j c_{ij} = 1$	$a(b - 1)$	$2ab(r-1) \frac{1}{a} \sum_i \log\left(\frac{1}{b} \sum_j \hat{b}_j \hat{c}_{ij}\right)$
$H_5 : a_i c_{ij} = 1$	$b(a - 1)$	$2ab(r-1) \frac{1}{b} \sum_j \log\left(\frac{1}{a} \sum_i \hat{a}_i \hat{c}_{ij}\right)$
$H_6 : a_i b_j c_{ij} = 1$	$ab - 1$	$2ab(r-1) \log\left(\frac{1}{ab} \sum_{i,j} \hat{a}_i \hat{b}_j \hat{c}_{ij}\right)$

closer approximations to the distribution of Bartlett's test statistic. This can be directly carried over to obtain closer approximations on the distribution of the test statistics M_4 and M_5. Since M_6 is exactly Bartlett's test statistic, the application of these results is direct.

Hartley's work has resulted in tables [10] which, for our application, give the .01 and .05 percentage points of M as a function of the auxiliary quantities k and c_1. These exact percentage points are those which are labeled (a) in the above-mentioned tables. Bartlett has shown that M/C (C a suitably chosen constant) is a better approximation to a chi-square variate than using M alone. On the other hand, Box has shown that a still better approximation than M/C can be obtained by taking $f_2 M/[f_1(b^* - M)]$ to have an F-distribution with f_1 and f_2 degrees of freedom, where f_1, f_2, and b^* are suitably defined. Table 2 summarizes the auxiliary constants for the above-mentioned approximations with respect to M_4, M_5, and M_6.

It is still necessary to find a way of approximating the small-sample distribution of M_1, M_2, and M_3. The scale adjustment C for Bartlett's test was derived so that $E(M/C) = f + O(\nu^{-2})$. It would appear that an adjustment made on any monotonic function of M should also do as well. In particular, since $M_i = 2ab(r-1) \log \Lambda_i$ and $E(\Lambda_i)$ can be found exactly, a scale factor C_i can be found by equating the expected value of $\Lambda_i = \exp M_i/2ab(r-1)$ to that of $\exp C_i \chi^2/2ab(r-1)$, i.e.,

$$E(\Lambda_i) = E\left\{\exp \frac{C_i \chi^2}{2ab(r-1)}\right\} = \left\{1 - \frac{C_i}{ab(r-1)}\right\}^{-f_i/2}, \qquad i = 1, 2, 3 . \tag{3.13}$$

Therefore

$$C_i = ab(r-1)[1 - E(\Lambda_i)^{-2/f_i}], \qquad i = 1, 2, 3 , \tag{3.14}$$

and from (3.10) we have, under the null hypothesis,

TABLE 2

FACTORS FOR M_i ($i = 4, 5, 6$)

Statistic	Degrees of freedom	Bartlett, $\chi^2 = M/C$	Hartley	Box, $F = \frac{f_2 M}{f_1(b^* - M)}$ [a]
M_4	$f_1 = a(b-1)$	$C_4 = 1 + \frac{b+1}{6b(r-1)}$	$c_1 = \frac{b^2-1}{2b(r-1)}$ $k = b$	$A_1 = \frac{b+1}{6b(r-1)}$
M_5	$f_1 = b(a-1)$	$C_5 = 1 + \frac{a+1}{6a(r-1)}$	$c_1 = \frac{a^2-1}{2a(r-1)}$ $k = a$	$A_1 = \frac{a+1}{6a(r-1)}$
M_6	$f_1 = ab - 1$	$C_6 = 1 + \frac{ab+1}{6ab(r-1)}$	$c_1 = \frac{(ab)^2-1}{2ab(r-1)}$ $k = ab$	$A_1 = \frac{ab+1}{6ab(r-1)}$

[a] The quantities f_2 and b^* are defined by $f_2 = (f_1 + 2)/A_1^2$, $b^* = f_2/(1 - A_1 + 2/f_2)$.

$$(3.15) \qquad \begin{aligned} E(\Lambda_1) &= E(\hat{a}_i) = g(a-1)^b\, g(-1)^{b(a-1)}, \\ E(\Lambda_2) &= E(\hat{b}_j) = g(b-1)^a\, g(-1)^{a(b-1)}, \\ E(\Lambda_3) &= E(\hat{c}_{ij}) = g[(a-1)(b-1)]g[-(a-1)]^{b-1} \\ &\qquad \cdot g[-(b-1)]^{a-1} g(1)^{(a-1)(b-1)}. \end{aligned}$$

Thus C_i can be evaluated exactly.

An approximation to C_i can be found by using the Barnes asymptotic expansion for $\log \Gamma(x+y)$, i.e.,

$$\log \Gamma(x+y) = (x + y - \tfrac{1}{2})\log x - x + \tfrac{1}{2}\log 2\pi + \sum_{n=1}^{m} \frac{(-1)^{n+1} B_{n+1}(y)}{n(n+1)x^n} + O(x^{-m-1}),$$

where $B_{n+1}(y)$ is the Bernoulli polynomial of order 1 and degree $n+1$. Hence one can approximate $g(p)^\alpha$ to $O((r-1)^{-3})$ by

$$(3.16) \qquad g(p)^\alpha = (r-1)^{\alpha p/ab} \exp\left\{\frac{\alpha p(p-ab)}{2(r-1)(ab)^2}\left[1 - \frac{(2p-ab)}{6(r-1)ab}\right] + O((r-\ )^{-3})\right\}.$$

Substituting (3.16) in (3.15) results in

$$(3.17) \qquad \begin{aligned} E(\Lambda_1) &= \exp\left\{\frac{a-1}{2(r-1)ab}\left[1 + \frac{3ab - 2a + 4}{6(r-1)ab}\right] + O((r-1)^{-3})\right\}, \\ E(\Lambda_2) &= \exp\left\{\frac{b-1}{2(r-1)ab}\left[1 + \frac{3ab - 2b + 4}{6(r-1)ab}\right] + O((r-1)^{-3})\right\}, \\ E(\Lambda_3) &= \exp\left\{\frac{(a-1)(b-1)}{2(r-1)ab}\left[1 + \frac{ab + 4a + 4b - 8}{6(r-1)ab}\right] + O((r-1)^{-3})\right\}, \end{aligned}$$

and finally, substituting (3.17) in (3.14), and using terms to $O((r-1)^{-3})$ gives

$$(3.18) \qquad C_1 = 1 + \frac{3ab - 2a + 1}{6(r-1)ab}, \qquad C_2 = 1 + \frac{3ab - 2b + 1}{6(r-1)ab},$$

$$C_3 = 1 + \frac{ab + 4a + 4b - 11}{6(r-1)ab}.$$

These values of C_i are correct to $O((r-1)^{-2})$ and should be sufficiently accurate for most applications. As a check on our procedure, when $b = 1$ in M_1 we have exactly Bartlett's test for testing the homogeneity of a independent variance estimates, each with $\nu = 2(r-1)$ degrees of freedom. With $b = 1$, $C_1 = 1 + (a+1)/(3a\nu)$, which is Bartlett's scale factor for equal degrees of freedom.

3.5 Confidence Intervals for Ratios of a_i and b_j. In many situations, it will be necessary to calculate confidence intervals for the ratios of main effects, $a_i/a_{i'}$, $b_j/b_{j'}$. Approximate confidence intervals can be obtained by considering

$$(3.19) \qquad (\hat{a}_i/\hat{a}_{i'})^b = \prod_{j=1}^{b} V_{ij} \Big/ \prod_{j=1}^{b} V_{i'j} \,, \qquad i \neq i' \,.$$

Since $2V_{ij}/ma_ib_jc_{ij} = \chi^2_{ij}$ with $\nu = 2(r-1)$ degrees of freedom, we have

$$(3.20) \qquad (\hat{a}_i/\hat{a}_{i'})^b = (a_i/a_{i'})^b \prod_{j=1}^{b} \chi^2_{ij} \Big/ \prod_{j=1}^{b} \chi^2_{i'j} = (a_i/a_{i'})^b \prod_{j=1}^{b} F_i \,,$$

where $F_j = \chi^2_{ij}/\chi^2_{i'j}$ is an F-statistic in which the numerator and denominator have ν degrees of freedom.

Let $z_j = \frac{1}{2}\log F_j$, and rewrite (3.20) in the form

$$\frac{1}{2}\log\left(\frac{\hat{a}_i/\hat{a}_{i'}}{a_i/a_{i'}}\right) = \bar{z} \,,$$

where $\bar{z} = \sum_{j=1}^{b} z_j/b$. The odd-order cumulants of z_j are equal to zero, i.e., $K_{2m+1}(z) = 0$, and the even-order cumulants can be approximated by (see Wishart, [11])

$$K_{2m}(z) \cong \frac{(2m-2)!}{(\nu-1)^{2m-1}} \,, \qquad m = 1, 2, \cdots \,.$$

Thus the first four cumulants of $\bar{z}$ are

$$K_1(\bar{z}) = 0 \,, \qquad K_2(\bar{z}) \cong [b(\nu-1)]^{-1} \,,$$
$$K_3(\bar{z}) = 0 \,, \qquad K_4(\bar{z}) \cong 2[b(\nu-1)]^{-3} \,.$$

Hence the standardized measures of skewness and kurtosis are

$$\beta_1(\bar{z}) = \frac{K_3^2}{K_2^3} = 0 \,, \qquad \beta_2(\bar{z}) \cong 3 + \frac{K_4}{K_2^2} = 3 + \frac{2}{b(\nu-1)} \,.$$

For practical applications, it seems sufficient to approximate the percentage points of $\bar{z}$ by equating the first four moments to an appropriate Pearson-type curve. If μ_α represents the 100α percentage point of the standardized variable $\sqrt{b(\nu-1)}\,\bar{z}$, then a two-sided confidence interval for $a_i/a_{i'}$ with confidence coefficient $1-2\alpha$ will be of the form $Z_\alpha^{-1}(\hat{a}_i/\hat{a}_{i'})$, $Z_\alpha(\hat{a}_i/\hat{a}_{i'})$, where $Z_\alpha = \exp[2\mu_\alpha/\sqrt{b(\nu-1)}\,]$.

Tables 3 and 4 give values of Z_α for $\alpha = .005$ and $.025$, respectively. These

TABLE 3

UPPER 0.5% POINTS OF $Z = \exp 2\bar{z}$

Number of terms in geom. ave. (m)	Degrees of freedom $\nu = 2(r-1)$				
	$\nu = 2$	$\nu = 4$	$\nu = 6$	$\nu = 8$	$\nu = 10$
2	54.7	8.99	5.34	4.08	3.44
3	24.5	5.89	3.89	3.12	2.72
4	15.5	4.59	3.23	2.67	2.38
5	11.3	3.89	2.84	2.41	2.16
6	8.99	3.44	2.58	2.23	2.02

For larger values of m and ν, use $Z_{.005} = \exp \frac{5.1516}{\sqrt{m(\nu - 1)}}$.

TABLE 4

UPPER 2.5% POINTS OF $Z = \exp 2\bar{z}$

Number of terms in geom. ave. (m)	Degrees of freedom $\nu = 2(r-1)$				
	$\nu = 2$	$\nu = 4$	$\nu = 6$	$\nu = 8$	$\nu = 10$
2	16.7	5.04	3.48	2.87	2.53
3	9.95	3.72	2.77	2.35	2.13
4	7.24	3.12	2.40	2.10	1.92
5	5.88	2.77	2.19	1.94	1.79
6	5.04	2.53	2.05	1.83	1.70

For larger values of m and ν, use $Z_{.025} = \exp \frac{3.9200}{\sqrt{m(\nu - 1)}}$.

tables enable one to calculate two-sided confidence intervals for $(a_i/a_{i'})$ with confidence coefficients .99 and .95. Similar results hold for confidence intervals on ratios of b_j. For placing confidence intervals on ratios of a_i, the row entry is $m = b$; for confidence intervals on ratios of b_j, the row entry is $m = a$. The values of μ_α used in the computation of Z_α were obtained by using linear interpolation from Table 42 of Pearson and Hartley [10].

4. Acknowledgment

The author wishes to thank Dr. E. L. Crow for reading an earlier draft and making several constructive suggestions.

REFERENCES

[1] Epstein, B., and Sobel, M. Life testing. *J. Amer. Stat. Assoc.*, 1953, **48**, 486–502.

[2] Epstein, B., and Sobel, M. Some theorems relevant to life testing from an exponential distribution. *Ann. Math. Stat.*, 1954, **25**, 373–81.

[3] Gupta, S. S., and Sobel, M. On the distribution of a statistic based on ordered uniform chance variables. *Ann. Math. Stat.*, 1958, **29**, 274–81.

[4] Birnbaum, Z. W., and Saunders, S. C. A statistical model for life-length of materials. *J. Amer. Stat. Assoc.*, 1958, **53**, 151–60.

[5] Weibull, W. A statistical representation of fatigue failures in solids. *Trans. Roy. Inst. Tech. (Stockholm)*, 1949 (No. 27), 1–50.

[6] Lieblein, J., and Zelen, M. Statistical investigation of the fatigue life of deep-groove ball bearings. *J. Res. N. B. S.*, 1956, **57**, 273–316.

[7] Bartlett, M. S. Properties of sufficiency and statistical tests. *Proc. Roy. Soc. (London)*, Ser. A, 1937, **160**, 268–82.

[8] Box, G. E. P. A general distribution theory for a class of likelihood criteria. *Biometrika*, 1949, **36**, 317–46.

[9] Hartley, H. O. Testing the homogeneity of a set of variances. *Biometrika*, 1940, **31**, 249–55.

[10] Pearson, E. S., and Hartley, H. O. *Biometrika Tables for Statisticians*, Vol. 1. Cambridge: Cambridge Univ. Press, 1954.

[11] Wishart, J. The cumulants of the z and of the logarithmic χ^2 and t distributions. *Biometrika*, 1947, **34**, 170–78.